Mobile Robots Motion Planning
New Challenges

Edited by
Xing-Jian Jing

Mobile Robots Motion Planning

Edited by Xing-Jian Jing

This is a reprint of the book published in 2008 by Intech

Republished by InTech

Janeza Trdine 9, 51000 Rijeka, Croatia

Technical Editor Teodora Smiljanic

Cover Designer Dino Smrekar

Additional hard copies can be obtained from orders@intechopen.com

Mobile Robots Motion Planning, Edited by Xing-Jian Jing
p. cm.
ISBN 978-953-7619-01-5

For EU product safety concerns:
IN TECH d.o.o., Prolaz Marije Krucifikse Kozulić 3, 51000 Rijeka, Croatia,
info@intechopen.com or visit our website at intechopen.com.

Preface

As an important extension of the motor capability of human being, mobile robots are now expected to implement various tasks and have become important tools in all kinds of application fields, such as manufacturing plants, warehouses, nurse/medical service, tour guider, resource or space exploration, NANO manipulation, national defence and so on. In the family of mobile robots, humanoid robots, robot pets or toys and robot assistants etc have now been becoming some new research and application trends, and more and more potential applications of mobile robots are emerging. In all these research and applications, a reliable and effective motion planning method plays a considerably important role in the successful completion of the corresponding tasks and purposes. The motion control problem of mobile robots, for example robot-cup competition, is also regarded as an effective platform in many institutes to study, test and demonstrate computationally intelligent methods and advanced control theories. For these reasons, the motion planning problem of mobile robots has been extensively studied in the field of robotics, and many noticeable methods have been proposed for different purposes.

The motion planning of mobile robots relies greatly on the known information about the involved environment perceived by sensors and the motion constraints of the robotic kinematics and dynamics. If the environment is static, well structured and completely known, there is usually less difficulty in the motion planning problems, which can be solved by using many existing path planning methods. However, when the environment is partially or totally unknown, unstructured, or dynamic changing, the demand of high autonomy for a mobile robot in such an environment will produce a great challenge for the motion planning task. For example, what sensors or energy supply should be adopted, how to model the environment with limited and noised environmental information from sensors, how to build the on line decision-making system of the robot, and how to find a satisfactory or optimal solution in real time which satisfies both the kinematical or dynamic constraints of the robot and the desired goal of the task, etc. Furthermore, the flexible and friendly interaction between mobile robots and environment including human being is more and more in great demand in many practical applications. All these provide great challenges to robotic techniques including sensor or video signal processing and communication, pattern recognition, online intelligent decision making, robust motion control, mechanical structure and sensor device design etc. These problems are all the hot research topics covered by current robotic literature, and lots of efforts have been made to cope with these challenges.

In this book, new results or developments from different research backgrounds and application fields are put together to provide a wide and useful viewpoint on these headed research problems mentioned above, focused on the motion planning problem of mobile robots. These results cover a large range of the problems that are frequently encountered in the motion planning of mobile robots both in theoretical methods and practical applications including obstacle avoidance methods, navigation and localization techniques, environmental modelling or map building methods, and vision signal processing etc. Different methods such as potential fields, reactive behaviours, neural-fuzzy based methods, motion control methods and so on are studied. Through this book and its references, the reader will definitely be able to get a thorough overview on the current research results for this specific topic in robotics.

The book is intended for the readers who are interested and active in the field of robotics and especially for those who want to study and develop their own methods in motion/path planning or control for an intelligent robotic system.

Editor

Xing-Jian Jing
University of Sheffield
United Kingdom
X.J.Jing@sheffield.ac.uk

Contents

1

Local Autonomous Robot Navigation using Potential Fields

Miguel A. Padilla Castañeda, Jesús Savage, Adalberto Hernández and Fernando Arámbula Cosío
University
Country

Chapter Abstract

The potential fields method for autonomous robot navigation consists essentially in the assignment of an attractive potential to the goal point and a repulsive potential to each of the obstacles in the environment. Several implementations of potential fields for autonomous robot navigation have been reported. The most simple implementation considers a known environment where fixed potentials can be assigned to the goal and the obstacles. When the obstacles are unknown the potential fields have to be adapted as the robot advances, and detects new obstacles. The implementation of the potential fields method with one attraction potential assigned to the goal point and fixed repulsion points assigned to the obstacles, has the important limitation that for some obstacle configurations it may not be possible to produce appropriate resultant forces to avoid the obstacles. Recently the use of several adjustable attraction points, and the progressive insertion of repulsion points as obstacles are detected online, have proved to be a viable method to avoid large obstacles using potential fields in environments with unknown obstacles. In this chapter we present the main characteristics of the different approaches to implement local robot navigation algorithms using potential fields for known and partially known environments. Different strategies to escape from local minima, that occur when the attraction and repulsion forces cancel each other, are also considered.

1. Introduction: The Potential Fields Method for Obstacle Avoidance

The local autonomous robot navigation problem consists of the calculation of a viable path between two points, an starting and a target point. The local navigation approach should produce an optimum (usually shortest) path, avoiding the obstacles present in the working environment. In general, the obstacles and the target could be static or dynamic. The obstacles could also be known *a priory* (e.g. the different walls in a building) or could be unknown (e.g. persons walking nearby the robot). In this chapter are presented the following aspects of a potential fields scheme for autonomous robot navigation: The potential and force field functions; The use of single or multiple attraction points; The construction of an objective function for field optimization; The field optimization approach in known and unknown environments. In the last section of the chapter we present hybrid approaches to recover from local minima of the potential field. During the chapter we have only considered potential fields defined in cartesian space, where attractive or repulsive potentials are a function of the position of the target or the obstacle. Recently, potential fields defined in a 2D trajectory space, using the path curvature and longitudinal robot velocity, have been reported (Shimoda et al., 2005).

1.1 Previous works on artificial potential fields for autonomous robot navigation

Artificial potential fields for autonomous robot navigation were first proposed by Khatib (1990). The main idea is to generate attraction and repulsion forces within the working environment of the robot to guide it to the target. The target point has an attractive influence on the robot and each obstacle tends to push away the robot, in order to avoid collisions. Potential field methods provide an elegant solution to the path finding problem. Since the path is the result of the interaction of appropriate force fields, the path finding problem becomes a search for optimum field configurations instead of the direct construction (e.g. using rules) of an optimum path. Different approaches have been taken to calculate appropriate field configurations.

Vadakkepat et al. (2000) report the development of a genetic algorithm (GA) for autonomous robot navigation based on artificial potential fields. Repulsion forces are assigned to obstacles in the environment and attraction forces are assigned to the target point. The GA adjusts the constants in the force functions. Multiobjective optimisation is performed on 3 functions which measure each: error to the target point, number of collisions along a candidate path, and total path length. This scheme requires *a priory* knowledge of the obstacle positions in order the evaluate the number of collisions through each candidate path. Kun Hsiang et al. (1999), report the development of an autonomous robot navigation scheme based on potential fields and the chamfer distance transform for global path planning in a known environment, and a local fuzzy logic controller to avoid trap situations. Simulation and experimental results on a real AGV are reported for a simple (4 obstacles) and known environment. McFetridge and Ibrahim (1998) report the development of a robot navigation scheme based on artificial potential fields and fuzzy rules. The main contribution of the work consists in the use of a variable for the evaluation of the importance of each obstacle in the path of the robot. Simulation results on a very simple environment (one obstacle) show that use of the importance variable produces smoother and shorter trajectories. Ge and Cui (2002) describe a motion planning scheme for mobile robots in dynamic environments, with moving obstacles and target point. They use potential field functions which have terms that measure the relative velocity between the robot and the target or obstacle.

The main disadvantage of artificial potential field methods is its susceptibility to local minima (Borenstein and Koren, 1991), (Grefenstette and Schultz, 1994). Since the objective function for path evaluation is usually a multimodal function of a large number of variables. Additionally, in the majority of works on artificial potential fields for robot navigation, a single attraction point has been used. This approach can be unable to produce the resultant forces required to avoid a large or several, closely spaced, obstacles (Koren and Borenstein, 1991). An scheme based on a fixed target attraction point and several, moving, auxiliary attractions points was reported in Arámbula and Padilla (2004). Multiple auxiliary attractions points with adjustable position and force intensity enable navigation around large obstacles, as well as through closely spaced obstacles, at the cost of increased

complexity of the field optimisation. A GA has been successfully used to optimise potential fields with a large number of unknown obstacles and four auxiliary attraction points. The approach is fast enough for on-line control of a mobile robot.
In the following section we present different potential and force field functions which have been used for robot navigation. In section 3 we present the main characteristics of potential fields with one, as well as several attraction points. In section 4 we present the basics of field optimization: objective function construction; function optimization in known and unknown environments. In section 5 we introduce a hybrid method to avoid local minima during field optimization.

2. Potential field and force field functions

The first formulation of artificial potential fields for autonomous robot navigation was proposed by Khatib (1990). Since then other potential fields formulation have been proposed (Canny 1990, Barraquand 1992, Guldner 1997, Ge 2000, Arámbula 2004).
In general, the robot is represented as a particle under the influence of an scalar potential field U, defined as:

$$U = U_{att} + U_{rep} \tag{1}$$

where U_{att} and U_{rep} are the attractive and repulsive potentials respectively.
The attraction influence tends to pull the robot towards the target position, while repulsion tends to push the robot away from the obstacles. The vector field of artificial forces $\mathbf{F}(\mathbf{q})$ is given by the gradient of U:

$$\mathbf{F}(\mathbf{q}) = -\nabla U_{att} + \nabla U_{rep} \tag{2}$$

where ∇U is the gradient vector of U at robot position $\mathbf{q}(x, y)$ in a two dimensional map. In this manner, $\mathbf{F}$ is defined as the sum of two vectors $\mathbf{F}_{att}(\mathbf{q}) = -\nabla U_{att}$ and $\mathbf{F}_{rep}(\mathbf{q}) = \nabla U_{rep}$, as shown in eq. 3.

$$\mathbf{F}(\mathbf{q}) = \mathbf{F}_{att}(\mathbf{q}) + \mathbf{F}_{rep}(\mathbf{q}) \tag{3}$$

2.1 Artificial Potential Fields Formulation

The most commonly used form of potential field functions proposed by Kathib (1990) is defined as:

Attraction potential field

$$U_{att} = \frac{1}{2}\xi\, d^2 \tag{4}$$

where $d = \left| \mathbf{q} - \mathbf{q}_a \right|$; $\mathbf{q}$ is the current position of the robot; $\mathbf{q}_a$ is the position of an attraction point; and ξ is an adjustable constant.

Repulsion potential field

$$U_{rep} = \begin{cases} \frac{1}{2}\eta\left(\frac{1}{d}-\frac{1}{d_0}\right)^2 & d \le d_0 \\ 0 & d > d_0 \end{cases} \tag{5}$$

where $d=|\mathbf{q}-\mathbf{q}_o|$ for the robot position **q** and the obstacle position $\mathbf{q}_{o,}$ d_0 is the influence distance of the force and η is an adjustable constant.
The corresponding force functions are:
Attraction force

$$\mathbf{F}_{att}(\mathbf{q}) = -\nabla U_{att} = -\xi(\mathbf{q}-\mathbf{q}_a) \tag{6}$$

where **q** is again the robot position, $\mathbf{q}_a$ the position of the attraction point and ξ is an adjustable constant.
Repulsion force

$$\mathbf{F_{rep}}(\mathbf{q}) = \nabla U_{rep} = \begin{cases} \eta(\frac{1}{d}-\frac{1}{d_0})\frac{(\mathbf{q}-\mathbf{q_0})}{d^3} & d \le d_0 \\ 0 & d > d_0 \end{cases} \tag{7}$$

where $d=|\mathbf{q}-\mathbf{q}_o|$ for the robot position **q** and the obstacle position $\mathbf{q}_{o,}$ d_0 is the influence distance of the force and η is an adjustable constant.
The above formulation is popular due to its mathematical elegance and its simplicity; unfortunately, it suffers of oscillations and local minima under some obstacle configurations could cause problems, such: trap situations due to local minima, oscillations in narrow passages or impossibility of passing between closely spaced obstacles.
Some different potential fields have been reported in the past in order to solve these problems. Ge and Cui (2000) proposed a modified formulation of Eq. 5 and Eq. 7 for repulsion forces for solving the problem of having a non-reachable target when it is placed nearby obstacles due to the fact that as the robot approaches the goal near an obstacle, the attraction force decrease and becomes drastically smaller than the increasing repulsion force. The modified repulsion potential takes the form of:

$$U_{rep} = \begin{cases} \frac{1}{2}\eta\left(\frac{1}{d}-\frac{1}{d_0}\right)^2 |\mathbf{q} - \mathbf{q_{goal}}|^n & d \le d_0 \\ 0 & d > d_0 \end{cases} \tag{8}$$

The term $\left|\mathbf{q}-\mathbf{q_{goal}}\right|$ is the distance between the robot and the goal position. The introduction of this term ensures that the total potential $U = U_{att} + U_{rep}$ arrives at its global minimum 0, if and only if $\mathbf{q} = \mathbf{q_{goal}}$. The corresponding repulsion force is given by:

$$\mathbf{F_{rep}}(\mathbf{q}) = \nabla U_{rep} = \begin{cases} \mathrm{F_{rep1}}\,\mathbf{n}_{OR} + \mathrm{F_{rep2}}\,\mathbf{n}_{RG} & d \le d_0 \\ 0 & d > d_0 \end{cases} \tag{9}$$

where

$$\mathbf{F_{rep1}} = \eta\left(\frac{1}{d} - \frac{1}{d_0}\right)\frac{\left|\mathbf{q}-\mathbf{q}_{goal}\right|^n}{d^2} \tag{10}$$

$$\mathbf{F_{rep2}} = \frac{\eta^2}{2}\left(\frac{1}{d} - \frac{1}{d_0}\right)^2\left|\mathbf{q}-\mathbf{q}_{goal}\right|^{n-1} \tag{11}$$

$\mathbf{n}_{OR} = \nabla d$ and $\mathbf{n}_{RG} = -\nabla\left|\mathbf{q}-\mathbf{q}_{goal}\right|$ are two unit vectors pointing from obstacle to the robot and from the robot to the goal, respectively. In this way, $F_{rep1}\mathbf{n}_{OR}$ repulses the robot away from the obstacle, while $F_{rep2}\mathbf{n}_{RG}$ attracts the robot toward the goal. Although this approach solves the problem of nonreachable goals which are nearby, still suffers of local minima at some obstacle configurations and combinations of η and ξ.

Arámbula and Padilla (2004) modified equations 6 and 7 experimentally in order to amplify the effect of repulsion in obstacles and designed a potential field scheme with movable and adjustable, in real time, auxiliary attraction points in order to reduce the risk of the robot to being trapped in local minima. The modified artificial attraction force $\mathbf{F_{att}}$ used for the target point and for each of the auxiliary attraction points is:

$$\mathbf{F}_{att}(\mathbf{q}) = -\nabla U_{att} = -\xi(\mathbf{q}-\mathbf{q}_a)\frac{1}{\left|\mathbf{q}-\mathbf{q}_a\right|} \tag{12}$$

The aim of normalization of Eq. 6 is to produce an attraction force independent of the distance between the robot and the target point (Eq.12). The artificial repulsion force F_{rep} is defined as:

$$\mathbf{F}_{rep}(\mathbf{q}) = \nabla U_{rep} = \begin{cases} \eta \cdot sqrt\left(\frac{1}{d} - \frac{1}{d_0}\right)\frac{(\mathbf{q}-\mathbf{q}_o)}{d^3} & d \le d_0 \\ 0 & d > d_0 \end{cases} \tag{13}$$

As the robot gets closer to an obstacle, the repulsion force of the closest obstacle points grows in the opposite direction of the robot trajectory. If the robot distance to an obstacle point is higher than d_0, that obstacle position has no effect on the robot. An steep repulsion force function is needed in order to enable navigation through narrow passages, however it was observed that taking the square root of $(1/d - 1/d_0)$ in Eq.13 provides a light increase of the repulsion forces at mid distances (as shown in Fig.1) enabling, in turn, a safer obstacle avoidance. The constant η is also adjustable in real time as the robot moves by a genetic algorithm as is explained in section 4.

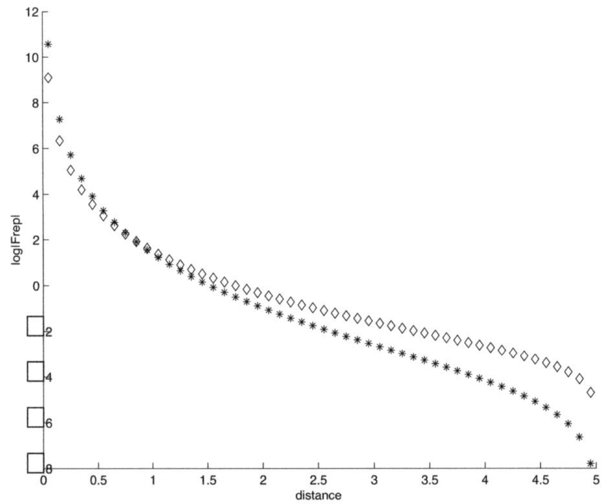

Figure 1. Plot of the magnitude of equation 9 (diamonds) and the same equation without taking the square root of (1/d-1/d_0) (stars)

2.2 Distance Fields as Potential Fields

Canny and Lin (1990) and Barraquand et al (1992) used a similar approach based on distance functions for building the potential field. Canny and Lin (1990) used the Euclidean distance field as a non-negative continuous and differentiable function defined as:

$$U = U_{att} = \min_i \eta(D_i(O_i, x)) \tag{14}$$

where $D_i(O_i, x)$ is the shortest Euclidean distance between an obstacle O_i and the position x of the robot and η is and adjustment constant. In this manner, the potential tends to zero as the robot approaches the obstacles, so the robot moves along the skeleton of the distance field that represents the path of maximum attraction potential. Unfortunately, under certain obstacle configurations the resulting potential field may contain local maxima, specially if the robot is near obstacle concavities.

Barraquand et al (1992) used a simple algorithm that computes the potential U as a grid where at the goal position x_{goal} is setting up the value of *0* and then progressively by region growing incrementing in 1 the value of the free obstacle neighbors and infinity in obstacle positions. Then the navigating path is found by tracking the flow of the negative

gradient vector field $-\nabla U$ starting from the initial robot position x_{init}. The idea behind this approach is to produce a free of local minima potential field.
Fig. 2 shows an example of an obstacle workspace and the potential fields produced by the calculation of the distance fields of both approaches.

(a) (b) (c)

Figure 2. Example of a workspace with distance field as potential field. a) Obstacle configuration; b) Skeleton of the potential field produced by Canny and Lin (1990); c) Skeleton of the potential field produced by Barraquand et al (1992)

2.3 Harmonic Artificial Potential Fields

Some authors have proposed the use of harmonic functions for building artificial potential fields which satisfies the $\nabla^T \nabla U = 0$ in order to avoid the problem of local minima (Connolly 1990, Utkin 1991, Guldner 1997). The generalized harmonic potential of a point charge q is:

$$U(r) = \begin{cases} \dfrac{q}{r^{n-2}} & n = 1,3,4,\ldots \\ q \ln \dfrac{1}{r} & n = 2 \end{cases} \tag{15}$$

for $r > 0$ in $\Re^n$ where $r = \sqrt{\sum_n x_i^2}, x = (x_1, x_2, \ldots, x_n)$ and the gradient $-\nabla U$ described by:

$$-\nabla U(r) = \begin{cases} (n-2)\dfrac{q}{r^{n-1}}\vec{e}_r & n = 1,3,4,\ldots \\ \dfrac{q}{r}\vec{e}_r & n = 2 \end{cases} \tag{16}$$

where $\vec{e}_r$ denotes a unit vector in radial direction.

In particular, Guldner et al (1997) introduced *the harmonic dipole potential* based on electrostatics, where points on the workspace represent *point charges* within a *security zone* inside ellipsoidal gradients. For a single obstacle, they defined the gradient of the harmonic potential field for a dipole charge as a *security circle* with radius R with a unit charge at the target point in the origin of the circle and a positive obstacle charge $q < 1$ defined as:

$$q = \frac{R}{R + D} \tag{17}$$

where D is the distance between the two charges. For multiple obstacles, independent security zones are determined for each obstacle in a transformed space and mapped in the original space without overlapping. When computing the navigation path, the method only considered the closest obstacle to the robot at each time and requires to switch obstacle potentials when the robot cross between security zones; in order to avoid discontinuities when switching potentials between obstacles, the resultant potential near the border of two zones is calculated by the weighted contribution of the obstacles, where the weight depends on the distance to the *security borders* of the obstacles.

2.4 Physical Fields as Artificial Potential Fields

Physical analogies for potential fields for robot navigation have been reported in the past for electrostatics (Guldner et al 1997), incompressible fluids dynamics (Keymeulen et al 1994), gaseous substance diffusion (Schmidt and Azarm 1992), mechanical stress (Masoud et al 1994) and steady-state heat transfer (Wang and Chirikjian 2000). For example, Wang and Chirikjian (2000) used temperature as the artificial potential field because in heat transfer the heat flux points in the direction of a negative temperature gradient; temperature monotonically decreases on the path from any point to the sink. In the analogy, the goal is treated as the sink that pulls the heat in and the obstacles as zero or very low thermal conductivity. With this approach the temperature is characterized as the harmonic field without local minima of the form:

$$\nabla \cdot (K \nabla T) = q \tag{18}$$

$$\int_{\Omega} q \mathrm{dV} = 0 \tag{19}$$

$$\left(\frac{\partial T}{\partial n} \right)_{\Gamma} = 0 \tag{20}$$

$$\underline{f} = -K \nabla T \tag{21}$$

where T is the temperature over the workspace, q indicates the heat sources and sink, K is the thermal conductivity which is a function of space coordinates, Ω is the configuration space where the robot moves and Γ is the boundary of this configuration space, n expresses the unit normal vector and f is the heat flux. Numerical solution is obtained from finite difference or finite element methods.

3. Attraction point configurations

In order to avoid trap situations or oscillations in the presence of large or closely spaced obstacles (Koren and Borenstein, 1991), in a map modelled as a two dimensional grid, several auxiliary attraction points can be placed around the goal cell (Fig. 3). Each attraction force $\mathbf{F}^i_{att}$ located at cell c_i, depends on the corresponding value of ξ_i (Eq. 6), which needs to be adjusted by an optimization algorithm as described in the next section. The effect of auxiliary attraction points has been evaluated in two modalities (Arámbula and Padilla, 2004): (1) auxiliary points placed at a fixed distance (of 15 cells) from the goal cell; and (2) auxiliary points placed at a variable distance (between 0 and 15 cells), which is adjusted automatically with a GA. Results from both approaches are shown in section 4. The use of auxiliary attraction points with a force strength and position automatically adjusted with a GA, allows for the generation of resultant force vectors which enable the robot to avoid large obstacles, as shown in Fig. 4.

3.1. Multiple attraction points

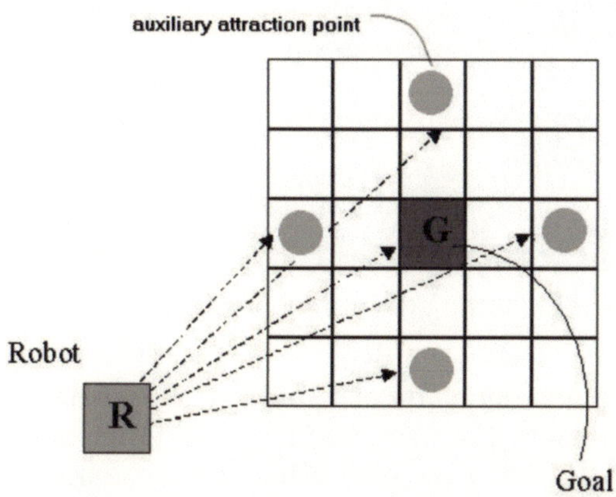

Figure 3. Attraction field composed of 5 attraction cells with adjustable position and force intensity

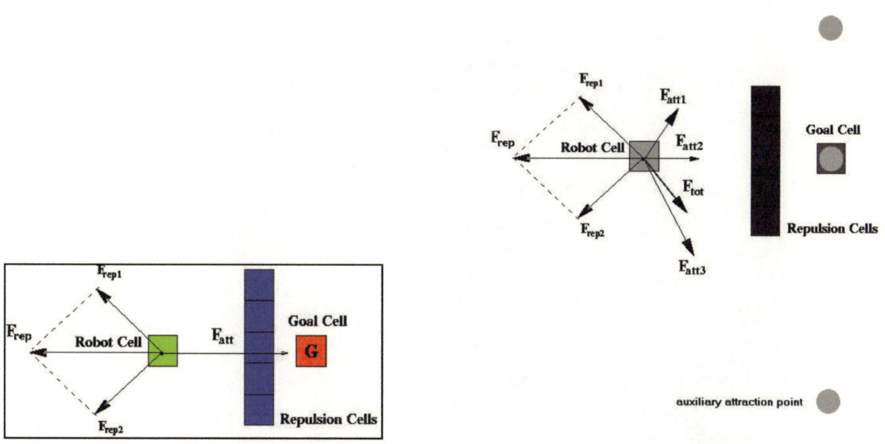

a) b)

Figure 4. a) A large obstacle which can not be avoided with one attraction point only; b) use of auxiliary attraction points of varying force intensity and position allow for the generation of resultant forces which guide the robot around the obstacle

4. Potential field optimization for obstacle avoidance

4.1 Pre-calculated potential fields

When the environment where a robot navigates is of the type of an office or a house, and it is known in advance, then the objects and walls can be represented using polygons.

Figure 6. Representation of the testing environment using polygons

Each polygon consists of a clockwise ordered list of its vertices. Representing the obstacles as polygons makes easier the definition of forbidden areas, which are areas which are not allowed for the robot to enter. They are built by growing the polygons that represent the objects by a distance greater than the radius of the robot, to consider it as a point and not as a dimensioned object (Lozano Pérez 1979). It is possible to create the configuration space in this way when the robot has a round shape. Fig. 6. shows a representation of a polygonal testing environment example testing environment. From the polygonal representation it is found the free space where the robot can navigate with this approach, which is formed by a set of equally spaced cells in which there are not obstacles, as it is shown in Fig. 7.

Figure 7. Representation of the free space using cells

For each cell it is calculated the repulsion forces that each of the obstacles generates, they are added and the resulting force is obtained, Fig. 8 shows the repulsion force map for the environment.

Figure 8. Repulsion force map for the environment shown in Fig. 6

By calculating in advance the repulsion force map liberates the robot's processors to perform other tasks, then knowing the destination the attraction force is calculated in each of the cells and added to the repulsion force calculated before. Figure 9 shows the attraction and repulsions force map, in which a robot navigates from the upper left corner to the lower right one.

Figure 9. Attraction and repulsions force map

The use of this kind of repulsion and attraction force maps improves the performance of the robot, because it is not necessary to calculate for each of the robot positions the repulsion forces on-line.

4.2 Optimization approaches

4.2.1 Objective functions

An objective function for robot navigation should measure the optimality of a path between two points. The main criteria to determine the optimality are: minimum travel distance, and safe obstacle avoidance throughout the path. Then the objective function should provide optimum values (minima or maxima) for the shortest travel paths, with maximum distances to each obstacle in the path. Objective functions are usually constructed by the system developer, according to the navigation conditions: known or unknown obstacles; one or several attraction points; navigation map. To illustrate we present two objective functions which have been succesfully used for local obstacle avoidance.
As mentioned in the introduction Kun Hsiang et al. (1999), reported the development of an autonomous robot navigation scheme based on potential fields and the chamfer distance transform for global path planning in a known environment, and a local fuzzy logic controller to avoid trap situations. The chamfer distance transform produces a matrix where each entry is the distance to the closest obstacle, these distances are used to calculate the repulsion forces exerted by the obstacles on the robot. The attraction force of the goal point is a constant with a user defined magnitude. A fuzzy logic controller based on two objective functions was developed to avoid trap situations where the robot is not able to avoid an obstacle using only the potential field functions. The objective functions measure: the angle between the repulsive force of the closest obstacle and the resultant force (Ec. 22); the distance to the closest obstacle (Eq. 23). The controller tries to maximize the distance to the obstacles. An stop condition is used when the robot reaches the goal.

$$\varphi=\theta_{obs} - \theta \tag{22}$$

where:
θ_{obs} is the angle of the repulsive force;
θ is the angle of the resultant force.

$$diff=M(xi+1, yi+1)- M(xi,yi) \tag{23}$$

where:
M(xi+1, yi+1) is the distance to the closest obstacle at the next position;
M(xi,yi) is the distance to the closest obstacle at the current position.
In Arambula and Padilla (2004) is reported an objective function to evaluate force field configurations which correspond to an optimum robot position (i.e. positions closer to the goal cell which also avoid obstacles). The objective function value of each candidate force field configuration is evaluated with two criteria: minimisation of the error distance E between the robot and the goal cell; and maximisation of the distance d_{min} to the closest obstacle cell. Equation 24 shows the objective function, which produces optimum (minimum) values for minimum E, and maximum d_{min}

$$f(\mathbf{q}) = \begin{cases} \sqrt{E/2} \cdot e^{-d_{\min}} & d_{\min} > 0 \\ 2000 & d_{\min} = 0 \end{cases} \tag{24}$$

where:
d_{min} is the distance to the closest obstacle cell

$$E = |q_{rx} - q_{gx}| + |q_{ry} - q_{gy}|$$

$\mathbf{q}_r$ is a candidate cell for the new robot position;
$\mathbf{q}_g$ is the goal cell;
The construction of the objective function (f) favors robot paths that run away from the obstacles and result in decreasing distance to the goal cell. The case where d_{min}= 0 (which corresponds to a collision) is severely penalised. In Fig. 5a is shown the plot of Eq.24 for: $0<=E<=44$ and $0.1<= d_{min}<=5$.

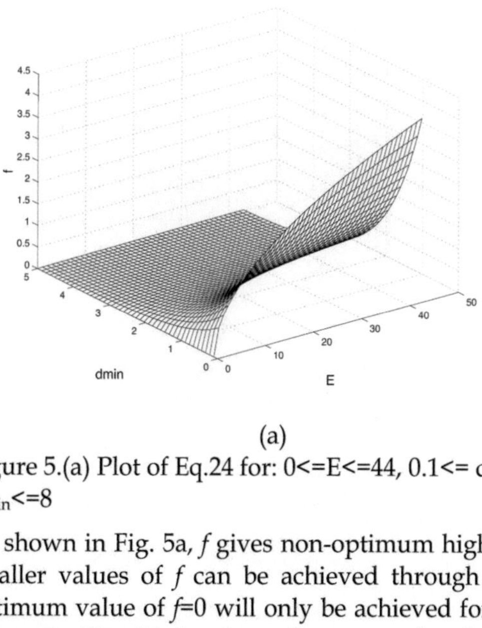

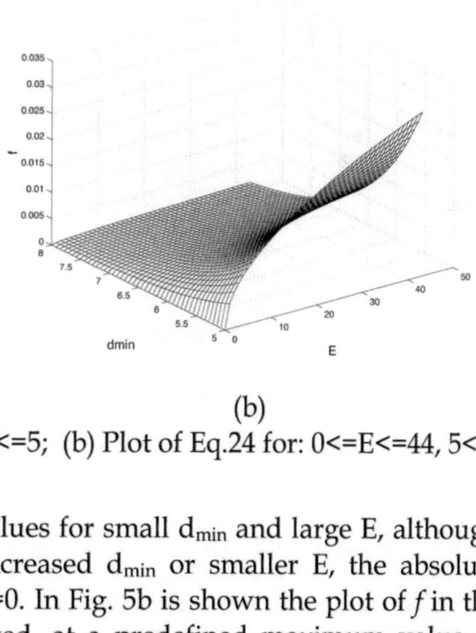

(a) (b)

Figure 5.(a) Plot of Eq.24 for: $0<=E<=44$, $0.1<= d_{min}<=5$; (b) Plot of Eq.24 for: $0<=E<=44$, $5<= d_{min}<=8$

As shown in Fig. 5a, f gives non-optimum high values for small d_{min} and large E, although smaller values of f can be achieved through increased d_{min} or smaller E, the absolute optimum value of f=0 will only be achieved for E=0. In Fig. 5b is shown the plot of f in the range: $0<=E<=44$, $5<=d_{min}<=8$; as can be observed, at a predefined maximum value of d_{min}=5, f still shows an slope which guarantees that optimum values correspond to decreasing E.

4.2.2 Adaptive potential fields

If the robot navigates in an environment with unknown obstacles it is necessary to detect and avoid obstacles as the robot moves towards the goal. In Arámbula and Padilla (2004) was reported an scheme for online obstacle detection. The robot is represented as a particle

R that moves in the configuration space C, modelled as a two dimensional grid, where each cell c_i inside C can be occupied by the robot, the goal or the obstacles. There is also an associated obstacle map M of the same size of C. The obstacle map is initially empty, and it is filled at the positions of the obstacles detected by the robot, as it moves inside C. The goal cell, and 4 auxiliary attraction points exert an attraction force on R given by Eq. 12, while each of the detected obstacle cells exerts repulsion forces given by Eq. 13. For obstacle detection, a $5x5$ grid simulates the robot sensors. When R moves, the positions of the sensors in the mask are updated and used to calculate the distance $d_{\min}$ to the closest detected obstacle (Fig. 10a). A predefined distance is assigned to obstacles outside of the detection mask, as shown in Fig. 10b.

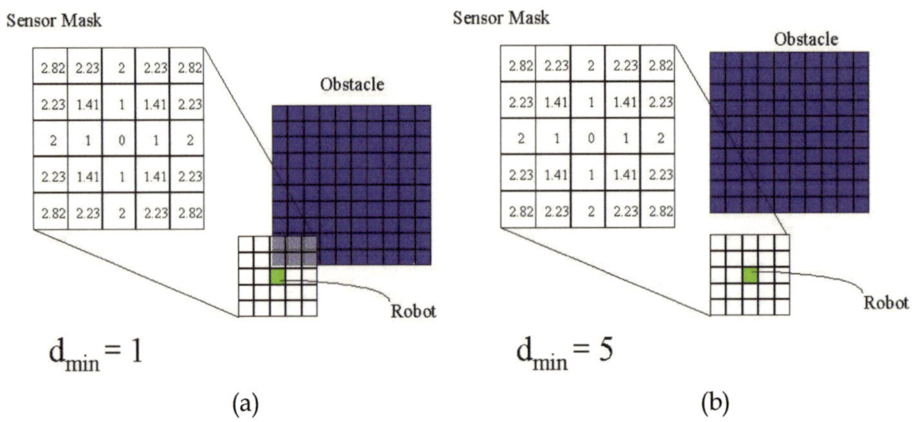

Figure 10. Examples of obstacle sensing. a) The robot detects an obstacle at $d_{\min} = 1$; b) The robot does not detect any obstacle and sets $d_{\min}$ to a predefined value of 5

In order to avoid trap situations or oscillations in the presence of large or closely spaced obstacles (Koren and Borenstein, 1991), 4 auxiliary attraction points have been placed around the goal cell (Fig. 3). Each attraction force $\mathbf{F}_{att}^{i}$ located at cell c_i, depends on the corresponding value of ξ_i, which is automatically adjusted by a genetic algorithm described in the next section. The effect of auxiliary attraction points was evaluated in two modalities: (1) auxiliary points placed at a fixed distance (of 15 cells) from the goal cell; and (2) auxiliary points placed at a variable distance (between 0 and 15 cells), which is adjusted automatically by the GA. Results from both approaches are reported in section 4.3. Use of auxiliary attraction points with a force strength and position automatically adjusted by the GA, allows for the generation of resultant force vectors which enable the robot to avoid large obstacles, as shown in Fig. 12.

4.2.2.1 Adaptive field optimization using genetic algorithms

Genetic algorithms are an efficient technique to optimise difficult functions in large search spaces. By testing populations of solutions represented as strings (called chromosomes) in an iterative process, a GA is able to find a near optimal solution in a robust manner, with the ability to produce a "best guess" from incomplete or noisy data (Goldberg, 1989) . A GA was used to optimise the values of the variables (ξ_i) of 5 attraction points and the values of the variables (η_j) of up to 155 obstacle cells. Each variable has a range of {0, 1000} and was binary coded with 20 bits of resolution in order to maintain a large number of values for the repulsion and attraction forces. A chromosome is formed by concatenation of the 160 binary coded variables. As mentioned above two modalities of the approach were evaluated: (1) with auxiliary attraction points placed at fixed positions, and (2) with auxiliary attraction points placed at variable distance from the goal cell. To implement modality (2), four additional binary variables in the range {0, 15} and coded with 4 bits each, are included in the chromosomes.

The GA searches for optimum values of ξ_i and η_j in a given binary string (chromosome) which move the robot to a position such that *f* (Eq. 24) has a minimum value. Only those η_j which correspond to obstacle cells detected by the robot are used to calculate the force fields given by Eq. 13, the rest of the repulsion weights in the string is ignored. At each generation of the GA, every chromosome in the current population is decoded and the value of Eqs. 12, and 13 is calculated, with this values is calculated the resultant force and the corresponding robot position. This robot position is evaluated with Eq. 24 and assigned a selection probability based on its objective function value (smaller values of the objective function correspond to higher selection probabilities). Each chromosome in the current population is assigned a number of copies with probability P_s using stochastic universal sampling (SUS) for selection and the ranking method to assign probabilities (Chipperfield et. al, 1995). Single point crossover is applied to the copies (offspring) with a probability of 0.6, mutation is applied to each string with a probability of 0.01 per bit. Finally, the next generation of the GA is formed using fitness based reinsertion with a generation gap of 0.8. This process continues until the robot reaches the goal cell or 200 generations (robot steps) are completed. Below is shown the pseudocode of the GA for robot navigation.

```
Pop= Random initialisation of 50 binary chromosomes
Step_count=0
While step_count<200
        Calculate Fatt (Eq.8), Frep (Eq.9), and the next robot position for each chromosome
        in Pop;
        Calculate f (Eq.10) for each robot position ;
        Assign a probability of selection (Ps) to each chromosome using the ranking
        method;
        Assign copies to each chromosome using SUS with probability of selection Ps;
        Mutate and cross the copies (offspring);
        Reinsert offspring in Pop with a generation gap of 0.8;
        Calculate f for Pop;
        Select best chromosome and move the robot to the corresponding position;
```

```
Increment step_count;
If(d=0)
        finish
end
```

4.3 Potential Field Optimization in a Partially known environment: Experiments and results

The genetic algorithm described above was implemented in Matlab using the GA toolbox developed at the University of Sheffield (Chipperfield et. al, 1995). A cell map of 40x40 cells simulating a five-room floor was used for evaluation. Random obstacle distributions were used, as shown in Fig. 11.

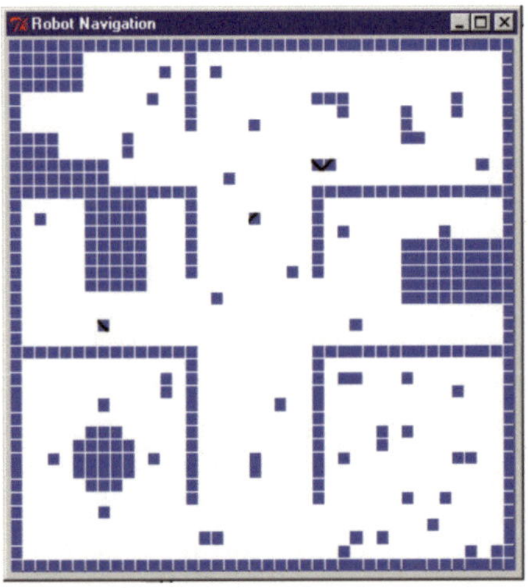

Figure 11. Cell map simulating a five-room floor with random obstacle

Ten experiments were performed, the start and goal positions for each experiment are shown in table 1, the origin is placed at the top-left corner of the cell map. Two intermediate goal points have been used to guide the robot through the corridor corner as well as through the door of the appropriate room. The positions of the intermediate goal points are also shown in table 1. The robot travels from the start position to each successive intermediate goal point and to the final goal point.

Two modalities of the navigation algorithm were evaluated: (1) with auxiliary attraction point placed at fixed positions, and (2) with auxiliary attraction points placed at variable distance from each goal cell. In table 2 are shown the results of the 20 experiments performed, the first column shows the experiment number corresponding to table 1. Columns two and three show respectively, the total distance traveled by the robot (measured in cells), and the deviation (as a percentage) from the optimum shortest path. Auxiliary attraction points were placed at a fixed distance of five cells from each goal position. Columns four and five show respectively, the total distance traveled by the robot and the deviation percentage, for auxiliary attraction points placed at a variable distance, which is automatically adjusted by the GA.

Exp. No	(start)-(goal)	intermediate goal 1	intermediate goal 2
1	(34, 9)-(11, 3)	(20,10)	(15,10)
2	(34, 9)-(13, 14)	(20,10)	(15,21)
3	(34, 9)-(3, 26)	(20,10)	(15,38)
4	(34, 9)-(36, 27)	(20,10)	(25,38)
5	(34, 9)-(37, 14)	(20,10)	(25,22)
6	(34, 4)-(3, 6)	(20,10)	(15,10)
7	(34, 4)-(3, 14)	(20,10)	(15,21)
8	(34, 4)-(12, 26)	(20,10)	(15,38)
9	(34, 4)-(30, 30)	(20,10)	(25,38)
10	(34, 4)-(38, 22)	(20,10)	(25,22)

Table 1. Start-goal and intermediate goal positions of each experiment

Exp.No.	Total distance 1 (cells)	Deviation from optimum 1 (%)	Total distance 2 (cells)	Deviation from optimum 2 (%)
1	34	17.2	48	65.5
2	44	29.4	34	0
3	collision	Collision	69	6.1
4	68	21.4	81	44.6
5	46	15.0	65	62.5
6	40	21.2	43	30.3
7	49	11.4	48	9.0
8	70	27.2	81	47.3
9	75	41.5	70	32.0
10	48	17.0	72	75.6
		Average: 22.3%		Average: 37.3%

Table 2. Experiment results: **Total distance 1**, and **Deviation from** optimum **1** obtained with auxiliary attraction points placed at fixed distance (five cells) from the goal; **Total distance 2**, and **Deviation from optimum 2** obtained with auxiliary attraction points placed at variable distance from the goal

From the results shown in table 2, the average deviation from the optimum path length is larger (37% vs. 22%) for the second approach, this is most likely because we have a larger and more complex search space which results in a higher probability of suboptimal points being chosen by the GA. However the second approach was able to produce a feasible path without collisions for all the experiments. In contrast the first approach (using fixed auxiliary attraction points) was not able to reach the goal for experiment 3. In Fig. 12 are shown five paths produced by the second approach. The average time for path completion on a Pentium III PC at 750MHz is 115s with an average path length of 56 cells (i.e.2.05 s/step).

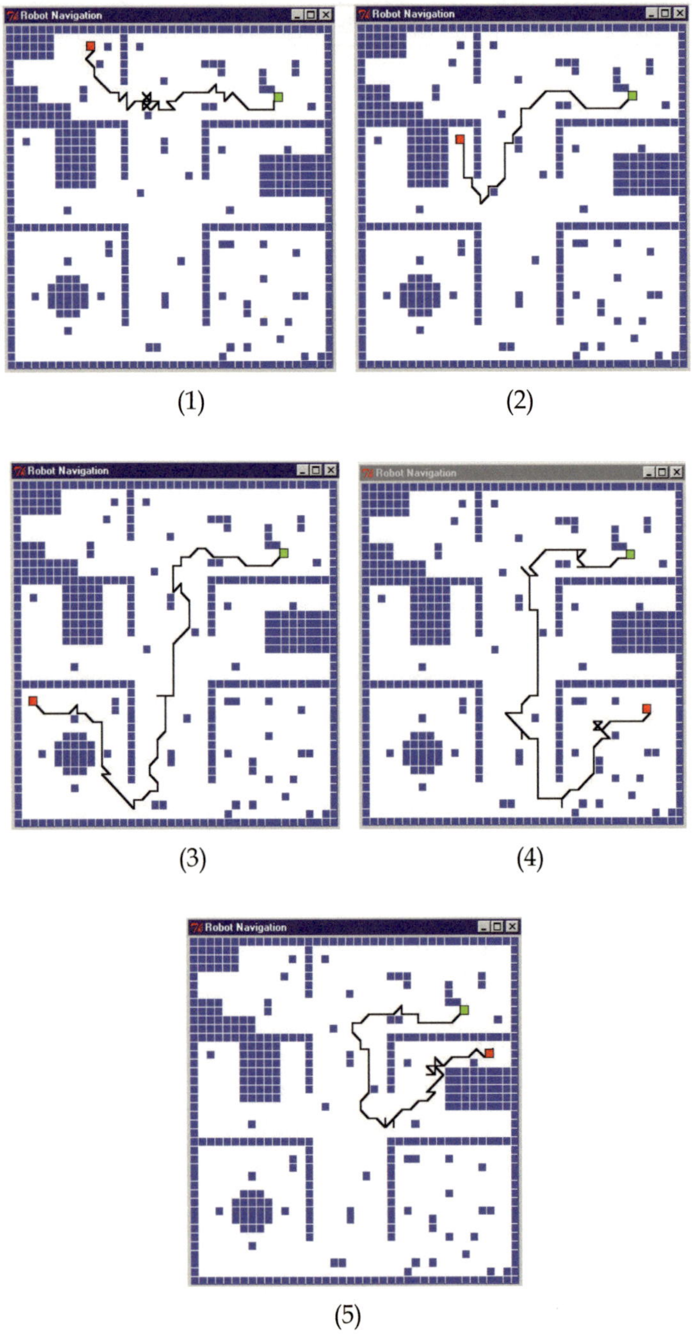

Figure 12. Paths produced by the navigation algorithm, using auxiliary attraction points placed at variable distance from the goal cell. Start-goal positions are as given in table 1

5. Hybrid Approaches to Recover from Local Minima

Hybrid approaches can be used to modify a potential field configuration in which a local minimum has been detected, for example Fig. 13, shows a robot that found an obstacle in the middle of the path between the origin and the goal and it is oscillating back and forth, due to the repulsion and attraction forces. First the repulsion forces repealed the robot from the obstacle, and when the robot is a little far away from it, the attraction force pushed it back to the obstacle, and then the repulsion force acts again repeating the whole process.
The potential field configuration can be modified by the addition of attraction forces that allow the robot to exit the local minima. By using the position of the known obstacle, additional attraction forces are added in places that will take the robot out of the local minimum. Usually additional attraction points are added in some of the vertices of the obstacles, as is shown in Fig. 14.

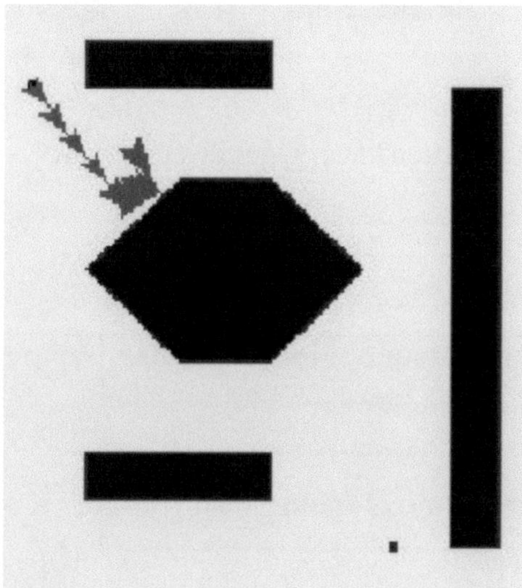

Figure 13. The robot is stuck in a local minimum

Basically the hybrid approach finds the obstacle in which the robot got stuck, then using its vertices $V=(v_1,v_2,...,v_N)$ it selects the vertices v_i, $v_{i+1},...,v_{k-1},v_k$, where v_i is the closest vertex from the stuck point, v_{i+1} is the clockwise vertex from v_i and v_k is the closest vertex to the goal. Using these selected vertices the approach places a new goal to reach at v_{i+1} disabling the original goal, after the goal in v_{i+1} is reached a new goal is issued at the next selected vertex and so on until v_k is reached. Finally the original goal is set again. In the Fig. 14 we can see that four additional attraction forces where added to the space to take the robot away from the local minimum.
There are cases in which this approach does not work because there are obstacles so large that can generate several local minima in which the robot can get stuck again. In this case another approach is to have a robotics behavioral architecture that consists of several behaviors in parallel (Arkin 1998), each of the behaviors generates an output according to the readings of the sensors connected to them and its internal state. Then a referee selects the output of one of the behaviors according to a selection mechanism and sends it to the robot's actuators. Figure 15 shows this type of architecture with two behaviors, one with potential fields and the other with an state machine.

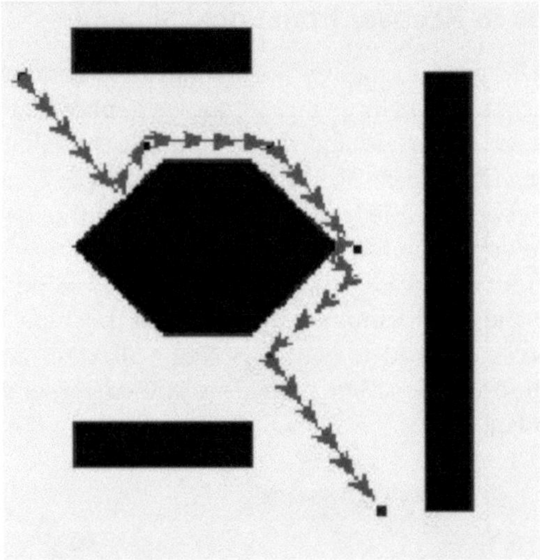

Figure 14. Four additional attraction forces are added to the environment to take the robot out of the local minima

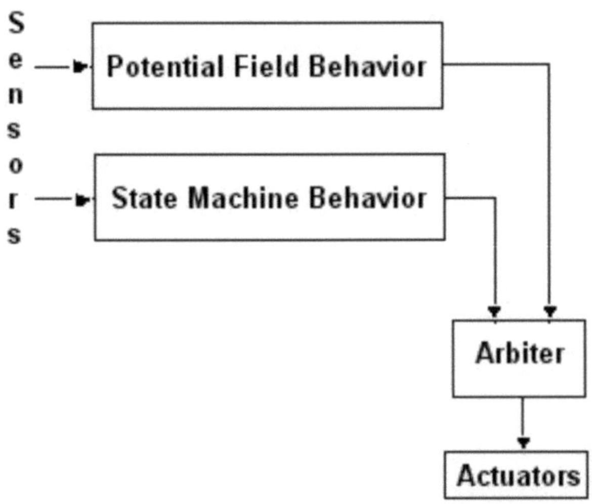

Figure 15. Behavior architecture used to control the movements of a robot

The function of the state machine behavior is to detect when the robot gets stuck in a local minima and take it out of it. After it takes the robot out of the local minima the referee selects again the potential field behavior. Figure 16 shows the behavior that the robot follows to avoid an obstacle. When the robots senses an obstacle in the left or in the right it will go backward first and then turn to the right or to the left accordingly, if it finds the obstacle in front of it, it goes backward then turns to the left 90 degrees, goes forward and then turns to the right and forward again. This simple behavior allows the robot to avoid local minima.

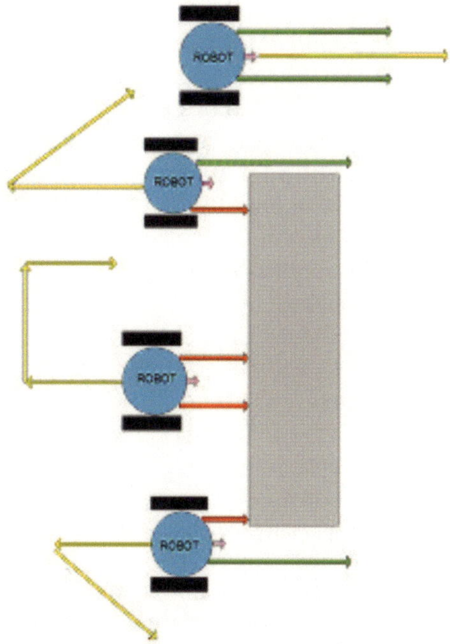

Figure 16. Robot behavior to take a robot out of a local minimum

6. References

Arámbula Cosío F. and Padilla Castañeda M.A. (2004), Autonomous robot navigation using adaptive potential fields, *Mathematical and Computer Modelling*, Volume 40, Issue 9-10, 1141-1156 .

Arkin R.C. (1998), Behavior-Based Robotics, Cambridge, MA: The MIT Press.

Borenstein J. and Koren Y. (1991), The vector field histogram-fast obstacle avoidance for mobile robots, *IEEE Transactions on Robotics and Automation 7*, 278-288.

Canny J.F. and Lin M.C. (1990), An opportunistic global path planner, *IEEE Int. Conf. on Robotics and Automation*, 1554-1559.

Chipperfield A., Fleming P., Pohlheim H. and Fonseca C. (1995), *Genetic Algorithm Toolbox for Use with* MATLAB *User's Guide,* Automatic Control and Systems Engineering, University of Sheffield, U.K.

Connolly C.I., Burns J.B., and Weiss R (1990), Path planning using Laplace's equation, *Proc. lEEEConf. on Robotics and Automation,* pp 2102-2106, Cincinnati, OH.

Ge S.S. And Cui Y.J. (2002), "Dynamic motion planning for mobile robots using potential field method", Autonomous Robots, 13, 207-222.

Ge S.S. and Cui Y.J. (2000), New Potential Functions for Mobile Robot Path Planning, *IEEE Transactions on Robotics and Automation*, vol. 16, No. 5, pp.615-620.

Goldberg D.E. (1989), *Genetic Algorithms in Search, Optimisation, and Machine Learning,* Addison-Wesley, MA.

Grefenstette J. and Schultz A.C. (1994), An evolutionary approach to learning in robots, In *Proceedings of the Machine Learning Workshop on Robot Learning, Int. Conf. on Robot Learning,* pp. 65-72, New Brunswick, N J.

Guldner J., Utkin V., Hashimoto H. (1997), Robot Obstacle Avoidance in n-Dimensional Space Using Planar Harmonic Artificial Potential Fields, *Journal of Dynamic Systems, Measurement, and Control*, vol. 119, pp. 160-166

Keymeulen D. and Decuyper J. (1994), The fluid dynamics applied to mobile robot motion: the stream field method, *IEEE Int. Conf. on Robotics and Automation*, 378-385.

Khatib O. (1990), Real-time obstacle avoidance for manipulators and mobile robots, In *Autonomous Robot Vehicles,* (Edited by I.J. Cox and G.T. Wilfong), pp. 396-404, Springer-Verlag.

Koren Y., Borenstein J., Potential field methods and their inherent limitations for mobile robot navigation. *In: Proceedings of the IEEE Int. Conf. on Robotics and Automation,* 1398-1404 (1991).

Kun-Hsiang Wu, Chin-Hsing Chen, and Juing-Ming Ko, Path planning and prototype design of an AGV. *Mathematical and Computer Modelling*, 30, 147-167 (1999).

Lozano-Pérez T. (1979), An algorithm for planning collision-free paths among polyhedral obstacles, *Communications of the ACM*, vol.~22 pp.~560--570.

Masoud A.A., Masoud S.A., and Bayoumi M.M. (1994), Robot navigation using a pressure generated mechanical stress field: the biharmonic potential approach, *IEEE Int. Conf. on Robotics and Automation*, 124-129.

McFetridge L. and Ibrahim M.Y. (1998), New technique of mobile robot navigation using a hybrid adaptive fuzzypotential field approach, *Computers Ind. Engng.* 35 (3-4), 471-474.

Schmidt G.K. and Azarm K. (1992), Mobile robot navigation in a dynamic world using an unsteady diffusion equation strategy, *IEEE/RSJ Int. Conf. on Intelligent Robots and Systems*, 642-647.

Shimoda S., Kuroda Y., Iagnemma K., (2005), Potential field navigation of high speed unmanned ground vehicles on uneven terrain, *IEEE Int. Conf. on Robotics and Automation*, Barcelona, Spain, 2828-2833.

Utkin V.I., Drakunov S., Hashimoto H., and Harashima F. (1991), Robot path obstacle avoidance control via sliding mode approach, *Proc. IEEE/RSJ Int. Workshop on Intelligent Robots and Systems,* pp. 1287-1290, Osaka, Japan.

Vadakkepat P., Kay Chen Tan, Wang Ming-Liang, Evolutionary artificial potential fields and their application in real time robot path planning. *Proc. of the 2000 Congress on Evolutionary Computation,* pp. 256-263, July (2000).

Barraquand, J., Langlois, B., Latombe, J.C., Numerical Potential Field Techniques for Robot Path Planning, *IEEE Transactions on Systems, Man and Cybernetics,* Vol. 22, No. 2, pp. 224-241, March/April, 1992.

Wang, Y., Chirikjian, G., A New Potential Field Method for Robot Path Planning, *Proc. IEEE Int. Conf. on Robotics and Automation*, pp. 977-982, 2000.

2

Foundations of Parameterized Trajectories-based Space Transformations for Obstacle Avoidance

J.L. Blanco, J. González and J.A. Fernández-Madrigal
University of Málaga
Spain

1. Introduction

It is essential for any approach to motion planning to account for some spatial representation of obstacles where collision-free paths could be found efficiently. This problem has been extensively studied by the robotics community and has traditionally led to two different research areas. On the one hand we have motion planning approaches, where an optimal path is computed for a known scenario and a target location. The Configuration Space (C-Space) (Lozano-Perez, 1983) has been successfully employed as representation in this scope. In C-Space the robot can be represented as a single point in the high-dimensionality space of its degrees of freedom. On the other hand, some navigation approaches deal with unknown or dynamic scenarios, where motion commands must be periodically computed in real-time during navigation (that is, there is no planning). Under these approaches, called *reactive* or obstacle avoidance, the navigator procedure can be conveniently seen as a mapping between sensor readings and motor action (Arkin, 1998). Although reactive methods are quite efficient and have simple implementations, many of them do not work properly in practical applications since they often rely on too restrictive assumptions, like a point or circular representation of robots or considering movements in any direction, that is, ignoring kinematic restrictions. C-Space is not an appropriate space representation for reactive methods due to its complexity, which prohibits real-time execution. Hence simplifications of C-Space have been proposed specifically for reactive methods. Finally, combinations of the two above approaches have also been proposed (Khatib et al., 1997; Lamiraux et al., 2004; Quinlan and Khatib, 1993), which usually start computing a planned path based on a known static map, and then deform it dynamically to avoid collision with unexpected obstacles. These hybrid approaches successfully solve the navigation problem in many situations, but purely reactive methods are still required for partially known or highly dynamic scenarios, where an a priori planned path may need excessive deformation to be successfully constructed by a hybrid method.
In this work we address purely reactive methods exclusively, concretely, the problem of reactively driving a kinematically-constrained, any-shape mobile robots in a planar scenario. This problem requires finding movements that approach the target location while avoiding obstacles and fulfilling the robot kinematic restrictions. Our main contribution is related to the process for detecting free-space around the robot, which is the basis for a reactive

navigator to decide the best instantaneous motor action. For this task, existing methods consider certain families of simple paths for measuring obstacle distances (which is equivalent to sampling the free-space). These families of paths, that we will call *path models*, must be considered not as planned paths but as artifacts for taking nearby obstacles into account. All existent reactive methods use path models that are an extension of the robot short-term action, as illustrated in Fig. 1: for holonomic robots that can freely move at any direction, straight lines are used, while for non-holonomic robots virtually all methods employ circular arcs.

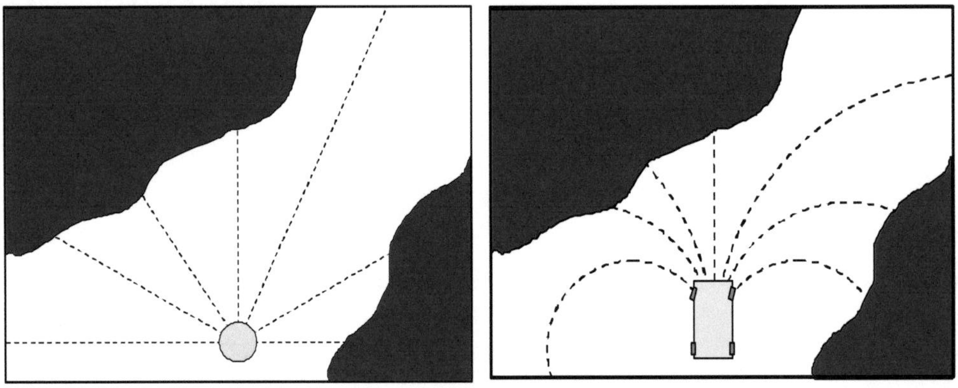

(a) (b)

Figure 1. (a) Holonomic robots can move following straight lines without restrictions, while (b) realistic non-holonomic robot can only move following sequences of circular arcs

We claim that straight and circular paths, used in previous reactive methods, are just two out from the infinity of path models that can be followed by a robot in a memory-less system, that is, reactively. It is clear that considering other path models is more appropriate to sample the free-space than using the classic straight or circular models only. We shed light into this issue through the example in Fig. 2, where a robot (reactively) looks for possible movements. If we employ a single circular path model for sampling obstacles as in Fig. 2(a), it is very likely that the obstacle avoidance method overlooks many good potential movements - notice that any reactive method must decide according solely to the information that path models provide about obstacles. In contrast, using a diversity of path models, as the example shown in Fig. 2(b), makes much easier to find better collision free movements. This is one of the distinctive features of our approach: the capability to handle a variety of path models simultaneously.

A fundamental point in the process of using path models to sample obstacles is that *not any arbitrary* path model is suitable for this purpose, since it must assure that the robot kinematic constrains are fulfilled while still being able of following the paths in a memory-less fashion, i.e. by a reactive method. It is worth discussing the properties of trajectories that fulfill this condition, called *consistent reactive trajectories* in Section 2.2, since it is an important reflection that cannot be found in previous works.

To motivate the discussion, consider the robot in Fig. 3(a), which must decide its next movement from a family of circular arcs, each one giving a prediction for the distance-to-obstacles. Since reactive navigation is a discrete time process, the decision will be taken iteratively, in a timely fashion, though at each time step the family of paths will be considered starting at the *current pose* of the robot. The central issue here is that, implicitly, it is assumed that if the robot chooses one path at some instant of time, at the next time step it will have the possibility of *continuing along the same trajectory*. Otherwise, the prediction of distance-to-obstacles would be useless since foreseen trajectories can not be actually followed. In the case of circular arcs, this property indeed holds, as illustrated in the example in Fig. 3(b). The main contribution of the present work is a detailed formalization of this and other properties that need to hold for a path model being applicable to obstacle avoidance.

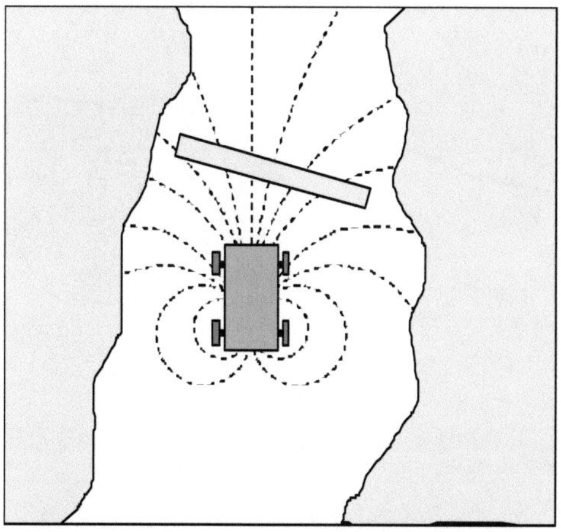

(a)

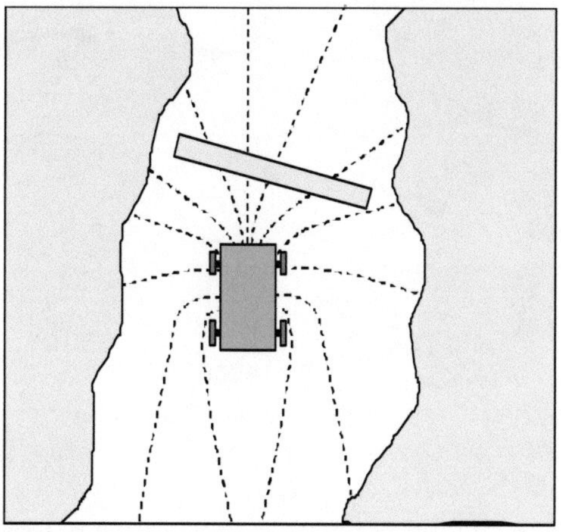

(b)

Figure 2. Reactive methods take obstacles into account through a family of paths, typically circular arcs (a). However, we claim that other possibilities may be useful for finding good collision-free movements, as the path family shown in (b)

As detailed in previous works (Blanco et al., 2006; Blanco et al., 2008), we decouple the problems of kinematic restrictions and obstacle avoidance by using path models to transform kinematic-compliant paths and real-world obstacles into a lower complexity space, a Trajectory Parameter Space (TP-Space for short). The transformation is defined in such a way that the robot can be considered as a free-flying-point in the TP-Space since its shape and kinematic restrictions are already embedded into the transformation. We can then entrust the obstacle avoidance task to any standard holonomic method operating in the transformed space.

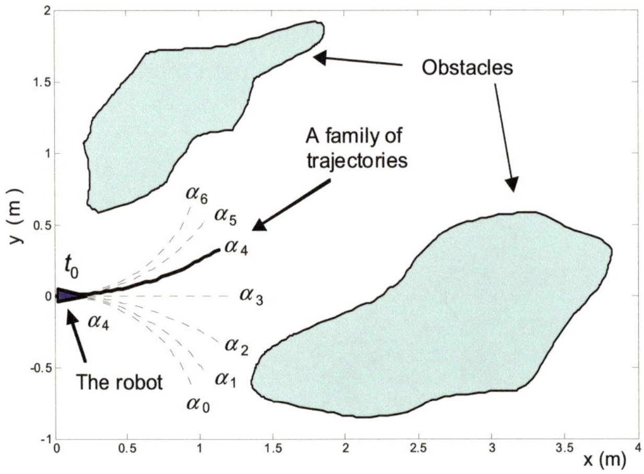

(a)

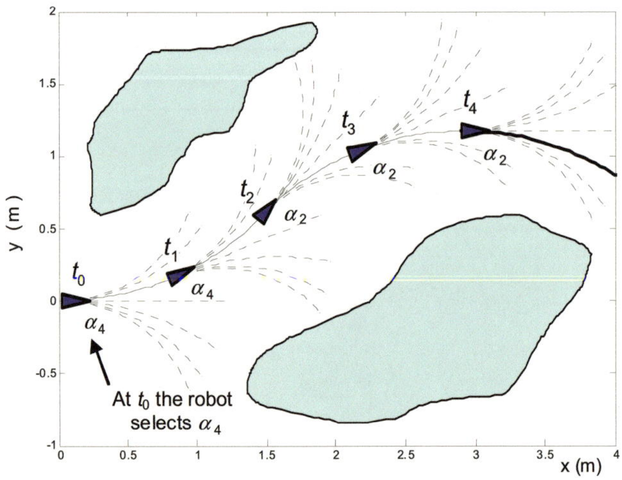

(b)

Figure 3. A schematic representation of the process involved in reactive navigation. At each time step, the robot employs a family of paths to sample the obstacles in the environment, and then chooses the most convenient action according to that information. It must be highlighted the important implicit assumption in the process, that the robot will be able to continue trajectories chosen at previous time steps. Since this does not hold in general for all path models, we develop in this work a template for path models that are proven to fulfill this requirement

This idea was firstly introduced by Minguez and Montano in (Minguez et al., 2002), and has subsequently evolved in a series of works (Minguez et al., 2006; Minguez and Montano, 2008). Our framework can be seen as an extension of (Minguez et al., 2002) since multiple space transformations can be defined instead of just the one corresponding to circular arcs. We allow any number of space transformations by generalizing path models through Parameterized Trajectory Generators (PTGs), which are described in detail in subsequent sections. For further details on how our framework can be integrated into a real navigation system, and for extensive experimental results from both simulations and real robots, the reader is referred to our previous works (Blanco et al., 2006; Blanco et al., 2008).

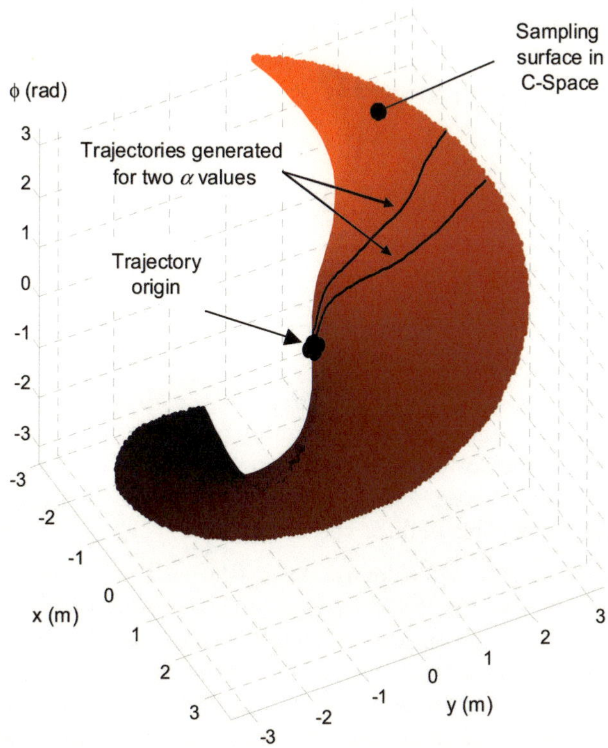

Figure 4. The simultaneous representation of all the trajectories of a path family in C-Space generates a 3D surface which comprises all the potential poses the robot can reach using the path family

2. Space Transformations for Obstacle Avoidance

2.1 Overview

Although not always put explicitly, any reactive navigation algorithm relies on the calculation of distance-to-obstacles to provide the robot with information for choosing the next movement. To the best of our knowledge, all previous (reactive) works make an implicit assumption that has never been questioned: distance-to-obstacles (i.e. collision distances) are computed by means of a single fixed path model: either straight or circular, commonly depending on the robot being holonomic or not. Distances are then taken along those 2D paths, though robot paths are actually defined as continuous sequences of locations and orientations, that is, as three-dimensional curves in C-Space - refer to Fig. 4. We propose instead to define distance-to-obstacles directly in C-Space, as described next.

If all the paths from a given path model are represented in C-Space simultaneously we obtain a 3D surface, as the example in Fig. 4. We will refer to these surfaces as *sampling surfaces*, since distance-to-obstacle can be computed by measuring the distance from the origin to the intersection of those surfaces with C-Obstacles. Next we can *straighten out* the surface into a lower dimensionality space where obstacle avoidance becomes easier, that is, a TP-Space. In this process the topology of the surface is not modified. Since we are proposing a diversity of path models to be used simultaneously, we will have different associated sampling surfaces in C-Space to compute distance-to-obstacles. The whole process is illustrated in Fig. 5.

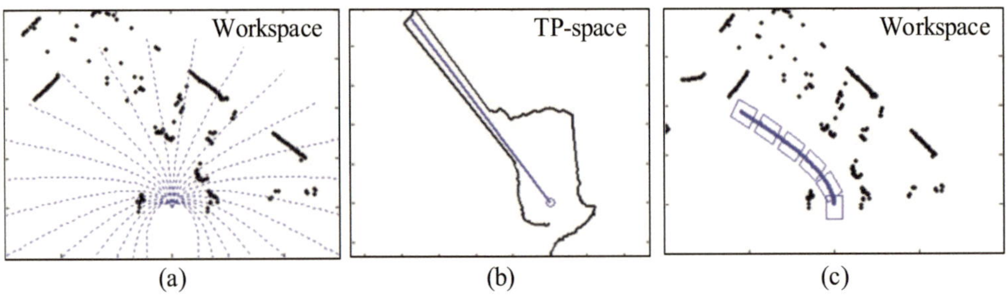

Figure 5. The process to apply simple obstacle avoidance methods to any-shape, non-holonomic robots comprises these steps: (a) A family of path is used to sample distance-to-obstacles, which gives us the obstacles in the transformed space (TP-Space), where (b) the obstacle avoidance method chooses a preferred direction. This straight line in TP-Space actually corresponds to a feasible path, as shown in (c)

We define a TP-Space as any two-dimensional space where each point corresponds to a robot pose in a sampling surface. It is convenient to consider points in a TP-Space by their polar coordinates: an angular component α and a distance d. In this way the angular coordinate has a closed range of possible values. The mapping between a TP-Space and a sampling surface is carried out by selecting an individual trajectory out from the family using the coordinate, while d establishes the distance of the pose along that selected trajectory.

This idea of applying obstacle avoidance in a transformed space was introduced in (Minguez et al., 2002), where the authors employed the Euclidean distance in the 2D plane, disregarding the robot orientation, as the distance measure for d. Alternatively, we measure distances through a non-Euclidean metric, directly along C-Space sampling surfaces. This has the advantage of taking into account robot turns, thus providing a more realistic estimate of how much a robot needs to move to follow a given trajectory. The region of interest in TP-space is a circle centered at the origin and of radius R_m (a constant that settles the collision avoidance maximum foresee range). We will refer to the TP-space domain as the 2D space $\mathbb{S}\times D$, with $\mathbb{S}=]-\pi,\pi]$ and $D=[0,R_m]$. Note as well that the transformation is applied at each iteration of the navigation process, thus for all our derivations the robot is always at the origin.

2.2 Definitions

We define a 2D robot trajectory for a given parameter value as:

$$\mathbf{q}(\alpha,t)=\begin{bmatrix} x(\alpha,t) \\ y(\alpha,t) \\ \phi(\alpha,t) \end{bmatrix} \qquad ,t\geq 0 \tag{1}$$

Since we address PTGs for realistic robots subject to non-holonomic constrains, trajectories are defined as the integration of their time derivative $\dot{\mathbf{q}}(\alpha,t)$, that is,

$$\mathbf{q}(\alpha,t)=\int_0^t \dot{\mathbf{q}}(\alpha,\tau)\,d\tau \tag{2}$$

where it applies the initial condition q(, 0) = 0 for any . Note from Eq. (4) that we define the transformed space in terms of distance d rather than time t, in which the kinematic equations are naturally defined. The reason for this change of variable is that we are interested in the geometry of paths, which remains unmodified if the velocity vector u(·) is multiplied by any positive scalar, an operation equivalent to modifying the speed of the robot dynamically. For example, it is common in navigation algorithms to adapt the robot velocities to the clearness of its surroundings.
Therefore, we define a PTG as:

$$PTG(\alpha,d)\triangleq\mathbf{q}\left(\alpha,\mu_\alpha^{-1}(d)\right) \tag{3}$$

where the function $\mu_\alpha^{-1}(d)$, mapping distances *d* to times *t*, is not relevant at this point and will be introduced later on. Thus, a PTG is a mapping of TP-Space points to a subset of C-Space:

$$\begin{array}{rccc} PTG: & \mathbb{S}\times D & \mapsto & \mathbb{R}^2\times\mathbb{S} \\ & (\alpha,d) & \mapsto & \mathbf{q} \end{array} \tag{4}$$

In the common case of car-like or differentially-driven robots, the derivatives in Eq. (2) are given by the same set of kinematic equations:

$$\begin{aligned} \dot{\mathbf{q}}(\alpha,t) &= \mathbf{J}(\mathbf{q}(\alpha,t))\mathbf{u}(\alpha,t) \\ &= \begin{bmatrix} \cos\phi(\alpha,t) & 0 \\ \sin\phi(\alpha,t) & 0 \\ 0 & 1 \end{bmatrix} \begin{bmatrix} v(\alpha,t) \\ \omega(\alpha,t) \end{bmatrix} \end{aligned} \tag{5}$$

Here u is the vector comprising the linear (v) and angular (ω) velocities of the robot at each instant of time t and for each value of the PTG parameter. The freedom for designing different PTGs is therefore bound up with the availability of different implementations of the actuation vector $\mathbf{u}(\alpha,t)$.

In despite of the fact *any* function $\mathbf{u}(\cdot)$ represents a kinematically valid path for a robot, which follows from Eq. (5) by definition, the present work is built upon the realization that not any arbitrary function leads to *valid* space transformations for obstacle avoidance methods. We specify next when such a transformation is valid for our purposes.

Definition. A space transformation between C-space and TP-space is *valid* when it fulfils the following conditions:

- **C1**. It generates *consistent reactive trajectories*. All path models are not applicable to reactive navigation because of the memoryless nature of the movement decision process, as discussed in section 1.
- **C2**. It is *WS-bijective*. For each WS location (x,y), at most one trajectory can exist taking the robot to it, regardless the orientation. Otherwise, the target position would be seen at two different directions (straight lines) in TP-Space – recall that a PTG maps straight lines of the TP-Space into trajectories of the C-Space.
- **C3**. It is *continuous*. Together with the last restriction, this condition assures that transformations do not modify the topology of the real workspace around the robot.

These three conditions hold for the case of paths that are circular arcs. The main contribution of the present work is the following theorem, which proves that a broader variety of *valid* PTGs indeed exist and is suitable to reactive navigation.

Theorem 1. *A sufficient, but not necessary, condition for a PTG to be valid is that its velocity vector is of the form:*

$$\mathbf{u}(\alpha,t)=\begin{bmatrix} v_m \cdot f_v\left(a\alpha+b\phi(\alpha,t)\right) \\ \omega_m \cdot \left(a\alpha+b\phi(\alpha,t)\right) \end{bmatrix} \tag{6}$$

where v_m *and* ω_m *settle the desired maximum linear and angular velocities in absolute value, respectively,* $f_v(\alpha,t)$ *is any Lipschitz continuous function which evaluates to non-zero over the whole domain, and* a, b *are arbitrary constants with the restrictions* $0<|a/b|\leq 1$ *and* $b<0$. *Furthermore, the velocity vector becomes fully defined by just specifying its value for* $t=0$.

The following section is devoted to a detailed analysis of PTGs in this form and to prove our claim of them always are valid in the sense that they fulfill all the conditions listed above.

3. Proofs

We start by defining the function $\mu_\alpha(t)$ as the distance traveled by the robot along a given trajectory α in C-space from the origin and until the instant *t*, that is:

$$\mu_\alpha(t)=\int_0^t\left\|\frac{\partial\mathbf{q}(\alpha,t)}{\partial t}\right\|d\tau \tag{7}$$

where the norm could be the Euclidean distance, though we will employ here a custom metric introduced in (Blanco et al., 2008), which accounts for robot turns through a constant ρ that roughly represents the robot radius, leading to:

$$\mu_\alpha(t) = \int_0^t \left(v(\alpha,t)^2 + \rho^2 \omega(\alpha,t)^2 \right)^{\frac{1}{2}} d\tau \tag{8}$$

Then, we can state the following lemma about the existence of $\mu_\alpha^{-1}(d)$, required in Eq. (3) for the definition of PTGs.

Lemma 1. *The function* $\mu_\alpha : t \mapsto d$ *is continuous and its inverse* $\mu_\alpha^{-1} : d \mapsto t$ *is well-defined for any* $t \geq 0$.

Proof. The first part, proving the continuity of $\mu_\alpha(t)$ is trivial since the function is defined as an integral, therefore it is differentiable. Next, it can be seen that the function is strictly increasing due to its derivative being the norm of $\dot{\mathbf{q}}$, which in general is non-negative, but given the hypothesis from theorem 1 of f_v evaluating to non-zero over all the domain, the case $\dot{\mathbf{q}} = 0$ can be ruled out. Being continuous and strictly-increasing $\mu_\alpha(t)$ becomes bijective for any $t \geq 0$ thus its inverse is well-defined.

An important feature of any valid PTG is that different values of α must generate unique trajectories (see condition C2), which is assured by the following lemma.

Lemma 2. *Provided* $b < 0$ *and* $0 < |a/b| \leq 1$, *then each value* $\alpha \in \mathbb{S}$ *determines a unique trajectory passing through the origin with its heading tending to* $-\alpha a / b$ *as* $t \to \infty$.

Proof. Since $\dot{\mathbf{q}}(\alpha,t)$ is Lipschitz continuous, and given the initial conditions $\mathbf{q}(\alpha,t) = \mathbf{0}$ for any value of α, there exists only one trajectory for each α value (Evans and Gariepy, 1992), which is determined by the value of $\dot{\mathbf{q}}(\alpha,0) = \left[v(\alpha,0) \quad \omega(\alpha,0) \right]^T$. From the hypotheses of theorem 1 it easily follows $\omega(\alpha_1,0) \neq \omega(\alpha_2,0)$ for any $\alpha_1 \neq \alpha_2$ as long as a $a \neq 0$ (refer to Eq. (6)), thus the uniqueness of each trajectory is assured.

Regarding the limit of the robot heading $\phi(\alpha,t)$, we can solve the differential equation of the kinematic model in Eq. (5) for this term, that is:

$$\begin{aligned} \dot{\phi}(\alpha,t) &= \omega(\alpha,t) \\ &= \omega_m \cdot \left(a\alpha + b\phi(\alpha,t) \right) \end{aligned} \tag{9}$$

which can be straightforwardly solved giving us:

$$\phi(\alpha,t) = -\alpha \frac{a}{b} \left(1 - e^{\omega_m b t} \right) \tag{10}$$

The parameter b determines the evolution of the heading over time. Since the robot heading must be bounded to the domain of $\mathbb{S}$, we discard the values $b > 0$. The case $b = 0$ must be avoided as well since in that case Eq. (10) is not defined. Therefore, for the valid values $b < 0$, the heading converges to:

$$\lim_{t\to\infty} \phi(\alpha,t) = -\alpha\frac{a}{b} \tag{11}$$

Notice the condition $0 < |a/b| \leq 1$ assures $\phi(\alpha,t)$ will always remain within its valid domain $\mathbb{S}$.

We address next the fundamental property of generated paths being consistent reactive trajectories - as stated by condition C1. The geometrical meaning of this property was discussed in section 1, recall Fig. 3 , and is now stated formally as follows.

Lemma 3. *For any* $\alpha \in \mathbb{S}$ *and* $t_0 \geq 0$, *there exists a function* $A(\alpha,t_0) : \mathbb{S}\times\mathbb{R}^{0+} \mapsto \mathbb{S}$ *such as:*

$$\mathbf{q}(\alpha,t_0+t) = \mathbf{q}(\alpha,t_0) \oplus \mathbf{q}(\alpha',t) \qquad ,\forall t \geq 0 \tag{12}$$

with $\alpha' = A(\alpha,t_0)$ *and where the* $\oplus$ *operator stands for pose composition (Smith et al., 1988).*

Proof. It can be trivially shown that this statement holds for $t = 0$, when Eq. (12) becomes:

$$\begin{aligned}\mathbf{q}(\alpha,t_0) &= \mathbf{q}(\alpha,t_0) \oplus \mathbf{q}(\alpha',0) \\ &= \mathbf{q}(\alpha,t_0) \oplus \mathbf{0} \\ &= \mathbf{q}(\alpha,t_0)\end{aligned} \tag{13}$$

Now, since both trajectories $\mathbf{q}(\alpha,t_0+t)$ and $\mathbf{q}(\alpha,t_0) \oplus \mathbf{q}(\alpha',t)$ pass through a common point in C-space at $t = 0$, it is enough to prove that their derivatives $\dot{\mathbf{q}}$ are identical at that instant for Lemma 2 to imply that both trajectories coincide for any $t > 0$. Taking into account the change of coordinates introduced by the pose composition operator, the condition of both time derivatives $\dot{\mathbf{q}}(\cdot)$ must coincide amounts to their velocity vectors $\mathbf{u}(\cdot)$ being identical at $t = 0$, that is, we must prove:

$$\mathbf{u}(\alpha',0) = \mathbf{u}(\alpha,t_0) \tag{14}$$

By noticing from Eq. (6) that $\mathbf{u}$ is a function of the term $a\alpha + b\phi(\alpha,t)$, the above condition can be rewritten as:

$$a \underbrace{\alpha'}_{A(\alpha,t_0)} + b\underbrace{\phi(\alpha',0)}_{0} = a\alpha + b\phi(\alpha,t_0)$$
$$aA(\alpha,t_0) = a\alpha + b\phi(\alpha,t_0) \tag{15}$$
$$A(\alpha,t_0) = \alpha + \frac{b}{a}\phi(\alpha,t_0)$$

where $\phi(\alpha,t_0)$ is given by the closed form expression in Eq. (11).

It is interesting to highlight that the resulting expression for $\alpha' = A(\alpha,t)$ indicates that α' is well-behaved, in the sense that it never exceed the limits $]-\pi,\pi]$. It also reveals that all trajectories eventually become a straight path, as can be seen by taking the limit:

$$\lim_{t\to\infty} A(\alpha,t_0) = \alpha - \frac{a}{b}\alpha\frac{b}{a} = 0 \tag{16}$$

where the fact that $\alpha = 0$ generates a straight trajectory follows from the PTG design equations in theorem 1. Note how the final part of all the trajectories being identical to one of them aligns perfectly with our goal of *consistent reactive trajectories* (condition C1).

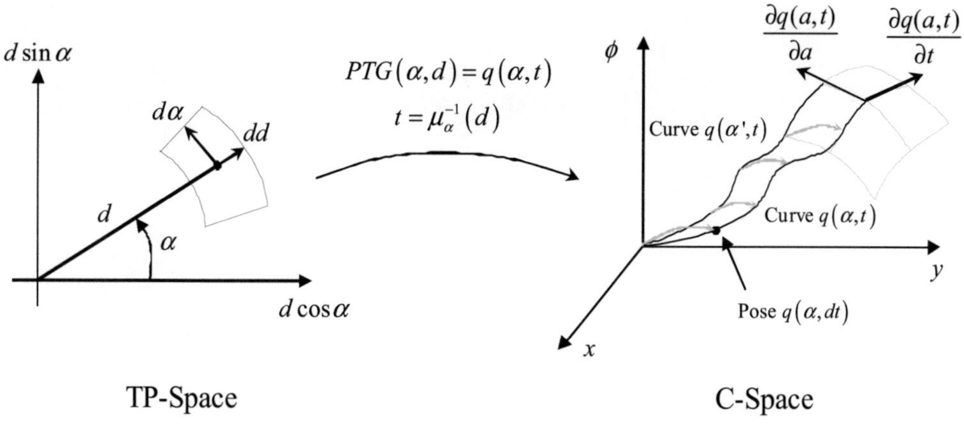

Figure 6. A schematic representation of the mapping a PTG performs between TP-Space points and C-Space robot poses. We represent the infinitesimal elements used in the proof of lemma 4. Basically, the idea represented here is that the curve for the trajectory $\mathbf{q}(\alpha',t)$ matches precisely trajectory $\mathbf{q}(\alpha,t)$ if the coordinate origin of the former is changed to $\mathbf{q}(\alpha,dt)$ for some α' infinitesimally close to α

Finally, the last requisite of a valid PTG (condition C3) is to generate *continuous* sampling surfaces in C-space, that is, the function $PTG(\alpha,d)$ must be continuous.

Lemma 4. *Given the hypotheses of theorem 1,* $PTG(\alpha,d)$ *is a 2-manifold of C-space with boundaries, and is continuous and derivable over the whole domain* $(\alpha,d)\in\mathbb{S}\times D$.

Proof. Firstly, due to lemma 1 it is enough to prove the continuity and differentiability of $\mathbf{q}(\alpha,t)$, since the mapping between distances d and times t conserves those properties of $\mathbf{q}$.

We show next that $\mathbf{q}(\alpha,t)$ has well-defined derivatives, which in turns implies it is continuous. For the case of $\frac{\partial\mathbf{q}(\alpha,t)}{\partial t}$ the proof is trivial since by definition this derivative is given by Eq. (5).

The derivation of $\frac{\partial\mathbf{q}(\alpha,t)}{\partial\alpha}$ is more involved. It is illustrative to keep Fig. 6 as a reference through the following derivations to clarify the geometrical meaning of each term. Let dt be an infinitesimal increment in time, and (α,t) some fixed point in the domain of TP-space. Then, using lemma 3.3 we can rewrite $\mathbf{q}(\alpha,t+dt)$ as:

$$\mathbf{q}(\alpha,t+dt)=\mathbf{q}(\alpha,dt)\oplus\mathbf{q}(\alpha',t) \tag{17}$$

where α' is given by:

$$\begin{aligned} A(\alpha,t) &= \alpha+\frac{b}{a}\phi(\alpha,dt) \\ &= \alpha+\underbrace{\frac{b}{a}\omega(\alpha,0)\,dt}_{d\alpha} \\ &= a+d\alpha \end{aligned} \tag{18}$$

Making use of the definition of pose composition operators (Smith et al., 1988) we can rearrange Eq. (17) as follows:

$$\mathbf{q}(\alpha',t)=\mathbf{q}(\alpha,t+dt)\ominus\mathbf{q}(\alpha,dt) \tag{19}$$

The geometrical meaning of this operation is that, as illustrated in Fig. 6, the curve $\mathbf{q}(\alpha',t)$ matches the curve $\mathbf{q}(\alpha,t)$ if translated and rotated to the pose $\mathbf{q}(\alpha,dt)$. As a result, this means that infinitesimal changes $d\alpha$ in a pair (α,t) leads to infinitesimal changes in $\mathbf{q}(\alpha,t)$ that can be written down as:

$$\mathbf{q}(\alpha+d\alpha,t)=\mathbf{q}(\alpha,t)\oplus\mathbf{J}(\mathbf{q}(\alpha,t))\mathbf{u}(\alpha,t)\,dt\ominus\begin{bmatrix} v(\alpha,0)\,dt \\ 0 \\ \omega(\alpha,0)\,dt \end{bmatrix} \quad (20)$$

which follows from Eq. (19) and the definition of $\mathbf{q}$ as an integral of the velocity vector $\mathbf{u}$. Since the pose composition $\oplus$ and inverse composition $\ominus$ operators are both continuous and differentiable, it follows that the derivative $\frac{\partial \mathbf{q}}{\partial \alpha}$ is well-defined for any point (α,t). Finally, given $\mathbf{q}(\alpha,t)$ is differentiable and so is $PTG(\alpha,d)$ at the whole domain of (α,d), the surface generated by a PTG can be seen as a 2-manifold with boundaries (Spanier, 1981).

4. Related work

In this section we review the most well-known space representations that have been employed in mobile robot motion planning and collision avoidance, and put them in contrast with our approach.

The C-Space has been extensively used in many fields, including robotic manipulators (Lozano-Perez, 1987), maneuver planning (Latombe, 1991), and mobile robot motion planning (Murphy, 2000). The complexity derived from its high dimensionality makes C-Space not applicable to real-time reactive navigation.

A first simplification for dealing with C-Space more efficiently is to assume a circular robot. Thus, C-Obstacles are no longer dependent on the robot orientation and the C-Space reduces to a planar space, the Workspace (WS). This space is employed by the well known potential field methods, like the VFF (Borenstein and Koren, 1989), VFH (Borenstein and Koren, 1991), and others (Haddad et al., 1998, Balch, 1993). Other reported methodologies are based on neural nets (Pal and Kar, 1995) and, more recently, the Nearness-Diagram (ND) approach (Minguez and Montano, 2004), which relies on a divide-and-conquer strategy that defines a set of different states according to the arrangement of nearby obstacles. All these methods deal with circular shaped robots, a too restrictive assumption for many real-life situations. For example, if a robotic wheelchair were assumed to be circular, it would never pass through a narrow doorway.

Most of the approaches that deal with any-shape robots and take into account their kinematic restrictions propose working with another less limiting simplification of C-Space: the velocity space (Arras et al., 2002; Feiten et al., 1994; Ramírez and Zeghloul, 2001; Schlegel, 1998; Simmons, 1996), or V-Space for short. For mobile robots of our interest, V-Space represents the space of the potential linear and angular robot velocities, hence the next movement can be chosen as a point in V-Space that results in constant curvature paths (i.e. circular paths). A common feature in many V-Space methods is the inclusion of a dynamic window (Fox et al., 1997), which restricts the range of reachable velocities to that compatible with the robot maximum acceleration. An important limitation of these methods is that, although many obstacles may be sensed, not all of them are exploited: only those ones falling into the robot dynamic window for the next step are considered for choosing the instantaneous motion command. It is clear that better paths would be found if more comprehensive obstacle information were taken into account, which indeed implies looking ahead for more than one step, as our approach does. In addition to the utilization of a dynamic window, most V-Space approaches use only the family of circular paths to sample the free-space, which entails the risk of not detecting many free-space areas. There are some exceptions (Ramírez and Zeghloul, 2001; Xu and Yang, 2002) that make use of straight paths, but this model is not appropriate for most actual mobile robots. Only these two path models have been reported in the reactive collision avoidance literature. While a generic path can only be described in the three-dimensional C-Space (2D position plus heading), poses along a circular path can be defined through two parameters: the path curvature and the distance along the arc. Upon this parameterization, a TP-Space was proposed in (Minguez et al., 2006) as an elegant and mathematically sound alternative to V-Space: if the navigation is carried out in the 2D TP-Space, the robot can be treated as a free-flying-point. That work demonstrates that navigation in a parameterized space allows us to decouple the problems of kinematic restrictions and obstacle avoidance. However, this approach has never been extended neither to cope with other path models apart from the circular one, nor to a number of different transformations, which are the contributions of the present work.

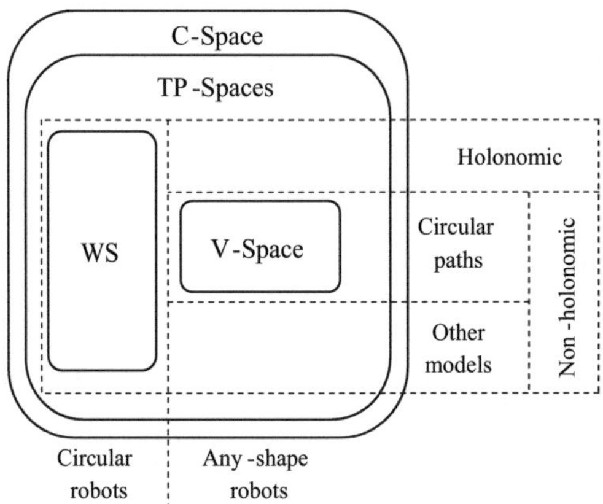

Figure 7. A summary with the classification of space representations used in obstacle avoidance

To further clarify the relationship between the different existing representation spaces, please refer to Fig. 7, where TP-Spaces appear as a generalization of spaces such as WS and V-Space. However, it must be remarked that C-Space is the most general space representation, but at the price of an elevated computational cost due to its high dimensionality.

5. Conclusions

In this work we have reviewed existing methods for obstacle avoidance and reactive navigation, and discussed how space transformations can be employed to extend their applicability to kinematically-constrained and any-shape mobile robots, making use of a clear and useful separation of the problems of robot shape and kinematic restrictions, and collision avoidance. We have developed a generalized kinematics abstraction mechanism which allows us using a variety of path models (PTGs) to obtain a better sampling of the whole C-Space from which more and better collision-free paths towards the target location can be found. We have settled the conditions that a PTG must hold to lead to a valid space transformation, and then we have introduced a PTG template which has been proven to always fulfill all the requirements. However, it must be remarked that theorem 1 determines a sufficient, but not necessary condition, thus indeed more valid PTGs may exist out of the given template (a prominent example are circular arcs). Finally, we should highlight that the applicability of PTGs is not limited to purely reactive navigation frameworks, hence their integration with hybrid planned-reactive approaches reveals as a promising research topic for the future.

6. References

Arkin, R. (1998). *Behavior-Based Robotics*. MIT Press.

Arras, K., Persson, J., Tomatis, N., and Siegwart, R. (2002). *Real-time obstacle avoidance for polygonal robots with a reduceddynamic window*. Vol. 3.

Balch, T. (1993). Avoiding the past: a simple but effective strategy for reactivenavigation. In *Proceedings of the IEEE International Conference on Robotics and Automation*, pages 678–685.

Blanco, J., Gonzalez, J., and Fernández-Madrigal, J. (2006). The Trajectory Parameter Space (TP-Space): A New Space Representation for Non-Holonomic Mobile Robot Reactive Navigation. *In Proceedings of the IEEE/RSJ International Conference on Intelligent Robots and Systems*, pages 1195–1200.

Blanco, J., Gonzalez, J., and Fernández-Madrigal, J. (2008). *Extending Obstacle Avoidance Methods through Multiple Parameter-Space Transformations. Autonomous Robots*, 24(1):29–48.

Borenstein, J. and Koren, Y. (1989). Real-time obstacle avoidance for fact mobile robots. *IEEE Transactions on Systems, Man and Cybernetics*, 19(5):1179–1187.

Borenstein, J. and Koren, Y. (1991). The vector field histogram-fast obstacle avoidance for mobilerobots. *IEEE Transactions on Robotics and Automation*, 7(3):278–288.

Evans, L. and Gariepy, R. (1992). *Measure Theory and Fine Properties of Functions*. CRC Press.

Feiten, W., Bauer, R., and Lawitzky, G. (1994). *Robust obstacle avoidance in unknown and cramped environments*. pages 2412–2417.

Fox, D., Burgard, W., and Thrun, S. (1997). The dynamic window approach to collision avoidance. *IEEE Robotics & Automation Magazine*, 4(1):23–33.

Haddad, H., Khatib, M., Lacroix, S., and Chatila, R. (1998). Reactive navigation in outdoor environments using potential fields. In *Proceedings of the IEEE International Conference on Robotics and Automation*, volume 2.

Khatib, M., Jaouni, H., Chatila, R., Laumond, J., and LAAS-CNRS, T. (1997). Dynamic path modification for car-like nonholonomic mobile robots. In *Proceedings of the IEEE International Conference on Robotics and Automation*, volume 4.

Lamiraux, F., Bonnafous, D., and Lefebvre, O. (2004). Reactive path deformation for nonholonomic mobile robots. *IEEE Transactions on Robotics*, 20(6):967–977.

Latombe, J. (1991). *Robot Motion Planning*. Kluwer Academic Publishers.

Lozano-Perez, T. (1983). Spatial Planning: A Configuration Space Approach. *IEEE Transactions on Computers*, 32(2):108–120.

Lozano-Perez, T. (1987). A simple motion-planning algorithm for general robot manipulators. *IEEE Journal of Robotics and Automation*, 3(3):224–238.

Minguez, J. and Montano, L. (2004). Nearness diagram (ND) navigation: collision avoidance in troublesome scenarios. *IEEE Transactions on Robotics and Automation*, 20(1):45–59.

Minguez, J. and Montano, L. (2008). Extending Reactive Collision Avoidance Methods to Consider any Vehicle Shape and the Kinematics and Dynamic Constraints. *IEEE Transactions on Robotics*.

Minguez, J., Montano, L., and Santos-Victor, J. (2002). Reactive navigation for non-holonomic robots using the ego-kinematic space. In *Proceedings of the IEEE International Conference on Robotics and Automation*, volume 3.

Minguez, J., Montano, L., and Santos-Victor, J. (2006). Abstracting Vehicle Shape and Kinematic Constraints from Obstacle Avoidance Methods. *Autonomous Robots*, 20(1):43–59.

Murphy, R. (2000). *Introduction to Ai Robotics*. MIT Press.

Pal, P. and Kar, A. (1995). Mobile robot navigation using a neural net. In *Proceedings of the IEEE International Conference on Robotics and Automation*, volume 2.

Quinlan, S. and Khatib, O. (1993). Elastic bands: connecting path planning and control. In *Proceedings of the IEEE International Conference on Robotics and Automation*, pages 802–807.

Ramírez, G. and Zeghloul, S. (2001). Collision-free path planning for nonholonomic mobile robots using a new obstacle representation in the velocity space. *Robotica*, 19(05):543–555.

Schlegel, C. (1998). *Fast local obstacle avoidance under kinematic and dynamicconstraints for a mobile robot*. volume 1.

Simmons, R. (1996). The curvature-velocity method for local obstacle avoidance. In *Proceedings of the IEEE International Conference on Robotics and Automation*, volume 4.

Smith, R., Self, M., and Cheeseman, P. (1988). A stochastic map for uncertain spatial relationships. *The fourth international symposium on Robotics Research*, pages 467–474.

Spanier, E. (1981). *Algebraic Topology*. Springer.

Xu, H. and Yang, S. (2002). Real-time collision-free motion planning of nonholonomic robots using a neural dynamics based approach. In *Proceedings of the IEEE International Conference on Robotics and Automation*, volume 3.

3

Text Detection and Pose Estimation for a Reading Robot

Marius Bulacu[1], Nobuo Ezaki[2] and Lambert Schomaker[1]
[1] *Dept. of Artificial Intelligence, University of Groningen,* [2] *Information and Control Engineering Dept., Toba National College of Maritime Technology*
[1]*The Netherlands,* [2]*Japan*

1. Introduction

This chapter presents two basic processes needed in a vision system for a mobile robot which is capable of reading the textual content encountered in its environment. We describe the general system outline and we present the structure and the experimental evaluation of the system modules that we built until the present. These two modules are designed to solve two essential problems facing our robotic reading system: text detection from scene images and text-pose estimation necessary for navigation guidance. Finding text in a natural image is the first problem that must be addressed and we propose four text-detection methods based on *connected components* (CoCos). We tested the effectiveness of these methods on the ICDAR 2003 Robust Reading Competition data. After text detection, for maneuvering the robot, we must estimate the orientation of the text surface with respect the viewing axis of the camera mounted on the robot. We propose an active-vision method for estimating the pose of a planar text surface using the *edge direction distribution* (EDD) as input to a neural network. We develop and evaluate a mathematical model to analyze how the EDD changes under canonical rotations and orthographic projection. We collected a set of camera-captured images with text in front-parallel view and, by applying single-axis synthetic rotations on these images, we obtain the data necessary to train and test the neural network. Further work will be directed at integrating our text detection and pose estimation modules within a complete robotic vision system.

2. Reading Robot

Our main research effort is concentrated on developing a vision system for an autonomous robot that will be able to detect and read the text encountered in its environment. A reading robot is an interesting proof of concept and building it is challenging as it raises many essential computer-vision problems requiring real-time solutions. Provided with text reading capabilities, mobile robots can capture this information intended for human visual communication and use it for navigation and task execution.

Our robot is essentially a computer on wheels with a controllable camera on top (equipped also with sonar sensors and odometry). Camera-based text reading in 3D space is a more defiant problem then classical optical character recognition (OCR) used for processing scanned documents. Two major aspects are different and play a very important role: the text areas must be first found in the image because text may be anywhere in the scene (*text detection*) and, secondly, the orientation of the text surface with respect to the camera viewing axis needs to be inferred (*pose estimation*) as it will be different from case to case. Our aim is to solve these problems by exploiting the robot's ability to move. We explore therefore, in a robotic setting, the role of active vision in the machine recognition of text in 3D (robotic OCR).

We can identify 4 important modules that need to be integrated in the vision system of the reading robot:

- text detection
- text-pose estimation
- robot/camera motion
- character classification.

The present chapter focuses on the first two modules of the system: the text detection module that finds the location of text in a natural scene image and the pose estimation module that computes the orientation of the text surface with respect to the viewing axis of the camera mounted on the robot. Once this information is known, the robot can be maneuvered and the camera can zoom-in to obtain (if possible) a high-resolution front-parallel view of the text, which, in principle, would give the best final OCR result.

Text detection is essentially a segmentation problem and, as such, it entails a known difficulty well established in the pattern recognition community. A perfect solution for this problem is hard to find. We adopt a connected-component-based approach for text detection. Connected components (CoCos) have the advantage of offering a quick access to the objects in the image. For a given text instance in a scene, the characters are usually similar in size and placed in a horizontal string. Using rules regarding size, aspect ratio and relative positioning can reduce the indiscriminate number of CoCos extracted from an image to obtain the final candidate text areas. In this chapter, we propose and evaluate four text detection methods using CoCos.

For tackling the problem of text-pose estimation we adopt a texture-based approach. The texture feature that we shall use is the angular distribution of directions in the text region extracted from the edges. This distribution changes systematically with the rotation angle and we develop a mathematical model to describe this trend. We then show how the rotation angle of the text surface can be recovered back from the edge-direction distribution (EDD) using a feed-forward neural network.

Here we consider, for simplicity, only single-axis rotations starting from front-parallel views. We impose this severe constraint in order to obtain a basic initial working system, perfectible in the future. Because robot motion is confined to the horizontal plane, only the rotation angle β of text around the vertical axis (Y) can be used for repositioning (see fig. 7). Initial experiments will be conducted in a simplified environment to test the basic robot functionality. In the totally unconstrained case, view normalization for the other rotations, around X and Z, will have to be performed in software.

This chapter is organized as follows. In section 3, we give a general overview of the field, providing pointers to literature and commenting on prospective applications. In section 4, we propose four CoCo-based methods for text detection and analyze their strengths and weaknesses. The next section presents an overall view of general pose-estimation methods and relates our approach to the more generic problem of shape-from-texture. In section 6, we present our texture-based method for text-pose estimation: we describe the extraction of the EDD, our underlying mathematical model and the use of the neural network. In the results section, first we evaluate the performance of the CoCo-based text detection methods when used individually and in combination. We then numerically check the validity of the theoretical EDD transform model and we evaluate the performance of the proposed text-pose estimation method in terms of angular error. A discussion of the strengths and weaknesses of our approach follows and conclusions end the chapter.

As regards our general approach to the problem of reading robotic systems, the following remarks are in place here. There is no formal mathematical solution to this complex problem. Therefore, a synthesizing approach has been followed, on the basis of what is known as 'best practice' in image processing, on the basis of geometric modeling and by using empirical evaluation.

3. Text Recognition in Real Images

Text detection and recognition in still images and video receives constant research attention, the references pointing to a number of systems (Clark & Mirmehdi, 2002b; Lienhart & Wernicke, 2002; Gao et al., 2001; Yang et al., 2001; Lopresti & Zhou, 2000; Li et al., 2000; Zhong et al., 2000; Wu et al., 1999) and a recent overview of camera-based document image analysis (Doermann et al., 2003). Two major categories of text have been identified: *scene text* incidentally picked up as part of the recorded scenery and *overlay text* added on the image by post hoc editing. Overlay text appears mostly in video and is carefully directed to carry information. It is assumed to appear in front-parallel view and with clear contrast, being less problematic to detect and recognize. Robust recognition of scene text is a more difficult problem and the robot has to confront it.

Automatic text detection and reading in natural images has many potential applications. To mention only a few: Intelligent transport systems (e.g. automatic reading of traffic signs or car license plates); office space with pervasive computing (e.g. intelligent cameras might be watching over a desk and automatically respond to commands to process the captured text); image retrieval (images can extracted from large multimedia databases using their text content).

Intelligent wearable cameras for visually impaired persons represent another particularly interesting application and an important research subject (Kang & Lee, 2002; Zandifar et al., 2002). The number of visually impaired persons is increasing every year due to eye disease, traffic accidents etc (e.g. in Japan alone there are about 200,000 people with acquired blindness). Such a support system, using a portable computer, a controllable camera and a speech synthesizer, can help an unaccompanied blind person by providing auditory information about the textual content of a scene (e.g. street or shop name, restaurant menu etc). This type of application shares many points in common with our robotic research theme: the acquired scene images are complex and their textual content has high variability in pose and in character size, color, font and contrast. Text-pose estimation, which in our case is primarily used for robot navigation, might also play a role, if coupled with acoustic feedback, in helping a blind person.

Text detection methods have been broadly classified in two categories: *texture based methods* and *connected-component based methods*. The methods in the first category use text texture for detection. To the casual observer, text areas have a distinctive general appearance in natural scenes. They exhibit a significant content of high frequencies, considerable variation in the gray levels, high density of edges, oriented in multiple directions, and these attributes are

uniform over a larger area. These rich textural features are exploited for text detection in combination with pattern-recognition techniques (k-means clustering, neural networks).
We opted to investigate the CoCo-based text detection methods for our robotic application because of their relative simplicity and effectiveness (Liu et al., 1998; Matsuo et al., 2002; Yamaguchi et al., 2003).

4. Text Detection Methods

In this section, we describe four connected-component based methods for detecting text in natural scene images. All four methods have the following processing steps:

- image preprocessing
- image binarization
- CoCo extraction
- CoCo selection using heuristic rules.

Only the image preprocessing step will be different from one method to another. Image binarization, CoCo extraction and selection will always be performed in similar fashion for all four proposed methods.
Using CoCos for detecting text in natural images raises a number of problems. The basic question is: "In what space are the pixels connected?" The simplest example concerns bitonal (B/W) images, where the concept of connectedness is straightforward. For gray-scale images, things already start to be complicated and binarization is needed for CoCo extraction. However, there are several ways in which a meaningful connectedness in the image can be imposed on the basis of local features such as color and texture. Assuming that such features can be transformed into a single scalar per pixel, a suitable binarization method may provide the basis for the CoCo extraction.
An inappropriate threshold might wipe away all or part of the text present in the image. A decision is also needed on whether to use a local or a global binarization method. For a defined class of images, local binarization methods can be adapted to perform significantly better than global methods (Trier & Jain, 1995). More research is needed in this direction, so we report here only on our results obtained using a global binarization method. Ideally, text characters should be extracted as individual CoCos. It is common knowledge that, unfortunately, on many occasions, this is not the case. Often, a CoCo might contain only a part of a broken character or several characters lumped together. This is an important difficulty and the different image preprocessing methods used represent an effort to confront this inherent problem. In the end, however, quite a significant number of errors still remain.

4.1 Binarization method

For binarization we use Otsu's classical method (Otsu, 1979). It is a simple, popular and quite effective global binarization method. The same threshold is used for the entire image.
Otsu's method automatically selects the binarization threshold that optimally partitions the gray-level histogram of the image into two separate subhistograms. The threshold T is selected such that the combined within-class variance σ_w^2 of the thresholded foreground and background pixels is minimized. This is also equivalent to maximizing the between-class variance σ_b^2 for the two classes of pixels:

$$\begin{aligned} T &= argmin(\sigma_w^2) = argmin(\omega_1\sigma_1^2 + \omega_2\sigma_2^2) \\ &= argmax(\sigma_b^2) = argmax[\omega_1(\mu_1 - \mu)^2 + \omega_2(\mu_2 - \mu)^2] \\ &= argmax[\omega_1\omega_2(\mu_1 - \mu_2)^2] \end{aligned} \tag{1}$$

where μ is the average value for the entire gray-level histogram, ω_1, ω_2 are the integrals of the two subhistograms (i.e. the proportions of pixels in the two classes after thresholding), σ_1, σ_2 are the standard deviations of the two subhistograms and μ_1, μ_2 are their average values. In programs, the third expression is usually implemented as it allows for an elegant and fast recursive computation.

4.2 Extraction of small characters using mathematical morphology operations

The first method we propose targets the small characters and it is based on mathematical morphology operations. We use a modified top-hat processing. In general, top-hat contrast enhancement is performed by calculating the difference between the original image and the opening image (Gu et al., 1997). As a consequence, the top-hat operation is applicable when the pixels of the text characters have higher values than the background. Additionally, in (Gu et al., 1997), the difference between the closing image and the original image is also used for text detection when character pixels have lower values than the background. This method is very effective, however it becomes computationally expensive if a large filter is used in order to extract large characters.

We developed an invariant method applicable to small characters. We use a disk filter with a radius of 3 pixels and we take the difference between the closing image and the opening image. The filtered images are binarized and then CoCos are extracted.

Top-hat image processing emphasizes the thin structures present in the image (thinner than the diameter of the structural filter used). As such, this method is only applicable for small characters (less than about 30 pixels in height). Besides text characters, other thin structures present in the image will also be detected (e.g. thin window frames).

This method detects connected text areas containing several small characters. As western text consists of characters that are usually horizontally placed, we take horizontally long areas ($1 < width / height < 25$) from the output image as the final candidate text regions (see figure 1).

4.3 Three methods for extracting large characters

We propose three extraction methods for large characters (more than about 30 pixels in height). The first two are based on Sobel edge detection, an image processing technique presented in detail in section 6.1. The third text extraction method is based on RGB color information. Fig. 2 shows how the three methods act on a sample image.

Each method extracts connected components that represent candidate text areas. Decision rules based on the sizes and relative positioning of these areas are afterwards used to prune the number of possibilities and reduce the large number of false hits.

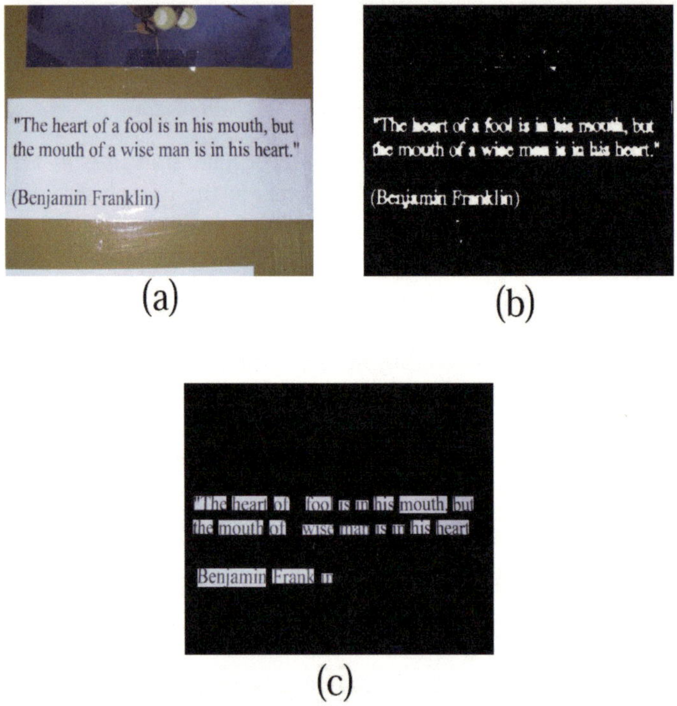

Figure 1. Extraction of small characters using morphological operations: a) original image, b) difference between closing and opening, c) extracted area

- **Character extraction from the edge image**

In this method, Sobel edge detection is applied on each color channel of the RGB image. The three edge images are then combined into a single output image by taking the maximum of the three gradient values corresponding to each pixel. The output image is binarized using Otsu's method and then CoCos are extracted.

This method fails when the edges of several characters are lumped together into a single large CoCo that is eliminated by the selection rules. This often happens when the text characters are close to each other or when the background is not uniform (see fig. 3).

- **Character extraction from the reverse edge image**

This method is complementary to the previous one; the binary edge image is reversed before connected component extraction. It will be effective only when characters are surrounded by continuous connected edges and the inner ink area is not broken (as in the case of boldface characters).

- **Color-based character extraction**

The three methods proposed until now use morphological and edge information for text detection. However, color information is also important, because, generally, text has almost the same color for a given instance encountered in the scene. The first step is to simplify the color space and we reduce it to 8 colors by the following procedure. We apply Otsu binarization independently on the three RGB color channels. Each pixel can now have only $2^3 = 8$ possible combinations of color values. We separate the 8 binary images and then we extract and select CoCos on each one independently. This method makes evident that global thresholding is not always appropriate and text characters can be lost.

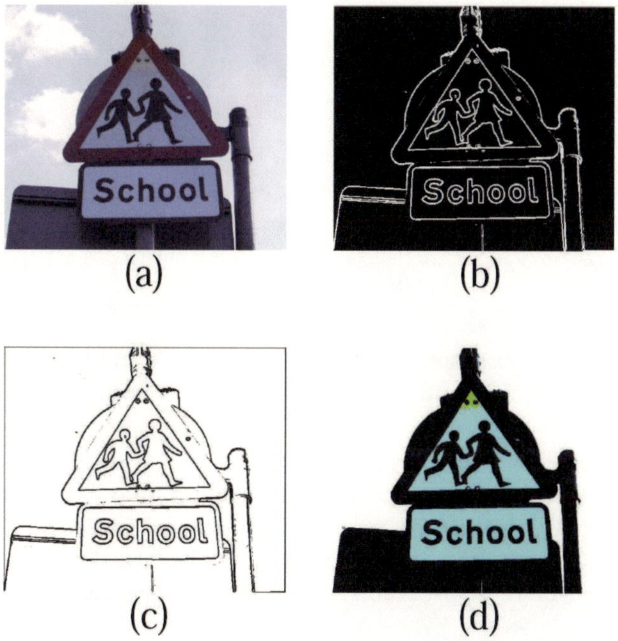

Figure 2. A "good" example: a) original image, b) edge image, c) reverse edge image, d) 8-color image

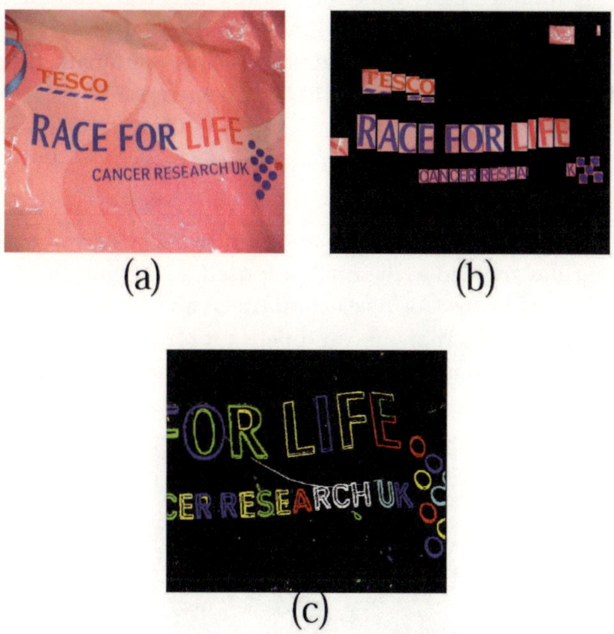

Figure 3. Several characters (namely 'R', 'C', 'H') are lumped into a single CoCo because their edges are connected by a background structure: a) original image, b) extracted area, c) close-up view of the problematic CoCo

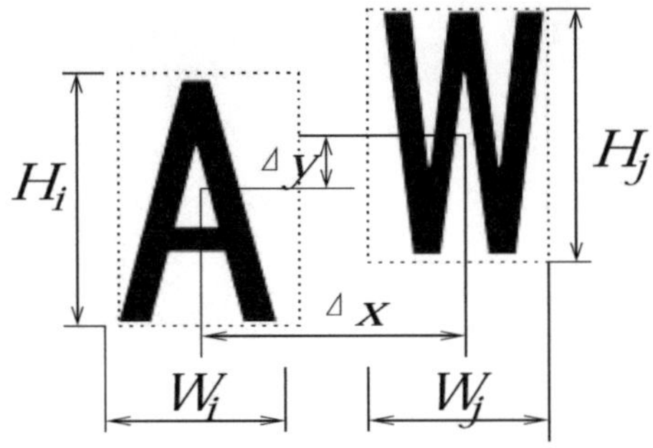

Figure 4. Size and relative positioning of extracted CoCos. W_i and H_i are the width and height of an extracted CoCo; Δx and Δy are the distances between their centers of gravity

4.4 Connected component selection rules

It can be noticed that, up to now, the proposed methods are very general in nature and not specific to text detection. As expected, many of the extracted CoCos do not actually contain text characters. At this point, rules are used to filter out the false detections (see fig. 4).
We impose constraints on the aspect ratio and area to decrease the number of non-character candidates:

$$0.1 < \frac{W_i}{H_i} < 2,\ 50 < W_i H_i \quad (2)$$

An important observation is that, generally, text characters do not appear alone, but together with other characters of similar dimensions and usually regularly placed in a horizontal string. We use the following rules to further eliminate from all the detected CoCos those that do not actually correspond to text characters:

$$0.5 < \frac{H_i}{H_j} < 2,\ \Delta y < 0.2\,max(H_i, H_j),\ \Delta x < 2\,max(W_i, W_j) \quad (3)$$

The system goes through all possible combinations of two CoCos and only those complying with all the selection rules succeed to the final proposed text region (see fig. 5).
The actual thresholds used in the CoCo selection rules can be further heuristically tuned for the individual methods. This remains an open problem. The proposed values work reasonably well on the test dataset (containing western characters). The selection rules do not completely eliminate all the erroneous non-text CoCos (see fig. 6).

Figure 5. Final result for the example given in fig. 2

(a) (b)

Figure 6. The windows of a building comply with CoCo selection rules and they are falsely detected as text characters

5. Approaches to Text-Pose Estimation

After text detection, the orientation of the text surface must be determined in order to provide navigation guidance to the mobile robot.

A very effective solution to text-pose estimation is based on finding vanishing points of text lines (Clark & Mirmehdi, 2002a; Myers et al., 2001). This type of knowledge-based approach has to impose restrictions on text layout (a minimum number of lines must be present, of sufficient length, with consistent paragraph justification) and the search for vanishing points is computationally expensive.

We adopt a different approach that can best be described as a simple shape-from-texture model. Determining the orientation (pose) and curvature (shape) of 3D surfaces from image texture information is a core vision problem. The proposed solutions make assumptions regarding the texture (isotropic (Garding, 1993) or homogenous (Malik & Rosenholtz, 1997)) and type of image projection (perspective (Garding, 1995; Clerc & Mallat, 1999) or orthographic (Super & Bovik, 1995)).

However, text texture does not have texels, it is homogeneous only in a stochastic sense and also, as we shall see, strongly directional, being a difficult candidate for the classical shape-from-texture algorithms.

We assume that text lies on a planar surface and we consider only single axis rotations. In this case, the general shape-from-texture problem reduces to determining the *slant angle* (the angle between the normal and the viewing axis Z) for rotations around X and Y (see fig. 7). For rotations around Z, the text surface remains parallel to the image plane and only text skew must de determined (a problem aptly addressed in the document analysis field).

General shape-from-texture algorithms rely on differential distortions in the local spatial frequency spectra of neighboring image patches. In contrast, we will use the edge-direction distribution (EDD) as a general texture signature for the entire text region and we will recover the rotation angle from it using a neural network. Realizing that a lot of information is disregarded prematurely, we consider our method presented here as a quick and helpful way of providing navigation guidance to a mobile robot, rather then a broad and generic solution.

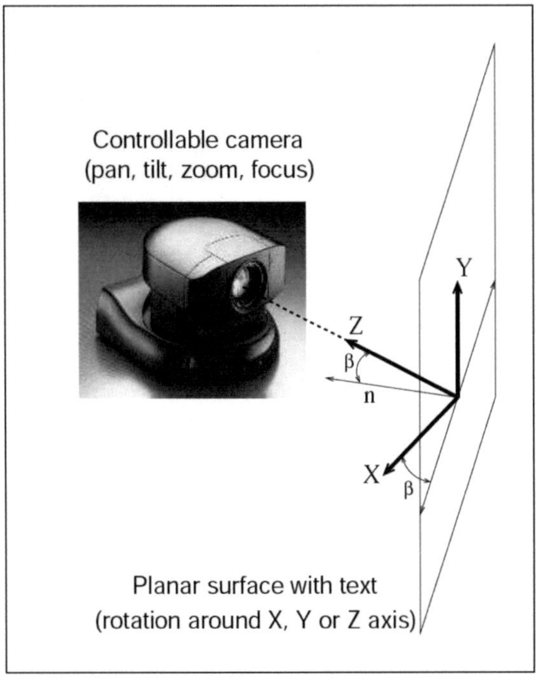

Figure 7. Experimental setup for text-pose estimation

6. Our Text-Pose Estimation Method

As demonstrated in (Clark and Mirmehdi, 2002b), the EDD and measures derived from it, such as symmetry and spread over directions, are very effective features for text detection as well. Here we will explore its use for pose estimation.

6.1 Extraction of the edge-direction distribution

As mentioned, one very important texture descriptor is the probability distribution of edge directions in the text area. EDD extraction starts with the classical Sobel edge-detection method, which is also used for two of the CoCo-based text detection methods described section 4.3 of this chapter.

Two orthogonal Sobel kernels S_x and S_y (eq. 4) are convolved with the image I (in eq. 5, $\otimes$ represents the convolution operator). The responses G_x and G_y represent the strengths of the local gradients along the x and y directions and G is their resultant total gradient (eq. 5). The orientation angle ϕ' of the gradient vector G measured from the horizontal (gradient phase) can be computed as in eq. 6. A final correction of 90° (eq. 6) is necessary to go from gradient-direction ϕ' to edge-direction ϕ, which is a more intuitive measure.

As the convolution runs aver the image, we build an angle histogram of the edge-directions. We count into the histogram bins only the pixels where G surpasses a chosen threshold (10% in our implementation). This makes sure that we take into consideration only the strong edge regions and not the quasi-uniform larger areas. In the end, the edge-direction histogram is normalized to a probability distribution $p(\phi)$.

$$S_x = \begin{bmatrix} -1 & 0 & 1 \\ -2 & 0 & 2 \\ -1 & 0 & 1 \end{bmatrix}, S_y = \begin{bmatrix} 1 & 2 & 1 \\ 0 & 0 & 0 \\ -1 & -2 & -1 \end{bmatrix} \tag{4}$$

$$G_x = S_x \otimes I, G_y = S_y \otimes I, G = \sqrt{G_x^2 + G_y^2} \tag{5}$$

$$\phi' = arctan(\frac{G_y}{G_x}), \phi = \phi' + \frac{\pi}{2} \tag{6}$$

The edge-direction distribution $p(\phi)$ is the texture feature that we shall use in the sequel for pose-estimation. The EDD is built mainly on phase information, which is known to be important for vision.

6.2 Text rotation in 3D and transform model for the edge-direction distribution

In this subsection, we analyze how the edge-direction distribution changes with the rotation angle. We shall consider only single axis rotations of a planar text surface under orthographic projection (a similar analysis under perspective projection would be mathematically more unwieldy).

- **Rotation around X axis**

Consider a needle OA of length l_0 initially contained in the front-parallel plane XOY and oriented at angle ϕ_0 with respect to the horizontal. We rotate it by angle $\alpha \in$ (-90°, +90°) around X axis to the new position OA' and then we project it back onto the front-parallel plane to OB (see fig. 8a). The projection OB will be of length l ($l < l_0$) and oriented at angle ϕ ($\phi < \phi_0$) with respect to the horizontal. The initial needle OA and its projection OB will appear at rescaled dimensions in the image. The projection equations are:

$$l_x = l\,cos\phi = l_0\,cos\phi_0 \tag{7}$$

$$l_y = l\,sin\phi = l_0\,sin\phi_0\,cos\alpha \tag{8}$$

Forward and backward relations for needle length and orientation are:

$$l = l_0\sqrt{1 - sin^2\phi_0\,sin^2\alpha}, \; l_0 = l\frac{\sqrt{1 - cos^2\phi\,sin^2\alpha}}{cos\alpha} \tag{9}$$

$$\phi = arctan(tan\phi_0\,cos\alpha), \; \phi_0 = arctan(\frac{tan\phi}{cos\alpha}) \tag{10}$$

If we consider that the needle actually stands for a small edge fragment, we can now describe how the text EDD changes from the initial $p_0(\phi_0)$ to $p_\alpha(\phi)$ after rotation. Two elements need to be taken into account: the length change $l_0 \to l$ and the angle change $\phi_0 \to \phi$. We express the new distribution as:

$$h(\phi) = p_0(\phi_0)\frac{l}{l_0}\frac{d\phi_0}{d\phi} \tag{11}$$

where $h(\phi)$ are some intermediary values. A renormalization of these values is necessary in order to obtain a proper final probability distribution that adds up to 1.

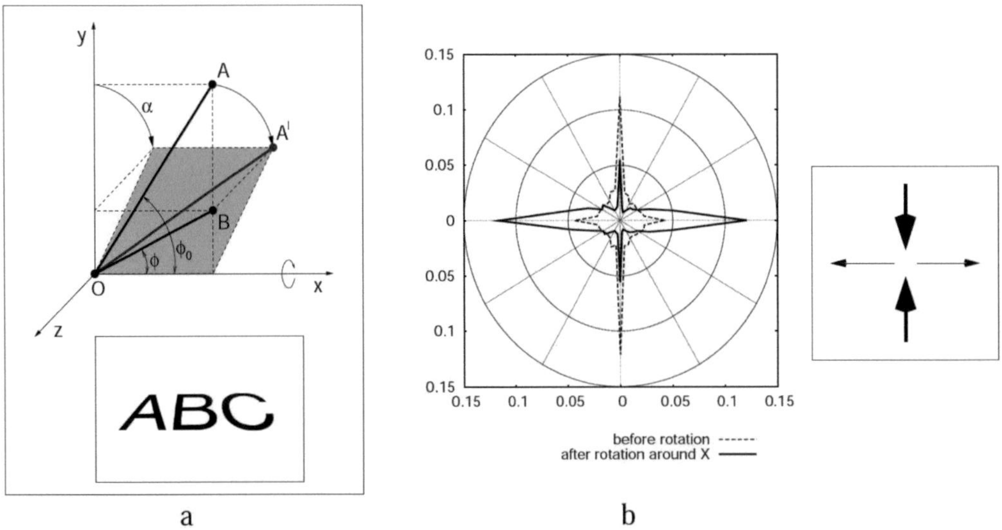

Figure 8. a) Text rotation around X axis. b) EDD change after rotation around X axis by 50°

Therefore, the EDD transform model that we propose is:

$$p_\alpha(\phi) = \frac{h_\alpha(\phi)}{\sum_\phi h_\alpha(\phi)} \text{ with } h_\alpha(\phi) = \frac{cos^2\alpha}{(1 - cos^2\phi sin^2\alpha)^{3/2}} p_0(arctan(\frac{tan\phi}{cos\alpha})). \tag{12}$$

In equation 12, the intermediary values h undergo renormalization. The expression for h is obtained from equation 11 after evaluating the lengths ratio and the angle derivative.

Unfortunately, the model cannot be formally developed beyond this point, making the numerical analysis our only option. This is the reason why we formulate equation 12 using discrete sums.

The EDD $p_\alpha(\phi)$ corresponding to rotated text cannot be expressed in closed form as a function of the rotation angle α and the base EDD $p_0(\phi_0)$ corresponding to front-parallel text.

Qualitatively, after rotation around X axis, text appears compressed vertically. This foreshortening effect is reflected in the EDD (see fig. 8b): the horizontal component of the distribution increases at the expense of the vertical one. These changes in EDD are more pronounced at larger angles and this makes possible recovering the rotation angle α.

- **Rotation around Y axis**

We apply a similar analysis considering a rotation of angle $\beta \in$ (-90°, +90°) around Y axis (see fig. 9a). The projection equations are:

$$l_x = l\cos\phi = l_0 \cos\phi_0 \cos\beta \tag{13}$$

$$l_y = l\sin\phi = l_0 \sin\phi_0 \tag{14}$$

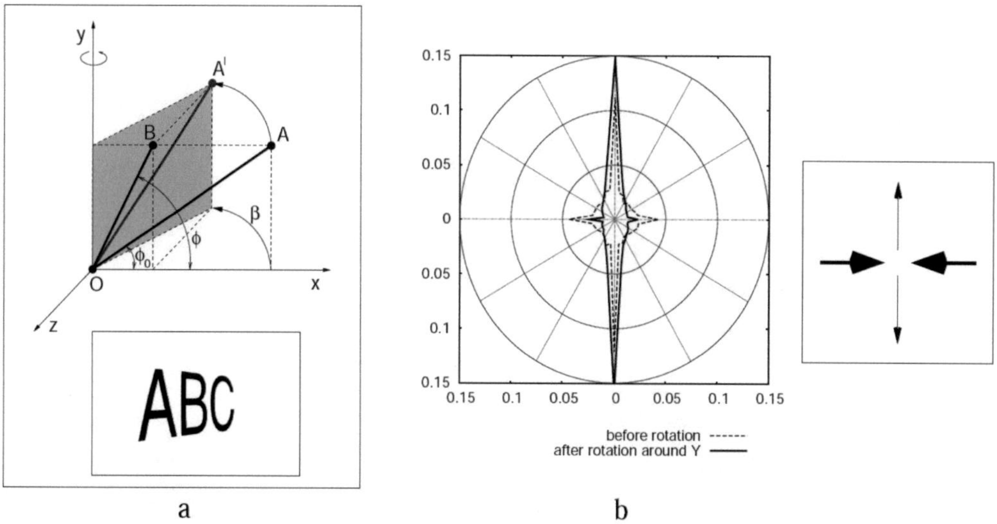

Figure 9. a) Text rotation around Y axis. b) EDD change after rotation around Y axis by 50°

Forward and backward relations for needle length and orientation are:

$$l = l_0\sqrt{1-\cos^2\phi_0 \sin^2\beta},\ l_0 = l\frac{\sqrt{1-\sin^2\phi \sin^2\beta}}{\cos\beta} \tag{15}$$

$$\phi = \arctan(\frac{\tan\phi_0}{\cos\beta}),\ \phi_0 = \arctan(\tan\phi\cos\beta) \tag{16}$$

Applying equation 11, the EDD transform model becomes:

$$p_\beta(\phi) = \frac{h_\beta(\phi)}{\sum_\phi h_\beta(\phi)} \quad \text{with } h_\beta(\phi) = \frac{\cos^2\beta}{(1-\sin^2\phi\sin^2\beta)^{3/2}} p_0(\arctan(\tan\phi\cos\beta)) \tag{17}$$

where h are intermediary values that undergo renormalization.

Here again, $p_\beta(\phi)$ (corresponding to rotated text) cannot be expressed in closed form as a function of the rotation angle β and the base EDD $p_0(\phi_0)$ (corresponding to front-parallel text).

Qualitatively, after rotation around Y axis, text appears compressed horizontally. This foreshortening effect is reflected in the EDD (see fig. 9b): the vertical component of the distribution increases at the expense of the horizontal one. The rotation angle β can be recovered because the changes in EDD are more pronounced at larger angles.

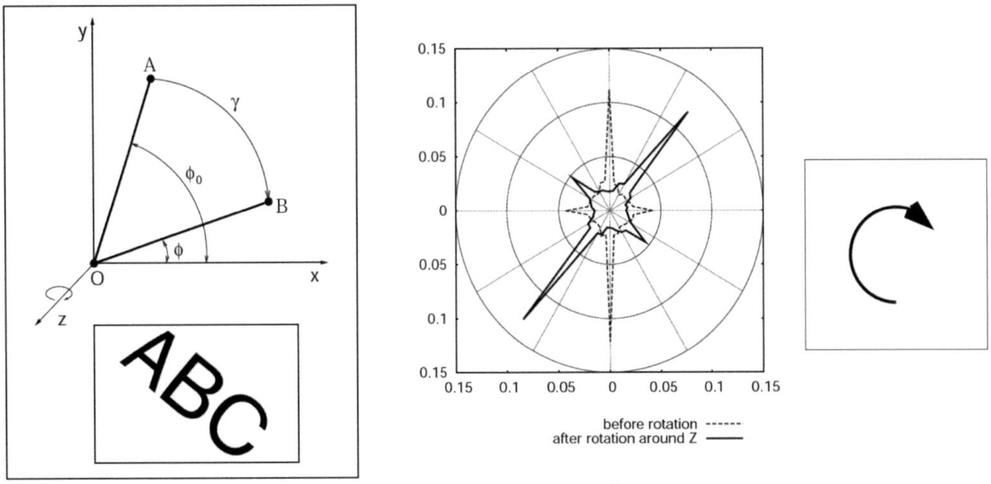

a b

Figure 10. a) Text rotation around Z axis. b) EDD change after rotation around Z axis by 40°

- **Rotation around Z axis**

In this case, text rotation by angle $\gamma \in$ (0°, 360°) simply results in a rotation of the EDD (considered in polar form) by the same angle:

$$\phi = \phi_0 + \gamma,\ \phi_0 = \phi - \gamma \tag{18}$$

$$l = l_0,\ l_0 = l \tag{19}$$

$$p_\gamma(\phi) = p_0(\phi - \gamma) \tag{20}$$

An example showing how the EDD changes for rotations around Z axis is given in fig. 10b.

6.3 The neural network

Very early in our attempts to recover the rotation angle using multilinear regression, we obtained correlation coefficients larger than 0.85 between the cosine squared of the rotation angle and the probability values in the EDD. But an obvious and more appropriate choice is to use a neural network to extract the nonlinear inverse relationship between the EDD and the rotation angle. The ground truth data needed to train and test the neural network is obtained using synthetic rotations starting from front-parallel views.

However, in trying to recover the rotation angle directly from the EDD, two problems appear: font dependence of the base EDD and quadrant ambiguity. One of the very important underlying assumptions is that the base EDD (i.e. that corresponding to the front-parallel view) is almost the same for all machine-print text. Otherwise, a change in the EDD due to font will be wrongly interpreted as a rotation. This assumption is not true: the EDD is actually different for different fonts. This font dependence of the EDD is in fact what we, very successfully, exploited in solving the problem of identifying people based on their handwriting (Bulacu et al., 2003; Bulacu & Schomaker, 2003; Schomaker et al., 2003) (an interesting biometrics method enjoying renewed interest for its forensic applicability).

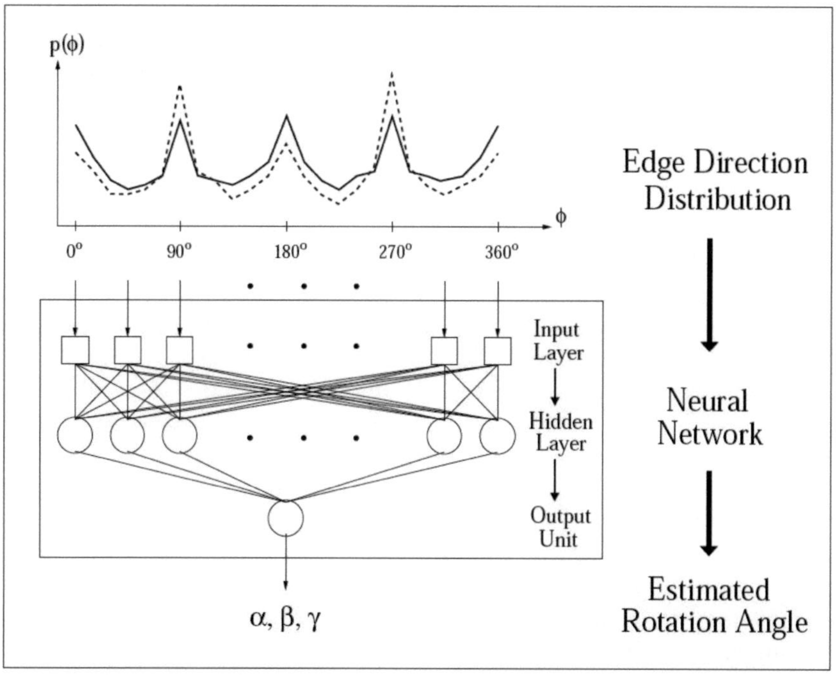

Figure 11. Text-pose estimation method. The neural network has one input unit for every EDD bin (36 in our implementation). The rotation angle is given by the output unit of the network. The difference between two successive EDDs is used as input for rotations around X and Y axes. The EDD itself is used as input for rotations around Z axis

The second problem is quadrant ambiguity for rotations around the X and Y axes: under orthographic projection, text looks the same under rotation of $+\alpha$ and $-\alpha$ ($+\beta$ and $-\beta$). The EDD cannot distinguish between the two situations and this can also be confirmed by observing that the functions depending on the rotation angle appearing in equations 12 and 17 are even.

In order to eliminate this problem, the idea is to consider in the analysis two images rather a single one, the second image being rotated at a fixed small angle δ from to the first. For a chosen δ, in one quadrant, the second image will be closer to the front-parallel view than the first. In the other quadrant, the situation will be reversed. The difference between the two EDDs extracted from the two images will clearly reflect this situation and the neural network has an easy job in inferring it from the training data. Using the difference between two EDDs diminishes also the font-dependence problem, which unfortunately cannot be completely eliminated resulting in inevitable final prediction errors.

The robot, therefore, will need - for rotations around Y axis - to make a small exploratory movement, always to the same side (say e.g. to the right) in order to alleviate the ambiguity. Two implicit assumptions are tacitly adopted here: tracking (the robot needs to look at the same text area) and satisfactory control of the rotation angle δ. We anticipate to solve these constraints using camera (auto)focus and wheel odometry.

For rotations around Z axis the quadrant ambiguity cannot be eliminated. While usually the vertical component of text is stronger than the horizontal one in machine-print, this difference is not reliable enough to obtain accurate predictions based on it. The EDD is almost symmetric to rotations of 90° around Z axis and consequently our solution can only encompass one quadrant. In this case, two images are not needed, the EDD from a single image suffices to determine the rotation angle. This problem is more effectively addressed in the document analysis field. We present our neural network solution only to have a unitary treatment throughout.

7. Experimental Results

7.1 Text detection results

For evaluating the performance of the proposed text detection methods, we used the dataset made available with the occasion of the ICDAR 2003 Robust Reading Competition (Lucas et al., 2003). The images are organized in three sections: Sample, Trial and Competition. Only the first two are publicly available, the third set of images being kept separate by the competition organizers to have a completely objective evaluation. The Trial directory has two subdirectories Trial-Train and Trial-Test. The Trial-Train images should be used to train and tune the algorithms.

As we do not use machine learning in our text detection methods, we included all the images in Trial-Test and Trial-Train for evaluation. This difficult dataset contains a total of 504 realistic images with textual content.

We used a similar evaluation method as that of the ICDAR2003 competition. It is based on the notions of precision and recall. Precision p is defined as the number of correct estimates C divided by the total number of estimates E:

$$p = \frac{C}{E} \tag{21}$$

Recall r is defined as the number of correct estimates C divided by the total number of targets T:

$$r = \frac{C}{T} \tag{22}$$

For a given image, we calculate precision and recall as the ratio between two image areas (expressed in terms of number of pixels). E is the area proposed by our algorithm, T is the manually labeled text area and C is their intersection. We then compute the average precision and recall aver all the images in the dataset.

There is usually a trade-off between precision and recall for a given algorithm. It is therefore necessary to combine them into a single final measure of quality f:

$$f = \frac{1}{\alpha / p + (1-\alpha)/r} \tag{23}$$

The parameter α was set to 0.5, giving equal weights to precision and recall in the combined measure f.

Our results on the ICDAR 2003 dataset are shown in table 1. The edge-based text detection method obtained top overall performance. In this context, we note that, at ICDAR 2003 (Lucas et al., 2003), the results for the winner of the competition were precision = 55%, recall = 46% and f = 50%.

The morphological method did not obtain good overall results because the dataset contains relative large text characters. Consequently, we selected, from the ICDAR 2003 dataset, a group of 55 images that contain only small characters. We evaluated the efficacy of the morphological method on these images and obtained precision = 38%, recall = 55% and f = 47%. We tested also the edge based method on these images and obtained precision = 26%, recall = 48% and f = 37%. The morphological method seems to be more effective for small characters.

Table 2 shows the results obtained by combining methods. Fusion is performed by ORing the results of the individual methods. By collecting all the candidate areas given by the different methods, we reduce the risk of missing a text instance. This is confirmed also by the high recall rate obtained when all methods are combined using OR. The increase in recall is outbalanced by the decrease in precision. However, for the same f value, the method with the highest recall rate is preferable.

In principle, it is naturally the job of the character recognizer to reject many of the false text detections based on its knowledge of character shape. The motivation for combining four text-detection methods is to have a high final recall rate.

Method	Precision	Recall	f
Edge (E)	60%	64%	62%
Edge reverse (R)	62%	39%	50%
8 colors (8)	56%	43%	49%
Morphology (M)	41%	16%	28%

Table 1. Results for the individual text extraction methods

Method	Precision	Recall	f
E + 8	54%	69%	62%
E + R	56%	70%	63%
E + M	55%	68%	62%
E + R + 8	51%	73%	62%
E + R + 8 + M	48%	76%	62%

Table 2. Results obtained after fusing methods using OR

7.2 Text-pose estimation results

For evaluating the text-pose estimation method another dataset of images was needed. We used a Sony Evi D-31 PAL controllable camera to collect 165 images containing text in front-parallel view. The images contain only text and the background is uniform. They are gray-scale (8 bits/pixel) and have a resolution of 748x556. We strived to obtain sufficient variability in the dataset: 10 different fonts, appearing at different sizes in the images, from a single word to a whole paragraph per image.

In order to test our text-pose estimation method, single-axis synthetic rotations are applied to these images using our own custom-built rotation engine. The number of bins in the EDD was set to N = 36. This was found to give a sufficiently fine description of text texture (10°/bin).

First we verify the validity of our EDD transform model and then we train a neural network to predict the rotation angle and evaluate its performance in terms of angular error.

- **Verification of the theoretical model**

From every image in the dataset, we extract the base EDD corresponding to the front-parallel view. We then randomly select a rotation angle and we theoretically compute (using equations 12, 17, 20) what the EDD should be for the rotated image (forward transform). We then apply the rotation on the image and we directly extract the EDD corresponding to the new pose. We compare the theoretically predicted EDD with the empirically extracted EDD to check the validity of our theoretical model.
An appropriate distance measure is the Bhattacharyya distance:

$$dist(f,g)=1-\sum_{i=1}^{N}\sqrt{f_i g_i} \tag{24}$$

where f and g are the two EDDs. The distance varies between 0 and 1 and we express it in percentages to have an intuitive measure. If the distance is null, the two distributions are identical.

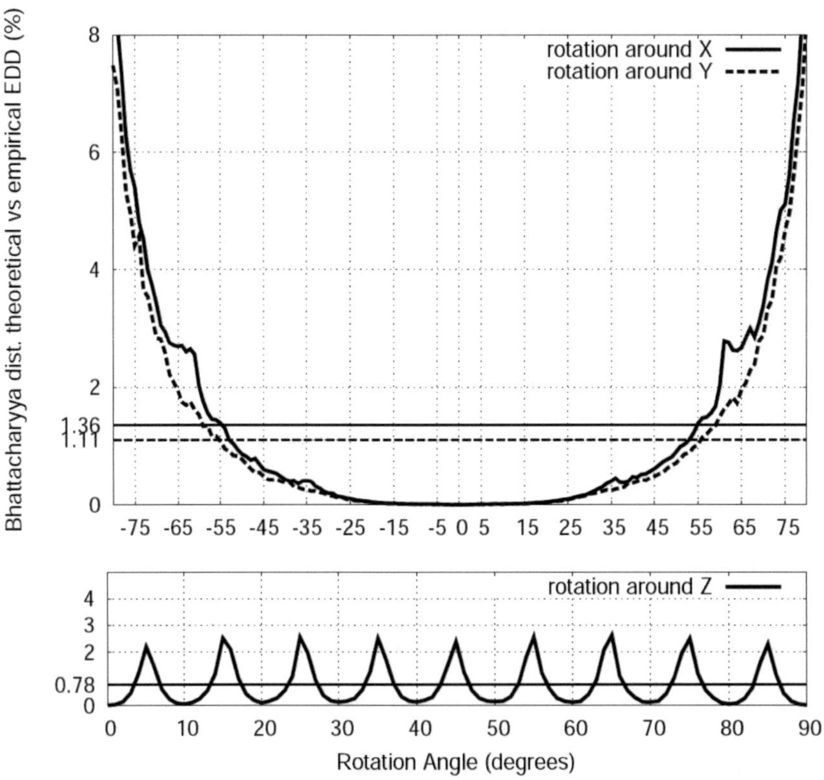

Figure 12. Verification of theoretical model: Bhattacharyya distance between theoretical and measured EDD (in percentages). The horizontal lines represent average values (from table 3 column 2)

We applied 400 random rotations on every image around each axis. The average distance is around 1% (see table 3) and in fig. 12 we show its dependence on the rotation angle.
For rotations around X and Y axes, the error increases with the rotation angle. At larger angles, text is so compressed that letters fuse together in a single lump and our mathematical model no longer correctly describes the changes in the EDD. For rotations around Z, the error is small and does not have a systematic trend, but we can observe a sampling artifact: the error shows an oscillatory behavior as the probability flows from one bin to another of the EDD.

Rotation around	Theoretical Model Error (percentages)	Angle Prediction Error (degrees)
X axis (pitch)	1.36%	3.8°
Y axis (yaw)	1.11%	6.6°
Z axis (roll)	0.78%	2.9°

Table 3. Correlation between theoretical model and empirical data (column 2). Overall angle prediction error (column 3)

- **Evaluation of the angle prediction method**

In order to predict the rotation angle from the EDD (inverse transform), we use a standard feed-forward neural network (3 layers, fully connected between layers, nonlinear transfer functions in the hidden layer). The network architecture is 36x10x1 (see fig. 11). The training method is Rprop (Riedmiller & Braun, 1993), a more effective variant of backpropagation algorithm.
From the beginning, we split the data into 100 images for training and 65 for testing. Every image is then rotated 400 times at randomly chosen angles (40000 training examples, 26000 testing examples). For rotations around X and Y axes, two rotated images are in fact generated with a slight pose difference between them $\delta = 10°$. The network is trained to predict the rotation angle (of the second image for example) using the difference between the two EDDs. For rotations around Z axis, a single EDD is used, but rotations are limited to one quadrant.
Fig. 13 shows how the method performs on two typical examples.
On the test data, we compute the root mean square (RMS) error between the predicted and the real rotation angle. The average angular prediction error is given in table 3. The method demonstrates good performance (3°- 7°angular error).
In fig. 14 we show the dependence of the angular error on the rotation angle. As expected, it can be observed again that the error increases at larger angles for rotations around X and Y axes.
Another interesting observation is that the prediction error for rotations around Y axis is larger than that for rotations around X axis. So we performed the following simple test: we first rotated all the images by 90° around Z and subsequently we applied all the regular analysis. The angular error for rotations around X axis snaps into the range of errors for rotations around Y axis and the reverse (see fig. 14), proving to be an inherent property of the data.
The explanation is that the vertical component of text is more reliable than the horizontal one and, as it is most affected by rotations around X axis, the prediction is more accurate in

this case. Unfortunately, rotations around Y axis represent the case of most interest for our robotic application.
For rotations around Z axis, an important observation is that for angles γ close to 0° and 90° the error increases as confusion appears (especially for uppercase characters) between the vertical and the horizontal components, which are the most prominent in the EDD. This is the reason why we opted for a single quadrant solution for this type of rotation.
The method becomes unreliable for small characters (less than 20 pixels in height or width) as the EDD cannot be consistently extracted. We found that the method works well if more than 10 characters are present in the image.

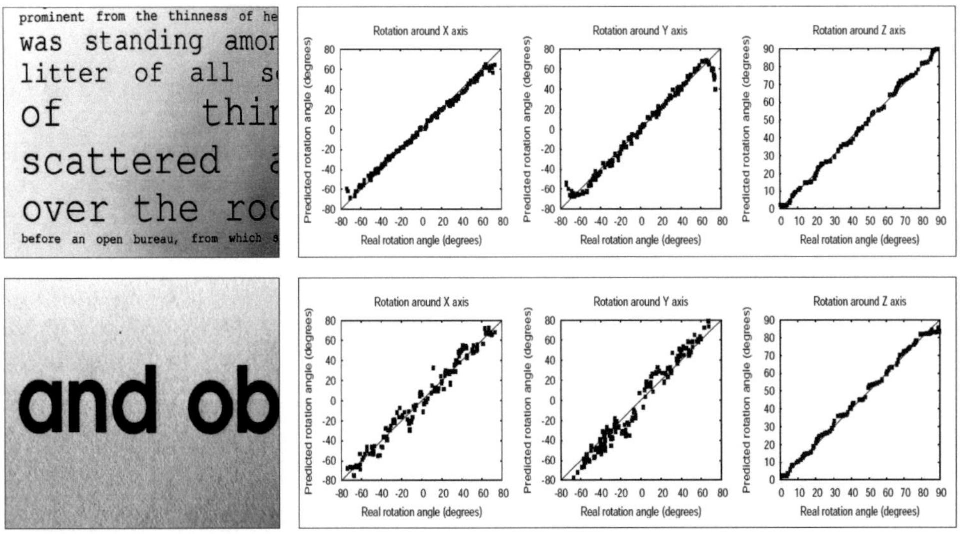

Figure 13. Typical performance: "good" example up, "bad" example down. Angular transfer functions are given for rotations around the X, Y, Z axis, from left to right panel. Ideally all the experimental points would be placed exactly on the diagonal for perfect predictions

In a qualitative evaluation, we found that the proposed method works also on-line in combination with our controllable camera. The neural network, trained and tested off-line on synthetic rotations, estimates reasonably well text-pose during on-line operation under real rotations. The errors are, nevertheless, relatively larger.
It is important to note at this point that the proposed algorithm is lightweight, on average 70 msec being necessary on a 3.0 GHz processor to extract the EDDs from 2 images and run the neural network on their difference to predict the rotation angle. Therefore, using the robot's ability to make small exploratory movements seems like an attractive idea for solving the pose-estimation problem.

8. Discussion

One very important advantage of using CoCos for text detection is that they naturally allow the analysis to take place across scales. In this approach, scale does not represent such a problematic issue because the CoCo extraction process is scale independent. CoCos give a prompt, but rather imperfect, hold to the structures present in the image and CoCo selection is an important complementary step. As the results indicate, further improvement is needed for our text detection module.

For text-pose estimation, we decided to base our analysis on orthographic projection, while most shape-from-texture methods rely on perspective effects. We consider this approach to be more robust, as perspective effects diminish if, after text detection, the camera zooms into the text area (long focal length, small field of view). The proposed texture-based method for text-pose estimation does not impose constraints on text layout. It works even when text lines are not present or they are very short, or when only a few characters are available. We found that Greek fonts can be handled surprisingly well by the same neural network.

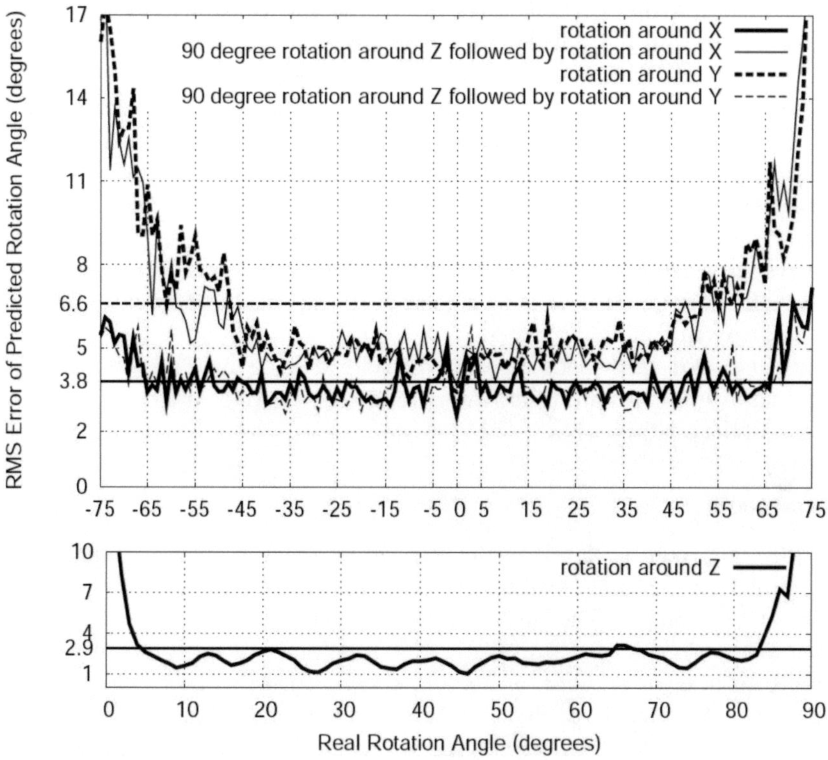

Figure 14. Prediction results: angular error (in degrees). Horizontal lines represent average values (from table 3 column 3)

We treated here only canonical rotations. The method can be directly extended to two-axis rotations, but our experiments are far from conclusive at the moment. We have not addressed free three-axis rotations. A particularly difficult instance is, for example, slanted text and text that is rotating about the surface normal. We will also consider in our future analysis second order moments of the EDD.

Our commitment to a single feature makes our approach limited in the end. However we believe that we have a promising starting point and an effective algorithm to implement on the robot for planning ballistic "text-hunting" movements.

9. Conclusions

In this chapter, we described the text detection and the pose estimation modules of a vision system for a reading robot.

Four connected-component-based methods for text detection have been implemented and evaluated. The most effective proves to be the sequence: Sobel edge detection, Otsu binarization, connected component extraction and rule-based connected component selection. A high recall rate can be achieved by collecting all the candidate text areas proposed by the four individual methods (recall = 76%, precision = 48%, f = 62%).

We also presented here a method for estimating the orientation of planar text surfaces using the edge-direction distribution (EDD) in combination with a neural network. We considered single-axis rotations and we developed a mathematical model to analyze how the EDD changes with the rotation angle under orthographic projection. We numerically verified the validity of our underlying mathematical model. In order to solve the quadrant ambiguity and improve performance, for rotations around X and Y axes, we consider a pair of images with a slight rotation difference between them. The change in the EDD is extracted and sent to a feed-forward neural network that predicts the rotation angle corresponding to the last image in the pair. For rotations around Z axis, a single EDD is used, the solution being applicable only to rotations in the first quadrant. The method has been tested off-line with single-axis synthetic rotations and shows good performance. In on-line operation, with real rotations, the errors are relatively larger.

Though limited in scope, the methods proposed here are elegant, quite simple and very fast. Our future work will concentrate on integrating the described modules in the complete vision system of the reading robot.

10. References

Bulacu, M. & Schomaker, L. (2003) Writer style from oriented edge fragments, *Proc. of 10th Int. Conf. on Computer Analysis of Images and Patterns (CAIP 2003): LNCS 2756*, pp. 460-469, Groningen, The Netherlands, Springer

Bulacu, M.; Schomaker, L. & Vuurpijl, L. (2003). Writer identification using edge-based directional features, *Proc. of 7th Int. Conf. on Document Analysis and Recognition (ICDAR 2003)*, Vol. II, pp. 937-941, Edinburgh, Scotland, IEEE Computer Society

Clark, P. & Mirmehdi, M. (2002)a. On the recovery of oriented documents from single images, *Proc. of Advanced Concepts for Intelligent Vision Systems (ACIVS 2002)*, pp. 190-197, Ghent, Belgium

Clark, P. & Mirmehdi, M. (2002)b. Recognizing text in real scenes, *International Journal on Document Analysis and Recognition*, Vol. 4, No. 4, pp. 243-257

Clerc, M., Mallat, S. (1999). Shape from texture and shading with wavelets, *Dynamical Systems, Control, Coding, Computer Vision, Progress in Systems and Control Theory*, Vol. 25, pp. 393-417

Doermann, D.; Liang, J. & Li, H. (2003). Progress in camera-based document image analysis, *Proc. of 7th Int. Conf. on Document Analysis and Recognition (ICDAR 2003)*, Vol. I, pp. 606-616, Edinburgh, Scotland, IEEE Press

Gao, J.; Yang, J.; Zhang, Y. & Waibel, A. (2001). Text detection and translation from natural scenes, *Technical Report CMU-CS-01-139*, Computer Science Department, Carnegie Mellon University, Pittsburgh, USA

Garding, J. (1993). Shape from texture and contour by weak isotropy, *J. of Artificial Intelligence*, Vol. 64, No. 2, pp. 243-297

Garding, J. (1995). Surface orientation and curvature from differential texture distortion, *Proc. of the IEEE Int. Conf. on Computer Vision (ICCV '95)*, pp. 733-739

Gu, L.; Tanaka, N.; Kaneko, T. & Haralick, R. (1997). The extraction of characters from cover images using mathematical morphology, *Transaction of The Institute of Electronics, Information and Communication Engineers of Japan*, Vol. J80-D-II, No. 10, pp. 2696-2704

Kang, S. & Lee, S.W. (2002). Object detection and classification for outdoor walking guidance system, *Proc. of 2nd Int. Workshop Biologically Motivated Computer Vision (BMCV 2002): LNCS 2525*, pp. 601-610, Tuebingen, Germany

Lienhart, R. & Wernicke, A. (2002). Localizing and segmenting text in images, videos and web pages, *IEEE Trans. on Circuits and Systems for Video Technology*, Vol. 12, No. 4, pp. 256-268

Li, H.; Doermann, D. & Kia, O. (2000). Automatic text detection and tracking in digital videos, *IEEE Trans. on Image Processing*, Vol. 9, No. 1, pp. 147-156

Liu, Y.; Yamamura, T.; Ohnishi, N. & Sugie, N. (1998). Extraction of character string regions from a scene image, *Transaction of The Institute of Electronics, Information and Communication Engineers of Japan*, Vol. J81-D-II, No. 4, pp. 641-650

Lopresti, D. & Zhou, J. (2000). Locating and recognizing text in www images, *Information Retrieval*, Vol. 2, No. 2/3, pp. 177-206

Lucas, S.M.; Panaretos, A.; Sosa, L.; Tang, A.; Wong, S. & Young, R. (2003). ICDAR 2003 robust reading competitions, *Proc. of 7th Int. Conf. on Document Analysis and Recognition (ICDAR 2003)*, Vol. II, pp. 682-687, Edinburgh, Scotland, IEEE Press

Malik, J. & Rosenholtz, R. (1997). Computing local surface orientation and shape from texture for curved surfaces, *Int. J. Computer Vision*, Vol. 23, No. 2, pp. 149-168

Matsuo, K.; Ueda, K. & Michio, U. (2002). Extraction of character string from scene image by binarizing local target area, *Transaction of The Institute of Electrical Engineers of Japan*, Vol. 122-C, No. 2, pp. 232-241

Myers, G.K.; Bolles, R.C.; Luong, Q.T. & Herson, J.A. (2001). Recognition of text in 3-d scenes, *Proc. of 4th Symposium on Document Image Understanding Technology*, Columbia, Maryland , USA

Otsu, N. (1979). A threshold selection method from gray-level histogram, *IEEE Trans. Systems, Man and Cybernetics*, Vol. 9, pp. 62-69

Riedmiller, M. & Braun, H. (1993). A direct adaptive method for faster backpropagation learning: The Rprop algorithm, *Proc. of the IEEE Int. Conf. on Neural Networks (ICNN)*, pp. 586-591, San Francisco, USA

Schomaker, L.; Bulacu, M. & van Erp, M. (2003). Sparse-parametric writer identification using heterogeneous feature groups, *Proc. of Int. Conf. on Image Processing (ICIP 2003)*, Vol. I, pp. 545-548, Barcelona, Spain

Super, B.J. & Bovik, A.C. (1995). Shape from texture using local spectral moments, *IEEE Trans on PAMI*, Vol. 17, No. 4, pp. 333-343

Trier, O.D. & Jain, A.K. (1995). Goal-directed evaluation of binarization methods, *IEEE Trans on PAMI*, Vol. 17, No. 12, pp. 1191-1201

Wu, V.; Manmatha, R. & Riseman, E.M. (1999). Textfinder: An automatic system to detect and recognize text in images, *IEEE Trans. on PAMI*, Vol. 21, No. 11, pp. 1224-1229

Yamaguchi, T.; Nakano, Y.; Maruyama, M.; Miyao, H. & Hananoi, T. (2003). Digit classification on signboards for telephone number recognition, *Proc. of 7th Int. Conf. on Document Analysis and Recognition (ICDAR 2003)*, Vol. I, pp. 359-363, Edinburgh, Scotland, IEEE Press

Yang, J.; Gao, J.; Zang, Y.; Chen, X. & Waibel, A. (2001). An automatic sign recognition and translation system, *Proceedings of the Workshop on Perceptive User Interfaces (PUI'01)*

Zandifar, A.; Duraiswami, R.; Chahine, A. & Davis, L. (2002). A video based interface to textual information for the visually impaired, *Proc. of 4th Int. Conf. on Multimodal Interfaces (ICMI 2002)*, pp. 325-330, Pittsburgh, USA

Zhong, Y.; Zhang, H. & Jain, A.K. (2000). Automatic caption localization in compressed video, *IEEE Trans. on PAMI*, Vol. 22, No. 4, pp. 385-392

4

Robust Vision-only Mobile Robot Navigation with Topological Maps

Toon Goedemé and Luc Van Gool
De Nayer Technical University, VISICS, Catholic University of Leuven
Belgium

1. Introduction

In this work we present a novel system for autonomous mobile robot navigation. With only an omnidirectional camera as sensor, this system is able to build automatically and robust accurate topologically organised environment maps of a complex, natural environment. It can localise itself using that map at each moment, including both at startup (kidnapped robot) or using knowledge of former localisations. The topological nature of the map is similar to the intuitive maps humans use, is memory-efficient and enables fast and simple path planning towards a specified goal. We developed a real-time visual servoing technique to steer the system along the computed path.
The key technology making this all possible is the novel *fast wide baseline feature matching*, which yields an efficient description of the scene, with a focus on man-made environments.

1.1 Application

Figure 1. Left: the robotic wheelchair platform. Right: the omnidirectional camera, composed by a colour camera and an hyperbolic mirror

This chapter describes a total navigation solution for mobile robots. It enables a mobile robot to efficiently localise itself and navigate in a large man-made environment, which can be

indoor, outdoor or a combination of both. For instance, the inside of a house, an entire university campus or even a small city lie in the possibilities.
Because of reliability problems of other sensors like e.g. GPS, why we aim at a *vision-only* solution to navigation. Vision is, in comparison with these other sensors, much more informative. Moreover, cameras are quite compact and increasingly cheap. We observe also that many biological species, in particular migratory birds, use mainly their visual sensors for navigation. We chose to use an omnidirectional camera as visual sensor, because of its wide field of view and thus rich information content of the images acquired with. For the time being, we added a range sensing device for obstacle detection, but this is to be replaced by an omnidirectional vision range estimator under development.
Our method works with *natural* environments. That means that the environment does not have to be modified for navigation in any way. Indeed, adding artificial markers to every room in a house or to an entire city doesn't seem feasible nor desirable.
In contrast to classical navigation methods, we chose a *topological* representation of the environment, rather than a metrical one, because of its resemblance to the intuitive system humans use for navigation, its flexibility, wide usability, memory-efficiency and ease for map building and path planning.
The targeted application of this research is the visual guidance of electric wheelchairs for severely disabled people. More in particular, the target group are people not able to give detailed steering commands to navigate around in their homes and local city neighbourhoods. If it is possible for them to perform complicated navigational tasks by only giving simple commands, their autonomy can be greatly enhanced. For most of them such an increase of mobility and independence from other people is very welcome.
Our test platform and camera are shown in fig. 1.

1.2 Method overview

An overview of the navigation method presented is given in fig. 2. The system can be subdivided in three parts: map building, localisation and locomotion.
The map building stage has to be gone through only once, to train the system in a new environment. The mobile system is lead through all parts of the environment, while it takes images at a constant rate (in our set-up one per second). Later, this large set of omnidirectional images is automatically analysed and converted into a topological map of the environment, which is stored in the system's memory and will be used when the system is actually in use.
The next stage is localisation. When the system is powered up somewhere in the environment, it takes a new image with its camera. This image is rapidly compared with all the images in the environment map, and an hypothesis is formed about the present location of the mobile robot. This hypothesis is refined using Bayes' rule as soon as the robot starts to move and new images come in.
When the present location of the robot is known and a goal position is communicated by the user to the robot, a path can be planned towards that goal using the map. The planned route is specified as a sequence of training images, serving as a reference for what the robot should subsequently see if on course. This path is executed by means of a visual servoing algorithm: each time a visual homing procedure is executed towards the location where the next path image is taken.

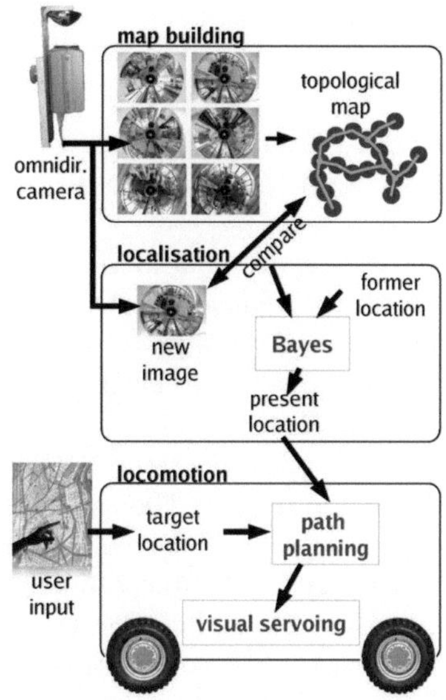

Figure 2. Overview of the navigation method

The remainder of this chapter is organised as follows. The next section gives an overview of the related work. In section 3, our core image analysis and matching technique is explained: fast wide baseline matching. The sections thereafter describe the different stages of our approach. Section 4 discusses the map building process, section 5 explains the localisation method, section 6 describes the path planning, and section 7 details the visual servoing algorithm. We conclude with an overview of experimental results (section 8) and a conclusion (section 9).

2 Related Work

2.1 Image comparison

A good image comparison method is of utmost importance in a vision-based navigation approach. Global methods compute a measure using all the pixels of the entire image. Although these methods are fast, they cannot cope with e.g. occlusions and severe viewpoint changes. On the other hand, techniques that work at a local scale, extracting and recognising *local features*, can be made robust to these effects. The traditional disadvantage of these local techniques is time complexity. In our approach, we combine novel global and local approaches resulting in fast and accurate image comparison.

2.1.1 Global techniques

Many researchers use global image comparison techniques. Straightforward *global methods* like histogram-based matching, used by (Ulrich & Nourbakhsh, 2000) don't seem distinctive enough for our application. Another popular technique is the use of an eigenspace decomposition of the training images (Jogan & Leonardis, 1999), which yields a compact database. However, these methods proved not useful in general situations because they are not robust enough against occlusions and illumination changes. That is why (Bischof et al.,

2001) developed a PCA-based image comparison that is robust against partial occlusions, respectively varying illumination.

2.1.2 Local techniques

A solution to be able to cope with partial occlusions is comparing local regions in the images. The big question is how to detect these local features, also known as visual landmarks.

A simple solution to do this is by adding artificial markers to strategically chosen places in the world. In this project we use *natural landmarks*, extracted from the scene itself, without modifications. Moreover, the extraction of these landmarks must be automatic and robust against changes in viewpoint and illumination to ensure the detection of these landmarks under as many circumstances as possible.

Many researchers proposed algorithms for natural landmark detection. Mostly, local regions are defined around interest points in the images. The characterisation of these local regions with descriptor vectors enables the regions to be compared across images. Differences between approaches lie in the way in which interest points, local image regions, and descriptor vectors are extracted. An early example is the work of (Schmid et al., 1997), where geometric invariance was still under image rotations only. Scaling was handled by using circular regions of several sizes. (Lowe, 1999) extended these ideas to real scale-invariance. More general affine invariance has been achieved in the work of Baumberg (Baumberg, 2000°), Tuytelaars & Van Gool (Tuytelaars et al., 1999; Tuytelaars & Van Gool, 2000), Matas (Matas et al.,2002), and Mikolajczyk & Schmid (Mikolajczyk & Schmid, 2002).

Although these methods are capable to find high quality correspondences, most of them are too slow to use in a real-time mobile robot algorithm. That is why we propose a much faster alternative, as explained in section 3.

2.2 Map structure

Many researchers proposed different ways to represent the environment perceived by vision sensors. We can order all possible map organisations by metrical detail: from dense 3D over sparse 3D to topological maps. We believe that the outer topological end of this spectrum offers the top opportunities.

One approach is building dense 3D models out of the incoming visual data (Pollefeys et al., 2004; Nistér et al., 2004). Such approach has some disadvantages. It is computationally and memory demanding, and fails to model planar and less-textured parts of the environment such as walls. Nevertheless, these structures are omnipresent in our application, and collisions need to be avoided.

One way to reduce the computational burden is to make abstraction of the visual data. Instead of modelling a dense 3D model containing billions of voxels, a sparse 3D model is built containing only special features, called *visual landmarks*.

Examples of researchers solving the navigation problem with sparse 3D maps of natural landmarks are (Se et al., 2001) and (Davison, 2003). They position natural features in a metrical frame, which is as big as the entire mapped environment. Although less than the dense 3D variant, these methods are still computationally demanding for large environments since their complexity is quadratic in the number of features in the model. Also, for larger models the metric error accumulates, so that feature positions are drifting away.

As a matter of fact, the need for explicit 3D maps in navigation is questionable. One step further in the abstraction of environment information is the introduction of topological maps. The psychological experiments of (Bülthoff et al., 1998) show that people rely more on a topological map than a metrical one for their navigation. In these topological maps, locally places are described as a configuration of natural landmarks. These places form the nodes of the graph-like map, and are interconnected by traversable paths. Other researchers (Vale & Ribeiro, 2003; Ulrich & Nourbakhsh, 2000; Kosecká & Yang, 2004) also chose for topological maps, mainly because they scale better to real-world applications than metrical, deterministic representations, given the complexity of unstructured environments. Other advantages are the ease of path planning in such a map and the absence of drift.

2.3 Toplogical map building

Vale (Vale & Ribeiro, 2003) developed a clustering-based method for automatic building of a topological environment map out of a set of images. Unfortunately, the latter method is only suited for image comparison techniques which are a metric function, and does not give correct results if *self-similarities* are present in the environment, i.e. places that are different but look similar.

Very popular are various *probabilistic* approaches of the topological map building problem. (Ranganathan et al., 2005) for instance use Bayesian inference to find the topological structure that explains best a set of panoramic observations, while (Shatkay & Kaelbling, 1997) fit hidden Markov models to the data. If the state transition model of this HMM is extended with robot action data, the latter can be modeled using a partially observable Markov decision process or POMDP, as in (Koenig & Simmons, 1996; Tapus & Siegwart, 2005). (Zivkovic et al., 2005) solve the map building problem using graph cuts.

In contrast to these global topology fitting approaches, an alternative way is detecting *loop closings*. During a ride through the environment, sensor data is recorded. Because it is known that the driven path is traversable, an initial topological representation consists of one long edge between start and end node. Now, extra links are created where a certain place is revisited, i.e. an equivalent sensor reading occurs twice in the sequence. This is called a loop closing. A correct topological map results if all loop closing links are added.

Also in loop closing, probabilistic methods are introduced to cope with the uncertainty of link hypotheses and avoid links at self-similarities. (Chen & Wang, 2005), for instance, use Bayesian inference. (Beevers & Huang, 2005) recently introduced Dempster-Shafer probability theory into loop closing, which has the advantage that ignorance can be modelled and no prior knowledge is needed. Their approach is promising, but limited to simple sensors and environments. In this chapter, we present a new framework for loop closing using rich visual sensors in natural complex environments, which is also based on Dempster-Shafer mathematics but uses it differently.

2.4 Visual Servoing

As explained in section 1.2, the execution of a path using such a topological environment map boils down to a series of visual servoing operations between places defined by images.

(Cartwright & Collett, 1987) proposed the so-called bearing-only 'snapshot' model, inspired by the visual homing behaviour of insects such as bees and ants. Their proposed algorithm consists of the construction of a home vector, computed as the average of landmark displacement vectors. (Franz et al., 1998) analysed the computational foundations of this method and derived its error and convergence properties. They conclude that every visual homing method based solely on bearing angles of landmarks like this one, inevitably depends on basic assumptions such as equal landmark distances, isotropic landmark distribution or the availability of an external compass reference. Unfortunately, because none of these assumptions generally hold in our targeted application we propose an alternative approach.

If both image dimensions are taken into account, not limiting the available information to the bearing angle, the most obvious choice is working via epipolar geometry estimation (e.g. Tuytelaars et al., 1999; Basri et al., 1993). Unfortunately, in many cases this problem is ill conditioned. A workaround for planar scenes is presented by (Sagüés et al., 2005), who opted for the estimation of homographies. (Svoboda et al., 1998) proved that motion estimation with omnidirectional images is much better conditioned compared to perspective cameras. That is why we chose a method based on omnidirectional epipolar geometry. Other work in this field is the research of (Mariottini et al., 2005), who split the homing procedure in a rotation phase and a translation phase, which can not be used in our application because of the non-smooth robot motion.

3. Fast wide baseline matching

The novel technique we use for image comparison is *fast wide baseline matching*. This key technique enables extraction of natural landmarks and image comparison for our map building, localisation and visual servoing algorithms.

We use a combination of two different kinds of wide baseline features, namely a rotation reduced and colour enhanced form of Lowe's *SIFT* features (Lowe, 1999), and the *invariant column segments* we developed (Goedemé et al., 2004). These techniques extract local regions in each image, and describe these regions with a vector of measures which are invariant to image deformations and illumination changes. Across different images, similar regions can be found by comparing these descriptors. This makes it possible to find correspondences between images taken from very different positions, or under different lighting conditions. The crux of the matter is that the extraction of these regions can be done beforehand on each image separately, rather than during the matching. Database images can be processed off-line, so that the images themselves do not have to be available at the time of matching with another image.

3.1 Camera motion constraint

The camera we use is a catadioptric system, consisting of an upward looking camera with a hyperboloidal mirror mounted above it. The result is a field of view of 360° in horizontal direction and more than 180° in vertical direction. The disadvantage is that these images contain severe distortions, as seen for instance in fig. 5.

We presume the robot to move on one horizontal plane. The optical axis of the camera is oriented vertically. In other words, allowed movements consist of translations in the plane and rotation around a vertical axis. Figure 3 shows an illustration on this.

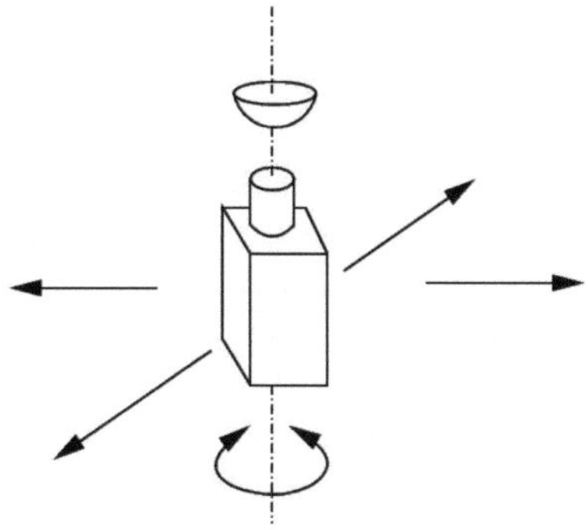

Figure 3. Illustration of the allowed movements of the camera

3.2 Rotation reduced and colour enhanced SIFT

David Lowe presented the *Scale Invariant Feature Transform* (Lowe, 1999), which finds interest points around local extrema in a scale-space of difference-of-Gaussian (DoG) images. The latter tend to correspond to blobs which contrast with their background. A dominant gradient orientation and scale factor define an image patch around each interest point so that a local image descriptor can be found as a histogram of normalised gradient orientations. SIFT features are invariant to rotation and scaling, and robust to other transformations.

A reduced form of these SIFT features for use on mobile robots is proposed by (Ledwich & Williams, 2004). They used the fact that rotational invariance is not needed for a camera with a motion constraint as in fig. 3. Elimination of the rotational normalisation and rotational part of the descriptor yields a somewhat less complex feature extraction and more robust feature matching performance.

Because the original SIFT algorithm works on greyscale images, some mismatches occur at similar objects in different colours. That is why we propose an outlier filtering stage using a colour descriptor of the feature patch based on global colour moments, introduced by Mindru *et al.* [32]. We chose three colour descriptors: C_{RB}, C_{RG} and C_{GB}, with

$$C_{PQ} = \frac{\int PQ\,\mathrm{d}\Omega \int \mathrm{d}\Omega}{\int P\,\mathrm{d}\Omega \int Q\,\mathrm{d}\Omega}, \tag{1}$$

where $P,Q \in \{R,G,B\}$, i.e. the red, green, and blue colour bands, centralised around their means. After matching, the correspondences with Euclidean distance between the colour description vectors above a fixed threshold are discarded.

We tested these algorithms on the image pair in fig. 5. With the original SIFT algorithm, the first 13 matches are correct. Using our rotation reduced and colour enhanced algorithm, we see that up to 25 correct matches are found without including erroneous ones.

3.3 Invariant column segments

We developed wide baseline features which are specially suited for mobile robot navigation. There we exploited the special camera motion (section 3.1) and the fact that man-made environments contain many vertical structures. Examples are walls, doors, and furniture.

These don't have to be planar, so cylindrical elements like pillars do comply too. Vertical lines in the world always project to radial lines in the omnidirectional image for the constrained camera motions.

Here, these new wide baseline features are described for the use on omnidirectional images. More details and how they can be used on perspective camera images are described in (Goedemé et al, 2004). The extraction process of the wide baseline features starts as illustrated in figure 4. We stress that every step is invariant to changes in viewpoint and illumination. Along every line through the centre of the original image (left), we look for points having a local maximum gradient value (centre). Every consecutive pair of gradient maxima along the line defines the begin and end of a new invariant column segment (right).

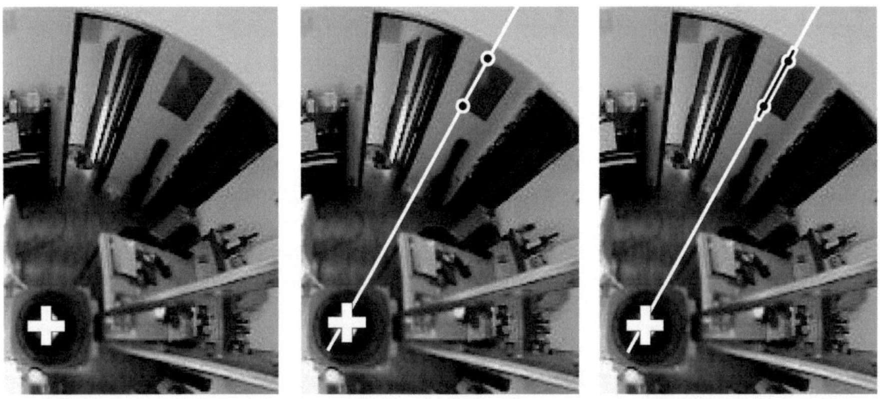

Figure 4. Illustration of the invariant column segment extraction algorithm: (left) part of the original image, the white cross identifies the projection centre, (centre) local maxima of the gradient for one radius, (right) one pair of maxima defines a column segment

We characterise the extracted column segments with a descriptor that holds information about colour and intensity properties of the segment. This 10-element vector includes:

- Three *colour invariants*. To include colour information in the descriptor vector, we compute the colour invariants, based on generalised colour moments (equation 1), over the column segment. To include information about the close neighbourhood of the segment, the line segment is expanded on both sides with a constant fraction of the segment length (in our experiments 0.2). Figure 4 (right) shows this.

- Seven *intensity invariants*. To characterise the intensity profile along the column segment, the best features to use are those obtained through the Karhunen-Lòeve transform (PCA). But because all the data is not known beforehand this is not practical. As is well known, the Fourier coefficients can sometimes offer a close approximation of the KL coefficients. In our method, because it is computationally less intensive and gives real output values, we choose to use the seven first coefficients of the *discrete cosine transform* (DCT), instead of Fourier.

In many cases there are horizontally constant elements in the scene. This leads to many very resembling column segments next to each other. To avoid matching over and over again very similar line segments, we first do a clustering of the line segments in each image. As a clustering measure we use the Mahalanobis distance of the descriptor vectors, extended with the horizontal distance between the line segments. In each cluster a prototype segment is chosen for use in the matching procedure.

3.4 Matching

These two kinds of local wide baseline features are very suited to quickly find correspondences between two widely separated images. A correspondence pair is established if for both features the other is the closest to it in the feature space, for the entire data set. Also, this match must be at least a fixed ratio better than the second best match. To be able to cope with different ranges of the elements of the descriptor vectors, distances are computed using the Mahalanobis measure (where we assume the cross-correlations to be zero):

$$d_{ij} = \sqrt{\sum_k \frac{(x_{ik} - x_{jk})^2}{\sigma_k^2}} \tag{2}$$

To speed up the matching, a Kd-tree of the reference image data is built.
Fig. 5 shows the matching results on a pair of omnidirectional images. As seen in these examples, the SIFT features and the column segments are complementary, which pleads for the combined use of the two. The computing time required to extract features in two 320x240 images and find correspondences between them is about 800 ms for the enhanced SIFT features and only 300 ms for the vertical column segments (on a 800 MHz laptop). Typically 30 to 50 correspondences are found.

Figure 5. A pair of omnidirectional images with colour-coded corresponding column segments (radial lines) and SIFT features (circles with tail)

For the description of a feature only the descriptors are used in the end, and not the underlying pixel data. As a result, the memory requirements for storing the reference images of entire environments can be kept limited.

4. Map Building

The navigation approach proposed is able to automatically construct a topological world representation out of a sequence of training images. During a training tour through the entire environment, omnidirectional images are taken at regular time intervals. The order of the training images is known. Section 4.1 describes the map structure targeted. In section 4.2, the image comparison technique based on fast wide baseline features is described which is used by the actual map building algorithm, presented in section 4.3.

4.1 Topological Maps

To be of use in the following parts of the navigation method, the topological map must describe all 'places' in the environment and the possible connections between these places. The topology of the world, being the maze of streets in a city or the structure of a house, must be reflected in the world model. The question remains what exactly is meant with such a *place* and how to delimit it. In a building, a place can be defined as a separate room. But then, what to do with long corridors, and outdoors with city streets?

That is why we define a place with regard to the needs of the localisation and locomotion algorithms. To be able to get a sufficiently detailed localisation output, the sampling of places must be dense enough. For the locomotion algorithm, which performs visual homing between two places at a time, the distance between these places must be not too big to ensure errorless motion. On the other hand, a compact topological map with fewer places requires less memory and enables faster localisation and path planning.

We discuss the image comparison method used in the map building algorithm before deciding on this place definition, as this comparison will lie at its basis.

4.2 Image Comparison Measure

The main goal of this section is to determine for each arbitrary pair of images a certain similarity measure, which tells how visually similar the two images are. Our image comparison approach consists of two levels, a global and a local comparison of the images. We first compare two images with a coarse but fast global technique. After that, a relatively slower comparison with more precision based on local features only has to be carried out on the survivours of the first stage.

4.2.1 Global colour similarity measure

To achieve a fast global image similarity measure between two images, we compute the same moments we used for the local features (equation 1) over the entire image. These moments are invariant to illumination changes, i.e. offset and scaling in each colour band. The Euclidean distance between two sets of these image colour descriptors gives a visual dissimilarity measure between two images.

With this dissimilarity measure, we can clearly see for instance the difference of images taken in different rooms. Because images taken in the same room but at different positions have approximately the same colour scheme, a second dissimilarity measure based on local features is needed to distinguish them.

4.2.2 Local measure based on matches

First, we search for feature matches between the two images, using the techniques described in section 3. The dissimilarity measure is taken to be inversely proportional to the *number of matches*, relative to the average number of features found in the images. Also the difference in relative configuration of the matches is taken into account. Therefore, we first compute a global angular alignment of the images by computing the average angle difference of the matches. The dissimilarity measure D_m is now also made proportional to the average angle difference of the features after this global alignment:

$$D_m = \frac{1}{N} \cdot \frac{n_1 + n_2}{2} \cdot \frac{\sum |\theta_i|}{N} \tag{3}$$

$$= \frac{(n_1 + n_2) \sum |\theta_i|}{2N^2}, \tag{4}$$

,where N corresponds to the number of matches found, n_i the number of extracted features in image i, and θ the angle difference for one match after global alignment.

4.2.3 Combined dissimilarity measure

We combine these two measures: only those pairs of images who have a colour dissimilarity under a predefined threshold are candidates for computing a matching dissimilarity.
This combined *visual* distance between two images is related to the physical distance between the corresponding viewpoints, but is certainly not a linear measure for it. As a matter of fact, the disparity and appearance difference of features is also related to the distance of the corresponding natural landmark to the cameras. Therefore, in large spaces (halls, market squares), a certain visual distance will be corresponding to a much larger physical distance, compared to the same visual distance between two images in a small space.
With this visual distance, the place definition problem in section 4.1 can be addressed on the basis of a constant visual distance between places instead of a constant physical distance.

4.3 Map Building Algorithm

We apply the mathematical theory of Demster and Shafer (Dempster, 1967; Shafer, 1976) on the topological map building problem posed. Out of a series of omnidirectional images, acquired at constant rate during a tour through the environment. Firstly, these images are clustered into *places*. Then, loop closing hypotheses are formulated between visually similar places of which evidence is collected using Dempster-Shafer theory. Once decisions are made about these hypotheses, the correct topology of the world is known.
This technique makes it possible to cope with *self-similar* environments. Places that look alike but are different will more likely get their link hypothesis rejected.

4.3.1 Image clustering

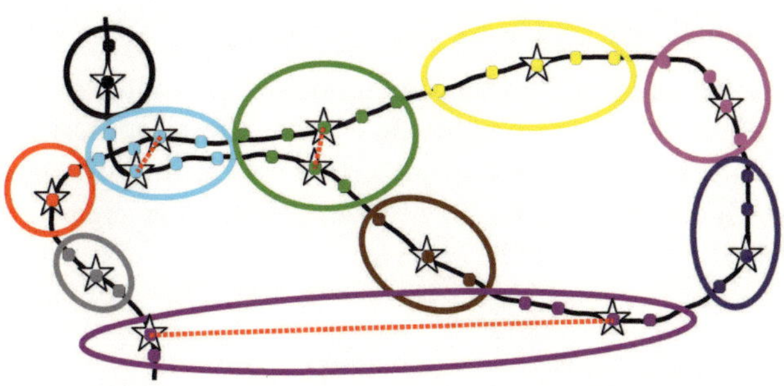

Figure 6. Example for the image clustering and hypothesis formulation algorithms. Dots are image positions, black is exploration path, clusters are visualised with ellipses, prototypes of (sub)clusters with a star. Hypotheses are denoted by a dotted red line

The dots in the sketch figure of 6 denote places where images are taken. Because they were taken at constant time intervals and the robot did not drive at a constant speed, they are not evenly spread. We perform agglomerative clustering with complete linkage based on the combined visual distance (see section 4.2.3) on all the images, yielding the ellipse shaped clusters in fig. 6. The black line shows the exploration path as driven by the robot.

4.3.2 Hypothesis formulation

As can be seen in the lower part of fig. 6, not all image groups nicely cover one distinct place. This is due to *self-similarities*, or distinct places in the environment that are different but look alike and thus yield a small visual distance between them.

For each of the clusters, we can define one or more subclusters. Images within one cluster which are linked by exploration path connections are grouped together. For each of these *subclusters* a prototype image is chosen as the *medoid*[1] based on the visual distance, denoted as a star in the figure.

For each pair of these subclusters within the same cluster, we define a *loop closing hypothesis* H, which states that if H=*true*, the two subclusters describe the same physical place and must be merged together. We will use Dempster-Shafer theory to collect evidence about each of these hypotheses.

4.3.3 Dempster-Shafer evidence collection

For each of the hypotheses defined in the previous step, a decision must be made if it was correct or wrong. Figure 7 illustrates four possibilities for one hypothesis. We observe that a hypothesis has more chance to be true if there are more hypotheses in the neighbourhood, like in case *a* and *b*. If no neighbouring hypotheses are present (*c*,*d*), no more evidence can be found and no decision can be made based on this data.

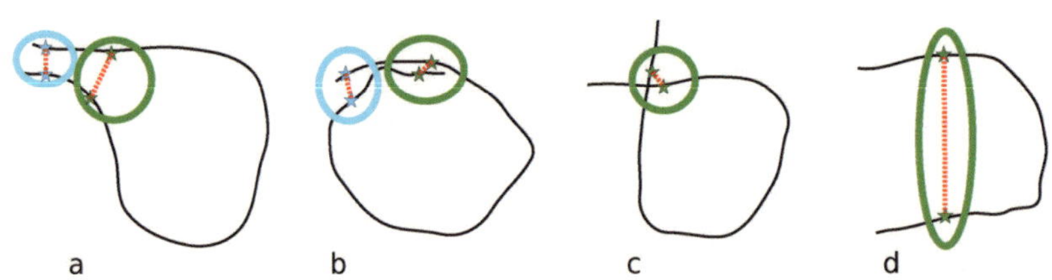

Figure 7. Four topological possibilities for one hypothesis

We conclude that for a certain hypothesis, a neighbouring hypothesis adds evidence to it. It is clear that, the further away this neighbour is from the hypothesis, the less certain the given evidence is. We chose to model this subjective uncertainty by means of the ignorance notion in Dempster-Shafer theory. That is why we define an *ignorance function* ξ containing the distance between two hypotheses H_a and H_b:

$$\xi(H_a, H_b) = \begin{cases} 1 - \sin\left(\frac{d_H(H_a,H_b)\pi}{2d_{th}}\right) & (d_H \leq d_{th}) \\ 0 & (d_H > d_{th}) \end{cases} \tag{8}$$

[1]The medoid of a cluster is computed analogous to the centroid, but using the median instead of the average.

where d_{th} is a distance threshold and d_H is the sum of the distances between the two pairs of prototypes of both hypotheses, measured in number of exploration images.
To gather *aleatory* evidence, we look at the visual similarity of both subcluster prototypes, normalised by the standard deviation of the intra-subcluster visual similarities. The visual similarity is the inverse of the visual distance, defined in equation 3.
Each neighbouring hypothesis H_b yields the following set of Dempster-Shafer masses, to be combined with the masses of the hypothesis H_a itself:

$$\begin{aligned} m(\{\emptyset\}) &= 0 \\ m(\{H_a\}) &= s_V(H_b)\xi(H_a, H_b) \\ m(\{\neg H_a\}) &= (1 - s_V(H_b))\xi(H_a, H_b) \\ m(\{H_a, \neg H_a\}) &= 1 - \xi(H_a, H_b) \end{aligned} \tag{9}$$

Hypothesis masses are initialised with the visual similarity of its subcluster prototypes and an initial ignorance value (0.25 in our experiments), which models its influenceability by neighbours.

4.4 Hypothesis decision

After combination of each hypothesis's mass set with the evidence given by neighbouring hypotheses (up to a maximum distance d_{th}), a decision must be made if this hypothesis was correct and thus if the subclusters must be united into one place or not.
Unfortunately, as stated above, only positive evidence can be collected, because we can not gather more information about totally isolated hypotheses (like *c* and *d* in fig. 7). This is not too bad, because of different reasons. Firstly, the chance for correct, but isolated hypothesis (case *c*) is low in typical cases. Also, adding erroneous loop closings (*c* and *d*) will yield an incorrect topological map, whereas leaving them out will keep the map useful for navigation, but a bit less complete. Of course, new data about these places can be aqcuired later, during navigation.
It is important to remind oneself that the computed Dempster-Shafer masses can not directly be interpreted as probabilities. That is why we compute the support and plausibility of each hypothesis after evidence collection. Because these values define a confidence interval for the real probability, a hypothesis can be accepted if the lower bound (the support) is greater than a threshold.
After this decision, a final topological map can be built. Subclusters connected with accepted hypotheses are merged into one place, and a new medoid is computed as prototype of it. For hypotheses that are not accepted, two distinct places should be constructed.

5. Localisation

When the system has learnt a topological map of an environment, this map can be used for a variety of navigational tasks, firstly *localisation*. For each arbitrary new position in the known environment, the system can find out *where* it is. The output of this localisation algorithm is a *location*, which is—opposed to other methods like GPS—not expressed as a metric coordinate, but as one of the topological places defined earlier in the formerly explained map building stage.
The training set doesn't need to cover every imaginable position in the environment. A relatively sparse coverage is sufficient to localise every possible position. That is because the

image comparison method we developed is based on wide baseline techniques and hence can recognise scene items from substantially different viewpoints.
Actually, two localisation modes exist. When starting up the system, there is no *a priori* information on the location. Every location is equally probable. This is called *global localisation,* alias the *kidnapped robot problem*. Traditionally, this is known to be a hard problem in robot localisation. In contrast, if there is knowledge about a former localisation not too long ago, the locations in the proximity of that former location have a higher probability than others further away. This is called *location updating*.
We propose a system that is able to cope with both localisation modes. A probabilistic approach is taken. Instead of making a hard decision about the location, a probability value is given to each location at each time instant. The Bayesian approach we follow is explained in the next subsection.

5.1 Bayesian Filtering

Define $x \in X$ a place of the topological map. Z is the collection of all omnidirectional images z, so that $z(x)$ corresponds to the training observation at place x. At a certain time instant t, the system acquires a new image z_t. The goal of the localisation algorithm is to reveal the place x_t where this image was taken.
We define the *Belief function* $Bel(x,t)$ as the probability of being at place x at time t, given all previous observations. So,

$$Bel(x,t) = P(x_t|z_t, z_{t-1}, \ldots, z_0), \tag{10}$$

for all $x \in X$. In the *kidnapped robot* case, there is no knowledge about previous observations hence $Bel(x,t_0)$ is initialised equal for all x.
Using Bayes' rule, we find:

$$Bel(x,t) = \frac{P(z_t|x_t, z_{t-1}, \ldots, z_0)P(x_t|z_{t-1}, \ldots, z_0)}{P(z_t|z_{t-1}, \ldots, z_0)}. \tag{11}$$

Because the denominator of this fraction is not dependent on x, we replace it by the normalising constant η. If we know the current location of the system, we assume that future locations do not depend on past locations. This property is called the *Markov Assumption*. Using it, together with the probabilistic sum rule, equation 11 yields:

$$Bel(x,t) = \eta\, P(z_t|x_t) \sum_{x_{t-1} \in X} [P(x_t|x_{t-1})Bel(x, t-1)] \tag{12}$$

This allows us to calculate the belief recursively based on two variables: the *next state density* or *motion model* $P(x_t|x_{t-1})$ and the *sensor model* $P(z_t|x_{t-1})$.

5.2 Motion Model

The motion model $P(x_t|x_{t-1})$ explicits the probability of a transition from one place x_{t-1} to another x_t. It seems logical to assume that a transition in one time instant between places that are far from each other is less probable than between places close to each other. We model this effect with a Gaussian:

$$P(x_t|x_{t-1}) = \frac{1}{\beta_x} e^{-dist(x_{t-1},x_t)/\sigma_x{}^2}. \tag{13}$$

In this equation, the function $dist(x_1,x_2)$ corresponds to a measurement of the distance between the two places. We approximate it as the minimum number of place transitions needed to go from x_1 to x_2 on the topological map, computed with the Dijkstra algorithm (Dijkstra, 1959). In equation 13, β_x is a normalisation constant, and σ_x^2 is the variance of the distances, measured on the map data. Once the topological map is known, the complete motion model can be computed off-line for usage during localisation.

5.3 Sensor Model

The entity $P(z_t | x_{t-1})$, called the sensor model, is the probability of acquiring a certain observation z_t if the location x_{t-1} is known. This is related to the visual dissimilarity of that observation and the training observation at location . The probability of acquiring an image at a certain place that differs much from the training image taken at that place has a low probability. We model this sensor model also by a Gaussian:

$$P(z_t|x_t) = \frac{1}{\beta_z} e^{-diss(z_t,z(x_t))/\sigma_z}. \tag{14}$$

This time, the function $diss(z_1,z_2)$ refers to the visual dissimilarity explained in section 4.2.3. Unlike the motion model, the sensor model cannot be computed beforehand. It depends on the newly incoming query image data. Every location update step the visual dissimilarities of the query image with many database images must be computed. This validates our efforts to make the computation of the visual dissimilarity measure as fast as possible.

6. Path planning

With the method of the previous section, at each time instant the most probable location of the robot can be found, from which a path to a goal can be determined. How the user of the system, for instance the wheelchair patient, gives the instruction to go towards a certain goal is highly dependent on the situation. For every disabled person, for instance, an individual interface must be designed adapted to his/her possibilities.

We assume a certain goal is expressed as a certain place of the topological map, e.g. as a voice command <<*Kitchen!* >>. From the present pose, computed by the localisation algorithm, a path can be easily found towards it using Dijkstra's algorithm (Dijkstra, 1959). This path is expressed as a series of topological places which are traversed.

7. Visual servoing

The algorithm described in this section makes the robot move along a path, computed by the previous section. Such a path is given as a sparse set of prototype images of places. The physical distance between two consecutive path images is variable (1 to 5 metres in our tests), but the visual distance is constant, such that there are enough local feature matches as needed by this algorithm.

It is easy to see that following such a sparse visual path boils down to a succession of *visual homing* operations. First, the robot is driven towards the place where the first image on the path is taken. When arrived, it is driven towards the next path image, and so on. Because a

smooth path is desired for the application, the motion must be continuous without stops at path image positions.
We tackle this problem by estimating locally the spatial structure of the wide baseline features using epipolar geometry. Hence, at this point we bring in some 3D information. This may seem at odds with our topological approach, but the depth maps are very sparse and only calculated locally so that errors are kept local, don't suffer from error build-up, and are efficient to compute.

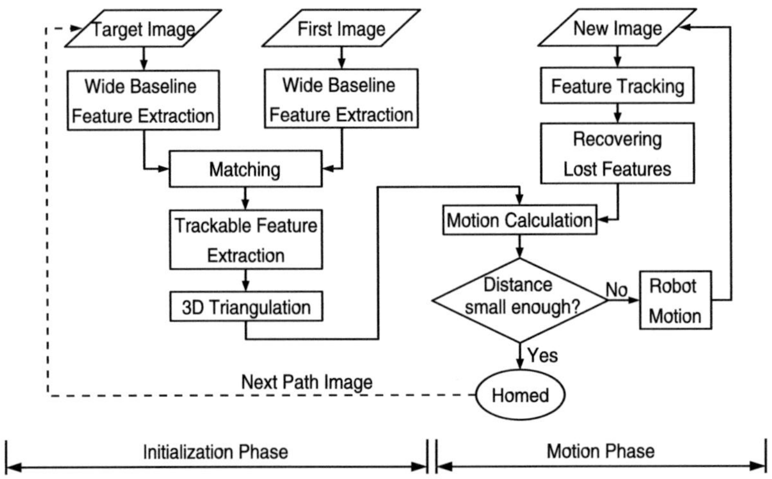

Figure 8. Flowchart of the proposed algorithm for visual servoing along a path

Fig. 8 offers an overview of the proposed method. Each of the *visual homing* operations is performed in two phases, an initialisation phase (section 7.1) and an iterated motion (section 7.2) phase.

7.1 Initialisation phase

From each position within the reach of the next path image (the *target* image), a visual homing procedure can be started. Our approach first establishes wide baseline local feature correspondences between the present and the target image, as described in section 3. That information is used to compute the epipolar geometry, which enables us to construct a local map containing the feature world positions, and to compute the initial homing vector.

7.1.1 Epipolar geometry estimation

Our calibrated single-viewpoint omnidirectional camera is composed of a hyperbolic mirror and a perspective camera. As imaging model, we use the model proposed by (Svoboda & Pajdla, 1998). This enables the computation of the epipolar geometry based on 8 point correspondences. In (Svoboda, 1999), Svoboda describes a way to robustly estimate the *essential matrix E*, when there are outliers in the correspondence set. The essential matrix is the equivalent of the fundamental matrix in the case of known internal camera calibration, .
Svoboda's so-called *generate-and-select* algorithm to estimate *E* is based on repeatedly solving an overdetermined system built from the correspondences that have a low *'outlierness'* and evaluating the *quality measure* of the resulting essential matrix. Because our tests with this

method did not yield satisfactory results, we implemented an alternative method based on the well-known Random Sample Consensus (RANSAC (Fischler & Bolles, 1981)) paradigm.

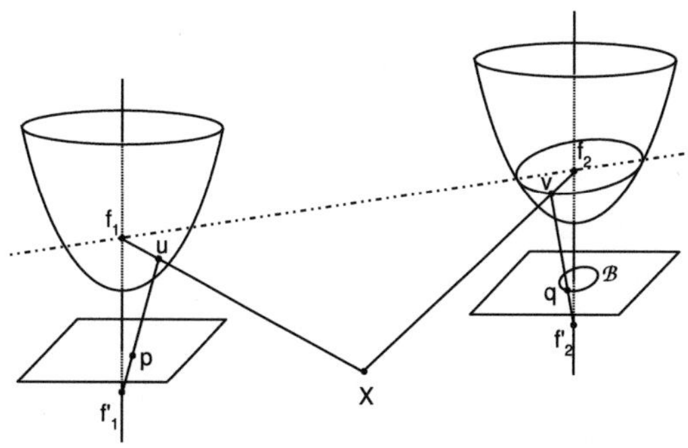

Figure 9. Projection model for a pair of omnidirectional images

The set-up is sketched in fig. 9. One visual feature with world coordinates **X** is projected via point **u** on the first mirror to point **p** in the image plane of the first camera. In the second camera, the mirror point is called **v** and the image plane point **q**. For each of the correspondences, the mirror points **u** and **v** can be computed as

$$\mathbf{u} = \mathcal{F}(K^{-1}\mathbf{p})K^{-1}\mathbf{p} + \mathbf{t}_C, \tag{15}$$

with $t_C=[0,0,-2e]^T$ and

$$\mathcal{F}(\mathbf{x}) = \frac{b^2(ex_1 + a\|\mathbf{x}\|)}{b^2x_1^2 - a^2x_2^2 - a^2x_3^2}. \tag{16}$$

In these equations K is the internal calibration matrix of the camera, and a, b and e are the parameters of the hyperbolic mirror.
If E is the *essential matrix,* for all correspondences $\boldsymbol{v}^T E\boldsymbol{u}=\boldsymbol{0}$. This yields for each correspondence pair one linear equation in the coefficients of $E=[e_{ij}]$.
For each random sample of 8 correspondences, an E matrix can be calculated. This is repeatedly done and for each E matrix candidate the inliers are counted. A correspondence is regarded an inlier if the second image point **q** lies within a predefined distance from the epipolar *ellipse,* defined by the first image point **q**. This epipolar ellipse B with equation $\boldsymbol{x}^T B\boldsymbol{x}=\boldsymbol{0}$ is computed with $B=$

$$\begin{bmatrix} -4t^2a^2e^2 + r^2b^4 & rsb^4 & rtb^2(-2e^2+b^2) \\ rsb^4 & -4t^2a^2e^2 + s^2b^4 & stb^2(-2e^2+b^2) \\ rtb^2(-2e^2+b^2) & stb^2(-2e^2+b^2) & t^2b^4 \end{bmatrix} \tag{17}$$

From the one essential matrix E with the maximal number of inliers the motion between the cameras can be computed using the SVD based method proposed by (Hartley,1992). If more than one E-matrix is found with the same maximum number of inliers, the one is chosen with the best (i.e. smallest) quality measure , where is the ith singular value of the matrix E. Out of this relative camera motion, a first estimate of the homing vector is derived. During the motion phase this homing vector is refined.

7.1.2 Local feature map estimation

In order to start up the succession of tracking iterations, an estimate of the local map must be made. In our approach the local feature map contains the 3D world positions of the visual features, centred at the starting position of the visual homing operation. These 3D positions are easily computed by triangulation.

We only use two images, the first and the target image, for this triangulation. This has two reasons. Firstly, these two have the widest baseline and therefore triangulation is best conditioned. Our wide baseline matches between these two images are also more plentiful and less influenced by noise than the tracked features.

7.2 Motion phase

Then, the robot is put into motion in the direction of the homing vector and an image sequence is recorded. We rely on lower-level collision detection, obstacle avoidance and trajectory planning algorithms to drive safely (Demeester et al., 2003). In each new incoming image the visual features are tracked. Robustness to tracking errors (caused by e.g. occlusions) is achieved by reprojecting lost features from their 3D positions back into the image. These tracking results enable the calculation of the present location and from that the homing vector towards which the robot is steered.

When the (relative) distance to the target is small enough, the entire homing procedure is repeated with the next image on the sparse visual path as target. If the path ends, the robot is stopped at a position close to the position where the last path image was taken. This yields a smooth trajectory along a sparsely defined visual path.

7.2.1 Feature tracking

The corresponding features found between the first image and the target image in the previous step, also have to be found in the incoming images during driving. This can be done very reliably performing every time wide baseline matching with the first or target image, or both. Although our methods are relatively fast this is still too time-consuming for a driving robot.

Because the incoming images are part of a smooth continuous sequence, a better solution is *tracking*. In the image sequence, visual features move only a little from one image to the next, which enables to find the new feature position in a small search space.

A widely used tracker is the KLT tracker of (Shi & Tomasi, 1994). KLT starts by identifying interest points (corners), which then are tracked in a series of images. The basic principle of KLT is that the definition of corners to be tracked is exactly the one that guarantees optimal tracking. A point is selected if the matrix

$$\begin{bmatrix} g_x^2 & g_x g_y \\ g_x g_y & g_y^2 \end{bmatrix}, \qquad (18)$$

containing the partial derivatives of the image intensity function over an $N \times N$ neighbourhood, has large eigenvalues. Tracking is then based on a Newton-Raphson style minimisation procedure using a purely translational model. This algorithm works surprisingly fast: we were able to track 100 feature points at 10 frames per second in 320×240 images on a 1 GHz laptop.

Because the well trackable points are not necessarily coinciding with the anchor points of the wide baseline features to be tracked, the best trackable point in a small window around such an anchor point is selected. In the assumption of local planarity we can always find back the corresponding point in the target image via the relative reference system offered by the wide baseline feature.

7.2.2 Recovering lost features

The main advantage of working with this calibrated system is that we can recover features that were lost during tracking. This avoids the problem of losing all features by the end of the homing manoeuvre, a weakness of our previous approach (Goedemé et al., 2005). This feature recovery technique is inspired by the work of (Davison, 2003), but is faster because we do not work with probability ellipses.

In the initialisation phase, all features are described by a local intensity histogram, so that they can be recognised after being lost during tracking. Each time a feature is successfully tracked, this histogram is updated.

When tracking, some features are lost due to invisibility because of e.g. occlusion. Because our local map contains the 3D positions of each feature, and the last robot position in that map is known, we can reproject the 3D feature in the image. Svoboda shows that the world point X_C (i.e. the point X expressed in the camera reference frame) is projected on point p in the image:

$$\mathbf{p} = \frac{K}{2e}(\lambda \mathbf{X}_C - \mathbf{t}_C), \tag{19}$$

wherein λ is the largest solution of

$$\lambda = \frac{b^2(-e)\mathbf{X}_{C3} \pm a\|\mathbf{X}_C\|}{b^2\mathbf{X}_{C3}^2 - a^2\mathbf{X}_{C1}^2 - a^2\mathbf{X}_{C2}^2}. \tag{20}$$

Based on the histogram descriptor, all trackable features in a window around the reprojected point $\mathbf{p}$ are compared to the original feature. When the histogram distance is under a fixed threshold, the feature is found back and tracked further in the next steps.

7.2.3 Motion computation

When in a new image the feature positions are computed by tracking or backprojection, the camera position (and thus the robot position) in the general coordinate system can be found based on these measurements.

It is shown that the position of a camera can be computed when for three points the 3D positions and the image coordinates are known. This problem is know as the *three point perspective pose estimation problem*. An overview of the proposed algorithms to solve it is given by (Haralick et al., 1994). We chose the method of Grunert, and adapted it for our omnidirectional case.

Also in this part of the algorithm we use RANSAC to obtain a robust estimation of the camera position. Repeatedly the inliers belonging to the motion computed on a three-point sample are counted, and the motion with the greatest number of inliers is kept.

7.2.4 Robot motion

In subsection 7.1.1, it is explained how the position and orientation of the target can be extracted from the computed epipolar geometry. Together with the present pose results of the last subsection, a homing vector can easily be computed. This command is communicated to the locomotion subsystem. When the homing is towards the last image in a path, also the relative distance and the target orientation w.r.t. the present orientation is given, so that the locomotion subsystem can steer the robot to stop at the desired position. This is needed for e.g. docking at a table.

8. Experiments

8.1 Test platform

We have implemented the proposed algorithm, using our modified electric wheelchair 'Sharioto'. A picture of it is shown in the left of fig. 1. It is a standard electric wheelchair that has been equipped with an omnidirectional vision sensor (consisting of a Sony firewire colour camera and a Neovision hyperbolic mirror, right in fig. 1). The image processing is performed on a 1 GHz laptop.

8.2 Map building

The wheelchair was guided around in a large environment, while taking images. The environment was a large part of our office floor, containing both indoor and outdoor locations. This experiment yielded a database of 545 colour images with a resolution of 320×240 pixels. The total distance travelled was approximately 450 *m*. During a second run 123 images were recorded to test the localisation. A map and some of these images are shown in fig. 10.

After place clustering with a fixed place size threshold (in our experiments 0.5), this resulted in a set of 53 clusters. Using the Dempster-Shafer based evidence collection, 6 of 41 link hypotheses were rejected, as shown in fig. 11. Fig. 12 shows the resulting 59 place prototypes along with the accepted interconnections.

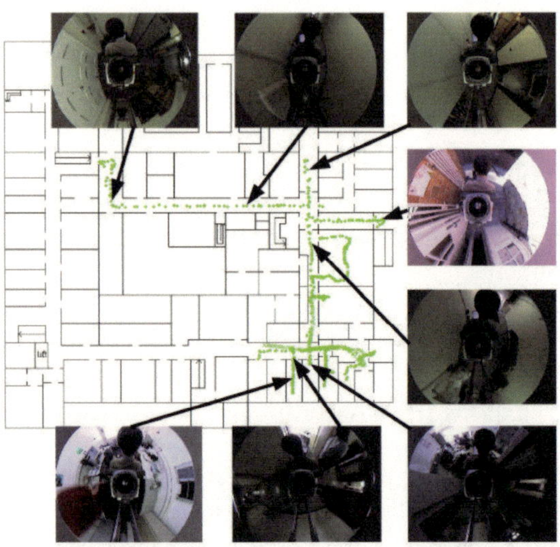

Figure 10. A map of the test environment with image positions and some of the images

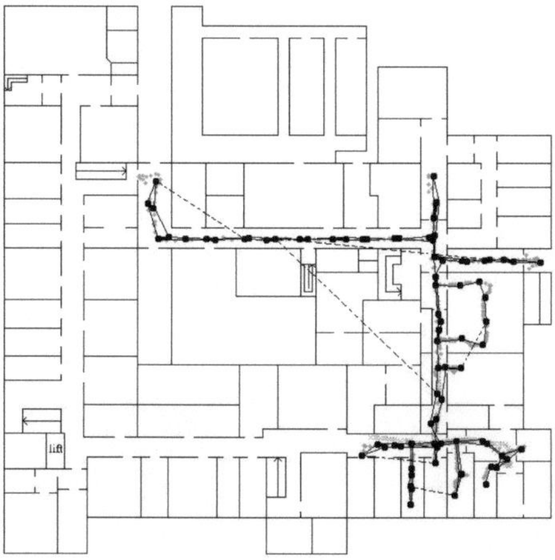

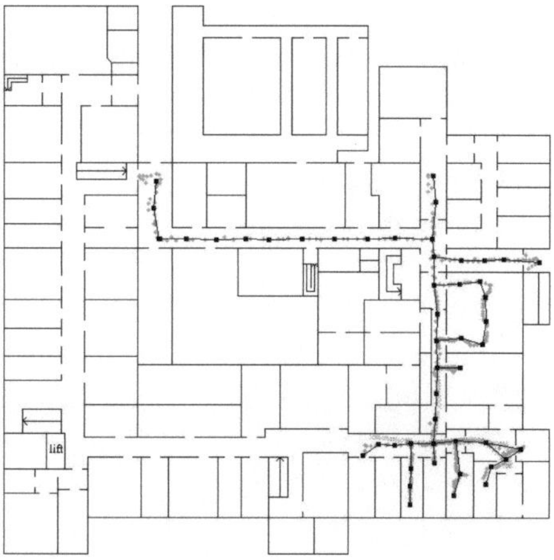

Figures 11 (left) and 12 (right). Left: topological loop closing, accepted hypotheses are shown in thick black lines, rejected in dashed thin black lines. Right: the resulting topological map, locations of the place prototypes with interconnections

Instead of keeping all the images in memory, the database is now reduced to only the descriptor sets of each prototype image. In our experiment, the memory needed for the database was reduced from 275 MB to 1.68 MB.

8.3 Localisation

From this map, the motion model is computed offline as explained in section 5.2. Now, for the separate test set, the accuracy of the localisation algorithm is tested. A typical experiment is illustrated in fig. 13.

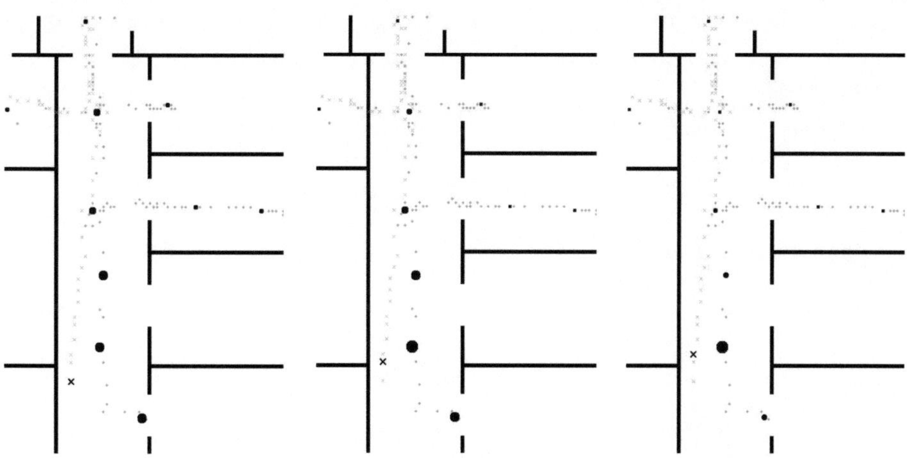

Figure 13. Three belief update cycles in a typical localisation experiment. The black x denotes the location of the new image. Place prototypes with a higher belief value are visualised as larger black circles

In total, for 78% of the trials the maximum of the belief function was located at the closest place at the first iteration, after the second and third belief update this percentage raised to 89% and 97%.

8.4 Visual servoing

8.4.1 Initialisation phase

During the initialisation phase of one visual homing step, correspondences between the present and target image are found and the epipolar geometry is computed. This is shown in fig. 14.

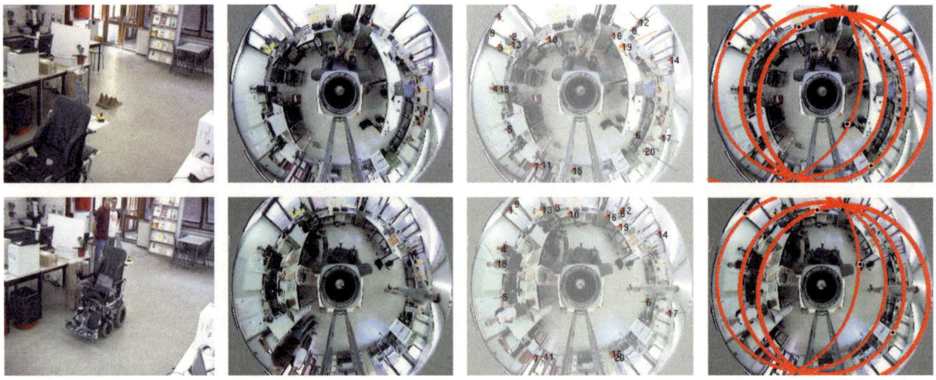

Figure 14. Results of the initialisation phase. Top row: target, bottom row: start. From left to right, the robot position, omnidirectional image, visual correspondences and epipolar geometry are shown

To test the correctness of the initial homing vector, we took images with the robot positioned at a grid with a cell size of 1 meter. The resulting homing vectors towards one of these images (taken at (6,3)) are shown in fig. 15. This proves the fact that even if the images are situated more than 6 metres apart the algorithm works thanks to the use of *wide baseline* correspondences.

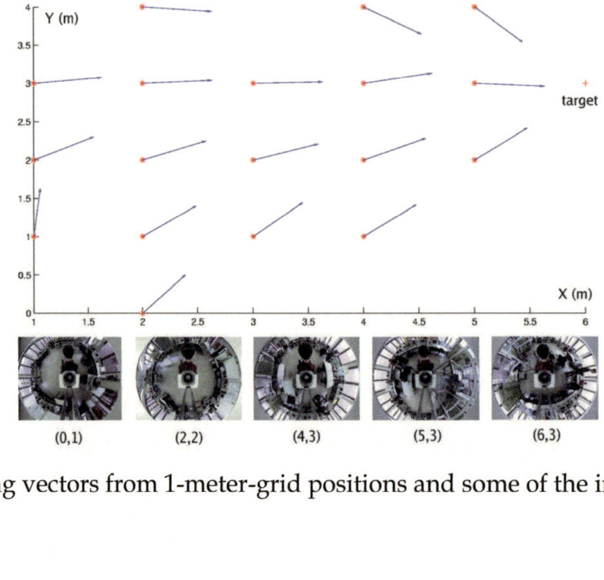

Figure 15. Homing vectors from 1-meter-grid positions and some of the images

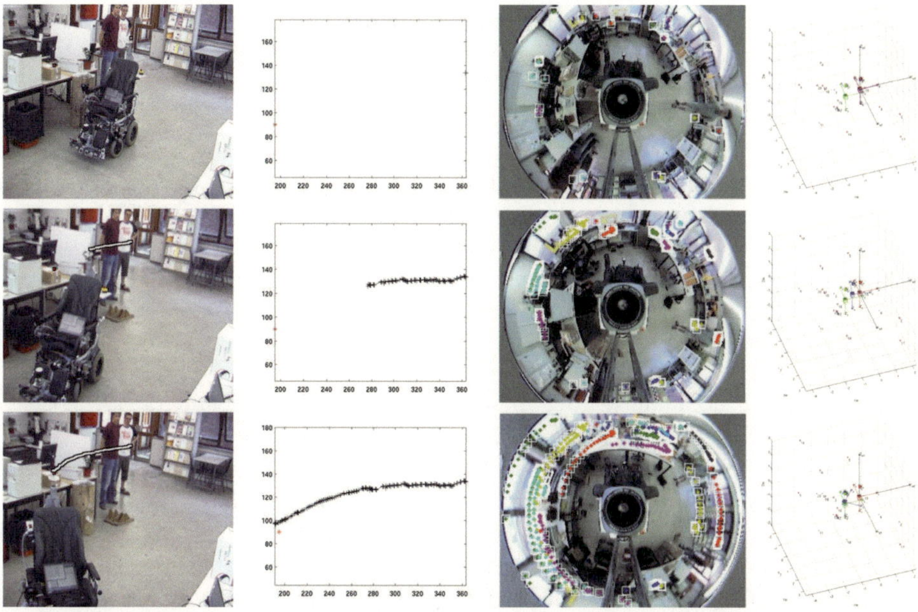

Figure 16. Three snapshots during the motion phase: in the beginning (left), half (centre) and at the end (right) of the homing motion. The first row shows the external camera image with tracked robot position. The second row shows the computed world robot positions [cm]. The third row shows the colour-coded feature tracks. The bottom row shows the sparse 3D feature map (encircled features are not lost)

8.4.2 Motion phase

We present a typical experiment in fig. 16. During the motion, the top of the camera system was tracked in a video sequence from a fixed camera. This video sequence, along with the homography computed from some images taken with the robot at reference positions, permits calculation of metrical robot position ground truth data.

Repeated similar experiments showed an average homing accuracy of 11 cm, with a standard deviation of 5 cm, after a homing distance of around 3 m.

8.4.3 Timing results

The algorithm runs surprisingly fast on the rather slow hardware we used: the initialisation for a new target lasts only 958 ms, while afterwards every 387 ms a new homing vector is computed. For a wheelchair driving at a cautious speed, it is possible to keep on driving while initialising a new target. This resulted in a smooth trajectory without stops or sudden velocity changes.

9. Conclusion

This chapter describes and demonstrates a novel approach for a service robot to navigate autonomously in a large, natural complex environment. The only sensor is an omnidirectional colour camera. As environment representation, a topological map is chosen. This is more flexible and less memory demanding than metric 3D maps. Moreover, it does

not show error build-up and enables fast path planning. As natural landmarks, we use two kinds of fast wide baseline features which we developed and adapted for this task. Because these features can be recognised even if the viewpoint is substantially different, a limited number of images suffice to describe a large environment.
Experiments show that our system is able to build autonomously a map of a natural environment it drives through. The localisation ability, with and without knowledge of previous locations, is demonstrated. With this map, a path towards each desired location can be computed efficiently. Experiments with a robotic wheelchair show the feasibility of executing such a path as a succession of visual servoing steps.

10. References

Baumberg, A. (2000). Reliable feature matching across widely separated views, *Computer Vision and Pattern Recognition*, pp. 774-781, Hilton Head, South Carolina.

Basri, R.; Rivlin, E. & Shimshoni, I. (1998), Visual homing: Surfing on the epipoles, *in IEEE International Conference on Computer Vision ICCV'98*, pp. 863-869, Bombay.

Beevers, K. & Huang W.(2005). Loop Closing in Topological Maps, *ICRA*, Barcelona, Spain.

Bischof, H.; Wildenauer, H. & Leonardis, A. (2001). Illumination insensitive eigenspaces, *In Proc. ICCV01*, pages 233-238, IEEE Computer Society, Vancouver, Canada.

Bülthoff, H.H.; van Veen, H.; Distler, H.K. & Braun, S.J. (1998), Navigating through a virtual city: Using virtual reality technology to study human action and perception, *Future Generation Computer Systems*, 14, pp. 231-242.

Cartwright, B. & Collett, T. (1987). *Landmark Maps for Honeybees*, Biological Cybernetics, 57, pp. 85-93.

Chen Ch. & Wang, H. (2005). Appearance-based topological Bayesian inference for loop-closing detection in cross-country environment, *IROS*, pp. 322-327, Edmonton.

Davison, A. (2003). Real-time simultaneous localisation and mapping with a single camera, *Intl. Conf. on Computer Vision*, Nice, France.

Dempster, A. P. (1967). *Upper and Lower Probabilities Induced by a Multivalued Mapping*, The Annals of Statistics (28), pp. 325–339, 1967.

Demeester, E.; Nuttin, M.; Vanhooydonck, D. & Van Brussel, H. (2003). Fine Motion Planning for Shared Wheelchair Control: Requirements and Preliminary Experiments, *International Conference on Advanced Robotics*, pp. 1278-1283, Coimbra, Portugal.

Dijkstra, E. W. (1959). *A note on two problems in connection with graphs*, Numerische Mathematik, 1: 269-271, 1959.

Fischler, M. & Bolles, R. (1981). *Random Sample Consensus: A Paradigm for Model Fitting with Applications to Image Analysis*, Comm. of the ACM, Vol 24, pp 381-395, 1981.

Franz, M.; Schölkopf, B.; Mallot, H. & Bülthoff, H. (1998). *Where did I take that snapshot? Scene-based homing by image matching*, Biological Cybernetics, 79, pp. 191-202, 1998.

Goedemé, T.; Tuytelaars, T. & Van Gool, L. (2004). Fast Wide Baseline Matching with Constrained Camera Position, *Computer Vision and Pattern Recognition*, Washington, DC, pp. 24-29.

Goedemé, T.; Tuytelaars, T.; Vanacker, G.; Nuttin, M. & Van Gool, L. (2005). Feature Based Omnidirectional Sparse Visual Path Following, *IEEE/RSJ International Conference on Intelligent Robots and Systems, IROS 2005*, Edmonton, Canada.

Haralick, R.; Lee, C.; Ottenberg, K. & Nölle, M. (1994). *Review and analysis of Solutions of the Three Point Perspective Pose Estimation Problem*, International Journal of Computer Vision, 13, 3, pp. 331-356, 1994.

Hartley, R. (1992). Estimation of relative camera positions for uncalibrated cameras, *2nd European Conference on Computer Vision*, pp. 579-587, Springer-Verlag, LNCS 588.

Jogan, M. & Leonardis, A. (1999). Panoramic eigenimages for spatial localisation, *In Proceedings 8th International Conference on Computer Analysis of Images and Patterns(1689)*, Solina, Franc and Leonardis, Ales, Eds., pages 558-567, Ljubljana.

Koenig, S. & Simmons, R. (1996). Unsupervised learning of probabilistic models for robot navigation, *Proceedings of ICRA, 1996.*

Kosecká, J. & Yang, X. (2004). Global Localization and Relative Pose Estimation Based on Scale-Invariant Features, *ICPR* (4), pp. 319-322, 2004.

Ledwich, L. & Williams, S. (2004). Reduced SIFT Features For Image Retrieval and Indoor Localisation, *Australasian Conf. on Robotics and Automation ACRA*, Canberra, Australia.

Lowe, D. (1999). Object Recognition from Local Scale-Invariant Features, *International Conference on Computer Vision*, pp. 1150-1157, Corfu, Greece.

Matas, J.; Chum, O.; Urban, M. & Pajdla, T. (2002). Robust wide baseline stereo from maximally stable extremal regions, *British Machine Vision Conference*, Cardiff, Wales, pp. 384-396.

Mariottini, G.; Alunno, E.; Piazzi, J. & Prattichizzo, D. (2005). Epipole-Based Visual Servoing with Central Catadioptric Camera, *IEEE ICRA05*, Barcelona, Spain.

Mikolajczyk, K. & Schmid, C. (2002). An affine invariant interest point detector, *ECCV*, vol. 1, 128–142, Copenhagen, Denmark.

Mindru, F.; Moons, T. & Van Gool, L. (1999). Recognizing color patters irrespective of viewpoint and illumination, *Computer Vision and Pattern Recognition*, vol. 1, pp. 368-373.

Nistér, D.; Naroditsky, O. & Bergen, J. (2004). Visual Odometry, *Conference on Computer Vision and Pattern Recognition*, Washington, DC.

Nuttin, M., Demeester, E.; Vanhooydonck, D. & Van Brussel, H. (2001). *Shared autonomy for wheelchair control: Attempts to assess the user's autonomy*, in Autonome Mobile Systeme, pp. 127-133, 2001.

Pollefeys, M.; Van Gool, L.; Vergauwen, M.; Verbiest, F.;. Cornelis, K; Tops, J.; Koch, R. (2004). Visual modeling with a hand-held camera, *International Journal of Computer Vision* 59(3), 207-232, 2004.

Ranganathan, A.; Menegatti E. & Dellaert, F., (2005). *Bayesian Inference in the Space of Topological Maps*, IEEE Transactions on Robotics, 2005.

Sagüés, C. & Guerrero, J. (2005). Visual correction for mobile robot homing, *Robotics and Autonomous Systems*, Vol. 50, no. 1, pp 41-49, 2005.

Schmid, C.; Mohr, R.; Bauckhage, C. (1997). *Local Grey-value Invariants for Image Retrieval,* International Journal on Pattern Analysis an Machine Intelligence, Vol. 19, no. 5, pp. 872-877, 1997.

Se, S.; Lowe, D. & Little, J. (2001) Local and Global Localization for Mobile Robots using Visual Landmarks, *In proceedings of the IEEE/RSJ International Conference on Intelligent Robots and Systems (IROS '01)*, Hawaii, USA, 2001.

Shafer, G. (1976). *A Mathematical Theory of Evidence*, Princeton University Press, 1976.

Shatkay, H. & Kaelbling, L. P. (1997) Learning Topological Maps with Weak Local Odometric Information, *IJCAI* (2), pp. 920-929, 1997.

Shi, J. & Tomasi, C. (1994) Good Features to Track, *Computer Vision and Pattern Recognition*, Seattle, pp. 593-600, 1994.

Svoboda, T.; Pajdla, T. and Hlavá, V. (1998) Motion Estimation Using Panoramic Cameras, *Conf. on Intelligent Vehicles*, Stuttgart, pp. 335-340, 1998.

Svoboda, T. (1999) *Central Panoramic Cameras, Design, Geometry, Egomotion*, PhD Thesis, Czech Technical University, 1999.

Tapus, A. & Siegwart, R. (2005) Incremental Robot Mapping with Fingerprints of Places, *IROS*, Edmonton, Canada.

Tuytelaars, T.; Van Gool, L.; D'haene, L. & Koch, R. (1999) Matching of Affinely Invariant Regions for Visual Servoing, *Intl. Conf. on Robotics and Automation*, pp. 1601-1606, 1999.

Tuytelaars, T. & Van Gool, L. (2000) Wide baseline stereo based on local, affinely invariant regions, *British Machine Vision Conference*, pp. 412-422, Bristol, UK.

Ulrich, I. & Nourbakhsh, I.(2000)Appearance-Based Place Recognition for Topological Localisation, *IEEE International Conference on Robotics and Automatisation*, pp. 1023-1029, San Francisco, CA, April 2000,

Vale, A.; Ribeiro, M. I. (2003) Environment Mapping as a Topological Representation, *Proceedings of the 11th International Conference on Advanced Robotics - ICAR2003*, Universidade de Coimbra, Portugal, June 30 - July 1- 3, 2003.

Zivkovic, Z.; Bakker, B. & Kröse, B. (2005), Hierarchical Map Building Using Visual Landmarks and Geometric Constraints, in proceedings of the International conference on Intelligent Robots and Systems (IROS), pp. 7-12, Edmonton, Canada.

5

A Practical Approach for Motion Planning of Wheeled Mobile Robots

Luis Gracia and Josep Tornero
Technical University of Valencia
Spain

1. Introduction

Wheeled Mobile Robots (WMR) have been widely studied in the past fifteen years. Due to kinematic constraints, many WMR are not integrable (non-holonomic). Therefore, standard techniques developed for robot manipulators are not directly applicable. In particular, the motion planning of WMR is still a relevant issue. Examples of motion planning for WMR are available in the literature (Latombe, 1991; Laumond et al., 1997; Gracia & Tornero, 2003; Borenstein & Koren, 1989). On the other hand, the singularity of WMR kinematics must be avoided since it implies slip or impossible control actions (Gracia & Tornero, 2007a). In the same way, in the vicinity of singularities there is high amplification of active joints' error or high values for active joints. Therefore, the aim of the present research is to develop a practical approach for motion planning of WMR based on avoiding singularities. The chapter is organized as follows. Section 2 presents the kinematics of WMR considering four types of wheels: fixed, centered orientable (hereinafter orientable), castor and Swedish. Afterwards, section 3 discusses the possibilities for motion planning and develops a cost index based on singularity. To illustrate the applications of the proposed motion planning an industrial forklift is considered and several simulation results are shown. Finally, section 4 points out the more outstanding contributions of this research.

2. Kinematics of Wheeled Mobile Robots

Firstly it will be introduced some terminology. Assuming horizontal movement, the position of the WMR body is completely specified by 3 scalar variables (e.g. x, y, θ), referred to in (Campion et al., 1996) as WMR posture, $\mathbf{p}$ in vector form. Its first-order time derivative is called WMR velocity vector $\dot{\mathbf{p}}$ and separately (v_x, v_y, ω) WMR velocities (Muir & Neuman, 1987). Similarly, for each wheel, wheel velocity vector and wheel velocities are defined.

2.1 Kinematic models of wheels

The kinematic modeling of a wheel is used as a previous stage for modeling the whole WMR (Gracia & Tornero, 2007a; Campion et al., 1996; Muir & Neuman, 1987; Alexander & Maddocks, 1989). Here, the four common wheels will be considered: *fixed*, *orientable*, *castor* and *Swedish*. As it is easy to obtain their equations using a vector approach, e.g. see (Gracia

& Tornero, 2007a) among many other possibilities, the detailed development will be omitted. The matrix equation of the off-centered orientable wheel or *castor* wheel is:

$$\mathbf{v}_{\text{slip } i} = \begin{pmatrix} \cos(\beta_i + \delta_i) & \sin(\beta_i + \delta_i) & l_i \sin(\beta_i + \delta_i - \alpha_i) - d_i \cos\delta_i & -d_i \cos\delta_i & 0 \\ -\sin(\beta_i + \delta_i) & \cos(\beta_i + \delta_i) & l_i \cos(\beta_i + \delta_i - \alpha_i) + d_i \sin\delta_i & d_i \sin\delta_i & r_i \end{pmatrix} \begin{pmatrix} \dot{\mathbf{p}} \\ \dot{\beta}_i \\ \dot{\varphi}_i \end{pmatrix}, \tag{1}$$

where it has been used the parameters of Fig. 1 (a) and the variables of Table 1.

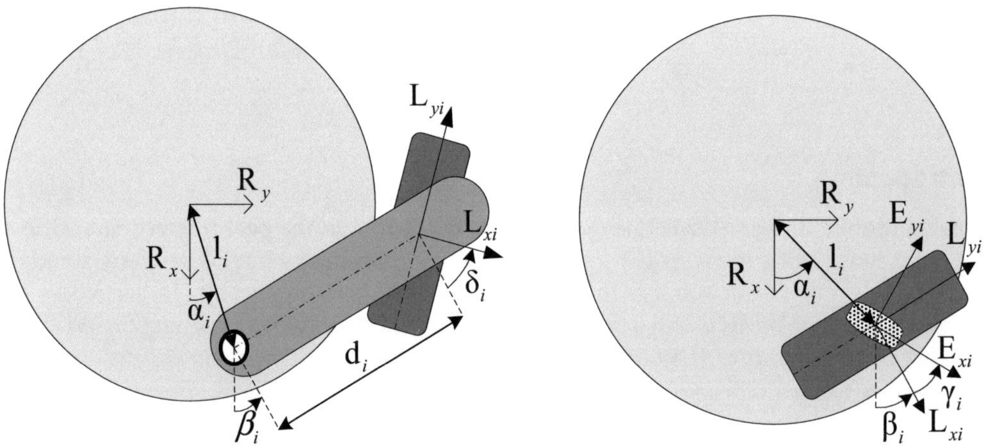

Figure 1. *Castor* wheel parameters: $l_i, d_i, \alpha_i, \beta_i, \delta_i$ *Swedish* wheel parameters: $l_i, \alpha_i, \beta_i, \gamma_i$

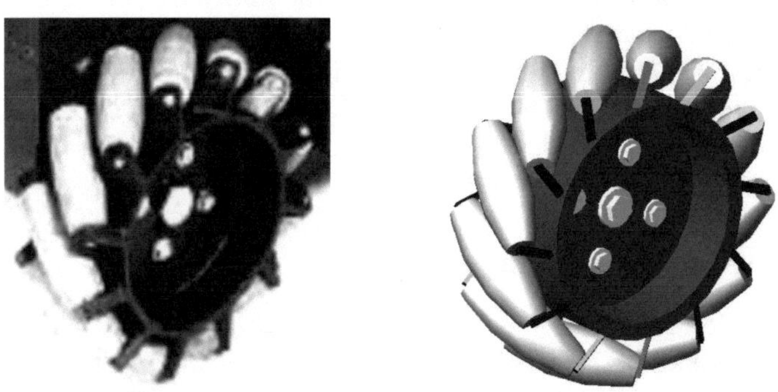

Figure 2. *Swedish* wheel (also called *Mecanum, Ilon* or *universal*) with rollers at 45°

The equation of the *orientable* wheel can be obtained from (1) with $d_i = \delta_i = 0$:

$$\mathbf{v}_{\text{slip } i} = \begin{pmatrix} \cos\beta_i & \sin\beta_i & l_i \sin(\beta_i - \alpha_i) & 0 \\ -\sin\beta_i & \cos\beta_i & l_i \cos(\beta_i - \alpha_i) & r_i \end{pmatrix} \begin{pmatrix} \dot{\mathbf{p}} \\ \dot{\varphi}_i \end{pmatrix}. \tag{2}$$

The previous equation is also valid for *fixed* wheels, where the angle β_i is constant.

The matrix equation of the *Swedish* wheel (see Fig. 2) is (3) where it has been used the parameters of Fig. 1 (b) and the variables and constants of Table 1.

Symbol	Description
R	Frame attached to the robot body with the Z-axis perpendicular to the floor surface
$\bar{R}$	Frame attached to the floor and instantaneously coincident with the robot frame R. This frame allows to avoid the dependency on a global stationary frame (Muir & Neuman, 1987)
(L_i, E_i)	Frames attached to the wheel *i* and to the roller of the *Swedish* wheel *i*, with the X-axes coincident with their rotation axle
$\dot{\mathbf{p}}$	WMR velocity vector in coordinate frame $\bar{R}$, equivalent to $({}^{\bar{R}}v_{Rx}\ {}^{\bar{R}}v_{Ry}\ {}^{\bar{R}}\omega_R)^T$ or $(v_x\ v_y\ \omega)^T$
$\mathbf{v}_{\text{slip}\,i}$	Sliding velocity vector of the wheel in coordinate frame L_i (E_i for *Swedish* wheels)
$(\dot{\beta}_i, \dot{\varphi}_i)$	Angular velocity of the steering link and rotation velocity of the wheel in L_{xi}-axis
$\dot{\varphi}_{ri}$	Rotation velocity of the rollers in E_{xi}-axis (it is usually a free wheel velocity)
(r_i, r_{ri})	Wheel equivalent radius and roller radius

Table 1. Frames, variables and constants

$$\mathbf{v}_{\text{slip}\,i} = \begin{pmatrix} \cos(\beta_i+\gamma_i) & \sin(\beta_i+\gamma_i) & l_i\sin(\beta_i+\gamma_i-\alpha_i) & r_i\sin\gamma_i & 0 \\ -\sin(\beta_i+\gamma_i) & \cos(\beta_i+\gamma_i) & l_i\cos(\beta_i+\gamma_i-\alpha_i) & r_i\cos\gamma_i & r_{ri} \end{pmatrix} \begin{pmatrix} \dot{\mathbf{p}} \\ \dot{\varphi}_i \\ \dot{\varphi}_{ri} \end{pmatrix} \tag{3}$$

2.2 Kinematic models of wheeled mobile robots

Once the type of WMR wheels and their equations are established, a compound kinematic equation for the WMR may be defined. Using (1), (2), and (3) we can obtain:

$$\mathbf{v}_{\text{slip}} = \begin{pmatrix} \mathbf{v}_{\text{slip}\,1} \\ \vdots \\ \mathbf{v}_{\text{slip}\,N} \end{pmatrix} = \begin{pmatrix} \mathbf{A}_{p1} & \mathbf{A}_{w1} & \cdots & 0 \\ \vdots & \vdots & \ddots & \vdots \\ \mathbf{A}_{pN} & 0 & \cdots & \mathbf{A}_{wN} \end{pmatrix} \begin{pmatrix} \dot{\mathbf{p}} \\ \dot{\mathbf{q}}_{w1} \\ \vdots \\ \dot{\mathbf{q}}_{wN} \end{pmatrix} = \begin{pmatrix} \mathbf{A}_p & \mathbf{A}_w \end{pmatrix} \begin{pmatrix} \dot{\mathbf{p}} \\ \dot{\mathbf{q}}_w \end{pmatrix} = \mathbf{A}\,\dot{\mathbf{q}}\,, \tag{4}$$

where N is the number of wheels; $\mathbf{v}_{\text{slip}}$ is the composite sliding velocity vector; $\dot{\mathbf{q}}_{wi}$ is a vector with all the wheel velocities of wheel *i*; $\dot{\mathbf{q}}_w$ is the composite vector of all the wheel velocities; $\dot{\mathbf{q}}$ is the vector of all the velocities; {$\mathbf{A}_{pi}$, $\mathbf{A}_{wi}$} are the multiplying matrices obtained from (1), (2), and (3); {$\mathbf{A}_p$, $\mathbf{A}_w$} are the composite multiplying matrices; and $\mathbf{A}$ is the WMR kinematic matrix. Under the *no-slip condition*, the kinematic solution for velocity vector $\dot{\mathbf{q}}$ results in:

$$\mathbf{A}\cdot\dot{\mathbf{q}} = \mathbf{0} \tag{5}$$

$$\dot{\mathbf{q}} \in \mathcal{N}(\mathbf{A}) \rightarrow \dot{\mathbf{q}} = \mathbf{B} \cdot \boldsymbol{\eta}, \tag{6}$$

where matrix **B** forms a basis of $\mathcal{N}(\mathbf{A})$, $\boldsymbol{\eta}$ is an m-dimensional vector representing WMR mobility, and m is the WMR mobility degree given by the nullity of **A**:

$$m = \dim(\boldsymbol{\eta}) = \dim(\mathcal{N}(\mathbf{A})) = \dim(\dot{\mathbf{q}}) - \operatorname{rank}(\mathbf{A}) = k - g\ . \tag{7}$$

In order to use variables with physical meaning, the mobility vector $\boldsymbol{\eta}$ should be replaced with a set of freely *assigned* velocities. Depending on whether wheel velocities or WMR velocities are chosen, a forward or inverse kinematic model is obtained. If a mix of both types of velocities is chosen a mixed solution is achieved. In order to check if an m-set of velocities $\dot{\mathbf{q}}_{\mathrm{a}}$ can be assigned, it must be verified that the determinant of the submatrix they define in (5) is non zero, that is:

$$\begin{pmatrix} \dot{\mathbf{q}}_{\mathrm{na}} \\ \dot{\mathbf{q}}_{\mathrm{a}} \end{pmatrix} = \begin{pmatrix} \mathbf{B}_{\mathrm{na}} \\ \mathbf{B}_{\mathrm{a}} \end{pmatrix} \cdot \boldsymbol{\eta} \tag{8}$$

$$\text{if } |\mathbf{B}_{\mathrm{a}}| \neq 0 \rightarrow \dot{\mathbf{q}}_{\mathrm{na}} = \mathbf{B}_{\mathrm{na}} \cdot \mathbf{B}_{\mathrm{a}}^{-1} \cdot \dot{\mathbf{q}}_{\mathrm{a}}\ , \tag{9}$$

where $\dot{\mathbf{q}}_{\mathrm{na}}$ are the remaining non-assigned velocities of $\dot{\mathbf{q}}$.
Alternatively to the previous procedure, based on the null space concept, it is possible to apply another method based on separating the m assigned velocities in (5):

$$\mathbf{A}_{\mathrm{na}}\ \dot{\mathbf{q}}_{\mathrm{na}} = -\mathbf{A}_{\mathrm{a}}\ \dot{\mathbf{q}}_{\mathrm{a}}\ . \tag{10}$$

To check if an m-set of velocities could be assigned $\dot{\mathbf{q}}_{\mathrm{a}}$, it must be verified that matrix $\mathbf{A}_{\mathrm{na}}$ is, in general, of full rank g:

$$\operatorname{rank}(\mathbf{A}_{\mathrm{na}}) = \operatorname{rank}(\mathbf{A}) = g\ . \tag{11}$$

Therefore, the singularity of a kinematic model is given by $|\mathbf{B}_{\mathrm{a}}| = 0$ in (9) or alternatively when matrix $\mathbf{A}_{\mathrm{na}}$ in (10) loses its full rank g. In (Gracia & Tornero, 2007a) it is characterized the singularity of WMR with a generic geometric approach.
On the other hand, (Gracia & Tornero, 2007b) consider a kinematic solution with redundant information (dim($\dot{\mathbf{q}}_{\mathrm{a}}$)>$m$) applying weighted left pseudoinverse to (10):

$$\dot{\mathbf{q}}_{\mathrm{na}} = -\left(\mathbf{A}_{\mathrm{na}}^{\mathrm{T}}\ \left(\sqrt{\boldsymbol{\mu}_{\mathrm{na}}}\right)^{\mathrm{T}} \sqrt{\boldsymbol{\mu}_{\mathrm{na}}}\ \mathbf{A}_{\mathrm{na}}\right)^{-1} \mathbf{A}_{\mathrm{na}}^{\mathrm{T}} \left(\sqrt{\boldsymbol{\mu}_{\mathrm{na}}}\right)^{\mathrm{T}} \sqrt{\boldsymbol{\mu}_{\mathrm{a}}}\ \mathbf{A}_{\mathrm{a}}\ \dot{\mathbf{q}}_{\mathrm{a}}\ . \tag{12}$$

where $\left(\sqrt{\boldsymbol{\mu}_{\mathrm{na}}}, \sqrt{\boldsymbol{\mu}_{\mathrm{a}}}\right)$ are the pre-multiplying weight matrices in (10) and, again, singularity arises when matrix $\mathbf{A}_{\mathrm{na}}$ loses its full rank or equivalently when $\left|\mathbf{A}_{\mathrm{na}}^{\mathrm{T}}\ \mathbf{A}_{\mathrm{na}}\right| = 0$.
When singularity arises for an m-set of assigned velocities there are two approaches:
- *Loss of degrees of mobility:* in order to avoid incompatibility the assigned velocities are coordinated properly, what implies a loss of degrees of mobility.

- *Kinematics Incompatibility:* no type of coordination for the assigned velocities is considered, so the kinematic incompatibility is not solved. If the assigned velocities are wheel velocities (forward kinematics), slip (due to the incompatibility, not because of accelerations) is inevitable. If they are WMR velocities (inverse kinematics), impossible (infinite) control action values are obtained.

In the same way, the singularity of a redundant forward kinematics (12) would produce an infinite error in the estimation of the WMR velocity vector. Therefore, it is obtained the following criterion: *singularity* (i.e. mobility degree loss, slip, impossible control actions, or infinite error in the estimation) *has to be avoided*. Moreover, *nearness to singularity is neither desirable* since it implies: high amplification of wheel velocities' error (redundant and non-redundant forward kinematics) or high values for wheel velocities (non-redundant inverse kinematics). If the singularity depends on the steering angles of *orientable* or *castor* wheels the previous criterion is a *planning criterion,* i.e. the upper level planner (path generator) has to develop paths not close to singularities, otherwise it becomes *design criterion.*

3. Motion Planning

Given a starting and ending configuration of a given WMR, a motion planning problem consists of automatically computing a collision-free path. This gives rise to the famous *piano mover problem,* i.e. any solution appears as a path in the admissible (i.e. collision-free) *configuration space.* Many papers have proposed general, exact, approximate, efficient ... methods in order to represent and explore this admissible configuration space: e.g. cellular decomposition, polygon representation, etc. (see (Latombe, 1991) for a synthesis of these approaches). One classical approach is based on *tree graphs* whose leafs are the WMR posture and whose branches are the paths from one posture to another. Then, the planner checks, during the construction of the tree graph, if the goal has been achieved. In order to avoid the high computational cost of the tree-graph method, it was developed the *roadmap* technique that builds a graph whose nodes are collision-free configurations and whose edges denote the presence of collision-free paths between two configurations. The roadmaps tend to capture both the coverage and connectivity of the configuration space and replace the concept of deterministic completeness by the concept of probabilistic completeness.

However, numerous classical methods work only when the WMR is holonomic and not when there is some *non-holonomic constraint* between its configuration parameters. In order to overcome this, in (Laumond et al., 1997) it is developed a planner that firstly generates a collision-free path ignoring the non-holonomic constraints and afterwards the path is transformed into one that is feasible with respect to these constraints.

On the other hand, other planners are *specific for one task,* e.g. in (Gracia & Tornero, 2003) it is presented a planner for parallel parking based on a geometric characterization for collision avoidance. Moreover, other types of approaches do not *explicitly* generate collision-free paths; instead, they integrate the WMR motion planning with the WMR control using tools like *fuzzy,* neural networks, reactive architecture, etc. For example, in (Borenstein & Koren, 1989) it is used artificial potential fields: the WMR is attracted by the objective configuration and repelled by the obstacles. If a time value is associated to each point of the path it becomes a trajectory; otherwise, a forward constant velocity is usually used across the path.

3.1 Proposed cost index

This research introduces a *cost index* based on kinematics singularity that is useful for many types of planners (based on tree graphs, roadmaps, etc.), since it allows to choose the path with minimum cost index among several possible collision-free paths. In the cost index it will be weighted the nearness to singularity of forward and inverse kinematics. This will allow avoiding singularity and nearness to singularity, i.e. high amplification of the WMR velocities' error or high values for wheel velocities. Similarly to robotic manipulators, the singularity of *inverse kinematic models* can be deal with a null velocity on the path at the singularity point, which is equivalent to a loss of degrees of mobility. It implies to stop the WMR in order to reorientate it and/or its wheels, as it is pointed out in (Gracia & Tornero, 2008) for the five types of WMR classified according to (Campion et al., 1996). This may be appropriate when there is not much space available (e.g. for parking maneuvers) but not in a general case, since it involves an important waste of time. Therefore, this option will not be considered here. The nearness to singularity of *forward kinematic models* produces high amplification of the WMR velocities' error, what implies a tracking error if the assigned wheel velocities are actuated wheel velocities or an estimation error if they are sensed wheel velocities. Both types of forward models will be considered in the cost index. Therefore, it is proposed the following *cost index*:

$$J = \sum_{i=1}^{\mathrm{N}} \left(\frac{1}{f_1\left(|\mathbf{B}_{\mathrm{a}}|_{\mathrm{inv}\, i}\right)} + \frac{f_2(\mathrm{N}-i+1)}{f_3\left(|\mathbf{B}_{\mathrm{a}}|_{\mathrm{fwd\ act}\, i}\right)} + \frac{f_4(\mathrm{N}-i+1)}{f_5\left(|\mathbf{A}_{\mathrm{na}}^{\mathrm{T}}\,\mathbf{A}_{\mathrm{na}}|_{\mathrm{fwd\ sensed}\, i}\right)} + f_6\left(\dot{\boldsymbol{\beta}}_{\mathrm{o}\, i}\right) \right) + f_7(D), \quad (13)$$

where N is the number of branches/edges of the path in the tree-graph/roadmap; f_i is a generic non-linear function; $|\mathbf{B}_{\mathrm{a}}|_{\mathrm{inv}\ i}$ is the singularity of the inverse kinematic model; $|\mathbf{B}_{\mathrm{a}}|_{\mathrm{fwd\ act}\ i}$ is de singularity of the forward models with actuated wheel velocities as assigned; $|\mathbf{A}_{\mathrm{na}}^{\mathrm{T}}\,\mathbf{A}_{\mathrm{na}}|_{\mathrm{fwd\ sensed}\ i}$ is de singularity of the forward model with redundant sensed wheel velocities; $\dot{\boldsymbol{\beta}}_{\mathrm{o}\, i}$ is the steering velocity vector of all the *orientable* wheels; and D is the length or distance of the collision-free path. Note that, the singularity of forward models has been multiplied by f_j (N – i +1) since the tracking/estimation error of the initial branches/edges is more important because it is propagated across the whole path. However, in order to limit the uncertainty of the estimation other global or local position sensors are required. Note also that, it has been introduced the steering velocities of the *orientable* wheels because they are not present in the velocity vector $\dot{\mathbf{q}}$, see (2).

3.2 Example of motion planning

The cost index of previous subsection will be obtained for the case of the industrial forklift of Fig. 3, which is equivalent to the tricycle WMR, where the origin of R (tracking point) has been located at the middle point of the *fixed* wheels. The traction of this industrial forklift is given by both *fixed* wheels, which are *properly coordinated* through a differential mechanism depending on the steering angle of the *orientable* wheel. Moreover, this WMR has three encoders measuring the rotation of both *fixed* wheels and the steering angle of the *orientable* wheel.

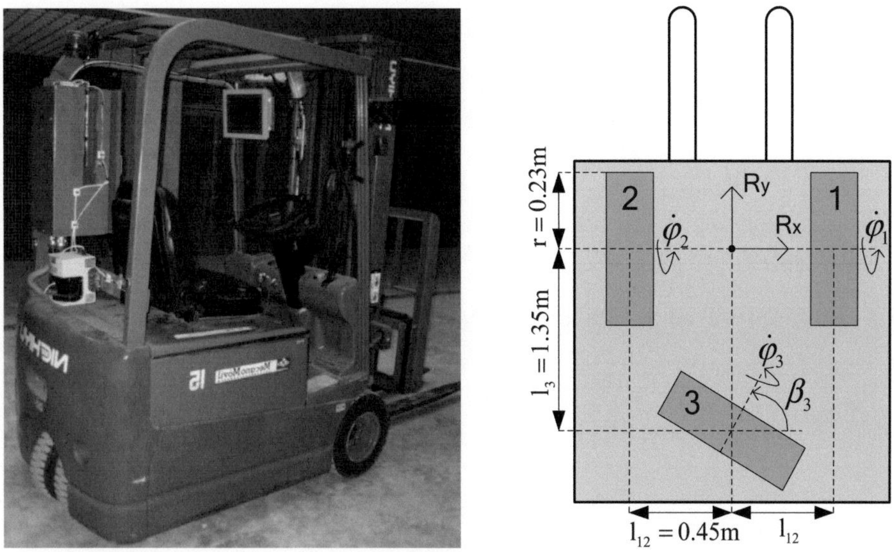

Figure 3. Industrial forklift Nichiyu FBT15 series 65 and equivalent tricycle representation

The composite equation (4) of this WMR results:

$$\mathbf{v}_{\text{slip}} = \begin{pmatrix} \mathbf{v}_{\text{slip 1}} \\ \mathbf{v}_{\text{slip 2}} \\ \mathbf{v}_{\text{slip 3}} \end{pmatrix} = \left(\begin{array}{ccc|ccc} 1 & 0 & 0 & 0 & 0 & 0 \\ 0 & 1 & l_{12} & r & 0 & 0 \\ \hline 1 & 0 & 0 & 0 & 0 & 0 \\ 0 & 1 & -l_{12} & 0 & r & 0 \\ \hline \cos\beta_3 & \sin\beta_3 & l_3\cos\beta_3 & 0 & 0 & 0 \\ -\sin\beta_3 & \cos\beta_3 & -l_3\sin\beta_3 & 0 & 0 & r \end{array}\right) \begin{pmatrix} {}^{\bar{R}}\dot{\mathbf{p}} \\ \dot{\varphi}_1 \\ \dot{\varphi}_2 \\ \dot{\varphi}_3 \end{pmatrix}. \tag{14}$$

Under the no-slip condition, a kinematic solution (8) is:

$$\begin{pmatrix} \dot{\mathbf{p}} \\ \dot{\varphi}_1 \\ \dot{\varphi}_2 \\ \dot{\varphi}_3 \end{pmatrix} = \begin{pmatrix} 0 \\ -l_3\cos\beta_3 \\ \sin\beta_3 \\ (L\sin\beta_3 - l_3\cos\beta_3)/r \\ (L\sin\beta_3 + l_3\cos\beta_3)/r \\ l_3/r \end{pmatrix} \eta. \tag{15}$$

For the redundant forward kinematics, (10) is particularized to:

$$\begin{pmatrix} 1 & 0 & 0 \\ 0 & 1 & l_{12} \\ 0 & 1 & -l_{12} \\ \cos\beta_3 & \sin\beta_3 & l_3\cos\beta_3 \end{pmatrix} \dot{\mathbf{p}} = -\begin{pmatrix} 0 & 0 \\ r & 0 \\ 0 & r \\ 0 & 0 \end{pmatrix} \begin{pmatrix} \dot{\varphi}_1 \\ \dot{\varphi}_2 \end{pmatrix} \rightarrow \mathbf{A}_{\text{na p}}\,\dot{\mathbf{p}} = -\begin{pmatrix} 0 & 0 \\ r & 0 \\ 0 & r \\ 0 & 0 \end{pmatrix} \begin{pmatrix} \dot{\varphi}_1 \\ \dot{\varphi}_2 \end{pmatrix}. \tag{16}$$

where it has been considered together the first and third equation of (14), and the last equation (used only to compute $\dot{\varphi}_3$) has been obviated. Therefore, the kinematics singularity is given by:

$$
\begin{aligned}
&\eta = v_y \rightarrow |\mathbf{B}_a|_{inv} = -l_3\cos\beta_3 = 0 \rightarrow \beta_3 = \pm 90^{o} \\
&\eta = \dot{\varphi}_1 \rightarrow |\mathbf{B}_a|_{fwd\ act1} = (l_{12}\sin\beta_3 - l_3\cos\beta_3)/r = 0 \rightarrow \beta_3 = \begin{cases} \text{atan}(l_3/l_{12}) \\ \text{atan}(l_3/l_{12}) + 180^{o} \end{cases} \\
&\eta = \dot{\varphi}_2 \rightarrow |\mathbf{B}_a|_{fwd\ act2} = (l_{12}\sin\beta_3 + l_3\cos\beta_3)/r = 0 \rightarrow \beta_3 = \begin{cases} -\text{atan}(l_3/l_{12}) \\ -\text{atan}(l_3/l_{12}) + 180' \end{cases} \\
&\dot{\mathbf{q}}_a = (\dot{\varphi}_1, \dot{\varphi}_2) \rightarrow \left|\mathbf{A}_{na\ p}^{T}\ \mathbf{A}_{na\ p}\right|_{fwd\ sensed} = 0 \rightarrow \text{No solution, never singular.}
\end{aligned} \tag{17}
$$

The cost index (13) will be particularized to:

$$
J = \sum_{i=1}^{N} \left(\frac{K_1}{|\mathbf{B}_a|_{inv\ i}^2} + \frac{K_2(N-i+1)}{\max\left(|\mathbf{B}_a|_{fwd\ act1\ i}^2, M\right)} + \frac{K_3(N-i+1)}{\max\left(|\mathbf{B}_a|_{fwd\ act2\ i}^2, M\right)} + \frac{K_4(N-i+1)}{\left|A_{na\ p}^{T} A_{na\ p}\right|_{fwd\ sensed\ i}^2} \right) + \\
+ \sum_{i=1}^{N} \left(K_5 \left(\dot{\beta}_{3\,i}\right)^2 \right) + D, \tag{18}
$$

where K_j is the weight of each term in the cost index and M is a kind of singularity saturation in order to not reduce in excess the WMR maneuverability. Note that the industrial forklift has one degree of mobility (m = 1), i.e. one instantaneous degree of freedom, that allows to specify a forward tracking velocity v_y. It has another non-instantaneous degree of freedom through the angle β_3 of the *orientable* wheel that allows turning. Therefore, this WMR can track 2-dimensional paths. In order to obtain simulation results, it will be considered the *tree graph technique* together with the previous cost index. It will be used a constant forward velocity on the path, e.g. v_y = 1 m/s, and the following motion equations between *leaves*/samples:

$$
\begin{aligned}
x_{k+1} &= x_k + (v_y/\omega)(\sin(\theta_k + \omega T) - \sin\theta_k) \\
y_{k+1} &= y_k - (v_y/\omega)(\cos(\theta_k + \omega T) - \cos\theta_k) \qquad \theta_{k+1} = \theta_k + \omega T,
\end{aligned} \tag{19}
$$

where T is the sample time, and it has been considered a constant forward motion v_y and a constant turning motion ω between samples. If the WMR angular velocity ω is null, it must be used the following equations:

$$
x_{k+1} = x_k + v_y T\cos\theta_k \qquad y_{k+1} = y_k + v_y T\sin\theta_k \qquad \theta_{k+1} = \theta_k. \tag{20}
$$

Note that the distance D of each collision-free path results $v_y \cdot T \cdot N$. For the construction of the tree graph it will be considered three possible steering velocities for the *orientable* wheel: $\{-\dot{\beta}_{3\ max}, 0, \dot{\beta}_{3\ max}\}$. During the construction of the tree graph it will be verified if the goal has been achieved within a tolerance. The parameters used for the simulations results of Fig. 4 are: v_y = 1 m/s, T = 0.5 s, N = 22, $\dot{\beta}_{3\ max} = 0.4$ rad/s, M = 0.01, $K_1 = K_5 = 18$, $K_2 = K_3 = K_4 = 3$; and it has been considered two rectangular obstacles that represent two warehouse shelves. The goal WMR posture **p** in the three examples of Fig. 4 are: (5.5, 0, 0); (2, 0, 180º); and (5, 0, any) respectively. The continuous thick line is the path with minimum cost index; the dashed thick line is the path with minimum distance; and the continuous thin lines are *some* (a sample) of the collision-free paths.

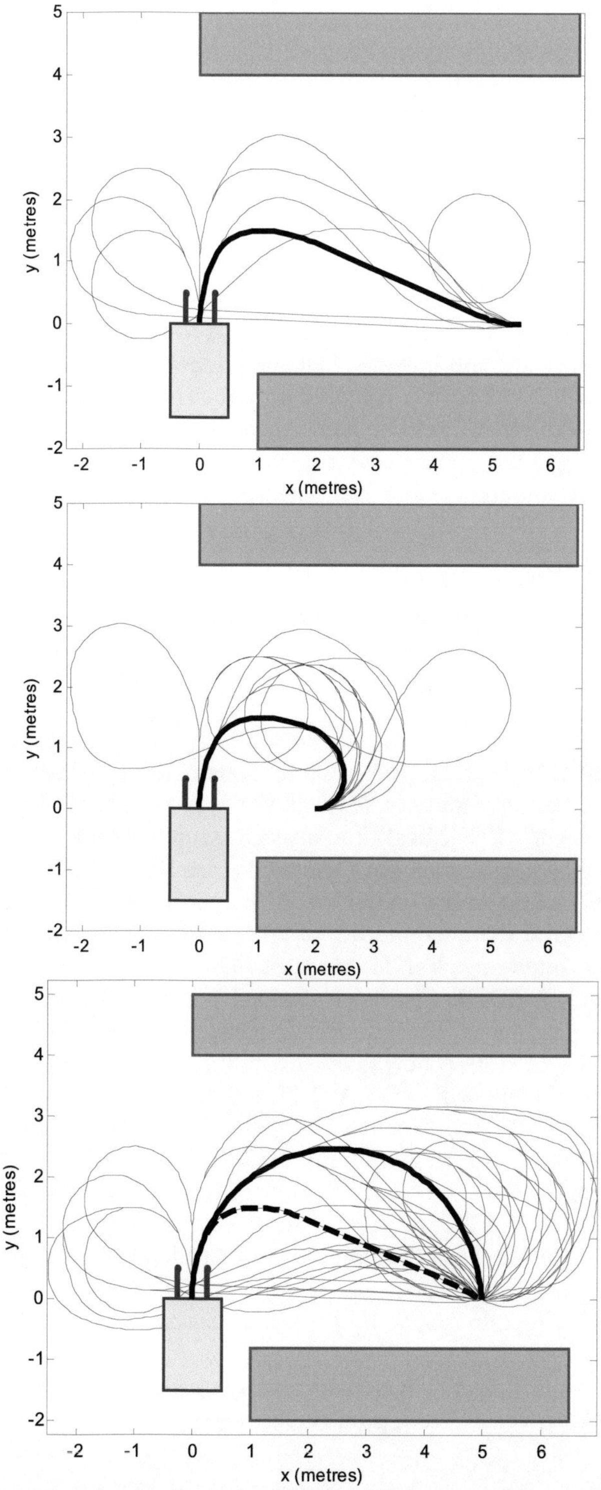

Figure 4. Simulation examples for the industrial forklift in a warehouse environment

4. Conclusion

In the previous work (Gracia & Tornero, 2007a) the authors had characterized the singularity of WMR kinematics. In this chapter it has been shown how to use WMR singularity or nearness to WMR singularity for motion planning. In particular, it has been proposed a cost index that assesses the nearness to singularity of forward and inverse kinematic models. This cost index can be used straightforward for many planning techniques (tree graphs, roadmaps, etc.) in order to choose one path among several possible collision-free paths. Therefore, the chosen path would avoid not only slip and impossible control actions (i.e. the singularity of forward and inverse kinematic models) but also high amplification of wheel velocities' error and high values for wheel velocities (i.e. the nearness to the singularity of forward and inverse kinematic models). To illustrate the applications of the proposed approach it has been considered an industrial forklift that is equivalent to the tricycle WMR. Finally, several results have been shown for this WMR in a simulated environment. It is suggested as further work to integrate the presented motion planning with other classical techniques like artificial potential fields, fuzzy planners, etc.

5. Acknowledgement

This work was supported in part by the Spanish Government: Research Projects DPI2004-07417-C04-01 and BIA2005-09377-C03-02

6. References

Alexander, J. C. & Maddocks, J. H. (1989). On the kinematics of wheeled mobile robots. *The International Journal of Robotics Research*, Vol. 8, No. 5, pp. 15–27.

Borenstein, J. & Koren, Y. (1989). Real-time obstacle avoidance for fast mobile robots. *IEEE Transactions on Systems Man and Cybernetics*, Vol. 19, pp. 1179-1187.

Campion, G.; Bastin, G. & D'Andrea-Novel, B. (1996). Structural properties and classification of kinematic and dynamic models of wheeled mobile robots. *IEEE Transactions on Robotics and Automation*, Vol. 12, No. 1, pp. 47–61.

Gracia, L. & Tornero, J. (2003). Geometric parallel parking planner for car-like vehicles, *Proceedings of the Industrial Simulation Conference*, pp. 357-361, Valencia, Spain.

Gracia, L. & Tornero, J. (2007a). A new geometric approach to characterize the singularity of wheeled mobile robots. *Robotica*, Vol. 25, No. 5, pp. 627-638.

Gracia, L. & Tornero, J. (2007b). Kinematic modeling of wheeled mobile robots with slip. *Advanced Robotics*, Vol. 21, No. 11, pp. 1253-1279.

Gracia, L. & Tornero, J. (2008). Kinematic control of wheeled mobile robots. *Latin American Applied Research*, Vol. 38, No. 1, pp. 7-16.

Latombe, J.-C. (1991). *Robot Motion Planning*, Kluwer Academic Publishers, chapter 7.

Laumond, J.-P.; Jacobs, P. E.; Taïx, M. & Murray, M. (1997). A motion planner for nonholonomic mobile robots. *IEEE Transactions on Robotics and Automation*, Vol. 10, No. 5, pp. 577-593.

Muir, P. F. & Neuman, C. P. (1987). Kinematic modeling of wheeled mobile robots. *Journal of Robotic Systems*, Vol. 4, No. 2, pp. 281–329.

6

SOVEREIGN: An Autonomous Neural System for Incrementally Learning to Navigate Towards a Rewarded Goal

William Gnadt and Stephen Grossberg
Boston University
USA

1. Introduction

1.1. Three Basic Design Themes

This chapter describes the SOVEREIGN (Self-Organizing, Vision, Expectation, Recognition, Emotion, Intelligent, Goal-oriented Navigation) neural model to clarify how an animal, or animat, can learn to reach valued goal objects through planned sequences of navigational movements. The SOVEREIGN model embodies a self-organizing control system that attempts to learn and perform such behaviors autonomously. As the name SOVEREIGN indicates, this control system unifies visual, recognition, cognitive, cognitive-emotional, and motor competences. We believe that this is the first neural model that embodies and coordinates such a wide range of behavioral competences in an autonomous self-organizing control system that can operate in real time. These results have been briefly reported in Gnadt and Grossberg (2005a, 2005b, 2006) and in more detail in Gnadt and Grossberg (2008).

SOVEREIGN contributes to three large themes about how the brain works. The first theme concerns how brains learn to balance between reactive and planned behaviors. During initial exploration of a novel environment, many reactive movements occur in response to unexpected and unfamiliar environmental cues (Leonard and McNaughton, 1990). These movements may initially appear to be locally random, as an animal orients toward and approaches many local stimuli. As such an animal becomes familiar with its surroundings, it learns to discriminate between objects likely to yield a reward and those that lead to punishment. Such approach-avoidance behavior is often learned via a perception-cognition-emotion-action cycle in which an action and its consequences elicit sensory cues that are associated with them. Rewards and punishments affect the likelihood that the same actions will be repeated in the future. When objects are out of direct sensory range, multiple reactive exploratory movements may be needed to reach them. Eventually, reactive exploratory behaviors are replaced by more efficient planned sequential trajectories within a familiar environment. One of the main goals of SOVEREIGN is to explain how erratic reactive exploratory behaviors can give rise to organized planned behaviors, and how both reactive and planned behaviors may remain balanced so that planned behaviors can be carried out where appropriate, without losing the ability to respond quickly to novel reactive challenges.

The second design theme illustrates the hypothesis that advanced brains are organized into parallel processing streams with complementary properties (Grossberg, 2000a). Each stream's properties are related to those of a complementary stream much as a lock fits its key, or two pieces of a puzzle fit together. The mechanisms that enable each stream to compute one set of properties prevent it from computing a complementary set of properties. As a result, each of these streams exhibits complementary strengths and weaknesses. How, then, do these complementary properties get synthesized into a consistent behavioral experience? It is proposed how interactions between these processing streams overcome their complementary deficiencies and generate behavioral properties that realize the unity of conscious experiences. In this sense, pairs of complementary streams are the functional units because only through their interactions can key behavioral properties be competently computed. SOVEREIGN clarifies how these complementary properties interact together to control goal-orienting sequences of navigational behaviors. For example, it is well-known that there are What and Where (or Where/How) cortical processing streams (Goodale and Milner, 1992; Mishkin, Ungerleider and Macko, 1983; Ungerleider and Mishkin, 1982). In particular, key properties of the What and Where cortical processing streams seem to be complementary.

A third design theme underlying the SOVEREIGN model is that brains use homologous circuits to compute navigational and hand/arm movements. In other words, movements of the body and of the hand/arms are controlled by circuits that share many properties. This proposed homology clarifies how navigational and arm movements can be coordinated when a body moves with respect to a goal object with the intention of grasping or otherwise manipulating it using the hand/arm system.

A considerable body of neural modeling of arm movement trajectory control (e.g., the VITE model: Bullock and Grossberg, 1988; Bullock, Cisek, and Grossberg, 1998) suggests that cortical arm movement control circuits compute a representation of where the arm wants to move (i.e., a target position) and compare this with an outflow representation of where the arm is now (i.e., the present position) by computing a difference vector between target position and present position representations. The difference vector represents the direction and distance that the arm needs to move to realize its goal position. Basal ganglia volitional signals of various kinds, such as a GO signal, translate the difference vector into a motor trajectory of variable speed. Additional cortical, spinal, and cerebellar circuitry is needed to ensure that the brain generates the forces that are needed to actually carry out such a commanded trajectory (e.g., the FLETE model: Bullock and Grossberg, 1991; Contreras-Vidal, Grossberg, and Bullock, 1997).

A key difference between navigation and hand/arm movement control concerns how present position is calculated. Because the arm is attached to the body, present position of the arm can be directly computed using outflow, or corollary discharge, movement commands that explicitly code the commanded arm position. In contrast, when a body moves with respect to the world, no such immediately available present position command is available. This difference requires more elaborate brain machinery to compute present position of the body in the world during navigational movements. The brain needs to use a variety of sensory cues, both proprioceptive and visual, to create a representation of present position that can be compared with representations of target position, so that a difference vector and volitional commands can move the body towards desired goal objects. In summary, both navigational movement in the world and movement of limbs with respect to the body use a difference vector computational strategy.

1.2. What SOVEREIGN Does

SOVEREIGN's perceptual competences include on-line, albeit simplified, visual representations of a 3D virtual reality environment in which the model controls navigation. SOVEREIGN computes, in parallel, both visual form and motion information about the world. As in the brain, the visual form of objects is computed within the What cortical processing stream, whereas visual motion is computed within the Where cortical processing stream. In this way, the brain can process both what objects are and where and how to track and act upon them.

SOVEREIGN uses the visual form information to incrementally learn spatially-invariant and size-invariant object recognition categories to recognize visually perceived objects in the world. These recognition categories, in turn, learn to read out top-down attentive expectations of the visual objects that they code. Object categories in the What stream are spatially-invariant and size-invariant to prevent a combinatorial explosion from occurring in which each position and size of an object would need its own representation. In contrast, the Where stream represents the spatial locations of these objects. In particular, visual motion information is used to guide reactive orienting movements and attention shifts to locations at which changes occur in SOVEREIGN's visual world. What-Where inter-stream interactions are needed to enable both recognition and acquisition of desired goal objects. These parallel streams help SOVEREIGN to balance between reactive and planned behaviors, in a manner that is further discussed below.

SOVEREIGN also includes cognitive processes, notably mechanisms to temporarily store sequences of events in working memory, and to learn sequential plans, or chunks, of these sequences with which to predict and control future planned behaviors. Parallel object and spatial working memories and sequential chunking networks are modeled. The object working memory and chunking network are in the model's What stream, and the spatial working memory and chunking network are in its Where stream. SOVEREIGN clarifies how these parallel cognitive processes cooperate to acquire desired goal objects that can only be reached through a sequence of actions, and to disambiguate sequential navigational decisions in contexts where only one of them would be insufficient.

Cognitive-emotional mechanisms include the role of rewards and punishments in shaping goal-oriented behaviors. In particular, reinforcement learning can influence which learned cognitive chunks will be attended and selected to elicit behaviors that acquire desired goals within a familiar environment. Learned interactions between cognitive and emotional representations, notably motivationally-mediated attention, play an important role in this context-sensitive selection process.

The SOVEREIGN model thus contributes solutions to three key problems: (1) How an animal, or animat that embodies biologically-inspired designs, learns to balance between reactive and planned behaviors in a task-appropriate way. (2) How plans are learned during erratic reactive behaviors in such a way that, after learning, they can be read out fluently at the correct times and in the correct spatial contexts. (3) How, in particular, an animat coordinates its reactive and planned behaviors so that its perception-cognition-emotion-action cycles of exploration, real-time vision, learned recognition, sequential working memory storage, learning of sequential plans, reinforcement learning, and planned action sequences are activated when needed as the animat navigates novel and familiar environments.

2. SOVEREIGN Model

2.1. Approach-Orienting Navigation and Learning in a 3D Virtual Reality Environment

The SOVEREIGN model simulates these processes for an animat that experiences a 3D visual world in a virtual reality environment. This world is the relatively simple spatial format of a plus-maze (Munn, 1950). Although simple, this environment tests in a clear way many of the critical competences that the animat needs to achieve. Much of the problem's difficulty arises because an animat may navigate the maze in different ways, including different speeds and directions of movement, on successive learning trials. Despite this variability of each experience of the maze, the animal can learn to navigate the maze to achieve valued goals in an efficient way. For our purposes, it is sufficient to assume that a learning trial starts after placing the animat in the maze, at the end of one arm. The goal location, in one of the other three arms, is baited with a cue that the animal finds rewarding. By shrouding the top of the maze, only route-based visual and motor cues can be used for navigation (O'Keefe and Nadel, 1978). Thus the model does not attempt to explain how spatial navigation, as supported by hippocampal place cells, is achieved. Only one visual cue is assumed to be visible at a time, at the end of each maze arm, from any location within the maze. A schematic diagram of the experimental setup appears in Figure 1a.

A sequence of images from the 3D virtual reality simulation during reactive approach toward a visual cue appears in Figure 1b. As the animat approaches the choice point, a competitive struggle occurs between the salience of form and motion signals. Suppose that the form signals have led to previous object category learning and have led to positive reinforcement that increases their motivational salience. Such motivational salience enhances the strength of the form representation through attentional feedback. Then the form signals may more effectively compete with the motion signals to determine the animat's momentary behavior. If the form cues win the competition, then the animat can continue to carry out an approach movement that is consistent with its recognition. If the motion signals win the competition, then they may trigger a reactive head-orienting movement to the right or left at a choice point, revealing another source of form signals at the end of an adjacent corridor. The outcome of this form-motion competition is sensitive to navigational variations that change from trial to trial. The sequence of visual scenes that are processed during a typical head-orienting behavior is illustrated in Figure 1c. An alternation of approach and orienting movements is characteristic of the animat's exploration of a novel environment.

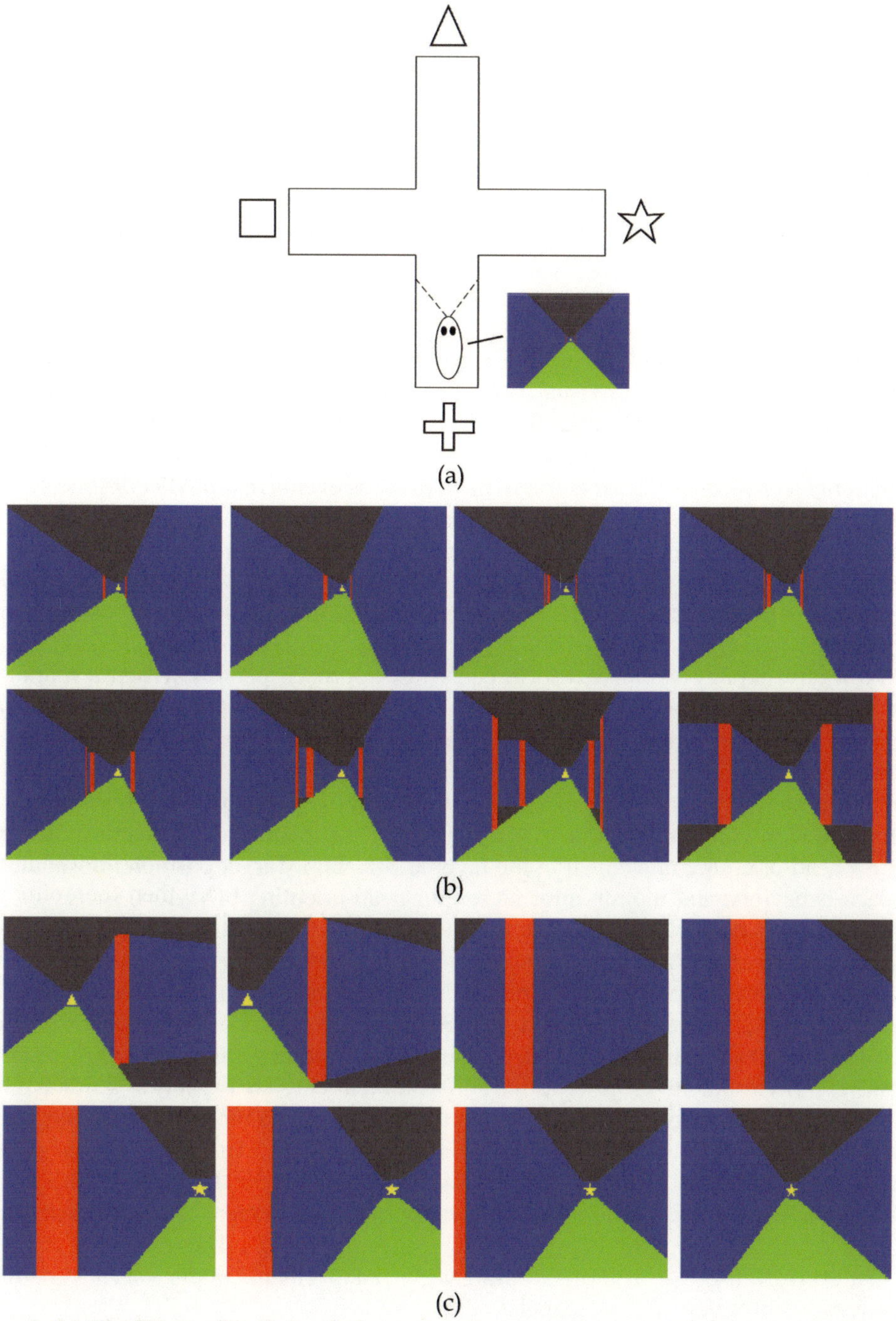

Figure 1. (a) The 3D graphical simulation generates perspective-views from any location within the plus-maze. (b) Snapshots from the 3D virtual reality simulation depict changes in the scene during reactive homing toward the triangle cue. (c) During reactive approach to the triangle cue, visual motion signals trigger a reactive head orienting movement to bring the star cue into view

2.2. Parallel Visual and Motor Working Memory and Chunking Networks

The animat's control system is split into a number of subsystems shown in the macrocircuit of Figure 2. The primary input is via the visual system. The Visual Form and Motion System processes visual cues within the What and Where cortical streams, respectively. The What cortical stream learns position-invariant and size-invariant object categories via on-line incremental learning. The Where stream computes measures of the relative location of visual cues from the animat. In particular, the distance and direction of the animat from a prescribed visual cue are used to cause approach movements towards that visual cue, or from memory. Motion cues result from the animat's self-motion, and determine whether the animat will make a left or right turn, and how big a turn, instead of continuing to approach a target cue.

These visual form and motion signals compete for control of the animat's approach-orienting behaviors within the Visual Form and Motion System. Learned visual categories can be amplified in strength, and thereby more probably attended, by feedback from motivational centers, called Drive Representations, through learned reinforcement-motivational feedback loops that embody their value as events that predict desired rewards. For this to happen, the invariant object categories are amplified by motivational signals that draw attention to them, and amplify, in turn, the approach commands corresponding to that object's position relative to the animat. Such a motivational amplification requires What-Where inter-stream interactions between position-invariant and position-variant information.

When a motivationally-modulated form cue wins, approach persists. When a motion cue wins, an orienting movement often begins. When motion cues are balanced in strength relative to the present gaze direction, the net left vs. right orienting signal is zero, after opponent competition between the opposite directions takes place. A form cue can then win with high probability. However, a suitably strong left/right motion cue difference can win the form-motion competition and direct the Motor Approach and Orienting System to initiate a head-orienting movement in the favored direction. Target position information for approach behaviors, and motion information for head-orienting behaviors, is relayed from the Visual Form and Motion System to the Motor Approach and Orienting System (Figure 2), where they direct body-approach movements or head-and-body orienting movements.

How does the animat know where a target cue is with respect to its current position? As noted in Figure 2, proprioceptive and vestibular signals provide the ground truth upon which the animat's location is estimated relative to its starting point, and with respect to targets in its environment. Proprioceptive and vestibular cues are capable of guiding animat navigation in a familiar environment even in the dark, and can modify movements quickly to cope with uneven or slippery terrains. Visual cues are also used during navigation to estimate body and head position and displacement relative to the animat. These visual signals are associated with, and adaptively calibrated with respect to, the representations that are activated by proprioceptive and vestibular motor signals. These multiple sources of information work together to more accurately guide movements under varying conditions than any one source of positional signals could.

Estimates of spatial displacement compute the NET body displacement and head rotation from a starting point. Sequences of such approach-orienting displacements represent a path that can command an animat to move from a starting location to a goal location in a maze.

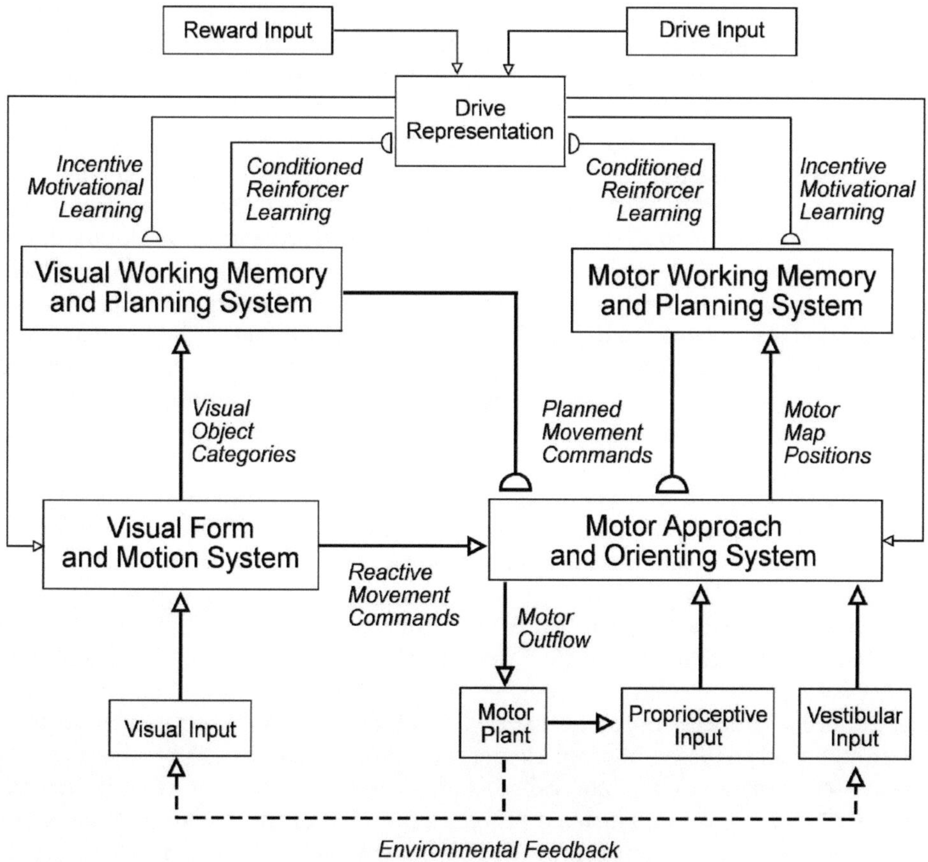

Figure 2. The key interactions between components of the SOVEREIGN model are shown in this flow diagram. See text for details

The Visual Working Memory and Planning System temporarily stores sequences of visual object categories in short-term working memory. It also categorizes, or chunks, sequences of stored object categories. Chunks are learned that are sensitive to object sequences of variable lengths. These visual list chunks can learn to activate motor commands in the Motor Approach and Orienting System via top-down learning. The motor commands encode approach-orient movements. The Visual Working Memory and Planning System operates in parallel with a Motor Working Memory and Planning System that temporarily stores sequences of motor commands in working memory. It also categorizes, or chunks, sequences of stored motor commands. These motor list chunks can also learn to activate approach-orient commands within the Motor Approach and Orienting System.

Why are visual and motor list chunks both needed? Together these parallel visual and motor working memories can disambiguate decisions that only one of them, acting alone, may find ambiguous. For example, the sequences of approach distances and head turns in two different environments may be the same, but their sequence of visual cues may be different. In a different environment, the sequence of visual cues may be the same, but their sequences of motor actions may differ. The visual and motor working memories induce the learning of list chunks, or sequential planning cells, that are sensitive to their respective object and action sequences, and can read out a prediction of the next motor command. The sequence that can disambiguate two different environments will typically win over one that cannot.

Rewards and punishments can modulate animat learning and determine what visual representations are attended and what motor plans are executed. Upon receiving reward, the active chunks are associated with active drives and actions. Drive inputs represent the animat's internal motivational state, and reward inputs represent valued inputs from the external environment. Both types of inputs combine in Drive Representations, which are most highly activated when a drive input representing a strong internal need combines with either a primary reward or a conditioned reinforcer input from the Visual Form and Motion System (Figure 2). After such a combination of cognitive and emotional learning occurs, when the animat sees a familiar sensory cue under a prescribed motivational state, it can recall a motivationally-compatible plan to reach the site of previous reward. Repeated, random exploration of the environment hereby effects a gradual transition from reactive to more efficient, planned control that leads the animat to its various motivated goals. Due to the selective role of motivational feedback, the animat is capable of learning to carry out different plans to satisfy different motivational goals even in response to the same sensory cues.

Visual and motor list chunks may learn to activate different approach-orient commands under different motivational states. How can a single chunk give rise to multiple responses? How this occurs can be seen by noting that emotional centers are often organized into *opponent* affective processes, such as fear and relief, and that oppositely valenced rewards can be conditioned to these opponent channels (Grossberg, 1984b, 2000b). These opponent-processing emotional circuits are called *gated dipoles*. In such a circuit, habituative transmitters "gate", or multiply, signal processing in each of the channels of the opponent "dipole." The response amplitude and sensitivity to external reinforcing inputs and internal drive inputs of these opponent-processing emotional circuits are calibrated by an arousal level and chemical transmitters that slowly inactivate, or habituate, in an activity-dependent way.

Sensory and cognitive representations, no less than emotional representations, can be organized into opponent channels with habituative ON and OFF cells. Unexpected events can trigger a burst, or sudden increment, of nonspecific arousal. When such an arousal burst is received on top of the baseline tonic arousal input of a normal dipole, it can cause an antagonistic rebound of activity in the OFF channel. In other words, the sensory, cognitive, or emotional hypothesis that is represented in a dipole's activity can be disconfirmed by an unexpected event. An unexpected event can hereby reset ongoing processing and lead to a shift of attention. SOVEREIGN expands the gated dipole mechanism into a gated multipole, which can select between multiple opponent drive channels. Each channel, whether representing an exploratory or consummatory drive state, can be associated with a particular learned response.

SOVEREIGN embodies a system synthesis and further development of biologically-derived neural networks that have been mathematically and computationally characterized elsewhere. These include LAMINART and FORMOTION models for form and motion processing (Berzhanskaya, Grossberg, and Mingolla, 2007; Cao and Grossberg, 2005; Grossberg, Mingolla, and Viswanathan, 2001; Grossberg and Yazdanbakhsh, 2005; Raizada and Grossberg, 2003), ART fast incremental learning classifiers (Carpenter, et. al., 1992), STORE working memories (Bradski, Carpenter, and Grossberg, 1994), Masking Field sequence chunking networks (Cohen and Grossberg, 1986, 1987; Grossberg and Myers, 2000; Grossberg and Pearson, 2007), Gated Dipole opponent processes (Grossberg, 1980, 1984a; Grossberg and Seidman, 2006), CogEM cognitive-emotional circuits for reinforcement learning (Grossberg and Merrill, 1992, 1996; Grossberg, 2000), Spectral Timing circuits for adaptively timed learning (Grossberg and Merrill, 1992; Fiala, Bullock, and Grossberg, 1996), and volitional (GO) and endogenous (ERG, Endogenous Random Generator) gates to release consummatory and exploratory behaviors, respectively (Bullock and Grossberg, 1988; Gaudiano and Grossberg, 1991; Pribe, Grossberg, and Cohen, 1997). We are not aware of any other autonomous agent that has yet integrated this range of self-organizing biological competences.

2.3. Reactive Exploration

The following sequence illustrates the functional flow of the visual input system during reactive exploration in the plus-maze of Figure 1. North designates the vertical direction, with South, East and West following accordingly. For definiteness, assume that the animat is placed into the maze and that all extra-maze cues are suppressed. Furthermore, the animat is motivated under both an exploratory and a hunger drive. The drive and reward inputs to the Drive Representation and then into the Visual and Motor Working Memory and Planning Systems are shown in Figure 2. The exploratory drive is assumed to be excited by an Endogenous Random Generator, or ERG, which is an internal arousal source. Such a source is active when the animat is placed into a new environment. The exploratory drive is inhibited by consummatory drive activity that can support realization of a valued goal. The animat receives a reward (e.g., food) upon reaching the goal location, which is located at the end of the West arm. We show how reactive visual signals during exploration eventually lead the animat toward the goal location, and reinforcement signals strengthen the association between stored plan items and the current motivational state. A step-by-step description of the model under reactive visual guidance follows.

Suppose that, by chance, the animat starts the maze shifted to the left of the corridor, with its head facing slightly to the right of the visual cue (Figure 3a). The left shift reduces the distance to motion cues on the left side of the maze. Because of this positional bias, motion signals within the Visual Form and Motion System (Figure 2) will receive a strong leftward bias. These assumptions are used to demonstrate an exploratory trial which ensures that the animat makes its first head-orienting movement toward the goal location. During the experimental trial, the animat moves forward (Figure 3b), turns left (Figure 3c) and approaches the goal location (Figure 3d) under reactive control.

Movement is organized into orienting and approach movements. In particular, a visually-activated motor command from the Visual Form and Motion System triggers a Motor Outflow command (Figure 2) that specifies a head-orienting angle to align the head with the triangle target. The resulting signals activate the Motor Plant (Figure 2), which converts the movement command into a physical displacement. A head-orienting movement towards the triangle target is thereby initiated. The head turn continues until the NET head-orienting displacement equals the commanded displacement angle.

When the animat faces the triangle cue, a Motor Outflow command from the Visual Form and Motion System activates the Motor Plant (Figure 2) to initiate an approach movement toward the triangle cue. When the Motor Plant converts the commanded approach movement into a physical displacement, the animat's body is passively aligned with the head during an approach movement to maintain a stable posture. Such dynamic stability control is assumed to be present, but is beyond the scope of this work.

During the approach movement, the Motor Approach and Orienting System continues to compute the NET head and body displacement toward the visual target cue. In the absence of competing cues, the body-approach movement could continue until the animat reaches the cue. However, the Visual Form and Motion System processes both form and motion signals while the animat continues to move. A sufficiently strong motion signal in the model's visual periphery can win a competition between Parvo form target locations and Magno motion cues. If a motion cue wins, then it can terminate the approach movement and trigger a reactive head-orienting movement away from the visual target cue.

As noted above, when the animat starts in a position that is shifted to the left side of the corridor, as in Figure 3a, motion signals in the left visual hemifield are stronger than those in the right hemifield. Left vs. right motion signals accumulate in the Visual Form and Motion System. When the left motion signal is sufficiently strong relative to the right motion cue and the form signal, a reactive head-orienting command is sent to the Motor Approach and Orienting System.

As the animat carries out these movements, it learns an invariant object category, or chunk, for the triangle visual cue. Top-down signals from the Visual Working Memory and Planning System (Figure 2) corresponding to the Triangle chunk learn the NET body approach and orienting movements computed by the Motor Approach and Orienting System. The triangle cue hereby learns to predict the Forward-Left body movement. The Forward-Left body movements are also stored in the Motor Working Memory and Planning System.

After the animat turns left, invariant preprocessing and learned ART categorization within the Visual Form and Motion System encode a 3D representation of the square cue. This 3D representation is stored in the Visual Working Memory and Planning System (Figure 2), while the NET body displacement in the Motor Approach and Orienting System is reset to prepare for the next movement. Then the cycle of computing the NET head and body displacements begin again, as the animat navigates toward the square cue.

The square cue is at the rewarded location. When the animat reaches this location, it receives a reward, such as food. The active hunger drive representation is then associated with the currently active plan chunks stored in both the Visual Working Memory and Planning System and the Motor Working Memory and Planning System (Figure 2). In particular, the visual Triangle-Square list chunk is learned and associated with the hunger drive representation. In addition, signals from the Triangle-Square chunk learn the NET body orienting and approach movement computed by the Motor Approach and Orienting System, and thereby learns to predict the Forward body movement that brings the animat to the square cue after it turns left in the West arm of the maze. Figure 4 summarizes this sequence of events.

One additional point should be made: All animat behaviors are motivated by some Drive Representation (Figure 2). During initial exploratory activities, an exploratory drive is active. As learning occurs, the exploratory drive is supplanted by the consummatory motivational sources that correspond to the reward; e.g., the hunger drive when the animat is rewarded by food. These processes are now described in greater detail.

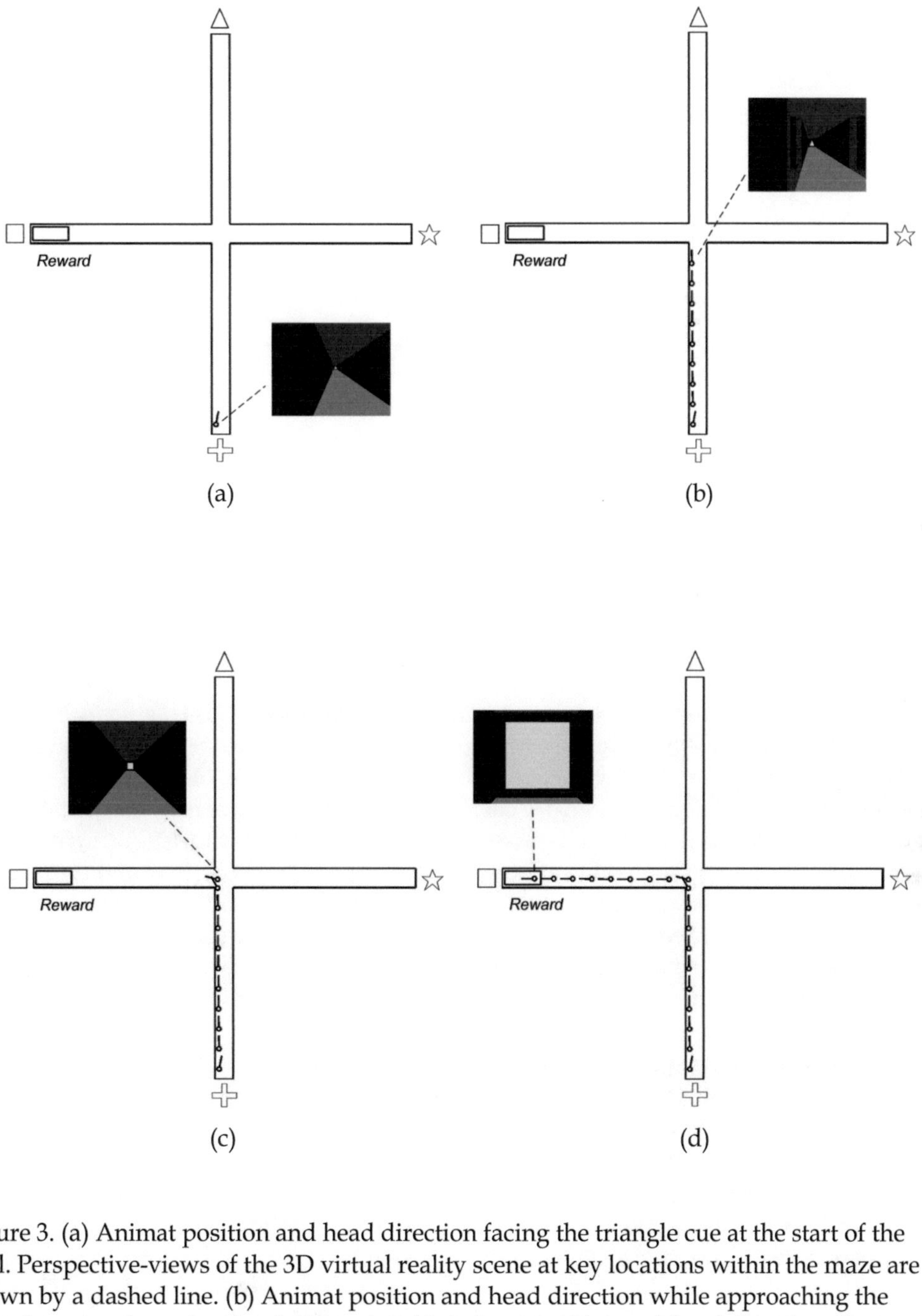

Figure 3. (a) Animat position and head direction facing the triangle cue at the start of the trial. Perspective-views of the 3D virtual reality scene at key locations within the maze are shown by a dashed line. (b) Animat position and head direction while approaching the triangle cue and nearing the choice point. (c) Animat position and head direction after a head orienting turn toward the square cue. (d) Animat position and head direction after reaching the goal location at the square cue

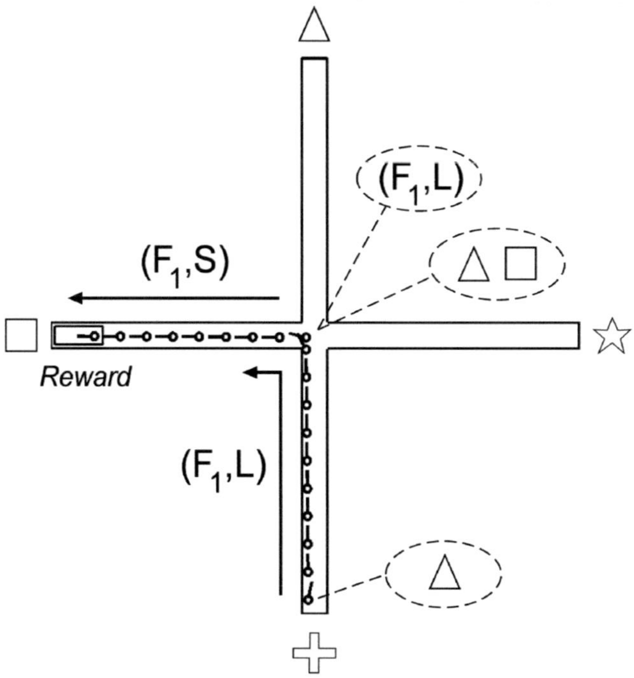

Figure 4. An initial maze trial in which the animat is under reactive visual guidance is shown in this diagram. An approach movement toward the triangle cue is interrupted by motion signals to the left. After a reactive head orienting movement, the square cue comes into view. After approaching the square cue, the rewarded location is reached and adaptive weights are adjusted to strengthen the association between the forward-left-forward sequence and the current motivational state. The arrows and symbols (F_1,L) and (F_1,S), along with the triangle and triangle-square symbols in the dotted ellipses, summarize that a forward-left movement sequence with a forward distance of F_1 is associated with the Triangle list category, and a forward-straight movement also with a forward distance of F_1 is associated with the Triangle-Square list category

2.4. Visual Form and Motion System

The Visual Form and Motion System processes signals from the What Parvo cortical processing stream and the Where Magno cortical processing stream (Figure 5). This separation of functionality endows the animat with three major capabilities. First, the animat can utilize target object recognition and cognitive-emotional conditioning circuitry to learn, choose, and execute motivationally–compatible movements within an overall plan. Second, the animat can use form information to localize visual references, or beacons, to measure its progress over varying terrain. Finally, the animat can process motion boundaries generated during movement toward a choice point within a maze. As the animat nears a choice point, its field–of–view and the intensity of boundary-derived motion signals increase, which can trigger a reactive head–orienting movement. The visual system also drives several important control signals within the model, as described below.

The visual environment is simulated in a virtual reality environment by rendering 3D chromatic scenes as 2D "snapshots" at regular intervals during head-orienting and body-approach movements. As indicated in Figure 1, the visual environment is simplified in SOVEREIGN, which focuses on the various learning and navigational aspects of sequential goal-oriented navigation. A visual target object is separated from the background by a two-stage Figure-Ground Separation module that is within the Parvo stream (Figure 5, left stream). At present, the first processing stage is accomplished in a simple way by using visual targets that are yellow (Figure 1), or have the grayscale corresponding to yellow, and are thereby selected from the background. The second processing stage selects object boundaries via convolution with a 2D Laplacian-of-Gaussian filter. Future model developments will include more sophisticated neural models for 3D vision and figure-ground separation (Cao and Grossberg, 2005; Fang and Grossberg, 2007; Grossberg and Yazdanbakhsh, 2005; Kelly and Grossberg, 2000).

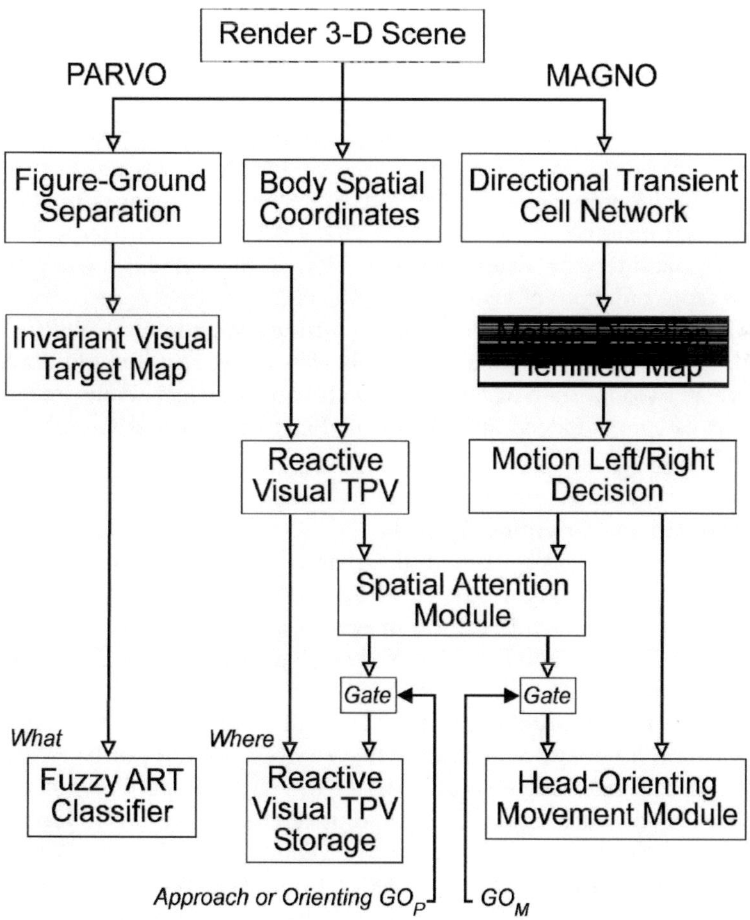

Figure 5. The Visual Form and Motion System flow diagram depicts the stages of visual processing in the model. See text for details

When an object falls within the visual field and it is separated from its background, the coordinates of its centroid, in the 2D image plane, are computed (cf., Russell, 1932) and passed to the Reactive Visual TPV module (Figure 5). The Reactive Visual TPV module converts the centroid from image plane coordinates to head-centered spatial coordinates by using the perspective transformation (Schilling, 1990). The Body Spatial Coordinates module computes the angle between the head and body, before combining this information with the target coordinates in the Reactive Visual TPV module to compute the body-centered distance and angle coordinates of the visual target. The Reactive Visual TPV module updates the Reactive Visual TPV Storage module until the volitional Approach and Orienting GO signal (GO_P) releases a head-orienting or body-approach movement (Figure 5). The head-orienting movement brings the visual target to the center of gaze. Such a transformation into body-centered coordinates can be learned by using a more elaborate network (Greve et. al., 1993; Grossberg et. al., 1993; Guenther et al., 1994).

The left path of the Parvo stream in Figure 5 is devoted to learning a size-invariant and position-invariant object category representation of a visual target within the Invariant Visual Target Map. In order to accomplish this, the figure-ground-separated visual target undergoes a log-polar transformation followed by Gaussian coarse-coding (Baloch and Waxman, 1991; Bradski and Grossberg, 1995). The log-polar transform computes a representation of the visual target object that is size-invariant and position-invariant. This invariant map representation of the target is then transformed into an object category, leading to further compression and invariance under modest changes in object shape, by using unsupervised incremental learning by a Fuzzy ART classifier (Carpenter et al., 1991). The Fuzzy ART classifier and Reactive Visual TPV Storage modules comprise What and Where cortical representations of visual target objects.

The Fuzzy ART classifier can be generalized in a future version of SOVEREIGN to enable learning of 3D target objects from one of multiple views. This requires additional processing stages to learn individual view categories which can be associatively linked to a view-invariant object category (Baloch and Waxman, 1991; Bradski and Grossberg, 1995; Fazl, Grossberg, and Mingolla, 2007).

2.5 Motor Approach and Orienting System

As noted above, the Motor Approach and Orienting System directs body-approach and head-orienting movements (Figure 2). Cumulative estimates of each approach-orienting movement that is processed within the Motor Approach and Orienting System are stored in the Motor Working Memory and Planning System (Figure 2). This section summarizes how these estimates are computed.

The Motor Approach and Orienting System flow diagram is shown in Figure 6. Target position information originates from one of two sources. First, it can be received from the body-centered distance and angle coordinates of the visual target object computed in the Reactive Visual TPV module (Figure 5). Second, it can be received from learned top-down signals from the processing stage that computes Motivated WHAT and WHERE Decisions (Figure 6). These decisions comprise responses from the animat's learned experience which are compatible with the current motivational state. An approximate measure of head-orienting and body displacement is computed by the NET module (Figure 6). Target position information flows from the Reactive Visual TPV to the Reactive Visual TPV Storage module. The NET activity is subtracted from the Stored TPV via learned weights to compute the Reactive Difference Vector, or DV, which represents the angle and distance to move. The learned weights from the NET activity are necessary to calibrate the DV activity. Similarly, learned top-down commands from the Motivated WHAT and WHERE Decisions activate the Planned DV, where NET movement signals are subtracted, yielding a planned angle and distance to move. Calibration of planned commands is accomplished entirely by the top-down adaptive weights. The Reactive DV and Planned DV are the first motor control stages which can elicit head-orienting and body-approach movements.

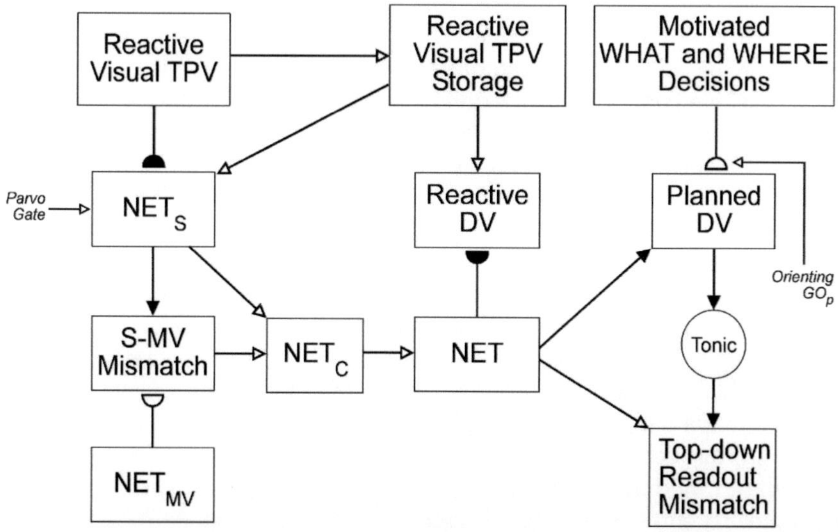

Figure 6. The Motor Approach and Orienting System flow diagram depicts the control hierarchy which generates motor outflow commands. See text for details

The NET estimates of head-orienting and body-approach displacement requires multiple stages of processing to be computed (Figure 6). NET estimates during navigation replace the outflow present position estimates that are computed during hand/arm movements. The NET_S module (Figure 6) calculates this displacement using target positions computed by the Visual Form and Motion System (Figure 5). Body-centered spatial coordinates are denoted by an "s" subscript. The NET_S field activity encodes the net body movement toward a target in spatial coordinates by subtracting the Reactive Visual TPV activity from the Reactive Visual TPV Storage module activity.

Initially, target position information flows from the Reactive Visual TPV to the Reactive Visual TPV Storage module and a short burst of learning zeros the difference at the NET_S module. As the animat moves toward a target, updates to the Reactive Visual TPV Storage module cease and the Reactive Visual TPV decreases, thereby allowing the NET_S module activity to grow. Vestibular and proprioceptive feedback signals are integrated into distances by the NET_{MV} module (Figure 6).

Learning at the output of the NET_{MV} module calibrates the vestibulo-motor signals relative to the visual signals at the S-MV Mismatch module. This adaptive process uses a slow learning rate while visual signals are available from the Visual Form and Motion System (Figure 5). The resulting activity at the S-MV Mismatch module serves as a correction factor which can account for the animat's progress either without visual feedback (e.g., in the dark) or over uneven (e.g., slippery) terrain. When the NET_S and NET_{MV} module activity are identical, then the correction factor is zero and the S-MV Mismatch module activity is also zero.

The NET module combines signals from the NET_S and S-MV Mismatch modules into a robust sensory-motor representation of body displacement. The NET_S module is only active when Parvo signals are present in the Invariant Visual Target Map module (Figure 5). Learned weights from the NET module inhibit activity of the Reactive DV (Figure 6). When the animat has reached the target under visual guidance, the Reactive Visual TPV reaches zero and these adaptive weights are updated, thereby inhibiting the Reactive DV, and stopping the movement. On future trials, the Reactive DV module can be driven to zero by a calibrated level of activity in the NET module, regardless of whether visual input is available.

While under reactive control, visual target coordinates flow into the Reactive Visual TPV Storage module and activate the Reactive DV module, which initiates head and body movements. The Approach or Orienting GO_P control signals are activated when the Reactive or Planned DV command is released under volitional control. The activation of the Approach or Orienting GO_P allows the DV signals to initiate a head-orienting or body-approach movement. Updates to the Reactive Visual TPV Storage module (Figure 6) continue until the Approach or Orienting GO_P is activated. However, under planned control, Motivated WHAT and WHERE Decisions (Figure 6) learn to read out planned head-orienting and body-approach movements. Top-down commands are computed in the Planned DV module, which can initiate head and body movements in response to motivationally-compatible plan items. As then plan unfolds, NET increases until the Planned DV approaches zero, thus terminating a planned movement. The Top-down Readout Mismatch module compares the activity of the learned top-down command and the NET module. A sufficiently large discrepancy between these fields can elicit a control signal to select a different top-down signal from the Motivated WHAT and WHERE Decisions. For instance, the control signal is released when a planned response is interrupted by a strong motion signal which activates the Head-Orienting Movement module (Figure 5) and the animat turns away from the planned response direction.

3. End-to-end SOVEREIGN Simulation

One key SOVEREIGN simulation is demonstrated herein. In the Motivated Choice Experiment, the animat learns the route to two different goal locations under two different motivational states. The simulation summary contains the following types of information: (1) Explanation of the experimental setup; (2) movement trajectories; (3) a step-by-step description of model dynamics; (4) snapshots of visual input at key moments; (5) multi-trial learning; and (6) summary of the model properties demonstrated.

3.1 Motivated Choice Experiment

In this classic example of spatial learning, the animat learns the route to two different goal locations under two different motivational states. Specifically, the Forward-Left-Forward sequence when hungry leads to a food reward, whereas the Forward-Right-Forward sequence leads to a water reward when thirsty. Upon reaching the end of the goal arm, the animat is rewarded and long-term memory weights are updated using a slow learning rate. For the first five training trials of this sequence, the animat has a high hunger drive and is rewarded with food at the end of each trial. For the next five trials, the animat is thirsty and is rewarded with water at the end of each trial. The eleventh and final trial retests the response under the hunger drive to demonstrate that learning under each drive is preserved.

The diagram shown in Figure 7a shows the position, actions and local views seen by the animat during this training trial. Similarly, the diagram in Figure 7b shows the Visual and Motor Working Memory and Planning System plan chunks which are stored during this experimental trial. This simulation is similar to that presented in earlier examples. However, a summary is offered here for completeness. The animat starts this trial shifted to the left within the corridor, thereby increasing Magno signals in the left hemifield. The hungry animat learns to categorize the triangle cue and updates the Visual Working Memory and Planning System. It approaches the triangle cue under reactive control. As it nears the choice point, Magno signals trigger a head-orienting movement to the left bringing the square cue into view. The Triangle chunk is associated with the exploratory drive and can now sample the Forward-Left movement. The animat then learns to categorize the square cue. The Visual and Motor Working Memory and Planning System are updated and the animat approaches the square cue under reactive control. Upon reaching the food reward, all active chunks are associated with the hunger drive and the Forward-Left and Triangle-Square chunks sample the Forward-Straight movement command. The initial training trial is complete.

The diagram in Figure 7c shows the learned plan chunks and their associated motor responses which are gradually strengthened during training. After several trials, the hungry animat starts this trial centered in the corridor, yielding Magno signals which are balanced between left and right. After learning to categorize the triangle cue and updating the Visual Working Memory and Planning System, the Triangle chunk can read out the command to go Forward-Left via top-down signals. The planned command overrides reactive signals and the animat moves forward and turns left. Both the approach speed and Parvo gain are increased because the previously rewarded plan element has been reactivated. After the turn is complete, the Triangle chunk is again associated with the exploratory drive. This learning is triggered by the exploratory learning signal in the absence of explicit reward, and is activated after a head turn is completed. After learning to categorize the square cue, the Visual and Motor Working Memory and Planning System are updated and the Forward-Straight command is directly read out via top-down Motivated WHAT and WHERE Decision signals. After approaching under planned control, the animat is rewarded with food and the active chunks are associated with the hunger drive. The Forward-Left and Triangle-Square chunks can sample the Forward-Straight movement command. The test trial is complete.

The diagram in Figure 8a shows the position, actions and local views seen by the animat during this experimental test trial. Similarly, the diagram in Figure 8b shows the Visual and Motor Working Memory and Planning System plan chunks which are stored during this experimental trial. After several learning trials, the thirsty animat starts this trial centered in the corridor, yielding Magno signals which are balanced between the left and right sides of the visual field. After learning to categorize the triangle cue and updating the Visual Working Memory and Planning System, the Triangle chunk is able to read out the command to go Forward-Right via top-down Motivated WHAT and WHERE Decision signals. The planned command overrides reactive signals and the animat moves forward and turns right. Both the approach speed and Parvo gain are increased because the previously rewarded plan element has been reactivated. After the turn is complete, the Triangle chunk is again associated with the exploratory drive. After learning to categorize the star cue, the Visual and Motor Working Memory and Planning System are updated and the Forward-Straight command is directly read out via top-down signals. After approaching under planned control, the animat is rewarded with water and the active chunks are associated with the thirst drive. The Forward-Right and Triangle-Star chunks can sample the Forward-Straight movement command. The test trial and this sequence of experiments are complete.

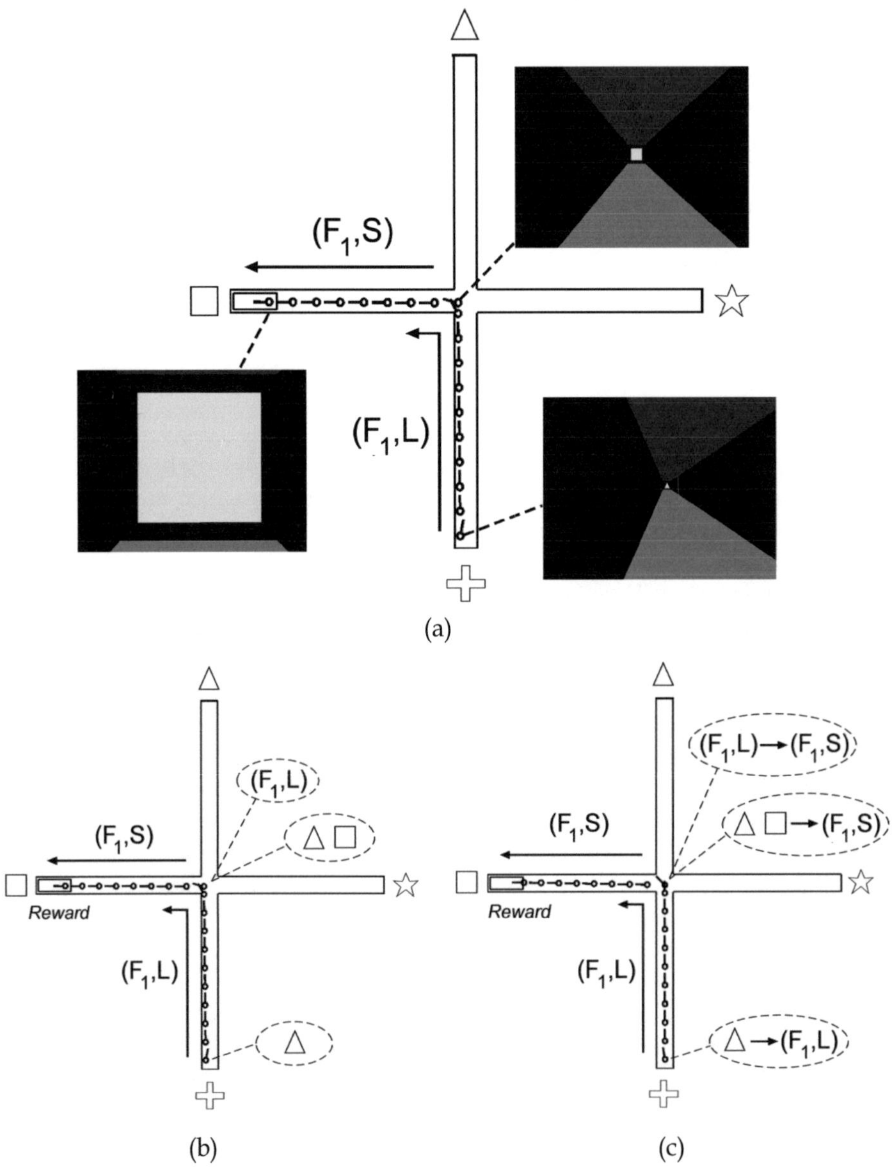

Figure 7. (a) Perspective views are shown for selected points during maze exploration toward the goal location in the left arm. (b) Each ellipse graphically depicts the short-term memory chunks represented in both the Visual and Motor Working Memory and Planning System during exploratory learning. (c) Each ellipse graphically depicts the short-term memory chunks and associated motor responses in both the Visual and Motor Working Memory and Planning System after exploratory learning

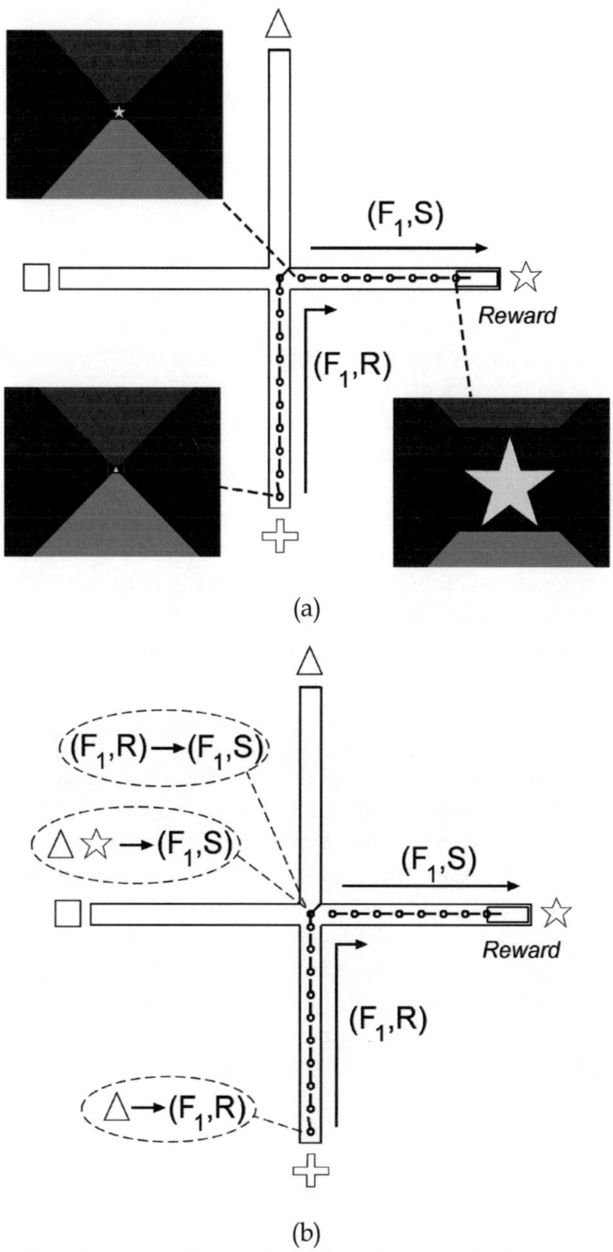

Figure 8. (a) Perspective views are shown for selected points during maze exploration under planned control. (b) Each ellipse graphically depicts the short-term memory chunks and associated motor responses in both the Visual and Motor Working Memory and Planning System after exploratory learning

4. General Discussion and Conclusions

The SOVEREIGN architecture embodies a number of design principles whose mechanistic instantiation as neural circuits (Figure 2) enable incremental learning of planned action sequences to carry out route-based navigation towards a rewarded goal. SOVEREIGN circuits are based on neural models that have elsewhere been used to explain and predict many behavioral and brain data. Here the emphasis is on designing a neuromorphic controller that emphasizes behavioral competence.

The model has several notable strengths relative to other available models, including the following ones: First, it provides an end-to-end model that includes on-line vision, visual recognition learning and categorization, working memory storage of sequences of visual and motor categories, learning of sequential cognitive and motor plans, cognitive-emotional interactions whereby reinforcement learning can select plans that can attain a currently valued goal, and balancing of visually reactive exploratory vs. planned movement decisions, based upon the relative salience of bottom-up and top-down information through time. Second, unlike various other models (e.g., Barto and Sutton, 1981; Dayan, 1987; Schmajuk, 1990), no explicit spatial goal gradient, proportional to the spatial distance from the goal, is required to guide goal-oriented sequential behavior in SOVEREIGN. Third, list chunks provide a compact context-sensitive code for learning plans to navigate a large number of different routes. Fourth, reliable, single–trial learning of a maze can occur if the animat happens to find the goal location during an exploratory trial. Fifth, the animat can respond to the same sequence of visual or motor events in different ways to achieve different goals when different drives are prepotent.

Whereas the detailed circuit realizations that are currently used in SOVEREIGN will doubtless be modified and further developed in the future, it embodies design principles that may need to be incorporated, in some form, into future autonomous adaptive controllers of navigational behaviors. One weakness in the current version of SOVEREIGN is that its navigational behaviors are all route-based. The model does not yet include mechanisms of spatial navigation (O'Keefe and Dostrovsky, 1971; O'Keefe and Nadel, 1978) such as the role of hippocampal place fields, head-direction cells, and the theta rhythm (e.g., Burgess et al., 1995). Such a development would require an understanding of how place fields form, notably the role of entorhinal grid cells in their formation (e.g., Hafting, Fyhn, Molden, Moser, and Moser, 2005), which other modeling research is currently investigating (e.g., Fuhs and Touretzky, 2006; Gorchetchnikov and Grossberg, 2007).

5. Acknowledgements

This work was supported in part by the National Science Foundation (SBE-0354378).

6. References

Baloch, A. A. and Waxman, A. M. (1991). Visual learning, adaptive expectations, and behavioral conditioning of the mobile robot MAVIN. *Neural Networks, 4*, 271–302.

Barto, A. G. and Sutton, R. S. (1981). Landmark learning: An illustration of associative search. *Biological Cybernetics, 42*, 1–8.

Berzhanskaya, J., Grossberg, S. and Mingolla, E. (2007). Laminar cortical dynamics of visual form and motion interactions during coherent object motion perception. *Spatial Vision*, in press.

Bradski, G., Carpenter, G.A., and Grossberg, S. (1994). STORE working memory networks for storage and recall of arbitrary temporal sequences. *Biological Cybernetics, 71*, 469-480.

Bradski, G. and Grossberg, S. (1995). Fast learning VIEWNET architectures for recognizing 3-D objects from multiple 2-D views. *Neural Networks, 8*, 1053-1080.

Bullock, D., Cisek, P. and Grossberg, S. (1998). Cortical networks for control of voluntary arm movements under variable force conditions. *Cerebral Cortex, 8*, 48-62.

Bullock, D. and Grossberg, S. (1988). Neural dynamics of planned arm movements: Emergent invariants and speed-accuracy properties during trajectory formation. *Psychological Review, 95*, 49-90.

Bullock, D. and Grossberg, S. (1991). Adaptive neural networks for control of movement trajectories invariant under speed and force rescaling. *Human Movement Science, 10*, 3-53.

Burgess, N., Recce, M., and O'Keefe, J. (1995). Hippocampus - Spatial models. In *The Handbook of Brain Theory*. Bradford Books / MIT Press, Cambridge, MA.

Cao, Y. and Grossberg, S. (2005). A laminar cortical model of stereopsis and 3D surface perception: Closure and da Vinci stereopsis. *Spatial Vision, 18*, 515-578.

Carpenter, G.A., Grossberg, S., Markuzon, N., Reynolds, J.H., and Rosen, D.B. (1992). Fuzzy ARTMAP: A neural network architecture for incremental supervised learning of analog multidimensional maps. *IEEE Transactions on Neural Networks, 3*, 698-713.

Carpenter, G. A., Grossberg, S., and Rosen, D. B. (1991). Fuzzy ART: Fast stable learning and categorization of analog patterns by an adaptive resonance system. *Neural Networks, 4*, 759–771.

Cohen, M.A. and Grossberg, S. (1986). Neural dynamics of speech and language coding: Developmental programs, perceptual grouping, and competition for short-term memory. *Human Neurobiology, 5*, 1-22.

Cohen, M. A. and Grossberg, S. (1987). Masking fields: A massively parallel neural architecture for learning, recognizing, and predicting multiple groupings of patterned data. In Grossberg, S. (Ed.), *Neural Networks and Natural Intelligence*, Chapter 7, pp. 317–367. The MIT Press, Cambridge, MA.

Contreras-Vidal, J.L., Grossberg, S., and Bullock, D. (1997). A neural model of cerebellar learning for arm movement control: Cortico-spino-cerebellar dynamics. *Learning and Memory, 3*, 475-502.

Dayan, P. (1987). Navigating through temporal difference. In *Neural Information Processing Systems, 3*, pp. 464–470. Morgan Kaufmann Publishers, San Mateo, CA.

Fang, L. and Grossberg, S. (2007). From stereogram to surface: How the brain sees the world in depth. Technical Report CAS/CNS TR-06-013, Boston University. Submitted for publication.

Fazl, A., Grossberg, S., and Mingolla, E. (2007). View-invariant object category learning, recognition and search: How spatial and object attention are coordinated using surface-based attentional shrouds. Technical Report CAS/CNS TR-07-011, Boston University. Submitted to *Cognitive Psychology*, April, 2007.

Fiala, J.C., Bullock, D., Grossberg, S. (1996). Metabotropic glutamate receptor activation in cerebellar Purkinje cells as substrate for adaptive timing of the classically conditioned eye blink response. *Journal of Neuroscience, 16,* 3760-3774.

Fuhs, M.C. and Touretzky, D.S. (2006). A spin glass model of path integration in rat medial entorhinal cortex. *Journal of Neuroscience, 26,* 4266 - 4276.

Gaudiano P. and Grossberg S. (1991). Vector associative maps: Unsupervised real-time error-based learning and control of movement trajectories. *Neural Networks, 4,* 147-183.

Gnadt, W. and Grossberg, S. (2005a). SOVEREIGN: *A Self-Organizing, Vision, Expectation, Recognition, Emotion, Intelligent, Goal-oriented Navigation system.* In Douglas Blank and Lisa Meeden (Eds.), Developmental Robotics: Papers from the 2005 Spring Symposium, March 21-23, pp. 106-110. American Association for Artificial Intelligence, Menlo Park, California.

Gnadt, W. and Grossberg S. (2005b). *SOVEREIGN: A Self-Organizing, Vision, Expectation, Recognition, Emotion, Intelligent, Goal-oriented Navigation system.* Dynamical Neuroscience XIII: Computational Cognitive Neuroscience satellite symposium of the annual Society for Neuroscience meeting, November 10-11, 2005.

Gnadt, W. and Grossberg S. (2006). *SOVEREIGN: A Self-Organizing, Vision, Expectation, Recognition, Emotion, Intelligent, Goal-oriented Navigation system.* Tenth International Conference on Cognitive and Neural Systems, Boston, Massachusetts, May 2006.

Gnadt, W. and Grossberg, S. (2008). SOVEREIGN: An autonomous neural system for incrementally learning to navigate towards a rewarded goal. *Neural Networks,* in press.

Goodale, M. A. and Milner, D. (1992). Separate visual pathways for perception and action. *Trends in Neurosciences, 15,* 20–25.

Gorchetchnikov, A. and Grossberg, S. (2007). Space, time, and learning in the hippocampus: How fine spatial and temporal scales are expanded into population codes for behavioral control. Neural Networks, 20, 182 - 193.

Greve, D., Grossberg, S., Guenther, F., and Bullock, D. (1993). Neural representations for sensory-motor control, I: Head-centered 3-D target positions from opponent eye commands. *Acta Psychologica, 82,* 115–138.

Grossberg, S. (1980). How does a brain build a cognitive code? *Psychological Review, 87,* 1-51.

Grossberg, S. (1984a). Some psychophysiological and pharmacological correlates of a developmental, cognitive and motivational theory. In R. Karrer, J. Cohen, and P. Tueting (Eds.), *Brain and information: Event related potentials.* New York: New York Academy of Sciences, pp.58-142.

Grossberg, S. (1984b). Some normal and abnormal behavioral syndromes due to transmitter gating of opponent processes. *Biological Psychiatry, 19,* 1075–1118.

Grossberg, S. (2000a). The complementary brain: Unifying brain dynamics and modularity. *Trends in Cognitive Sciences, 4,* 233-246.

Grossberg, S. (2000b). The imbalanced brain: From normal behavior to schizophrenia. *Biological Psychiatry, 48,* 81-98.

Grossberg, S., Guenther, F., Bullock, D., and Greve, D. (1993). Neural representations for sensory-motor control, II: Learning a head-centered visuomotor representation of 3-D target position. *Neural Networks, 6,* 43–67.

Grossberg, S. and Merrill, J.W.L. (1992). A neural network model of adaptively timed reinforcement learning and hippocampal dynamics. *Cognitive Brain Research, 1,* 3-38.

Grossberg, S. and Merrill, J.W.L. (1996). The hippocampus and cerebellum in adaptively timed learning, recognition, and movement. *Journal of Cognitive Neuroscience, 8,* 257-277.

Grossberg, S., Mingolla, E. and Viswanathan, L. (2001). Neural dynamics of motion integration and segmentation within and across apertures. *Vision Research, 41,* 2521-2553.

Grossberg, S. and Myers, C.W. (2000). The resonant dynamics of speech perception: Interword integration and duration-dependent backward effects. *Psychological Review, 107,* 735-767.

Grossberg, S., & Pearson, L. (2007). Laminar cortical dynamics of cognitive and motor working memory, sequence learning and performance: Towards a unified theory of how the cerebral cortex works. Submitted for publication.

Grossberg, S. and Seidman, D. (2006). Neural dynamics of autistic behaviors: Cognitive, emotional, and timing substrates. *Psychological Review, 113,* 3, 483-525.

Grossberg, S. and Yazdanbakhsh, A. (2005). Laminar cortical dynamics of 3D surface perception: Stratification, transparency, and neon color spreading. *Vision Research, 45,* 1725-1743.

Guenther, F. H., Bullock, D., Greve, D., and Grossberg, S. (1994). Neural representations for sensory-motor control, III: Learning a body-centered representation of 3-D target position. *Journal of Cognitive Neuroscience, 6,* 341 - 358.

Hafting, T., Fyhn, M., Molden, S., Moser, M.H., and Moser, E.I. (2005). Microstructure of the spatial map in the entorhinal cortex, *Nature, 436,* 801 - 806.

Kelly, F.J. and Grossberg, S. (2000). Neural dynamics of 3-D surface perception: Figure-ground separation and lightness perception. *Perception and Psychophysics, 62,* 1596-1619.

Leonard, B. and McNaughton, B. L. (1990). Spatial representation in the rat: Conceptual, behavioral, and neurophysiological perspectives. In Kesner, R. P. and Olton, D. S. (Eds.), *Neurobiology of Comparative Cognition,* Chapter 13, pp. 363–422. Lawrence Erlbaum Associates, Hillsdale, NJ.

Mishkin, M., Ungerleider, L. G., and Macko, K. A. (1983). Object vision and spatial vision: Two cortical pathways. *Trends in Neurosciences, 6,* 414–417.

Munn, N. L. (1950). *Handbook of Psychological Research on the Rat.* Houghton Mifflin Co., Boston, MA.

O'Keefe, J.M. and Dostrovsky, J. (1971). The hippocampus as a spatial map. Preliminary evidence from unit activity in the freely-moving rat. *Brain Research, 34,* 171 - 175.

O'Keefe, J. and Nadel, L. (1978). *The Hippocampus as a Cognitive Map.* Oxford University Press, New York, NY.

Pribe, C., Grossberg, S., and Cohen, M.A. (1997). Neural control of interlimb oscillations, II: Biped and quadruped gaits and bifurcations. *Biological Cybernetics, 77,* 141-152.

Raizada, R. and Grossberg, S. (2003). Towards a theory of the laminar architecture of cerebral cortex: Computational clues from the visual system. *Cerebral Cortex, 13,* 100-113.

Russell, J. T. (1932). Depth discrimination in the rat. *Journal of Genetic Psychology, 40,* 136–159.

Schilling, R. J. (1990). *Fundamentals of Robotics: Analysis and Control.* Prentice Hall, Englewood Cliffs, NJ.

Schmajuk, N. A. (1990). Role of the hippocampus in temporal and spatial navigation: An adaptive neural network. *Behavioral Brain Research, 39,* 205–229.

Ungerleider, L. G. and Mishkin, M. (1982). Two cortical visual systems: Separation of appearance and location of objects. In Ingle, D. L., Goodale, M. A., and Mansfield, R. J. W. (Eds.), *Analysis of Visual Behavior,* pp. 549–586. The MIT Press, Cambridge, MA.

7

Stereo Matching and 3D Reconstruction via an Omnidirectional Stereo Sensor[1]

Lei He, Chuanjiang Luo, Feng Zhu and Yingming Hao
Shenyang Institute of Automation, Chinese Academy of Sciences
P.R. China

1. Introduction

A catadioptric vision system using diverse mirrors has been a popular means to get panoramic images(K.Nayar, 1997) which contains a full horizontal field of view (FOV). This wide view is ideal for three-dimensional vision tasks such as motion estimation, localization, obstacle detection and mobile robots navigation. Omnidirectional stereo is a suitable sensing method for such tasks because it can acquire images and ranges of surrounding areas simultaneously. For omnidiretional stereo vision, an obvious method is to use two (or more) cameras instead of each conventional camera (K.Tan et al., 2004; J.Gluckman et al., 1998; H.Koyasu et al. 2002; A.Jagmohan et al. 2004). Such two-camera (or more-camera) stereo systems are relatively costly and complicated compared to single camera stereo systems. Omnidirectional stereo based on a double-lobed mirror and a single camera was developed (M.F.D. Southwell et al. 1996; T.L. Conroy & J.B. Moore, 1999; E. L. L. Cabral, et al. 2004; Sooyeong Yi & Narendra Ahuja, 1996) . A double lobed mirror is a coaxial mirror pair, where the centers of both mirrors are collinear with the camera axis, and the mirrors have a profile radially symmetric around this axis. This arrangement has the advantage to produce two panoramic views of the scene in a single image. But the disadvantage of this method is the relatively small baseline it provides. Since the two mirrors are so close together, the effective baseline for stereo calculation is quite small. We have developed a novel omnidirectional stereo vision optical device (OSVOD) based on a common perspective camera coupled with two hyperbolic mirrors, which are separately fixed inside a glass cylinder. As the separation between the two mirrors provides much enlarged baseline, in our system, the baseline length is about 200mm, the precision has improved correspondingly (Fig. 1).

[1] This work is supported by National Science Foundation of P.R. China (60575024).

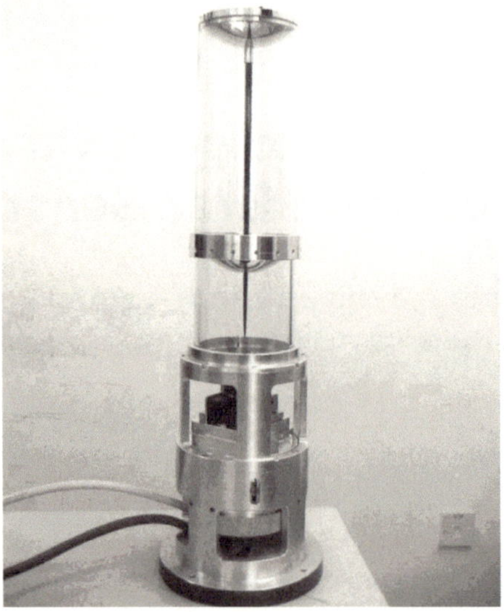

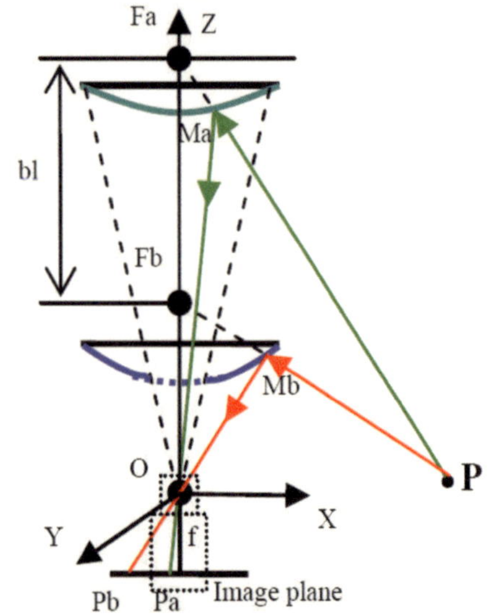

a b

Figure 1. a: The appearance of the stereo vision system. b: The configuration of the system

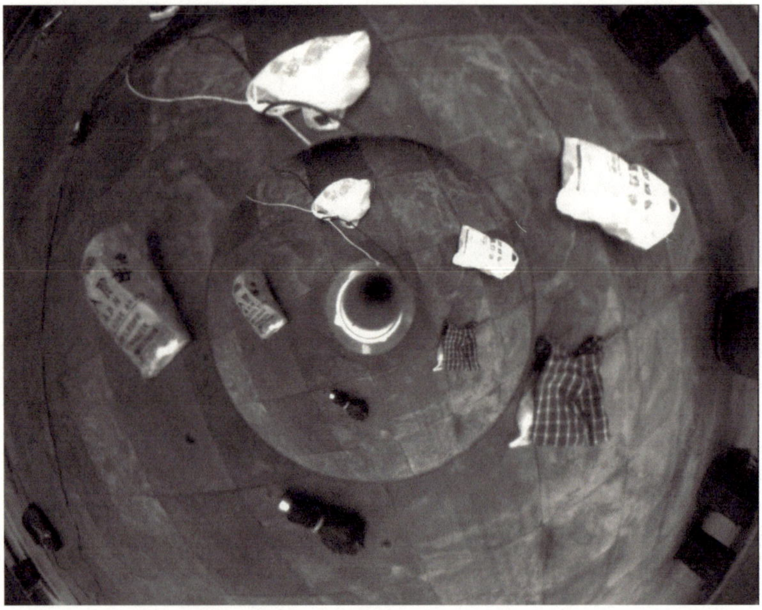

Figure 2. Real indoor scene captured for depth map reconstruction

The coaxial configuration of the camera and the two hyperbolic mirrors makes the epipolar line radially collinear, which makes the system free of the search process for complex epipolar curve in stereo matching (Fig. 2). The OSVOD is mounted on top of mobile robot looking downwards, at a height of approximately 0.75 meters above the ground plane. We also notice that G. Jang et al. (G. Jang, S. Kim & I. Kweon, 2005) proposed a wide-baseline catadioptric stereo system, in which the system design is similar to us but the system design and mirror types are different. Furthermore, for a mobile robot, our system is designed to obtain omnidirectional images that are different from above mentioned stereo systems. Note that our images are taken from a special angle of view to obtain the planform of scene, which aims to reconstruct the depth map that denotes the heigh information (vertical depth) but not the horizontal distance information in common stereo systems. Albeit the calculative precision of triangulation was improved theoretically due to the wider baseline, more complex and difficult stereo matching problem should be brought on because of the wider disparity space that exists and more serious image distortion. A major aim of this paper is to propose an integrated framework, which mainly focuses on stereo matching to enhance the performance for depth map regeneration. Since our primary goal is to propose a precise and suitable algorithm for stereo matching to satisfy the reliability requirement via an omnidirectional stereo vision system, we chiefly review previous stereo matching methods as follows. State of the art algorithms for dense stereo matching can be divided into two categories. One category is local methods, in which some kind of similarity measure over an area is calculated (Devernay & F. Faugeras, 1994). They work well in relatively textured areas in a very fast speed, while they cannot gain correct disparity map in textureless areas and areas with repetitive textures, which is an unavoidable problem in most situations. In (Sara, R., 2002) a method of finding the largest unambiguous component has been proposed, but the density of the disparity map varies greatly depend on the discriminability of the similarity measure in a given situation. The other one is global methods which are generally energy minimization approaches, these methods make explicit smoothness assumptions and try to find a global optimized solution of a predefined energy function that take into account both the matching similarities and smoothness assumptions. Most recent high-perfomance algorithms belong to energy minimization approaches (Pedro F. Felzenszwalb & Daniel P. Huttenlocher, 2006; Y. Boykov et al., 2001; V. Kolmogorov & R. Zabih, 2001; Yedidia, J.S. et al., 2000) due to powerful new optimization algorithms such as graph cuts and loopy belief propagation. The results, especially in stereo, have been dramatic, according to the widely-used Middlebury stereo benchmarks (D. Scharstein & R. Szeliski, 2002), almost all the top-performing stereo methods rely on graph cuts or LBP. Moreover, these methods give substantially more accurate results than what were previously possible. Although numerous methods exist for stereo matching, as to our knowledge, most algorithms are presented and implemented using standard images, there are few algorithms specifically designed for single camera omnidirectional stereo. Sven Fleck et al. (Sven Fleck et al., 2005) completely use a common graph cut method to acquire a 3D model using a mobile robot that is equipped with a laser scanner and a panoramic camera. Other work in point based omnidirectional reconstruction on mobile robotics can be found in (R. Bunschoten & B. Kr¨ose, 2003), however this is not based on graph cuts that we are concerned in this paper. Considering the peculiarities of omnidirectional images, we adapt improved graph cut method, in which a new energy model is introduced for more general priors corresponding to more reasonable piecewise smoothness assumption since the well-known swap move algorithm can be applied to a wider class of energy functions (Y. Boykov et al., 2001). The proposed energy function is different from previous any other ones since the smooth item is based on three variables whereas others only consist of two variables. We also show the necessary modification to handle panoramic images, including deformed matching template, adaptable template scale to elaborate the date term. In the rest of the paper, we first necessarily introduce a full model of calibration in the system. In section 3, our improved optimization model is presented. We generalize our completed omnidirectional stereo matching framework in section 4. In section 5, experiments and their results are given. Finally, we conclude the paper.

2. Calibrating the System

2.1 Principle of Our Vision System

The system we have developed (Su & Zhu, 2005) is based on a common perspective camera coupled with two hyperbolic mirrors, which are separately fixed inside a glass cylinder (Fig.1a). The two hyperbolic mirrors share one focus which coincides with the camera center. A hole in the below mirror permits imaging via the mirror above. As the separation between the two mirrors provides much enlarged baseline, the precision of the system has been improved correspondingly. The coaxial configuration of the camera and the two hyperbolic mirrors makes the epipolar line radially collinear, thus making the system free of the search process for complex epiploar curve in stereo matching (Fig. 3).
To describe the triangulation for computing 3D coordinates of space points, we define the focal point O as the origin of our reference frame, z-axis parallel to the optical axis pointing above. Then mirrors can be represented as:

$$\frac{(z-c_i)^2}{a^2}-\frac{(x^2+y^2)}{b^2}=1, \quad (i=1,2) \tag{1}$$

Only the incident rays pointing to the focus $F_a(0,0,2c_a)$, $F_b(0,0,2c_b)$ will be reflected by the mirrors to pass through the focal point of the camera. The incident ray passing the space point $P(x,y,z)$ reaches the mirrors at points M_a and M_b, being projected onto the image at points $P_a(u_a,v_a,-f)$ and $P_b(u_b,v_b,-f)$ respectively. As P_a and P_b are known, M_a and M_b can be represented by:

$$\frac{x_{M_i}}{u_i}=\frac{y_{M_i}}{v_i}=\frac{z_{M_i}}{-f}, \quad (i=1,2) \tag{2}$$

Since point M_a and M_b are on the mirrors, they satisfy the equation of the mirrors. Their coordinates can be solved from equation group (1) and (2). Then the equation of rays F_aP and F_bP are:

$$\frac{x_p}{x_i}=\frac{y_p}{y_i}=\frac{z_p-2c_i}{z_i-2c_i}, \quad (i=1,2) \tag{3}$$

We can finally figure out coordinate of the space point P by solving the equation (3).

2.2 Overview of Omnidirectional Camera Calibration

In using the omnidirectional stereo vision system, its calibration is important, as in the case of conventional stereo systems (Luong & Faugeras, 1996; Zhang & Faugeras, 1997). We present a full model of the imaging process, which includes the rotation and translation

between the camera and mirror, and an algorithm to determine this relative position from observations of known points in a single image.
There have been many works on the calibration of omnidirectional cameras. Some of them are for estimating intrinsic parameters (Ying & Hu, 2004; Geyer & Daniilidis, 1999; Geyer Daniilidis, 2000; Kang, 2000). In (Geyer & Daniilidis, 1999; Geyer Daniilidis, 2000), Geyer & Daniilidis presented a geometric method using two or more sets of parallel lines in one image to determine the camera aspect ratio, a scale factor that is the product of the camera and mirror focal lengths, and the principal point. Kang (Kang, 2000) describes two methods. The first recovers the image center and mirror parabolic parameter from the image of the mirror's circular boundary in one image; of course, this method requires that the mirror's boundary be visible in the image. The second method uses minimization to recover skew in addition to Geyer's parameters. In this method the image measurements are point correspondences in multiple image pairs. Miousik & Pajdla developed methods of calibrating both intrinsic and extrinsic parameters (Miousik & Pajdla, 2003a; Miousik & Pajdla, 2003b). In (Geyer & Daniilidis, 2003), Geyer & Daniilidis developed a method for rectifying omnidirectional image pairs, generating a rectified pair of normal perspective images.
Because the advantages of single viewpoint cameras are only achieved if the mirror axis is aligned with the camera axis, these methods mentioned above all assume that these axes are parallel rather than determining the relative rotation between the mirror and camera. A more complete calibration procedure for a catadioptric camera which estimates the intrinsic camera parameters and the pose of the mirror related to the camera appeared at (Fabrizio et al., 2002), the author used the images of two known radius circles at two different planes in an omnidirectional camera structure to calibrate the intrinsic camera parameters and the camera pose with respect to the mirror. But this proposed technique cannot be easily generalized to all kinds of catadioptric sensors for it requires the two circles be visible on the mirror. Meanwhile, this technique calibrated the intrinsic parameters combined to extrinsic parameters, so there are eleven parameters (five intrinsic parameters and six extrinsic parameters) need to be determined. As the model of projection is nonlinear the computation of the system is so complex that the parameters cannot be determined with good precision.
Our calibration is performed within a general minimization framework, and easily accommodates any combination of mirror and camera. For single viewpoint combinations, the advantages of the single viewpoint can be exploited only if the camera and mirror are assumed to be properly aligned. So for these combinations, the simpler single viewpoint projection model, rather than the full model described here, should be adopted only if the misalignment between the mirror and camera is sufficiently small. In this case, the calibration algorithm that we describe is useful as a software verification of the alignment accuracy.
Our projection model and calibration algorithm separate the conventional camera intrinsics (e.g., focal length, principal point) from the relative position between the mirrors and the camera (i.e., the camera-to-mirrors coordinate transformation) to reduce computational complexity and improve the calibration precision. The conventional camera intrinsics can be determined using any existing method. For the experiments described here, we have used the method implemented in http://www.vision.caltech.edu/bouguetj/calib_doc/. Once the camera intrinsics are known, the camera-to-mirrors transformation can be determined by obtaining an image of calibration targets whose three-dimensional positions are known, and then minimizing the difference between coordinates of the targets and the locations

calculated from the targets' images through the projection model. Fig. 3 shows one example of calibration image used in our experiments. The locations of the three dimensional points have been surveyed with an accuracy of about one millimeter. If the inaccuracy of image point due to discrete distribution of pixels is taken into account, the total measuring error is about five millimeters.

2.3 Projection Model

Fig. 3 depicts the full imaging model of a perspective camera with two hyperbolic mirrors. There are three essentially coordinate systems.

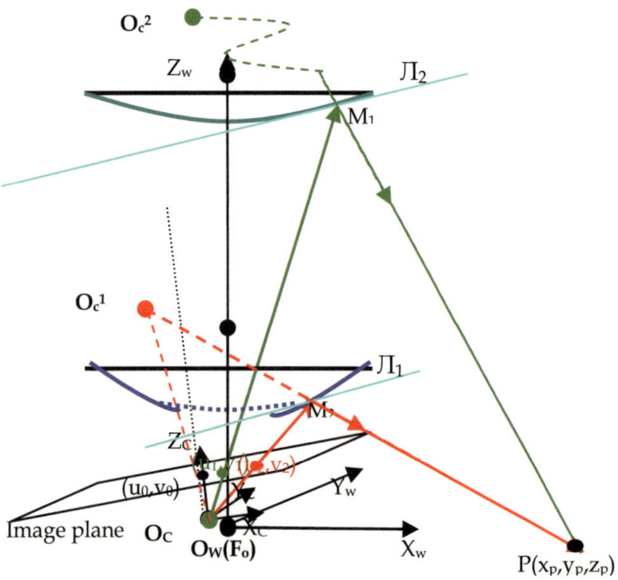

Figure 3. The projection model of the omnidirectional stereo vision system. There are transformations between the camera coordinate system and the mirror (or world) coordinate system

1. The camera coordinate system centered at the camera center O_c, the optical axis is aligned with the z-axis of the camera coordinate system;
2. The mirror system centered at common foci of the hyperbolic mirrors F_o, the mirrors axes is aligned with the z-axis of the mirror coordinate system (We assume that the axes of the mirrors are aligned well, and the common foci are coincident, from the mirrors manufacturing sheet we know it is reasonable);
3. The world system centered at O_w. The omnidirectional stereo vision system was placed on a plane desk. As both the base of vision system and desk surface are plane, the axis of the mirror is perpendicular to the base of the system and the surface of the desk feckly. We make the mirror system coincide with the world system to simplify the model and computations.

So the equations of hyperboloid of two sheets in the system centered at O_w are the same as equation (1). For a known world point $P(x_w,y_w,z_w)$ in the world (or mirror) coordinate system whose projected points in the image plane are also known, $q_1(u_1,v_1)$ and $q_2(u_2,v_2)$ are respectively projected by the upper mirror and bellow mirror. Then we get their coordinates in the camera coordinate system:

$$\begin{bmatrix} x_i^c \\ y_i^c \\ z_i^c \end{bmatrix} = \begin{bmatrix} (u_i-u_0)k_u \\ (v_0-v_i)k_v \\ f \end{bmatrix}, \quad (i=1,2) \tag{4}$$

Where f is the focal length; k_u and k_v are the pixel scale factors; u_0 and v_0 are the coordinates of the principal point, where the optical axis intersects the projection plane. They are intrinsic parameters of the perspective camera.

So the image points $P_c(x_i^c, y_i^c, z_i^c)$ of the camera coordinate system can be expressed relative to the mirror coordinate system as:

$$\begin{bmatrix} x_i^m \\ y_i^m \\ z_i^m \end{bmatrix} = R\begin{bmatrix} x_i^c \\ y_i^c \\ z_i^c \end{bmatrix} + t, \quad (i=1,2) \tag{5}$$

Where R is a 3×3 rotation matrix with three rotation angles around the x-axis (pitch α), y-axis (yaw β) and z-axis (title χ) of the mirror coordinate system respectively; $t=[t_x,t_y,t_z]$ is the translation vector. So the origin $O_c=[0,0,0]^T$ of the camera coordinate system can be expressed in the world coordinate system $O_m=[t_x,t_y,t_z]^T$, so the equations of lines O_cM_1 and $f=\{f_p|p\in P\}$ which intersect with the upper mirror and bellow mirror respectively at points M_1 and M_2, can be determined by solving simultaneous equations of the line O_cM_1 or O_cM_2 and the hyperboloid. Once the coordinates of the point M_1 and M_2 have been worked out, we can write out the equations of the tangent plane л1 and л2 which passes the upper and the bellow mirror at point M_1 and M_2 respectively. Then the symmetric points O_c^1 and O_c^2 of the origin of the camera coordinate system O_c relative to tangent plane л1 and л2 in the world coordinate system can be solved from the following simultaneous equations:

$$\begin{cases} \dfrac{x_{O_c}{}^i - tx}{a_i^2 x_{M_i}} = \dfrac{y_{O_c}{}^i - ty}{a_i^2 y_{M_i}} = \dfrac{z_{O_c}{}^i - tz}{-b_i^2 z_{M_i} + b_i^2 c_i} \\ a_i^2 x_{M_i}(tx + x_{O_c}{}^i) + a_i^2 y_{M_i}(ty + y_{O_c}{}^i) - (-b_i^2 z_{M_i} + b_i^2 c_i)(tz + z_{O_c}{}^i), \\ +2[-a_i^2 x_{M_i}{}^2 - a_i^2 y_{M_i}{}^2 - z_{M_i}(-b_i^2 z_{M_i} + b_i^2 c_i)] = 0 \end{cases} \quad (i=1,2) \tag{6}$$

Hitherto the incident ray $O_c^1 M_2$ and $O_c^2 M_1$ can be written out to determine the world point $P(x_w, y_w, z_w)$. Generally, the two lines are non-co-plane due to various parameter errors and measuring errors, we solve out the midpoint $G=(\hat{x}_w, \hat{y}_w, \hat{z}_w)^T$ of the common perpendicular of the two lines by

$$\begin{cases} [\overrightarrow{O_c^1 M_2} \times (\overrightarrow{O_c^1 M_2} \times \overrightarrow{O_c^2 M_1})] \bullet \overrightarrow{G_1 M_2} = 0 \\ \overrightarrow{G_1 M_1} = t\overrightarrow{G_1 O_c^2} \end{cases} \Rightarrow \overrightarrow{OG_1} \qquad \begin{cases} [\overrightarrow{O_c^2 M_1} \times (\overrightarrow{O_c^1 M_2} \times \overrightarrow{O_c^2 M_1})] \bullet \overrightarrow{G_2 M_1} = 0 \\ \overrightarrow{G_2 M_2} = t\overrightarrow{G_2 O_c^1} \end{cases} \Rightarrow \overrightarrow{OG_2} \qquad \overrightarrow{OG} = (\overrightarrow{OG_1} + \overrightarrow{OG_2})/2 \tag{7}$$

From all of them above, we finally come to the total expression to figure out the world point $G = (\hat{x}_w, \hat{y}_w, \hat{z}_w)^T$ from two image points respectively projected by the upper mirror and bellow mirror and six camera pose parameters left to be determined.

$$G\left(\alpha, \beta, \chi, t_x, t_y, t_z, u_1, v_1, u_2, v_2\right) = \begin{bmatrix} \hat{x}_w \\ \hat{y}_w \\ \hat{z}_w \end{bmatrix} \tag{8}$$

Equation (8) is a very complex nonlinear equation with high power and six unknown parameters to determine. The artificial neural network trained with sets of image points of the calibration targets is used to estimate the camera-to-mirror transformation.

Taking advantage of the ANN capability, which adjusts the initial input camera-to-mirror transformations step by step to minimize the error function, the real transformations parameters of the camera-to-mirror can be identified precisely.

2.4 Error Function

Considering the world points with known coordinates, placed onto a calibration pattern, at the same time, their coordinates can be calculated using the equation (8) from back-projection of their image points. The difference between the positions of the real world coordinates and the calculated coordinates is the calibration error of the model. Minimizing the above error by means of an iterative algorithm such as Levenberg-Marquardt BP algorithm, the camera-to-mirror transformation is calibrated. The initial values for such algorithm are of consequence. In our system, we could assume the transformation between cameras and mirrors is quite small, as the calculation error without considering the camera-to-mirror transformation is not significant thus using R=I and T=0 as the initial values is a reasonable choice.

We minimize the following squared error ε^2:

$$\varepsilon^2 = \sum_{i=1}^{n} \left\| P_i - G_i\left(\alpha, \beta, \chi, t_x, t_y, t_z, u_1^i, v_1^i, u_2^i, v_2^i\right) \right\|^2 \tag{9}$$

Where n is the number of the calibration points.

Because $G_i\left(\alpha,\beta,\chi,t_x,t_y,t_z,u_1^i,v_1^i,u_2^i,v_2^i\right)$ depends on the camera-to-mirror transformation, (9) is optimized with respect to the six camera-to-mirror parameters.

2.5 Calibration Result

The calibration was performed using a set of 81 points equally distributed on a desk with different heights from 0 to 122mm around the vision system.

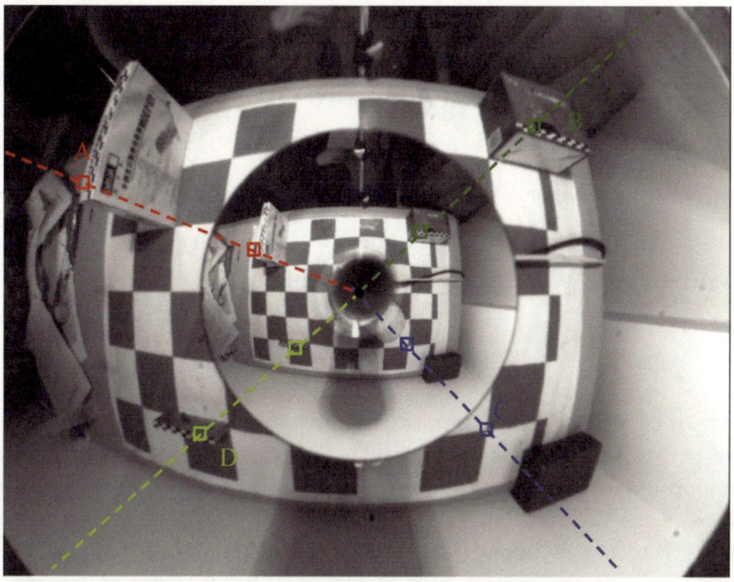

Figure 4. A calibration image used in our experiments. The coaxial configuration of the camera and the two hyperbolic mirrors makes the epipolar line radially collinear, which makes the system free of the search process for complex epipolar curve in stereo matching

The calibration results with real data are listed in Table 1.

	α	β	χ	t_x	t_y	t_z
value	-0.9539°	0.1366°	0.1436°	-0.0553mm	-0.1993mm	1.8717mm

Table 1. Calibration result with real data

The calibration error was estimated using a new set of 40 untrained points, the average square error of the set points is 34.24mm without considering the camera-to-mirror transformation. Then we calculate the error with the transformation values listed in Table 1, the average square error decrease to 12.57mm.

3. Improved Graph-cut Model

In this section, we first briefly introduce the prior work on graph cuts, after that our improved optimization model and corresponding algorithm are presented. Note that our work has been done about graph cuts mainly based on related excellent work in [14], [15], [20] and [21], in our paper, most general depiction is based on above papers.

3.1 Two-variable Smooth Model

In terms of the energy minimization, many early vision problems, which of course contain stereo matching, can be formulated in following energy model:

$$E(f)=E_{data}(f)+E_{smooth}(f). \tag{10}$$

In Eq. (10), the data term $E_{data}(f)$ represents some extent similarities between f and the observed data and typically,

$$E_{data}(f) = \sum_{p\in P} D_p(f_p). \tag{11}$$

Where, P denotes the set of image pixels that need to be assigned labels. The label assigned to pixel $p\in P$ is denoted by f_p, and f is the set of all assignments: $f = \{f_p \mid p \in P\}$. D_p measures how well the label f_p fits the pixel p in the observed image. The smooth term $E_{smooth}(f)$ usually represents the smoothness of f ,which is a critical issue and lots of functions have been proposed, typically it can be expressed as follows,

$$E_{smooth}(f) = \sum_{(p,q)\in N} V_{pq}(f_p,f_q). \tag{12}$$

Where N represents the set of neighboring pixel pairs, in this case, Vpq represents the smoothness of pixels p and q that are respectively denoted by fp and fq, thereby *Esmooth* reflects the smoothness of all the neighboring pixels because our prior knowledge tells us that the surfaces of objects invariably keep relatively smooth except for some discontinuity area. We call the smoothness model that based on this function two-variable smoothness model because E_{smooth} refers to two variables. Considering above description, the following energy function is commonly minimized in computer vision and graphical fields such as image restoration, image segmentation and stereo matching:

$$E(f)= \sum_{p\in P} D_p(f_p)+ \sum_{(p,q)\in N} V_{pq}(f_p,f_q). \tag{13}$$

Currently, there are two typical two-variable smoothness models that specify smooth assumption that exist. One is Potts model, in which, the smooth term is depicted by following expression,

$$V_{pq}(f_p,f_q)=C_{pq}\cdot\min(1,\left|f_p-f_q\right|) \tag{14}$$

where once any difference exists in label pairs, the penalty to this difference via smooth energy is bound to be the same amount, theoretically, the expectation of labels f should be constant since only the smoothness assumption is considered, however, the presence of the data term usually results in piecewise constant in fact. The other model is truncated convex priors,

$$V_{pq}(f_p,f_q)=C_{pq}\cdot\min(T,g(f_p-f_q)). \tag{15}$$

Generally, $g(x)$ is convex and symmetric, for linear truncated function, $g(x)=|x|$ and when $g(x)=x^2$, it is truncated quadratic function, this smoothness energy function tend to give a piecewise constant assumption theoretically on smoothness term, albeit practically the piecewise smooth results may come out owing to the data term. In this case, label set f is expected to be consisted of several pieces and in each piece the difference between the adjacent pixels in f is just a little, in other words, the neighboring pixels tend to vary smoothly. The truncation constant T is crucial to limit the penalty on smooth, otherwise, the penalty on neighboring pixel fp and fq might be unboundedly large, this is somewhat ridiculous since discontinuity always exist in most images and it may lead to oversmoothed label set f.

3.2 Three-variable Smooth Model

Our three-variable smoothness model is proposed to solve the problem that exists in two-variable smoothness one. Note that only two variables, which are usually labels of neighboring pixels, are considered in smoothness assumption. Apparently, only the zero order and first order derivatives can be obtained via only two adjacent pixels. Nevertheless, sometimes they are not enough to specify the smoothness of a surface although expecting to minimize the difference of neighboring pixels (lesser first order derivative of f) invariably gives a reasonable constraint to make the whole surface that denoted by f varies relatively smoothly. This constraint is reasonable customarily mainly because the less difference between neighboring pixels likely means the more smoothly f varies. And also the data term, which accounts for biggish proportion in the whole energy function, always tend to enshroud the imprecise smoothness model. However, the smoothness term needs to be more precise to obtain reliable results when the disparity space is large, which means the reconstructed scene surface needs to be more elaborated and when the data term can not contribute to the energy function well in some conditions such as weak textured and textureless area. Unfortunately, these two take place in our omnidirectional images. To represent the smoothness of f better, we propose a three-variable smoothness model in which Esmooth typically can be expressed in following form:

$$E_{smooth}(f)=\sum_{(p,q,r)\in N} V_{pqr}(f_p,f_q,f_r). \tag{16}$$

In Eq. (16), neighboring pixels *p*, *q* and *r* are series-wound orderly, *Esmooth* contains three variables so that it can represent the smoothness of *f* more appropriately than two-variable model does. Now we give the concrete expression:

$$V_{pq}(f_p,f_q,f_r)=C_{pqr}\cdot\min(T,g(|f_p+f_r-2f_q|+|f_p-f_r|)), \tag{17}$$

where g(x) can be defined as the same in Eq. (16), |fp+fr-fq| represents the second derivative of label fr. In this case, the labels between the neighboring pixels tend to vary consistently and that also means less curvature which can represent the smoothness of surface better. Without doubt, to vary steadily is also important, thereby, |fp-fr|, which denotes the offset between first pixel p and last pixel r, is added in Vpqr. Obviously, the labels are expected to vary both smoothly and consistently in three-variable smoothness model while in two-variable smoothness model, the labels are only emphasized on varying smoothly.

3.3 Graph cuts for 3-variable smooth model

As we know, it is NP-hard to optimize the energy functions in Eq. (14) and Eq. (15). To solve this problem, Boykov et al. (Y. Boykov et al., 2001) developed the expansion and swap algorithm in which when Vpq is Potts and truncated linear or quadratic, the energy function can be optimized approximately. Now we use swap algorithm to optimize our 3-variable smoothness model.

Before constructing graph for 3-variable smoothness model, in the first place, we necessarily prove this energy function is graph-representable which is equal to proving Esmooth(f) is regular. According to (V. Kolmogorov & R. Zabih, 2004), we only need to confirm all the projections of Vpqr of two variables are regular. To complete the proof of this conclusion, following three inequalities should be proved, for simplicity, we use V to represent Vpqr.

$$V(\alpha,\alpha,f_r)+V(\beta,\beta,f_r)\le V(\alpha,\beta,f_r)+V(\beta,\alpha,f_r), \tag{18}$$

$$V(\alpha,f_q,\alpha)+V(\beta,f_q,\beta)\le V(\alpha,f_q,\beta)+V(\beta,f_q,\alpha), \tag{19}$$

$$V(f_p,\alpha,\alpha)+V(f_p,\beta,\beta)\le V(f_p,\alpha,\beta)+V(f_p,\beta,\alpha), \tag{20}$$

For (18), because g(x) increase monotonously when x > 0 , combining Eq. (17), we just need to prove following inequality:

$$|\alpha-f_r| + |\beta-f_r| \le |\alpha-2\beta+f_r| + |\beta-2\alpha+f_r|. \tag{21}$$

Without loss of generality, supposing $\alpha<\beta$, three possible relationships might exist among α, β and fr, we discuss them respectively as follows. When $\alpha<f_r<\beta$, it yields $|\alpha-f_r|+|\beta-f_r|=\beta-\alpha$, while $|\alpha-2\beta+f_r|+|\beta-2\alpha+f_r|=3(\beta-\alpha)$. In this case, (18) is proper. When $fr<\alpha<\beta$, it yields $|\alpha-f_r|+|\beta-f_r|-|\alpha-2\beta+f_r|=-\beta+2\alpha-fr\le|\beta-2\alpha+f_r|$. So (18) is proper too. While $\alpha<\beta<f_r$, likewise the same conclusion can be acquired.

To prove (19) analogously, only following inequality should be fulfilled:

$$2|\alpha-f_q|+2|\beta-f_q|\le 2|\alpha+\beta-2f_q|+2|\alpha-\beta|. \tag{22}$$

Note that the inequality below is invariably true, $|x|+|y|\le|x-y|+|x+y|\le 2|x|+2|y|$. Thus (19) can be proved simply. Also we can prove (20) similarly, as (20) is very similar to (19). Now we have proved that our three-variable smoothness model is graph-representable definitely. We use swap algorithm in (Y. Boykov et al., 2001) analogously to optimize our energy model, since the integrated illustration about graph construction has been described in (V. Kolmogorov & R. Zabih, 2004), we only need to specify three possible cases in energy function Vpqr as follows. One case is that $f_p,f_q,f_r\in\{\alpha,\beta\}$, in this case, Vpqr is the typical regular function of three binary variables. The second case is only two of fp, fq, fr belong to $\{\alpha,\beta\}$, in this case, Vpqr is virtually a regular function of two binary variables. The last case is the simplest one in which only one of these three continuous labels belongs to $\{\alpha,\beta\}$,

Vpqr is a regular function of one binary variable. Note that above three cases nearly invariably exist in fact. We no longer describe the graph construction process since all the cases are amply described in (V. Kolmogorov & R. Zabih, 2004), although our graph construction process is slightly different when considering the actual model.

4. Stereo Matching

In this section, combining the improved model and corresponding graph cuts algorithm, we present detailed steps for omnidirectional stereo matching.

4.1 Handling Omnidirectional Images

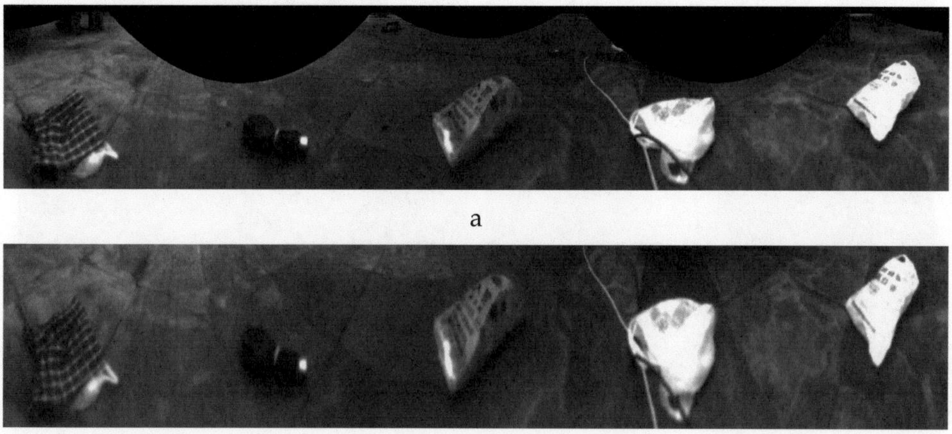

a

b

Figure 3. Unwrapped cylindrical images, of which a corresponding to outer circle image and b inner circle image

The images acquired by our OSVOD (Fig. 2) have some particularities in contrast to normal stereo pairs as follows, which may lead to poor results using traditional stereo matching methods: (1) The upper mirror and nether mirror have different focal length that the camera focal length has to compromise with the two, thus causing defocusing effect. As a result, similarity measures, such as SSD, take on much less discriminability. (2) In this close-quarter imaging, the object surface is always not frontal-parallel to the virtual retina of the camera, resulting in large foreshortening effect between the outer circle image and the inner circle image. (3) Quite a number of weak textured and textureless areas exist in our real indoor scene, more difficulties are bound to bring on in stereo matching. We choose this scene with an eye to the actual ground circumstance, which is usually apt to be weak textured and textureless. (4) The wide-baseline vision system can enhance the calculative precision, whereas, the pending disparity space is larger correspondingly, this increases the search range and ambiguous results tend to bring on. To solve these problems, our method consists of the following several steps: we first convert the raw image both to two cylindrical images and planform images corresponding to images via nether and upper mirrors respectively (Fig. 5 and Fig. 6). The vertical lines with the same abscissa in the cylindrical images are the same epipolar. We compute a similarity measurement for each disparity on every epipolar curve in the cylindrical images. The similarity measurement of a pixel pair is set as the average cost value of that computed from cylindrical images and that from planform images. This is to solve problem (2), as surface perpendicular to the ground tend to have good similarity measurement on the cylindrical images and surface parallel to the ground on the planform images. Second, we choose an appropriate similarity measure described below and also make necessary modification in the iteration of graph cuts to handle panoramic images, including deformed matching template, adaptable template scale. Finally we use improved 3-variable smoothness model via graph cuts to enhance the performance of our algorithm largely. These two steps are to solve problems (1), (3) and (4).

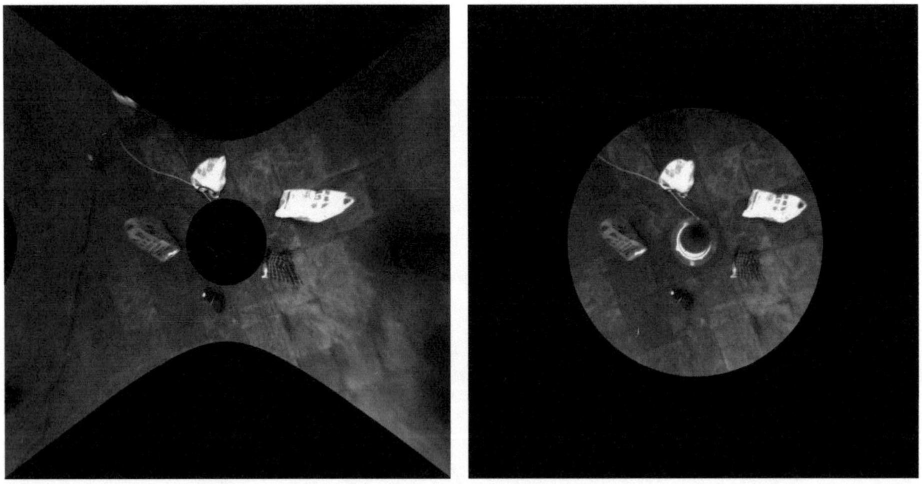

a b

Figure 4. Converted planform images, of which a corresponding to outer circle image and b inner circle image

4.2 MZNCC

The similarity measure we choose here is zero-mean normalized cross correlation (ZNCC), since it is invariant to intensity and contrast between two images. But directly using this measure would result in low discriminability. Chances exist that two templates with great difference in average gray-level or standard deviation which cannot be deemed as matched pair may have high ZNCC value. To avoid this possibility, we modified ZNCC (called MZNCC) by multiplying a window function as follows:

$$MZNCC(p,d)=\frac{\sum(I_a(i,j+d)-\mu_a)\cdot(I_b(i,j)-\mu_b)}{\sigma_a\cdot\sigma_b}\cdot w(|\mu_a-\mu_b|)\cdot w(\frac{\max(\sigma_a,\sigma_b)}{\min(\sigma_a,\sigma_b)}-1) \quad (23)$$

where $w(x)=\begin{cases}1, & x<\lambda \\ \lambda/x & x\geq\lambda\end{cases}$, μ_a and μ_b are the average grey-level of matching window, σ_a and σ_b are the standard deviation, d denotes the disparity of pixel p.

We define our texure level of each point following the notion of bandwidth of the bandpass filter. For a given pixel and a given template centred in the pixel, we slide the template one pixel at a time in the two opposite directions along the epipolar line and stop at the location the MZNCC value of the shifted template with the primary one decrease below a certain threshold for the first time. Let l be the distance between the two stop points, which is inverse proportional the texture level. The definition of texture intensity can be formalized as:

$$Tex(u,v)=\sum_{-r\leq(i,j)\leq r} \left(I(u+i,v+j)-\bar{I}\right)^2 \Big/ l^2 \tag{24}$$

Where r is the radius of the template. With the use of this defined texture intensity and two thresholds, the whole image can be divided into three regions: strong textured, weak textured and textureless regions. Unlike others straightforwardly use sum of intensity difference, we define the data term in our energy function in the form of MZNCC value multiplies a penalty coefficient Cp aim to assign different weights to different points based on the texture level. Generally, as the less reliability of the weak textured area and textureless one, we give following form of Cp.

$$C_p=\begin{cases}\mu_s, & Tex>t_s\\ \mu_w, & t_w\leq Tex\leq t_s.\\ \mu_l, & t_l<Tex<t_w\end{cases} \tag{25}$$

where $\mu_s>\mu_w>\mu_l$ and t_s,t_w,t_l represent corresponding thresholds to differentiate above three typical areas.

4.3 Template Rectification and Adaptive Scale

For certain corresponding pixel pairs, it is expected that the MZNCC value of the two templates centered at these two points are very close to 1. This expectation is well satisfied when the two image templates are the projections of a single surfaces and this surface is frontal-parallel to the imaging plane of the virtual camera. When larger image templates straddle depth discontinuities, possibly including occluded regions, the MZNCC value may decrease to a value much smaller than 1. Also, if the surface is not parallel to the imaging plane, especially as the ground plane in our scene perpendicular to the imaging plane, the foreshortening effect makes the two templates differ quite significantly, also reduce the MZNCC value to an unsatisfactory amount.

In this iterative framework of graph cuts, it is natural to estimate the appropriate template scale not to straddle depth discontinuities and rectify the template to compensate the foreshortening effect from current temporal result at each step. At each pixel in the image, we first use the largest template scale. We then compute the variance of the depth data in the template. If the variance exceeds an appropriately chosen threshold, it may be that the template scale is too large. Otherwise we continue to make use this template scale for MZNCC calculation.

After the scale is determined, it is ensured that the template corresponds to a single surface. We use the depth data to fit this surface to a plane in 3-D space, and then reproject this plane to the other image. Normally, the reprojected template is not a rectangle any more if the surface is not frontal-parallel. And we compute the MZNCC value between the rectangle template in the reference image and the rectified template in the other image. In this way, foreshortening effect is well compensated.

4.4 Graph Cuts

Considering the high-performance results in stereo matching, eventually, we use improved graph cuts based on three-variable smoothness model to optimize our energy function. Since the critical part of energy function has been discussed in section 3, now we briefly present the general form corresponding to our omnidirectional stereo. We use the appropriate energy function E in the form (10) to accommodate close-quarter measurement. $E_{data}(f)$ describes the MZNCC of corresponding pixels, and we also use different penalty coefficients to distinguish pixels with different texture.

$$E_{data}(f)=\sum(1-C_p \cdot MZNCC(p,f_p)), \tag{26}$$

E_{smooth} introduces a penalty for neighboring pixels having different disparity values.

$$E_{smooth}(f)=\sum_{(p,q,r)\in N} C_{pqr} \cdot \min(T, g(|f_p+f_r-2f_q|+|f_p-f_r|)). \tag{27}$$

Note that we do not take occlusion into account when considering no occlusions come forth since we invariably choose the unwrapped cylindrical image, which corresponds to the outer image reprojected by nether mirror, as the reference image in stereo. You see, as the upper mirror is higher than the nether one, any point that can be seen in outer image must be seen in inner image unless it exceeds the view area in the camera or the theoretical matching point can not exist. In this case, we call it falls in 'blind area', virtually this always comes true in our omnidirectional images and we invariably neglect the pixels in 'blind area'.

5. Experimental Results

In this section, we present our experimental results. In data term E_{data} ,Cp can be defined as follows:

$$C_p=\begin{cases} 1, & Tex>10 \\ 0.5 & 2\le Tex\le 10, \\ 0.25 & 0<Tex<2 \end{cases}$$

in view of E_{smooth} , Cpqr =0.01, T=100, g(x)=x2 .

To observe the experimental results legibly, we utilize a part of the real images (Fig. 7(a), 200*200) in Fig. 3 to test our algorithm. Fig.8 shows the experimental results via two typical 2-variable smoothness models and our 3-variable one. Fig. 7(b) gives the initial depth map by MZNCC via winner-take-all (WTA). Fig. 7(c) shows the depth map based on the Potts model. Fig. 7(d) denotes the results of traditional 2-variablated convex priors. Fig. 7(e) and Fig. 7(f) present our results based on 3-variable truncated convex priors model respectively. The 2-variable model, which results in smaller gaps in depth map, performs better than Potts model. Notice that our algorithm produces an answer which varies most smoothly and consistently in three models because our energy function can approximate the true scene better, especially in our omnidirectional images which possess some particularities we have mentioned above. Note that when introducing template rectification and adapt scale (f) performances better at the boundary than (e), however, due to the imprecise initial depth map, the amelioration is not very obvious but unspent. Unlike some standard images, obtaining precise groundtruth depth map for our real images is quite complicated, therefore, it is hard to describe our results quantitatively. Note that in all the experimental results, we neglect the 'blind area', in which we simply set a certain depth value.

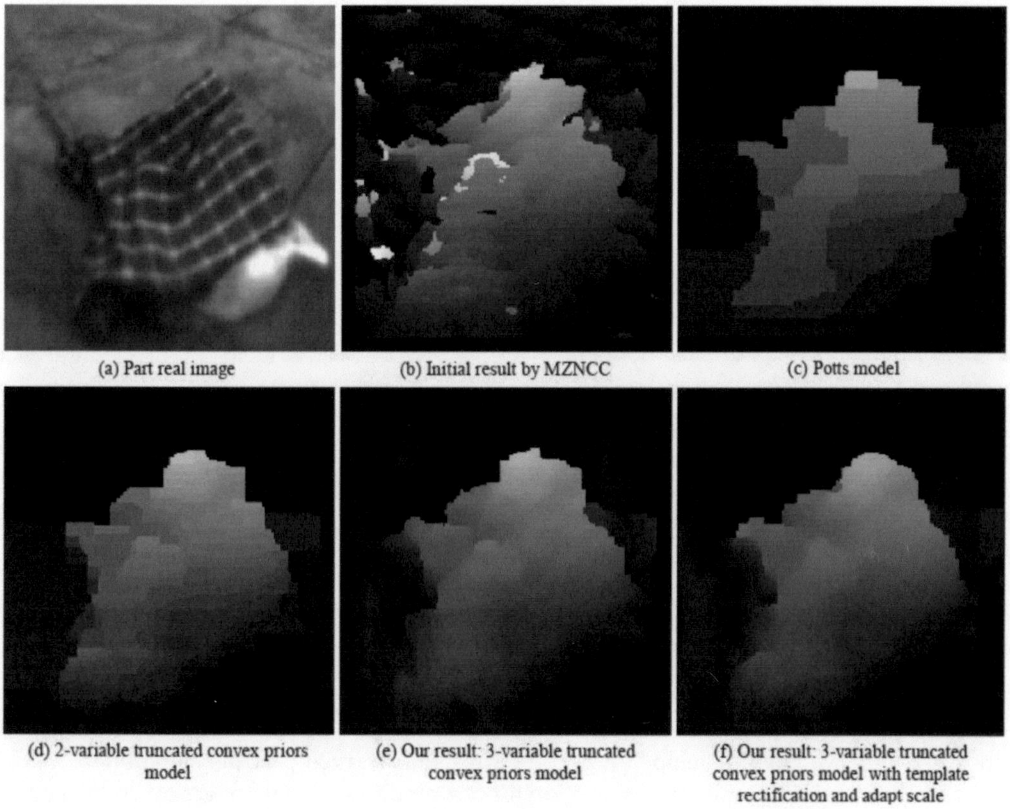

Figure 7. Depth maps of part omnidirectional images

Finally, we give the complete depth map in Fig. 8 and also the corresponding obstacle sketch map consisting of point clouds can be found in Fig. 9. Fig. 8 describes the stereo correspondence of images in Fig. 7 precisely and this preferable performance is reliable for depth map regeneration, even though the obstacle we placed on ground is somewhat low, which results in relatively difficulties in the reconstructed process. Fig. 9 shows an effective obstacle sketch map for a mobile robot. These point clouds denote existence of obstacle, we can see the resolution gets lower when the object is far from the center (maybe need to zoom in to see clearly), this was also mentioned in (G. Jang, S. Kim & I. Kweon, 2005). To get a reasonable obstacle sketch map, we necessarily set a small threshold (about 10mm in this paper) to judge if the corresponding point should be seen as an obstacle point since the error nearly invariably exists.

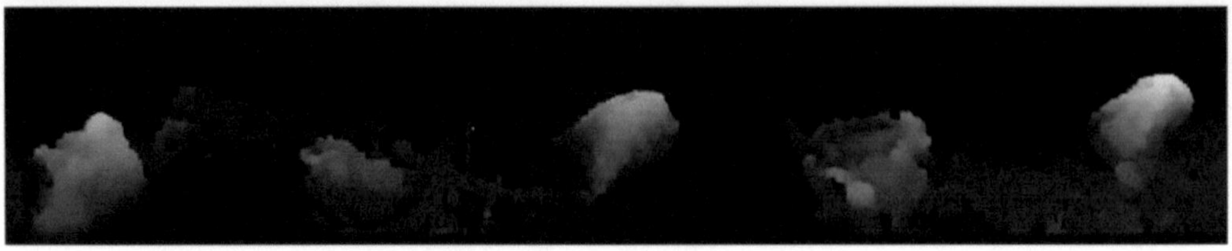

Figure 8. Final depth map (Fig.5(a) is reference image)

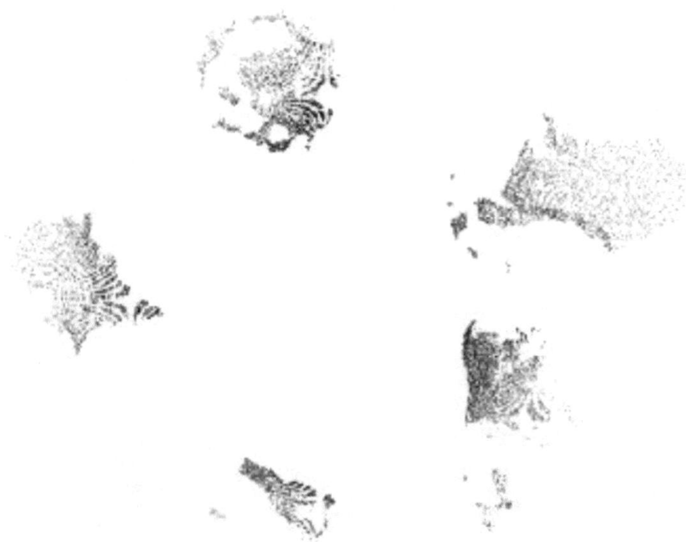

Figure 9. Final omnidirectional obstacle sketch map

6. Summary

In this paper, we have developed a graph-representable three-variable smoothness model for graph cuts to fit the smooth assumption for our omnidirectional images taken by a novel vision sensor. We further develop MZNCC as a suitable similarity measurement and also the necessary modification, including deformed matching template and adaptable scale. Experiments demonstrate the effectiveness of our algorithm, based on which, the regenerated obstacle map is finer for a mobile robot.

7. References

A. Jagmohan, M. Singh and N. Ahuja(2004). Dense Two View Stereo Matching Using Kernel Maximum Likelihood Estimation, *Proc. of ICPR'04*. pages 28-31, 2004.

D. Scharstein and R. Szeliski(2002). A Taxonomy and Evaluation of Dense Two-Frame Stereo Correspondence Algorithms. *International Journal of Computer Vision*. 2002

Devernay and F. Faugeras(1994). Computing Differential Properties of 3-D Shapes from Stereoscopic Images without 3-D Models. Proc. of CVPR'94. pages 208–213, 1994.

Fabrizio, J.; Tarel, J. & Benosman, R. (2002). Calibration of Panoramic Catadioptric Sensors Made Easier, *Proceedings of Workshop on Omnidirectional Vision*, pp. 45-52, Copenhagen, Denmark, Jun 2002

Geyer, C. & Daniilidis, K.(1999). Catadioptric Camera Calibration, *Proceedings of International Conference on Computer Vision*, Vol. 1, pp. 398-404, Kerkyra, Greece, Sep 1999

Geyer, C. & Daniilidis, K. (2000). Paracatadiopric Camera Calibration, *IEEE Trans. on Pattern Analysis and Machine Intelligence*, Vol. 24, No. 5, pp. 687-695

Geyer, C. & Daniilidis, K. (2003). Conformal Rectification of Omnidirectional Stereo Pairs, *Proceedings of IEEE Workshop on Omnidirectional Vision and Camera Networks*, pp. St. Louis, USA, Jun 2003

G. Jang, S. Kim and I. Kweon(2005). Single Camera Catadioptric Stereo System. *Proc. Of Workshop on Omnidirectional Vision, Camera Networks and Nonclassical cameras(OMNIVIS2005)*, 2005

H. Ishikawa(2003). Exact optimization for markov random fields with convex priors. *IEEE Trans. on PAMI.* 25(10):1333–1336, October 2003.

H. Koyasu, J. Miura and Y. Shirai(2002). Recognizing Moving Obstacles for Robot Navigation Using Real-time Omnidirectional Stereo Vision. *Journal of Robotics and Mechatronics*, 14 (2), pages 147-156, 2002.

J. Gluckman, S. K. Nayar and K. J. Thoresz(1998). Real-time Omnidirectional and Panoramic Stereo. Proc. of DARPA Image Understanding Workshop. pages 299-303, 1998.

E. L. L. Cabral, J. C. de Souza Junior and M. C. Hunold. Omnidirectional Stereo Vision with a Hyperbolic Double Lobed Mirror, Proc. of ICPR'04, 2004.

Kang, S.B. (2000). Catadioptric Self-calibration, *Proceedings of IEEE Conf. on Computer Vision and Pattern Recognition*, pp. 201-207, Hilton Head Island, South Carolina, Jun 2000

K. Nayar(2004). Catadioptric Omnidirectional Camera. *Proc. Of CVPR'97*, pages 482-488, 1997.

K. Tan, H. Hua and N. Ahuja(2004). Multiview Panoramic Cameras Using Mirror Pyramids, IEEE Trans. on PAMI. 26(6), 2004.

Luong, Q.T. & Faugeras, O.D. (1996). The Fundamental Matrix: Theory, Algorithms, and Stability Analysis. *Int. J. of Computer Vision*, Vol. 17, No. 1, pp. 43-76

M.F.D. Southwell, A. Basu and J. Reyda. Panoramic Stereo. *Proc. of ICPR'96*, pages 378-382, 1996.

Miousik, B. & Pajdla, T. (2003a). Estimation of Omnidirectional Camera Model from Epipolar Geometry, *Proceedings of 2003 IEEE Conf. on Computer Vision and Pattern Recognition*, Vol. 1, pp. 485-490, Madison, USA, Jun 2003

Miousik, B. & Pajdla, T. (2003b). Omnidirectional Camera Model and Epipolar Geometry Estimation by RANSAC with Bucketing, *Proceedings of Scandinavian Conf. on Image Analysis*, pp. 83-90, Goteborg, Sweden, Jun 2003

Pedro F. Felzenszwalb and Daniel P. Huttenlocher(2006). *International Journal of Computer Vision*, 70(1), 2006.

R. Bunschoten and B. Kr¨ose(2003). Robust scene reconstruction from an omnidirectional vision system. *IEEE Transactions on Robotics and Automation*. pages 351– 357, 2003.

Sara, R(2002). Finding the Largest Unambiguous Component of Stereo Matching. *Proc. of ECCV'02*, pages 900–914, 2002.

Sooyeong Yi and Narendra Ahuja(2006). An Omnidirectional Stereo System Using a Single Camera. *Proc. of ICPR '06*, 2006.

Su, L. & Zhu, F. (2005). Design of a Novel Stereo Vision Navigation System for Mobile Robots, *Proceedings of IEEE Conference on Robotics and Biomimetics*, No. 1, pp. 611-614, Hong Kong, Jun 2005

Sven Fleck, Florian Busch, Peter Biber, Wolfgang Straßer and Henrik Andreasson(2005). Omnidirectional 3D Modeling on a Mobile Robot using Graph Cuts. *Proc. of ICRA '05*. 2005.

T.L. Conroy and J.B. Moore(1999). Resolution Invariant Surfaces for Panoramic Vision Systems. *Proc. of ICCV '99*, pages 392-397, 1999.

V. Kolmogorov and R. Zabih(2001). Computing visual correspondence with occlusions using graph cuts. *Proc. Of ICCV '01*. pages 508-515, 2001.

V. Kolmogorov and R. Zabih(2004). What Energy Functions Can Be Minimized via Graph Cuts? *IEEE Trans. on PAMI*. 26(2), 2004.

Y. Boykov, O. Veksler and R. Zabih(2001). Fast approximate energy minimization via graph cuts. *IEEE Trans. on PAMI*. pages 1222–1239, 2001.

Yedidia, J.S., Freeman, W.T., Weiss, Y.(2000). Generalized belief propagation. *Proc. of NIPS '00*. pages 689-695, 2000.

Ying, X. & Hu, Z. (2004). Catadioptric Camera Calibration Using Geometric Invariants, *IEEE Trans. on Pattern Analysis and Machine Intelligence*, Vol. 26, No. 10, pp. 1260-1271

Zhang, Z. & Faugeras, O. (1997). An Effective Technique for Calibrating a Bincular Stereo Through Projective Reconstruction Using Both a Calibration Object and the Environment, *Journal of Computer Vision Research*, Vol. 1, No. 1, pp. 58-68

8

Motion Estimation of Moving Target using Multiple Images in Intelligent Space

TaeSeok Jin[1] and Hideki Hashimoto[2]
[1]*Dept. of Mechatronics Engineering, DongSeo University,*
[2]*Institute of Industrial Science, the University of Tokyo*
[1]*Korea,* [2]*Japan*

1. Introduction

During the past two decades, researchers in mobile robotics have dealt with different path planning methods. In most cases, the methods goal is to find free collision paths; which will meet the initial and final configurations to complete a mission. Some researchers have proposed methods where the robot's configuration is perfectly known at each instant during the planning and navigation stages(Moreno & Dapena, 2003) . This is not always possible. Dealing with uncertainty, in the planning stage, is essential when the position errors values approach values close to the allowed thresholds for the mission. Plans based on geometrical models, assuming null uncertainty, are clearly insufficient when the mobile robot has to coexist with humans or other kind of difficult situations. Thus, the use of planners, which not explicitly deal with uncertainty, is limited to simple situations, where the errors are less than the allowed thresholds to accomplish the missions (Bouilly & Siméon, 1996). In general, the basic requirements for the autonomous navigation of a mobile robot are environmental recognition, path planning, driving control and location estimation/correction capabilities (Nakamura, 1991, Haralick & Shapiro). The location estimation and correction capabilities are practically indispensable for the autonomous mobile robot to execute the given tasks efficiently. There are many factors involved in obtaining accurate location information while the mobile robot is moving (Sim & Dudek). To get reliable and precise location data, sensor fusion techniques (Ayache & Faugeras, 1989, Zhou & Sakane, 2001) have also been developed. When a CCD camera is utilized under good illumination conditions, certain patterns or shapes of objects are also effective for determining the location (Han et.al, 1999, Segvic & Ribaric, 2001). Similarly when a mobile robot is moving in a building, the walls, edges, and doors can be utilized for position estimation (Betke, 1994 David, 1989). Most researches (Choset, 2001, Sanisa, 2001, Philippe & Colle, 2001) focus on the indoor navigation of a mobile robot in a well-structured environment. In other words, beacons, doors, and corridor edges are utilized to estimate the current location of the mobile robot. However, in cases such as when a mobile robot is navigating under a deep sea or in a forest (Kim, et.al, 2001), there are no landmarks that can be utilized to determine the location.
This paper considers the situations where a mobile robot and a walking human coexist in a structured intelligent environment, such as assembly line in a factory. In these cases, one cannot utilize any landmarks or special features known *a priori* (Lallet & Lacroix, 1998,

Olson, 2000) to localize the mobile robot. The only thing that can be utilized for the localization is the information on a human captured by a CCD camera attached on top of the mobile robot. And an intelligent environment is used in order to solve these problems, a new scheme for the mobile robot localization using the information on the moving object has been developed. This situation might be considered as the inverse of tracking of an unknown moving object using a navigating robot with a camera whose location is precisely calibrated and stored all the time. The tracking problem has been already tackled in many researches (Nair & Aggarwal, 1989).

In this research, the data coming from the dead reckoning sensors are used to obtain the initial location of the mobile robot and it is corrected through the position estimation procedure using the information on the moving object/walking human. For the quantitative analysis of this approach, the position uncertainty of the mobile robot (Adam & Rivlin, 2000, Caglioti, 2001) is represented by an uncertainty ellipsoid that shows the directional uncertainty quantitatively. The trajectory of the moving object is transformed to the image frame and represented as a geometrical constraint equation that is used for the Kalman filtering process (Kalman, 1960, Sorenson, 1966) that estimates the position of the mobile robot to reduce the size of the uncertainty ellipsoid. And a mobile robot cooperates with multiple intelligent sensors, which are distributed in the environment. The distributed sensors recognize the mobile robot and the moving objects/walking human, and give control commands to the mobile robot. The mobile robot receives the necessary support for localization control from the environmental sensors. We aim to perform mobile robot localization without applying any burden to the human with a mobile robot that is simple in structure. We propose intelligent space (ISpace) as an intelligent environment with many intelligent sensors, and are building an environment where humans and mobile robots can now coexist. The mobile robot of this research is one of the physical agents for human support in ISpace (Lee & Hashimoto, 2001).

This paper is organized as follows. In Section 2, the concept of ISpace and the robot localization in intelligent space are explained. Section 3 describes the driving model of a mobile robot and the position estimation uncertainty. In Section 4, the image transformation relation, image projection of the walking human' trajectory, and the position correction technique using the Kalman filter are described. Section 5 explains the proposed control method is applied to ISpace. The simulations and the experiments of robot localization are performed and the effect of the proposed method is verified. Finally, the Conclusions and directions for future work are described in Section 6.

2. Robot Localization in Intelligent Space

2.1 Structure of ISpace

ISpace is a space where many intelligent devices are distributed throughout the whole of the space, as shown in Fig. 1. These intelligent devices have sensing, processing and networking functions, and are named distributed intelligent networked devices (DINDs). These devices observe the positions and behavior of both humans and robots coexisting in the ISpace. The information acquired by each DIND is shared among the DINDs through the network communication system. Based on the accumulated information, the environment as a system is able to understand the intention of humans. For supporting humans, the environment/system utilizes machines including computers and robots.

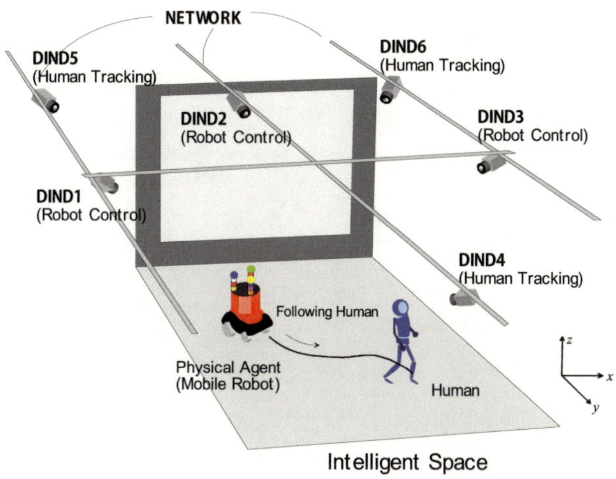

Figure 1. Structure of Intelligent environment by distributed cameras

2.2 ISpace System with Multi-camera

In ISpace, the CCD camera is adopted as the sensor for DINDs, and the tracking of moving objects, such as walking human, and mobile robot, is performed. There are two advantages in using CCD cameras. One is that the position measurement of the targets is a noncontact method. The other is that the human doesn't have to carry any special devices for the DINDs to be able to measure his position. Experiments were performed in the laboratory room, which is about 7 m in both width and depth, is used for the ISpace. The ISpace has a mobile robot as a human-following agent, six DINDs which can obtain the situation in the environment, and a projector and a screen which present suitable information to the human. Each module is connected through the network communication. Three DINDs are used in order to recognize the mobile robot and to generate the control commands. The other three DINDs are used to recognize the position of the human. DINDs are placed as shown in Fig. 1.

In the Intelligent Space, the DIND understands events occurring within the space and offers appropriate services for people by using devices such as robots, displays, and speakers. We can avoid the redundancy of function by sharing the information obtained by the DINDs and assigning each DIND to either recognize a walking human or a robot. The recognition and control performance in 3-D space were improved by locating each DIND for a walking human or robot using the position of triangular form. This intelligence, obtained spatially by the DINDs, connects the physical and digital spaces and enables the understanding of the human's motion.

Fig. 2 is a picture of the actual ISpace. The placement of the three DINDs for human recognition is optimized to expand the viewable area of the cameras so that the head and hands of the human can be recognized over a wide area (Akiyama, Lee, & Hashimoto, 2002). On the other hand, the placement of DINDs for the mobile robot has to be decided by trial and error. It is desirable that the DINDs for the mobile robot recognize the whole of the area covered by three DINDs for human recognition in order to achieve the human-following system and reliable mobile robot control. Thus, three DINDs are placed so that the area for human recognition is completely covered. Human walking information is extracted by background subtraction and by detecting the skin color of a face and hands on captured images.

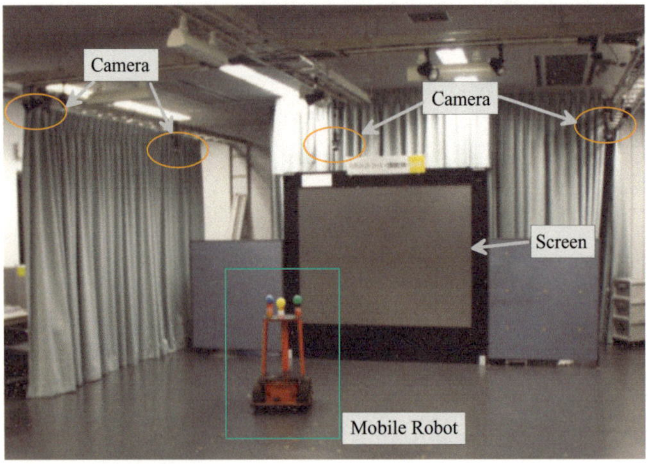

Figure 2. Experimental environment

A mobile robot is connected to the DIND network via wireless LAN, as shown in Fig. 3, and shares the resources of the DIND's. For recognizing the position of the robot, one color panels are installed around the mobile robot. The pattern of the color panel is recognized by the DIND and it estimates the posture and position of the robot by kinematics of robot projected onto an image plane. Since the height of the mobile robot is already known, the position of a mobile robot is reconstructed from one camera image.

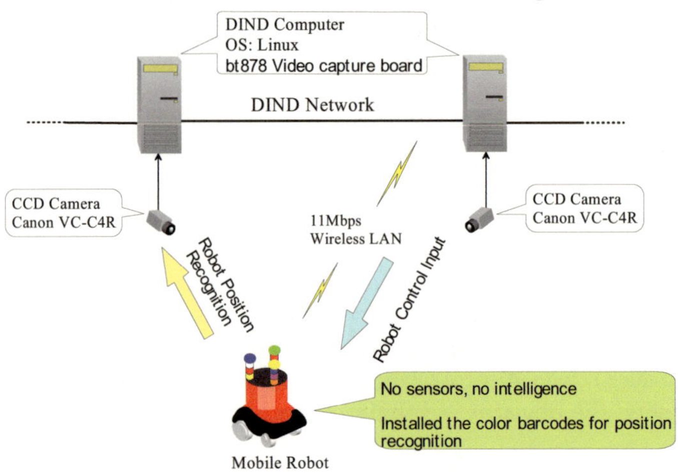

Figure 3. Mobile robot and network system in ISpace

3. Position Uncertainty Modeling

3.1 Ellipsoid point with uncertainty ellipse

The "ellipsoid point with uncertainty ellipse" is characterized by the coordinates of an ellipsoid point (the origin), distances *r*1 and *r*2, and the angle of orientation A. It formally describes the set of points on the ellipsoid that fall within or on the boundary of an ellipse with semi-major axis of length *r1* oriented at angle A (0 to 180o) measure d clockwise from the north and semi-minor axis of length *r2*. The distances are the geodesic distances over the ellipsoid, i.e., the minimum length of a path on the ellipsoid that joins the two points, as shown in Fig. 5(4).

The ellipsoid point can be used to indicate points on the Earth's surface or near the Earth's surface that have the same latitude and longitude. The confidence level with which the position of a target entity is included within this set of points is also specified along with this shape. The typical use of this shape is to indicate a point when its position is known only with a limited accuracy, but the geometrical contributions to uncertainty can be quantified.

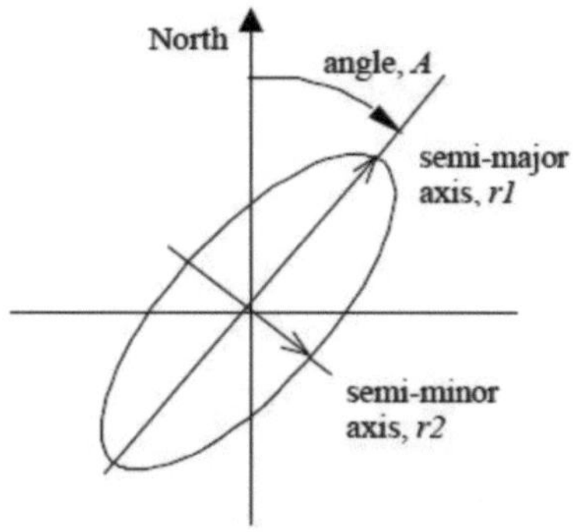

Figure 4. Description of an uncertainty ellipse

Mathematically, an ellipsoid E^t is a region of space defined as follows:

$$E^t := \{\mathbf{x} \in \mathbf{R}^n : (\tilde{\mathbf{x}}_o \mathbf{x}_m) P_t^{-1} (\tilde{\mathbf{x}}_o \mathbf{x}_m) \leq 1.0\}, \quad (1)$$

for some $\mathbf{x} \in \mathbf{R}^n$, where the subscripts **o** and **m** represent the observed point and the point in the map, respectively. Pt is a symmetric and positive definite matrix.

The aim of the ellipsoid method is to begin with a large ellipsoid (usually an n-dimensional sphere where n is a dimension of a state space) that contains the whole uncertainty set and to then generate a sequence of progressively smaller/larger ellipsoids, leading to an ellipsoid that fits the state uncertainty as tightly as possible.

The positional uncertainty of the object point in world coordinates is calculated by first applying the rotation between the robot and the world coordinates and then adding the uncertainty of the robot position. In order to determine the correspondence between the observation and the map, we integrate the observation result with the map to reduce the positional uncertainty of object points. For this purpose, we need to determine the correspondence between the object points in the observed data and those in the map. For each observed point, we identify the corresponding point on the map that satisfies the Eq. (1).

3.2 Modeling of Mobile Robot

The initial position of a mobile robot can be specified precisely. However, measurement error and slippage during the movement may cause the position estimation uncertainty to be large. This uncertainty increases with driving distance, and may eventually result in losing the location of the mobile robot.

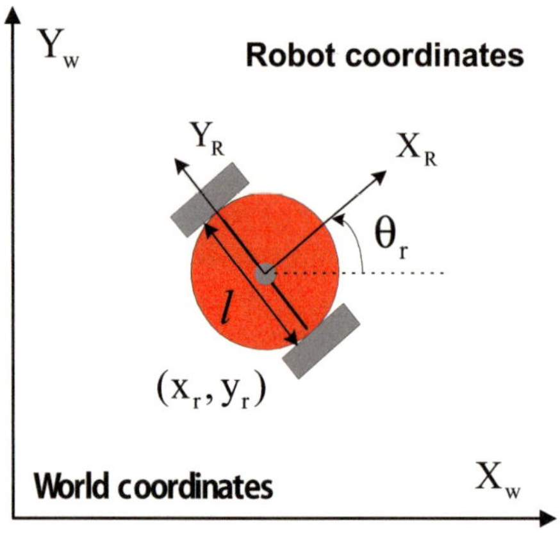

Figure 5. Geometrical model of a mobile robot

The robot's position will be represented by the vector of its spatial variables $\mathbf{x}(k)$, as a point in the Cartesian plane, with $x_r(k)$ and $y_r(k)$ coordinates and an orientation $\theta_r(k)$, $\mathbf{x}=[x_r, y_r, \theta_r]^T$. The simplified kinematic model proposed in (Adam, Rivlin & Shimshoni, 2000) describes how the robot's position changes in time, in relation to an initial position, in response to a $\mathbf{u}(k)$ control input formed by a $T(k)$ translation followed by a $\theta_r(k)$ rotation: $\mathbf{u}(k) = [T(k),\ \theta_r(k)]^T$. The state, for a given instant, is obtained from the state transition function $f(\mathbf{x}(k),\ \mathbf{u}(k))$, represented in Eq. (2)

$$\begin{aligned}\mathbf{x}(k+1) &= f(\mathbf{x}(k),\ \mathbf{u}(k)) + \mathbf{v}(k) \\ &= \begin{bmatrix} x_r(k) + T(k)\cos(\theta_r(k)) \\ y_r(k) + T(k)\sin(\theta_r(k)) \\ \theta_r(k) + \Delta\theta(k) \end{bmatrix} + \mathbf{v}(k)\end{aligned} \tag{2}$$

where $f(\mathbf{x}(k),\mathbf{u}(k))$, is a non-linear state transition function, $\mathbf{v}(k)$ is a noise source assumed to be zero-mean Gaussian with covariance $\mathbf{Q}(k) \rightarrow N(0,\ Q(k))$ and finally $\mathbf{u}(k)$ is the control input. The position uncertainty of the robot is modeled by means of a Gaussian distribution of probability centered in the vehicle position at a given moment. Eq. (2) allows to obtain the mean vector estimation in the $k + 1$ position. It is now necessary to estimate the covariance matrix in the same position. The first two moments, the mean and the covariance of the distribution function, which follow the spatial position relationship, must be determined. The covariances matrix related to the prediction, in the non-linear spatial relationship case, is obtained from the Taylor series expansion. Therefore, the estimated position of a mobile robot and $\hat{\mathbf{x}}(k\,|\,k)$ covariance matrix equation is shown in Eq. (3) and Eq. (4), respectively (Kalman, 1960).

$$\hat{\mathrm{x}}(k+1) = f(\mathrm{x}(k),\mathrm{u}(k)) \tag{3}$$

$$\mathbf{P}(k+1\,|\,k) = \nabla\mathbf{f}\,\mathbf{P}(k\,|\,k)\nabla\mathbf{f}^T + \mathbf{Q}(k), \tag{4}$$

where $\nabla \mathbf{f}$ is the state transition function Jacobian, obtained as the linearizing result around the estimated state. The state transition function Jacobian is described in Eq. (5)

$$\nabla \mathrm{f} = \begin{bmatrix} 1 & 0 & T(k)\sin(\hat{\theta}_r(k \mid k)) \\ 0 & 1 & T(k)\cos(\hat{\theta}_r(k \mid k)) \\ 0 & 0 & 1 \end{bmatrix}. \quad (5)$$

In order to estimate the vehicle position, odometry is not enough. The $\hat{x}(k \mid k)$ covariance matrix equation shown in Eq. (4) tends to grow continuously (Fig. 6). Using this covariance matrix, the position estimation uncertainty can be represented as a hyper-ellipsoid. That is, the uncertainty hyper-ellipsoid can be defined (Nakamura, 1991) from the SVD (singular value decomposition) of the covariance matrix. This SVD provides the principal axis by the left singular vectors and the length along the axis by the corresponding singular values. Fig. 6 illustrates the effectiveness of the uncertainty ellipsoid as an example. It shows that the uncertainty ellipsoid becomes larger with the movement of a mobile robot, and that the geometrical shape of the ellipsoid directly represents the position estimation uncertainty along a given axis.

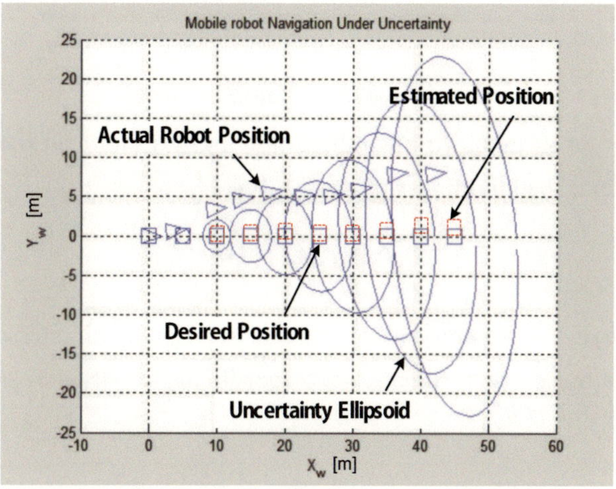

Figure 6. Uncertainty ellipsoids with the movement of a mobile robot

4. Position Estimation based DINDs

4.1 Image projection of walking human

During navigation, a mobile robot may need to re-locate its position. When there is a walking human that can be captured by the CCD camera of DINDs and the motion information on the walking human is available to the mobile robot, it may stop at its current position to improve the position estimation accuracy of itself by observing the walking human. The given object trajectory can be represented as a linear equation in the image frame, and using the current position estimation of the mobile robot, geometric constraint equations can be derived through coordinate transformation.

The derivation procedure of geometric constraint equations is going to be illustrated with an example shown in Fig. 7. The conventional pin-hole model (Haralick & Shapiro, 1993) is utilized to form a geometrical model of the camera. In Fig. 7, $(x_W \; y_W \; z_W)$ and (u, v) represent the reference coordinates and the image coordinates, respectively.

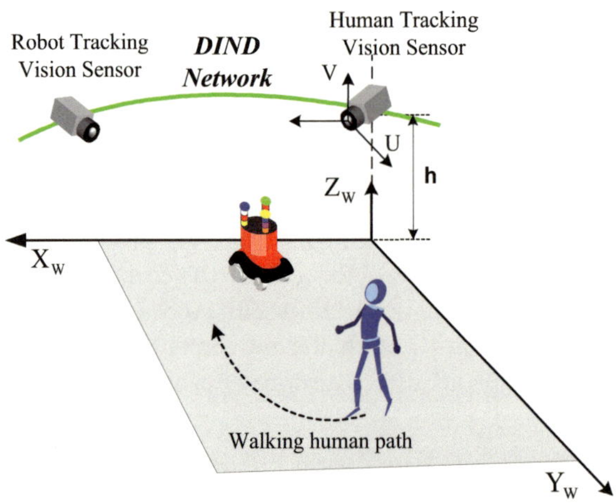

Figure 7. Coordinates for a walking object and a mobile robot

The walking human is assumed to have the following trajectory on the $x_W - y_W$ plane of the reference coordinates, without loss of generality:

$$f(x_W, y_W) = 0 \tag{6}$$

where $z_W (= z_0 \neq h)$ is also assumed to be constant and not equal to the camera height, h.

The walking human trajectory in the reference coordinates can then be transformed into the robot coordinates, as follows:

$$\begin{bmatrix} x_R \\ y_R \\ z_R \end{bmatrix} = \begin{bmatrix} \cos\hat{\theta}_r & \sin\hat{\theta}_r & 0 \\ -\sin\hat{\theta}_r & \cos\hat{\theta}_r & 0 \\ 0 & 0 & 1 \end{bmatrix} \begin{bmatrix} x_W - \hat{x}_r \\ y_W - \hat{y}_r \\ z_W \end{bmatrix} \tag{7}$$

where $[\hat{x}_r \quad \hat{y}_r \quad \hat{z}_r]^T$ represents the current estimated position of the mobile robot and $[x_W \quad y_W \quad z_W]^T$ represents the position of the walking human.

This point ($[x_R \quad y_R \quad z_R]^T$) is again mapped onto the image frame using the perspective projection, as follows (Haralick & Shapiro, 1993, Han et.al, 1999):

$$\begin{bmatrix} u \\ v \end{bmatrix} = \begin{bmatrix} \lambda \dfrac{y_R}{x_R} \\ \lambda \dfrac{z_R - h}{x_R} \end{bmatrix} \tag{8}$$

where λ represents the camera focal length, and $z = [u \quad v]^T$ is the position of the walking human on the image frame. Based on the equations (7) and (8), the geometric constraint equation can be generally represented as

$$f(z, \hat{x}) = 0 \tag{9}$$

where $\hat{x} = [x_r, y_r, \hat{z}_r]^T$ represents the current estimated position of the mobile robot.

4.2 Position Correction

The calculated position of the walking human in the image frame, based on the estimated robot position, has some discrepancy from the actual value. Utilizing this error, the practical position of the robot can be corrected recursively. To overcome vague input information, *i.e.*, the human position in the image frame includes noise and the position estimation of the robot has uncertain components, the Kalman filtering technique is adopted to form a robust observer (Kalman, 1960, Sorenson, 1966). The geometric constraint equations between the human image coordinates and the robot position are approximated to a linear system equation, and the Kalman filtering technique is applied to estimate the robot position.

It is assumed that i-th measured vector, *i.e.*, the position of the walking human, $\hat{z}_i$, includes noise with the following average and variance:

$$\hat{z}_i = z_i + v_i \tag{10}$$

where $E[v_i] = 0$ and $E[v_i v_i^T] = S_i$.

The nonlinear constraint equations are approximated to linear ones using the Taylor series expansion, ignoring the higher order nonlinear terms at the measured vector, $\hat{z}_i$, and the estimated position of the mobile robot, $\hat{x}_{i-1}$. That is,

$$\begin{aligned} f(z_i, \hat{x}) &= 0 \\ &\approx f(\hat{z}_i, \hat{x}_{i-1}) + \left.\frac{\partial f}{\partial z}\right|_{z=\hat{z}_i} (z_i - \hat{z}_i) + \left.\frac{\partial f}{\partial \hat{x}}\right|_{\hat{x}=\hat{x}_{x-1}} (\hat{x} - \hat{x}_{i-1}) \end{aligned} \tag{11}$$

where $\left.\dfrac{\partial f}{\partial z}\right|_{z=\hat{z}_i}$ and $\left.\dfrac{\partial f}{\partial \hat{x}}\right|_{\hat{x}=\hat{x}_{i-1}}$ represent the estimated value of $\dfrac{\partial f}{\partial z}$ and $\dfrac{\partial f}{\partial \hat{x}}$ at $\hat{z}_i$ and $\hat{x}_{i-1}$, respectively.

For a linear system, the equation (11) can be rearranged as the following matrix equation (Ayache & Faugeras, 1989).

$$y_i = M_i \hat{x} + u_i \tag{12}$$

where $y_i = -f(\hat{z}_i, \hat{x}_{i-1}) + \left.\frac{\partial f}{\partial \hat{x}}\right|_{\hat{x}=\hat{x}_{i-1}} \hat{x}_{i-1}$, $M_i = \left.\frac{\partial f}{\partial \hat{x}}\right|_{\hat{x}=\hat{x}_{i-1}}$ and $u_i = \left.\frac{\partial f}{\partial z}\right|_{z=\hat{z}_i} (z_i - \hat{z}_i)$.

In this equation, y_i becomes the new measured vector, M_i combines the measured vector and the robot position, $\hat{x}$, linearly, and u_i is the error for linearization of the measured vector with the following average and variance values (Nakamura, 1991).

$$E[u_i] = 0 \tag{13}$$

$$E[u_i u_i^T] = W_i = \frac{\partial \hat{f}}{\partial z} S_i \frac{\partial \hat{f}}{\partial z}^T \tag{14}$$

Since M_i and y_i are *a priori* given values, if we have the average and variance of u_i, we can obtain the optimal estimated value of $\hat{x}$ with the new variance. The Kalman filter provides the estimated value, $\hat{x}$, which minimizes the expected squared error norm, $E[(\hat{x}-x)^T(\hat{x}-x)]$ as the linear combination of the measured vectors, $\{y_i\}$, as follows:

$$\hat{x}_i = \hat{x}_{i-1} + K_i(y_i - M_i \hat{x}_{i-1}) \tag{15}$$

$$K_i = P_{i-1} M_i^T (M_i P_{i-1} M_i^T + W_i)^{-1} \tag{16}$$

$$P_i = (I - K_i M_i) P_{i-1} \tag{17}$$

where K_i represents the Kalman gain, P_i is the zero-mean-variance matrix of the estimated error by the ith measured vector, and $\hat{x}_i$ is the estimated robot position by the ith measured vector.

The initial robot position estimation and variance, $\hat{x}_0$ and P_0, can be obtained using the mobile robot driving model. Using the *n* image frames from the image coordinates of the moving object, the final robot position is recursively estimated as $\hat{x}_n$, with a variance of P_n.

5. Simulation and Experiments

5.1 Simulation

The simulations were performed for two different patterns of walking human motion: the parabolic motion and the sinusoidal motion. To make it realistic, the following camera parameters in Table 1 were utilized:

Parameter	value
Camera height (h)	$220\,cm$
Focal length (λ)	$1.25\,cm$
CCD size $(H)\times(V)$	$0.66\,cm \times 0.48\,cm$

Table 1. The parameters of camera system

The variances of the measured vectors were independent of each other; the empirical variances were obtained as follows:

$$S_i = \begin{bmatrix} 0.005^2 & 0 \\ 0 & 0.005^2 \end{bmatrix}. \tag{18}$$

The estimated initial position and the error variance came from a mobile robot driving model with 100 msec of control cycle, where the input errors of the wheel angle readings are limited to within 2% for Kalman filtering.

Case 1: a parabolic walking motion case

As the simplest example, an object was assumed to be moving along a parabolic curve, specified as the range, $X_w = 400 - 600[cm]$, $Y_w = -0.1(X_w - 500)^2 + 1200[cm]$, $Z_w = 100[cm]$, walking speed: 30 Cm/sec.

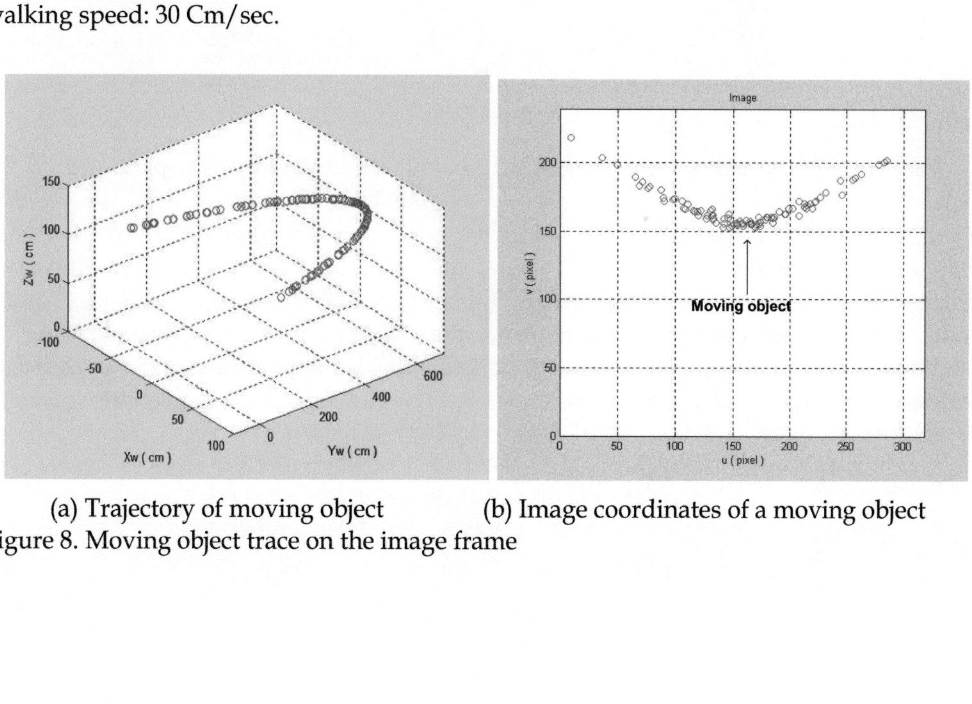

(a) Trajectory of moving object (b) Image coordinates of a moving object

Figure 8. Moving object trace on the image frame

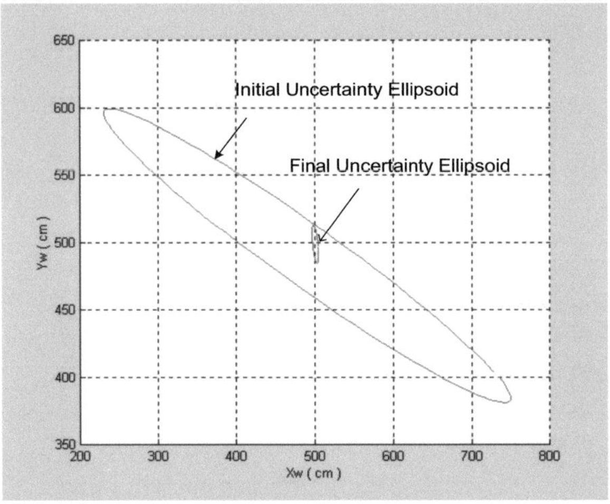

Figure 9. Improvement of position estimation uncertainty

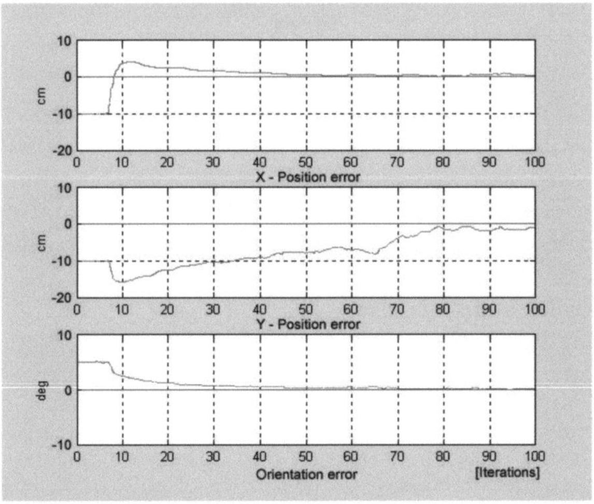

Figure 10. Position estimation errors

Fig. 8 (a)-(b) represents the moving path of the walking human in 3-D space and on the image frame respectively, and Fig. 9 shows the quantitative improvement of the position estimation uncertainty. Finally, Fig. 10 represents the position estimation error for each component of the robot position, (x,y,θ) when the parabolic walking motion is utilized for the localization. Note that since the walking human provides the useful information useful for the position correction on the x-y plane, the uncertainty ellipsoid shrunk along all directions, as shown in Fig. 9. Also notice that both of the x and y position errors of the robot converse to zero for the same reason, as shown in Fig. 10. The position estimations and zero-mean variances before and after the 100 times iterations are shown as:

Initial: $\hat{x}_0 = [490cm \;\; 480.5cm \;\; 95°]^T$ and $P_0 = \begin{bmatrix} 243.1162 & -89.8719 & -0.6576 \\ -89.8719 & 62.9118 & 0.2529 \\ -0.6576 & 0.2529 & 0.0105 \end{bmatrix}$

Final: $\hat{x}_{100} = [500.4cm \quad 499.6cm \quad 90.2°]^T$ and $P_{100} = \begin{bmatrix} 3.6294 & -2.1574 & 0.0080 \\ -2.1574 & 15.4273 & -0.0044 \\ 0.0080 & -0.0044 & 0 \end{bmatrix}$.

Note that the estimated position, $\hat{x}_{100} = [500.4cm \quad 499.6cm \quad 90.2°]^T$, converges to the real position, $x = [500cm \; 500cm \; 90°]^T$ precisely.

Case 2: a sinusoidal walking motion case

As a general case, the trajectory of a walking human was assumed to be a sinusoidal trajectory, represented as follows:

$X_w = 400 - 600[cm]$, $Y_w = -0.01(X_w - 500)^3 + 300[cm]$, $Z_w = 100[cm]$, walking speed: 30 Cm/sec.

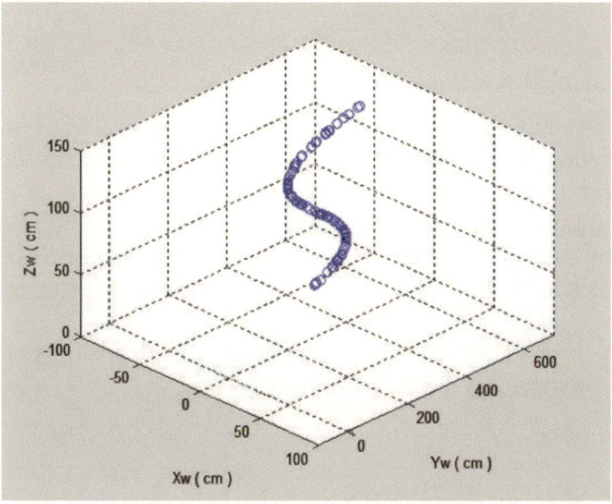

(a) Trajectory of moving object

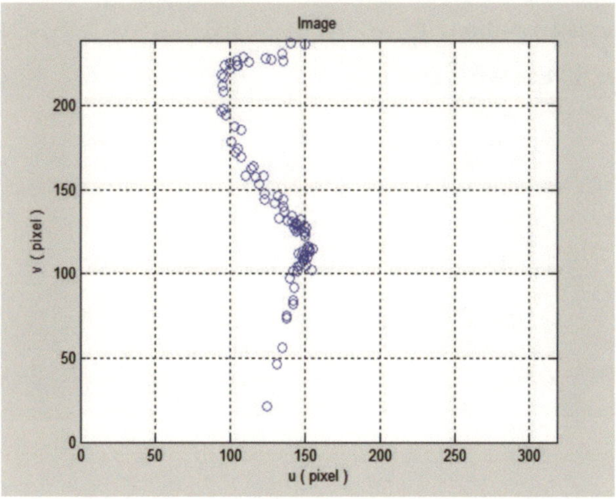

(b) Image coordinates of a moving object

Figure 11. Moving object trace on the image frame

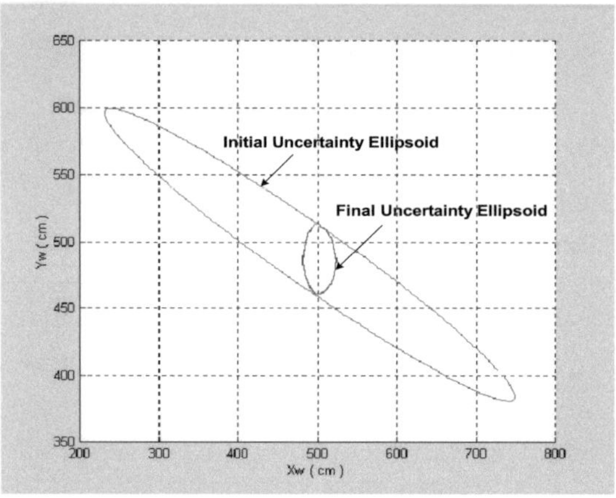

Figure 12. Improvement of position estimation uncertainty

Fig. 11 (a)-(b) represents the sinusoidal moving path of the walking human and the image coordinates of a moving object respectively, and Fig. 12 shows the uncertainty ellipsoid for moving object closer to the walking human tends to be smaller, as expected for visual information with updated estimation. Finally, Fig. 13 represents the position error and the orientation error for each component of the robot position respectively. These simulation results indicate that the Multi-visual estimation correspond well to actual objects in the ISpace.

Initial: $\hat{x}_0 = [490cm\ \ 480cm\ \ 95°]^T$ and $P_0 = \begin{bmatrix} 242.9162 & -89.7719 & -0.5576 \\ -89.7719 & 62.3118 & 0.1829 \\ -0.5576 & 0.1829 & 0.0015 \end{bmatrix}$

Final: $\hat{x}_{100} = [503.4cm\ \ 504.0cm\ \ 90.2°]^T$ and $P_{100} = \begin{bmatrix} 5.7384 & -2.2684 & 0.0180 \\ -2.2684 & 15.8873 & -0.0174 \\ 0.0180 & -0.0174 & 0 \end{bmatrix}$.

Note that the estimated position, $\hat{x}_{100} = [503.4cm\ \ 504.0cm\ \ 90.2°]^T$, converges to the real position, $x = [500cm\ 500cm\ 90°]^T$ precisely.

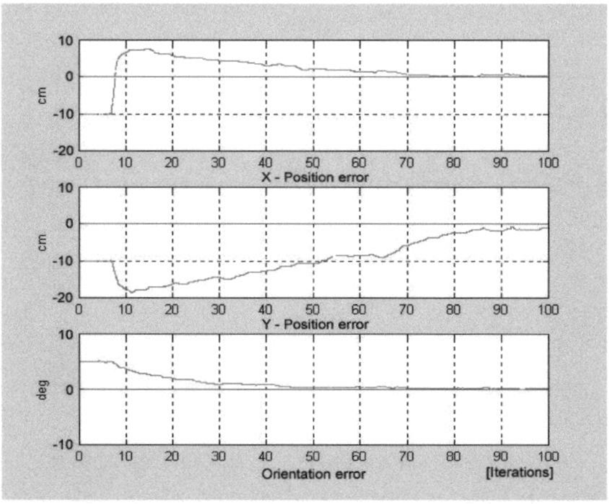

Figure 13. Position estimation errors

5.2 Experiments

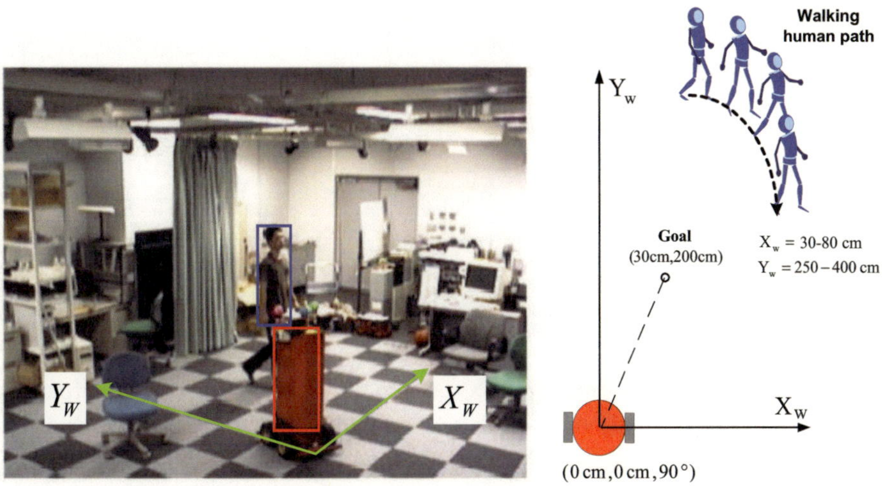

Figure 14. Experimental environment

The real experiments were performed at the ISpace as shown in Fig. 2. In this experiment, the robot is driven to follow walking human in the ISpace. The initial position of the mobile robot was set as (0 cm, 0 cm, 90°) and aimed towards the goal position, (30 cm, 200 cm, 90°), where the CCD camera of DIND tracks the walking human on the floor to correct its own position. The trajectory of the walking human was captured by a color CCD camera, VC-C4R, made by CANON Inc. and the trace was given as $X_w = 30 - 80[cm]$, $Y_w = 250 - 400[cm]$, and $Z_w = 100[cm]$ with the speed of 100 Cm/sec. The parameters used for the experiments were the same as for the simulations. When the robot arrived at the goal position, it captured 20 frames of images every 100 msec for the walking human to estimate and correct its own position utilizing the information on the walking human. The real position of the mobile robot was $[33cm \quad 208cm \quad 90°]^T$, and the estimated position of the mobile robot and the variance of the estimated error were given as

$$\hat{x}_0 = [28cm \quad 195cm \quad 87.9°]^T \text{ and } P_0 = \begin{bmatrix} 49.951 & -1.021 & -0.154 \\ -1.021 & 0.183 & 0.003 \\ -0.154 & 0.003 & 0.001 \end{bmatrix}.$$

After the 100 frames of observation, the estimated position of the mobile robot became very close to the real position, and the variance of the estimated error was greatly reduced, as shown below:

$$\hat{x}_{100} = [29.4cm \quad 196.4cm \quad 89.7°]^T \text{ and } P_{100} = \begin{bmatrix} 8.0406 & -0.1784 & 0.0252 \\ -0.1784 & 0.1660 & -0.0006 \\ 0.0252 & -0.0006 & 0.0001 \end{bmatrix}.$$

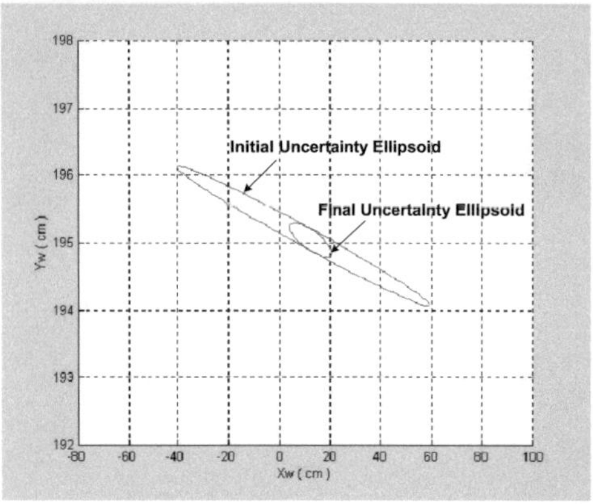

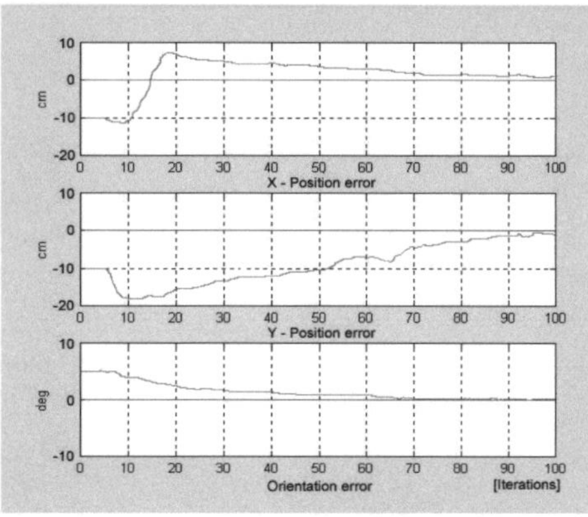

(a) Robot position estimation uncertainty (b) Position errors

Figure 15. Improvement of position estimation uncertainty

The robot position estimation uncertainty was represented as an uncertainty ellipsoid and it was very impressive that the size of the ellipsoid shrunk only along the normal direction of the walking human, as shown in Fig. 15 (a). Fig 15(b) shows the position errors of the mobile robot. As expected, we can see that the position error reduced significantly as the robot moves, depending on the motion it observes. The overall robot pose uncertainty for each cycle decreases over iterations, showing that by observing the scene repeatedly with DINDs, the estimation uncertainty reduces and hence a better robot pose is obtained. The robot orientation also decreases slightly for this reason, when walking motion has not acute angle turn.

These experimental results correspond to the experiment performed inside the 7 m x 7 m space. Evidently, in the experiment, the simulation error range appears low since the mobile robot and the walking human were assumed to be point objects, and the size of point was set to 30 cm x 100 cm. Moreover, the simulation result is better than the experimental result because of the following factors in the real experiment: nonlinear elements in the motion of walking human, the robot position error due to wheel slippage, rough surface, and sensor error.

As regards the additional experimental results, Fig. 16 presents the experimental results with respect to walking speed for case 1. For comparison purposes, the Kalman filter was also designed based on case 1, and its outputs are plotted in the figure. The walking human moved along a parabolic trajectory in the x-y axis, and the motion models had the same zero-mean variance. Fig. 16(a) shows the data plot of the localization error when only the constraints imposed by the visibility regions are used. As evident from the figure, the DIND system is unable to localize the mobile robot based on this information alone. Moreover, when the average walking speed is 100 cm/s, the localization accuracy is reasonably high and the robot is capable of quickly determining its absolute position in the environment.

Fig. 16(b) presents the experimental result when the walking speed is above 110 cm/s and the turns of human are included. When the walking human accelerates, the DIND4, DIND5, and DIND6 are used to track the accurate position and direction of the human. As compared to Fig. 16, the response of the position is reasonably smooth, whereas the orientation θ continues to be very noisy during the sampling step 30. This is due to distortions of the

images caused by the human's motion and walking speed. During the sampling step of 90 ms, the position error is at most 100 mm, and the orientation error is up to 5 degrees.
In future research efforts, it is necessary to examine the influence of the mobile robot, which maintains a flexible distance between the robot and the human. In the ISpace, since a human-walking trajectory is newly generated at every step, it is considered to be a function of time. Therefore, the application of tracking control is effective. However, although the target trajectory of a mobile robot is continuous and smooth when the usual tracking control is applied, the actual human-walking trajectory that is to be tracked by the robot is generally unstable. In such cases, stable human-following behavior may not be achieved by the usual tracking control.

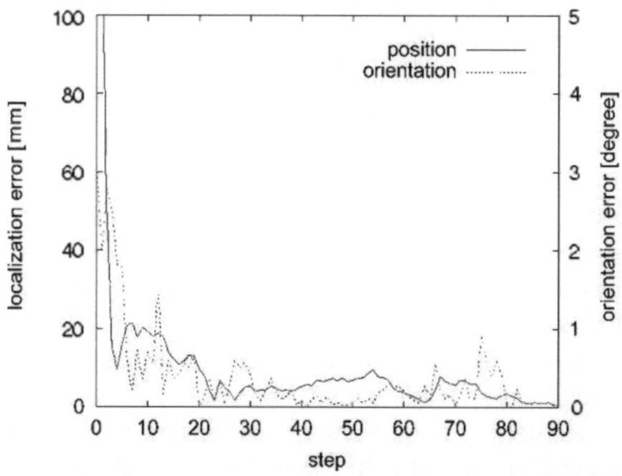

(a) Position and orientation error at an average speed of 100 cm/s

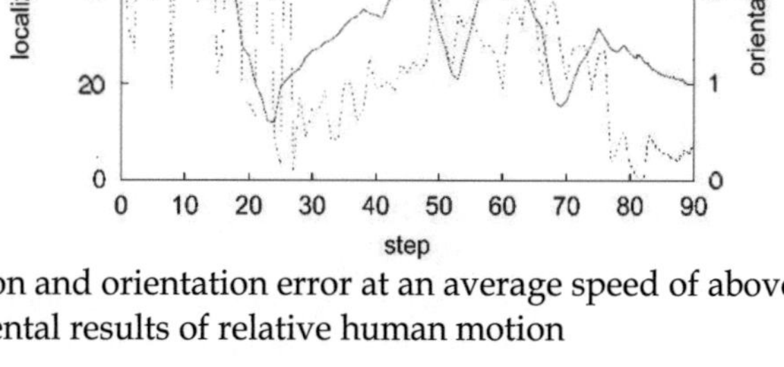

(b) Position and orientation error at an average speed of above 110 cm/s

Figure 16. Experimental results of relative human motion

6. Conclusion

In this paper, using the images of walking human, an absolute position estimation method for a human following robot in ISpace was presented. First, the position estimation uncertainty of the mobile robot is quantitatively represented by the uncertainty ellipsoid.

The real human position is transformed to geometric constraint equations in the image coordinates for a given robot position. The control algorithm using the linear constraint equations and the Kalman filtering technique was proposed for a mobile robot in order to estimate and correct its position recursively, and follow a walking human whose position is estimated incompletely.
Specifically, the pre-determined path of a walking human is projected onto an image frame and the geometrical constraint equations between the human's image coordinates and the estimated human following robot position are derived. Since the location is based on the estimated position of the mobile robot, there exists positional discrepancy between the estimated image coordinates and the real position of the walking human. Using this discrepancy, the position of the mobile robot is corrected recursively. Since the image coordinates of the human are subjected to noise, the Kalman filtering technique is adopted for robust estimation. Next, cooperation between the multiple DINDs was described. The position of the human and the mobile robot in ISpace was measured with DINDs. To control a mobile robot in a wide area, cooperation of the DINDs, effective communication and role assignment are required. Finally, simulations and an experiment into the human-following control of a mobile robot were performed using the proposed control algorithm. It was recognized that position estimation accuracy depends on the path of the walking human. The effectiveness of this algorithm is verified through real experiments.
Future studies will involve improving the estimation accuracy for the human following robot and applying this system to complex environments where many people, mobile robots and obstacles coexist. Real time image processing and camera calibration are needed to improve the estimation accuracy for the distance between the human and the mobile robot. Since the proposed algorithm absorbs the kinematic differences between humans and robots, any kind of mobile robot, including legged robots, can be used as human-following robots, as long as the robot is able to move at the speed of human walking. Moreover, it is necessary to survey the influence of the mobile robot which maintains a flexible distance between the robot and the human, and introduces the knowledge of cognitive science and social science.

7. Acknowledgments

This work was supported by the Korea Research Foundation Grant funded by the Korean Government(MOEHRD) (KRF-2007-331-D00152)

8. References

L. Moreno and E. Dapena. (2003). Path quality measure for sensor-based motion planning, *Robotics and Autonomous Systems*, vol. 44, pp. 131-150.

B. Bouilly and T. Siméon. (1996). A sensor-based motion planner for mobile robot navigation with uncertainty, in *Proc. of the International Workshop* on RUR'95, Springer, Berlin, pp. 235–247.

M. Betke *et al.* (1994). Mobile Robot Localization Using Landmarks, *Proc. of the IEEE/ RSJ/GI Int. Conf. on Intelligent Robots and Systems*, pp. 135~142.

J. David, Kreigman *et al.* (1989). Stereo vision and navigation in buildings for mobile robots, *IEEE Trans. Robotics and Automation*, vol. 5, no. 6, pp. 792~803.

J.-H. Lee, G. Appenzeller, and H. Hashimoto. (1997). An agent for intelligent spaces: Functions and roles of mobile robots in sensored, networked, thinking spaces, in *Proc. IEEE Conf. Intelligent Transportation Systems*, Boston, pp. 983–988.

Y. Nakamura. (1991). *Advanced Robotics: Redundancy and Optimization*, Addison-Wesley.

R. M. Haralick and Linda G. Shapiro. (1993). *Computer and Robot Vision*, Addison-Wesley.

N. Ayache and O. D. Faugeras. (1989). Maintaining Representations of the Environment of a Mobile Robot, *IEEE Trans. Robotics and Automation*, vol. 5, no. 6.

R. E. Kalman. (1960). New Approach to Linear Filtering and Prediction Problems, *Trans, ASME, J. Basic Eng*, Series 82D, pp. 35-45.

H. W. Sorenson. (1966). Kalman Filtering Techniques, *Advances in Control Systems Theory and Applications*, vol. 3, pp. 219-292.

M. Y. Han, B. K. Kim, K. H. Kim, and Jang M. Lee. (1999). Active Calibration of the Robot/Camera Pose Using the Circular Objects, in Korean, *Transactions on Control, Automation and Systems Engineering*, vol. 5, no. 3, pp. 314-323.

T. Akiyama, J.-H. Lee, and H. Hashimoto. (2002). Evaluation of CCD camera arrangement for positioning system in intelligent space, in *Proc. Seventh Int. Symp. Artificial Life and Robotics (AROB'02)*, pp. 310–315.

Dinesh Nair and Jagdishkumar K. Aggarwal. (1989). Moving Obstacle Detection From a Navigation Robot, *IEEE Trans. Robotics and Automation*, vol. 14, no. 3, pp. 404~416.

Anthony Lallet and Simon Lacroix. (1998). Toward Real-Time 2D Localization in Outdoor Environments, *Proc. of the 1998 IEEE International Conference on Robotics & Automation*, pp. 2827-2832.

Amit Adam, Ehud Rivlin, and Ilan Shimshoni. (2000). Computing the Sensory Uncertainty Field of a Vision-based Localization Sensor, *Proc. of the 2000 IEEE International Conference on Robotics & Automation*, pp. 2993-2999.

Robert Sim and Gregory Dudek. (1999). Learning Visual Landmarks for Pose Estimation, *Proc. of the 1999 IEEE International Conference on Robotics & Automation*, pp.1972-1978.

B.H. Kim, D.K. Roh, Jang M. Lee, M.H. Lee, K. Son, M.C. Lee, J.W. Choi, and S.H. Han. (2001). Localization of a Mobile Robot using Images of a Moving Target, *Proc. of the 2001 IEEE International Conference on Robotics & Automation*.

Vincenzo Caglioti. (2001). An Entropic Criterion for Minimum Uncertainty Sensing in Recognition and Localization Part II-A Case Study on Directional Distance Measurements, *IEEE Trans. On Systems, Man, and Cybernetics*, vol. 31, no. 2, pp. 197-214.

Clark F. Olson. (2000). Probabilistic Self-Localization for Mobile Robots, *IEEE Trans. On Robotics and Automation*, vol. 16, no. 1, pp.55-66.

Hongjun Zhou and Shigeyuki Sakane. (2001). Sensor Planning for Mobile Robot Localization Based on Probabilistic Inference Using Bayesian Network," *Proc. of the 4th IEEE International Symposium on Assembly and Task Planning*, pp. 7-12.

Muriel Selsis, Christophe Vieren, and Francois Cabestaing. (1995). Automatic Tracking and 3D Localization of Moving Objects by Active Contour Models, *Proc. of the 95' IEEE International Symposium on Intelligent Vehicles*, pp. 96-100.

Howie Choset and Keiji Nagatani. (2001). Topological Simultaneous Localization and Mapping (SLAM): Toward Exact Localization Without Explicit Localization, *IEEE Trans. On Robotics and Automation*, vol. 17, no. 2.

Sanisa Segvic and Slobodan Ribaric. (2001). Determining the Absolute Orientation in a Corridor Using Projective Geometry and Active Vision, *IEEE Trans. On Industrial Electronics*, vol. 48, no. 3, pp. 696-710.

Philippe Hoppernot and Etienne Colle. (2001). Localization and Control of a Rehabilitation Mobile Robot by Close Human-Machine Cooperation, *IEEE Trans. On Neural Systems and Rehabilitation Engineering*, vol. 9, no. 2, pp. 181-190.

J.-H. Lee and H. Hashimoto. (2002). Intelligent space -concept and contents, *Adv. Robot.*, vol. 16, no. 3, pp. 265–280.

9

Robot Tracking using the Particle Filter and SOM in Networked Robotic Space

TaeSeok Jin
Dept. of Mechatronics Engineering, DongSeo University
Korea

1. Introduction

Detection of moving objects has been utilized in industrial robotic systems, for example, in the recognition and monitoring of unmanned systems that also require compression of moving images (Senior, 2002, Bierlaire & Antonini, 2003). Trajectory prediction of moving objects is required for a mobile manipulator that aims at the control and observation of motion information such as object position, velocity, and acceleration. Prediction and estimation algorithms have generally been required for industrial robots. For a simple example, in a pick-and-place operation with a manipulator, the precise motion estimation of the object on the conveyor belt is a critical factor in stable grasping (Nummiaro, et.al, 2002, Allen, et.al, 1992). A well-structured environment, such as the moving-jig that carries the object on the conveyor belt and stops when the manipulator grasps the object, might obviate the motion estimation requirement. However, a well-structured environment limits the flexibility of the production system, requires skillful designers for the jig, and incurs a high maintenance expense; eventually it will disappear from automated production lines.
To overcome these problems, to tracking a moving object stably without stopping the motion, the trajectory prediction of the moving object on the conveyor belt is necessary (Ma, Kosecka & Sastry, 1999). The manipulator control system needs to estimate the most accurate position, velocity, and acceleration at any instance to capture the moving object safely without collision and to pick up the object stably without slippage. When the motion trajectory is not highly random and continuous, it can be modeled analytically to predict the near-future values based on previously measured data (Choo & Fleet, 2001). However, this kind of approach requires significant computational time for high-degrees-of-freedom motion, and its computational complexity increases rapidly when there are many modeling errors. In addition, performance is highly sensitive to the change of the environment. Those state-of-the-art techniques perform efficiently to trace the movement of one or two moving objects but the operational efficiency decreases dramatically when tracking the movement of many moving objects because systems implementing multiple hypotheses and multiple targets suffer from a combinatorial explosion, rendering those approaches computationally very expensive for real-time object tracking (Anderson & Moore, 1979).

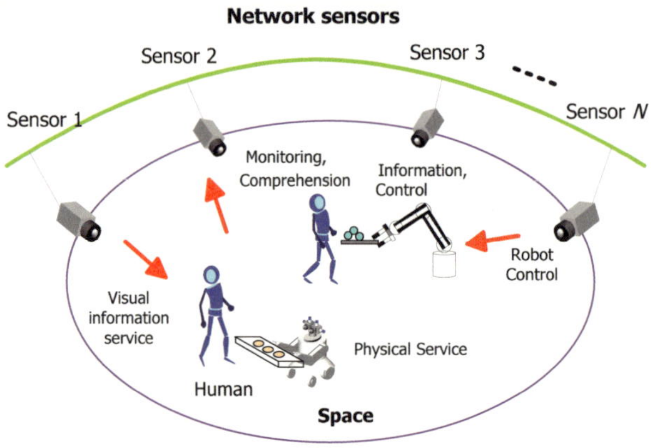

Figure 1. Intelligent environment by distributed cameras

It is necessary for the intelligent environment to acquire various information about humans and robots in the environment. When the environment does not know where humans and robots are respectively, the environment cannot give the enough service to the appropriate user as for the physical service especially. Therefore, it is considered that how to get the location information is the most necessary of all. The system with multiple color CCD cameras is utilized as one of the means to acquire the location information in an intelligent environment. It can achieve the human centered system because the environment acquires the location of human noncontactly and the equipment of the special devices isn't required for human. Moreover, camera has the advantage in wide monitoring area. It also leads to acquisition of details about objects and the behavior recognition according to image processing. Our intelligent environment is achieved by distributing small intelligent devices which don't affect the present living environment greatly.

2. Vision System in Robotic Space

2.1 Structure of Robotic Space

Fig. 2 shows the system configuration of distributed cameras in Intelligent Space. Since many autonomous cameras are distributed, this system is autonomous distributed system and has robustness and flexibility. Tracking and position estimation of objects is characterized as the basic function of each camera. Each camera must perform the basic function independently at least because over cooperation in basic level between cameras loses the robustness of autonomous distributed system. On the other hand, cooperation between many cameras is needed for accurate position estimation, control of the human following robot (Kitagawa, 1996), guiding robots beyond the monitoring area of one camera (Chen & Young, 2002), and so on. These are advanced functions of this system. This distributed camera system of Intelligent Space is separated into two parts as shown in Fig. 2. This paper will focus on the tracking of multiple objects in the basic function. Each camera has to perform the basic function independently of condition of other cameras, because of keeping the robustness and the flexibility of the system. On the other hand, cooperation between many cameras is needed for accurate position estimation, control of mobile robots to supporting human (Harris, 1995, Norlund & Eklundh, 1997), guiding robots beyond the monitoring area of one camera (Atsushi, et.al, 2002), and so on. These are advanced functions of this system.

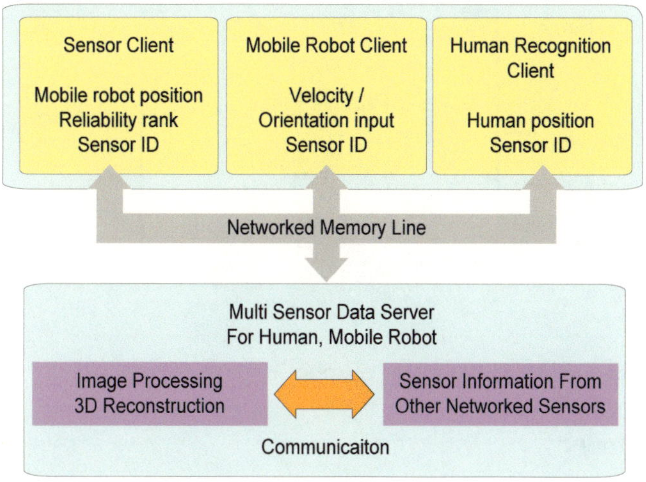

Figure 2. Configuration of distributed camera system

2.2 Previous Research for Tracking

Neural networks can be classified into two categories: supervised learning- and unsupervised learning methods. In most of the previous research, the supervised learning method was adopted to overcome the nonlinear properties (Roberts, 1995, Atsushi, et.al, 2002). Since the supervised learning method requires the relation between input and output (Chen & Young, 2002) at all times, it is not suitable for real-time trajectory estimation for which the input-output relation cannot be obtained instantaneously in the unstructured environment. Therefore, in this study, SOM (Self Organizing Map), that is, a type of unsupervised learning method, was selected to estimate the highly nonlinear trajectory that cannot be properly predicted by the Particle filter. Also, SOM is a data-sorting algorithm, which is necessary for real-time image processing since there is so much data to be processed. Among the most popular data-sorting algorithms, VQ (Vector Quantization), SOM, and LVQ (Learning Vector Quantization), SOM is selected to sort the data in this approach since it is capable of unsupervised learning. Since VQ is limited to the very special case of a zero neighborhood and LVQ requires preliminary information for classes, neither of them is suitable for the unsupervised learning of the moving trajectory. Fig. 3 shows the estimation and tracking system for this research. The input for the dynamic model comes from either the Particle filter or SOM according to the following decision equation:

$$Predicted value = k \cdot ParticleFilter_{out} + (1-k) \cdot SOM_{out} \tag{1}$$

where k=1 for $error \leq threshold$ and k=0 for $error > threshold$.

The threshold value is empirically determined based on the size of the estimated position error.

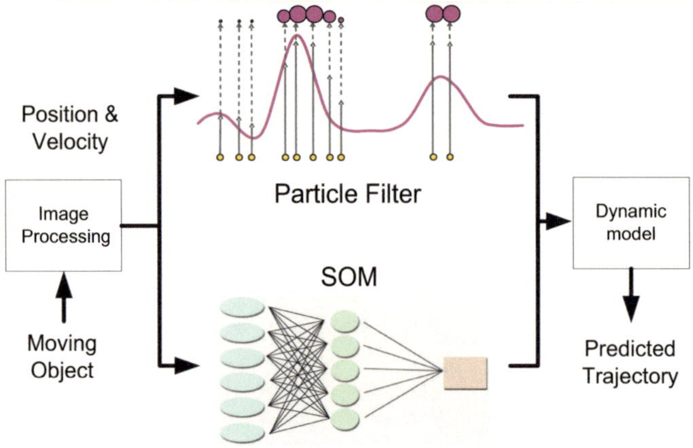

Figure 3. Estimation and tracking system of robotic space

3. Processing Flow

3.1 Extraction of Object

Classifying the moving-object pattern in the dynamically changing unstructured environment has not yet been tackled successfully (Wren, et.al, 1997). Therefore, in this research, the camera was fixed on a stable platform in order to capture static environment images. To estimate the states of the motion characteristics, the trajectory of the moving object was pre-recorded and analyzed. Fig. 4(a) and Fig. (b) represent the object images at (t-1) instance and (t) instance, respectively.

(a) (t-1) instance (b) (t) instance

Figure 4. Detected image of moving object at each instance

As recognized in the images, most parts of the CCD image correspond to the background. After eliminating the background, the difference between the two consecutive image frames can be obtained to estimate the moving-object motion. To compute the difference, either the absolute values of the two image frames or the assigned values can be used. The difference method is popular in image pre-processing for extracting desired information from the whole image frame, which can be expressed as

$$\text{Output}(x, y) = \text{Image1}(x, y) - \text{Image2}(x, y) \tag{2}$$

The difference image between Fig. 4(a) and Fig. 4(b) is represented in Fig. 5. When the difference image for the whole time interval can be obtained, the trajectory of the moving object can be calculated precisely.

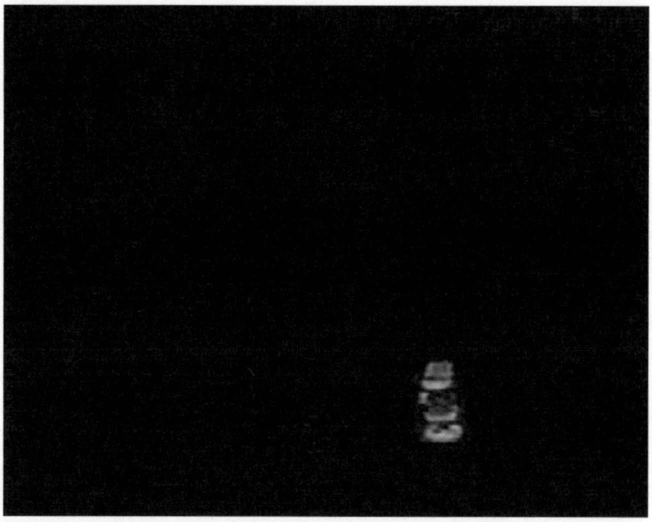

Figure 5. Difference image between (t) and (t-1) instance images

3.2 Target Regions Encoded in a State Vector

Particle filtering provides a robust tracking framework, as it models uncertainty. Particle filters are very flexible in that they not require any assumptions about the probability distributions of data. In order to track moving objects (e.g. pedestrians) in video sequences, a classical particle filter continuously looks throughout the 2D-image space to determine which image regions belong to which moving objects (target regions). For that a moving region can be encoded in a state vector. In the tracking problem the object identity must be maintained throughout the video sequences. The image features used therefore can involve low-level or high-level approaches (such as the colored-based image features, a subspace image decomposition or appearance models) to build a state vector. A target region over the 2D-image space can be represented for instance as follows:

$$\mathrm{r} = \{\mathrm{l}, \mathrm{s}, \mathrm{m}, \gamma\} \tag{3}$$

where l is the location of the region, s is the region size, m is its motion and γ is its direction. In the standard formulation of the particle filter algorithm, the location l, of the hypothesis, is fixed in the prediction stage using only the previous approximation of the state density. Moreover, the importance of using an adaptive-target model to tackle the problems such as the occlusions and large-scale changes has been largely recognized. For example, the update of the target model can be implemented by the equation

$$\bar{\mathrm{r}}_t = (1-\lambda)\bar{\mathrm{r}}_{t-1} + \lambda \mathrm{E}[\mathrm{r}_t] \tag{4}$$

where λ weights the contribution of the mean state to the target region. So, we update the target model during slowly changing image observations.

4. Tracking Moving Objects

4.1 State-Space over the Top-View Plan

In a practical particle filter (Ma, Kosecka & Sastry, 2001, Choo & Fleet, 2001) implementation, the prediction density is obtained by applying a dynamic model to the output of the previous time-step. This is appropriate when the hypothesis set approximation of the state density is accurate. But the random nature of the motion model induces some non-zero probability everywhere in state-space that the object is present at that point. The tracking error can be reduced by increasing the number of hypotheses (particles) with considerable influence on the computational complexity of the algorithm. However in the case of tracking pedestrians we propose to use the top-view information to refine the predictions and reduce the state-space, which permits an efficient discrete representation. In this top-view plan the displacements become Euclidean distances. The prediction can be defined according to the physical limitations of the pedestrians and their kinematics. In this paper we use a simpler dynamic model, where the actions of the pedestrians are modeled by incorporating internal (or personal) factors only. The displacements $\mathrm{M}^{t}_{topview}$ follows the expression

$$\mathrm{M}^{t}_{topview} = \mathrm{A}(\gamma_{topview})\mathrm{M}^{t-1}_{topview} + \mathrm{N} \tag{5}$$

where **A**(.) is the rotation matrix, $\gamma_{topview}$ is the rotation angle defined over top-view plan and follows a Gaussian function $g(\gamma_{topview};\sigma_{\gamma})$, and **N** is a stochastic component. This model proposes an anisotropic propagation of **M** : the highest probability is obtained by preserving the same direction. The evolution of a sample set is calculated by propagating each sample according to the dynamic model. So, that procedure generates the hypotheses.

4.2 Estimation of Region Size

The height and width of the robot can be obtained using geometric analysis.

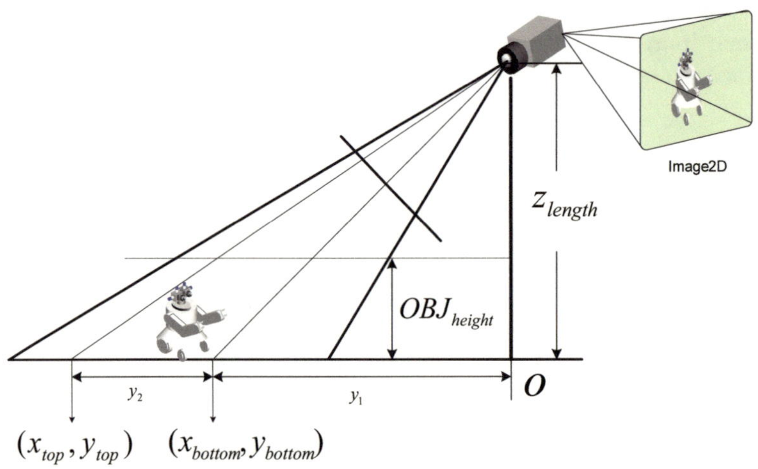

Figure 6. Approximation of Top-view plan by image plan with a monocular camera

As shown in Fig. 6, the distance y_1 from the lowest coordinates of the object to the origin is calculated using y_{bottom} by geometric analysis as,

$$y_1 = y_{bottom} - O \tag{10}$$

where O represents the origin.

In the same manner, y_{top} can be calculated by replacing y_{bottom} as y_{top} and $P_{y,bottom}$ as $P_{y,top}$. Therefore, the distance y_2 from the highest coordinates of the object to y_{bottom} is calculated as,

$$y_2 = y_{top} - y_{bottom} \tag{11}$$

When the coordinates, y_1 and y_2 are obtained, the height OBJ_{height} of the robot can be calculated as,

$$OBJ_{height} = \frac{z_{length} \times y_2}{(y_1 + y_2)} \tag{12}$$

from the similarity properties of triangles.
Following the same procedure, the width of the mobile robot can be obtained as follows: The real length $length_{pixel}$ per pixel is calculated as follow:

$$length_{pixel} = OBJ_{height} /(P_{y,top} - P_{y,bottom}) \tag{13}$$

Then, the width, OBJ_{width}, of the object is calculated as

$$OBJ_{width} = length_{pixel} \times (p_{x,right} - p_{x,left}) \tag{14}$$

5. Experiments

To compare the tracking performance of a mobile robot using the algorithms of the Particle filter and SOM, experiments of capturing a micro mouse with random motion by the mobile robot were performed. As can be recognized from Fig. 7, SOM based Particle Filter provided better performance to the mobile robot in capturing the random motion object than the other algorithms. As shown in Fig. 7, the mobile robot with SOM based Particle Filter has a smooth curve to capture the moving object. As the result, the capturing time for the moving object is the shortest with SOM based Particle Filter. Finally, as an application experiments were performed to show the tracking and capturing a mobile object in robotic space.
Fig. 8 shows the experimental results for tracking a moving object that is an 8x6[cm] red-colored mouse and has two wheels with random velocities in the range of 25-35[cm/sec]. First, mobile robot detects the moving object using an active camera. When a moving object is detected within view, robot tracks it following the proposed method.

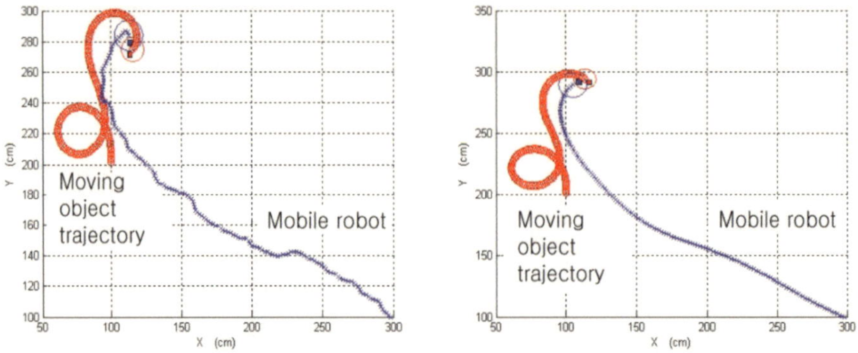

(a) SOM (b) SOM + Particle Filter

Figure 7. Tracking trajectory by SOM and SOM based particle filter

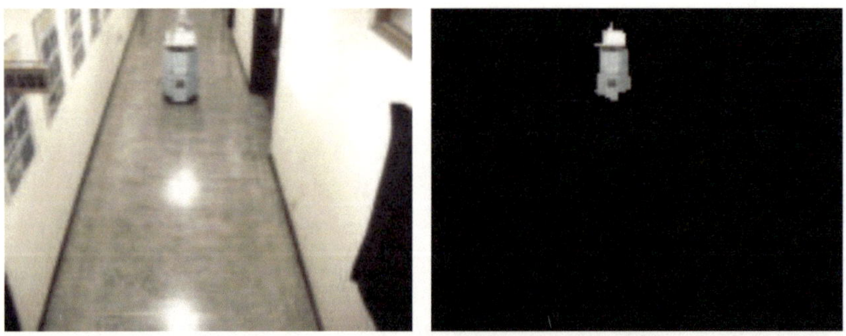

(a) Input Image (b) Extracted Image

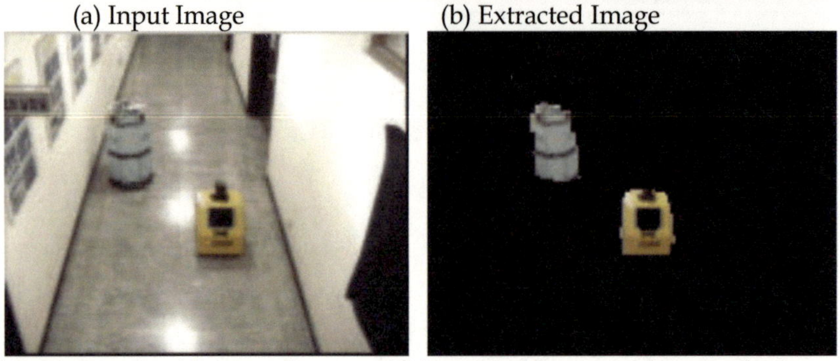

(c) Input Image with two robots (d) Extracted Image with two robots

Figure 8. Experimental results for tracking a moving object

6. Conclusion

In this paper, the proposed tracking method adds an adaptive appearance model based on color distributions to particle filtering. The color-based tracker can efficiently and successfully handle non-rigid and fast moving objects under different appearance changes. Moreover, as multiple hypotheses are processed, objects can be tracked well in cases of occlusion or clutter. This research proposes estimation and tracking scheme for a moving object using images captured by multi cameras. In the scheme, the state estimator has two algorithms: the particle filter that estimates the states for the linear approximated region, and SOM for the nonlinear region. The decision for the switchover is made based on the size of the position estimation error that becomes low enough for the linear region and becomes large enough for the nonlinear region. The effectiveness and superiority of the proposed algorithm was verified through experimental data and comparison. The adaptability of the algorithm was also observed during the experiments. For the sake of simplicity, this research was limited to the environment of a fixed-camera view. However, this can be expanded to the moving camera environment, where the input data might suffer from higher noises and uncertainties. As future research, selection of a precise learning pattern for SOM in order to improve the estimation accuracy and the recognition ratio, and development of an illumination robust image processing algorithm, remain.

7. Acknowledgments

This work was supported by the Korea Research Foundation Grant funded by the Korean Government(MOEHRD) (KRF-2007-331-D00152)

8. References

Senior, A. (2002). Tracking with Probabilistic Appearance Models. In *Proc. ECCV workshop on Performance Evaluation of Tracking and Surveillance Systems*, pp. 48-55.

Bierlaire, M., Antonini, G., Weber, M. (2003). Behavioural Dynamics for Pedestrians, in K. Axhausen (Ed.), *Moving through nets: the physical and social dimensions of travel*, Elsevier, pp.1-18.

Nummiaro, K., Koller-Meier, E., Van Gool, L.J. (2002). Object Tracking with an Adaptive Color-Based Particle Filter, *DAGM-Symposium Pattern Recognition*, pp.353-360.

P. K. Allen, A. Tmcenko, B. Yoshimi, and P. Michelman. (1992). Trajectory filtering and prediction for automated tracking and grasping of a moving object, *IEEE International Conference on Robotics and Automation*, pp.1850-1856.

Yi Ma, J. Kosecka, and S. S. Sastry. (1999). Vision guided navigation for a nonholonomic mobile robot, *IEEE Transaction on Robotics and Automation*, vol. 15, no. 3, pp. 521-536.

Choo, K., Fleet, D.J. (2001). People tracking using hybrid Monte Carlo filtering, In *Proc. Int. Conf. Computer Vision*, vol. II, pp.321-328.

Anderson, B., Moore, J. (1979). *Optimal Filtering*. Prentice-Hall, Englewood Cliffs.

Kitagawa, G. (1996). Monte Carlo Filter and Smoother for Non-Gaussian Nonlinear State Space Models, *Journal of Computational and Graphical Statistics*, Vol. 5, pp.1-25.

Yi-Yuan Chen and Kuu-young Young. (2002). An intelligent radar predictor for high-speed moving-target tracking, *TENCON '02. Proceedings. IEEE Region 10 Conference on Computers, Communications, Control and Power Engineering*, vol. 3, pp.1638 -1641.

J. M. Roberts, D. J. Mills, D. Charnley, and C. J. Harris. (1995). Improved Kalman filter initialization using neuro-fuzzy estimation, *Int'l. Conf. on Artificial Neural Networks*, pp. 329-334.

Norlund, P., Eklundh, J.O. (1997). Towards a Seeing Agent. *Proc. of First Int. Workshop on Cooperative Distributed Vision*, pp.93-120.

Atsushi, N., Hirokazu, K., Shinsaku, H., Seiji, I. (2002). Tracking Multiple People using Distributed Vision Systems. *Proc. of the 2002 IEEE Int. Conf. on Robotics & Automation*, pp.2974-2981.

Wren, C., Azarbayejani, A., Darrell, T., Pentland, A.: finder. (1997). Real-Time Tracking of the Human Body, *IEEE Transactions on Pattern Analysis and Machine Intelligence*, Vol. 19, pp.780-785.

Gardner, W.F., Lawton, D.T.(1996). Interactive model based vehicle tracking. *IEEE Transaction on Pattern Analysis and Machine Intelligence*, Vol.18, pp.1115-1121.

Swain, M.J., Ballard, D.H. (1991). Color indexing. *Int. J. of Computer Vision*, Vol.7, No.1, pp.11-32.

10

Artificial Coordinating Field based Motion Planning of Mobile Robots

Xing-Jian Jing[1,2*] and Yue-Chao Wang[2]
[1]*Automatic Control and Systems Engineering, University of Sheffield,*
[2] *State Key Laboratory of Robotics, Shenyang Institute of Automation, Chinese Academy of Sciences*
[1]*U.K.*, [2]*China*

1. Introduction

Motion planning of mobile robots in uncertain dynamic environments has been a hot topic in robotic literature. It requires a mobile robot to decide its motion behaviour on line using limited and noised information of the local environment from sensors. There are many methods having been proposed to deal with this problem (Salichs and Moreno 2000, Jing 2005). Noticeably, artificial potential field (APF) based methods have gained increasingly popularity among researchers due to its high safety, simplicity and elegance (Khatib 1986, Rimon and Koditschek 1991, Kant and Zucher 1988, Rimon and Koditschek 1992, Koren and Borenstein 1991, Guldner and Utkin 1995, Ge and Cui 2000, Prassler 1999, Noborio et al 1995, Krogh 1984, Satoh 1993, Louste and Liegeois 2000, Wong and Spetsakis 2000, Singh et al 1997, Tsourveloudis et al 2001, Masoud and Masoud 2000). However, when the involved environment is totally or partially unknown or even dynamically changing, local minima are usually encountered, where the robot is trapped and cannot move on. There may also be unnecessary oscillations on the planned trajectory between multiple obstacles (Koren and Borenstein 1991). These inhibit the practical applications of this methodology to a certain extent. To overcome these problems, there are already some methods having been proposed in literature. For example, Krogh (1984) proposed a generalized potential field, in which the strength of repulsion is directly proportional to the speed of approach and inversely proportional to the minimum avoidance time. Satoh (1993) proposed Laplace potential field, which requires the potential field to be harmonic, and satisfy the Laplace equation. In Louste and Liegeois (2000), the authors used viscous fluid field instead of conventional APF to achieve near optimal path planning. Moreover, electric-like fields (Wong and Spetsakis 2000), magnetic field (Singh et al 1997), electrostatic potential field (Tsourveloudis et al 2001) were all proposed for the navigation and motion planning problems. But all these methods either require some global environment information or only deal with navigation problems in static environments, and only a few take into consideration of the actual dynamic constraints of the mobile robot such as saturations of velocity and acceleration. Moreover, few of the existing potential fields can guarantee the safety and reachability of the mobile robot with consideration of the actual dynamic constraints in uncertain dynamic environments.

* The first author has been with the University of Sheffield since Oct 2005.

The reason for the drawbacks of the ACF as mentioned above is that, in our opinion, this simple "repulsive and attractive" information model of the environments in APF methods cannot completely and accurately reflect the actual states and real motion purpose of the mobile robot. Hence, it is difficult or even impossible to decide the optimal or satisfactory motion behaviour in some complicated situations just based on this simple information model of the environments of APF using only attractive and repulsive forces. In order to overcome the drawbacks of the conventional APFs, it does need to change the simple "repulsive and attractive" information model to another more appropriate information model of the environments, and make the new model be adaptable to motion purpose and relative states of the mobile robot with respect to obstacles.

Therefore, an artificial coordinating field (ACF) is proposed in this chapter. In order to overcome the drawbacks of APF, a special force vector called Coordinating Force is defined and added to the conventional APF, and the ACF is designed to be adaptable to the motion purpose and relative states of the mobile robot with respect to obstacles, which includes not only the information of relative positions of the robot with respect to an obstacle, but also the information of the relative velocity, maximum acceleration and velocity of the robot. Decision-making of the robot's behavior when avoiding an obstacle is based on a special variable, called coordinating factor λ, which is simple and in an optimal way. The safety and reachability of the proposed method are theoretically analyzed with some assumptions on the environments. Simulation results are given to illustrate our method.

2. Definition of the ACF

Our study is restricted to the 2-D planar case. Some notations are introduced as follows. The planar U can be denoted as a point set $U=\{p=(x,y)^T \mid x,y \in \mathbb{R}\}$, where point $(x,y)^T$ is a column vector, $(*)^T$ is the transpose of vector (*), $\mathbb{R}$ is the set of all the real numbers. ∂D denotes the boundary of a subset D in U. Without specialty, a bold italic symbol denotes a vector. e($\boldsymbol{A}$) denotes the unitary vector of a vector $\boldsymbol{A}$, *i.e.*, $e(\boldsymbol{A}) = \mathbf{A}/\|\mathbf{A}\|$, where $\|\boldsymbol{A}\|$ denotes the Euclidian norm of $\boldsymbol{A}$. Difference of two point is a vector, *e.g.*, $\boldsymbol{A}=q_1-q_2$, where $q_1, q_2 \in U$, the direction of $\boldsymbol{A}$ is from q_2 to q_1, *i.e.*, $e(\boldsymbol{A})=e(q_1-q_2)$. Moreover, "$a \rightarrow b$" denotes "$a$ is approaching to b nearly or very nearly". On the contrary, "$a>>b$" and "$a<<b$" denote that "a is much larger or smaller than b" respectively.

In addition, the mobile robot can be regarded as a point mass with weight M, its goal is denoted by q_d. An obstacle can be regarded as a point set O or O_i in U, where the subscript i is to distinguish different obstacles. The obstacle O_i may also be called obstacle i later on. The distance between two point set O_i and O_j is defined as $d(O_1, O_2) = \min_{p\in\partial O_1, q\in\partial O_2} \|p-q\|$.

Define a mapping g_O: $q \rightarrow \partial O$ such that $p = g_O(q) = \arg\min_{p\in\partial O}\|q-p\|$, where $p, q \in U$.

Obviously, p is the nearest point on the boundary of O to q. For an obstacle O_i, this mapping function is also written in short as $g_i(q)$.

The ACF is defined as a force vector field as follows (see Figure 1). The attractive field at the goal q_d of the mobile robot is defined as: $\forall q \in U$

$$\boldsymbol{F_a}(q) = K_a \cdot (q_d - q) \tag{1a}$$

where K_a is to be defined. For an obstacle O, define the ACF as: $\forall q \in U \setminus O$

$$\boldsymbol{F}_{cO}(q) = \boldsymbol{F}_{rO}(q) + \boldsymbol{F}_{nO}(q) \tag{1b}$$
$$\boldsymbol{F}_{rO}(q) = K_{rO} \cdot (q - g_O(q)) \tag{1c}$$
$$\boldsymbol{F}_{nO}(q) = K_{nO} \cdot \lambda \cdot \boldsymbol{T} \cdot (q - g_O(q)) \tag{1d}$$

where $\lambda \in \{1,0,-1\}$ is called coordinating factor, $\boldsymbol{T} = \begin{bmatrix} 0 & 1 \\ -1 & 0 \end{bmatrix}$; (1b) is the artificial repulsive-coordinating field of obstacle O, which is also called in short as artificial coordinating field (ACF) in this study; (1c) is the repulsive force vector, K_{rO} is to be defined; (1d) is the coordinating force vector, which is orthogonal to the repulsive force and whose direction is determined by λ, K_{nO} is to be defined. For different obstacle O_i, the aforementioned force vectors are rewritten in short as: $\boldsymbol{F}_{ci}$, $\boldsymbol{F}_{ri}$, $\boldsymbol{F}_{ni}$, respectively, and the corresponding parameters are rewritten as K_{ri}, K_{ni}, λ_i, respectively. If the repulsive force (1c) is substituted by the attractive force (1a), then the new artificial field is called artificial attractive-coordinating field. Moreover, we can also define the ACF in 3-dimensional space using a similar method as above.

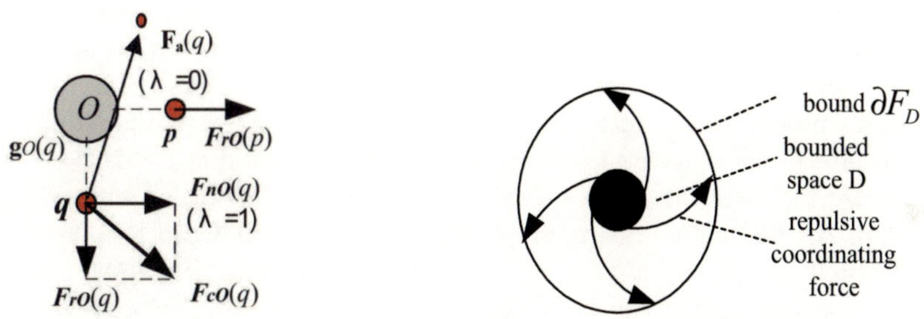

(a) Different forces in an ACF (b) A bounded repulsive ACF with λ =1

Figure 1. The repulsive force and attractive force in an ACF

At any time instant *t*, let the *x*-coordinate of the dynamic coordinates on the mobile robot with respect to an obstacle O be parallel to the coordinating force vector, and the y-coordinate be parallel to the repulsive force vector. Obviously, the ACF has two-dimensional orthogonal force vectors, thus the mobile robot has two DOF to be controlled in its dynamic coordinates when meeting an obstacle, this may help to realize some desired motion behavior. Compared with ACF, the conventional APF can only exert one-dimensional force to the mobile robot in the dynamic coordinates. Thus the mobile robot can only run away from the APF when meeting an obstacle, but not avoid the obstacle with intention. This may be a major reason that there are local minima in conventional APFs for uncertain dynamic environments. Especially, it is noted that the direction of the coordinating force vector in an ACF at any time is determined by λ. If let λ =0, then $\boldsymbol{F}_{cO}(q)$=0 (referring to the point *p* in Figure 1), and there is only a repulsive force at point *p* in this case, which is right the APF. Hence, APF is only a special case of ACF. Since more environmental information and motion purpose of the mobile robot can be represented in the ACF, the states of a mobile robot can be controlled for some special purposes by using the orthogonal forces in the ACF. Moreover, considering the motion planning problem in uncertain dynamic environments, only the distance between an obstacle and the mobile robot is near enough (*e.g.*, less than a constant R), for the obstacle to be detected by the robot's sensors. Therefore, the radius of the ACF of an obstacle should be less than the distance R around the boundary of the obstacle.

It shall be noted that similar but different work to our ACF defined above can be found in literature. Note in Masoud and Masoud (2000), two orthogonal fields were used, a scalar potential field in normal space and a circular field in tangent space around obstacles, to locally switch the robot from one trajectory to another in order to adapt the unknown changing environments. The idea of orthogonal fields is very similar to ours. The difference is that it needs to solve boundary value problems, and needs also some global environment information. The circular field is only used to shift the robot from one path to another when meeting unknown static obstacles, and the whole field is still a passive one as most existing fields, namely, it cannot be adaptable to the states and motion purpose of the robot in the local environment. Note also in Medio and Oriolo (1999), a vortex field was proposed, which is also a passive field, and has no repulsive force compared with APFs.

3. Properties and Designs of ACF

This section discusses the properties of the ACF and studies how to design the parameters of the ACF to achieve the desired performance in the motion-planning problem of mobile robots in uncertain dynamic environments. More notations are introduced as follows. The position of the mobile robot is denoted by q without specialty, the maximum velocity and acceleration of the mobile robot are V_{max} m/s and a_{max} m/s^2, and the radius within which an obstacle can be effectively detected by the sensors is R. Assume that $R >> 2(V_{max})^2/a_{max}$. The region covered by the circle at point $q=(x,y)^T$ with a radius R is called observable region denoted by $P(q)$. All the static and moving obstacles in $P(q)$ that can be detected by the sensors are denoted by sets O_s and O_d, respectively. For instance, if a static obstacle O_i is detected by the sensors, then it can be written as $O_i \in O_s$ or $i \in O_s$. Velocity vectors of the mobile robot and an obstacle O_i are denoted by $\boldsymbol{V_r}$ and $\boldsymbol{V_i}$, respectively, and the relative velocity of the mobile robot with respect to the obstacle O_i is $\boldsymbol{V_{ir}}=\boldsymbol{V_r}-\boldsymbol{V_i}$. In this section, we design the ACF based on the analysis of the dynamics of a point in ACFs.

3.1 Controllability of a mobile robot in the ACFs

For the motion-planning problem, there is an attractive field $\boldsymbol{F_a}(q)$ at the goal point $q_d=(x_d,y_d)^T$, and some ACFs $\boldsymbol{F_{ci}}(q)$ with respect to different obstacles O_i (i=1,2,…,N) in the planar U. Assume that $d(O_i,O_j)>0$, $d(q_d,O_i)>0$ for all i=1,2,…,N. For any time t, the dynamics of the mobile robot in the artificial fields can be written as:

$$M\ddot{q} = -K_f(\dot{q}-\dot{q}_d) + \mathbf{F_a}(q) + \sum_{i\in O_d \cup O_s}\mathbf{F_{ri}}(q) + \sum_{i\in O_d \cup O_s}\mathbf{F_{ni}}(q) \tag{2}$$

where, K_f>0 is a parameter to be defined, $\dot{q}_d$ is the desired velocity of the mobile robot. Equation (2) is called Planning Equation. The first term on the right side of the equality is to balance the dynamics of the mobile robot and control the velocity to a desired level stably. The last three terms stand for all the virtual forces received by the mobile robot, they can be rewritten as

$$\mathbf{F_{total}} = \mathbf{F_a}(q) + \sum_{i \in O_d \cup O_s} \mathbf{F_{ri}}(q) + \sum_{i \in O_d \cup O_s} \mathbf{F_{ni}}(q) .$$

Substitute (1) into (2), and transform (2) into state space equation form, we can have:

$$M\begin{bmatrix} \ddot{x} \\ \ddot{y} \\ \dot{x} \\ \dot{y} \end{bmatrix} = \begin{bmatrix} -K_f & 0 & \sum K_{ri} - K_a & -\sum K_{ni}\lambda \\ 0 & -K_f & \sum K_{ni}\lambda & \sum K_{ri} - K_a \\ 1 & 0 & 0 & 0 \\ 0 & 1 & 0 & 0 \end{bmatrix} \cdot \begin{bmatrix} \dot{x} \\ \dot{y} \\ x \\ y \end{bmatrix} + \begin{bmatrix} 1 & 0 \\ 0 & 1 \\ 0 & 0 \\ 0 & 0 \end{bmatrix} \cdot \mathbf{u} \tag{3}$$

where

$$\mathbf{u} = K_f \begin{bmatrix} \dot{x}_d \\ \dot{y}_d \end{bmatrix} + K_a \begin{bmatrix} x_d \\ y_d \end{bmatrix} - \sum_{i \in O_d \cup O_s} K_{ri}\, \mathrm{g_i}(q) + \sum_{i \in O_d \cup O_s} K_{ni}\lambda_i \cdot \mathbf{T} \cdot \mathrm{g_i}(q) \tag{4}$$

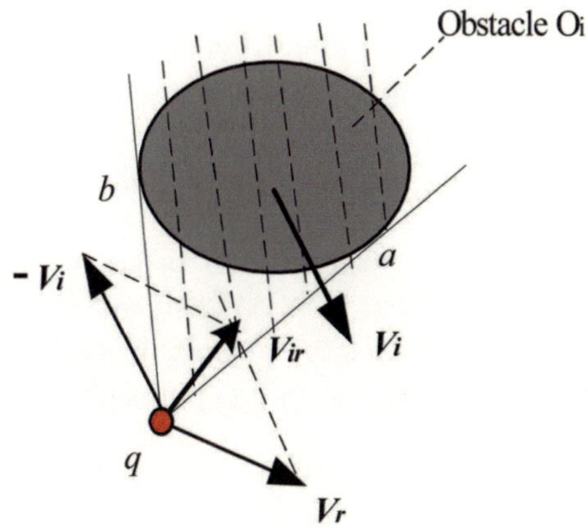

Figure 2. The mobile robot meets a moving obstacle

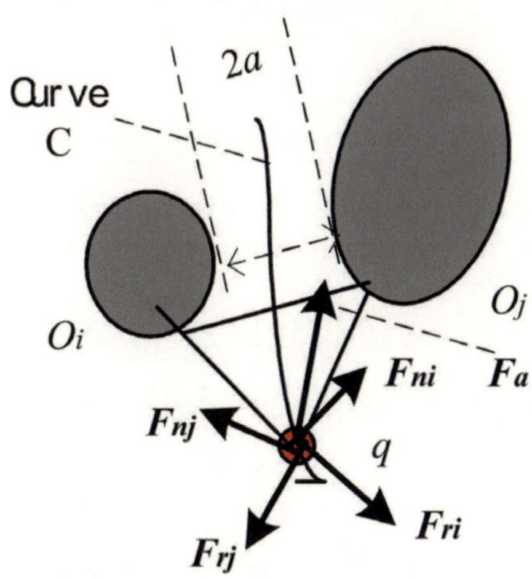

Figure 3. The mobile robot is passing a passage between two obstacles

It can be verified that, (3) is complete state controllability. From (4), the states of (3) is completely controlled by the variables K_a, K_{ri}, K_{ni} and λ_i of the ACFs. Hence, by properly choosing the variables of the ACFs, the desired motion behavior of the mobile robot can be obtained. Without loss of generality, let $\dot{q}_d=0$ in (2). Specially considering the APF case, *i.e.*, let $\lambda_i=0$ in (2-4). $\boldsymbol{F_{total}}$ will be zero in the local minima where the robot may be trapped. However, in ACFs, this is not the case. The local minima can be removed by properly choosing the coordinating factors and other variables of the ACFs such that $\boldsymbol{F_{total}}$ cannot be zero. This is further discussed in the following sections.

3.2 Adaptability of the ACFs

For different collision risk or different relative states of a mobile robot with respect to obstacles, the mobile robot should adopt different motion behavior or strategy according to an optimal evaluation. For this purpose, the ACFs should be adaptable to the collision risk or relative states of the mobile robot and can generate virtual forces of different magnitude and properties corresponding to different situations. This adaptability of ACFs can help the mobile robot to coordinate its motion behavior to avoid different type of obstacles and go to its goal in an optimal or a satisfactory way. To this aim, evaluations of the collision risk of a mobile robot are investigated with only the local information of the environments at first. And then the ACF is designed using these evaluation functions. Note that $\angle(*,*)$ denotes the angle of two vectors later on.

Assume that the mobile robot with velocity $\boldsymbol{V_r}$ and position q meets an obstacle O_i with velocity $\boldsymbol{V_i}$ at time t (See Figure 2). From point q, make two lines tangent to the boundary ∂O_i at point a and b, respectively. If the relative velocity $\boldsymbol{V_{ir}}=\boldsymbol{V_r}-\boldsymbol{V_i}$ is regarded as the velocity of the mobile robot, then the obstacle O_i can be regarded to be static. Let

$$\theta_{v1}=\angle(\mathbf{V}_{\mathbf{ir}},(\mathrm{e}(a-q)+\mathrm{e}(b-q))/2)\,,\ \theta_{v2}=\angle(\mathrm{e}(a-q),\mathrm{e}(b-q))/2\,.$$

It is easy to verify that, if $\theta_{v1}\le\theta_{v2}$ holds and $\boldsymbol{V_{ir}}$ is also kept unchanging, then the robot must be heading a collision with the obstacle. Let

$$E_{Vi}=\begin{cases}1 & \theta_{v1}\le\theta_{v2}\\ 0 & else\end{cases},$$

which is called velocity risk with respect to obstacle O_i. Obviously, $E_{Vi}=1$ implies a possible collision. Define the absolute collision risk with respect to O_i as:

$$E_i=k_{risk_1}/\|q-\mathrm{g}_i(q)\|+k_{risk_2}\cdot E_{Vi}\cdot(1+(\theta_{v2}-\theta_{v1})/\sup(\theta_{v2})+\|\boldsymbol{V_{ir}}\|/\sup(\|\boldsymbol{V_{ir}}\|))$$

where $0<k_{risk1}, k_{risk2}<1$. The total collision risk with respect to all the observable obstacles is

$$E=\sum_{i\in Od\cup Os}E_i\,.$$

Then define the relative collision risk with respect to obstacle O_i as:

$$E_{ri}=E_i/E$$

Using the evaluations of the collision risk with respect to obstacles above, we define the corresponding variables of the ACF as follows:

i. When the collision risk is increasing, the attractive force should be increased accordingly in order to attract the mobile robot to move towards its goal. However, too large magnitude of the attractive force may affect the safety of the robot in complicated situations. Hence, we let

$$K_a = k_a(1 + \min(M_a, E)) \tag{5}$$

where k_a>0, M_a>0 are constants.

ii. In order to guarantee the safety, the magnitude of the repulsive force should be proportional to the collision risk with respect to an obstacle O_i. Hence, we let

$$K_{ri} = k_{rc} \cdot (E_{ri} + \varepsilon_1) \cdot k_{ri} / \|q - g_i(q)\| \tag{6}$$

Where k_{rc}>0 is a constant, k_{ri} is to be defined, $0 \le \varepsilon_1 < 1$ is the minimal relative collision risk corresponding to different situations to be defined.

iii. As for the magnitude of the coordinating force, it is defined similarly to the repulsive force, and additionally it is defined to be limited:

$$K_{ni} = \min(M_n, k_{nc} \cdot (E_{ri} + \varepsilon_2) \cdot k_{ni} / \|q - g_i(q)\|) \tag{7}$$

where, k_{nc}>0 is a constant, M_n is the upper bound of K_{ni} and satisfies M_n>>sup($\|\boldsymbol{F_a}\|$), $0 \le \varepsilon_2 < 1$ is similar to ε_1, and k_{ni} is to be defined.

It should be noted that from (5-7), the magnitude of all the virtual forces has relation with the collision risk. Especially, the magnitude of an ACF is a function of the relative collision risk. The higher the collision risk with respect to an obstacle is, the larger is the force generated by the ACF of the obstacle. Hence, the robot dynamic is basically dominated by the obstacles with higher collision risks. This helps to guarantee the safety of the mobile robot and can pass the collision risk from one robot to another.

3.3 Safety of a mobile robot in the ACFs

The safety of a mobile robot not only has relation with the complexity of the environments but also is subjected to the dynamic constraints of the mobile robot. Obviously, if $\boldsymbol{V_{ir}}$.e($g_i(q)$ - q) ≤ 0 whenever d(q,O_i)→0, then there will be no collision to happen. For this reason, we let k_{ri} in (6) be:

$$k_{ri} = 1/\left(\mathrm{pos}(d(q, O_i) - (\mathrm{pos}(\boldsymbol{V_{ir}} \cdot \mathrm{e}(g_i(q) - q)))^2 / 2a_{\max})\right)^n \tag{8}$$

where $n \ge 1$, pos(x)=max(0,x) (this is directly used later on), d(q,O_i)=$\|q - g_i(q)\|$.

Based on the designs of the ACF above, the safety of the mobile robot can be guaranteed with some environment constraints. The main results are given as follows.

Proposition 1. Assume R>>2$(V_{max})^2/a_{max}$, and at time t the mobile robot is at point q satisfying d(q,O_i)<R. Let $\boldsymbol{F_{other}}$=$\boldsymbol{F_{total}}$-$\boldsymbol{F_{ri}}$. If $\mathbf{F_{other}} \cdot \mathrm{e}(g_i(q) - q) \ll \infty$, and the velocity and acceleration of the obstacle O_i satisfy $\|\boldsymbol{V_i}\| < V_{\max}$ and $\dot{\boldsymbol{V}}_i \cdot \mathrm{e}(g_i(q) - q) \ge 0$ whenever

$d(q, O_i) \to S_0$, then $\forall t, g_i(q) \neq q$, that is there will be no collision to happen between the mobile robot and the obstacle O_i, where $S_0 = (\mathrm{pos}(\boldsymbol{V}_{ir} \cdot e(g_i(q) - q)))^2 / 2a_{\max}$.

Proof. Let $v = \boldsymbol{V}_{ir} \cdot e(g_i(q) - q)$, which is the relative velocity of the mobile robot with respect to the nearest point of obstacle O_i. If $v \le 0$ then the safety is guaranteed. Consider the case $v > 0$ in the following. Since $\|\boldsymbol{V}_i\| < V_{\max}$, and $R >> 2(V_{max})^2/a_{max}$, we have $R >> S_0$. Hence, $d(q, O_i) > S_0$ usually holds at the beginning that the robot meets the obstacle. Whenever $d(q, O_i) \to S_0$, if $\dot{\boldsymbol{V}}_r \cdot e(g_i(q) - q) = -a_{\max}$ holds, and due to the assumptions on the obstacle O_i, we have $\dot{\boldsymbol{V}}_{ir} \cdot e(g_i(q) - q) = (\dot{\boldsymbol{V}}_r - \dot{\boldsymbol{V}}_i) \cdot e(g_i(q) - q) \le -a_{\max}$. Hence, if the robot can go with an acceleration $-a_{max}$ in the direction of $e(\mathbf{g}_i(q) - q)$, there must be $\boldsymbol{V}_{ir} \cdot e(g_i(q) - q) \le 0$ whenever $d(q, O_i) \to 0$, that is, there is no collision to happen between the mobile robot and the obstacle O_i. From (6) and (8), whenever $d(q, O_i) \to S_0$, then $k_{ri} \to \infty$, that is, $\boldsymbol{F}_{ri} \cdot e(g_i(q) - q) = -\infty$. If additionally $\boldsymbol{F}_{other} \cdot e(g_i(q) - q) << \infty$, then $\boldsymbol{F}_{total} \cdot e(q - g_i(q)) = \infty$, that is, $\boldsymbol{F}_{total} \cdot e(q - g_i(q)) - K_f(\dot{q} - \dot{q}_d) >> M.a_{\max}$. Then from (2), the mobile robot must go with an acceleration $-a_{max}$ in the direction of $e(g_i(q) - q)$. This completes the proof.

According to Proposition 1, we define **Environment Constraint 1**:

$$\forall i \in O_d,\ \|\boldsymbol{V}_i\| < V_{\max} \text{ and } \dot{\boldsymbol{V}}_i \cdot e(g_i(q) - q) \ge 0$$

$$\text{whenever } d(q, O_i) \to (\mathrm{pos}(\boldsymbol{V}_{ir} \cdot e(g_i(q) - q)))^2 / 2a_{\max}$$

In (8), the dynamic constraints of the mobile robot are considered in the design of the magnitude of the repulsive force. It can guarantee the safety of the mobile robot with the environment constraint 1 from Proposition 1. In most of the conventional APF, the repulsive force is only a function of the relative position of the mobile robot with respect to an obstacle. Hence it cannot guarantee the safety of the mobile robot in applications. From the results above, the following result is obvious.

Theorem 1. In static environments, the ACFs, based on the parameter designs in (6) and (8), can guarantee the safety of the mobile robot.

By using contradiction, it is easy to prove Theorem 1. In order to guarantee that the ACF can guarantee the safety of the mobile robot in a dynamic environment, we firstly prove the following proposition.

Proposition 2. Assume two moving obstacles O_i, O_j and the mobile robot are moving in a same line path at time t, the mobile robot is between O_i and O_j. Their velocities are $\boldsymbol{V}_i$, $\boldsymbol{V}_j$ and $\boldsymbol{V}_r$, respectively. And assume $e(\boldsymbol{V}_i) = -e(\boldsymbol{V}_j)$. Then when the two obstacles are approaching each other, *i.e.*, $d(O_i, O_j) \to \delta$ as $t \to \infty$, where δ is a small positive number, the robot will not collide with the obstacles based on the ACFs defined above.

Proof. Let $v_i = \boldsymbol{V}_{ir} \cdot e(g_i(q) - q)$, and $v_j = \boldsymbol{V}_{jr} \cdot e(\mathbf{g}_j(q) - q)$. Whenever $d(O_i, O_j) \to \delta$, we have $d(q, O_i) \to \mathrm{pos}(v_i)^2 / 2a_{\max}$, $d(q, O_j) \to \mathrm{pos}(v_j)^2 / 2a_{\max}$. According to (6,8), $\boldsymbol{F}_{ri}$ and $\boldsymbol{F}_{rj}$ are both very large for this case. According to (5,7), the attractive force and the coordinating force are both limited, and the coordinating force is orthogonal to the repulsive force at any time, thus

they both can be neglected to consider the safety problem. Then (2) can be rewritten as: $M\ddot{q} \approx -k_f \dot{q} + \boldsymbol{F}_{ri} - \boldsymbol{F}_{rj}$. Obviously, the mobile robot will track the balance point of the two repulsive fields in this case. From the assumptions of this proposition, the balance point of the repulsive forces on the line path is a safe point. This completes the proof.
According to Proposition 2, define the **Environment Constraint 2**:

$$d(O_i, O_j) > 0, \forall O_i, O_j \in O_s \cup O_d$$

In face, the effect of the coordinating forces in the case of Proposition 2 helps to guarantee the safety of the mobile robot, though it is not considered there. By far, we obtain the following result.
***Theorem* 2**. Assume the maximum velocity and acceleration of the mobile robot are V_{max} and a_{max}, respectively. The maximum radius of the sensors within which the obstacles can be effectively detected is $R>>2(V_{max})^2/a_{max}$, and assume the environment constraints 1-2 are satisfied. Then the mobile robot is safety in the ACFs using the designs in (5-8).
Proof. If there only one observable obstacle, it is the case in Proposition 1. Otherwise, any other case can be regarded as the typical case in Proposition 2 that the robot is moving between two obstacles. Hence, from Proposition 1 and 2, the mobile robot is safe in the ACFs.

3.4 Reachability of the ACFs

Reachability of the ACFs is the ability of the mobile robot using ACFs to reach its goal provided that there is a safe path from the starting point to the goal in the environment. This requires that there are no local minima in ACFs. In conventional APFs, the attractive force and repulsive force may be balanced at some points where local minima exist. These points are usually between multiple obstacles or on the opposite side of an obstacle with respect to the goal point. However, local minima at these points can be removed by properly using the coordinating forces in ACFs.

- **Using the Coordinating Force to Remove Local Minima**

Let k_{ni} in (7) be

$$k_{ni} = 1/\left(\text{pos}(\|q - \text{g}_i(q)\| - (\text{pos}(\mathbf{V}_{\mathbf{ir}} \cdot \text{e}(\text{g}_i(q) - q)))^2 / 2a_{\max})\right)^m \tag{9}$$

where $m>0$ is to be defined.
Obviously, it is easy to remove the local minima using the coordinating force if there is only one observable obstacle. As for the multiple obstacles case, the coordinating force should be designed to remove the local minima between any two obstacles in the observable region such that the mobile robot can go through the passage between any two obstacles satisfying the environment constraints 1-2.
***Definition* 1**. Curve C is an equi-repulsive-force curve between two obstacles O_i and O_j, if the following equation holds: $\forall p \in C, \|\mathbf{F}_{\mathbf{ri}}(p)\| = \|\mathbf{F}_{\mathbf{rj}}(p)\|$ (Referring to Figure 3).

***Lemma* 1**. Neglecting the effects of the attractive and coordinating forces, the mobile robot is moving on the curve C when passing a passage between two obstacles, and the following equations hold: (The proof is omitted)

$$\boldsymbol{V}_{ir} \approx \boldsymbol{V}_{jr}, d(q, O_i) \approx d(q, O_j), \ \mathrm{E_{ri}} \approx \mathrm{E_{rj}}.$$

***Proposition** 3*. Assume that the mobile robot is at point q on the curve C between obstacles O_i and O_j at time t, $d(O_i,O_j)= 2a > 0$, and the obstacles satisfy environment constraints 1-2. Neglecting the effect of the attractive force, in order for the mobile robot to pass the passage between O_i and O_j, the ACFs should satisfy the following minima-free conditions:
before passing the passage:

$$\lambda_l = \arg\min_{\lambda \in \{1,-1\}} (\angle(\lambda \cdot \mathbf{T} \cdot (g_l(q) - q), (-\mathbf{F}_{\mathbf{rj}}(q) - \mathbf{F}_{\mathbf{ri}}(q))/2)),$$

after passing the passage:

$$\lambda_l = \arg\min_{\lambda \in \{1,-1\}} (\angle(\lambda \cdot \mathbf{T} \cdot (g_l(q) - q), (\mathbf{F}_{\mathbf{rj}}(q) + \mathbf{F}_{\mathbf{ri}}(q))/2)) \text{ or } 0,$$

and

$$(k_{rc}/F(q,l)^n)\sqrt{R^2 - a^2} \le (k_{nc}/F(q,l)^m) \cdot a$$

where

$$F(q,l) = (\|q - g_l(q)\| - \mathrm{pos}(\boldsymbol{V}_{lr} \cdot \mathrm{e}(g_l(q) - q))^2 / 2a_{\max}), \ l \in \{i, j\}.$$

Proof. The repulsive and coordinating forces exerted on the mobile robot by obstacles O_i and O_j are

$$\boldsymbol{F}_r = \boldsymbol{F}_{ri} + \boldsymbol{F}_{rj} = \frac{k_{rc}(E_{ri} + \varepsilon_1)}{F(q,i)^n} \cdot \frac{q - g_i(q)}{\|q - g_i(q)\|} + \frac{k_{rc}(E_{rj} + \varepsilon_1)}{F(q,j)^n} \cdot \frac{q - g_j(q)}{\|q - g_j(q)\|},$$

$$\boldsymbol{F}_n = \boldsymbol{F}_{ni} + \boldsymbol{F}_{nj} = \frac{k_{nc}(E_{ri} + \varepsilon_2)}{F(q,i)^m} \cdot \frac{q - g_i(q)}{\|q - g_i(q)\|} + \frac{k_{nc}(E_{rj} + \varepsilon_2)}{F(q,j)^m} \cdot \frac{q - g_j(q)}{\|q - g_j(q)\|},$$

According to Lemma 1, in order for the robot to pass the passage along Curve C, the following equations hold: $F(q,i) \approx F(q,j)$, $E_{ri} \approx E_{rj}$. Let $\theta = \angle(\mathrm{e}(q - g_i(q)), \mathrm{e}(q - g_j(q)))$, and according to the cosine lemma in a triangle and the principle to choose λ given in the proposition, we have

$$\|\mathbf{F}_\mathbf{r}\| = 2 \cdot (\bar{k}_{rc}/F^n)^2 + 2 \cdot (\bar{k}_{rc}/F^n)^2 \cos\theta,$$

$$\|\mathbf{F}_\mathbf{n}\| = 2 \cdot (\bar{k}_{nc}/F^m)^2 + 2 \cdot (\bar{k}_{nc}/F^m)^2 \cos(\pi - \theta) = 2 \cdot (\bar{k}_{nc}/F^m)^2 - 2 \cdot (\bar{k}_{nc}/F^m)^2 \cos\theta.$$

where, $\bar{k}_{rc} = k_{rc}(E_{ri} + \varepsilon_1)$, $\bar{k}_{nc} = k_{nc}(E_{ri} + \varepsilon_2)$, $F = F(q,i)$. Neglecting $\boldsymbol{F}_a$, in order for the robot to pass the passage, only if $\|\boldsymbol{F}_r\| \le \|\boldsymbol{F}_n\|$ holds. Hence, we have

$$(\bar{k}_{rc}/F^n)^2 (1 + \cos\theta) \le (\bar{k}_{nc}/F^m)^2 (1 - \cos\theta),$$

(this inequality is denoted as p1). Due to the maximum detecting radius of the sensors is R, and min($\left\| g_i(q) - g_j(q) \right\|$)=$2a$, applying the cosine lemma to the minimum θ in the triangle (q,$g_i(q)$,$g_j(q)$), we can obtain $R^2+R^2-2R^2\cos\theta=(2a)^2$, which further yields: $\cos(\theta)_{max}=1-(\sqrt{2}a/R)^2$. Substitute it into (p1), and let $\varepsilon_1 = \varepsilon_2$, we obtain

$$(k_{rc}/F^n)\sqrt{R^2 - a^2} \le (k_{nc}/F^m)\cdot a.$$

This completes the proof.
Let m=n, according to Proposition 3 we have

$$k_{nc}/k_{rc} \ge \sqrt{(\tfrac{R}{a})^2 - 1}.$$

Note that the smaller a is, the larger is k_{nc}/k_{rc} in this case. It is consistent with the practical fact. If the robot is a circle with radius r, then the corresponding condition should be:

$$k_{nc} \ge k_{rc}\sqrt{\tfrac{R}{r}(\tfrac{R}{r} + 2)}.$$

This is verified in the simulations. Proposition 3 provides a theoretical view point to the design of the ACFs' parameters, though some assumptions are strict. In fact, the minima-free conditions in Proposition 3 are just sufficient, since the attractive force is neglected in the proof.

- **Online Decision Making Based on Coordinating Factors**

In order to remove the local minima between multiple obstacles in uncertain dynamic environments, the coordinating factors with respect to different obstacles should be properly decided on line such that the coordinating forces can provide actuating forces to the mobile robot to balance the repulsive forces. On the other hand, the wall-following behavior (Lumelshy and Skewis 1999) should be adopted when the mobile robot meets a large obstacle of even nonconvex shape such that the robot can follow the boundary of the obstacle to go until the robot can directly find in free space the direction in which the goal exists. In this case, the coordinating force is used directly as the actuating force for the robot to follow the "wall". For this purpose, the coordinating factors should also be properly decided on line.

The decision making of λ is based on the decision making of the local sub-goal in the observable region of the mobile robot. The local sub-goal (Xu et al 1998) is denoted by $\boldsymbol{e}_{ds}$, which should be an appropriate tradeoff between the collision-avoidance behavior and the going-to-goal behavior. In this study, the mobile robot is expected to avoid an obstacle along the shortest path in local environment. For a static obstacle, the wall-following behavior should be able to be generated; and for a moving obstacle, the robot is expected to run away from the trajectory of the obstacle as fast as possible. To these aims, the local sub-goal is decided as follows with respect to an obstacle O:

$$\text{If } \left\|V_O\right\| \le V_O: \ \boldsymbol{e}_{ds} = \mathrm{e}(\boldsymbol{F}_a) + \kappa\, \mathrm{e}(\boldsymbol{V}_r) \quad (10a)$$

$$\text{else} \quad \boldsymbol{e}_{ds} = -\mathrm{e}(\boldsymbol{V}_O) + \kappa\, \mathrm{e}(\boldsymbol{V}_r) \quad (10b)$$

where V_r and V_O are the velocities of the mobile robot and the obstacle, respectively, $\kappa>0$, V_O is a constant. Then the optimal decision-making of the coordinating factor with respect to the obstacle O is

$$\lambda=\arg(\underset{\lambda\in\{1,-1\}}{\angle}(\lambda\cdot\boldsymbol{T}\cdot(g_O(q)-q),\boldsymbol{e}_{ds})\leq 90^0) \tag{10c}$$

If the velocity of an obstacle is lower than a constant V_O, then it can be regarded as a static obstacle. (10a) is a tradeoff between the going-to-goal behavior and the collision-avoidance behavior, and (10b) provides such a sub-goal that the robot is expected to avoid a moving obstacle as fast as possible. The angle between the optimal direction of the coordinating force and the local sub-goal is less than 90^0 such that the coordinating force can provide an actuating force to the mobile robot. It should be noted that different coordinating factor may correspond to a different motion behavior, the desired motion behavior of the mobile robot is basically determined by the optimal decision-making of the coordinating factors. It can also be verified that the coordinating factor decided by (10c) is consistent with the minima-free conditions in Proposition 3.

- **Realization of the Wall-Following Behavior and no Local Minima**

In order to realize the wall-following behavior with respect to an obstacle, the coordinating factor should be kept constant once the mobile robot meets the obstacle. To show that (10c) can provide a consistent coordinating factor with respect to an obstacle, we have the following results.

Fact 1. Assume the boundary ∂O of an obstacle O is differentiable. q^* is a point outside of O. $\forall p\in\partial O$, $t(p)$ is a tangent line at this point. $e_r(t(p))$ is the unitary vector of the tangent line at point p in anticlockwise, and $e_l(t(p))$ is in clockwise. Then we have $\angle(\boldsymbol{e}_s(t(p)),\boldsymbol{e}_d)\leq 90^0$, where $\boldsymbol{e}_d$=e(q^*-p)+$\boldsymbol{e}_s(t(p))$, $s=r$ or l.

Define **Environment Constraint 3**:

All the obstacles are convex, and their boundaries are one-time differentiable.

Proposition 4. All the obstacles satisfy the environment constraints 1-3. Let $f_r=\boldsymbol{F}_a\mathrm{e}(\boldsymbol{F}_{ri})$, $f_n=\boldsymbol{F}_a\mathrm{e}(\boldsymbol{F}_{ni})$. If choose k_{nc} in (7) such that $\boldsymbol{F}_{ni}\mathrm{e}(\boldsymbol{F}_{ni})+f_n>0$, then wall-following behavior can be generated based on (10) once the robot meets the obstacle O_i in the case $f_r<0$.

Proof. Due to the obstacle O_i is convex, its boundary ∂O_i can be classified into two parts according to the sign of f_r. In the side of $f_r\geq 0$, the angle between the repulsive and attractive forces is less than 90^0, the mobile robot can run away from the ACF of the obstacle quickly. Hence the wall-following behavior is unnecessary in this case. And for the case $f_r<0$, the angle between the repulsive and attractive forces is more than 90^0. In the latter case, it can be regarded that f_r=-$\boldsymbol{F}_r\mathrm{e}(\boldsymbol{F}_r)$, and then the planning equation in (2) can be rewritten as:

$$M\,\ddot{q}+K_f\,\dot{q}=\boldsymbol{F}_a+\boldsymbol{F}_{ri}+\boldsymbol{F}_{ni}=\left[f_r\cdot e(\boldsymbol{F}_{ri})+\boldsymbol{F}_{ri}\right]+\left[f_n\cdot e(\boldsymbol{F}_{ni})+\boldsymbol{F}_{ni}\right]=f_n\cdot e(\boldsymbol{F}_{ni})+\boldsymbol{F}_{ni},$$

Obviously, the velocity of the mobile robot is basically determined by the coordinating force, and it finally converges to $\dot{q}=\left(f_n\cdot\mathrm{e}(\boldsymbol{F}_{ni})+\boldsymbol{F}_{ni}\right)/K_f$. If $\boldsymbol{F}_{ni}\mathrm{e}(\boldsymbol{F}_{ni})+f_n>0$ holds, then we have e($\dot{q}$)=e($\boldsymbol{F}_{ni}$). Now utilizing Fact 1, e($\boldsymbol{F}_{ni}$) can be regarded as the direction of the tangle line of the obstacle boundary, *i.e.*, $e_r(t(p))$ or $e_l(t(p))$. If $\|\boldsymbol{V}_i\|\leq V_O$, the sub-goal decided by (10a) is equivalent to $\boldsymbol{e}_d$ in Fact 1 with $q^*=q_d$, otherwise, $q^*=p-\boldsymbol{V}_i$. In both cases, we always have $\angle(e(\boldsymbol{F}_{ni}), \boldsymbol{e}_d) \le 90^0$. That is, if the current coordinating factor $\lambda = 1$ (or -1) with respect to obstacle O_i, then $\lambda = 1$ (or -1) still holds according to (10c) for the next time. The wall-following behavior is generated for this case. This completes the proof.

Further study can show that the wall-following behavior can also be realized for a non-convex obstacle based on (10), this was discussed in Jing and Wang (2004). The repulsive force may prevent the mobile robot to reach its goal if the goal point is very near to an obstacle. To overcome this problem, we let (Wong and Spetsakis 2000) (with respect to a static obstacle)

$$k_{rc} = k_1 \cdot \min\left(\|q - q_d\|^3, \|q - q_d\|^{k_2}\right) \tag{11}$$

where $k_1>0$, $0<k_2\le 1$ are both constants.

To achieve minima-free ACFs, define **Environment Constraint 4**:

For any two obstacles O_i, O_j, let $\boldsymbol{e}_d=(e(q_d-p)+\boldsymbol{e}_s(t(p)))/2$, $p\in \partial O_i \cap P(q)$. Considering the case $g_j(p) \in \partial O_j \cap P(q)$, if $s=l$, clock($e(g_j(p)-p),\boldsymbol{e}_d$)= 1 holds, and if $s=r$, clock($e(g_j(p)-p),\boldsymbol{e}_d$)= -1 holds. Where, clock($\boldsymbol{a},\boldsymbol{b}$) is defined as: it is -1 if $\boldsymbol{b}$ can be obtained by rotating $\boldsymbol{a}$ with the angle $\angle(\boldsymbol{a},\boldsymbol{b})$, otherwise, it is 1.

With the deliberate designs above, it can be seen that the reachibility of the robot can be guaranteed under the environmental contraints 1-4. Due to the attractive force, the mobile robot is always approaching an obstacle that is on the line jointing the current position of the mobile robot and its goal. If the mobile robot meets a static obstacle O_i, then according to Proposition 4, the wall-following behavior is generated. During the wall-following with respect to this static obstacle O_i, if the mobile robot meets another moving obstacle satisfying the environment constraints 1-4, then according to (10) and Proposition 3, the mobile robot either runs away from the trajectory of the moving obstacle as fast as possible, or passes a passage between two obstacles to avoid the moving obstacle. And then, the mobile robot comes back again to the state following the boundary of an obstacle that is on its desired shortest path to the goal. If the mobile robot meets a static obstacle O_j during it is following the boundary of the static obstacle O_i in anti-clockwise (*i.e.*, λ_i =1). Due to the environment constraint 4 and according to Proposition 4, we can have clock($e(\boldsymbol{F}_{rj}(q)),\boldsymbol{e}_d$)= 1 and $\boldsymbol{e}_d=(e(q_d-q)+e(\boldsymbol{V}_r))/2$. Utilizing (10) again, we have $\angle(\boldsymbol{F}_{nj}(q),\boldsymbol{e}_d) \le 90^0$, then clock($e(\boldsymbol{F}_{rj}),e(\boldsymbol{F}_{nj})$)=1 must hold in this case, *i.e.*, λ_j =-1 with respect to O_j. It is easy to verify that, the minima-free conditions in Proposition 3 are satisfied in this case. That is, the mobile robot can pass the passage between O_i and O_j, and the wall-following behavior can be kept during passing this passage. If the mobile robot follows the boundary of O_i in clockwise, the same conclusion can be made. After the mobile robot avoids obstacle O_i completely, it may meet another obstacle and then the similar process as above is carried out again due to the actuation of the attractive force, until it reaches the goal finally.

4. Motion Planning of the Mobile Robot in Uncertain Dynamic Environments

Assume the goal of the mobile robot is known. The motion planning problem can be written as: *To find the optimal* $\boldsymbol{u}(t)$*, i.e.,*

$$\mathbf{u} = \arg\min_{\mathbf{u}\in U_u}(J(\mathbf{u}))$$

(in the following algorithm, it is transformed to be the optimal decision making equation (10) for λ*, where* U_u *is the decision making space of* **u** *satisfying the dynamic constraints) such that the mobile robot can go safely from its stating point* q_0 *to its goal point* q_d*, i.e.,*

$$(1)\ \forall i,\ q \cap O_i = \Phi,\ (2)\ \exists T < \infty,\ \lim_{t\to T} q(t) = q_d\ .$$

Considering the dynamic constraints of mobile robots, the states of (3,4) should be subjected to the saturation constraints of the velocity and acceleration (Jing and Wang 2002). For (3,4), we use the following control law:
If there is no observable obstacle

$$\mathbf{u} = K_f \begin{bmatrix} \dot{x}_d \\ \dot{y}_d \end{bmatrix} \tag{12a}$$

where

$$\begin{bmatrix} \dot{x}_d \\ \dot{y}_d \end{bmatrix} = V_d \cdot \begin{bmatrix} x_d - x \\ y_d - y \end{bmatrix} \Big/ \left\| \begin{matrix} x_d - x \\ y_d - y \end{matrix} \right\|,$$

$$V_d = \min\left(\alpha \cdot V_{\max}, \sqrt{2 \cdot \left\| \begin{matrix} x_d - x \\ y_d - y \end{matrix} \right\| \cdot a_{\max} \cdot \beta}\right),$$

$$0<\alpha,\beta\le 1,\ K_a=0,\ K_{ri}=0, K_{ni}=0,\ (i=1,2\ldots),\ K_f>0.$$

If there are observable obstacles, let $\dot{q}_d = 0$ and

$$\boldsymbol{u} = K_a \begin{bmatrix} x_d \\ y_d \end{bmatrix} - \sum_{i\in Od\cup Os} K_{ri}\, g_i(q) + \sum_{i\in Od\cup Os} K_{ni} \lambda_i \cdot \boldsymbol{T} \cdot g_i(q) \tag{12b}$$

where, $K_a, K_{ri}, K_{ni}, \lambda_i$ are chosen according to (5-11). The outputs of the planning equation are the desired behavior for the mobile robot to take.

5. Simulations

The planning equation (3) is used in the simulations. The parameters of the mobile robot are set as follows: its radius is r=0.3m, the maximum acceleration a_{max}=0.5 m/s^2, the maximum velocity V_{max}=0.5m/s, the maximum detecting radius of the sensors is R=1.5m>2$(V_{max})^2/a_{\max}$.

The parameters of the ACF are chosen as follows:

Step1. For the parameters of the attractive force (in (5)), let k_a=1, M_a =4. If these parameters are set to be too large, they may affect the safety of the robot.

Step2. For the parameters of the repulsive force (in (6)(11)), let n=2 , ε_1=0.05, $k_{rc} = k_1 \cdot \min\left(\|q - q_d\|^3, \|q - q_d\|^{k_2}\right)$, k_1=0.5□k_2=0.5. The repulsive force should be much larger than the attractive force within the minimum safe radius predefined for the robot.

Step3. For the parameters of the coordinating force (in (7)), let ε_2 =0.05, M_n =200, k_{nc}=6k_r m=2 (in Proposition 3). Based on the parameters chosen for the repulsive force, these parameters for the coordinating force are chosen basically according to Proposition 3.

Step4. For collision risk, let k_{risk1}=0.1, k_{risk2}=0.9. They are chosen according to different inclinations.

Step5. For the decision making of the coordinating factor(in (10)), let V_O=0.2m/s, κ =2. The larger κ is, the larger is the impact of the current velocity on the sub-goal, which further affects the trajectory of the mobile robot.

Step6. For the parameters for the control law(in (12)), let α =1, β =0.5, K_f=10. The larger K_f is, the larger is the damp of the planning equation.

- **The ACFs can reduce oscillations**

Figure 4 is the result using the conventional APFs, and the results of the ACFs are given in Figure 5. The coordinating forces can exert an actuating force to the mobile robot with proper decision making of the coordinating factors, and all the virtual forces in ACFs are proportional to the collision risk. Hence, the ACFs can effectively reduce the oscillation on the trajectory between multiple obstacles. However, the oscillation of "S" shape exists on the trajectory planned by the conventional APFs based methods. It should also be noted that the velocity and acceleration planned for the mobile robot by ACFs are both satisfied with the dynamic constraints.

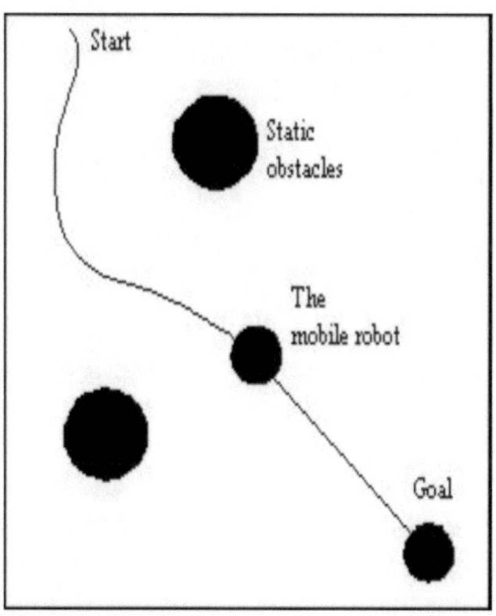

Figure 4. Results of the conventional APFs

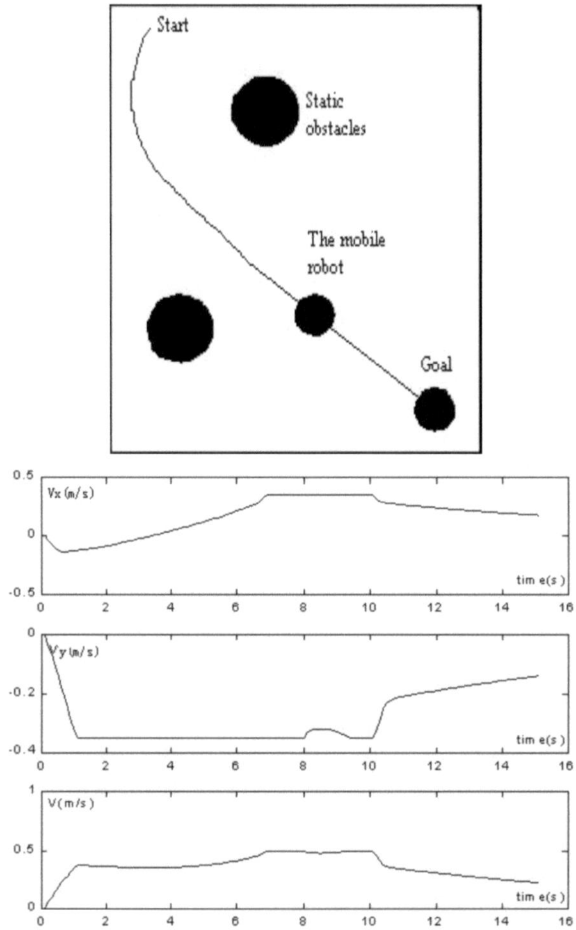

Figure 5. Results of the ACFs

- **The ACFs can improve the autonomy and intelligence of the mobile robot and remove local minima when meeting obstacles and other robots**

The moving obstacle is assumed to be a circle with radius 0.35m and velocity 0.35m/s. A simulation process is in Figure 6 (A-H). In Figure 6, Ri denotes a robot i, Oj denotes a dynamic unknown obstacle j, and others are static obstacles. A line between the current position of a robot and its goal indicates the desired direction. A ray on dynamic obstacle indicates its moving direction. In figure A and B, R3 meets R5, by anti-clockwise rotating they avoid collision with each other, and obviously it is ideal for a shorter collision-free path. In Figure A and B, R1 meets a large static obstacle, wall-following behavior is used. In Figure C, R1 passes a passage between two static obstacles, in conventional APF there may be local minima which will prevent the robot passing. When the robot meets dynamic obstacles, coordinating force can make the collision-avoidance behavior of the robot more intentionally and effectively. See it in Figure C, if no coordinating force, R1 might be pushed back, but in fact from its trajectory in Figure H we can note that a turning-left behavior occurred due to the coordinating force, which makes the motion more effective and rational. In other Figures, we can also see such effective and intelligent collision-avoidance behaviors.

More discussions about this subject can also be referred to Jing and Wang (2003), Jing et al (2004c) and Jing (2004).

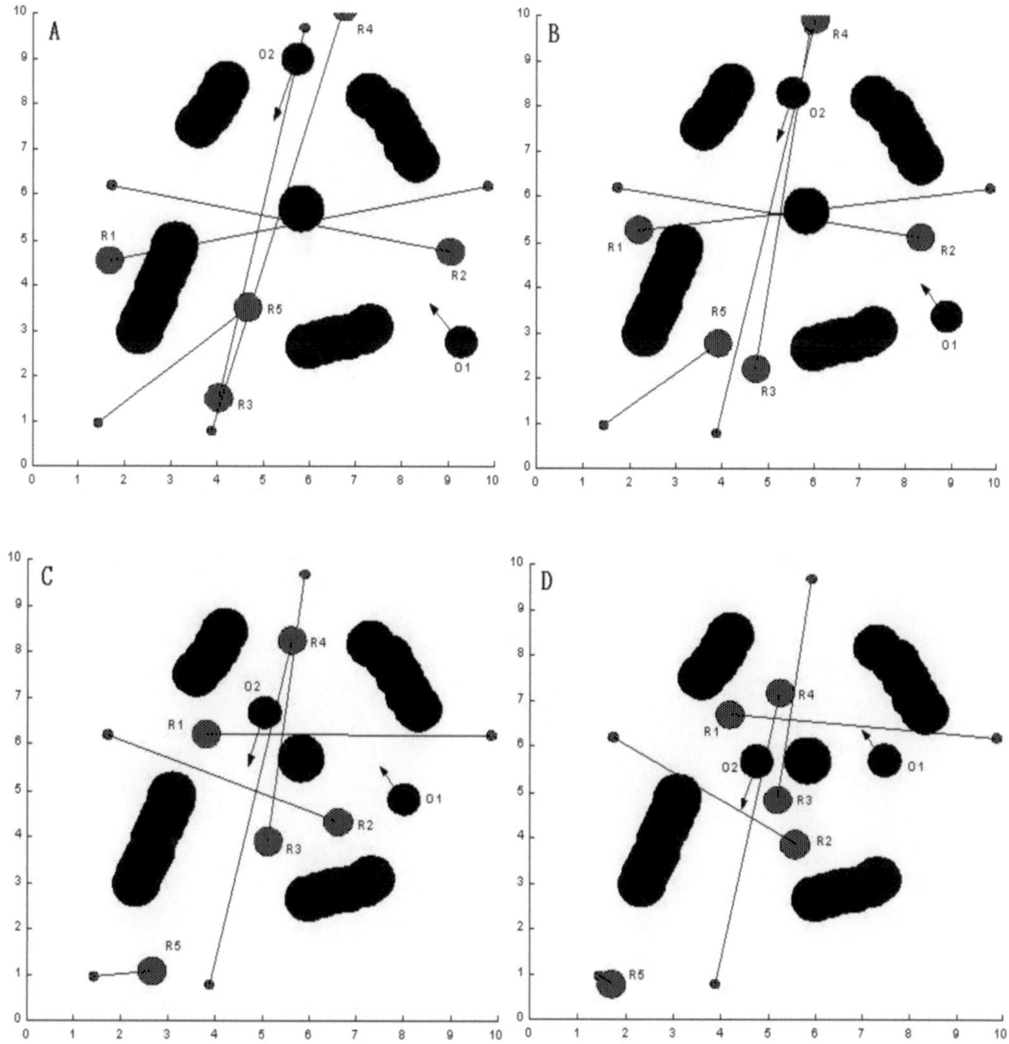

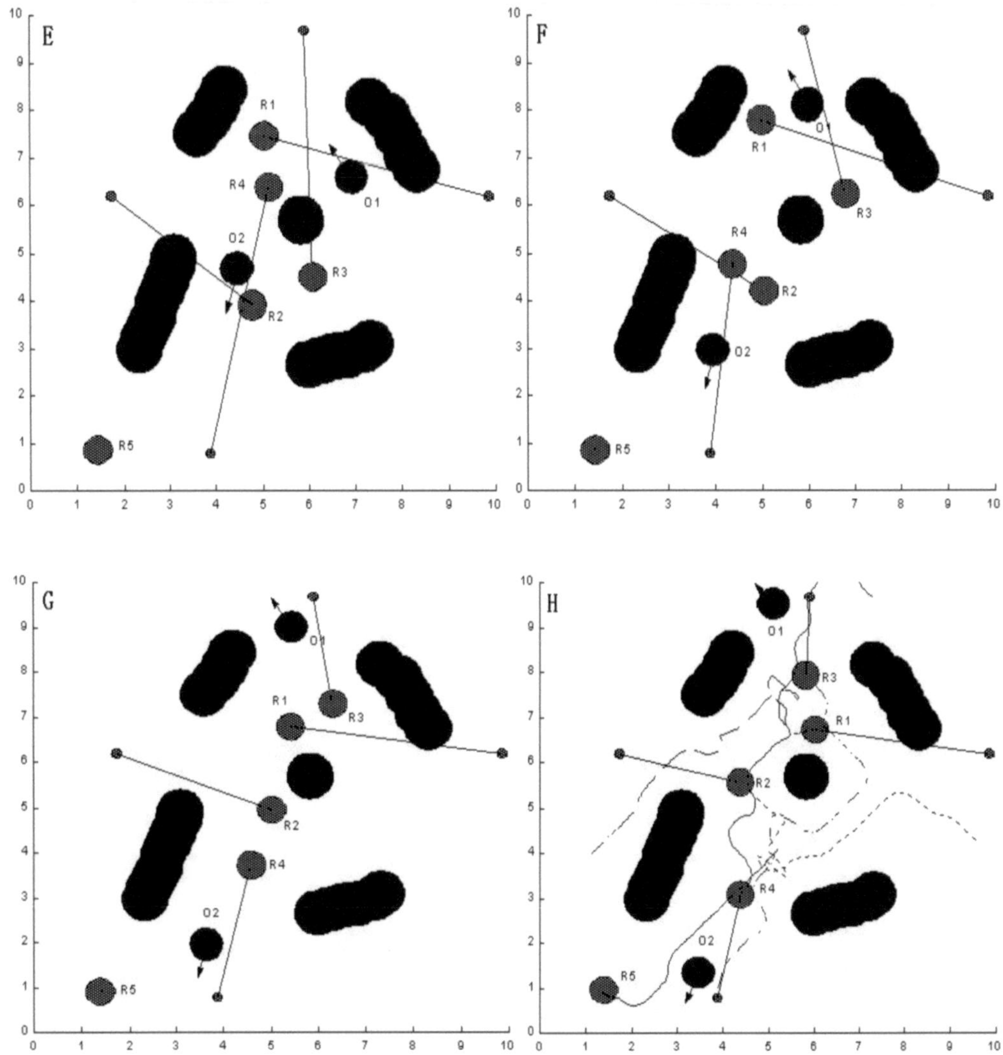

Figure 6. A simulation process (A-H)

6. Conclusions

In order to overcome some noticeable drawbacks of the traditional APF based methods such as local minima and oscillations on the planned trajectory, an artificial coordinating field was proposed recently (Jing et al 2002, 2003, 2004abc). This chapter provides a simple introduction for these newly developed results. A coordinating force is added to the conventional APF which is orthogonal to the repulsive force, and the field parameters are designed with consideration of the states and task of mobile robots under different enviromental situations. These enable the ACF to be more robust and effective for behavior decisions of mobile robots and adaptable to the change of environments when there are different intelligent and unintelligent obstacles. Local minima and unnecessary oscillation in planned trajectory can be avoided. More intelligent coordination between different mobile robots in obstacled environments can also be achieved.

There are three principles in designing the ACFs:

a. All the virtual forces are functions of the motion purpose and relative states of the mobile robot with respect to the local environments,
b. The repulsive force should satisfy the dynamic constraints of the mobile robot,
c. The coordinating force should satisfy the minima-free conditions.

Based on these designs, the ACFs are adaptable to environments and controllable for robots. The ACFs based motion planning can guarantee the safety and reachability under certain environment constraints. Since more information of the robot and environment can be represented, the ACFs are more robust. In the local dynamic coordinates defined in Section 2, the ACF has full-dimensional forces, instead of one-dimensional force in the conventional APF. This is the most important and fundamental difference between ACF and APF, and the conventional APF is just a special case of the ACF.

7. References

Ge S.S. and Cui Y.J (2000). New potential functions for mobile robot path planning. *IEEE trans. Robot. and Automat.* Oct. 16(5): 615-620

Guldner J and Utkin V. I. (1995) Sliding mode control for gradient tracking and robot navigation using artificial potential fields. *IEEE Trans. Robot. and Automat*, April, 11(2):247-253

Jing X.J. (2004), *Stability and planning of internet-based tele-robotic systems* (in Chinese). PhD thesis in the Graduate School of Chinese Academy of Sciences, pp 72-101

Jing X.J. (2005), Behavior dynamics based motion planning of mobile robots in uncertain dynamic environments, *Robotics and Autonomous Systems* 53, 99–123

Jing X. J., Wang Y.C (2002). Artificial Coordinating Field (ACF) Control of Mobile Robot for Collision Avoidance and Navigation. *Proceedings of international conference on Control and Automation*, Xiamen, China: Xiamen University Press, June, pp2478-2482

Jing X. J., Wang Y. (2003), Artificial Coordinating Field based Coordinating Collision Avoidance, in: *IEEE International Conference on Robotics, Intelligent Systems and Signal processing*, Changsha, Hunan, China, pp. 126–130.

Jing X. J. and Wang Y.C. (2004a), Local minima-free design of artificial coordinating fields, *Journal of Control Theory and Applications*, Vol 2, No 4, pp371-380

Jing X. J., Wang Y., Tan D. (2004b), Artificial Coordinating Field and its application to motion planning of robots in uncertain dynamic environments, *Science in China, Ser. E Eng. Mater. Sci.*(both in Chinese and English) 47 (5), 577–594.

Jing X. J., Wang Y., Tan D. (2004c), Artificial coordinating field based real-time collision avoidance planning of multiple mobile robots, *Contr. Theor. Appl.* (in Chinese) 21 (5), 757–764.

Kant K. and Zucher S (1988). Planning collision free trajectories in time-varying environments: a two level hierarchy. *Proc. IEEE Inter. Conf. Robotics and Automation*, pp1644-1649

Khatib O. (1986). Real time obstacle avoidance for manipulators and mobile robots. *International Journal of Robotics Research*, 5(1):90-98

Koren Y., Borenstein J. (1991). Potential field methods and their limitations for mobile robot navigation. *Proc. IEEE Inter. Conf. Robot. and Automat. Sacramento,* California: International House, April 7-12: 1398-1404

Krogh B (1984). A generalized potential field approach to obstacle avoidance control. *In SME Conf. Proc. Robotics Research: The Next Five Years and Beyond,* Bethlehem, Pennsylvania: International house, pp 950-955

Louste C. and Liegeois A (2000). Near optimal robust path planning for mobile robots: the viscous fluid method with friction. *Journal of intelligent and robotic systems,* 27: 99-112

Lumelshy V., Skewis T.(1999) Incorporating range sensing in the robot navigation function. *IEEE Trans. Syst. Man and Cyber.* 30: 2749-1296

Noborio H., Yoshiolka T., Hamaguchi T (1995). On-line deadlock-free path planning algorithms by means of a sensor-feedback tracing. *Proc. IEEE Inter. Conf. Syst. Man and Cyber.* Vancouver: International House, Oct. 22-25:1291-1296

Masoud S. A. and Masoud A. A. (2000) Constrained motion control using vector potential fields. *IEEE trans. on Systems, Man, and Cybernetics – Part A: Systems and Humans,* May, 30(3): 251-272

Medio C. D. and Oriolo G.(1999) Robot obstacle avoidance using vortex fields. In *Advance in robot kinematics,* S.Stifter and J.Lenatc(Eds), Springer-Verlag, Wien, pp227-235

Prassler E. (1999) Navigating a robotic wheelchair in a railway station during rush hour. *The international journal of robotics research,* 8(7): 711-727

Rimon E. and Koditschek D. (1991). The construction of analytic diffeomorphisms for exact robot navigation on star worlds. *Transactions of the American Mathematical Society,* 327(1):71-115.

Rimon E. and Koditschek D.E. (1992) Exact robot navigation using artificial potential functions. *IEEE trans. Robot. and Automat.* Vol.8, no.5, pp501-518.

Salichs M.A. and Moreno L. (2000). Navigation of mobile robots: open questions. *Robotica,* 18:227-234.

Satoh K.(1993) Deadlock-free motion planning using the laplace potential field. *Adv. Robotics,* 7(5):449-461

Singh L., Wen J., and Stephanou H (1997). Motion planning and dynamic control of a linked manipulator using modified magnetic fields. *Proc. IEEE int. conf. on robotics and automation,* Albuquerque, NM: International House, pp1142-1147

Tsourveloudis N. C., Valavanis K. P. and Hebert T (2001). Autonomous vehicle navigation utilizing electrostatic potential fields and fuzzy logic. IEEE *trans. On Robotics and Automation,* Aug, 17(4): 490-497

Wong B and Spetsakis M (2000) Scene reconstruction and robot navigation using dynamic fields. *Autonomous Robots,* 8: 71-86

Xu W.L., Tso S.K., Fung Y.H (1998). Fuzzy reactive control of a mobile robot incorporating a real/virtual target switching strategy. *Robotics and Autonomous Systems,* 23:171-186

11

Minimum-Energy Motion Planning for Differential-Driven Wheeled Mobile Robots

Chong Hui Kim[1] and Byung Kook Kim[2]

[1]*Agency for Defense Development,*

[2]*School of Electrical Engineering and Computer Science, KAIST*

Republic of Korea

1. Introduction

With the remarkable progresses in robotics, mobile robots can be used in many applications including exploration in unknown areas, search and rescue, reconnaissance, security, military, rehabilitation, cleaning, and personal service. Mobile robots should carry their own energy source such as batteries which have limited energy capacity. Hence their applications are limited by the finite amount of energy in the batteries they carry, since a new supply of energy while working is impossible, or at least too expensive to be realistic. ASIMO, Honda's humanoid robot, can walk for only approximately 30 min with its rechargeable battery backpack, which requires four hours to recharge (Aylett, 2002). The BEAR robot, designed to find, pick up, and rescue people in harm's way, can operate for approximately 30 min (Klein et al., 2006). However, its operation time is insufficient for complicated missions requiring longer operation time. Since operation times of mobile robots are mainly restricted by the limited energy capacity of the batteries, energy conservation has been a very important concern for mobile robots (Makimoto & Sakai, 2003; Mei et al., 2004; Spangelo & Egeland, 1992; Trzynadlowski, 1988; Zhang et al., 2003). Rybski et al. (Rybski et al., 2000) showed that power consumption is one of the major issues in their robot design in order to survive for a useful period of time.

Mobile robots usually consist of batteries, motors, motor drivers, and controllers. Energy conservation can be achieved in several ways, for example, using energy-efficient motors, improving the power efficiency of motor drivers, and finding better trajectories (Barili et al., 1995; Mei et al., 2004; Trzynadlowski, 1988; Weigui et al., 1995). Despite efficiency improvements in the motors and motor drivers (Kim et al., 2000; Leonhard, 1996), the operation time of mobile robots is still limited in their reliance on batteries which have finite energy. We performed experiments with mobile robot called Pioneer 3-DX (P3-DX) to measure the power consumption of components: two DC motors and one microcontroller which are major energy consumers. Result shows that the power consumption by the DC motors accounts for more than 70% of the total power. Since the motor speed is largely sensitive to torque variations, the energy dissipated by a DC motor in a mobile robot is critically dependent on its velocity profile. Hence energy-optimal motion planning can be achieved by determining the optimal velocity profile and by controlling the mobile robot to follow that trajectory, which results in the longest working time possible.

The total energy drawn from the batteries is converted to mechanical energy by driving motors, which is to induce mobile robot's motion with some losses such as armature heat dissipation by the armatures in the motors. The DC motor is most widely used to produce mechanical power from electric power. It converts electric power into mechanical power during acceleration and cruise. Moreover, during deceleration, mechanical energy can be converted back to electrical energy (Electro-Craft, 1977). However, the motor is not an ideal energy converter, due to losses caused by the armature resistance, the viscous friction, and many other loss components. Many researchers have concentrated on minimizing losses of a DC motor (Trzynadlowski, 1988; Angelo et al., 1999; Egami et al., 1990; El-satter et al., 1995; Kusko & Galler, 1983; Margaris et al., 1991; Sergaki et al., 2002; Tal, 1973). They developed cost function in terms of the energy loss components in a DC motor in order to conserve limited energy. The loss components in a DC motor include the armature resistance loss, field resistance loss, armature iron loss, friction and windage losses, stray losses, and brush contact loss. Since it is difficult to measure all the parameters of the loss components, its implementation is relatively complex. To overcome this problem, some researches considered only the armature resistance loss as a cost to be minimized (Trzynadlowski, 1988; Tal, 1973; Kwok & Lee, 1990). However, loss-minimization control is not the optimal in terms of the total energy drawn from the batteries.

Control of wheeled mobile robot (WMR) is generally divided into three categories (Divelbiss & Wen, 1997).

- *Path Planning*: To generate a path off-line connecting the desired initial and final configurations with or without obstacle avoidance.
- *Trajectory Generation*: To impose a velocity profile to convert the path to a trajectory.
- *Trajectory Tracking*: To make a stable control for mobile robots to follow the given trajectory.

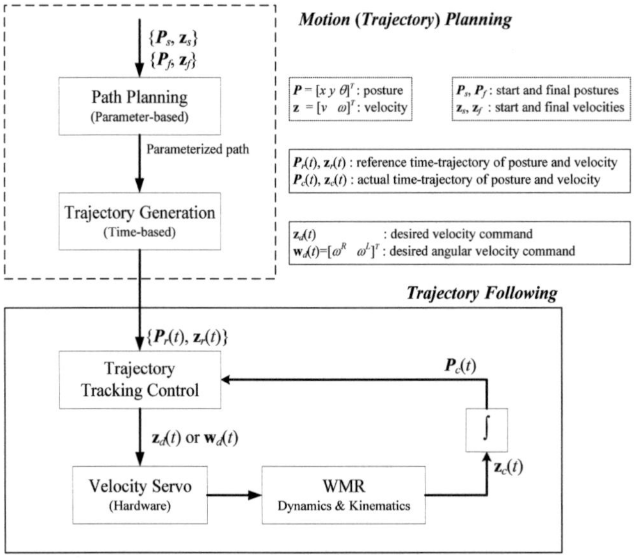

Figure 1. Traditional overall scheme of WMR control

Trajectory means a time-based profile of position and velocity from start to destination while paths are based on non-time parameters. Fig. 1 shows the overall control architecture

of WMR system (Choi, 2001). Finding a feasible trajectory is called trajectory planning or motion planning (Choset et al., 2005).
Trajectory planning (motion planning) is a difficult problem since it requires simultaneously solving the path planning and velocity planning (trajectory generation) problems (Fiorini & Shiller, 1998). Most of the paths of WMR consist of straight lines and arcs. The pioneering work by Dubins (Dubins, 1957) and then by Reeds and Shepp (Reeds & Shepp, 1990) showed that the shortest paths for car-like vehicle were made up of straight lines and circular arcs. Since these paths generate discontinuities of curvature at junctions between line and arc segment, a real robot would have to stop at each curvature discontinuity. Hence frequent stops and turnings cause unnecessary acceleration and deceleration that consume significant battery energy. In order to remove discontinuity at the line-arc transition points, several types of arcs have been proposed. Clothoid and cubic spirals provide smooth transitions (Kanayama & Miyake, 1985; Kanayama & Harman, 1989). However, these curves are described as functions of the path-length and it is hard to consider energy conservation and dynamics of WMR. Barili et al. described a method to control the travelling speed of mobile robot to save energy (Barili et al., 1995). They considered only straight lines and assumed constant acceleration rate. Mei et al. presented an experimental power model of mobile robots as a function of constant speed and discussed the energy efficiency of the three specific paths (Mei et al., 2004; Mei et al., 2006). They did not consider arcs and the energy consumption in the transient sections for acceleration and deceleration to reach a desired constant speed.
In this book chapter, we derive a minimum-energy trajectory for differential-driven WMR that minimizes the total energy drawn from the batteries, using the actual energy consumption from the batteries as a cost function. Since WMR mainly moves in a straight line and there is little, if any, rotation (Barili et al., 1995; Mei et al., 2005), first we investigate *minimum-energy translational trajectory generation problem* moving along a straight line. Next we also investigate *minimum-energy turning trajectory planning problem* moving along a curve since it needs turning trajectory as well as translational trajectory to do useful actions. To demonstrate energy efficiency of our trajectory planner, various simulations are performed and compared with loss-minimization control minimizing armature resistance loss. Actual experiments are also performed using a P3-DX mobile robot to validate practicality of our algorithm.
The remainder of the book chapter is organized as follows. Section 2 gives the kinematic and dynamic model of WMR and energy consumption model of WMR. In Section 3, we formulate the minimum-energy translational trajectory generation problem. Optimal control theory is used to find the optimal velocity profile in analytic form. Experimental environment setup to validate simulation results is also presented. In Section 4, we formulate the minimum-energy turning trajectory planning problem and suggest iterative search algorithm to find the optimal trajectory based on the observation of the cost function using the solution of Section 3. Finally, we conclude with remarks in Section 5.

2. WMR Model

2.1 Kinematic and Dynamic Model of WMR

It is well known that a WMR is a nonholonomic system. A full dynamical description of such nonholonomic mechanical system including the constraints and the internal dynamics can be found in (Campion et al., 1991). Yun (Yun, 1995; Yun & Sarkar, 1998) formulated a dynamic system with both holonomic and nonholonomic constraints resulting from rolling contacts into the standard control system form in state space. Kinematic and dynamic modeling of WMRs has been addressed by several researches. A systermatic procedure for kinematic model derivation can be found in (Alexander & Maddocks, 1989; Muir & Neuman, 1987). Campion et al. (Campion et al., 1996) have given a general and unifying presentation of the modeling issue of WMR with an arbitrary number of wheeles of various types and various motorizations. They have pointed out the structural properties of the kinematic and dynamic models taking into account the restriction to the robot mobility induced by constraints.

Unlike car-like robot (Jiang et al., 1996; Laumond et al., 1994; Laumond et al., 1998), we assumed that a WMR has a symmetric structure driven by two identical DC motors, as shown in Fig. 2. Define the posture (position x, y and orientation θ) as $\boldsymbol{P}(t)=\left[x(t) \quad y(t) \quad \theta(t)\right]^T$, the translational velocity of a WMR as v, and its rotational velocity as ω. Then the WMR's kinematics is defined by

$$\begin{bmatrix} \dot{x} \\ \dot{y} \\ \dot{\theta} \end{bmatrix} = \mathbf{T}_P \begin{bmatrix} v \\ \omega \end{bmatrix}, \mathbf{T}_P = \begin{bmatrix} \cos\theta & 0 \\ \sin\theta & 0 \\ 0 & 1 \end{bmatrix} \tag{1}$$

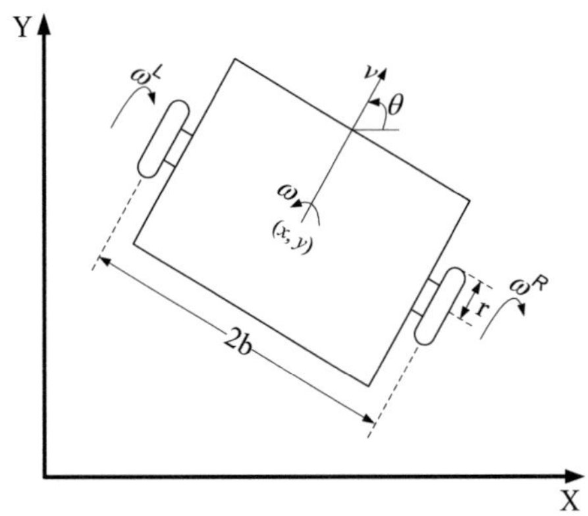

Figure 2. Structure of WMR

Assume that two identical DC motors have the same armature resistance R_a, back-emf constant K_b, and gear ratio n. To simplify dynamics, we ignore the inductance of the armature circuits because the electrical response is generally much faster than the mechanical response. Letting V_s be the battery voltage, the armature circuits of both motors are described as

$$R_a\mathbf{i} = V_s\mathbf{u} - K_b n\mathbf{w} \tag{2}$$

where $\mathbf{i} = \left[i^R \quad i^L\right]^T$ is the armature current vector, $\mathbf{w} = \left[\omega^R \quad \omega^L\right]^T$ is the angular velocity vector of the wheeles, and $\mathbf{u} = \left[u^R \quad u^L\right]^T$ is the normalized control input vector. Superscripts R and L correspond to right and left motors, respectively.

In addition, the dynamic relationship between angular velocity and motor current, considering inertia and viscous friction, becomes (Yun & Yamamoto, 1993)

$$\mathbf{J}\frac{d\mathbf{w}}{dt} + F_v \mathbf{w} = K_t n \mathbf{i} \tag{3}$$

where F_v is the viscous friction coefficient and equivalent inertia matrix of motors $\mathbf{J}$ is $\mathbf{J} = \mathbf{S}^T\mathbf{MS}$, which is 2x2 symmetric.
From Eqs. (2) and (3), we obtain the following differential equation.

$$\dot{\mathbf{w}} + \mathbf{Aw} = \mathbf{Bu} \tag{4}$$

where

$$\mathbf{A} = \begin{bmatrix} a_1 & a_2 \\ a_2 & a_1 \end{bmatrix} = \mathbf{J}^{-1}\left(F_v + \frac{K_t K_b n^2}{R_a} \right), \quad \mathbf{B} = \begin{bmatrix} b_1 & b_2 \\ b_2 & b_1 \end{bmatrix} = \mathbf{J}^{-1}\frac{V_s K_t n}{R_a}$$

Define a state vector as $\mathbf{z} = [v \quad \omega]^T$. Then v and ω are related to ω^R and ω^L by

$$\mathbf{z} = \begin{bmatrix} v \\ \omega \end{bmatrix} = \mathbf{T}_q \begin{bmatrix} \omega^R \\ \omega^L \end{bmatrix} = \mathbf{T}_q \mathbf{w}, \quad \mathbf{T}_q = \begin{bmatrix} r/2 & r/2 \\ r/2b & -r/2b \end{bmatrix} \tag{5}$$

Using the similarity transformation, from Eqs. (4) and (5), we obtain the following equation

$$\dot{\mathbf{z}} + \bar{\mathbf{A}}\mathbf{z} = \bar{\mathbf{B}}\mathbf{u} \tag{6}$$

where

$$\bar{\mathbf{A}} = \mathbf{T}_q \mathbf{A} \mathbf{T}_q^{-1} = \begin{bmatrix} \pi_v & 0 \\ 0 & \pi_\omega \end{bmatrix} = \begin{bmatrix} a_1 + a_2 & 0 \\ 0 & a_1 - a_2 \end{bmatrix}$$

$$\bar{\mathbf{B}} = \mathbf{T}_q \mathbf{B} = \begin{bmatrix} \beta_1 & \beta_1 \\ \beta_2 & -\beta_2 \end{bmatrix} = \begin{bmatrix} r(b_1 + b_2)/2 & r(b_1 + b_2)/2 \\ r(b_1 - b_2)/2b & -r(b_1 - b_2)/2b \end{bmatrix}$$

The overall dynamics of a WMR is shown in Fig. 3, where I_2 is the 2x2 unit matrix.

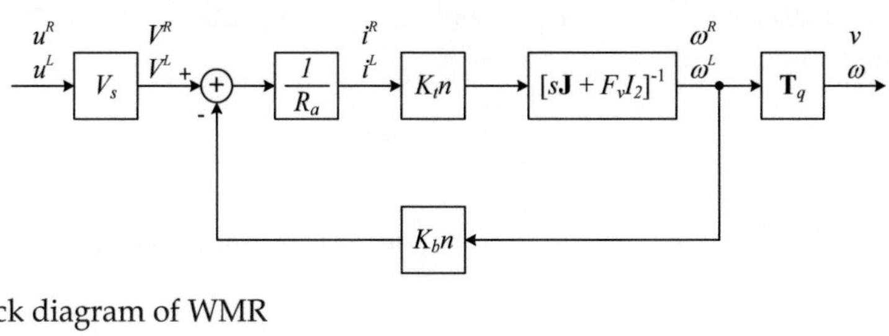

Figure 3. Block diagram of WMR

2.2 Energy Consumption of WMR

The energy drawn from the batteries is converted to mechanical energy to drive motors and losses such as the heat dissipation in the armature resistance. In a WMR, energy is dissipated by the internal resistance of batteries, amplifier resistance in motor drivers, armature resistance, and viscous friction of motors. Fig. 4 shows a simplified circuit diagram of a WMR system.

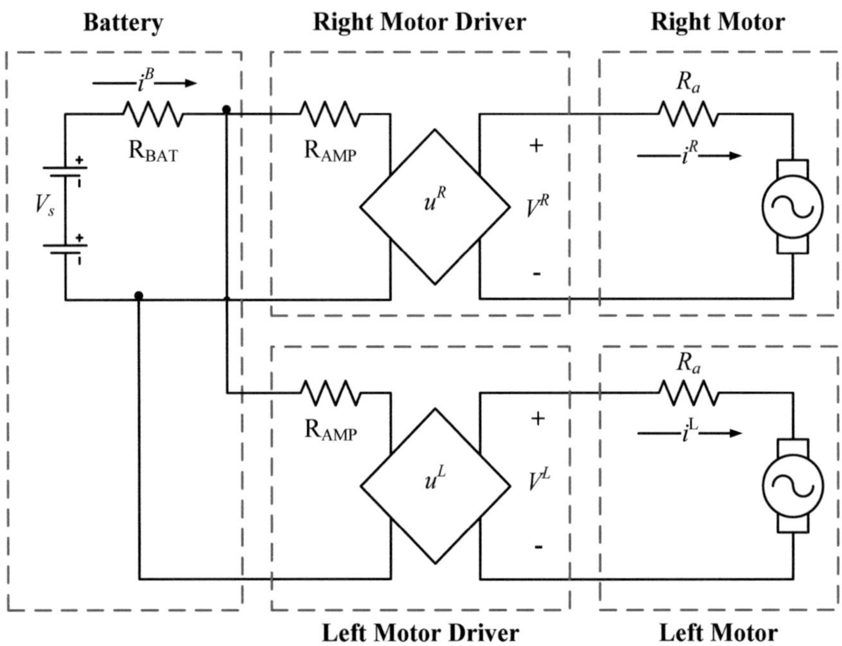

Figure 4. Circuit diagram of batteries, motor drivers, and motors of WMR

A pulse width modulated (PWM) controller is the preferred motor speed controller because little heat is generated and it is energy efficient compared to linear regulation (voltage control) of the motor. We assume that an H-bridge PWM amplifier is used as a motor driver, and this is modeled by its amplifier resistance R_{AMP} and PWM duty ratio u^R and u^L. In our robot system, P3-DX, internal resistance of battery (CF-12V7.2) is approximately 22mΩ and power consumption by the motor drivers is 0.2W. Since internal resistance of battery is much smaller compared with armature resistance of motor (710mΩ) and the power consumption by the motor drivers is much smaller than that of motors (several watts), they are ignored here. Hence the total energy supplied from the batteries to the WMR, E_W, is the cost function to be minimized and is defined as

$$E_W = \int \mathbf{i}^T \mathbf{V} dt = V_s \int \mathbf{i}^T \mathbf{u} dt \tag{7}$$

where $\mathbf{V} = \begin{bmatrix} V^R & V^L \end{bmatrix}^T$ is the input voltage applied to the motors from the batteries, and $\mathbf{u} = \mathbf{V}/V_s = \begin{bmatrix} u^R & u^L \end{bmatrix}^T$.

As there is a certain limit on a battery's output voltage, WMR systems have a voltage constraint on batteries:

$$-u^{\max} \le u^R \le u^{\max}, \; -u^{\max} \le u^L \le u^{\max} \tag{8}$$

From Eqs. (2) and (5), E_W can be written in terms of the velocity and the control input as

$$E_W = \int \left(k_1 \mathbf{u}^T \mathbf{u} - k_2 \mathbf{z}^T \mathbf{T}_q^{-T} \mathbf{u} \right) dt \tag{9}$$

where $k_1 = V_s^2 / R_a$ and $k_2 = K_b n V_s / R_a$.
From Eqs. (2) and (3), the cost fuction E_W becomes

$$E_W = R_a \int \mathbf{i}^T \mathbf{i} dt + F_v \frac{K_b}{K_t} \int \mathbf{z}^T \mathbf{T}_q^{-T} \mathbf{T}_q^{-1} \mathbf{z} dt + \frac{K_b}{K_t} \int \dot{\mathbf{z}}^T \mathbf{T}_q^{-T} \mathbf{J}^T \mathbf{T}_q^{-1} \mathbf{z} dt \tag{10}$$

Note that the first term, $E_R \left(= R_a \int \mathbf{i}^T \mathbf{i} dt \right)$, is the energy dissipated by the armature resistance in the motors and the cost function of loss-minimization control considering only the armature resistance loss. The second term, $E_F \left(= F_v \frac{K_b}{K_t} \int \mathbf{z}^T \mathbf{T}_q^{-T} \mathbf{T}_q^{-1} \mathbf{z} dt \right)$, corresponds to the velocity sensitive loss due to viscous friction. The last term, $E_K \left(= \frac{K_b}{K_t} \int \dot{\mathbf{z}}^T \mathbf{T}_q^{-T} \mathbf{J}^T \mathbf{T}_q^{-1} \mathbf{z} dt \right)$, is the kinetic energy stored in the WMR and will have zero average value when the velocity is constant or final velocity equal to the initial velocity. This means that the net contribution of the last term to the energy consumption is zero.

3. Minimum-Energy Translational Trajectory Generation

A mobile robot's path usually consists of straight lines and arcs. In the usual case, a mobile robot mainly moves in a straight line and there is little, if any, rotation (Barili et al., 1995; Mei et al., 2005). Since the energy consumption associated with rotational velocity changes is much smaller than the energy consumption associated with translational velocity changes, we investigate minimum-energy translational trajectory generation of a WMR moving along a straight line. Since the path of WMR is determined as a straight line, this problem is reduced to find velocity profile minimizing energy drawn from the batteries.

3.1 Problem Statement

The objective of optimal control is to determine the control variables minimizing the cost function for given constraints. Because the rotational velocity of WMR, ω, is zero under translational motion constraint, let $P(t) = \left[x(t) \;\; 0 \;\; 0 \right]^T$ be the posture and $\mathbf{z}(t) = \left[v(t) \;\; 0 \right]^T$ be the velocity at time t. Then the *minimum-energy translational trajectory generation problem* investigated in this section can be formulated as follows.

Problem: Given initial and final times t_0 and t_f, find the translational velocity $v(t)$ and the control input $u(t)$ which minimizes the cost function

$$E_W = \int_{t_0}^{t_f} \left(k_1 \mathbf{u}^T \mathbf{u} - k_2 \mathbf{z}^T \mathbf{T}_q^{-T} \mathbf{u} \right) dt$$

for the system described by Eq. (6) subject to

(1) initial and final postures: $\boldsymbol{P}(t_0)=[x_0 \quad 0 \quad 0]^T$ and $\boldsymbol{P}(t_f)=[x_f \quad 0 \quad 0]^T$,

(2) initial and final velocities: $\mathbf{z}(t_0)=[v_s \quad 0]^T$ and $\mathbf{z}(t_f)=[v_f \quad 0]^T$, and

(3) satisfying the batteries' voltage constraints, $u^{\max}$

As time is not critical, a fixed final time is used.

3.2 Minimum-Energy Translational Trajectory

Without loss of generality, we assume that the initial and final velocities are zero, and the initial posutre is zero. Then the *minimum-energy translational trajectory generation problem* can be written as

$$\text{minimize } E_W = \int_0^{t_f} \left(k_1 \mathbf{u}^T \mathbf{u} - k_2 \mathbf{z} \mathbf{T}_q^{-T} \mathbf{u}\right) dt \tag{11}$$

$$\text{subject to} \qquad \dot{\mathbf{z}} = -\overline{\mathbf{A}}\mathbf{z} + \overline{\mathbf{B}}\mathbf{u} \tag{12}$$

$$\mathbf{z}(0) = \mathbf{z}(t_f) = [0 \quad 0]^T \tag{13}$$

$$P_f = \int_0^{t_f} \mathbf{T}_P \mathbf{z} dt = \left[x_f \quad 0 \quad 0\right]^T \tag{14}$$

$$\begin{bmatrix} -u^{\max} \\ -u^{\max} \end{bmatrix} \le \mathbf{u} = \begin{bmatrix} u^R \\ u^L \end{bmatrix} \le \begin{bmatrix} u^{\max} \\ u^{\max} \end{bmatrix} \tag{15}$$

We used the Pontryagin's Maximum Principle to find the minimum-energy velocity profile that minimizes Eq. (11) while satisfying the constraints in Eqs. (13) – (15) for the system, with Eq. (12). Let the Largrange multiplier for the posture constraint, Eq. (14), be $\boldsymbol{\alpha} = \left[a_x \quad a_y \quad a_\theta\right]^T$. Defining the multipler function for Eq. (12) as, $\boldsymbol{\lambda} = [\lambda_v \quad \lambda_\omega]^T$, the Hamiltonian H is

$$H = k_1 \mathbf{u}^T \mathbf{u} - k_2 \mathbf{z}^T \mathbf{T}_q^{-T} \mathbf{u} - \boldsymbol{\alpha}^T \mathbf{T}_p \mathbf{z} + \boldsymbol{\alpha}^T P_f / t_f + \boldsymbol{\lambda}^T \left(-\overline{\mathbf{A}}\mathbf{z} + \overline{\mathbf{B}}\mathbf{u}\right) \tag{16}$$

The necessary conditions for the optimal velocity $\mathbf{z}^*$ and the control input $\mathbf{u}^*$ are

$$\partial H / \partial \mathbf{u} = 2k_1 \mathbf{u} - k_2 \mathbf{T}_q^{-1} \mathbf{z} + \overline{\mathbf{B}}^T \boldsymbol{\lambda} = 0 \tag{17}$$

$$\partial H / \partial \mathbf{z} = -k_2 \mathbf{T}_q^{-T} \mathbf{u} - \mathbf{T}_P^T \boldsymbol{\alpha} - \overline{\mathbf{A}}^T \boldsymbol{\lambda} = -\dot{\boldsymbol{\lambda}} \tag{18}$$

$$\partial H / \partial \boldsymbol{\lambda} = -\overline{\mathbf{A}}\mathbf{z} + \overline{\mathbf{B}}\mathbf{u} = \dot{\mathbf{z}} \tag{19}$$

From Eqs. (17) – (19), we obtain the following differential equation.

$$\ddot{\mathbf{z}} - \left(\overline{\mathbf{B}}\overline{\mathbf{B}}^T \overline{\mathbf{A}}^T \overline{\mathbf{B}}^{-T} \overline{\mathbf{B}}^{-1} \overline{\mathbf{A}} - \frac{k_2}{k_1} \overline{\mathbf{B}}\overline{\mathbf{B}}^T \mathbf{T}_q^{-T} \overline{\mathbf{B}}^{-1} \overline{\mathbf{A}} \right) \mathbf{z} + \frac{1}{2k_1} \overline{\mathbf{B}}\overline{\mathbf{B}}^T \mathbf{T}_P^T \boldsymbol{\alpha} = 0 \tag{20}$$

As $\bar{\mathbf{B}}\bar{\mathbf{B}}^T$ and $\bar{\mathbf{A}}$ are diagonal matrices, Eq. (20) is reduced to quadratic differential form as follows.

$$\ddot{\mathbf{z}} - \mathbf{Q}^T\mathbf{Q}\mathbf{z} + \mathbf{R}^T\mathbf{T}_p^T\boldsymbol{\alpha} = 0 \tag{21}$$

where $\mathbf{Q}^T\mathbf{Q} = \bar{\mathbf{A}}^T\bar{\mathbf{A}} - \frac{k_2}{k_1}\bar{\mathbf{B}}\bar{\mathbf{B}}^T\mathbf{T}_q^{-T}\bar{\mathbf{B}}^{-1}\bar{\mathbf{A}}$, $\mathbf{Q} = \begin{bmatrix} 1/\tau_v & 0 \\ 0 & 1/\tau_\omega \end{bmatrix}$, and $\mathbf{R} = \frac{\bar{\mathbf{B}}^T\bar{\mathbf{B}}}{2k_1} = \begin{bmatrix} \eta_v & 0 \\ 0 & \eta_\omega \end{bmatrix}$. Here $\tau_v = (J_1 + J_2)/\sqrt{F_v(F_v + K_tK_bn^2/R_a)}$ denotes the mechanical time constant for translation and $\tau_\omega = (J_1 - J_2)/\sqrt{F_v(F_v + K_tK_bn^2/R_a)}$ denotes the mechanical time constant for rotation of WMR. Since we ignore energy dissipation associated with rotational velocity changes and consider only a WMR moving along a straight line (i.e., rotational velocity is zero), the optimal velocity $\mathbf{z}^*$ becomes

$$\mathbf{z}^*(t) = \begin{bmatrix} v^*(t) \\ \omega^*(t) \end{bmatrix} = \begin{bmatrix} C_1e^{t/\tau_v} + C_2e^{-t/\tau_v} + K_v \\ 0 \end{bmatrix} \tag{22}$$

where

$$C_1 = \frac{e^{-t_f/\tau_v} - 1}{e^{t_f/\tau_v} - e^{-t_f/\tau_v}}K_v,\; C_2 = \frac{1 - e^{t_f/\tau_v}}{e^{t_f/\tau_v} - e^{-t_f/\tau_v}}K_v,\; K_v = \frac{x_f\left(e^{t_f/\tau_v} - e^{-t_f/\tau_v}\right)}{2\tau_v\left(2 - e^{t_f/\tau_v} - e^{-t_f/\tau_v}\right) + t_f\left(e^{t_f/\tau_v} - e^{-t_f/\tau_v}\right)}$$

To investigate the properties of the minimum-energy velocity profile, the minimum-energy translational velocity profile, Eq. (22), is shown in Fig. 5 as velocity per unit versus time per unit, where the reference velocity is taken as the x_f/t_f ratio and the reference time t/t_f for various $k = \tau_v/t_f$ (the ratio of translational mechanical time constant per displacement time) using the parameters shown in Table 1.

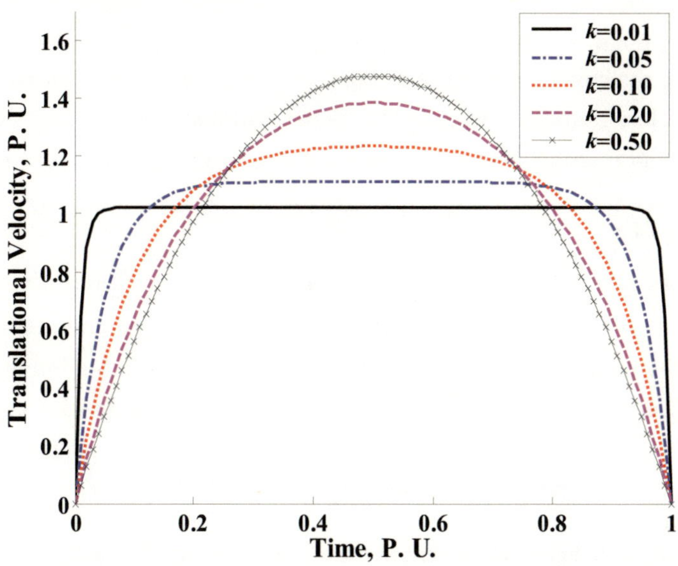

Figure 5. Minimum-Energy velocity profiles for incremental motion at various $k = \tau_v/t_f$

Parameter	Value	Parameter	Value
R_a	0.71Ω	K_t	0.023Nm/A
K_b	0.023V/(rad/s)	n	38.3
m_c	13.64Kg	m_ω	1.48Kg
V_s	12.0V	$u^{\max}$	1.0
F_v	0.039Nm/(rad/s)	r	0.095m
b	0.165m	$\mathbf{J}=\begin{bmatrix} J_1 & J_2 \\ J_2 & J_1 \end{bmatrix}$	$\begin{bmatrix} 0.0799 & 0.0017 \\ 0.0017 & 0.0799 \end{bmatrix}$

Table 1. Parameters of the WMR, P3-DX

This shows that for k close to zero the minimum-energy velocity profile resembles a widely used trapezoidal velocity profile, whereas for values of $k > 0.2$ the profile rapidly converges to a parabolic profile instead of the widely used trapezoidal profile. As shown in Fig. 5, the minimum-energy velocity profile has a symmetric form for the *minimum-energy translational trajectory generation problem*, of Eqs. (11) – (15) as follows.

$$v^*(t) = \frac{x_f}{\tau_v} \frac{\sinh(t_f/\tau_v) - \sinh((t_f - t)/\tau_v) - \sinh(t/\tau_v)}{2(1-\cosh(t_f/\tau_v)) + \frac{t_f}{\tau_v}\sinh(t_f/\tau_v)} \tag{23}$$

Eq. (23) means that the minimum-energy velocity profile depends on the ratio of the mechanical time constant τ_v and the displacement time t_f.

3.3 Simulations and Experiments

3.3.1 Simulations

Several simulations were performed to evaluate the energy saving of the minimum-energy control optimizing the cost function E_W of Eq. (11); these were compared with two results of other methods: loss-minmization control (Trzynadlowski, 1988; Tal, 1973; Kwok & Lee, 1990) optimizing energy loss due to armature resistance of a DC motor, $E_R\left(= R_a \int \mathbf{i}^T \mathbf{i} dt\right)$, and the fixed velocity profile of commonly used trapezoidal velocity profile optimizing the cost function E_W of Eq. (11).

Table 3.2 shows the simulation results of the energy saving for various displacements x_f and displacement time t_f. *Minimum-Energy* denotes the mnimum-energy control optimizing the cost fucntion E_W, *Loss-Minimization* denotes the loss-minimization control optimizing the cost function E_R, and *TRAPE* denotes the trapezoidal velocity profile optimizing the cost fuction E_W. Values in parenthesis represent percentage difference in the total energy drawn from the batteries with respect to that of minimum-energy control. It shows that minimum-energy control can save up to 8% of the energy drawn from the batteries compared with loss-minimization control and up to 6% compared with energy-optimal trapezoidal velocity profile. Because the minimum-energy velocity profile of Eq. (23) resembles a trapezoid for a sufficiently long displacement time, the energy-optimal trapezoidal velocity profile converges to minimum-energy velocity profile and is a near energy-optimal velocity profile for a longer displacement time. However, it expends more energy when frequent velocity changes are required due to obstacles.

Constraints		Total Energy Drawn from the Batteries E_W (J)		
t_f	x_f	*Minimum-Energy*	*Loss-Minimization*	*TRAPE*
2.0s	1.0m	7.26	7.38 (1.65%)	7.70 (6.06%)
5.0s	3.0m	19.07	20.26 (6.24%)	19.57 (2.62%)
10.0s	5.0m	24.26	26.22 (8.08%)	24.57 (1.27%)
20.0s	10.0m	46.56	49.38 (6.06%)	46.85 (0.62%)
30.0s	15.0m	68.92	71.91 (4.34%)	69.20 (0.41%)

Table 2. Comparison of energy saving for various t_f and x_f

Compared with loss-minimization control, minimum-energy control has a significant energy saving for a displacement time greater than 2s. For a further investigation, we performed a careful analysis of two optimization problems: minimum-energy control and loss-minimization control. Fig. 6 shows the simulations for various time constants τ_v that were performed for t_f = 10.0s and x_f = 5.0m. As the energy-optimal velocity profile depends on $k=\tau_v/t_f$, as shown in Fig. 5, the mechanical time constant affects the velocity profiles of the two optimization problems with different cost functions.

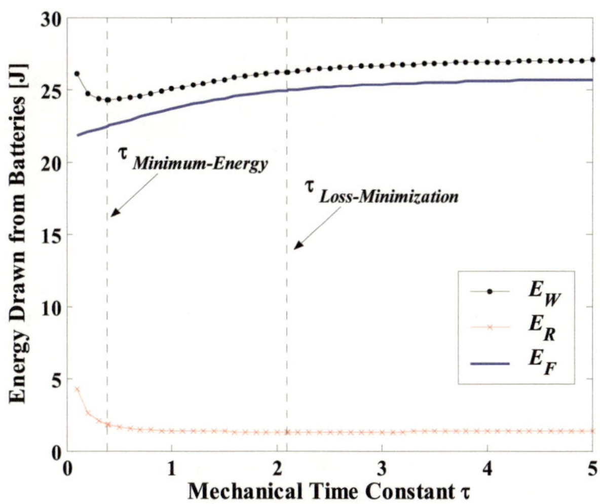

Figure 6. Cost function plot with respect to mechanical time constant τ

Applying the Pontryagin's Maximum Princple to optimize the cost function, the mechanical time constants are $\tau_v=(J_1+J_2)/\sqrt{F_v\left(F_v+K_tK_bn^2/R_a\right)}$ for minimum-energy control and $\tau_v=(J_1+J_2)/F_v$ for loss-minimization control. Fig. 6 shows the change of the cost function with respect to various mechanical time constant τ.

From Eq. (2), decreasing the armature current increases the value of the back-emf and the motor speed. Because the mechanical time constant of minimum-energy control less than that of loss-minimization control, the armature current in minimum-energy control quickly decreases, as shown in Fig. 7(c) during acceleration and deceleration. Hence minimum-energy control can accelerate and decelerate at a higher acceleration rate as shown in Fig. 7(a). Corresponding control inputs are shown in Fig. 7(b) and energy consumptions for each case are shown in Fig. 7(d).

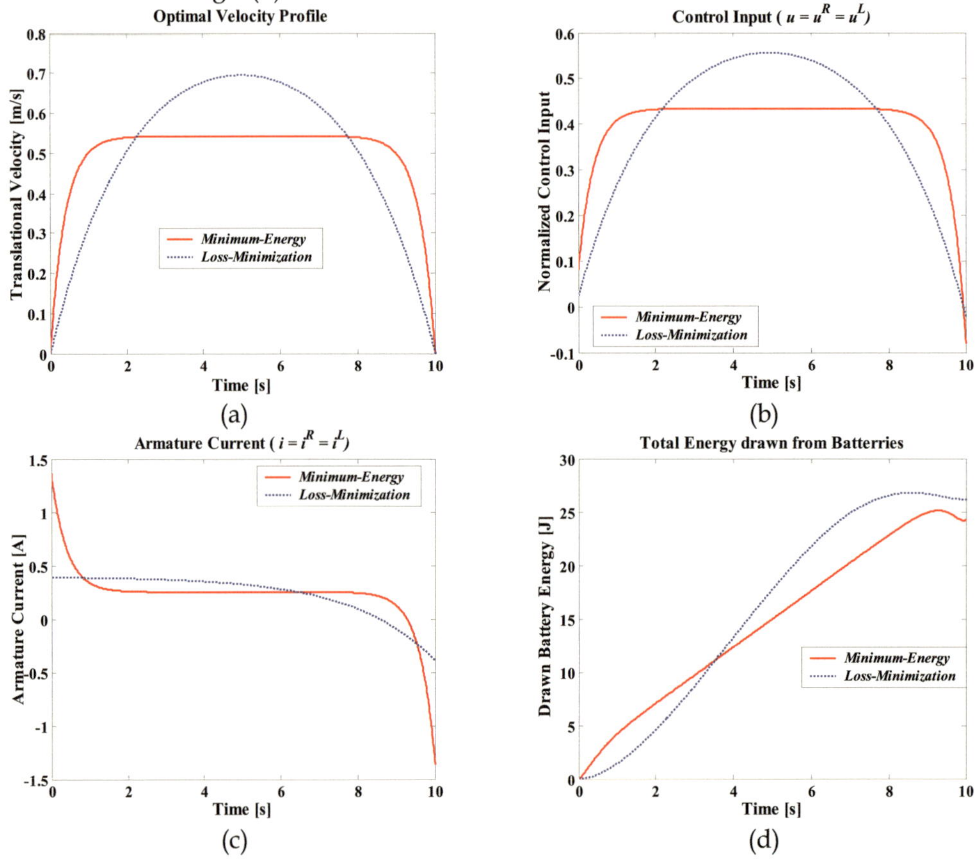

Figure 7. Simulations of minimum-energy control and loss-minimization control for t_f = 10.0s and x_f = 5.0m, (a) Optimal velocity profile, (b) Corresponding control inputs, (c) Armature current change, (d) Comparison of energy consumption

Table 3 shows the ratio of consumed energy for each energy component of Eq. (10) with respect to total energy drawn from the batteries during the entire process for minimum-energy control. Note that the kinetic energy acquired at start up is eventually lost to the whole process when the final velocity is equal to the initial velocity, as shown in Table 3.

t_f	x_f	E_W(%)	E_R (%)	E_F (%)	E_K (%)
2.0s	1.0m	7.26	2.30 (31.68%)	4.96 (68.32%)	0.00 (0.00%)
5.0s	3.0m	19.07	2.35 (12.32%)	16.72 (87.68%)	0.00 (0.00%)
10.0s	5.0m	24.26	1.82 (7.50%)	22.44 (92.50%)	0.00 (0.00%)
20.0s	10.0m	46.56	2.51 (5.39%)	44.05 (94.61%)	0.00 (0.00%)
30.0s	15.0m	68.92	3.26 (4.73%)	65.66 (95.27%)	0.00 (0.00%)

Table 3. Ratio of energy consumption of each energy component for minimum-energy control

Since most of the battery energy is dissipated by the armature resistance for a short displacement time, the minimum-energy control does not have significant energy savings for a short displacement time but shows significant energy savings for a long displacement time, as shown in Table 2.

Fig. 8 shows the power consumption for each energy component of minimum-energy control and loss-minimization control for the constraints given in Fig. 7. It shows that the minimum-energy control requires greater energy consumption than loss-minimization control during accleleration, whereas minimum-energy control consumes less energy after accleration. It means that even though the minimum-energy control requires larger energy consumption than loss-minimization control druing acceleration, it consumes less energy after acceleration. During deceleration a certain amount of energy is *regenerated* and stored in the batteries: 0.94J for minimum-energy control and 0.62J for loss-minimization control.

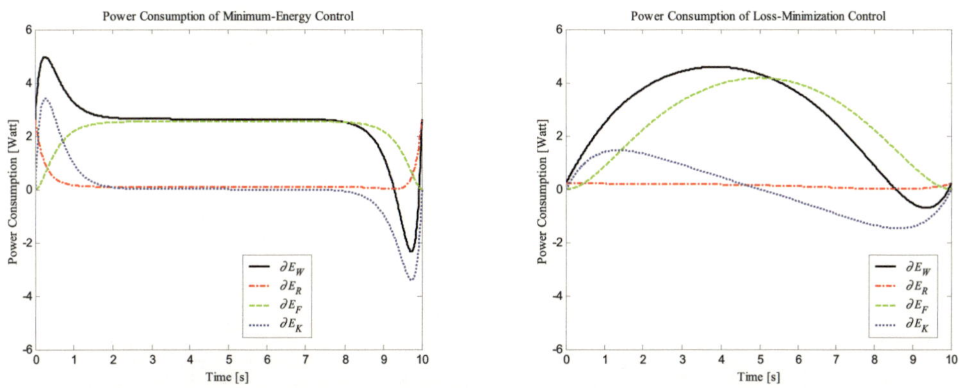

(a) (b)

Figure 8. Comparison of power consumption for each energy component, (a) Minimum-energy control, (b) Loss-minimization control

3.3.2 Experimental Environment Setup

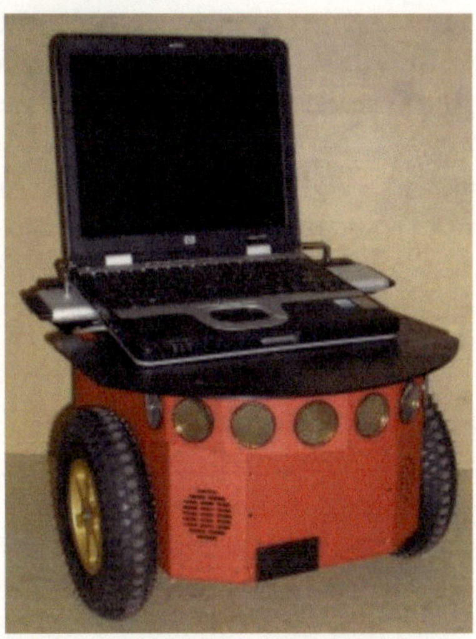

Figure 9. The Pioneer 3-DX robot with a laptop computer

To validate the energy saving of the proposed minimum-energy control, we performed experiments with an actual robot. We use a commercial mobile robot, P3-DX. Fig. 9 is a picture of P3-DX with a laptop computer.
The robot is powered by rechargeable batteries with 12V and has two DC motors with encoders driving two wheels. The maximum translational velocity is approximately 1.2m/s. A Renesas SH2-7144 RISC microcontroller is used to control motors and it communicates with PC client through RS232 serial port. The microcontroller is managed by an Advanced Robot Control and Operations Software (ActiveMedia, 2006).

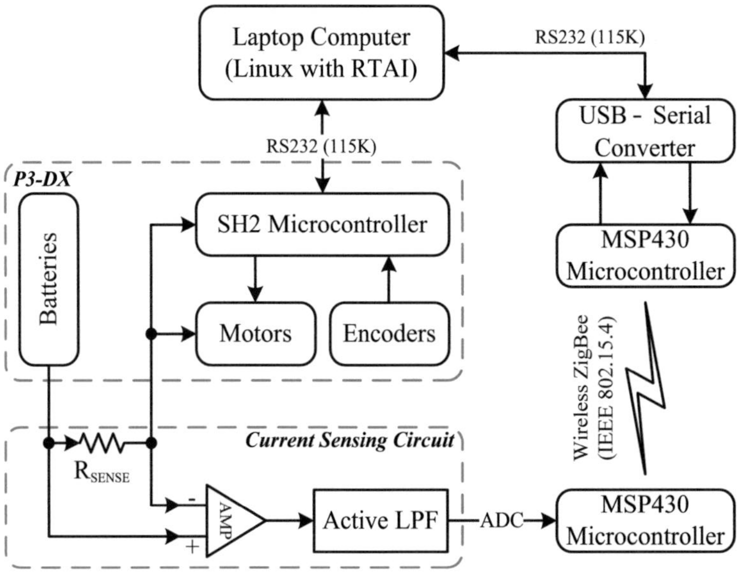

Figure 10. Experimental environment setup

Fig. 10 shows our experimental environment setup. The laptop computer is used to control the robot and to measure the current drawn from the batteries for calculating energy consumption. The robot is controlled by acceleration rate and desired velocity as control commands, and robot's velocity profile is piecewise linear. Since the velocity profiles of minimum-energy and loss-minimization control are nonlinear, we approximated them to piecewise linear velocity profile with 10ms sampling time. The laptop computer is connected to the robot through a serial port with 115Kbps baud rate, and sends a set of acceleration rate and desired velocity of approximated piecewise linear velocity profile to the robot every control period of 10ms, and receives a Standard Information Packet (ActiveMedia, 2006) including velocity and position from the robot every 10ms. Since it is difficult to control every 10ms in Windows or general Linux, we adopted Real-Time Application Interface (RTAI), one of Linux real-time extension, as an operating system of the laptop computer for real-time control (Lineo, 2000).
To measure the drawn energy from the batteries, we sense high side battery current using bi-directional current sensing circuit as shown in Fig. 10. We monitor the current through R_{SENSE} using LT1787 current sense amplifiers with 1.25V reference and filter output of amplifier to obtain average output with unity gain Sallen-Key 2nd order active low pass filter with 1KHz cut-off frequency and unity damping ratio. Then MSP430 microcontroller samples the filtered output with 200Hz sampling rate using 12-bit ADC and transmits

sampled array data to the laptop computer energy 10ms. Since the measured current includes current drawn by microcontroller as well as current drawn by motors, we subtract the measured current when the robot is in initial stop state to obtain the current drawn by motor driving. Fig. 11 shows the bi-directional battery current sensing hardware.

(a)

(b)

Figure 11. Battery current sensing hardware, (a) Bi-directional current sensing circuit, (b) MSP430 microcontroller with 12-bit ADC for data acquisition with USB-to-Serial converter

3.3.3 Experiments

We performed experiments for the constraints in Table 2 and compared with loss-minimization control. To calculate energy consumption, we calculated the armature current and the applied voltages of each motor using the ratio of the armature current between two motors since we can only measure the batteries' current of P3-DX. The ratio of the armature currents can be obtained from Eqs. (3) and (5) using measured velocity of WMR as follows.

$$\rho = \frac{i^R}{i^L} = \frac{(J_1 + J_2)\dot{v} + b(J_1 - J_2)\dot{\omega} + F_v(v + b\omega)}{(J_1 + J_2)\dot{v} + b(J_2 - J_1)\dot{\omega} + F_v(v - b\omega)} \tag{24}$$

Since battery current is $i^B = i^R + i^L$, the armature current of two motors are

$$i^R = \frac{\rho}{\rho + 1} i^B, \; i^L = \frac{1}{\rho + 1} i^B \tag{25}$$

and applied voltages of two motors is obtained from Eq. (2). Then we can calculate the drawn energy from the batteries, Eq. (7).

Figs. 12 and 13 show the experimental results that were performed for t_f = 10.0s and x_f = 5.0m compared with simulation results. Actual velocity of the robot follows well desired velocity. Since we ignore the armature inductance of the motor, armature current change and power consumption has slightly different change during accleration and deceleration. However, they show the similar overall response.

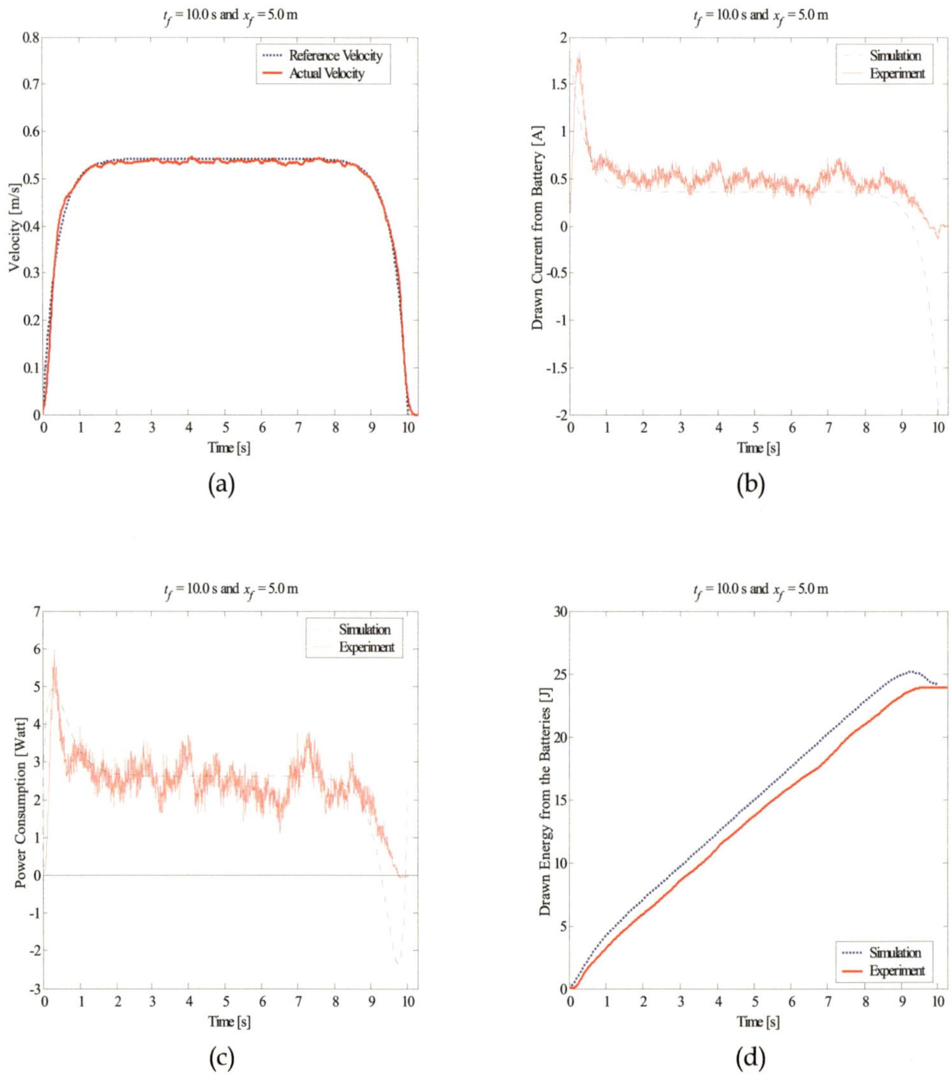

Figure 12. Experimental results of minimum-energy control for t_f = 10.0s and x_f = 5.0m, (a) Velocity profile, (b) Armature current change, (c) Power consumption, (d) Energy consumption

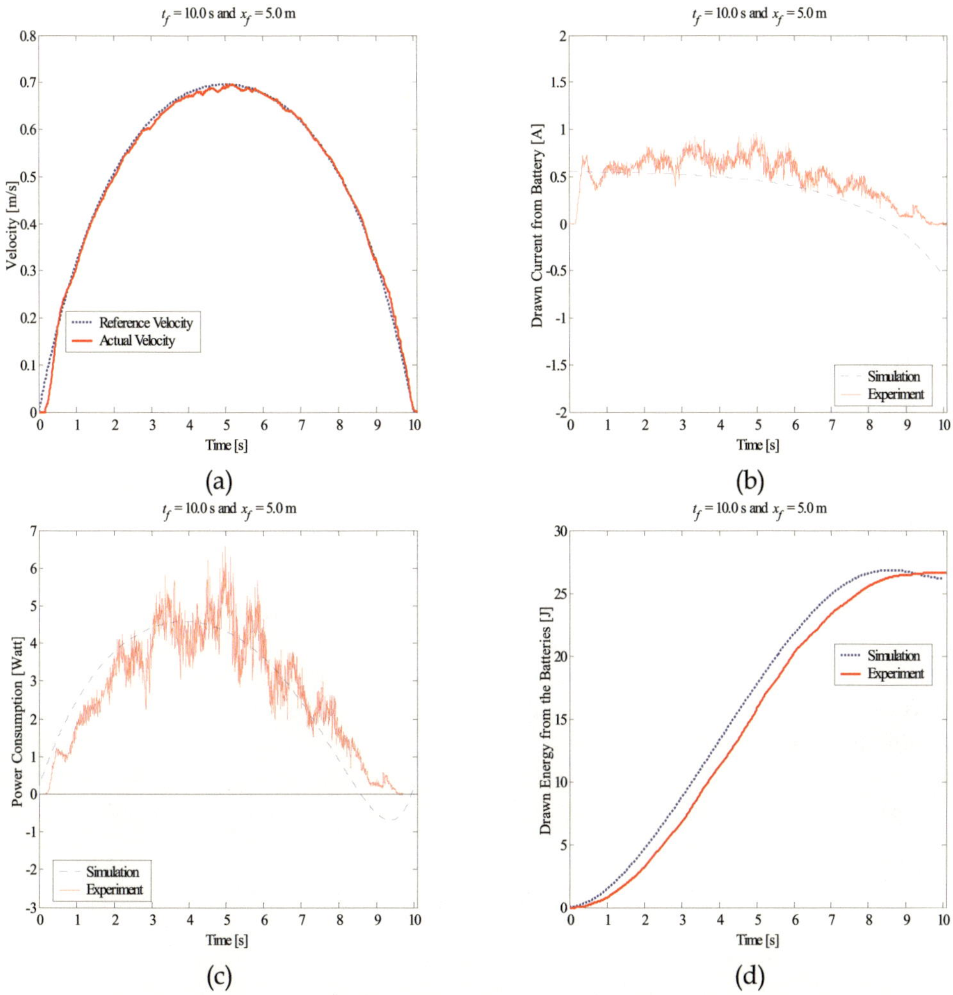

Figure 13. Experimental results of loss-minimization control for t_f = 10.0s and x_f = 5.0m, (a) Velocity profile, (b) Armature current change, (c) Power consumption, (d) Energy consumption

Table 4 shows the experimental results for energy savings for various displacements x_f and displacement time t_f. Values in parenthesis represent percentage difference in the total energy drawn from the batteries with respect to that of minimum-energy control. Experimental results revealed that the minimum-energy control can save up to 11% of the energy drawn from the batteries compared with loss-minimization control.

Since we ignore the inductance of the motors and there can be errors in modelling and measuring the energy drawn from the batteries for experiments is slightly different to that of simulations. However, we can see that the minimum-energy control can save the battery energy compared with loss-minimization control in both experiments and simulations. Table 4 also shows that the percent of energy savings difference between minimum-energy control and loss-minimization control has a similar tendency with that of simulation results in Table 2.

Constraints		Total Energy Drawn from the Batteries E_W (J)	
t_f	x_f	*Minimum-Energy*	*Loss-Minimization*
2.0s	1.0m	8.73	8.97 (2.75%)
5.0s	3.0m	20.42	22.63 (10.82%)
10.0s	5.0m	23.93	26.59 (11.12%)
20.0s	10.0m	45.27	50.16 (10.80%)
30.0s	15.0m	66.61	69.95 (5.01%)

Table 4. Comparison of experimental results of energy saving for various t_f and x_f

4. Minimum-Energy Turning Trajectory Planning

4.1 Problem Statement

In Section 3, we investigated minimum-energy translational trajectory generation of WMR moving along a straight line. To do useful actions, WMR needs rotational trajectory as well as translational trajectory.

According to the configurations of initial and final postures, we can consider two basic paths. The one is single corner path which consists of an approach heading angle followed by a departure heading angle that is along a line at some angle relative to the approach line. That is, it is unnecessary to change the sign of rotational velocity to reach final posture, and the other is double corner path which necessary to change the sign of rotational velocity to reach final posture as shown in Fig. 14. More complicated paths can be constructed by combining multiple single corner paths.

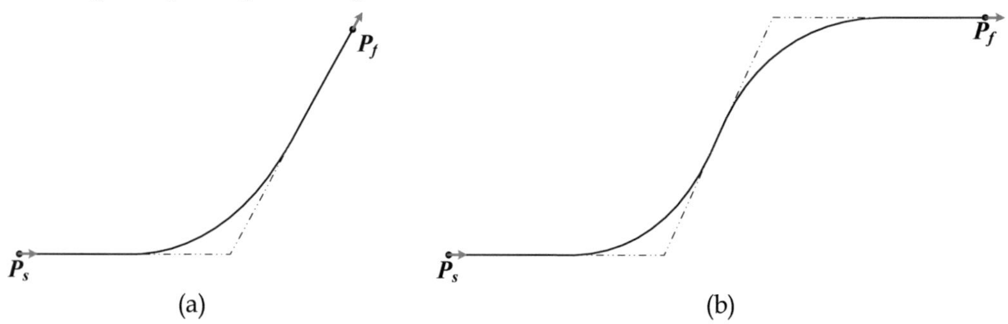

Figure 14. Classification of paths, (a) Single corner path, (b) Double corner path

For simplicity, we consider the single corner path. Futhermore, because of nonlinear and nonholonomic properties, a graphical approach will be used with the following definition about two sections to solve trajectory planning problem.

- Rotational section is a section where the rotational velocity of WMR is not zero, as a result, turning motion is caused.
- Translational section is a section where the rotational velocity is zero, as a result, linear motion is caused only.

Since the paths for single corner are expected to be made up with one rotational section and two translational sections surrounding the rotational section, we divide our trajectory planning algorithm into three sections. The first is *RS* (rotational section) which is focused on the required turning angle, and the others are *TSB* (translational section before rotation) and *TSA* (translational section after rotation) which are secondary procedure to satisfy the condition of positions.

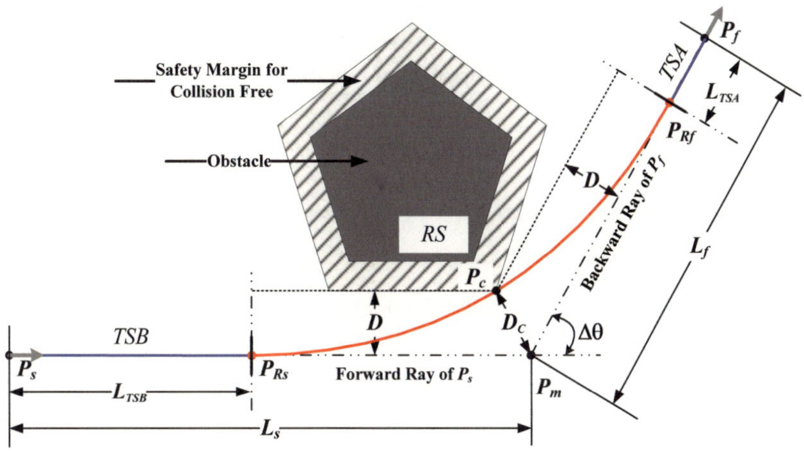

Figure 15. Sections in path and requirement of path-deviation

As shown in Fig. 15, let $\boldsymbol{P}_s = [x_s \quad y_s \quad \theta_s]^T$ and $\boldsymbol{P}_f = [x_f \quad y_f \quad \theta_f]^T$ be given initial and final postures, and $\boldsymbol{P}_m = [x_m \quad y_m \quad -]^T$ be the via point which is an intersection of the forward ray of $\boldsymbol{P}_s$ and the backward ray of $\boldsymbol{P}_f$, where '-' means that value is not used. Since there are limitations by obstacles or walls in real world, we consider the bound of path-deviation D (or deviation from the corner, $D_c = D/\cos(\Delta\theta/2)$ where $\Delta\theta = \theta_f - \theta_s$) as shown in Fig. 15, which limits path-deviation from the given configuration. In Fig. 15, path-deviation is given considering the safety margin to avoid collisions to the obstacles. Hence the path for single corner is divided into three sections: *TSB*, *RS*, and *TSA*.

Then the *minimum-energy turning trajectory planning problem* can be formulated as follows.

Problem: Given initial and final times t_0 and t_f, find the trajectory, that is path and velocity profiles, which minimizes the cost function

$$E_W = \int_{t_0}^{t_f} \left(k_1 \mathbf{u}^T \mathbf{u} - k_2 \mathbf{z}^T \mathbf{T}_q^{-T} \mathbf{u} \right) dt$$

for the system described by Eq. (6) subject to

(1) initial and final postures: $\boldsymbol{P}(t_0) = \boldsymbol{P}_s$ and $\boldsymbol{P}(t_f) = \boldsymbol{P}_f$,

(2) initial and final velocities: $\mathbf{z}(t_0) = \mathbf{z}_s$ and $\mathbf{z}(t_f) = \mathbf{z}_f$,

(3) satisfying the batteries' voltage constraints $u^{\max}$, and

(4) satisfying the path-deviation constraint D.

4.2 Minimum-Energy Turning Trajectory Planning

4.2.1 Overview of the Method

WMR's path is described by finite sequences of two straight lines for translational motions and an arc for rotational motion in between as shown in Fig. 15. The velocity of the WMR depends on planning path and the path of the WMR depends on planning velocity profile. Hence energy consumption depends on path and velocity, i.e., the trajectory. We define a trajectory of the WMR at time t as $\mathbf{q}(t)=\left[S(t) \quad \theta(t)\right]^T$, where $S(t)$ is the displacement profile of the WMR and $\theta(t)$ is the orientation profile of the WMR. Since velocity of the WMR is a time derivative of trajectory, trajectory has the following relationship.

$$\dot{\mathbf{q}}=\mathbf{z} \text{ or } \mathbf{q}=\int \mathbf{z}dt \tag{26}$$

In Fig. 15, let $\boldsymbol{P}_{Rs}$ and $\boldsymbol{P}_{Rf}$ be start and end postures of *RS*, and velocities at $\boldsymbol{P}_{Rs}$ and $\boldsymbol{P}_{Rf}$ be $\mathbf{z}_{Rs}$ and $\mathbf{z}_{Rf}$, respectively. Without loss of generality, we stipulate that initial posture is $\boldsymbol{P}_s=\left[0 \quad 0 \quad 0\right]^T$, and the initial and final velocities are $\mathbf{z}(t_0)=\mathbf{z}(t_f)=\left[0 \quad 0\right]^T$. We define the velocities at t_{Rs} and t_{Rf} are $\mathbf{z}(t_{Rs})=\left[v_{Rs} \quad 0\right]^T=\mathbf{z}_{Rs}$ and $\mathbf{z}(t_{Rf})=\left[v_{Rf} \quad 0\right]^T=\mathbf{z}_{Rf}$, respectively. Then the velocity profile has a shape as shown in Fig. 16.

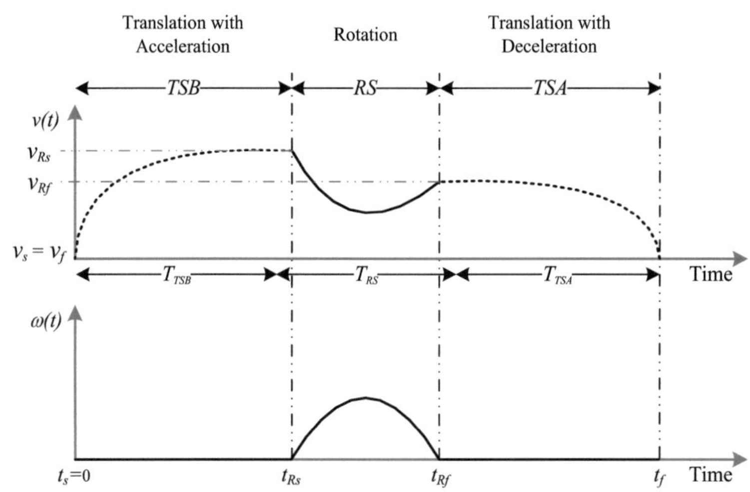

Figure 16. Possible shape of the velocity profile for single corner trajectory

The t_{Rs} and t_{Rf} denote initial and final time of *RS*. We define a time interval of *TSB* as $T_{TSB}=t_{Rs}-T_0$, a time interval of *RS* as $T_{RS}=t_{Rf}-t_{Rs}$, and a time interval of *TSA* as $T_{TSA}=t_f-t_{Rf}$.

WMR is a nonholonomic system. Since its position should be integrated along the curved trajectory by Eq. (1), there are no analytic expressions available. In our trajectory planning strategy, *RS* is planned first to turn the required turning angle $\Delta\theta=\theta_f-\theta_s$. To satisfy the condition of positions, *TSB* and *TSA* are planned to cover remaining distances L_{TSB} (along θ_s) and L_{TSA} (along θ_f) as shown in Fig. 17.

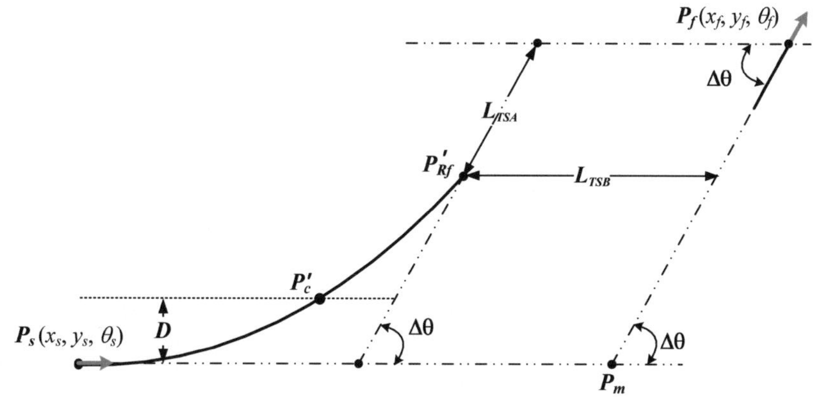

Figure 17. Calculation of L_{TSB}, L_{TSA}, and $P_c^{'}$

4.2.2 Rotational Section

In rotational section *RS*, let trajectories at t_{Rs} and t_{Rf} be the $\mathbf{q}(t_{Rs}) = [S_{Rs} \quad \theta_{Rs}]^T$ and $\mathbf{q}(t_{Rf}) = [S_{Rf} \quad \theta_{Rf}]^T$. Then the minimum-energy turning trajectory planning problem for *RS* can be written as follows.

Problem RS: Find a trajectory for $t_{Rs} \le t \le t_{Rf}$ which minimizes the cost function Eq. (9) subject to

(1) initial and final postures: $\mathbf{q}(t_{Rs}) = \mathbf{q}_{Rs}$ and $\mathbf{q}(t_{Rf}) = \mathbf{q}_{Rf}$,

(2) initial and final velocities: $\dot{\mathbf{q}}(t_{Rs}) = \mathbf{z}_{Rs}$ and $\dot{\mathbf{q}}(t_{Rf}) = \mathbf{z}_{Rf}$, and

(3) satisfying the path-deviation constraint *D*.

We used the Pontryagin's Maximum Principle to deal with the minimum-energy trajectory of *RS*. Defining the Largrange multiplier for the trajectory $\mathbf{q}(t)$ as $\boldsymbol{\alpha} = [a_S \quad a_\theta]^T$ and the multiplier function for Eq. (6) as $\boldsymbol{\lambda} = [\lambda_v \quad \lambda_\omega]^T$, the Hamiltonian is

$$H = k_1\mathbf{u}^T\mathbf{u} - k_2\mathbf{z}^T\mathbf{T}_q^{-T}\mathbf{u} - \boldsymbol{\alpha}^T\left(\mathbf{z} - \frac{\mathbf{q}_{Rf} - \mathbf{q}_{Rs}}{t_{Rf} - t_{Rs}}\right) + \boldsymbol{\lambda}^T\left(-\bar{\mathbf{A}}\mathbf{z} + \bar{\mathbf{B}}\mathbf{u}\right) \tag{27}$$

The necessary conditions for the optimal velocity $\mathbf{z}^*$ and the control input $\mathbf{u}^*$ are

$$\partial H / \partial \mathbf{u} = 2k_1\mathbf{u} - k_2\mathbf{T}_q^{-1}\mathbf{z} + \bar{\mathbf{B}}^T\boldsymbol{\lambda} = 0 \tag{28}$$

$$\partial H / \partial \mathbf{z} = -k_2\mathbf{T}_q^{-T}\mathbf{u} - \boldsymbol{\alpha} - \bar{\mathbf{A}}^T\boldsymbol{\lambda} = -\dot{\boldsymbol{\lambda}} \tag{29}$$

$$\partial H / \partial \boldsymbol{\lambda} = -\bar{\mathbf{A}}\mathbf{z} + \bar{\mathbf{B}}\mathbf{u} = \dot{\mathbf{z}} \tag{30}$$

From Eqs. (28) - (30), we obtain the following differential equation.

$$\ddot{\mathbf{z}} - \mathbf{Q}^T\mathbf{Q}\mathbf{z} + \mathbf{R}^T\mathbf{T}_P^T\boldsymbol{\alpha} = 0 \tag{31}$$

where $\mathbf{Q}^T\mathbf{Q} = \bar{\mathbf{A}}^T\bar{\mathbf{A}} - \frac{k_2}{k_1}\bar{\mathbf{B}}\bar{\mathbf{B}}^T\mathbf{T}_q^{-T}\bar{\mathbf{B}}^{-1}\bar{\mathbf{A}}$, $\mathbf{Q} = \begin{bmatrix} 1/\tau_v & 0 \\ 0 & 1/\tau_\omega \end{bmatrix}$, and $\mathbf{R} = \frac{\bar{\mathbf{B}}^T\bar{\mathbf{B}}}{2k_1} = \begin{bmatrix} \eta_v & 0 \\ 0 & \eta_\omega \end{bmatrix}$. Here $\tau_v = (J_1 + J_2)/\sqrt{F_v(F_v + K_t K_b n^2 / R_a)}$ denotes the mechanical time constant for translation and $\tau_\omega = (J_1 - J_2)/\sqrt{F_v(F_v + K_t K_b n^2 / R_a)}$ denotes the mechanical time constant for rotation of WMR.

Solving Eq. (31), the optimal velocity profile in RS, $\mathbf{z}^*_{RS}$, becomes

$$\mathbf{z}^*_{RS}(t - t_{Rs}) = \begin{bmatrix} C_1^v e^{(t-t_{Rs})/\tau_v} + C_2^v e^{-(t-t_{Rs})/\tau_v} + K_v \\ C_1^\omega e^{(t-t_{Rs})/\tau_\omega} + C_2^\omega e^{-(t-t_{Rs})/\tau_\omega} + K_\omega \end{bmatrix} \tag{32}$$

where

$$C_1^v = -\frac{v_{Rs}e^{-T_{RS}/\tau_v} - v_{Rf} - K_v\left(e^{-T_{RS}/\tau_v} - 1\right)}{e^{T_{RS}/\tau_v} - e^{-T_{RS}/\tau_v}},\ C_2^v = \frac{v_{Rs}e^{T_{RS}/\tau_v} - v_{Rf} - K_v\left(e^{T_{RS}/\tau_v} - 1\right)}{e^{T_{RS}/\tau_v} - e^{-T_{RS}/\tau_v}}$$

$$K_v = \frac{\left(S_{Rf} - S_{Rs}\right)\left(e^{T_{RS}/\tau_v} - e^{-T_{RS}/\tau_v}\right) + \tau_v\left(v_{Rs} + v_{Rf}\right)\left(2 - e^{T_{RS}/\tau_v} - e^{-T_{RS}/\tau_v}\right)}{2\tau_v\left(2 - e^{T_{RS}/\tau_v} - e^{-T_{RS}/\tau_v}\right) + T_{RS}\left(e^{T_{RS}/\tau_v} - e^{-T_{RS}/\tau_v}\right)}$$

$$C_1^\omega = \frac{e^{-T_{RS}/\tau_\omega} - 1}{e^{T_{RS}/\tau_\omega} - e^{-T_{RS}/\tau_\omega}}K_\omega,\ C_2^\omega = -\frac{e^{T_{RS}/\tau_\omega} - 1}{e^{T_{RS}/\tau_\omega} - e^{-T_{RS}/\tau_\omega}}K_\omega,$$

$$K_\omega = \frac{\left(\theta_f - \theta_s\right)\left(e^{T_{RS}/\tau_\omega} - e^{-T_{RS}/\tau_\omega}\right)}{2\tau_\omega\left(2 - e^{T_{RS}/\tau_\omega} - e^{-T_{RS}/\tau_\omega}\right) + T_{RS}\left(e^{T_{RS}/\tau_\omega} - e^{-T_{RS}/\tau_\omega}\right)}$$

Since the path of WMR depends on the velocity profile, the displacement length of RS, $\Delta S_{RS} = S_{Rf} - S_{Rs}$, is an unknown parameter. From Eq. (32), the trajectory of RS is determined by four unknown variables of RS: time interval T_{RS}, initial velocity v_{Rs}, final velocotiy v_{Rf}, and the displacement length ΔS_{RS}. To consider the path-deviation requirement, we calculate the path of WMR in RS with respect to the $\mathbf{P}_s$ from the planned trajectory of Eq. (32) using the integral of $\mathbf{z}$, Eq. (26). Let $\mathbf{P}'_{Rf} = \begin{bmatrix} x'_{Rf} & y'_{Rf} & \theta'_{Rf} \end{bmatrix}^T$ be the final posture of RS with respect to the $\mathbf{P}_s$, $\mathbf{P}_c = \begin{bmatrix} x_c & y_c & - \end{bmatrix}^T$ be the corner point, and $\mathbf{P}'_c = \begin{bmatrix} x'_c & y'_c & \theta'_c \end{bmatrix}^T$ be the point which the y-coordinate of the planned RS path is equal to the y-coordinate of the corner point $\mathbf{P}_c$. Corner point $\mathbf{P}_c$ can be obtained from the path-deviation requirement D, the required turning angle $\Delta\theta = \theta_f - \theta_s$, and via point $\mathbf{P}_m$ as follows.

$$\mathbf{P}_c = \begin{bmatrix} x_m - D \cdot \tan\Delta\theta & D & - \end{bmatrix}^T \tag{33}$$

To connect the path of RS planned with respect to $\mathbf{P}_s$ with those of TSB and TSA, we calculate the remaining distances $\mathbf{L}_{TSB}$ for TSB and $\mathbf{L}_{TSA}$ for TSA from $\mathbf{P}'_{Rf}$ and $\mathbf{P}_f$ as follows.

$$L_{TSB} = \begin{cases} x_f - x'_{Rf} & , \Delta\theta = \pi/2 \\ x_f - \dfrac{y_f - y'_{Rf}}{\tan\Delta\theta} - x'_{Rf} & , otherwise \end{cases} \quad \text{and } \boldsymbol{L}_{TSA} = \frac{y_f - y'_{Rf}}{\sin\Delta\theta} \tag{34}$$

Then the postures of $\boldsymbol{P}_{Rs}$ and $\boldsymbol{P}_{Rf}$ (See Fig. 15) are

$$\boldsymbol{P}_{Rs} = \begin{bmatrix} x_{Rs} \\ y_{Rs} \\ \theta_{Rs} \end{bmatrix} = \begin{bmatrix} L_{TSB} \\ y_s \\ \theta_s \end{bmatrix} \text{ and } \boldsymbol{P}_{Rf} = \begin{bmatrix} x_{Rf} \\ y_{Rf} \\ \theta_{Rf} \end{bmatrix} = \begin{bmatrix} x'_{Rf} + L_{TSB} \\ y'_{Rf} \\ \theta_f \end{bmatrix} \tag{35}$$

To obtain a feasible trajectory of *RS*, the following conditions should be satisfied.

$$x'_c + L_{TSB} \geq x_c,\ L_{TSB} > 0, \text{ and } L_{TSA} > 0 \tag{36}$$

4.2.3 Two Translational Sections

After planning the rotational section *RS*, we obtain the remaining distances L_{TSB} and L_{TSA} of Eq. (34) to plan the trajectories of *TSB* and *TSA*. To obtain energy-optimal trajectories of *TSB* and *TSA*, let trajectories at t_0 and t_f be the $\mathbf{q}(t_0) = [S_s \ \ \theta_s]^T$ and $\mathbf{q}(t_f) = [S_f \ \ \theta_f]^T$. Then the minimum-energy trajectory planning problem for *TSB* and *TSA* can be written as follows.

Problem TSB: Find a trajectory for $t_0 \leq t \leq t_{Rs}$ which minimizes the cost function Eq. (9) subject to

(1) initial and final postures: $\mathbf{q}(t_0) = \mathbf{q}_s$ and $\mathbf{q}(t_{Rs}) = \mathbf{q}_{Rs}$, and

(2) initial and final velocities: $\dot{\mathbf{q}}(t_0) = \mathbf{z}_s$ and $\dot{\mathbf{q}}(t_{Rs}) = \mathbf{z}_{Rs}$.

Problem TSA: Find a trajectory for $t_{Rf} \leq t \leq t_f$ which minimizes the cost function Eq. (9) subject to

(1) initial and final postures: $\mathbf{q}(t_{Rf}) = \mathbf{q}_{Rf}$ and $\mathbf{q}(t_f) = \mathbf{q}_f$, and

(2) initial and final velocities: $\dot{\mathbf{q}}(t_{Rf}) = \mathbf{z}_{Rf}$ and $\dot{\mathbf{q}}(t_f) = \mathbf{z}_f$.

Applying the same process in *RS*, the optimal velocity profiles of *TSB* and *TSA*, $\mathbf{z}^*_{TSB}$ and $\mathbf{z}^*_{TSA}$ become

$$\mathbf{z}^*_{TSB}(t - t_0) = \begin{bmatrix} C_1^{TSB} e^{(t-t_0)/\tau_v} + C_2^{TSB} e^{-(t-t_0)/\tau_v} + K_v^{TSB} \\ 0 \end{bmatrix} \tag{37}$$

$$\mathbf{z}^*_{TSA}(t - t_{Rf}) = \begin{bmatrix} C_1^{TSA} e^{(t-t_{Rf})/\tau_v} + C_2^{TSA} e^{-(t-t_{Rf})/\tau_v} + K_v^{TSA} \\ 0 \end{bmatrix} \tag{38}$$

where

$$C_1^{TSB} = -\frac{v_s e^{-T_{TSB}/\tau_v} - v_{Rs} - K_v^{TSB}\left(e^{-T_{TSB}/\tau_v} - 1\right)}{e^{T_{TSB}/\tau_v} - e^{-T_{TSB}/\tau_v}}, \quad C_2^{TSB} = \frac{v_s e^{T_{TSB}/\tau_v} - v_{Rs} - K_v^{TSB}\left(e^{T_{TSB}/\tau_v} - 1\right)}{e^{T_{TSB}/\tau_v} - e^{-T_{TSB}/\tau_v}}$$

$$K_v^{TSB} = \frac{\left(S_{Rs} - S_s\right)\left(e^{T_{TSB}/\tau_v} - e^{-T_{TSB}/\tau_v}\right) + \tau_v\left(v_s + v_{Rs}\right)\left(2 - e^{T_{TSB}/\tau_v} - e^{-T_{TSB}/\tau_v}\right)}{2\tau_v\left(2 - e^{T_{TSB}/\tau_v} - e^{-T_{TSB}/\tau_v}\right) + T_{TSB}\left(e^{T_{TSB}/\tau_v} - e^{-T_{TSB}/\tau_v}\right)}$$

$$C_1^{TSA} = -\frac{v_{Rf} e^{-T_{TSA}/\tau_v} - v_f - K_v^{TSA}\left(e^{-T_{TSA}/\tau_v} - 1\right)}{e^{T_{TSA}/\tau_v} - e^{-T_{TSA}/\tau_v}}, \quad C_2^{TSA} = \frac{v_{Rf} e^{T_{TSA}/\tau_v} - v_f - K_v^{TSA}\left(e^{T_{TSA}/\tau_v} - 1\right)}{e^{T_{TSA}/\tau_v} - e^{-T_{TSA}/\tau_v}}$$

$$K_v^{TSA} = \frac{\left(S_f - S_{Rf}\right)\left(e^{T_{TSA}/\tau_v} - e^{-T_{TSA}/\tau_v}\right) + \tau_v\left(v_{Rf} + v_f\right)\left(2 - e^{T_{TSA}/\tau_v} - e^{-T_{TSA}/\tau_v}\right)}{2\tau_v\left(2 - e^{T_{TSA}/\tau_v} - e^{-T_{TSA}/\tau_v}\right) + T_{TSA}\left(e^{T_{TSA}/\tau_v} - e^{-T_{TSA}/\tau_v}\right)}$$

Note that $\Delta S_{TSB} = S_{Rs} - S_s$ and $\Delta S_{TSA} = S_f - S_{Rf}$ are the displacement lengths of *TSB* and *TSA*, respectively. From Eqs. (37) and (38), the trajectories of *TSB* and *TSA* are determined by two unknown variables of *TSB* and *TSA*: time interval T_{TSB} of *TSB* and time interval T_{TSA} of *TSA*.

4.2.4 Trajectory Optimization

Since $t_f - t_0 = T_{TSB} + T_{RS} + T_{TSA}$, from Sections 4.2.2 and 4.2.3, the minimum-energy turning trajectory is determined by five variables: T_{RS}, v_{Rs}, v_{Rf}, ΔS_{RS}, and T_{TSB}. Since there is no analytic expression for the posture function of WMR, we performed numerous simulations to analyze the convexity of the cost function. Fig. 18 is the one of numerous simulations of the cost function using full search.

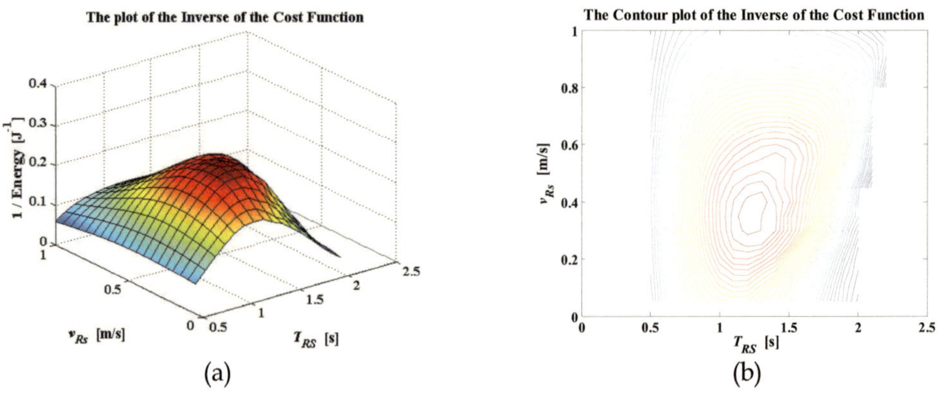

Figure 18. Plot of the cost function of feasible solution for t_f = 2.5s and P_f = [0.50m 0.75m 90°]^T, (a) The plot of the inverse of the cost function, (b) Contour plot of the inverse of the cost function

It shows that the cost function is convex with respect to T_{RS} and v_{Rs}. Numerous simulations also showed that the cost function is convex with respect to v_{Rf} and T_{TSB} also. Hence we constructed an iterative search with quintuple loops with variables to find the minimum-energy turning trajectory.

Fig. 19 is the overall flowchart of iterative search with quintuple loops to find minimum-energy turning trajectory. Fig. 20 contains partial detailed flowcharts of Fig. 19. Figs. 20(a) –

20(d) are iterative search loops to plan the trajectory of RS satisfying the feasibility conditions of Eq. (36). In Fig. 20(d), maximum displacement length of RS is $S_{RS}^{\max} = L_s + L_f$ as shown in Fig. 15. Combining the solutions of minimum-energy trajectories for the required cornering motions, we can get the overall minimum-energy turning trajectory.

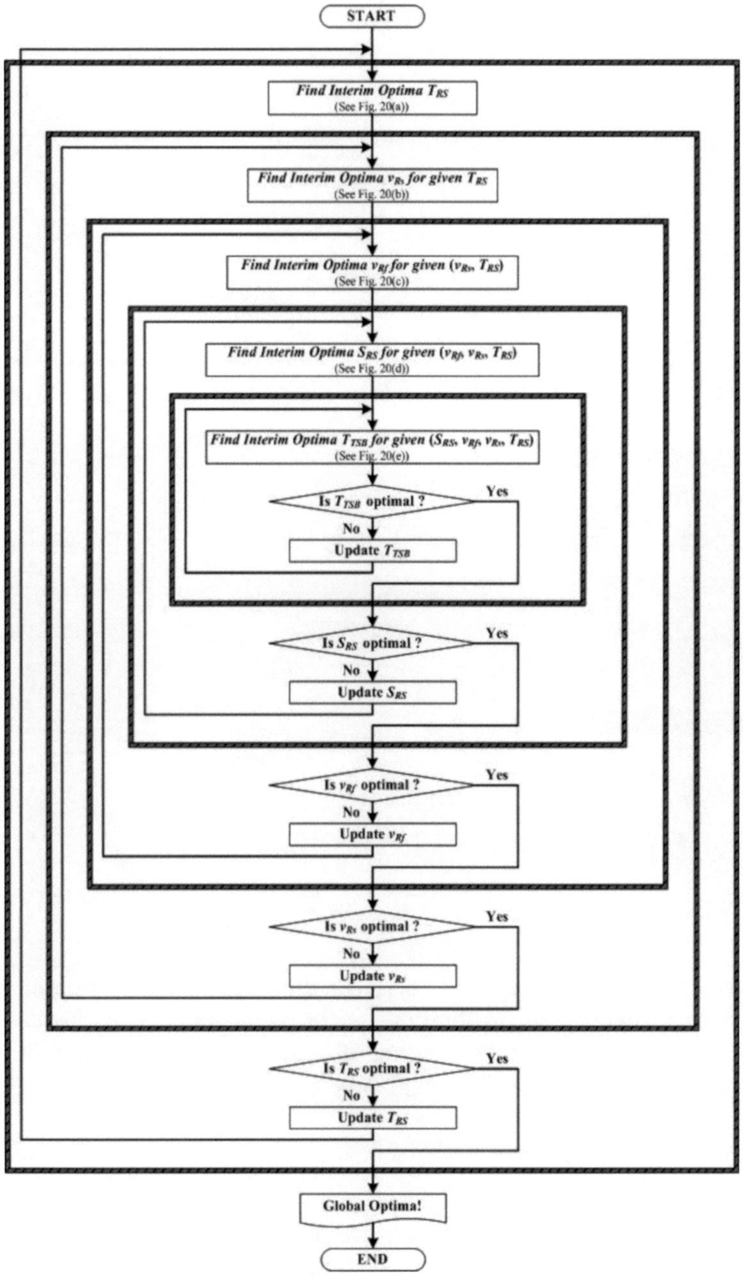

Figure 19. Overall flowchart of itertaive serach with quintuple loops to find minimum-energy turning trajectory

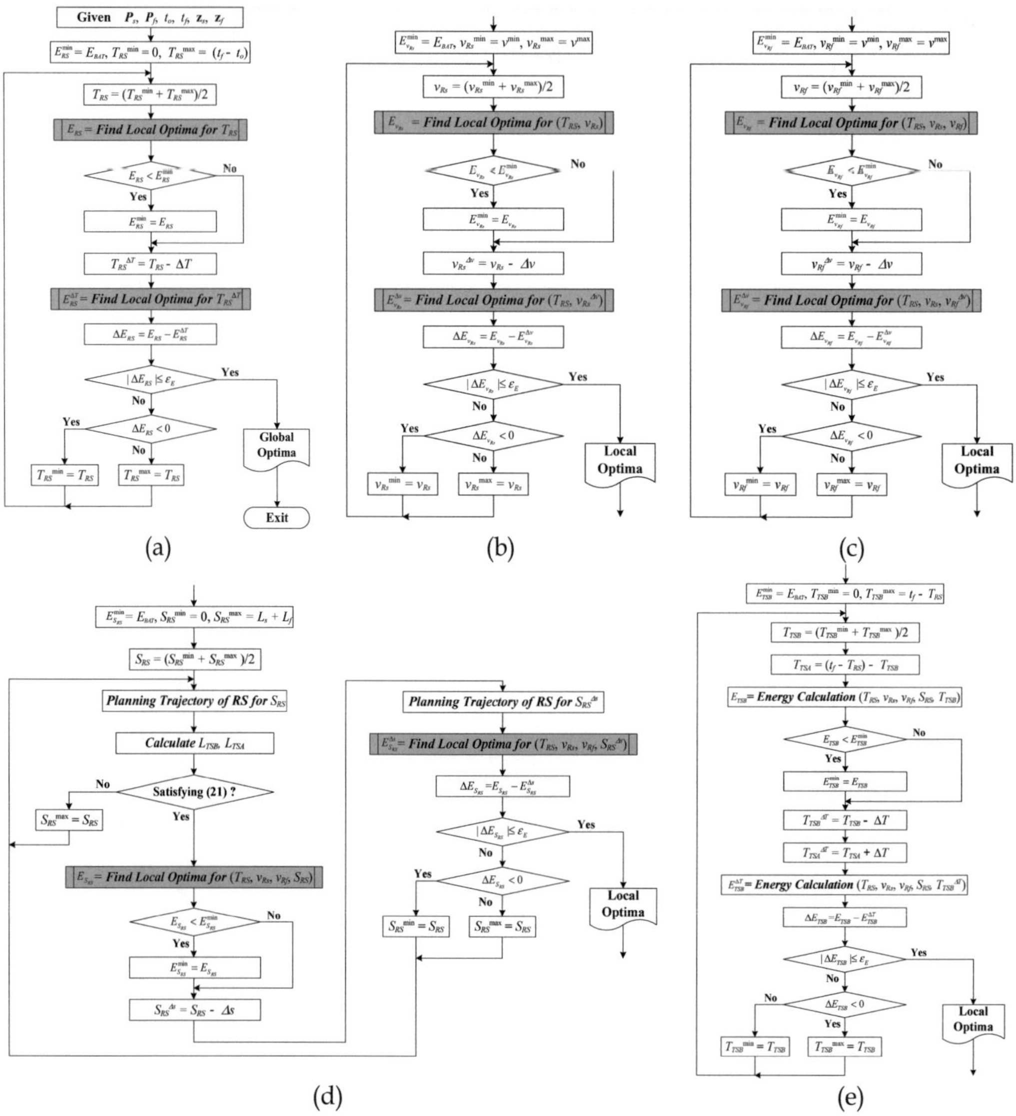

Figure 20. Partial flowchart of iterative serach with quintuple loops to find minimum-energy turning trajectory, (a) Search loop for finding T_{RS}, (b) Search loop for finding v_{RS}, (c) Search loop for finding v_{Rf}, (d) Search loop for finding ΔS_{RS} satisfying feasibility conditions of RS, Eq. (36), (e) Search loop for finding T_{TSB}

4.3 Simulations and Experiments

4.3.1 Simulations

A number of simulations were performed to evaluate the energy savings of the minimum-energy turning trajectory minimizing the cost function E_W of Eq. (9) and compared to the trajectory using loss-minimization control, which optimizes the energy loss due to armature resistance of the DC motor, E_R.

Tables 5 and 6 show the simulations results for the variables of planned turning trajectory by the minimum-energy control and loss-minimization control, respectively. For a sufficient cruise motion in *TSB* and *TSA*, based on observation of the property of the minimum-energy velocity profile shown in Fig. 5, we set the final time to t_f = 15.0s greater than $20\tau_v$ (≈ 7.8s). Since there are many interations to find minimum-energy turning trajectory, we implemented the algorithm in C language and simulation takes much time about 4 ~ 6 hours for simulation constraints of Tables 5 and 6 using a PC with Intel Core2 Duo 2.13GHz processor.

Constraints				Optimized Variables of Turning Trajectory								
$\boldsymbol{P}_f$			D	T_{RS}	v_{RS}	v_{Rf}	ΔS_{RS}	T_{TSB}	T_{TSA}	ΔS_{TSB}	ΔS_{TSA}	E_W
[(m)	(m)	(°)]	(m)	(s)	(m/s)	(m/s)	(m)	(s)	(s)	(m)	(m)	(J)
2.50	2.00	90	0.1	2.201	0.305	0.305	0.669	7.212	5.587	2.091	1.592	12.33
			0.2	4.698	0.300	0.309	1.256	5.968	4.334	1.726	1.222	11.34
2.50	1.50	120	0.1	2.109	0.347	0.314	0.556	8.726	4.165	2.944	1.320	15.58
			0.2	3.398	0.319	0.319	1.011	8.311	3.291	2.586	0.949	13.43

Table 5. Simluation results of minimum-energy turning trajectory

Constraints				Optimized Variables of Turning Trajectory								
$\boldsymbol{P}_f$			D	T_{RS}	v_{RS}	v_{Rf}	ΔS_{RS}	T_{TSB}	T_{TSA}	ΔS_{TSB}	ΔS_{TSA}	E_W
[(m)	(m)	(°)]	(m)	(s)	(m/s)	(m/s)	(m)	(s)	(s)	(m)	(m)	(J)
2.50	2.00	90	0.1	2.703	0.323	0.319	0.747	6.736	5.562	2.050	1.552	12.87
			0.2	3.724	0.370	0.348	1.384	6.310	4.966	1.652	1.168	12.21
2.50	1.50	120	0.1	2.607	0.319	0.309	0.616	7.854	4.539	2.915	1.286	16.46
			0.2	3.223	0.381	0.338	1.115	7.873	3.904	2.516	0.910	14.23

Table 6. Simluation results of loss-minimization turning trajectory

The results of energy savings for various simulations are summarized in Table 7. It shows that the minimum-energy turning trajectory can save up to 8% of the energy drawn from the batteries compared with loss-minimization turning trajectory.

$\boldsymbol{P}_f$			D	Total Energy Drawn from the Batteries		
[(m)	(m)	(°)]	(m)	*Minimum-Energy*	*Loss-Minimization*	*Energy Saving*
2.50	2.00	90	0.1	12.33J	12.87J	4.38%
			0.2	11.34J	12.21J	7.67%
2.50	1.50	120	0.1	15.58J	16.46J	5.65%
			0.2	13.43J	14.23J	5.96%

Table 7. Comparison of energy savings of minimum-energy turning trajectory planning and loss-minimization turning trajectory planning

Fig. 21 shows a typical resultant trajectory that were performed for t_f = 15.0s, P_f = [2.50m 1.50m 120°]^T, and $D = 0.2\text{m}$. The mechanical time constant affects the velocity profiles of the two optimization problems with different cost functions. Applying the Pontryagin's Maximum Principle to optimize the cost function of loss-minimization control, we obtain the mechanical time constants for loss-minimization control as follows: $\tau_v = (J_1 + J_2)/F_v$ for translational motion and $\tau_\omega = (J_1 - J_2)/F_v$ for rotational motion.

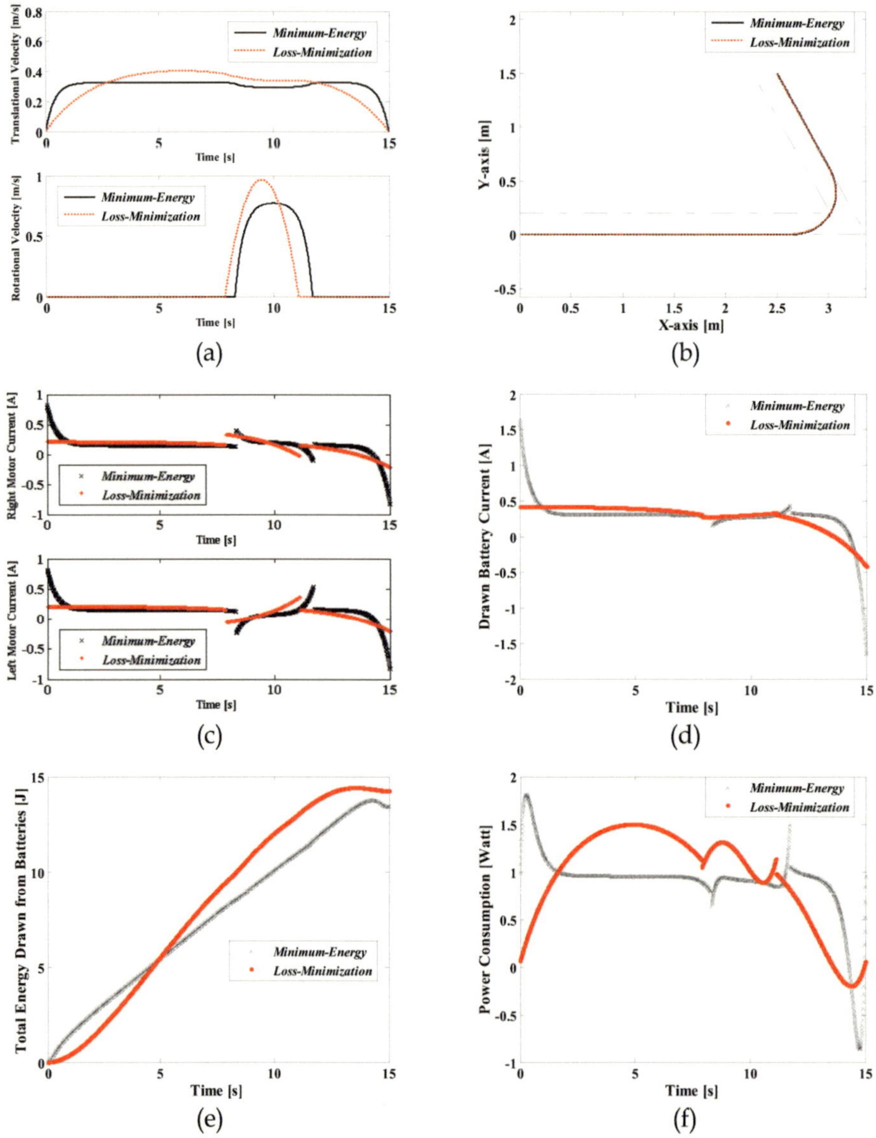

Figure 21. Simulations of minimum-energy control and loss-minimnization control turning trajectory for t_f = 15.0s, P_f = [2.50m 1.50m 120°]^T, and D=0.2m, (a) Optimal velocity profile, (b) Optimal planned path, (c) Armature current change, (d) Corresponding drawn battery current, (e) Comparison of energy consumption, (f) Corresponding power consumption

From Eq. (2), decreasing armature current increases the value of the back-emf and the motor speed. Because the mechanical time constants of minimum-energy control are less than those of loss-minimization control, the armature current in minimum-energy control quickly decreases during acceleration and deceleration, as shown in Fig. 21(d). Hence we can see that the minimum-energy control accelerates and decelerates more quickly than the loss-minimization control as shown in Fig. 21(a) and the minimum-energy control gives a cruise or cruise-like motion to WMR in *TSB* and *TSA* as we expected. However, the loss-minimization control accelerates and decelerates in whole time since its mechanical time constant is much greater than that of minimum-energy control. Fig. 21(f) shows that although the minimum-energy control requires larger energy consumption than the loss-minimization control during acceleration, it consumes less energy after acceleration. Also note that during deceleration a certain amount of energy is *regenerated* and stored into the batteries in both minimum-energy control and loss-minimization control. However, amount of regenerated energy in the loss-minimization control is much smaller than that of the minimum-energy control since deceleration rate of loss-minimization control is much smaller than that of minimum-energy control due to larger mechanical time constants.

4.3.2 Experiments

To validate the energy savings of the proposed minimum-energy turning trajectory, we performed experiments with P3-DX and compared with the loss-minimization control for the constraints of simulations. Figs. 22 and 23 show typical experimental results that were performed to compare with simulation results of Fig. 21.

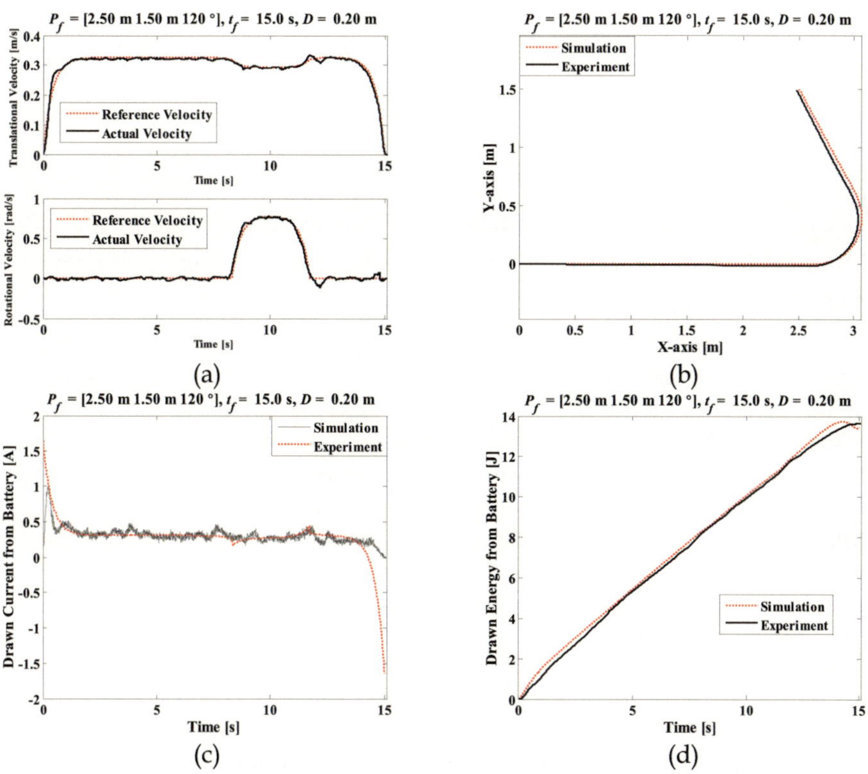

Figure 22. Experimental result of minimum-energy turning trajectory for t_f = 15.0s, $\boldsymbol{P}_f$ = [2.50m 1.50m 120°]^T, and D=0.2m, (a) Actual robot velocities, (b) Actual robot path, (c) Drawn battery current, (d) Energy Consumption

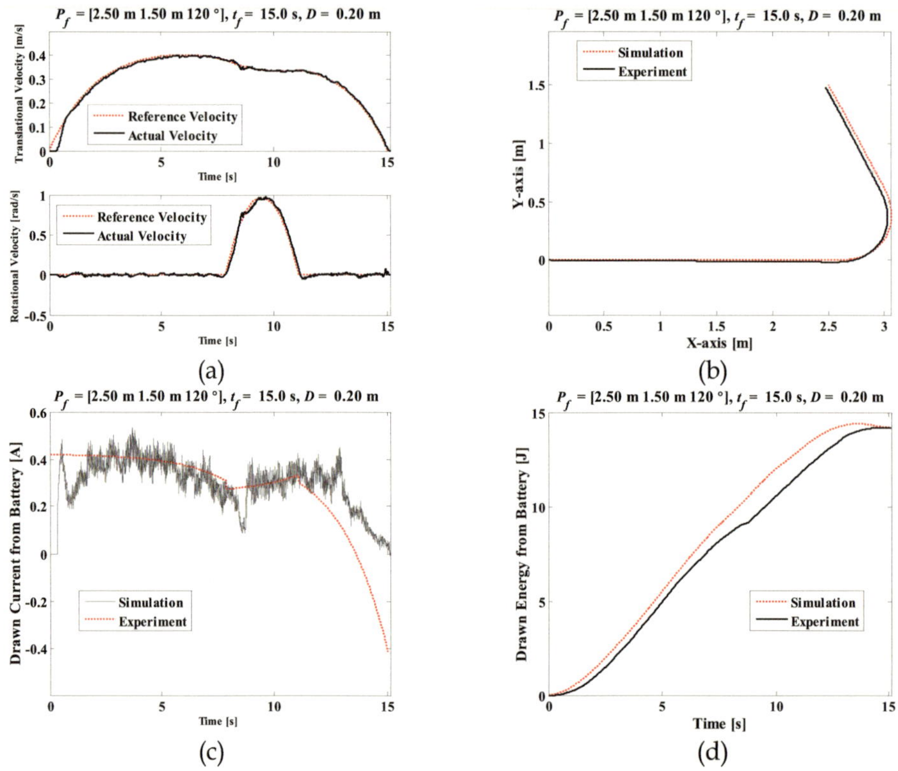

Figure 23. Experimental result of loss-minimization turning trajectory for t_f = 15.0s, $\boldsymbol{P}_f$ = $[2.50\text{m } 1.50\text{m } 120°]^T$, and D=0.2m, (a) Actual robot velocities, (b) Actual robot path, (c) Drawn battery current, (d) Energy consumption

Actual velocity of the robot follows well reference velocity as shown in Figs. 22(a) and 23(a). Since there are some errors in kinematic parameters and velocity tracking, actual trajectory is slightly different to reference trajectory as shown in Figs. 22(b) and 23(b), and there is a final position error about 20mm ~ 30mm. However, experimental results show similar response with simulation results. Figs. 22(c) and 23(c) show the drawn battery current. Since we ignore the armature inductance of the motor, drawn battery current has slightly different change during acceleration and deceleration. However, they show the similar overall response.

Table 8 shows the total energy drawn from the batteries of experiments. Values in parenthesis represent the final position errors. Experimental results revealed that the minimum-energy turning trajectory can save up to 9% of the energy drawn from the batteries compared with loss-minimization turning trajectory.

Since we ignored the inductance of the motors and there can be errors in modeling and measuring, the energy drawn from the batteries is slightly different to that for simulations. However, we can see that the minimum-energy turning trajectory can save the battery energy compared with loss-minimization turning trajectory in both expereiments and simulations. Table 8 also shows that the the percent of energy saving difference in experiments has a similar tendency that of simulations.

P_f			D	Total Energy Drawn from the Batteries		
[(m)	(m)	(°)]	(m)	*Minimum-Energy*	*Loss-Minimization*	*Energy Saving*
2.50	2.00	90	0.1	12.05J (24mm)	13.10J (17mm)	8.71%
			0.2	11.59J (23mm)	12.36J (24mm)	6.64%
2.50	1.50	120	0.1	15.34J (25mm)	16.71J (25mm)	8.93%
			0.2	13.63J (25mm)	14.21J (31mm)	4.26%

Table 8. Comparison of experimental results of minimum-energy turning trajectory planning and loss-minimization turning trajectory planning

5. Conclusion

In this book chapter, we derived the minimum-energy trajectory for WMR considering practical energy drawn from the batteries. First we investigated the *minimum-energy translational trajectory generation* moving along a straight line. Using the Pontryagin's Maximum Principle, the energy-optimal velocity profile that minimizes total energy drawn from the batteries is found to be a reasonable complex analytic form. The minimum-energy velocity profile is shown to depend on the ratio of the mechanical time constant and displacement time. Simluations show that minimum-energy control can give significant energy savings, up to 8% compared with loss-minmization control and up to 6% compared with the widely used trapezoidal velocity profile, minimizing the total energy drawn from the batteries. The experimental results also showed that the proposed minimum-energy control can save the battery energy up to 11% compared with loss-minimization control.

Since WMR also needs turning trajectory as well as translational trajectory to do useful actions, we also investigated the *minimum-energy turning trajectory planning* for WMR. To overcome nonholonomic and nonlinear properties of a WMR, we divided our trajectory into three sections. The first is *RS*, which is focused on the rotational motion with the required turning angle, and the others are *TSB* and *TSA*, which are adjoining procedures focused on translational motion. Energy optimal trajectory for each section was obtained using the Pontryagin's Maximum Principle. To combine three sections and find the minimum-energy trajectory, since there is no closed-form solution combining three solutions, we suggested an iterative search method with quintuple loops based on observations of the cost function. Since iterative search is composed of quintuple loops and there are many iterations loops to calculate robot's position due to its nonholonomic property, we implemented the algorithm in C language and simulations took several hours using a PC with Intel Core2 Duo 2.13GHz. Simulation results showed that the minimum-energy turning trajectory can save the battery energy up to 8% compared with loss-minimization turning trajectory. The experimental results also revealed that the minimum-energy turning trajectory can save up to 9% of the energy drawn from the batteries compared with loss-minimization turning trajectory.

As a further works, it remains a problem to solve about on-line trajectory planning for the overall real-time control of WMR. Also it is necessary to design a trajectory tracking controller reducing velocity tracking and posture errors for more accurate motion control for actual robots.

6. References

ActiveMedia Robotics. (2006). *Pioneer 3 operations manual*, Ver. 3

Alexander, J.C. & Maddocks, J.H. (1989). On the kinematics of wheeled mobile robots, *International Journal of Robotics Research*, Vol. 8, No. 5, pp. 15-27, ISBN 0-387-97240-4

Angelo, C.D. ; Bossio, G. & Garcia, G. (1999). Loss minimization in DC motor drives, *Proceedings of the International Conference on Electric Machines and Drives*, pp. 701-703, ISBN 0-7803-5293-9, Seattle, USA, May 1999

Aylett, R. (2002). *Robots: Bringing intelligent machines to life*, Barrons's Educational Series Inc., ISBN 0-7641-5541-5, New York

Barili, A. ; Ceresa, M. & Parisi, C. (1995). Energy-saving motion control for an autonomous mobile robot, *Proceedings of the IEEE International Symposium on Industrial Electronics*, pp. 674-676, ISBN 0-7803-2683-0, Athens, Greece, Jul. 1995

Campion, G. ; d'Andrea-Novel, B. & Bastin, G. (1991). Modeling and state feedback control of nonholonomic mechanical systems, *Proceedings of the IEEE International Conference on Decision and Control*, pp. 1184-1189, Brighton, England, Dec. 1991

Campion, G. ; Bastin, G. & d'Andrea-Novel, B. (1996). Structural properties and classification of kinematic and dynamic models of wheeled mobile robots, *IEEE Transactions on Robotics and Automation*, Vol. 12, No. 1, pp. 47-62, ISSN 1042-296x

Choi, J.S. (2001). *Trajectory planning and following for mobile robots with current and voltage constraints*, Ph.D's Thesis, KAIST, Daejeon, Korea.

Choset, H. ; Lynch, K.M. ; Hutchinson, S. ; Kantor, G. ; Burgard, W. ; Kavraki, L.E. & Thrun, S. (2005). Principles of robot motion, The MIT Press, ISBN 0-262-03327-5, Cambridge, MA

Divelbiss, A.W. & Wen, J. (1997). Trajectory tracking control of a car-trailer system, *IEEE Transactions on Control Systems Technology*, Vol. 5, No.3, pp. 269-278

Dubins, L.E. (1957). On curves of minimal length with a constraint on average curvature and with prescribed initial and terminal positions and tangents, *American Journal of Mathematics*, Vol. 79, No. 3, pp. 497-516

Egami, T. ; Morita, H. & Tsuchiya, T. (1990). Efficiency optimized model reference adaptive control system for a DC motor, *IEEE Transactions on Industrial Electronics*, Vol. 37, No. 1, pp. 28-33, ISSN 0278-0046

El-satter, A.A. ; Washsh, S. ; Zaki, A.M. & Amer, S.I. (1995). Efficiency-optimized speed control system for a separately-excited DC motor, *Proceedings of the IEEE International Conference on Industrial Electronics, Control, and Instrumentation*, pp. 417-422, Orlando, USA, Nov. 1995

Electro-Craft Corporation. (1977). *DC motors, speed control, servo systems: An engineering handbook*, Pergamon Press, ISBN 0-0802-1715-x, New York

Fiorini, P. & Shiller, Z. (1998). Motion planning in dynamic environments using velocity obstacles, *International Journal of Robotics Research*, Vol. 17, No. 7, pp. 760-772, ISSN 0278-3649

Jiang, J. ; Seneviratne, L.D. & Earles, S.W.E. (1996). Path planning for car-like robots using global analysis and local evaluation, *Proceedings of the IEEE International Conference on Emerging Technologies and Factory Automation*, Vol. 10, No. 5, pp. 577-593, ISBN 0-7803-3685-2, Kauai, USA, Nov. 1996

Kanayama, Y. & Miyake, N. (1985). Trajectory generation for mobile robots, *Proceedings of the International Symposium of Robotics Research*, pp. 333-340, ISBN 0-2620-6101-5, Gouvieux, France, Oct. 1985

Kanayama, Y. & Harman, B.I. (1989). Smooth local path planning for autonomous vehicles, *Proceedings of the IEEE International Conference on Robotics and Automation*, pp. 1265-1270, ISBN 0-8186-1938-4, Scottsdale, USA, May 1989

Kim, J. ; Yeom, H. ; Park, F.C. ; Park, Y.I. & Kim, Y. (2000). On the energy efficiency of CVT-based mobile robots, *Proceedings of the IEEE International Conference on Robotics and Automation*, pp. 1539-1544, ISBN 0-7803-5886-4, San Francisco, USA, Apr. 2000

Klein, J.T. ; Larsen, C. & Nichol, J. (2006). The design of the TransferBot: A robotic assistant for patient transfers and repositioning, *International Journal of Assistive Robotics and Mechatronics*, Vol. 7, No. 3, pp. 54-69, ISSN 1975-0153

Kusko, A. & Galler, D. (1983), Control means for minimization of losses in AC and DC motor drives, *IEEE Transactions on Industrial Applications*, Vol. 19, No. 4, pp. 561-570, ISSN 0885-8993

Kwok, S.T. & Lee, C.K. (1990). Optimal velocity profile design in incremental servo motor systems based on a digital signal processor, *Proceedings of the Annual conference of the IEEE Industrial Electronics Society*, pp. 262-266, ISBN 0-87942-600-4, Pacific Grove, USA, Nov. 1990

Laumond, J.P. ; Jacobs, P.E. & Murry, R.M. (1994). A motion planner for nonholonomic mobile robots, *IEEE Transactions on Robotics and Automation*, Vol. 10, No. 5, pp. 577-593, ISSN 1042-296x

Laumond, J.P. ; Nissoux, C. & Vendittelli, M. (1998). Obstacle distances and visibility for car-like robots moving forward, *Proceedings of the IEEE International Conference on Robotics and Automation*, pp. 33-39, ISBN 0-7803-4300-x, Leuven, Belgium, May 1998

Leonhard, W. (1996). *Control of electrical drives*, Springer-Verlag, ISBN 3-5405-9380-2, Berlin, Germany

Lineo, Inc. (2000). *DIAPM RTAI Programming Guide 1.0*, Italy

Makimoto, T. & Sakai, Y. (2003). Evolution of low power electronics and its future applications, *Proceedings of the International Symposium on Low Power Electronics and Design*, pp. 2-5, ISBN 1-58113-682-x, Seoul, Korea, Aug. 2003

Margaris, N. ; Goutal, T.; Doulgeri, Z. & Paschali, A. (1991). Loss minimization in DC drives, *IEEE Transactions on Industrial Electronics*, Vol. 38, No. 5, pp. 328-336

Mei, Y. ; Lu, Y.-H. ; Hu, Y.C. & Lee, C.S.G. (2004). Energy-efficient motion planning for mobile robots, *Proceedings of the IEEE International Conference on Robotics and Automation*, pp. 4344-4349, ISBN 0-7803-8232-3, New Orleans, USA, Apr. 2004

Mei, Y. ; Lu, Y.-H. ; Hu, Y.C. & Lee, C.S.G. (2005). Deployment strategy for mobile robots with energy and timing constraints, *Proceedings of the IEEE International Conference on Robotics and Automation*, pp. 2816-2821, ISBN 0-7803-8914-x, Barcelona, Spain, Apr. 2005

Mei, Y. ; Lu, Y.-H. ; Hu, Y.C. & Lee, C.S.G. (2006). Deployment of mobile robots with energy and timing constraints, *IEEE Transactions on Robotics and Automation*, Vol. 22, No. 3, pp. 507-522

Muir, P.F. & Neuman, C.P. (1987). Kinematic modeling of wheeled mobile robots, *Journal of Robotic Systems*, No. 2, pp. 281-329, ISSN 0741-2223

Reeds, J.A. & Shepp, R.A. (1990). Optimal paths for a car that goes both forward and backward, *Pacific Journal of Mathematics*, Vol. 145, No. 2, pp. 367-393, ISSN 0030-8730

Rybski, P.E. ; Papankolopoulos, N.P. ; Stoeter, S.A. ; Krantz, D.G. ; Yesin, K.B. ; Gini, M. ; Voyles, R. ; Hougen, D.F. ; Nelson, B. & Erickson, M.D. (2000). Enlisting rangers and scouts for reconnaissance and surveillance, *IEEE Robotics and Automation Magazine*, Vol. 7, No. 4, pp. 14-24, ISSN 1070-9932

Sergaki, E.S. ; Stavrakakis, G.S. & Pouliezos, A.D. (2002). Optimal robot speed trajectory by minimization of the actuator motor electromechanical losses, *Journal of Intelligent and Robotic Systems*, Vol. 33, No. 2, pp. 187-207, ISSN 0921-0296

Spangelo, I. & Egeland, O. (1992). Generation of energy-optimal trajectories for an autonomous underwater vehicle, *Proceedings of the IEEE International Conference on Robotics and Automation*, pp. 2107-2112, ISBN 0-8186-2720-4, Nice, France, May 1992

Tal, J. (1973). The optimal design of incremental motion servo systems, *Proceedings of the Annual Symposium on Incremental Motion Control Systems and Devices*, , pp. 72-76, Urbana, USA

Trzynadlowski, A.M. (1988). Energy optimization of a certain class of incremental motion DC drives, *IEEE Transactions on Industrial Electronics*, Vol. 35, No. 1, pp. 60-66

Weigui, W. ; Huitang, C. & Peng-Yung, W. (1999). Optimal motion planning for a wheeled mobile robot, *Proceedings of the IEEE International Conference on Robotics and Automation*, pp. 41-46, ISBN 0-7803-5180-0, Detroit, USA, May 1999

Yun, X. and Yamamoto, Y. (1993). Internal dynamics of a wheeled mobile robot, *Proceedings of the IEEE/RSJ International Conference on Intelligent Robots and Systems*, Vol. 2, pp. 1288-1294, ISBN 0-7803-0823-9, Yokohama, Japan, Jul. 1993

Yun, X. (1995). State space representation of holonomic and nonholonomic constraints resulting from rolling contacts, *Proceedings of the IEEE International Conference on Robotics and Automation*, pp. 2690-2694, ISBN 0-7803-1965-6, Nagoya, Japan, May 1995

Yun, X. & Sarkar, N. (1998). Unified formulation of robotic systems with holonomic and nonholonomic constraints, *IEEE Transactions on Robotics and Automation*, Vol. 14, No. 4, pp. 640-650, ISSN 1042-296x

Zhang, G. ; Schmidhofer, A. & Schmid, A. (2003). Efficiency optimization at DC drives for small electrical vehicles, *Proceedings of the IEEE International Conference on Industrial Technology*, pp. 1150-1155, ISBN 0-7803-7852-0, Maribor, Slovenia, Dec. 2003

12

Performance Evaluation of Potential Field based Distributed Motion Planning Methods for Robot Collectives

Leng-Feng Lee and Venkat N. Krovi
University at Buffalo
USA

1. Introduction

In the past decade, ongoing revolutions in computing effectiveness and miniaturization of processors/sensors/actuators have facilitated the transition of research focus from an individual mobile robot to networked distributed teams of mobile robots. The development of effective motion-planning methods for such collectives is critical to realizing the full potential of the group in numerous applications ranging from reconnaissance, foraging, herding to cooperative payload transport.

While considerable literature exists for motion planning of individual mobile agents, the renewed challenge lies in creating motion plans for the entire team while incorporating notions such as cooperation. The "formation" paradigm has emerged as a convenient mechanism for abstraction and coordination with approaches ranging from leader-following (Wang, 1991; Desai et al., 2001), virtual structures (Lewis and Tan, 1997; Beard et al., 2001) and virtual leaders (Leonard and Fiorelli, 2001; Ogren et al., 2002). The group control problem now reduces to a well-known single-agent control problem from which the other agents derive their control laws but requires communication of some coordination information. Early implementations involved the kinematic specification of the followers' motion-plans as a "prescribed motions" relative to a team-leader without the ability to affect the dynamics of the leader. Subsequent approaches have incorporated some form of "formation-feedback" from the members to the overall group using natural or artificially introduced dynamics within the constraints. The formation paradigm has evolved to allow prescription of parameterized formation maneuvers and group feedback (Egerstedt and Hu, 2001; Young et al., 2001; Ogren et al., 2002). From these seemingly disparate approaches, a dynamic system-theoretic perspective has emerged for examining the decentralized multi-agent "behavioral control" in the context of "formations" (Lawton et al., 2000; Egerstedt and Hu, 2001; Leonard and Fiorelli, 2001; Young et al., 2001; Ogren et al., 2002; Olfati-Saber and Murray, 2006). "Behavioral" control laws, derived implicitly as gradients of limited-range artificial potentials, can be implemented in a decentralized manner while permitting a Lyapunov-based analysis of formation maintenance.

Various variants of the Artificial Potential Field (APF) framework have been leveraged in implementing such behavioral motion-planning/control of robot collectives due to their seeming ease of formulation, decentralization and scalability. However, we note that while stability guarantees (typically asymptotic) may be obtained, APF approaches are unable to guarantee strict formation maintenance. Such strict formation maintenance is critical in applications such as cooperative payload transport by collectives (Bhatt et al., 2008) or in distributed sensor deployment applications where the robots are to form some geometric pattern and maintain it while moving about in the world (Young et al., 2001).

We note that the group of independent mobile robots moving together in formation and coupled together by constraint dynamics can alternatively be viewed as a constrained mechanical system. The computation of motion plans for such collectives in a potential field may also be viewed as simulating the forward dynamics of a constrained multi-body mechanical system. By doing so, we would like to link (and leverage) the extensive literature on formulation and implementation of computational simulation of multibody systems (Haug, 1989; Shabana, 1989; Schiehlen, 1990; García de Jalón and Bayo, 1994; Ascher and Petzold, 1998) to the problem of motion planning of mobile robot collectives.

In this chapter, *we evaluate the formation maintenance performance of several formulations* developed by analogy to the approaches used for constrained mechanical systems. These include: (i) a direct Lagrangian multiplier elimination approach (to serve as the benchmark); (ii) a penalty-formulation approach which is the most popular implementation; and (iii) a constraint manifold projection approach. We note that the instabilities introduced in the form of the "formulation stiffness" at the algorithm development stage have the potential to hinder the subsequent control and requires a careful quantitative examination (Ascher et al., 1997). Hence, we compare and contrast the various approaches on the basis of modular formulation, distributed computation and relative computational efficiency and accuracy. These aspects are studied in the context of the motion-planning of a group of point-mass mobile robots which are constrained together by means of rheonomous holonomic constraints.

The rest of the paper is organized as follows: Section 2 presents a brief discussion of various candidate formulations of forward dynamics approaches for constrained multibody systems. In Section 3, the dynamic model of the system of point-mass robots moving in plane is introduced and the candidate methods are evaluated from viewpoint of distribution of computation. Section 4 discusses the standardized test arena and the performance evaluation metric which is then used in Section 5 to compare and contrast the methods. Section 6 presents a brief discussion and concluding remarks.

2. Forward Dynamics Formulations for Constrained Mechanical Systems

In this section we briefly review some of the available alternative formulations for developing the forward dynamics simulations in constrained mechanical systems. At the outset, we note that suitable selection of a set of *configuration coordinates* is of particular importance due to its impact both on the ease of formulation and the subsequent computational efficiency. We make use of expanded sets of *dependent Cartesian coordinates* linked together by *holonomic constraints* as being most appropriate for modular composition and general-purpose analysis.

The overall dynamics can be formulated as a system of ODEs whose solutions are required to satisfy additional holonomic (algebraic) constraint equations as Lagrangian equations of the first kind (Arnold, 1989):

$$\dot{\underset{\sim}{q}} = \underset{\sim}{v} \tag{1}$$

$$M\left(\underset{\sim}{q}\right)\dot{\underset{\sim}{v}} = \underset{\sim}{f}\left(\underset{\sim}{q},\underset{\sim}{v},\underset{\sim}{u},t\right) - A^T\left(\underset{\sim}{q}\right)\underset{\sim}{\lambda} \tag{2}$$

$$\underset{\sim}{C}\left(\underset{\sim}{q},t\right) = \underset{\sim}{0} \tag{3}$$

where

$\underset{\sim}{q}$ is the n-dimensional vector of generalized coordinates.

$\underset{\sim}{v}$ is the n-dimensional vector of generalized velocities.

$M\left(\underset{\sim}{q}\right)$ is the $n \times n$ dimensional inertia matrix.

$\underset{\sim}{f}\left(\underset{\sim}{q},\underset{\sim}{v},\underset{\sim}{u},t\right)$ is the n-dimensional vector of external forces.

$\underset{\sim}{u}$ is the vector of actuator forces/torques.

$\underset{\sim}{C}\left(\underset{\sim}{q},t\right)$ is the m-dimensional vector of holonomic constraints.

$A\left(\underset{\sim}{q}\right) = \partial \underset{\sim}{C} / \partial \underset{\sim}{q}$ is the $m \times n$ dimensional constraint Jacobian matrix.

$\underset{\sim}{\lambda}$ is the m-dimensional vector of Lagrange multipliers.

The solution of resulting system of index-3 Differential Algebraic Equations (DAEs) by direct finite difference discretization is not possible using explicit discretization methods. We adopt a *converted ODE approach,* wherein all the algebraic position and velocity level constraints are differentiated and represented at the acceleration level to obtain an augmented index-1 DAE (in terms of both, the unknown accelerations and the unknown multipliers). Differentiating the position constraints in Eq.(3), with respect to time, yields the velocity-level constraints:

$$\dot{\underset{\sim}{C}} = A\underset{\sim}{v} = \underset{\sim}{0} \tag{4}$$

Further differentiation with respect to time yields the acceleration level constraints as:

$$\ddot{\underset{\sim}{C}} = A\dot{\underset{\sim}{v}} + \dot{A}\underset{\sim}{v} = \underset{\sim}{0} \tag{5}$$

Thus, Eq. (2) can then be written together with Eq. (5) as an index-1 DAE as:

$$\begin{bmatrix} M & A^T \\ A & 0 \end{bmatrix}_{(n+m)\times(n+m)} \begin{bmatrix} \dot{\underset{\sim}{v}} \\ \underset{\sim}{\lambda} \end{bmatrix}_{(n+m)\times 1} = \begin{bmatrix} \underset{\sim}{f} \\ -\dot{A}\underset{\sim}{v} \end{bmatrix}_{(n+m)\times 1} \tag{6}$$

In a typical forward dynamics simulation setting, the index-1 DAE systems resulting from the converted ODE approach are then converted into final system of first-order ODEs by: (a) direct Lagrange multiplier elimination; (b) penalty-formulation; or (c) constraint manifold projection.

2.1 Direct Lagrange Multiplier Elimination

In this approach, a simultaneous solution of the augmented linear system of Eq. (6) is obtained at each time step. While an explicit inversion of the augmented system may be avoided by adopting a Gaussian elimination method, the overall approach may still be denoted as:

$$\begin{bmatrix} \dot{\underset{\sim}{v}} \\ \underset{\sim}{\lambda} \end{bmatrix} = \begin{bmatrix} M & A^T \\ A & 0 \end{bmatrix}^{-1} \begin{bmatrix} \underset{\sim}{f} \\ -\dot{A}\underset{\sim}{v} \end{bmatrix} = \begin{bmatrix} \underset{\sim}{f_1}\left(\underset{\sim}{q},\underset{\sim}{v}\right) \\ \underset{\sim}{f_2}\left(\underset{\sim}{q},\underset{\sim}{v}\right) \end{bmatrix} \tag{7}$$

Thus, the overall system may now be written as a system of first order ODEs as:

$$\dot{\underset{\sim}{x}}_{2n\times 1} = \begin{bmatrix} \dot{\underset{\sim}{q}}_{n\times 1} \\ \ddot{\underset{\sim}{q}}_{n\times 1} \end{bmatrix} = \begin{bmatrix} \underset{\sim}{v} \\ \underset{\sim}{f_1}\left(\underset{\sim}{q},\underset{\sim}{v}\right) \end{bmatrix} \tag{8}$$

which may then be integrated using standard numerical solvers. The main advantage is its conceptual simplicity and simultaneous determination of the accelerations and the Lagrange multipliers by solving a linear system of equations. However, this is a centralized approach and does not scale up very well.

2.2 Penalty-Formulation

In penalty-based approaches the holonomic constraints are relaxed and replaced by linear/non-linear virtual springs and dampers, thereby incorporating the constraint equations as a dynamical system penalized by a large factor. The Lagrange multipliers are *approximated* using a virtual spring type law (based on the extent of the constraint violation and assumed spring stiffness) and eliminated from the list of $n+m$ unknowns leaving behind a system of $2n$ first order ODEs. While the sole initial drawback may appear to be restricted to the numerical ill-conditioning due to selection of large penalty factors, it is important to note that penalty approaches only *approximate* the true constraint forces and can create unanticipated problems (as will be discussed later). This individual multiplier values can be explicitly calculated as $\lambda_i = K_{P_i} C_i + K_{D_i} \dot{C}_i$ where K_{P_i} is the spring constant, K_{D_i} is the damping constant and $C_i\left(\underset{\sim}{q}\right)$ is the constraint violation in the direction of the respective λ_i. By substituting the value of $\underset{\sim}{\lambda}$ in Eq. (2), the final ODE system can be written as:

$$\dot{\underset{\sim}{x}}_{2n\times 1} = \begin{bmatrix} \dot{\underset{\sim}{q}}_{n\times 1} \\ \ddot{\underset{\sim}{q}}_{n\times 1} \end{bmatrix} = \begin{bmatrix} \underset{\sim}{v} \\ M^{-1}\left(\underset{\sim}{f} - A^T\left(K_P\underset{\sim}{C} + K_D\dot{\underset{\sim}{C}}\right)\right) \end{bmatrix} \tag{9}$$

where $K_P = diag\left[K_{P_i}\right]$ and $K_D = diag\left[K_{D_i}\right]$.

2.3 Constraint Manifold Projection

This approach seeks to take the dynamical equations with constraint-reactions into the *tangent and cotangent* subspace. The rheonomous holonomic constraints, $\underset{\sim}{C}\left(\underset{\sim}{q},t\right)=\underset{\sim}{0}$, can be written in differential form as:

$$\left[\frac{\partial \underset{\sim}{C}}{\partial \underset{\sim}{q}}\right]\dot{\underset{\sim}{q}}+\left[\frac{\partial \underset{\sim}{C}}{\partial t}\right]=\underset{\sim}{0} \Leftrightarrow A\dot{\underset{\sim}{q}}=\underset{\sim}{a}\left(\underset{\sim}{q}\right) \tag{10}$$

Let $S\left(\underset{\sim}{q}\right)$ be an $n\times(n-m)$ dimensional full rank matrix whose column space is in the null space of A i.e. $AS=\underset{\sim}{0}$. The orthogonal subspace is spanned by the so-called constraint vectors (forming the rows of the matrix A) while the tangent subspace *complements* this orthogonal subspace in the overall generalized velocity vector space. All *feasible* dependent velocities, $\dot{\underset{\sim}{q}}$, of a constrained multibody system necessarily belong to this tangent space, appropriately called the *space of feasible motions.* This space is spanned by the columns of S) and is parameterized by an $(n-m)$-dimensional vector of independent velocities, $\underset{\sim}{\nu}(t)$, yielding the expression for the feasible dependent velocities as:

$$\dot{\underset{\sim}{q}}=\underset{\sim}{v}=S\underset{\sim}{\nu}(t)+\underset{\sim}{\eta}\left(\underset{\sim}{q}\right) \tag{11}$$

where $\underset{\sim}{\eta}\left(\underset{\sim}{q}\right)$ is the particular solution of (10). Differentiating this further we get:

$$\dot{\underset{\sim}{v}}=S\dot{\underset{\sim}{\nu}}+\dot{S}\underset{\sim}{\nu}+\dot{\underset{\sim}{\eta}}=S\dot{\underset{\sim}{\nu}}+\underset{\sim}{\gamma}\left(\underset{\sim}{q},\underset{\sim}{\nu}\right) \tag{12}$$

where $\underset{\sim}{\gamma}\left(\underset{\sim}{q},\underset{\sim}{\nu}\right)=\dot{S}\underset{\sim}{\nu}+\dot{\underset{\sim}{\eta}}$ needs to be calculated numerically which has potential of introducing errors. In order to avoid this situation, we adopted the method in (Yun and Sarkar, 1998).

Such a projection process works out well in a Riemannian setting (where the notion of orthogonal complement subspaces exists). Special care needs to be exercised when treating configuration spaces such as $SE(2)$ or $SE(3)$. A family of projections exists depending on selection of dependent/independent velocities. However, once a projection is selected, the dynamic equations of motion can now be projected on to the instantaneous feasible motion directions, to obtain the so-called constraint-reaction-free equations of motion. Pre-multiplying both sides of Eq. (2) by S^T and noting that $S^T A^T=\underset{\sim}{0}$ we get:

$$S^T M \dot{\underset{\sim}{v}} = S^T \underset{\sim}{f} \tag{13}$$

By substituting $\dot{v}$ from Eq. (12) into Eq. (13) and solving for $\dot{v}$ we get:

$$\dot{\underset{\sim}{\nu}} = -\left(S^T M S\right)^{-1}\left(S^T M \underset{\sim}{\gamma} - S^T \underset{\sim}{f}\right) \tag{14}$$

The resulting overall system of ODEs may be expressed in state-space form as:

$$\dot{\underset{\sim}{x}}_{(2n-m)\times 1} = \begin{bmatrix} \dot{\underset{\sim}{q}}_{n\times 1} \\ \dot{\underset{\sim}{\nu}}_{(n-m)\times 1} \end{bmatrix} = \begin{bmatrix} S\underset{\sim}{\nu} + \underset{\sim}{\eta} \\ -\left(S^T M S\right)^{-1}\left(S^T M \underset{\sim}{\gamma} - S^T \underset{\sim}{f}\right) \end{bmatrix} \tag{15}$$

The final solution may be obtained either by numerically integrating a system of $2n-m$ first-order ODEs in the n dependent velocities and $n-m$ independent accelerations.

2.4 Baumgarte Stabilization

The drawbacks of the Constraint Manifold Projection approach include: (i) the need to provide additional consistent initial conditions; and (ii) the mild instability of the differentiated constraints resulting in state-drift from the position-level constraint manifold. While the growth rate can be reduced by lowering the error tolerance and by using smaller step-sizes or greater numerical precision, this comes at the cost of longer and more expensive computations. Baumgarte stabilization (Baumgarte, 1983) involves the creation of an artificial first or second-order dynamical system which has the algebraic position-level constraint as its attractive equilibrium configuration. For example, when the holonomic constraints in Eq. (3) are approximated by a first order system of the form, we obtain:

$$\dot{\underset{\sim}{C}}\left(\underset{\sim}{q},t\right) + \sigma \underset{\sim}{C}\left(\underset{\sim}{q},t\right) = \underset{\sim}{0}, \quad \sigma > 0 \tag{16}$$

where σ is the rate of convergence. The equilibrium condition for this first order system is the constraint manifold $\underset{\sim}{C}\left(\underset{\sim}{q},t\right) = \underset{\sim}{0}$ and for any initial condition $\underset{\sim}{q}(0)$, which may not satisfy the holonomic constraint equation $\underset{\sim}{C}\left(\underset{\sim}{q}(0)\right) = \underset{\sim}{0}$, the above first order equation guarantees exponential convergence of $\underset{\sim}{C}\left(\underset{\sim}{q}(t)\right) = \underset{\sim}{0}$ to zero as the time t progresses. The rate of convergence will be determined by σ, which can be chosen based on specific application. Eq. (16) can be suitably modified as:

$$\left[\frac{\partial \underset{\sim}{C}}{\partial \underset{\sim}{q}}\right] \dot{\underset{\sim}{q}} = -\sigma \underset{\sim}{C} - \left[\frac{\partial \underset{\sim}{C}}{\partial t}\right] \Leftrightarrow A\dot{\underset{\sim}{q}} = \underset{\sim}{\bar{a}}\left(\underset{\sim}{q}\right) \tag{17}$$

and the rest of solution process remains unchanged. While Baumgarte's technique is very popular in the engineering application community, principally due to the resulting augmented ODE formulation, the practical selection of the parameters of the stabilization system depends both on the discretization methods and step-size and is widely regarded as an open research problem (Ascher et al., 1995).

3. Distributed Modeling of the N-Mobile Robot Collective

We consider a collective formed by N -robots, each with point-mass m_i operating in the horizontal plane with a configuration vector $\underset{\sim}{q}_i = [x_i, y_i]^T \in \mathbb{R}^2$ w.r.t an inertial frame $\{F\}$, as shown in Fig 1.

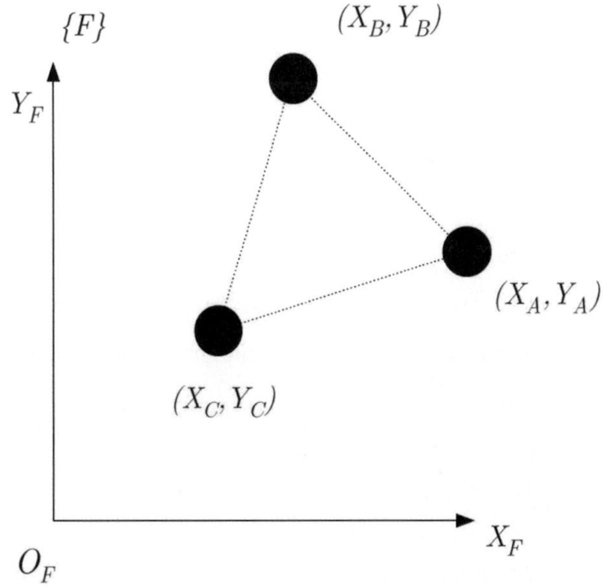

Figure 1. A robot collective form by $N = 3$ point mass robots operating in the horizontal plane w.r.t an inertial frame $\{F\}$

The governing EOM for each robot take the simple form $M_i \ddot{\underset{\sim}{q}}_i = \underset{\sim}{u}$, where $M_i = m_i I_{2\times2}$ $\forall i = 1,\cdots,N$. The equations of the overall collective moving in formation can be written in an index-3 DAE form as:

$$\begin{aligned} \dot{\underset{\sim}{q}} &= \underset{\sim}{v} \\ M(\underset{\sim}{q})\ddot{\underset{\sim}{q}} + \underset{\sim}{V}(\underset{\sim}{q}, \dot{\underset{\sim}{q}}) + \underset{\sim}{G}(\underset{\sim}{q}) &= E(\underset{\sim}{q})\underset{\sim}{u} - A^T \underset{\sim}{\lambda} \\ \underset{\sim}{C}(\underset{\sim}{q}) &= \underset{\sim}{0} \end{aligned} \tag{18}$$

where

$$\underset{\sim}{q}=\begin{bmatrix}\underset{\sim}{q}_1\\ \vdots\\ \underset{\sim}{q}_N\end{bmatrix};\ M\left(\underset{\sim}{q}\right)=\begin{bmatrix}[M_1]_{2\times2} & \cdots & \underset{\sim}{0}\\ \vdots & \ddots & \vdots\\ \underset{\sim}{0} & \cdots & [M_N]_{2\times2}\end{bmatrix};\ \underset{\sim}{u}=-k\left(\nabla_{\underset{\sim}{q}}V_p\right)^T;\ E\left(\underset{\sim}{q}\right)=I_{2N\times2N};\ G\left(\underset{\sim}{q}\right)=\underset{\sim}{0};$$

$$V\left(\underset{\sim}{q},\dot{\underset{\sim}{q}}\right)=\underset{\sim}{0}$$

We will consider the case where a *rigid formation is desired.* The $2N-3$ constraint equations (Olfati-Saber and Murray, 2002) that maintain the rigidity are obtained from the requirement that each robot tries to maintain a desired distance with the others:

$$C\left(\underset{\sim}{q}\right)=\left[C_{ij}\left(\underset{\sim}{q}\right)\right]_{(2N-3)\times1}=\underset{\sim}{0} \tag{19}$$

Equation (18) represents the centralized form of the governing equations using artificial potentials. We now consider the possibility of distributing the motion-planning computations between the multiple agents. Further details are available in Lee (2004).

3.1 The Penalty Formulation

Noting that the state vector $\underset{\sim}{q}=\left[\underset{\sim}{q}_A^T\ \ \underset{\sim}{q}_B^T\ \ \underset{\sim}{q}_C^T\right]^T$ has state variables belonging to each of the robots A,B and C, the distributed model may be obtained in state-space form as:

$$\left[\underset{\sim}{\dot{x}}_i\right]_{4\times1}=\begin{bmatrix}\dot{\underset{\sim}{q}}_i\\ \ddot{\underset{\sim}{q}}_i\end{bmatrix}=\begin{bmatrix}\underset{\sim}{v}_i\\ M_i^{-1}\left(E_i\underset{\sim}{u}_i-A_i^T\left(K_{P_i}\underset{\sim}{C}_i+K_{D_i}\dot{\underset{\sim}{C}}_i\right)\right)\end{bmatrix};\quad \forall i=A,B,C \tag{20}$$

where K_{P_i}, K_{D_i} are the compliance and damping matrices and $\underset{\sim}{C}_i$ represents the extent of the constraint violation as pertinent to robot i.
The three dynamic sub-systems, shown in Eq. (20), can be simulated in a distributed manner if at every time step: (i) either the information pertaining to $\underset{\sim}{C}_i\left(\underset{\sim}{q}\right)$, the extent of the constraint violation, is made available explicitly or (ii) computed by exchanging state information between the robots. The sole coupling between the two sub-parts is due to the Lagrange multipliers, which are now explicitly calculated using the virtual spring. While this is shown for a "three robot system", the process generalizes easily for "N-robot" system.

3.2 Constraint Manifold Projection

We examine this approach as an appropriate alternative to the penalty formulation where again our emphasis is on distribution of the motion planning computations to be performed by the individual robots. Noting that the state vector may be written as $\underset{\sim}{q}=\left[\underset{\sim}{q}_A^T\ \ \underset{\sim}{q}_B^T\ \ \underset{\sim}{q}_C^T\right]^T$, the projected dynamics equations may be partitioned as:

$$\begin{bmatrix} S_A^T & S_B^T & S_C^T \end{bmatrix} \begin{bmatrix} M_A & 0 & 0 \\ 0 & M_B & 0 \\ 0 & 0 & M_C \end{bmatrix} \begin{bmatrix} S_A^T \\ S_B^T \\ S_C^T \end{bmatrix} \dot{\underset{\sim}{v}} + \begin{bmatrix} S_A^T & S_B^T & S_C^T \end{bmatrix} \begin{bmatrix} M_A & 0 & 0 \\ 0 & M_B & 0 \\ 0 & 0 & M_C \end{bmatrix} \begin{bmatrix} \underset{\sim}{\gamma}_A^T \\ \underset{\sim}{\gamma}_B^T \\ \underset{\sim}{\gamma}_C^T \end{bmatrix} = \begin{bmatrix} S_A^T & S_B^T & S_C^T \end{bmatrix} \begin{bmatrix} I_{6\times 6} \end{bmatrix} \begin{bmatrix} \underset{\sim}{u}_A^T \\ \underset{\sim}{u}_B^T \\ \underset{\sim}{u}_C^T \end{bmatrix} \tag{21}$$

Thus, it is now possible to calculate the state vectors forming separately as:

$$\dot{\underset{\sim}{x}}_i = \begin{bmatrix} \dot{\underset{\sim}{q}}_i \\ \dot{\underset{\sim}{\nu}}_i \end{bmatrix} = \begin{bmatrix} S_i \underset{\sim}{\nu} + \eta_i \\ -\left(S^T M S\right)^{-1} \left(S_i^T \left(M_i \underset{\sim}{\gamma}_i - E_i \underset{\sim}{u}_i\right)\right) \end{bmatrix}; \ \forall i = A,B,C \tag{22}$$

By examining Eq. (21), we note that the overall system can be evaluated in a distributed manner if states $\underset{\sim}{q}_i$ and $\underset{\sim}{\nu}_i$ are made available. Each independent sub-part can now be numerically integrated on a mobile robot thereby permitting independent operation. At each time-instant, the complete state of the system needs to be exchanged between the robots. The coupling between the various sub-parts is due to the existence of the $\left(S^T MS\right)^{-1}$. This matrix inverse needs to be computed on each and every processor (although we note that the explicit calculation of the inverse is typically avoided by using an optimal equation solver). Alternatively, state information from the slave processors could be collected by a central processor at each time instant, the $\left(S^T MS\right)^{-1}$ computed and the result subsequently broadcasted to all robots.

4. The Standardized Test Arena

In order to compare the performance of various methods for motion planning of robot collectives within a potential-field framework, we developed a standardized test course. A graphical user interface (GUI) is used to locate the positions of the initial robot configurations, the obstacles and the target. As shown in Figure 2(a). Then an APF is developed in the form of a navigation function (Rimon and Koditschek, 1992) to ensure a unique minimum. This is shown as a 3D plot in Figure 2(b) and as a contour plot in Figure 2(c).

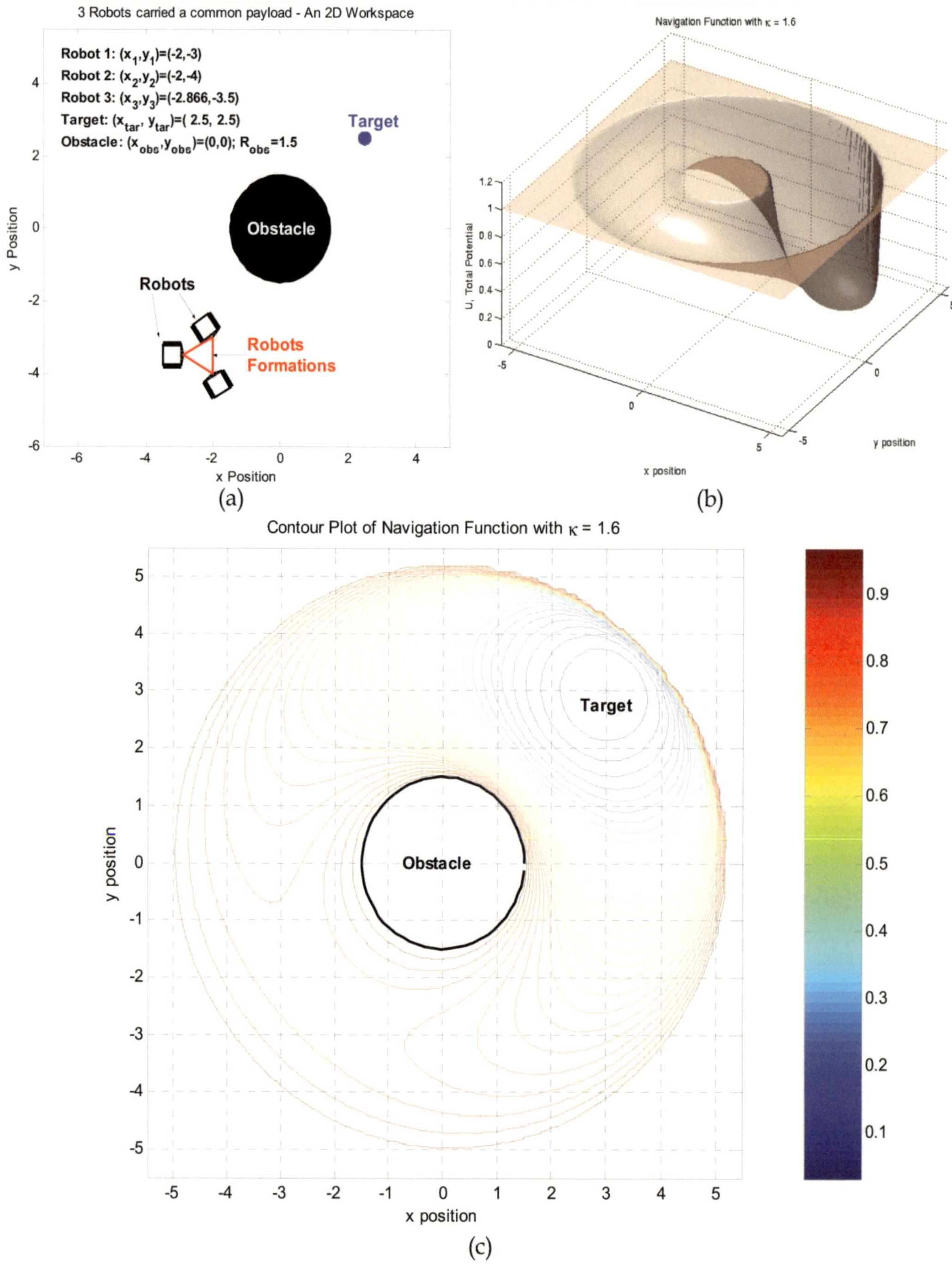

Figure 2. Standard test course for performance measurement (a) Formation, with environment and target; (b) 3-d plot; and (c) contour plot of the generated navigation function

5. Performance Evaluation

In what follows, we treat the results from the forward dynamic simulations with a fixed time step as generating the motion plans for the robot collective. In Section 3, we examined how both the penalty- and projection-based formulations for motion planning of N -robot collective (for $N = 3,\ 10$), that could be distributed to run on separate processors (requiring only the exchange of state information at every time instant). This is implemented using MATLAB. Figure 3(a) shows the corresponding simulation result of the 3-robot collective while Figure 3(b) shows the results from a larger 10-robot collective.

We then study the performance of the various formulations in the context of accumulated individual constraint errors as well as the overall formation error for a fixed time step and additionally study the effects of varying the time-step. The formation error is computed as $\left\| \underset{\sim}{C}\left(\underset{\sim}{q}\right) \right\|$ and corresponds to the structural error used by Egerstedt and Hu (2001) and Olfati-Saber and Murray (2006). A number of simulations with different values for fixed time-steps (ranging from 1e-2 to 1e-4) were performed by Lee (2004). However, only the resulting formation errors from running the two methods for a fixed step size of 1e-3 seconds are shown in Figure 4. Each method has independent parameters that could potentially affect the performance of the method – the virtual spring/damper parameters $\left(K_{P_i}, K_{D_i}\right)$ in the penalty formulation and the stabilization factor (σ in the constraint manifold projection method). The effects of these parameters are studied in greater detail in Lee (2004).

Figure 4(a) shows the results from the benchmark formulation using direct Lagrange elimination method. In Figure 4(c), we note that the selection of the value of the independent parameter σ only plays a minor role since regardless of the selected value the constraint error remains near about 1e-6 which is in agreement with benchmark problem. In contrast, in Figure 4(b) we see that for small values of the spring stiffness, considerable constraint error results which decreases as K_P is increased. While this constraint error reduces to the order of 1e-3 as the spring stiffness is increased to 500, formation maintenance never reaches the levels observed for the projection-based method.

We also performed a similar simulation studies (with fixed time-step of 1e-3 seconds) with 10 point mass mobile robots in order to test the scalability of the adopted approaches, the corresponding simulation result is shown in Figure 3(b). The results shown in Figure 5 follow the same general trend observed in Figure 4. However, the distinction between the three methods (as manifested in the total formation error) is far more pronounced.

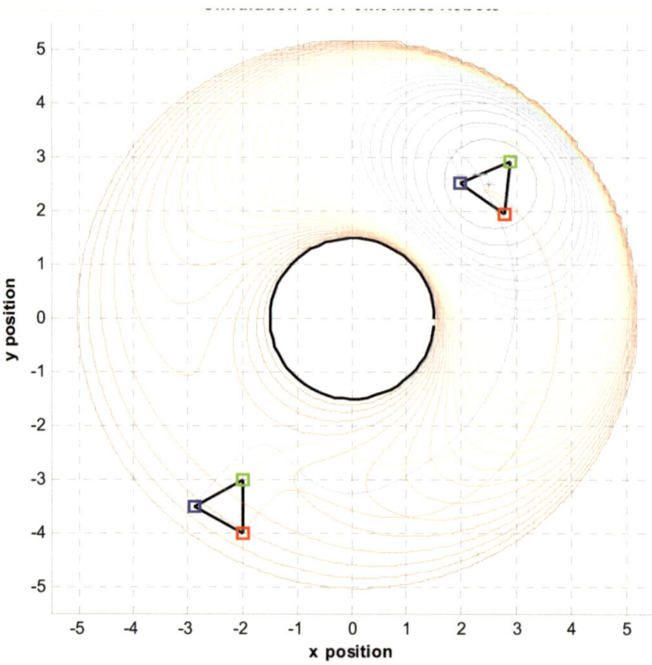

(a)

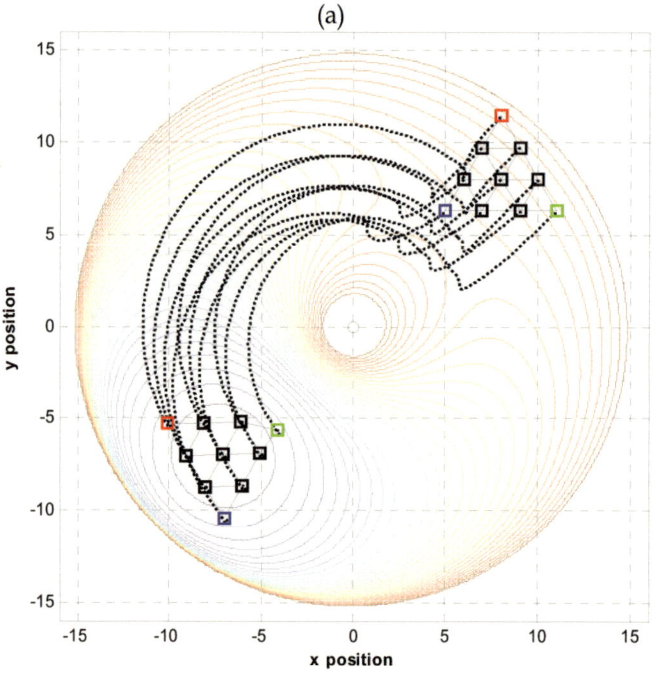

(b)

Figure 3. (a) The simulation result showing the three robots in a triangular formation move from their initial position to the target position while maintaining formation; and (b) The simulation result of 10 robots forming an interconnected triangular formation in a workspace with one obstacle

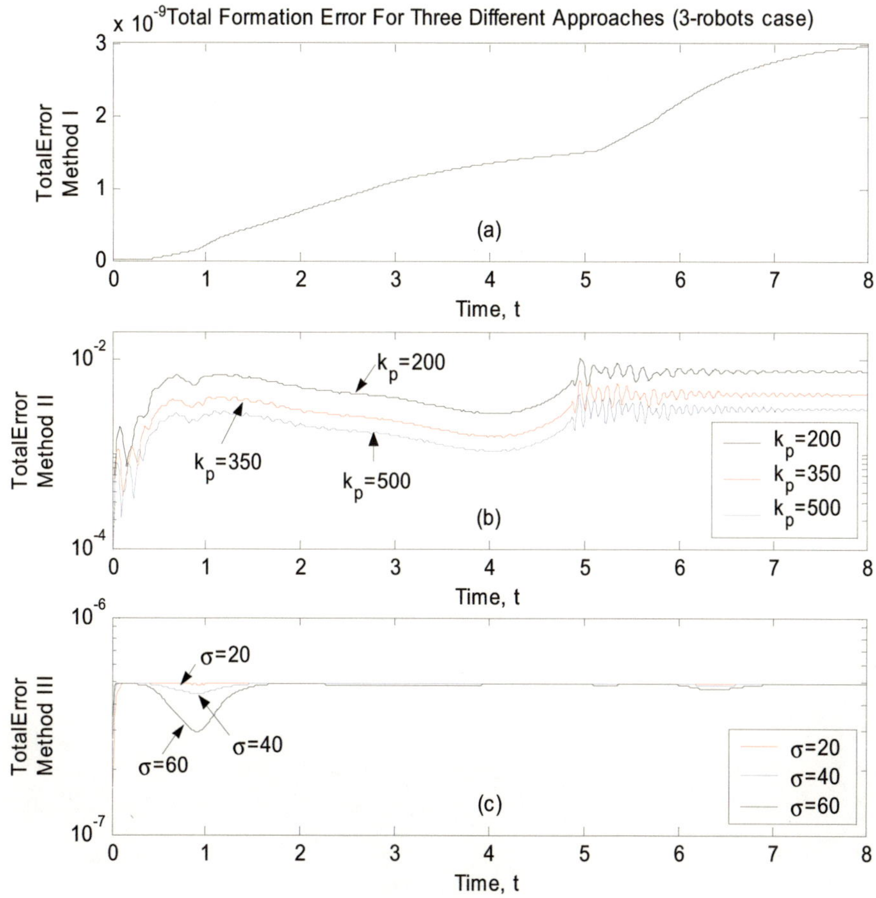

Figure 4. Constraint error for numerical integration with fixed time-step (1e-3 sec) for 3 point mass robots case, using (a) Method I: Direct Lagrange Elimination (Benchmark); (b) Method II: Penalty formulation; and (c) Method III: Constraint manifold projection approaches

Many have noted the various advantages of penalty formulation including: automated treatment of appearing/ disappearing constraints, robustness near singularities, in addition to the natural decoupling offered by the formulation. However, the Lagrange multipliers only form a part of the complete picture regarding the constraint forces. They represent the magnitude-type contribution while the other (and perhaps most important) part is the directional information that is embedded in the constraint Jacobian. The imperfect approximation of the Lagrange multipliers, coupled with the (artificial) relaxation of the constraints can over time lead to alternate configurations thereby indirectly affecting the directions of constraint vectors. Hence, not withstanding the small magnitudes of the constraint violations, the incorrect projection of the Lagrange multipliers would: (i) yield seemingly correct but non-physical results; (ii) and additionally act as a continuous source of disturbance. Schiehlen *et al.* (2000) noted very similar results when a similar comparison was performed in the context of distributed dynamic simulation by coupling two or more minimal local subsystem with explicit (force-coupled) or implicit (DAE approach) enforcement of holonomic constraints.

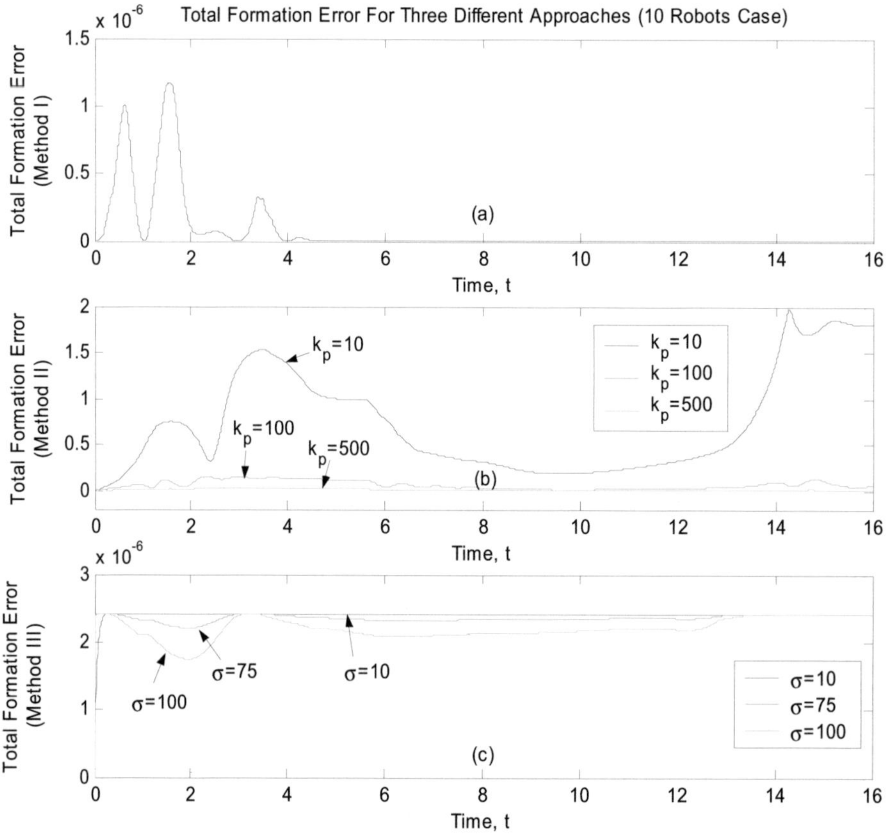

Figure 5. Constraint error for numerical integration with fixed time-step (1e-3 sec) for 10 point mass robots case, using (a) Method I: Direct Lagrange Elimination (Benchmark); (b) Method II: Penalty formulation; and (c) Method III: Constraint manifold projection approaches

6. Discussion & Summary

In this research, we examined aspects of the development and performance-evaluation of two alternate methods for distributed motion-planning for robot collectives within an artificial potential framework. These approaches arise by drawing the analogy to formulation methods in use for modular and distributed forward dynamics simulations of constrained mechanical systems. (Similar situations may also be encountered in other arenas where the governing equations take the form of sets of ODEs coupled together by algebraic constraints and solution of the combined system of DAEs needs to be found).

Our preliminary results (examined in the context of distributed motion planning of 3-robot collective and 10-robot collective discussed in the previous section) indicate that a global unified view of the evaluation of the computational complexity of the simulation is advisable. Specifically, at an algorithmic development level, the penalty-formulation within an APF framework provides a seemingly natural method for decoupling and distributing the computation, reduced computational complexity and an elegant Lyapunov-based setting to prove stability results. However, this is typically at the cost of formation maintenance - the projection-based approach does not distribute as well and is computationally more expensive per time-step. However, in the overall picture, this approach generates motion plans with smaller formation errors for a specified time-step and would have overall computational advantages over using the penalty formulation with a much smaller time-step.

7. Acknowledgements

We gratefully acknowledge the support from The Research Foundation of State University of New York and National Science Foundation CAREER Award (IIS-0347653) for this research effort.

8. References

Arnold, V. (1989). *Mathematical Methods of Classical Mechanics*, Springer-Verlag, New York.

Ascher, U., Chin, H., Petzold, L. & Reich, S. (1995). Stabilization of Constrained Mechanical Systems with DAEs and Invariant Manifolds. *Mechanical Structures & Machines*, Vol.23, 135-158.

Ascher, U., Pai, D. K. & Cloutier, B. (1997). Forward Dynamics, Elimination Methods, and Formulation Stiffness in Robot Simulation. *The International Journal of Robotics Research*, Vol.16, No.6, 749-758.

Ascher, U. & Petzold, L. (1998). *Computer Methods for Ordinary Differential Equations and Differential-Algebraic Equations*, SIAM: Society for Industrial and Applied Mathematics, Philadelphia, PA.

Baumgarte, J. (1983). A New Method of Stabilization for Holonomic Constraints. *ASME Journal of Applied Mechanics*, Vol.50, 869-870.

Beard, R. W., Lawton, J. & Hadaegh, F. Y. (2001). A Feedback Architecture for Formation Control. *IEEE Transactions on Control Systems Technology*, Vol.9, No.6, 777-790.

Bhatt, R. M., Tang, C. P. & Krovi, V. N. (2008). Formation Optimization for A Fleet of Wheeled Mobile Robots - A Geometric Approach. *Robotics and Autonomous Systems*. doi:10.1016/j.robot.2006.12.012

Desai, J. P., Ostrowski, J. P. & Kumar, V. (2001). Modeling and Control of Formations of Nonholonomic Mobile Robots. *IEEE Transactions on Robotics and Automation*, Vol.17, No.6, 905-908.

Egerstedt, M. & Hu, X. (2001). Formation Constrained Multi-Agent Control. *IEEE Transactions on Robotics and Automation*, Vol.17, No.6, 947-951.

García de Jalón, J. & Bayo, E. (1994). *Kinematic and Dynamic Simulation of Multibody Systems: The Real-Time Challenge*, Springer-Verlag, New York.

Haug, E. J. (1989). *Computer Aided Kinematics and Dynamics of Mechanical Systems: Basic Methods*, Allyn and Bacon Series in Engineering, Prentice Hall.

Lawton, J., Beard, R. & Young, B. (2003). A Decentralized Approach To Formation Maneuvers. *IEEE Transactions on Robotics and Automation*, Vol.19, No.6, 933-941.

Lee, L.-F. (2004). Decentralized Motion Planning within an Artificial Potential Framework (APF) For Cooperative Payload Transport by Multi-Robots Collectives. *M.S. Thesis*. Mechanical and Aerospace Engineering. University at Buffalo, Buffalo, NY.

Leonard, N. E. & Fiorelli, E. (2001). Virtual Leaders, Artificial Potentials and Coordinated Control of Groups, *Proceedings of IEEE Conference on Decision and Control*, pp. 2968-2973, Orlando, FL.

Lewis, M. A. & Tan, K.-H. (1997). High Precision Formation Control of Mobile Robots Using Virtual Structures. *Autonomous Robots*, Vol.4, 387-403.

Ogren, P., Egerstedt, M. & Hu., X. (2002). A Control Lyapunov Function Approach to Multi-Agent Coordination. *IEEE Transactions on Robotics and Automation*, Vol.18, No.5, 847-851.

Olfati-Saber, R. & Murray, R. M. (2002). Graph Rigidity and Distributed Formation Stabilization of Multi-Vehicle Systems, *Proceedings of 41st IEEE Conference on Decision and Control*, pp. 2965-2971, Las Vegas, Nevada.

Olfati-Saber, R. (2006). Flocking for Multi-Agent Dynamic Systems: Algorithms and Theory. *IEEE Transactions on Automatic Control*, Vol.51, No.3, 401-420.

Rimon, E. & Koditschek, D. E. (1992). Exact Robot Navigation Using Artificial Potential Functions. *Robotics and Automation, IEEE Transactions on*, Vol.8, No.5, 501-518.

Schiehlen, W. (1990). *Multibody Systems Handbook*, Springer-Verlag, Berlin.

Schiehlen, W., A., R. & Schirle, T. (2000). Force Coupling Versus Differential Algebraic Description of Constrained Multibody Systems. *Multibody System Dynamics*, Vol.4, No.4, 317-340.

Shabana, A. A. (1989). *Dynamics of Multibody Systems*, Wiley, New York.

Wang, P. K. C. (1991). Navigation Strategies for Multiple Autonomous Mobile Robots Moving in Formation. *Journal of Robotic Systems*, Vol.8, No.2, 177-195.

Young, B., Beard, R. W. & Kelsey, J. M. (2001). A Control Scheme for Improving Multi-Vehicle Formation Maneuvers, *Proceedings of American Control Conference*, pp. 704-709, Arlington, VA.

Yun, X. & Sarkar, N. (1998). A Unified Formulation of Robotic Systems with Holonomic and Non Holonomic Constraints. *IEEE Transactions on Robotics and Automation*, Vol.14, No.4, 640-650.

13

Motion Planning of Intelligent Explorer for Asteroid Exploration Mission

Takashi Kubota, Tatsuaki Hashimoto and Jun'ichiro Kawaguchi
Japan Aerospace Exploration Agency
Japan

1. Introduction

In-situ observations of minor bodies like asteroids or comets are scientifically very important because their sizes are too small to have high internal pressures and temperatures, which means they should hold the early chemistry of the solar system. In recent years, some rendezvous or sample-return missions to small body have received a lot of attention in the world. To date, the missions of NEAR (Farquhar, 2001), Deep Space 1 (Rayman et al., 2000), Deep Impact (Chiu et al., 2000), and Stardust (Atkins, et al., 2000) have been successfully performed, while MUSES-C (Kawaguchi et al., 2000) and Rosetta (Wittmann, et al. 1999) are currently in operation. NEAR spacecraft was successfully put into the orbit of the asteroid 433 Eros in February 2000. After precise remote-sensing observations, NEAR spacecraft succeeded in hard-landing on the surface of EROS in February 2001. In Japan, meanwhile, ISAS (Institute of Space and Astronautical Science) launched an asteroid sample and return spacecraft MUSES-C toward a near Earth asteroid 1998SF36 in 2003 and performed soft landing on the asteroid in 2005.

In deep space missions, ground based operation is very limited due to the communication delay and low bit-rate communication. Therefore, autonomy is required for deep space exploration. On the other hand, because little information on the target asteroid is known in advance, robotics technology is used for the spacecraft to approach, rendezvous with, and land on the asteroid safely. Various kinds of advanced and intelligent robotics technologies (Kubota et al. 2001) have been developed and used for navigation and guidance of the explorer to touch down and collect samples. This chapter describes the outline of the sample return mission MUSES-C, descent and touch-down scenario, vision based navigation scheme, sensor based motion planning, autonomous functions, and flight results in detail.

This chapter is structured as follows. Section 2 describes the mission purpose and the configuration of MUSES-C spacecraft. In Section 3, navigation sensors are explained. In Section 4 discusses the strategy for autonomous approach and landing. Autonomous descent scheme based on navigation sensors is introduced. In Section 5, a vision based navigation scheme is presented. In Section 6, flight results in MUSES-C mission is presented. Finally, Section 7 is for discussions and conclusions.

2. MUSES-C Mission

2.1 Outline of MUSES-C Mission

ISAS launched the MUSES-C spacecraft toward the asteroid in May 2003, which is named "Hayabusa". The MUSES-C project (Kawaguchi et al., 1998) is aiming at demonstrating four key technologies required for the future sample and return missions from extra-terrestrial bodies. Those technologies are : 1) solar electrical propulsion with ion thrusters in an interplanetary space, as a primary propulsive means, 2) autonomous optical guidance and navigation, 3) automated sampling mechanism, and 4) direct hyperbolic reentry of the recovery capsule to the ground.

The target body of the MUSES-C spacecraft is a near Earth asteroid 1998SF36 which is named "Itokawa". The launch date was May 9th in 2003 and the arrival at 1998SF36 in September 2005. Leaving the asteroid in December 2005, the spacecraft returns to the Earth in June 2010. The mission duration from launch to the Earth return is about 7 years. In MUSES-C mission, the spacecraft Hayabusa has stayed for about four months around the asteroid and both mapping and sampling operations were carried out during that short period. Figure 1 shows the illustration of MUSES-C mission.

Hayabusa was launched via the ISAS medium class launch vehicle M-V. The mass of the spacecraft is about 500[kg] including chemical and ion engine propellant of 130[kg]. The solar cell is a tri-junction one and the solar panel generates approximately 1.8[kW]. During the flight, the distance from the earth is shorter than 2[AU].

Figure 1. MUSES-C Mission (Ikeshita/MEF/JAXA)

2.2 MUSES-C Spacecraft

In deep space, it is difficult to operate a spacecraft on a real-time basis remotely from the earth mainly due to the communication delay. So autonomous navigation and guidance are required for descent and touch-down to the asteroid. For this purpose, MUSES-C has some navigation sensors and onboard image processing system (Hashimoto et al., 2000). The rendezvous and touch down to the asteroid, whose size, shape, surface condition are unknown, requires intelligent and advanced navigation, guidance and control system. Figure 2 shows the overview and the configuration of the MUSES-C spacecraft.

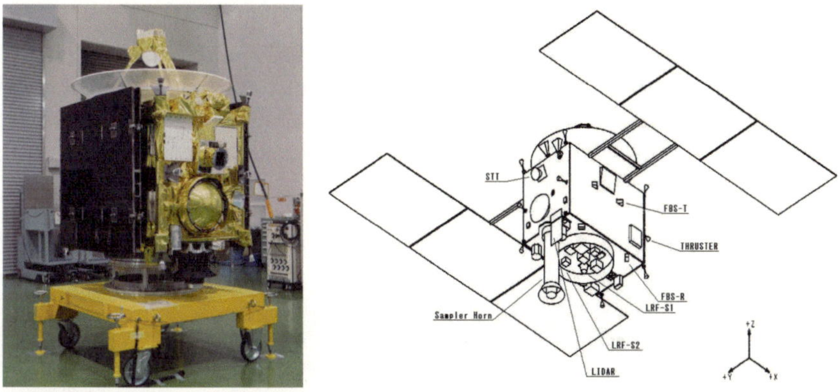

Figure 2. Overview and Configuration of MUSES-C Spacecraft

2.3 AOCS and GNC System

When the spacecraft was designed, the exact size, the shape, and the surface condition of the target asteroid were unknown. The GNC system was designed so that it could cope with various situations within the severe weight and power restrictions for the spacecraft. Figure 3 shows the AOCS and GNC system of Hayabusa spacecraft. TSAS (Two axis Sun Aspect Sensor), STT (Star Tracker) and IRU (Inertial Reference Unit) are combined to determine the spacecraft attitude. ACM (AcceleroMeter) is used to accurately measure the velocity increment gained by RCS (Reaction Control System) firings. RW (Reaction Wheel) and RCS thrusters are used for attitude and position control. Twelve thrusters were installed on the spacecraft and this arrangement allows the control of translational and rotational motion independently.

Some navigation sensors (Hashimoto et al., 2003) for descent and touchdown are equipped on Hayabusa spacecraft. The spacecraft has two kinds of optical navigation cameras. The narrow angle camera (ONC-T) is used for mapping (Maruya et al., 2006) and multiple scientific observations of the asteroid from the Home Position. The wide angle cameras (ONC-W1 and ONC-W2) are both used for onboard navigation, though W2 is used only when high-phase angle observation. Control electronics for ONCs (ONC-E) equipped with a RISC processor and a Gate Array image processor has some image processing functions such as image compression, center-finding of bright object, correlation tracking, feature terrain extraction, etc. Measurement of the altitude is performed with LIDAR (Light radio Detecting And Ranging). LIDAR covers the measurement range from 50[m] to 50[km]. For sampling the surface materials, cancellation of the relative horizontal speed is essential while touch down. To accomplish this requirement, the spacecraft drops a Target Marker (TM) that can act as a navigation aid by posing as an artificial landmark on the surface. The position of TM is estimated by using ONC. Since it is not so easy to detect TM from several tens of meters altitude, TM is equipped with reflexive reflector and ONC has a flash lamp (FLA) whose radiation is synchronized with camera exposure. Laser Range Finder (LRF) is used at lower altitude. LRF has four beams that are canted with 30 [deg] and can measure the range from 7[m] to 120[m]. LRF can provide the height and attitude information with respect to the surface. Four sets of Fan Beam Sensors (FBS) are equipped onboard as alarm sensors to detect some potential obstacles that may hit the solar cell panels.

The GNC logic is implemented in AOCU (Attitude and Orbit Control Unit), where a high performance microprocessor is equipped. Figure 4 shows the block diagram of GNC functions. The core of onboard navigation system is an extended Kalman filter. The filter outputs the estimated position and velocity relative to Itokawa. The state dynamics for the Kalman filter employs orbit dynamics model around Itokawa. Simple gravity field model is included in the dynamics. The observations for spacecraft position come from ONC, LIDAR and LRF.

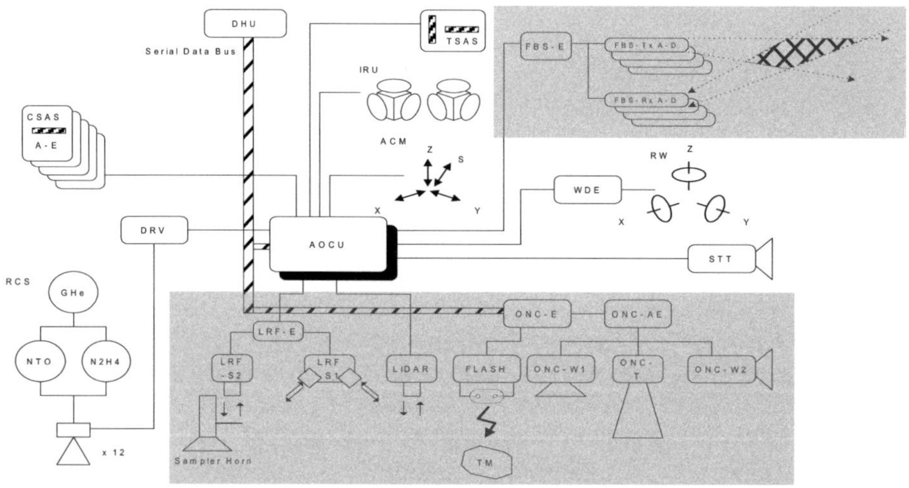

Figure 3. AOCS and GNC System of Hayabusa Spacecraft

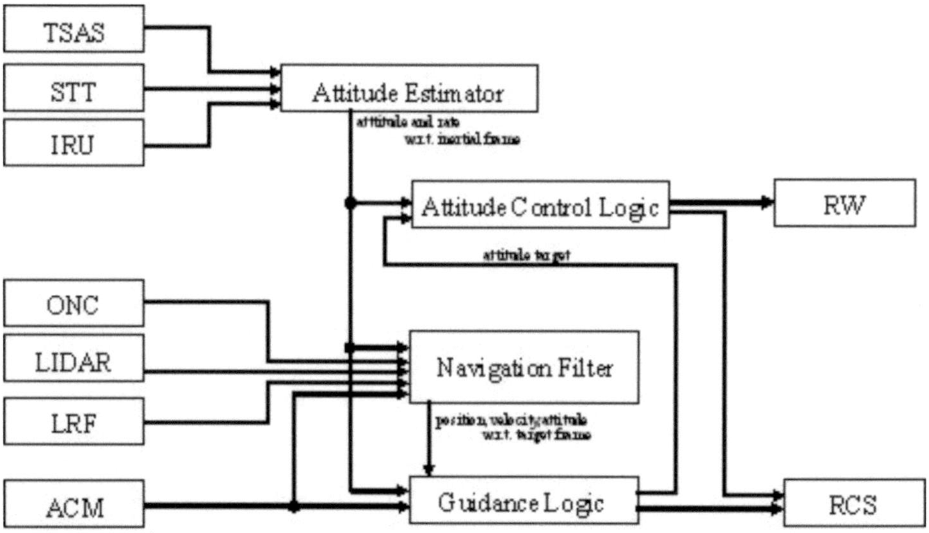

Figure 4. Block Diagram of AOCS/GNC System

3. Navigation Sensors

For autonomous descent and touchdown, various kinds of navigation sensors are used. The specifications of the main sensors for navigation are described as follows.

3.1 ONC-W

Hayabusa spacecraft has one telescopic camera ONC-T and two wide FOV cameras: ONC-W1 and W2. ONC-W1 whose FOV aligned to -Z axis of the spacecraft is used for on-board navigation. ONC-W2 has the FOV of -Y direction, which is used for terminator observation phase. The FOV of ONC-W1 is 60deg x 60deg and the resolution is 1000(H) x 1024(V). The overview of ONC-W1 is shown in Figure 5.

3.2 LIDAR

LIDAR is a pulse laser radar which measures the traveling time of the pulse between the spacecraft and the asteroid surface. A Photo of the prototype model is shown in Figure 5. Since the magnitude of received signal will change about 10^6 orders between 50km and 50m, LIDAR has automatic gain control function of APD. Transmitting pulse can be synchronized with external signal such as AOCS timing. This function is not only for precise range measurement but also synchronization with the exposure of ONC-T or NIRS, which helps the alignment measurement of both sensors. To minimize the weigh of optics, the reflecting mirror is made of SiC.

Figure 5. Overviews of ONC-W (Left) and LIDAR (Right)

3.3 LRF

LRF(Laser Range Finder) consists of four beams sensors for navigation(LRF-S1), one beam sensor for touchdown detection(LRF-S2), and an electronics circuit (LRF-E). Photos of LRF-S1 and LRF-S2 are shown in Fig.6. LRF detects the range to the surface with the phase deference between AM-modulated transmitting and receiving laser light. LRF-S1 has four beams canted 30deg from vertical direction and AOCU can calculate relative attitude and position to the surface using four beam range information. The target of LRF-S2 is the side surface of the sampler horn and it detects the change of the length of the horn which means that the horn is collide with the surface. LRF has single electronics and S1 and S2 are switched by commands when used.

3.4 FBS

FBS (Fan Beam Sensors) are sensors for detecting obstacles bigger than 10cm. A pair of a transmitter (FBS-T) and a receiver (FBS-R) forms a three-dimensional detection area shown in Fig.7. Four pairs of FBS cover almost half of the area beneath the spacecraft's solar cell panels as shown in Fig.8.

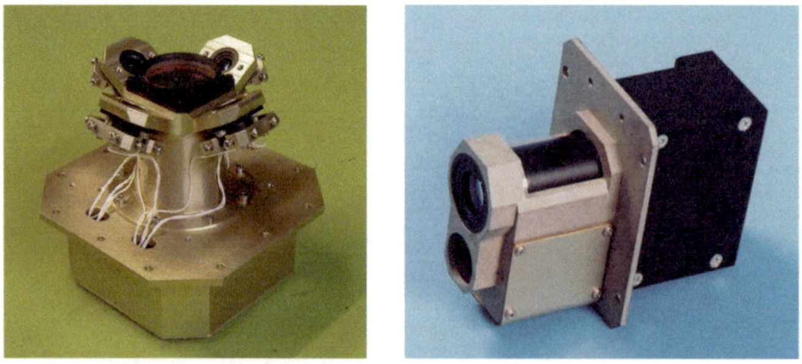

Figure 6. Overviews of LRF-S1 and LRF-S2

Figure 7. Overviews of FBS-T and FBS-R

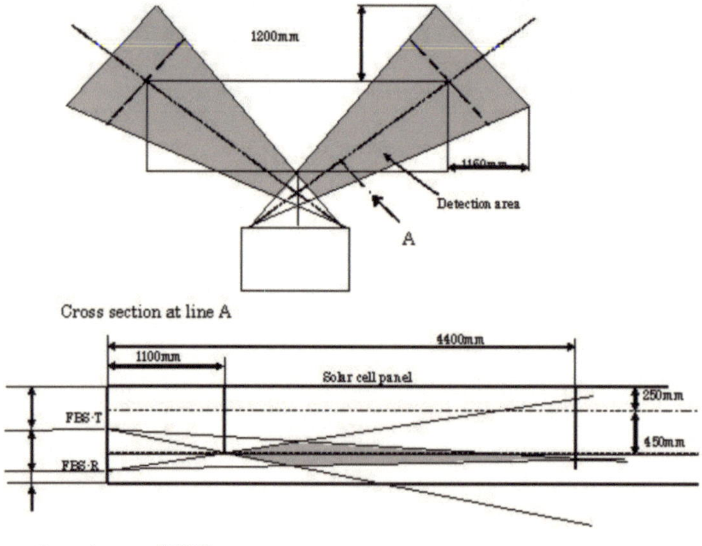

Figure 8. Detection Area of FBS

4. Autonomous Landing

For landing on an unknown body safely, it is necessary to obtain the terrain information of a planetary surface around a landing point. It is also important to guide a spacecraft to the landing point without hitting rocks or big stones. In the touch down phase, cancellation of the horizontal speed relative to the surface of the landing site is essential. Hayabusa spacecraft uses a new method to land on the surface of an unknown body autonomously, using optical cameras and laser altimeter (Hashimoto et al., 2002). The strategy for autonomous descent and touch-down consists of the following phases. The autonomous descent and landing sequence is illustrated in Fig.9.

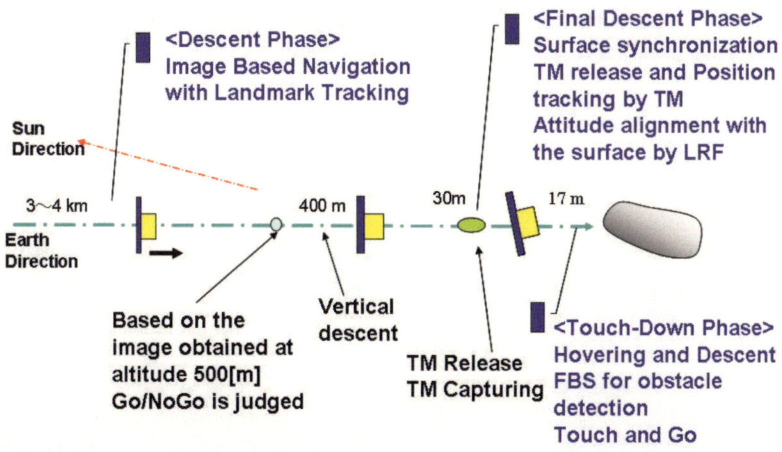

Figure 9. Scenario for final descent and touchdown

4.1 Descent Phase

While the whole of the asteroid image is in the field of view (FOV) of the optical navigation camera (ONC), the 3D navigation scheme (Hashimoto et al. 2001) based on the center-finding of the asteroid is used. After a part of the asteroid goes out of the FOV (about 1 km altitude), the spacecraft will nominally descent with only vertical velocity control because it cannot detect the direction of the asteroid center. Therefore, navigation accuracy depends upon the initial position and velocity, which are determined in the Home Position.

For an experiment, Navigation, Guidance, and Control (NGC) system has also a natural terrain tracking function. That is, characteristic features like craters on the surface are extracted from images and tracked autonomously. If some tracked features are recognized to be unsuitable for tracking, new appropriate features are extracted automatically. The line-of-sight vectors to the extracted features provide relative position to the surface. Since the locations of the features are unknown in the asteroid-fixed coordinate system, the spacecraft measures only the deviation of the vectors, that is, it can obtain relative velocity to the surface.

4.2 Final Descent Phase

The sampling method in MUSES-C mission is a so-called touch-and-go way. That means the spacecraft shoots a small bullet to the surface just after the touch-down is detected, collects ejected fragments with a sampler horn, and lifts off before one of solar cell panels hits the

surface. Therefore, the control of the vertical velocity and the cancellation of the horizontal speed are essential for both successful sampling and the spacecraft safety. To meet these requirements, TM is released from the spacecraft at the altitude of about 30[m]. At the altitude of about 30[m], ONC tries to capture TM, that would be placed near the target landing point. To ensure the visibility of TM, the surface of TM is covered with reflexive reflector and ONC provides the differential image taken by flash-on and flash-off.
After TM is successfully captured, the relative navigation logic is initiated to obtain the position with respect to TM and the local horizon, which is calculated based on the asteroid model. The spacecraft moves to the position right above the TM, and then the attitude of the spacecraft is aligned to the local horizon determined from Laser Range Finder (LRF) measurements. The spacecraft is guided to the landing point and stays there until the relative velocity is stabilized within a limit. The introduction of an artificial landmark drastically reduces the computational load and uncertainty of image processing, even though the function of natural terrain tracking (Misu et al., 1999) remains as an experiment and backup.

4.3 Touchdown Phase

After the alignment, the spacecraft starts descending again and touches down the asteroid surface to collect samples. During the touchdown descent, some potential obstacles are checked with Fan Beam Sensors (FBS). If any obstacle is detected, the touchdown and sampling sequence is terminated and emergency assent is initiated. When the touchdown is detected, the spacecraft collects the sample as soon as possible and then lifts off. Figure 10 shows the sensors used for touchdown detection (Uo et al., 2006). Before the touchdown descent, AOCU changes the sensor from LRF-S1 to LRF-S2, which can sense the distance between LRF and the target on the horn and also sense the brightness of the target on the horn. LRF-S2, ACM and IRU are used for touchdown detection.

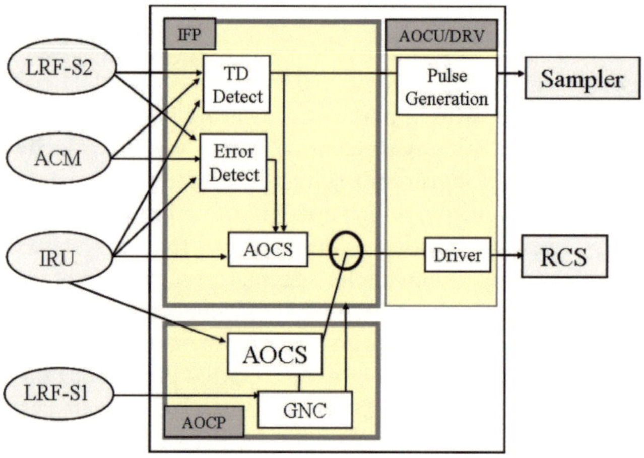

Figure 10. Touchdown Detection Scheme

4.4 Sampling Phase

A sample collection technique is what the MUSES-C spacecraft demonstrates first in the world. Different from the large planets, the asteroid is a very small object whose gravity field is too little for any sampler to dig and drill the surface. Nevertheless, the spacecraft has to cope even with the hard surface such as rocks, while it is requested to function for soft surface like sands as well. Therefore, a novel sample collection system is introduced as shown in Fig.11. The introduced method is the combination of the Shooting Projectile and the Fragment Catcher. The basic idea is retrieving fragments from the surface ejected by the projectile shot. And a key in the mechanism is the use of the catcher whose inlet surface covers the shot area that is concealed from the main body of the spacecraft, so that the fragments and dusts cannot hit the spacecraft at all. The spacecraft extends a mast whose tip end is equipped with a gun shooting a projectile of 10[g] at the speed of 300[m/sec]. A tiny hole that opens above a flange relieves the high-pressured gas after the shot. It has deceleration device inside that absorbs the fragments/projectile kinetic energy.

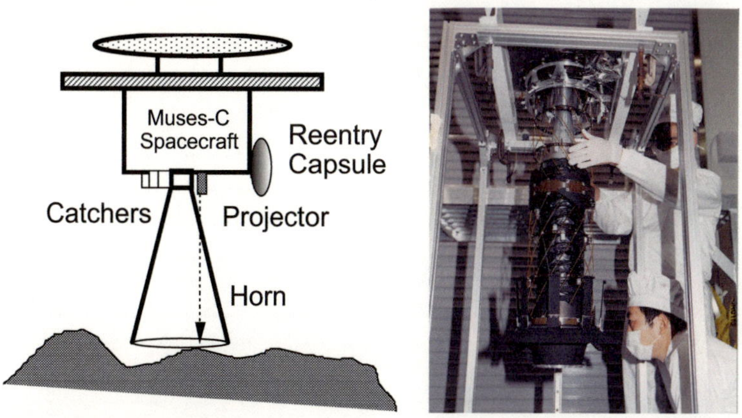

Figure 11. Sample Collection System

4.5 Descent and Touchdown Sequence

Figure 12 shows the operation sequence below the height of 500[m]. First, onboard navigation system is initialized by ground command. Initial position and velocity is calculated using GCP navigation. The onboard guidance logic is then initiated, and the descending begins. The GNC system keeps the constant descending velocity. In this case, the descending velocity is set at 0.1[m/s]. When the spacecraft reaches the height of 100[m], the spacecraft checks if the "continue" command is sent from the ground. If the continue command is not received by this time, the spacecraft interrupts descending and return to Home Position. At a height of 40[m], the wire that ties target marker on the spacecraft is cut. Right after that, the spacecraft decreases the descending velocity. Thus the target marker leaves the spacecraft and continues to fall on to the surface at a speed of about 10[cm/s]. The spacecraft begins slow free fall. During the free fall, the spacecraft examines if the LRF data is valid, and the consistency between the LRF data and the navigation solution generated using LIDAR data. If the consistency is within an expected range, the navigation filter begins to use LRF data instead of LIDAR data. Using the LRF based navigation solutions, the spacecraft hovers at a height of 17[m]. In this hovering point, the spacecraft waits the target marker to reach the surface. ONC is changed to the TMT mode after the TM separation. When the target marker image is acquired by ONC, after the interval to wait for the TM to reach the surface, relative position with respect to the target marker is determined combined with the LRF data. The Six DOF controller is activated after the navigation filter solution converges.

Several abort functions are implemented on the spacecraft. The spacecraft stops the sequence and returns to home position in the following events. These functions are automatically enabled and disabled according to the progress of the sequence.

1. Loss of LIDAR observation
2. Loss of LRF observation
3. Loss of ONC observation
4. Obstacle detection by FBS
5. Large error of attitude and attitude rate

Nominal Altitude	Nominal Velocity	Activity
500	0.05	Initialize navigation filter by ground command
	0.1	Initiate onboard guidance logic and start vertical descent
70	0.1	Check if "continue" command is set from the ground
40	0.1	Cut wire that ties Target Marker to the spacecraft
	0.06	Decrease spacecraft velocity from 0.1m/s->0.06m/s (TM continues to descend at 0.1 m/s)
35	0.06	Check if LRF data is consistent with current navigation solution
30	0.06	Initialize navigation filter for altitude keeping
17	0	Keep hovering altitude and wait TM reach the surface
17	0	Start TM image acquisition with ONC
17	0	Initialize final descent navigation filter Wait nav filter converges
17	0	Calculate attitude relative to local surface with LRF data
17	0	Start attitude alignment maneuver
17	0	Change LRF from navigation mode to TD mode
17	0.04	Execute descent dV of 0.04m/s for touchdown
0	< 0.1m/s	On touchdown detection, execute sampling sequence: projecter ignition escape dV

Figure 12. Operation Sequence

5. Vision based Navigation

The following image processing techniques are used for visual navigation in MUSES-C mission as shown in Fig.13.

5.1 WCT (Whole image Center Tracking)

In this mode, groups of adjoining pixels whose brightness beyond the specified threshold are extracted and the center address and the total number of the pixels of each group are calculated. Though less than nine groups are extracted as the specification, only one group, which has maximum number of pixels is usually used for center finding of the asteroid. When the image of the asteroid is divided into some portions with the illumination condition, Attitude and orbit control unit (AOCU) uses some relatively large groups and estimates real center of the asteroid.

5.2 TMT (Target Marker Tracking)

The function of this mode is basically the same as WCT mode except that TMT uses a differential image between flush-on and off. Different from the center finding of the whole asteroid image, the size of the extracted groups are expected to be a few pixels considering the distortion of the optics and the number of extracted groups must be one in order to track TM properly.

Artificial target marker is dropped down onto the asteroid surface to cancel the relative velocity. To use such a target marker, it is needed to develop a marker object with low restitution coefficient under the micro-gravity environment. To develop an object with low restitution coefficient, Japanese traditional "otedama" concept is introduced. "Otedama" is made of some amount of small beads inside a soft cover cloth. When "otedama" collides with other object, beads are expected to reduce the total collision energy as shown in Fig.14. To investigate the collision mechanism of "otedama", the dropping tower micro-G experiments were performed at MGLAB in Gifu in Japan. Experimental results show that the restitution coefficient values of "otedamas" mark below 0.1 (Kubota et al., 2007).

5.3 FWT (Fixed Window correlation Tracking)

This mode is prepared for an experiment and a backup when TM is not captured. Some tracking windows are designated on an image and the windows are used for templates of the tracking. Each window on the next coming image is correlated with corresponding template and the deviation of horizontal and vertical pixels between images are calculated. FWC is used to measure the relative velocity against the surface.

5.4 AWC (Auto Window Tracking)

Auto window tracking (AWC) (Misu et al., 1999) is also prepared for an experiment and a backup and the function of the correlation tracking is the same as FWC. AWC autonomously sets tracking windows. Firstly, some edges are extracted on the image, and the areas which contain a lot of edges are selected as characteristic terrain, and then tracking windows are set around the terrain. Though the developed algorithm is advanced and somehow complex, it seems more robust against terrain and illumination condition than FWC, because it uses featured windows, which can be easily tracked.

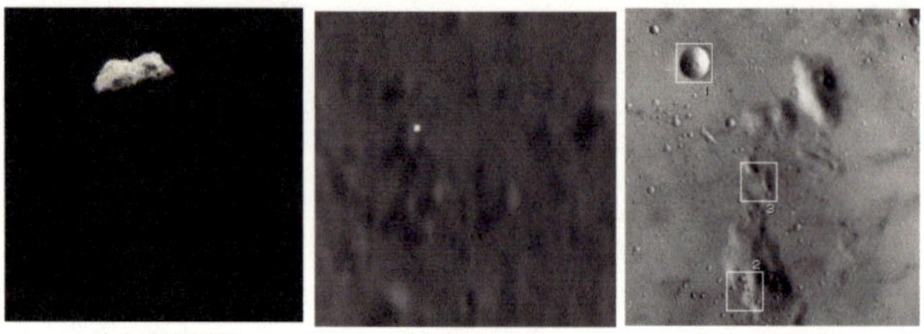

(a) WCT (b) TMT (c) AWC

Figure 13. Image Processing

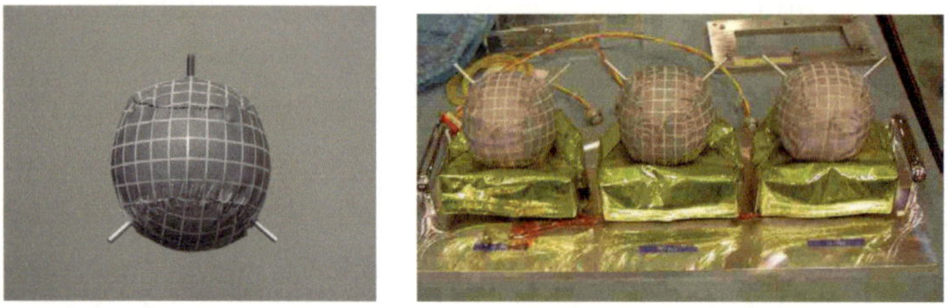

Figure 14. Target Markers

5. Navigation and Control Scheme

5.1 Navigation Scheme

To satisfy the stringent requirement on position and velocity estimation, a navigation filter that utilizes Kalman filter technique (Hashimoto et al., 2001) is adopted. The outputs of the navigation sensors are used to update the propagated states. The update gain is calculated so that the estimation error is minimized. The linearlized state dynamics and observation equations used in the position filter are:

$$\mathbf{x}_i=\varphi\mathbf{x}_{i-1}+\psi(d\mathbf{v}_{i-1}+d\mathbf{v}_G),\ \xi_i=\mathbf{G}(\mathbf{x}_{0i})\mathbf{x}_i \tag{1}$$

where: (suffix i is omitted)

$$\mathbf{x}=(\mathbf{r}_{SC/\#}{}^T\quad \dot{\mathbf{r}}_{SC/\#}{}^T)^T \tag{2}$$

$$\varphi=\begin{bmatrix}\mathbf{I}_{3X3} & dT\,\mathbf{I}_{3X3}\\ \mathbf{0}_{3X3} & \mathbf{I}_{3X3}\end{bmatrix}\quad \psi=\begin{bmatrix}(dT/2)\mathbf{I}_{3X3}\\ \mathbf{I}_{3X3}\end{bmatrix} \tag{3}$$

dT	:state integration interval
$d\mathbf{v}_{i-1}$	:velocity increment measured with ACM
$d\mathbf{v}_G$	:estimated gravitational effect

$$\xi=\xi^*-\mathbf{g}(\mathbf{x}_0)+\mathbf{G}(\mathbf{x}_0)\mathbf{x}_0\ \ (\mathbf{x}_0\text{: current best estimates}) \tag{4}$$

$$\xi^*=(L\quad n_X/n_Z\quad n_Y/n_Z)^T \tag{5}$$

$$\mathbf{G} = \partial \mathbb{g} / \partial \mathbb{x} \tag{6}$$

$$\mathbb{g}(\mathbb{x}) = \begin{pmatrix} \mathbb{r}_{LO/\#} - \mathbb{r}_{SC/\#} \\ (\mathbb{r}_{CO/\#} - \mathbb{r}_{SC/\#})_X / (\mathbb{r}_{CO/\#} - \mathbb{r}_{SC/\#})_Z \\ (\mathbb{r}_{CO/\#} - \mathbb{r}_{SC/\#})_Y / (\mathbb{r}_{CO/\#} - \mathbb{r}_{SC/\#})_Z \end{pmatrix} \tag{7}$$

L : Measurements of LIDAR or LRF
$\mathbb{m}$: Measurement vector of ONCs $=(n_X, n_Y, n_Z)^T$
$\mathbb{r}_{SC}$: Position of the spacecraft
$\mathbb{r}_{LO}$: Position of the range measurement point
$\mathbb{r}_{CO}$: Position of the camera target
$*_{/\#}$: Denotes a vector expressed in navigation reference frame:#
Note) Bold letter such as $\boldsymbol{\varphi}$ denotes a matrix, and $\mathbb{x}$ denotes a vector.

The dynamics and observation equations are processed in Kalman filter algorithm. The propagation of the covariance is executed as follows:

$$\mathbf{P}_i = \varphi \mathbf{P}_{i-1} \varphi^T + \psi \mathbf{Q}_{i-1} \psi^T \tag{8}$$

$$\mathbf{Q}_i = \begin{pmatrix} q_X & 0 & 0 \\ 0 & q_Y & 0 \\ 0 & 0 & q_Z \end{pmatrix} \tag{9}$$

$$q_{X,Y,Z} = k q_V + q_G \tag{10}$$

where:
k : equals 1 when delta V is executed. Otherwise zero
q_V : DeltaV measurement error
q_G : Velocity increment estimation error including gravity and dynamics model error

This filter implementation is applied through all the phases by properly selecting/switching the reference frame and for example, $\mathbb{r}_{CO}$ means the center of the asteroid in high altitude descent phase and TM in final descent phase. $\mathbb{r}_{LO}$ is approximated onboard, by calculating the intersection of laser beam and asteroid shape model.

5.2 Control Scheme

In the final descent and touchdown phase, various operations, such as surface synchronization, attitude alignment to local horizontal surface and stable hovering are required. Six degree-of-freedom variables, three for position and three for attitude, are independently controlled by the thruster control law that utilizes thruster switching curves defined on the phase plane. The outline of the switching curve is shown in Fig.15. In the fine control region, attitude is controlled by RW. Details of the control law have been presented in the paper (Yamashita et al., 2001).

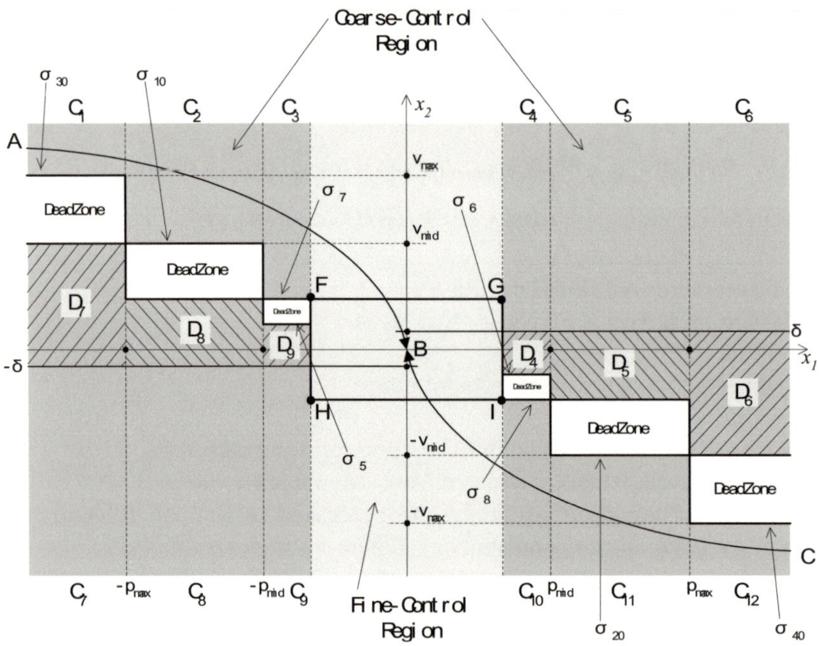

Figure 15. Thruster Switching Curve

6. Flight Results

6.1 Landing Site on Asteroid Itokawa

Figure 16 shows the picture of Itokawa taken by onboard optical navigation camera. The landing and sampling site was selected at the Joint Science Team meeting held in the end of October 2005, considering the scientific interest and the spacecraft safety. From the results on the global mapping of Itokawa from Home Position, it was found that most of the Itokawa surfaces were rocky or steep area, and "Muses-sea" was the only one candidate of landing points are shown by the circle in Fig.16.

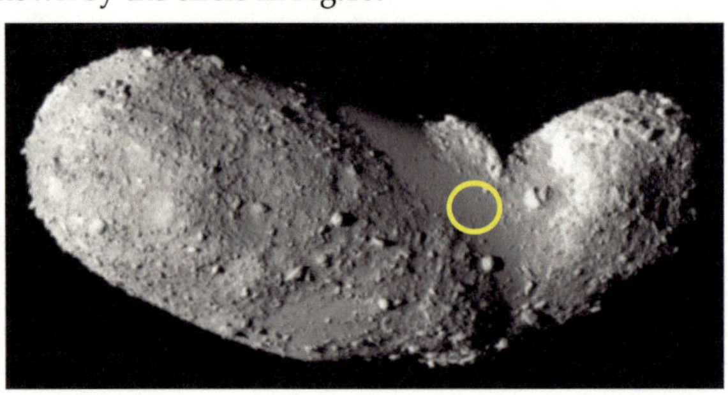

Figure 16. Picture of Itokawa and Landing Site "Muses-sea"

6.2 First Touchdown Results

The first landing for sampling was tried on November 20th 2005. The guidance and the navigation (Kubota et al., 2000) were all performed in order as almost planned. The guidance accuracy was within 30 meters horizontally in terms of the hovering point (Kubota et al., 2006). The target marker was released at about 40m altitude, because the velocity of target marker was smaller than nominal one and the spacecraft needed more time to get to the surface. The optical navigation camera could track the target marker properly. Figure 17 shows the low-altitude image, in which the shadow of Hayabusa spacecraft on the surface and the shinning released target marker could be seen.

After the obstacle had detected, the spacecraft continued descending because the attitude error was so large enough to prevent ascending the thruster firing. As a result, the spacecraft did unexpected touch-down without sampling sequence, and stayed on the surface for about 34 minutes until the forced ascent was commanded from the ground. The attitude of the spacecraft was controlled during the free-fall and then touchdowns were performed as shown in Fig.18. Therefore natural sample collections seemed to be conducted.

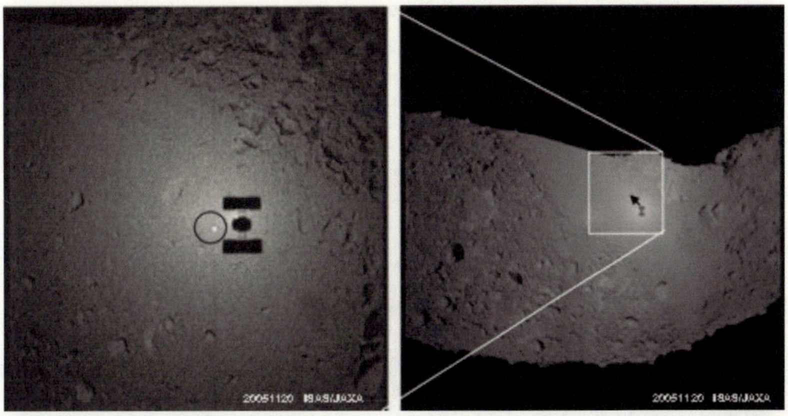

Figure 17. Navigation image taken during descent on 20th Nov. 2005 (Left: taken at 30m altitude, Right: taken at 200m altitude)

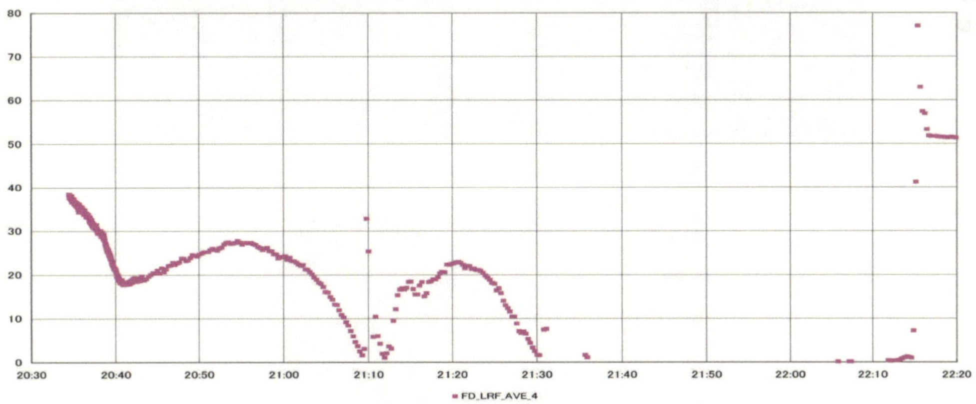

Figure 18. LRF data during bouncing and landing

6.3 Second Touchdown Results

The second and final landing was performed on November 26th 2005 (Yano et al., 2006). The descent path taken was almost same as that at the 1st touching-down attempt, toward the west part of the Muses-sea. As already one TM was in the Muses-sea, to avoid the confusion for image processing, a new marker was not released, this time. That is, TM was not used for the horizontal speed canceling, because GNC team had confidence to control the spacecraft remotely from the ground station. And also the obstacle detection was not set to be active, but to be monitored, since it seemed reporting too-sensitive signal in the 1st touchdown trial on 20th Nov 2005. The touch-down sequence was set so that the lift off must be only after the sampler horn deformation had detected and the sampling sequence had completed. In the second touchdown trial, the navigation and guidance to the aimed landing point was perfect. Figure 19 shows the image sequence taken by ONC-W1. In Fig.20, TM released on 20th Nov could be seen in the same position. Figure 21 shows the obtained LRF data. Terrain alignment was successfully performed at about 7[m] altitude. Touching-down speed was estimated about 10[cm/second]. When Hayabusa lifted-off, the +Z axis (High Gain Antenna axis) was 7 degrees off from the Sun direction as expected. The communication was established very well and every instrument aboard functioned normally. Though it seemed perfect landing and sampling, after the spacecraft returned to Home Position, it lost the attitude and could not communicate with the high-speed link to the ground stations. Therefore detailed data could not be obtained so far.

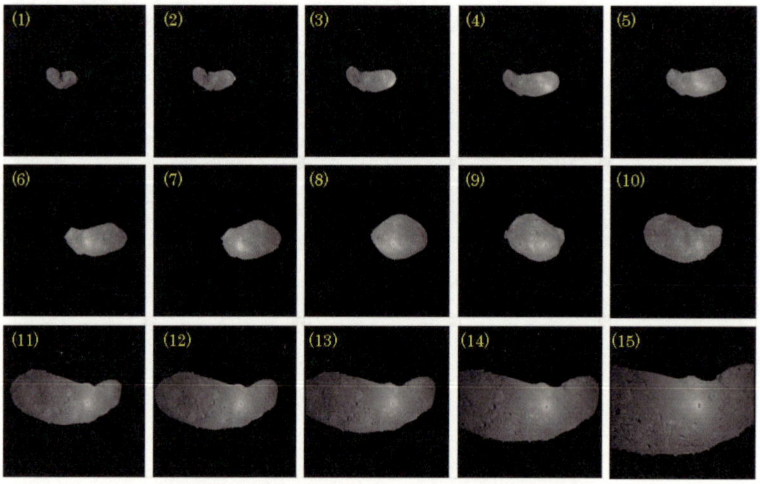

Figure 19. Image Sequence for 2nd Touchdown on 26th Nov. 2005

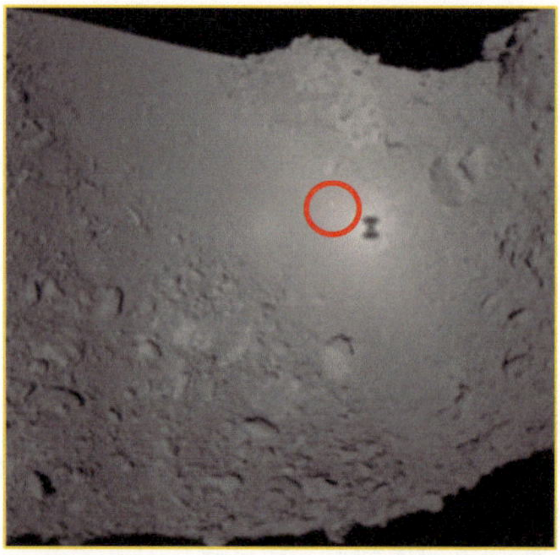

Figure 20. Target Marker released 1st Touchdown in 20th Nov. 2005

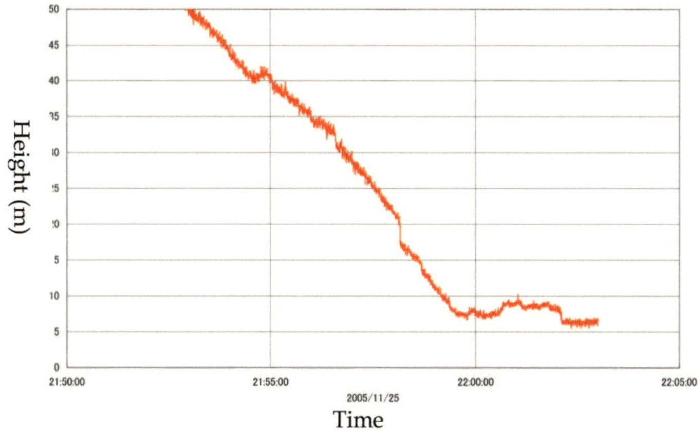

Figure 21. LRF data before touch-down on 26th Nov. 2005

7. Conclusion

This chapter has presented the motion planning of intelligent explorer in Japanese asteroid sample return mission. This chapter explained the navigation sensors and navigation strategy in the descent and touchdown phase for Hayabusa spacecraft. Target marker tracking and attitude alignment have been described in detail. And the flight data showed the effectiveness of the proposed and installed schemes. Hayabusa spacecraft succeeded in touchdown on the surface of Itokawa and lift-off the surface.

8. References

Farquhar, R. (2001). NEAR Shoemaker at Eros : Rendezvous, Orbital Operations, and a Soft Landing. *Advances in the Astronautical Sciences : Astrodynamics 2001*, Vol. 109, 2001, pp. 953-972.

Rayman, M.D. Varghese, P. H.Lehman, D. Livesay, L.L. (2000). Results from the Deep Space 1 Technology Validation Mission. *Acta Astronautica Journal*, vol.47, pp.475-487.

Chiu, M.C. Veverka, J. Reunolds, E.L. (2000). The Contour Mission - Status of Implementation. *IAA Int. Conf. on Low-Cost Planetary Missions*, No.IAA-L-0205.

Atkins, K.L. Martin, B.D. Vellinga, J. Price, R. (2000). STARDUST: Implementing a New Manage-to-Budget Paradigm. *IAA Int. Conf. on Low-Cost Planetary Missions*, No.IAA-L-0202.

Kawaguchi, J. Uesugi, K. Fujiwara, A. (2000). The MUSES-C Mission for the Sample Return - Its Technology Development Status and Readiness. *IAA Int. Conf. on Low-Cost Planetary Missions*, No. IAA-L-0306, 2000.

Wittmann, K. Feuerbacher, B. Ulamec, S. Rosenbauer, H. Bibring, JP. Moura, D. et al., (1999). Rosetta Lander in situ Characterization of a Comet Nucleus. *Acta Astronautica Journal*, Vol. 45, 1999, pp. 389-395.

Kubota, T. Sawai, S. Hashimoto, T. Kawaguchi, J. Fujiwara, A. (2001). Robotics Technology for Asteroid Sample Return Mission MUSES-C. *6th Int. Symposium on Artificial Intelligence, Robotics and Automation in Space*, AS016.

Kawaguchi, J. Uesugi, K. Fujiwara, A. Saitoh, H. (1998). The MUSES-C Mission Description and its Status. *3rd IAA Int. Conf. on Low-Cost Planetary Missions*, IAA-L-98-0505.

Hashimoto, T. Kubota, T. Mizuno, T. (2003). Light Weight Sensors for the Autonomous Asteroid Landing of MUSES-C Mission. *Acta Astronautica*, Vol.52, pp.381-388.

Maruya, M. Ohyama, H. Uo, M. Muranaka, N. Morita, H. Kubota, T. Hashimoto, T. Saito, J. Kawaguchi, J. (2006). Navigation Shape and Surface Topography Model of Itokawa. *AIAA/AAS Astrodynamics Specialist Conference and Exhibit*, 2006-6659.

Hashimoto, T. Kubota, T. Sawai, S. Uo M. (2002). Image-based Guidance, Navigation, and Control for MUSES-C Sample and Return Spacecraft. *Advances in the Astronautical Sciences*, Univelt Inc., Vol. 111, pp. 181-192.

Uo, M. Shirakawa, K. Hashimoto, T. Kubota, T. Kawaguchi, J. (2006). Hayabusa's Touching-down to Itokawa - Autonomous Guidance and Navigation. *AAS2006*, AAS-06-214.

Kubota, T. Sawai, S. Hashimoto, T. Kawaguchi, J. (2007). Collision Dynamics of a Visual Target Marker for Small-body Exploration, *Advanced Robotics*, Vol.21, No.14, pp.1635-1651.

Misu, T. Hashimoto, T. Ninomiya, K. (1999). Optical Guidance for Autonomous Landing of Spacecraft. *IEEE Trans. on Aerospace and Electronic Systems*, Vol.35, No.2, pp.459-473.

Hashimoto, T. Kubota, T. Kawaguchi, J. Uo, M. Baba, K. Yamashita, T. (2001a). Autonomous Descent and Touch-down via Optical Sensors. *Advances in the Astronautical Sciences*, Vol. 108, pp.469-480, Univelt Inc.

Yamashita, T. Uo, M. Hashimoto, T. (2001). Nonlinear Six-degree-of-freedom Control for Flexible Spacecraft. *IFAC Automatic Control in Aerospace*, pp.327-332.

Kubota, T. Hashimoto, T. Sawai, S. Kawaguchi, J. Ninomiya, K. Uo, M. Baba, K. (2000). An Autonomous Navigation and Guidance System for MUSES-C Asteroid Landing. *4th IAA Int. Conf. on Low-Cost Planetary Missions*, IAA-L-307.

Kubota, T. Hashimoto, T. Uo, M. Tsuno, K. Kawaguchi, J. (2006). Use of Laser and Optical Sensors in Terrain Alignment and Touchdowns. *16th AAS/AIAA Space Flight Mechanics Conference*, AAS06-216.

Yano, H. Kubota, T. Miyamoto, H. Okada, T. Scheeres, D. Takagi, Y. Yoshida, K. Abe, M. Abe, S. Barnouin-Jha, O. Fujiwara, A. Hasegawa, S. Hashimoto, T. Ishiguro, M. Kato, M. Kawaguchi, J. Mukai, T. Saito, J. Sasaki, S. Yoshikawa, M. (2006). Touchdown of the Hayabusa Spacecraft at the Muses Sea on Itokawa. *SCIENCE*, Vol.312, pp.1350-1353.

14

Modification of Kohonen Rule for Vehicle Path Planing by Behavioral Cloning

Ranka Kulić
Faculty of Maritime of Studies
Montenegro

Abstract

The problem of path generation for the autonomous vehicle in environments with infinite number obstacles is considered. Generally, the problem is known in the literature as the path planning. This chapter treated that problem using the algorithm, named MKBC, which is based on the behavioral cloning and Kohonen rule. In the behavioral cloning, the system learns from control traces of a human operator. Kohonen rule connected with the weighting coefficients, while the MKBC algorithm does not use the weighting values as values from the previous time, but permanentlly uses the training values as weighting values. That is something which enables an intelligent system to learn from the examples (operator's demonstrations) to control a vehicle in the process of the obstacles avoiding, like the human operator does. Like that, the very important MKBC characteristic is the symplicity. The MKBC simplicity is something which is so obviously, specialy according to the RBF neural network and the machine learnig algorithm which is used the previously. Following the MKBC given context the problem narrow passage avoiding and the goal position reaching fundamentally is observed. Namely, defining if - then rule, according to the named cases is treated as destroying of the consistency of the methodology. In that sense, using MKBC neural network the solution was found. A the end, the autonomous vehicle mathematical model which is given by nonlinear equations describing a 12 state dynamical system is used and in that case the MKBC algorithm is applied successfully. Eventually, as it has been illustrated the previously, the advantage of the entire methodology lies in the fact that a complete path of the vehicle can be defined off-line, without using sophisticated symbolical models of obstacles. These are facts that MKBC algorithm and the given methodology substantially differ from the others. In the next phase it is expected to confirm results in on - line simulation process.
Key words: vehicle path planning, behavioral cloning, cloning success, obstacle avoiding, machine learning, Kohonen rule, neural network, Shark dynamical model.

1. Introduction

In the last years en increasing interest in mobile robots has appeared, notably in aeronautical space exploration, automatized agriculture, collective mobile robot games, and so on (P. Vadakkepat, X. Peng et al., 2007). These application require the mobile robot to move in partially known environments with the high amount of uncertainty. Moving - obstacle avoidance with unknown obstacle trajectory remains remarkable challenge and has opened a research area in the control of the mobile robots, but it is treated in our research using the methodology which is connected with unmoving obstacles. Many approaches are available to study this research area. In its simplest form, the motion planning problem can be defined as follows (C. Latombe, 1991; J. Schwartz, M. Sharir, J. Hopcroft, 1987). Let B be the autonomous vehicle consisting of collection of rigid subparts having a total k degrees of freedom, and let B be free to move in two - or three - dimensional space V, avoiding obstacles whose geometry is known. For a given initial position S and a desired target position G of B, the task is to determine whether there exist a continuous obstacle - avoiding motion of B from S to G, and if so, to find such a motion. The simplest collision avoidance algorithm fall into the generate and test paradigms. A simple path from S to G, usually a straight line, is hypothesized and then it is tested for potential collisions between B and obstacles. If collision is detected, a new path is proposed using information about detected collision. This process repeats until no collision is detected. But in spite of its simplicity these methods have not found significant application. They have several fundamental drawbacks. One of these is inability to propose a radically different and better path from local information about potential collision. Another is that collection methods are based on a configuration space approach(J. Reif, 1987; C. Latombe, 1991; J. Schwartz, M. Sharir, J. Hopcroft, 1987). The configuration of rigid body is set of independent parameters that characterize the position of every point of it. For the vehicle B some regions represent illegal configuration space because there are obstacles. So the find path (the vehicle motion planning) approach means that the vehicle have to be shrank to dimension of a reference point and to grow obstacles, i.e. to compute forbidden regions for the reference point. Finding the path of the vehicle is in this way transformed in finding the path of the reference point, moving in configuration space and avoiding obstacles. For the vehicle B moving from position S to position G the desired path is the shortest path which takes into account all constraints of the position of the reference point of B. It is possible to obtain this path by generating an appropriate graph (visibility graph, connectivity graph,...) and finding a path from graph node S to graph node G. Fundamental problem arising during the implementation of these methods is concerned with the obstacle growing and graph searching. For both cases the problem complexity is very large. For example, while in the planar case the shortest path can be found in time that is in the worst case the quadratic in the number of obstacle vertices and edges, finding the shortest path between two points in three dimensions, which avoids a collections of polyhedral obstacles is NP - hard (J. Schwartz, M. Sharir, and J. Hopcroft,1987; J. Reif,1987). This is a specially very large problem if the world model changes. Other classes of approaches are developed as alternative to the traditional ones. A typical such approach (O. Khatib, 1986) regards the obstacles as the sources of repelling potential field, while the goal position G of the vehicle is considered as a strong attractor. The vehicle B follows potential gradient vector field. These approaches try to find the local minimum only. As the next, we can consider the direction which assumes problem solution capability of the vehicle motion planning based on the transfer of skill into controllers of the vehicle (D. Michie, R. Camacho, 1994; C. Sammut, S.Hurst, D. Kedzier, D. Michie, 1992).

2. Motion planning based on the behavioral cloning

Skill is often defined as an ability to perform a high quality sensory-motor coordination and control in real time. Humans exhibits such a skill as a result of training over a period of time. It would be especially useful if we can also provide systems with the capability of acquiring such a skill. In this sense two approaches have been known. The former treats the skill as something that could be acquired in a dialogue with an operator. In that process it is expected from the operator to describe skills he has been governed over the control of the vehicle. Here arises some difficulties because the skill is human subconscious action and so cannot be completely consciously and reliably described. An alternative approach is to start from the assumption that the skill can be reconstructed, using learning algorithms, from the manifestation trace of it (D. Michie, R. Camacho, 1994; C. Sammut, S.Hurst, D. Kedzier, D. Michie, 1992). Sammut, Hurst, Kedzier and Michie give a description of the solution belonging to the flight control area. Our idea (R. Kulic, Z. Vukic, 2006) is to enable an intelligent system to learn from the examples (operator's demonstrations) to control a vehicle avoiding obstacles, like the human operator does.
In section III the intelligent controller concept is given. In section IV the results in controller development are presented. In section V the conclusion is given and possibilities of further development are discussed.

3. Elements of concept controller development

3.1 Learning problem

Suppose that is given a data set giving living area and price of m houses in some place. How can is possible learn to predict the prices of other houses in that place, as function of living areas? To denote the input variables or input features (living areas in this case) x_{1i} is used. And y_i is used to denote the output or target variable which is need training to predict. The dataset $\{(x_i,y_i), i=1,..,m\}$ that will be used to learn is called a training set. If X denote the space input values and Y denote the space output values the goal is for given a training set to learn a function h: X→Y. The function h(X) is called a hypothesis and it have to be a good predictor for the corresponding values of Y. When the target variable is continuous, the learning problem is called a regression problem. When the target variable take on only small number of discrete values, the learning problem is called a classification problem. Lets consider a slightly richer dataset with a number of bedrooms in each house. The X is two - dimensional vector, with x_{1i}, x_{2i} features. In general, when designing learning problem, it is up to us to decide what features to choose. To perform learning it is have to decide how we are going to represent hypothesis h. A choice can be to approximate it as a linear function of $x = \{ x_{1i}, x_{2i} \dots x_{ni}\}$:

$$h(x) = \theta_0 + \theta_1 x_{1i} + \theta_2 x_{2i} + .. + \theta_n x_{ni}$$

The θ_i 's are the parameters (or weights) parameterizing the space of linear functions mapping from X to Y. To simplify notation it is introduced $x_{i0}=1$. So that

$$h(x) = \sum_{j=1}^{n} \theta_j x_{ji},$$

where n is the number of input variables , without x_0. For given training set, how it is possible learn the parameters θ_j? One reason is to make $h(X)$ close to Y, at least for the training example. To formalize this, it can be defined a function that measures for each value of the θ_j how close $h(x_i)$ to the corresponding y_i. So, the least –squares cost function is defined:

$$J(\theta) = \frac{1}{2}\sum_{i=1}^{m}(h(x_i) - y_i)^2 .$$

That function is a special case of a much broader family of algorithms. Now, it is needed to choose vector θ to minimize J(θ). To do so, lets us a search algorithm that starts with some initial guess for θ and repeatedly changes θ to make J(θ) smaller, until it is reached the value of θ that minimize J(θ). Lets consider the gradient descent algorithm, which starts with some initial θ, and repeatedly perform the update:

$$\theta_j = \theta_j - \alpha \frac{\partial}{\partial \theta_j} J(\theta).$$

This update is simultaneously performed for all values of j=0,…,n. The learning rate α repeatedly changes in the direction of steepest decrease of J(θ). For one training example we have:

$$\frac{\partial}{\partial \theta_j} J(\theta) = \frac{\partial}{\partial \theta_j}\frac{1}{n}(h(x) - y)^n = n\frac{1}{n}(h(x) - y)\frac{\partial}{\partial \theta_j}(h(x) - y) =$$

$$(h(x) - y)\frac{\partial}{\partial \theta_j}\left(\sum_{j=0}^{n}\theta_j x_{ji}\right) = (h(x) - y)x_j$$

For single training example that gives the rule:

$$\theta_j = \theta_j + \alpha(y - h(x))x_{j.}$$

This is called LMS (least mean squares) update rule or Widrow - Hoff learning rule. For instance the magnitude of the update is proportional to the error $(y_i - h(x_i))$. For m training example the corresponding update rule

Repeat until convergence exist {

$$\theta_j = \theta_j + \alpha\sum_{i=1}^{m}(y_j - h(x_j))x_{ji} \text{ for every j}$$

}.

The optimization problem have been given here for linear regression has only one global optima. Assuming the learning rate not too large this batch gradient descent method always converges to the global minimum.

3.2. Bihevioral cloning process

The idea of controller development by cloning the human operator (D. Michie, R. Camacho,1994) is illustrated in Figure 1.

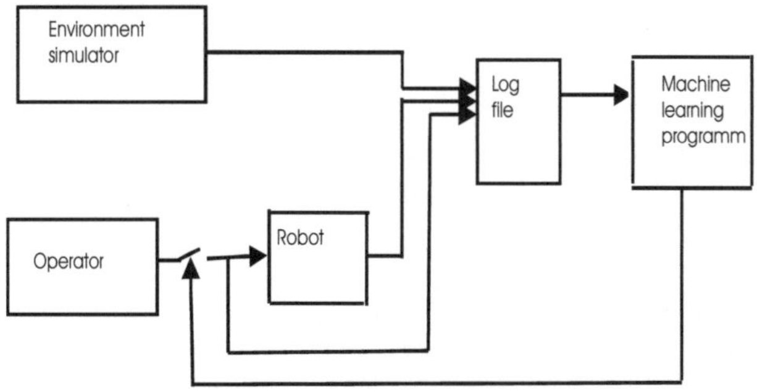

Figure 1. Behavioral cloning process. The autonomous underwater vehicle is named as the robot

In the obstacles avoiding problem, this idea could be interpreted in the following way. During one of the simulation phases, called the training phase, operator guides the vehicle avoiding unmoving obstacles located in its working space. During this phase variables that are evaluated as relevant are written into LOG FILE. In the second simulation phase, called the learning phase, the machine learning program, takes data from the LOG FILE, generates differential equations, that define the operator's trajectory. In the third phase, called the verifying phase, operator is excluded from the vehicle control process and the vehicle is controlled solely by a clone induced in the learning phase. This development of process phases are needed for the repeated changing of both problem domain representation and/or learning system regarding cloning success criterion. Using "several" vehicle models (problem domain representations) and "several" machine learning systems, we attempt to find an appropriate domain model and an appropriate machine learning system that will enable the vehicle to avoid obstacles according to cloning success criterion.

3.3 The vehicle kinematical model

The following kinematical model of the vehicle is used:

$$\psi(n)=\psi(n-1) + \Delta t\, r(n),$$

$$x(n) = x(n-1) + \Delta t\, v \cos(\psi(n)),$$

$$y(n) = y(n-1) + \Delta t\, v \sin(\psi(n)),$$

where: ψ is the heading angle of the vehicle (ψ=0 if the vehicle is oriented parallel to x-axis); r and v are control variables i.e. desired rotation speed and translation speed respectively; x, y are position coordinates, Δt is the sampling time and n is the time index. The vehicle is represented as a geometrical figure. Its dimensions are not neglected and we should point out that this is a very important fact. The selection of the vehicle model is inspired by conventional methodology that is used in control systems for a given path (R. Stojic, R. Kulic, M. Zivanovic, 1990).

3.4 Environment models

Environment model amounts to the distances of the vehicle gravity center from the goal position (dx_G and dy_G) and from the obstacles (d_i). dx_G and dy_G are calculated as:

$$dx_G = x - x_G,$$

$$dy_G = y - y_G\,,$$

where x_G, y_G are the goal position coordinates. Obstacles are represented by its characteristic values, as illustrated in Figure 2. Obstacle area is divided into sub-areas .

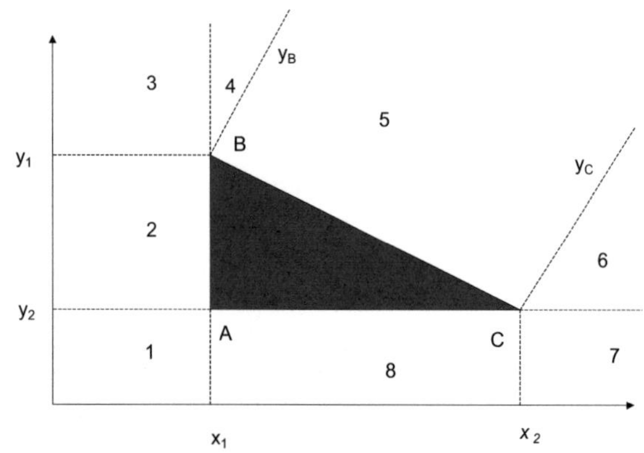

Figure 2. Triangular obstacle as the environment

A procedure, for the simulation purpose, calculating the vehicle distance d_i from i-th obstacle is explained for a triangle obstacle as:
SubArea-4:
if**$((x>=x_1)$**and**$(y>=y_B))$* ***then $d_i= ((x-x_1)^2 + (y-y_1)^2)^{1/2}$, SubArea5:
if**$((x>=x_1)$**and**$(y<y_B)$**and**$((y>=y_C)$* ***then
$d_i = \left| [y+[(y_1-y_2)/(x_2-x_1)]x + [(y_2-y_1)/(x_2-x_1)]x_1-y_1]/[[(y_1-y_2)/(x_2-x_1)]^2+1]^{1/2} \right|$,
y_B and y_C are lines that are normal onto the line BC at points B and C.

3.5 Cloning success criterion

Performance error is very important for the evaluation of the quality of clone that was constructed. Regarding the ideal case the goal concept and the approximation concept of the vehicle trajectory are identical and the performance error is equal to zero. Ideal trajectory in x-y plane without obstacles is, for example, a straight line between the start S and the goal G positions of the vehicle. Operator, in a training phase, mostly does not manage to realize this trajectory. Position error E_{xy} is based upon a distances $d_{op(i)}$ and $d_{cl(i)}$ of operator and clone trajectories, respectively, from the named straight line. Our problem is to avoid obstacle and so we can consider only E_{perf} as:

$$E_{xy} = \frac{\sum_{j=1}^{N} \frac{\left|d_{op}(j) - d_{cl}(j)\right|}{\max(d_{op}(j) - d_{cl}(j))}}{N} \qquad (1)$$

For avoiding n obstacles we have to find $(d)_m= \min\{ d_i ,i=1,..n\}$ in order to define:

$$E_{xy} = \frac{\sum_{j=1}^{N} \frac{\left|(d_{cl}(j))_m - (d_{cl}(j-1))_m\right|}{\max(d_{cl}(j))_m - (d_{cl}(j-1))_m}}{N-1} \quad (2).$$

Eventually, we can say that between two clones more successful is the one which produces lower performance error regarding equations (1) and (2).

3.6 The vehicle goal position as the obstacle

The vehicle goal position can be treated as the obstacle, but it has not been applid earlier in [12], [13]. In that sense the distance from the vehicle goal position and the vehicle gravity centre was taked into account, as d_{goal}. Now the attribute number is extended and ¨minimum distance¨ is determined by relation: $d_{min} = \min\{ d_{goal}, d_1, d_2, ...,d_k \}$, where d_1, d_2, ...,d_k are the mimial distance for avoiding K obstacles. Eventually, when we consider the vehicle goal position as the obstacle, we enable the vehicle path generation to become simple, but not to become optimal.

3.7 Learning systems

3.7.1 Radial basis function (RBF) of neural network

The model is commonly referred to as the radial basis function (RBF) network. The most important attribute that distinguishes the RBF network from earlier radial based models is its adaptive nature. It generally allows to utilize a relative small number of locally tuned units. RBF network were independently proposed by several authors (D. S. Broomhead, D. Lowe ,1988; S. Lee, R. Kill, 1988; M. Niranjan , F. Fallside,1988).The following is a description of the basic RBF architecture Figure 3. The RBF network has a feedforward structure consisting of a single hidden layer of Q locally tuned units, which are interconnected to an output layer of L linear units. All hidden units simultaneously receive R dimensional real -valued input vector p. Notice the absence of hidden layer weights. Each hidden unit output a_j is obtained by calculating the closeness of the input p to n dimensional parameter vector μ_j (IW in Figure 2). This parameter is associated with jth hidden units. The response characteristics of the jth hidden units are given by

$$a_j[p] = K\left(-\frac{\| p - \mu_j \|}{2\sigma_j^2}\right),$$

where K is a strictly positive radially symmetric function (kernel) with a unique maximum at its center μ_j and which drops off rapidly to zero away from the center. The parameter σ_j is the width of the receptive field of the input space for unit j. This means that a_j has an appreciable value only when distance $\| p - \mu_j \|$ is smaller than the width σ_j. A specially but commonly used RBF network assumes a Gaussian basis function for the hidden units, i.e.:

$$a_j[p] = \exp\left(-\frac{\| p - \mu_j \|^2}{2\sigma_j^2}\right),$$

where σ_j and μ_j, are the standard deviation and mean of the jth unit. The norm is the Euclidian norm. The output of the RBF network is the L dimensional vector a2, which is given by:

$$a2 = \sum_{j=1}^{L} LW_{ij} a_j \ .$$

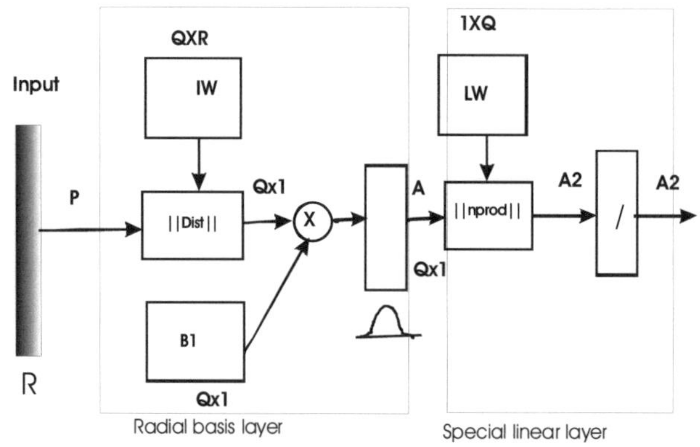

Figure 3. A radial basis function neural network consisting of a single hidden layer of locally tuned units that is fully interconnected to an output layer of linear units

RBF networks are best suited for approximating continuous real valued mappings f : $R^n \rightarrow R^L$, where n is sufficiently small. According to the previously named equations the RBF network may be as approximating a desired unction f(p).The degree of accuracy can be controlled by three parameters the number of basis functions to be used, their location and their width. The RBF networks are considered as universal approximators (T. Poggio, F. Girosi, 1990). The training of RBF network is addressed. .Consider training set of m labeled pairs $\{x_j, y_j)$ which are represent samples of a continuous multivariate function. The criterion function is an error function E to be minimized over the given training set. It is desired to develop a training method that minimizes E by updating the free parameters of the RBF. These parameters are σ_j, μ_j and w_{ij}. One of the first training methods that comes to mind is a fully supervised gradient descent methods over E, as it is given in section 3.1.

3.7.2 RBF neural network algorithm and behavioral cloning

The algorithm RBF neural network algorithm is given below for kinematical model which is given in 3.3.

for m:=1 to N do *% N is number of training examples*
begin
*dist[m,1] := (0.8326 / spread)^2 * (IW[m,1] – d_min)^2;* *% d_min = d_{min} ;*
a[m,1] := $e^{-dist[m,1]}$; *% IW[m,1] is the training examples set of the distance d_{min}*
end;
for j :=1 to N do
begin
*a2 := a2+ LW[1,j] * a[j,1];*
am := am + a[j,1];
end; *% LW[m,1] is the training examples set of the angle ψ*

if ((am > 0) or (am <0)) then a2 := a2/am;
psides := a2;

r := (psides –psi) / dt; % $$r = \frac{\psi_{des} - \psi}{\Delta t}$$, ***dt= Δt=0.05 s***

Control strategy was r = (ψ_{des} - ψ) / Δt, v=0.1, where ψ_{des} is the desired value of angle ψ.

3.7.3 Modification Kohonen rule and behavioral cloning - MKBC algorithm

Thus, as it is given in Figure 3, the neuron whose weight vector was closest to the input vector is updated to become even more closer. Suppose that the ith neuron wins, then the elements of the ith row of the input weighting matrix are adjusted as shown:

$${}_iLW^{1,1}(q) = {}_iLW^{1,1}(q-1) + \alpha(p(q) - {}_iLW^{1,1}(q-1)),$$

where p(q) is input vector, q is the time index and α must be specially tuned according to the domain. With regarding to the original Kohonen rule connected with the weighting coefficients, as the neural network, ***this algorithm does not use the weighting values as the values from the previous time, but permanentlly uses the training values as the weighting values. This enables an intelligent system to learn from the examples (operator's demonstrations) to control a vehicle in avoiding obstacles, like the human operator does.*** This seems, at least for this case, to be better according to the original idea of the Kohonen rule. The algorithm is given below.

```
for i:=1 to M do
begin
AS(1,1) :=0;
end;
 alfa(i,n);             % Must be specially tuned
                        % according to the domain;
                        % n is the time index;
for i:=1 to M do        % M is the number of inputs;
begin
for j :=1 to N  do      % N is the number of the
                        %training examples;
  begin

    LW[i,j,n] :=  alfa[i,n] * ( p[i,n0]  -  p[i,n]); % p[i,n] is the input
                                                     % vector;
                                                     % p(i,n0) are the training example set
   AS(i,n) :=  AS(i,n) + ( LW[i,j,n0] – LW(i,j,n))^2 ; % LW[i,j,n0] is the training examples set

  end;
  a(i,n) := sqrt (AS(i,n));                          % a(i,n) is the output
                                                     % vector;

end;
```

The algorithm means that by using N training examples the weight vector *LW [i,j,n0], {i=1,M; j=1,N}* can be formed[1]. Then *LW[i,j,n0]* need to be modified according to: 1) the actual values of the input vector[2] *p[i,n], {i=1,M}* and 2) the actual value of the tuned vector *α[i,n],{i=1,M},* where *n* is the time index. The resulting weighting vector is *LW[i,j,n]* and it has deviation according to the input vector *p[i,n].* The output vector is *a[i,n]* and it is connected with the weight vector as is illustrated by two last lines of the algorithm.

3.8 The MKBC algorithm and the motion planning

The MKBC algorithm is needed to be adapted in order to be used in the motion planning domain. The motion planning with the ***weight vector*** uses LW (Table 1), the training values of the learned variable which is the heading angle ψ (Section 3.3). Firstly, the algorithm requires the modification of the training vector LW. The modification executes, as is given below, using: 1) N training examples in IW and LW, where IW are the training values of the minimal distance d_{min} (section 3.6), 2) the current input value d_{min} and 3) tuning factor α . Secondly, the algorithm finds the desired output value using, in some sense, least - square cost function of the heading angle. The algorithm output value enables the obstacle avoidance and it is the desired value of the heading angle (ψ_{des}). The algorithm is given below.

```
AS :=0;
alfa = Kc – Ka * d_min;     % The values Kc and Ka must be tuned;

for j :=1 to N do            % N is the number of the training examples;
  begin

    LW[2,j] := alfa * ( IW [1,j] – dmin);
    AS := AS + ( LW[2,j] – LW[1,j])^2 ;

  end;
  psides := sqrt (AS);          % ψdes = √AS ;
                                % r := (psides –psi) / dt; , dt= Δt=0.05 s;
```

On the other hand, when the modified Kohonen rule with the weighting parameters is used, then the algorithm does not use the weighting values IW as the distance d_{min} from the previous time, but always uses IW as the training values of the distance d_{min} to determine the training values LW of the heading angle ψ . This enables an intelligent system to learn from the examples (operator's demonstrations) to control a vehicle in avoiding obstacles, like the human operator does. The very important MKBC characteristics are the operator cloning and symplicity, the simplicity specialy according to the RBF neural network. The coefficient α has values that are changed from time to time. Firstly, when it is tuned this factor enables the heading angle ψ to increase when the vehicle distance from the obstacles edges is small (i.e. $d_{min} \approx 0.08$). The performance error E_{xy} has a minimum for that α value. While α factor increases, the total time T of the autonomous vehicle moving without touching the obstacle also increases.

---

[1] *n0 means the training example set.*

[2] *The training value of the input vector is p[i,n0].*

4. Experiments and results

In order to form training instances the task was only to avoid obstacles. The robot start position was its goal position. The minimal number of robot traveling from the start to the goal was to be one. *It is the framework for selecting appropriate training instances. The idea is to combine this control strategy with control strategy without obstacles in order to form* a full controller.
When an obstacle was included, the robot trajectory for training scene is illustrated in Figure 4.

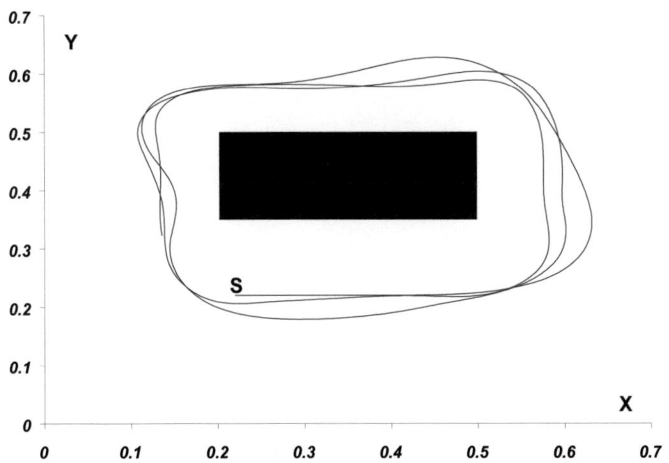

Figure 4. The robot trajectory used in order to gather training example

Control strategy was $r = (\psi_{desired} - \psi) / \Delta t$, v=0.1.

4.1 Experiments with RBF neural network

In the following is described application of RBF network in mobile robot motion planning. A relatively small corresponding training set is given in Table 1.

dex J	Distance (IW[j,1]) or d_{min}	Angle Ψ or LW[1,j]	Angle modified Ψ_m or LW[1,j]	Factor f_m
1	0.04	10.906	10.906	1
2	0.0494	13.8025	13.8025	1
3	0.0591	11.425	11.425	1
4	0.08	0.0035	81.025	23151
5	0.08	9.513	85.617	9
6	0.081	12.070	12.070	1
7	0.0894	12.692	12.692	1
8	0.0923	12.556	12.556	1
9	0.1006	7.6405	7.6405	1
10	0.12	0.6895	3.4475	5

Table 1. The original end the modified training examples according to the movement of the vehicle

The firstly by tuning spread factor (section 3.7.1 and 3.7.2) is tuned σ_j. For step k=1 spread =1, and then spread decreased in 0.01. Process was stopped for spread=0.002. The performance error E_{xy} has minima for that spread value. While spread factor decreased, total time T of the mobile robot moving without touching of obstacles increased: (spread=0.005, T=534 s; spread=0.003, T=600 s; spread=0.002, T=637 s). In order to decreased performance error for small the robot distance from obstacles edges ($d_{min} \approx 0.08$), LW[1,j] is multiplied by f_m [j] as it is given in Table 1 and in Figure 5.

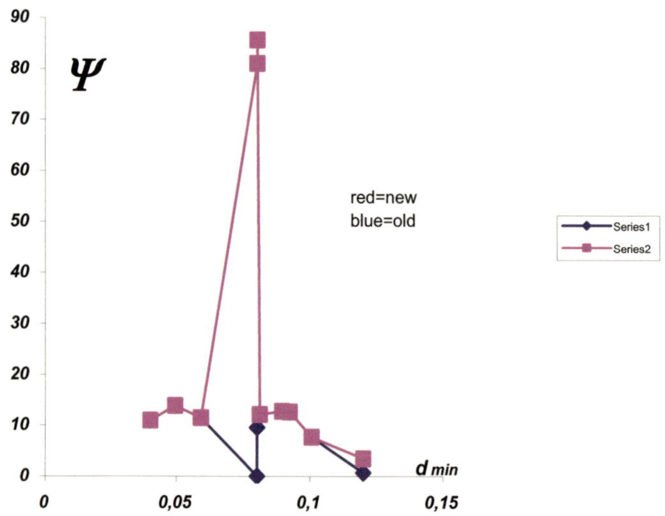

Figure 5. Modifying the LW_{ij} parameters or Ψ training instances for the RBF network in order to decrease performance error $\mathbf{E_{xy}}$

For f_m [4]=23151, f_m[5]=6 and another {f_m [j]=1,j=1,2,3,6,7,8,9,10}, it was T=1248 s. But for f_m [j] which have values as it was given in Table 1 performance error was E_{xy} = 0.00382 and robot moving was not time limited. Two scenes with five static obstacles is given in Figure 6. In that case we have situation when the mobile robot moves away from P1 obstacle to be close, until some critical distance and without touching, to obstacles P2,P3,P4 and P5. The robot safely avoids the obstacles touching. Changing f_m[j] repeatedly takes a step in the direction of steepest decrease of E_{xy}, like the cases described for function J in section 3.1. But some changes are inappropriate. Accordingly generally speaking we wished, but we could not manage, to synthesize the controller to guide the mobile robot to "oscillate" between obstacle P1 and set of obstacles P2,P3,P4 and P5. Exists the tendency of the clone to guide the robot in such a fashion that its distance from obstacle edges enhance gradually. The problem is when the controller solves turning to the left, i.e. it have to conclude that the robot distance, for example, from vertices of obstacle P4 is greater than from the robot distance from vertices of obstacle P5 . Until that the robot avoids obstacle P4 attempting to turn, not to the left, but to the right. It is reason that the robot more and more is closed to obstacles P2, P3, P4 and P5. For small distance of the robot from obstacle vertices we attempt to increase reaction time when turns around a square.

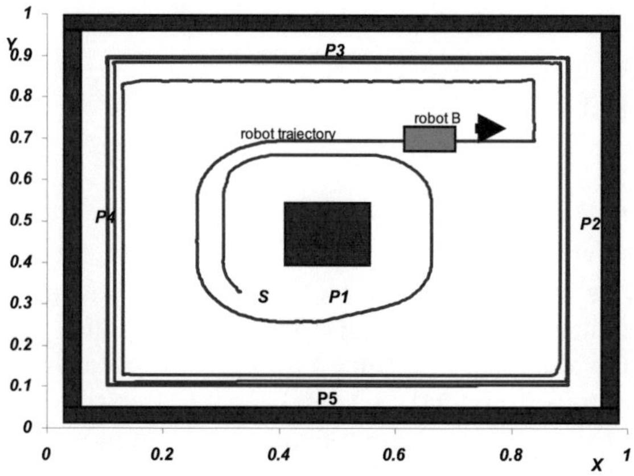

a)

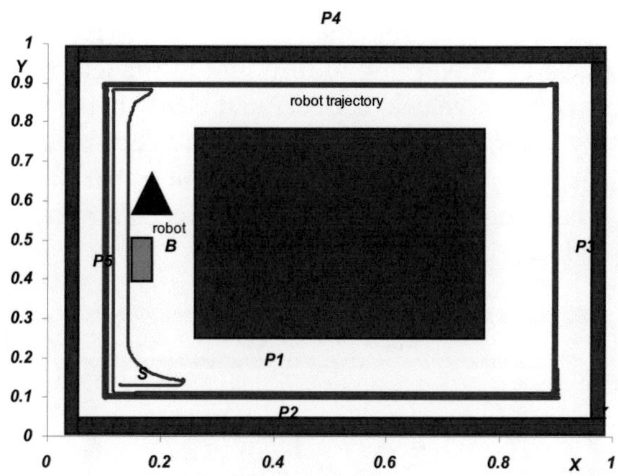

b)

Figure 6. The mobile robot trajectory in xy plane for five static obstacles . The mobile robot is controlled by RBF neural network controller

4.2 MKBC algorithm as advance according to RBF neural network algorithm

The training instances Table 1 (selected 10 from 1801) the previously were tuned using RBF neural network (Section 4.1), i.e. an appropriate f_m factor was selected. In order to form training instances the task was to avoid rectangle obstacle and achieve the minimal number of one vehicle travels from the start to the goal position. On the other hand, the goal is the number of instances to be as small as possible. The MKBC algorithm (section 3.8) was tested as it is illustrated by Figure 7 for $K_c = 1000$ *and* $K_a = 50$ *and*

$$r = \frac{3.1623 * \psi_{des} - 0.9 * \psi}{\Delta t}$$ control strategy, for $\Delta t = 0.05$.

The vehicle start position and obstacles are given by relations:
Start S = (0.15, 0.15),
Position 1 P1 = (0.3,0.3,0.7,0.7),
Position 2 P2 = (0.0, 0.95,1.0,1.0),
Position 3 P3 = (0.95,0.0,1.0,1.0),
Position 4 P4 = (0.0,0.0,1.0,0.05),
Position 5 P5 = (0.,0.0.05, 1.0), for *v =0.2 [m/s]*.
For training instances given in Table 1 and for f_m [4] = 23151, f_m[5] = 6 and another {f_m [i] = 1, j = 1,2,3,6,7,8,9,10}, as it is given in Table 1 the simulation time T using the exposed algorithm was greater then about 6000 [s]. LW[1,i] is multiplied by f_m [i] as is given in Table 1. For f_m [i] with values given in Table 1 performance error was E_{xy} = 0.00180 and the vehicle movement was time unlimited. So we take that f_m is exactly as it is given in Table 1. In this case we have situation when the autonomous vehicle moves away from P1 obstacle to be closed, until some critical distance and without touching obstacles ***P2,P3,P4 and P5.*** Here, the tendency of the clone exists to guide the vehicle in such a fashion that its distance from the obstacle edges enhance gradually. The problem appears when the controller solves the turn to the left. Namely, it should conclude that the vehicle distance, for example, from vertices of obstacle P4 is greater than from the vehicle distance from vertices of obstacle P5. The vehicle avoids obstacle P4 attempting to turn, not to the left, but instead to the right. This is the main reason why the vehicle is closer and closer to obstacles P2, P3, P4 and P5. Changing f_m[j] repeatedly takes a step in the direction of steepest decrease of E_{xy}, like in the cases described early. Like that, if we make the possibility that the coefficint α changes according to the d_{min} changing, we will going to induce a stroughly encahement with regard to the simulation time which is the touch free.

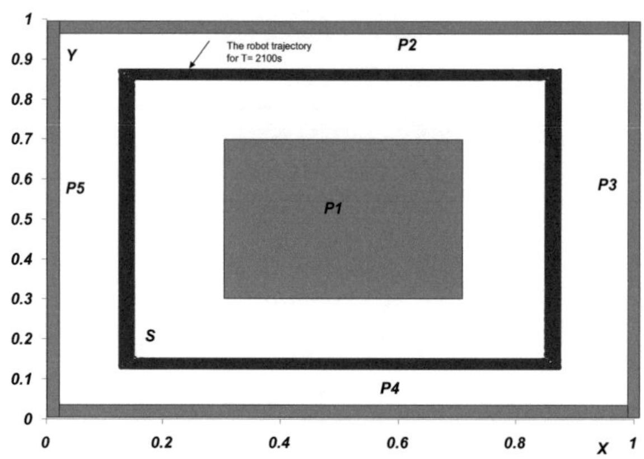

Figure 7. XY trajectory of the vehicle controlled by clone based on modified Kohonen rule

For K_c *=1000 and* K_a *=50* and for the named control strategy we have practically infinite the time which is touch of free. The simplicity and that is reason to accept the MKBC algorithm as an advancement according to the RBF neural network. At the end, Simulink implementation MKBC for the vehicle kinematical model is very simple and it is illustrated in Figure 8, but according to Matlab integration method constants K_{cte},K_α and $K_{psidesired}$ must be corrected as K_{cte}=500,K_α=100 , $K_{psidesired}$=0.7.

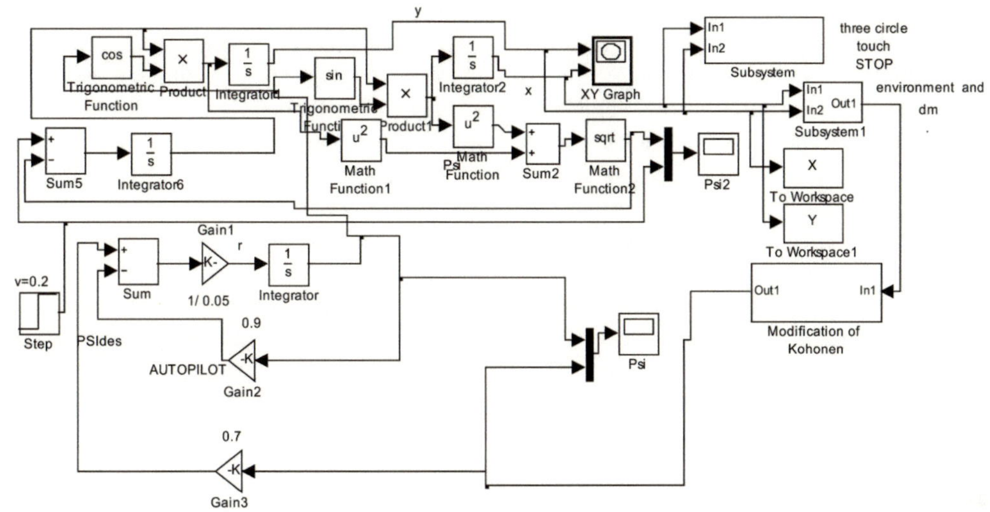

a) Simulink model

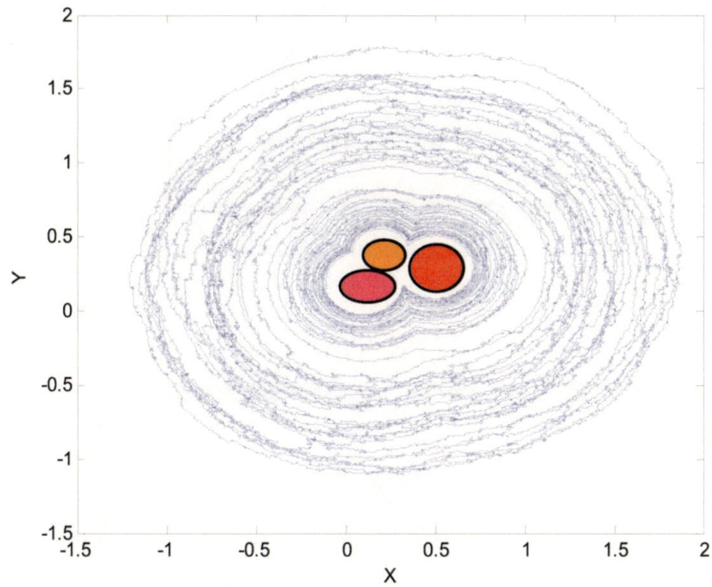

b) 2D trajectory for three obstacles when Kcte=500,K_α=100

Figure 8. Simulink model for MKBC algorithm and an appropriate result

4.3 The narrow passage avoiding by MKBC algorithm

In literature (R. Kulić, 2004) is treated the narrow passage avoiding. In order to avoid the robot traveling through narrow passages between obstacles is needed to define a rule regarding dimension of the autonomous vehicle. In this section MKBC algorithm, as an advancement according to the RBF, is used. That algorithm is applied in order to solve the situation which is given in Figure 9b) -9h). In different situations which are illustrated by

Figure 9b) - 9h) the vehicle moves from different start positions to different goal positions avoiding obstacles and narrow passages also. All experiments are quite successful. The narrow passages are avoided using the reason which is described the previously. In Figure 9b), for example, the autonomous vehicle moves away from P6 obstacle to be close P7, until some critical distance, and so on, to be close until some critical distance from obstacles P4, P3, P2, P1, ..., and always changes Ψ rapidly in order to be safely avoided the obstacle touch for long time. Changes Ψ rapidly is connected with the changing $f_m[j]$ also rapidly as it illustrated in Figure 5. It is the manner repeatedly takes a step in the direction of steepest decrease of E_{xy}. According to the different situations the vehicle managed to avoid obstacles again and again, and never did not have the chance to stop, i.e. that process is has not time limiting and we observed that all experiments are quite successful. It was found that the obstacle avoiding and also the narrow passage avoiding is enabled by the same MKBC algorithm or by the methodology which is the previously described and it means that the solution consistency is saved.

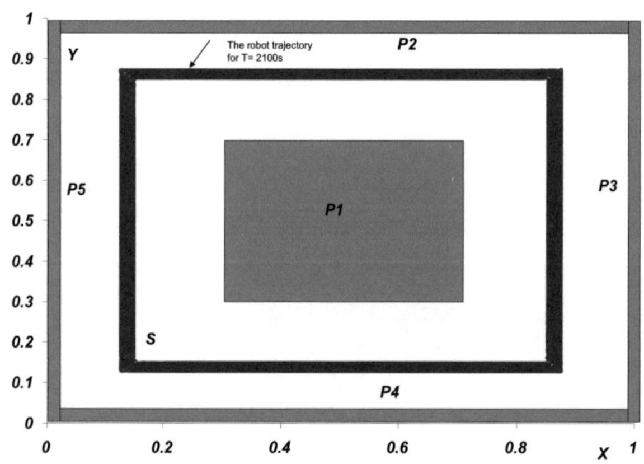

a)

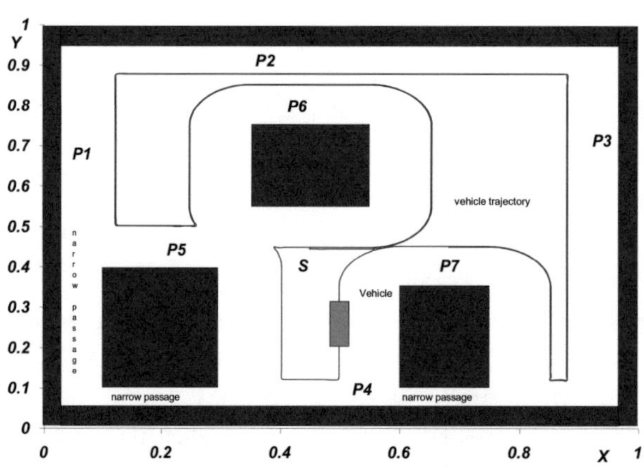

b)

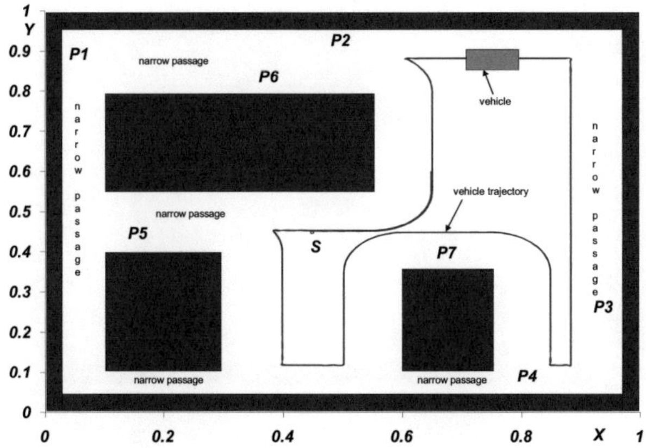
P1
P2
P3
P4
P5
P6
P7
S
narrow passage
narrow passage
narrow passage
narrow passage
narrow passage
narrow passage
narrow passage
vehicle
vehicle trajectory
Y
X
0
0.1
0.2
0.3
0.4
0.5
0.6
0.7
0.8
0.9
1
0
0.2
0.4
0.6
0.8
1

c)

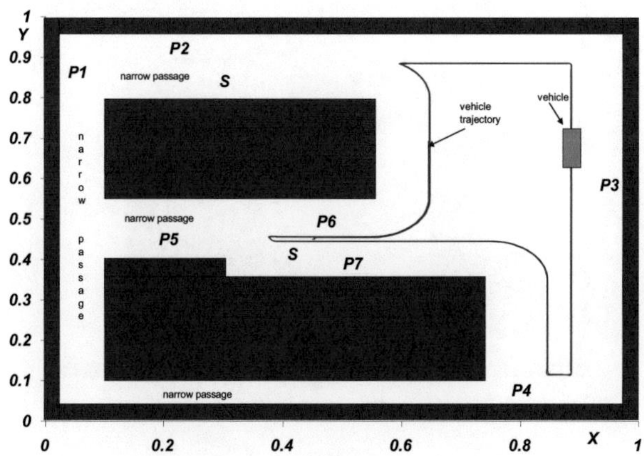
P1
P2
P3
P4
P5
P6
P7
S
S
narrow passage
narrow passage
narrow passage
narrow passage
vehicle
vehicle trajectory
Y
X
0
0.1
0.2
0.3
0.4
0.5
0.6
0.7
0.8
0.9
1
0
0.2
0.4
0.6
0.8
1

d)

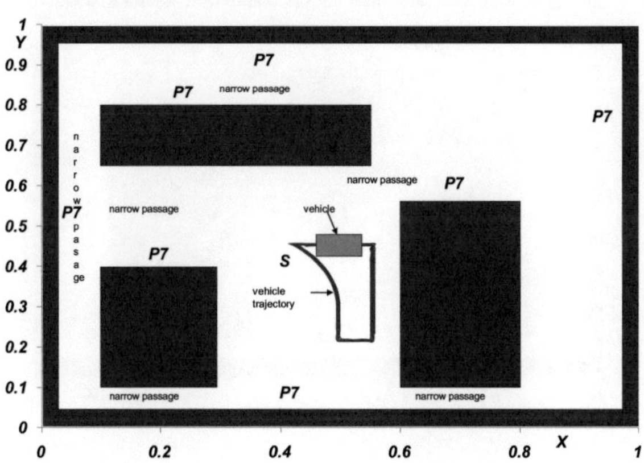
P7
P7
P7
P7
P7
P7
P7
S
narrow passage
narrow passage
narrow passage
narrow passage
narrow passage
narrow passage
vehicle
vehicle trajectory
Y
X
0
0.1
0.2
0.3
0.4
0.5
0.6
0.7
0.8
0.9
1
0
0.2
0.4
0.6
0.8
1

e)

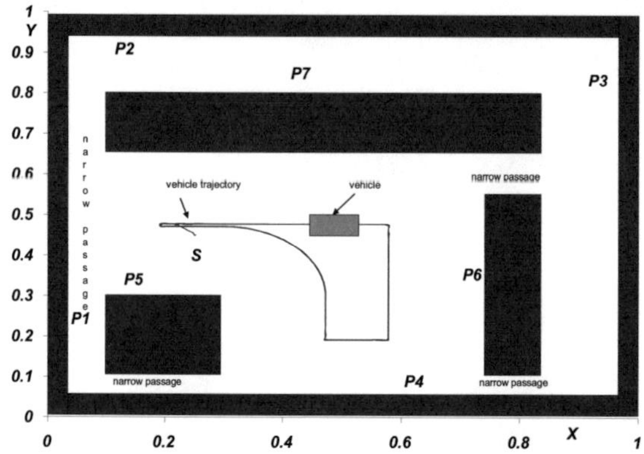

f)

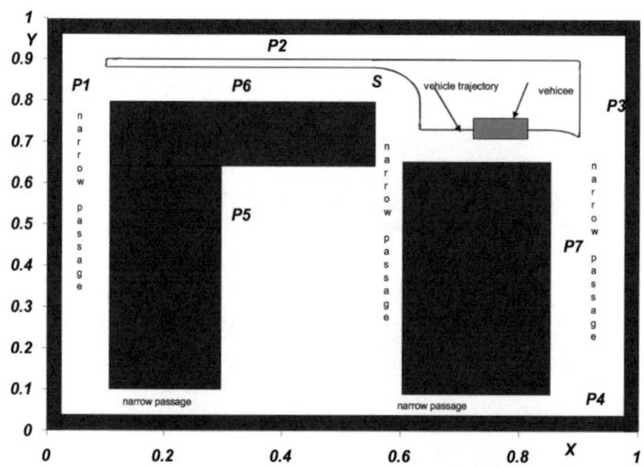

g)

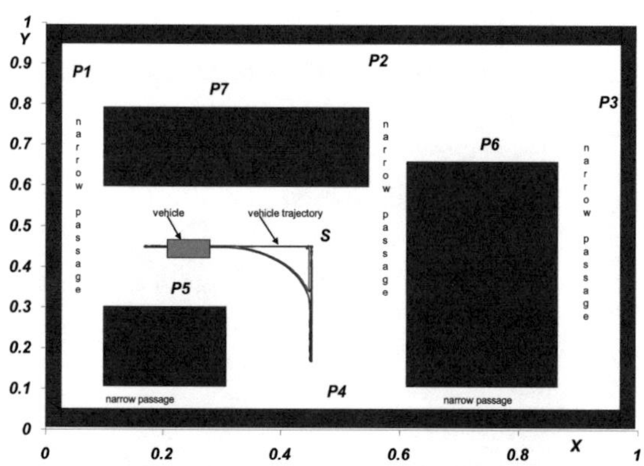

h)

Figure 9. The autonomous vehicle trajectory in xy plane for 5 and for 7 static obstacles

The previously, in literature (R. Kulic, 2006), it is concluded, according to the given machine learning system, to avoid the vehicle traveling through the narrow passages between obstacles it is needed to define a rule regarding their dimensions and now it could be added that rules can be escaped by changing in structure of the given learning system in sense that it is to be able to change Ψ rapidly.

4.4 The goal position reaching

We interested in the goal reaching as it is given in (R. Kulic, 2004) and many disadvantages of if – then rules are named according to this problem. Situation is more and more difficult when the number of obstacles are increasing.

4.4.1 The virtual obstacle based on line connecting the goal and the start position

In the next experiments the line which connect the goal and the start position serves to be formed a virtual obstacle as it is illustrated in Fig. 10. According to the Fig. 10 it is posible to form relations (3 - 8):

$$\begin{aligned} X1 &= XS + 0.05, Y1 = YS; \\ X2 &= XS, Y2 = YS + 0.05; \\ X3 &= XG - 0.05, Y3 = YG; \\ X4 &= XG, Y4 = Y4 - 0.05; \end{aligned} \tag{3}$$[3]

$$YP1 = Y1 + \frac{Y2 - Y1}{X2 - X1}(X - X1) \tag{4}$$

$$YP2 = Y2 + \frac{Y3 - Y2}{X3 - X2}(X - X2) \tag{5}$$

$$YP3 = Y3 + \frac{Y4 - Y3}{X4 - X3}(X - X3) \tag{6}$$

$$YP4 = Y1 + \frac{Y4 - Y1}{X4 - X1}(X - X1) \tag{7}$$

It is known that the normal distance of the point M(X,Y) from the line aX+bY+c=0 is:

$$d = \frac{|aX + bY + c|}{\sqrt{a^2 + b^2}} \tag{8}$$

A procedure, for the simulation purpose, calculating the vehicle distance from the virtual obstacle is given as: SubArea-1: ***if***((Y<=YP1)***and***(Y<=YP2)) ***then*** $d_v = ((X-X1)^2 + (Y-Y1)^2)^{1/2}$, SubArea2: ***if***((Y>=YP4)***and***(Y<YP2)***and***(Y<=YP3)) ***then*** $d_v = |Y+[(Y1-Y2)/(X2-X1)]X +[(Y2-Y1)/(X2-X1)]X1-Y1]| / [[(Y1-Y2)/(X2-X1)]^2+1]^{1/2}$,..... and so on for all of from the

[3] 0.05 *2 means that the virtual obstacle is not only the line but has dimensions which is randomly determined. Equations (3) – (8) define the line which is connected with two determined points.

eight areas. This approach enables the vehicle moving from the start to the goal position expanding distance set by defining: $(d)_m = \min\{d_v, d_i, i=1,..n\}$, using only the equation according to the obstacle avoiding.

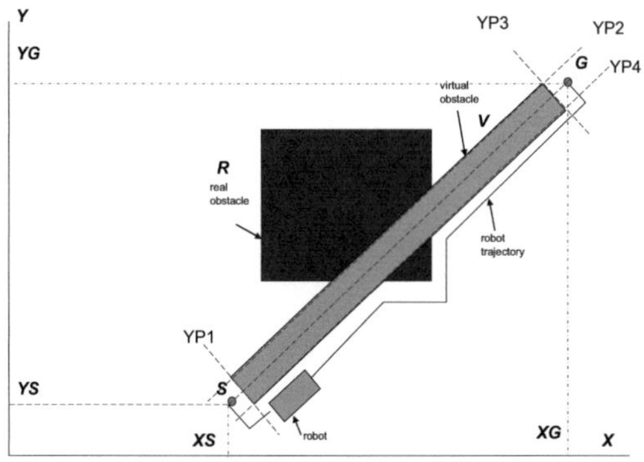

Figure 10. The virtual obstacle based on the start - the goal line

4.4.2 Experiments using the virtual obstacle

As the first, the methodology which is exposed in section 4.1 is tested using GoldHorn machine learning system and (A. Karalic,1991) and equation (9)

$$\psi_{desired} = \frac{-d_m + 0.13735}{0.0064} \tag{9}$$

and control strategy: $r = (\psi_{desired} - \psi) / \Delta t$, $v=0.1$. The result is given in Figure 11.

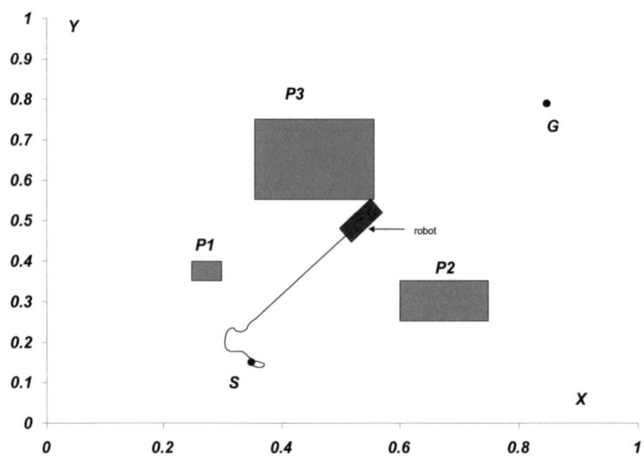

Figure 11. The virtual obstacle based on the start - the goal line and GoldHorn clone

The virtual obstacle is treated and according to all of obstacles obviously the touch is detected because reaction time of GoldHorn clone is inappropriate, as it was observed the previously.

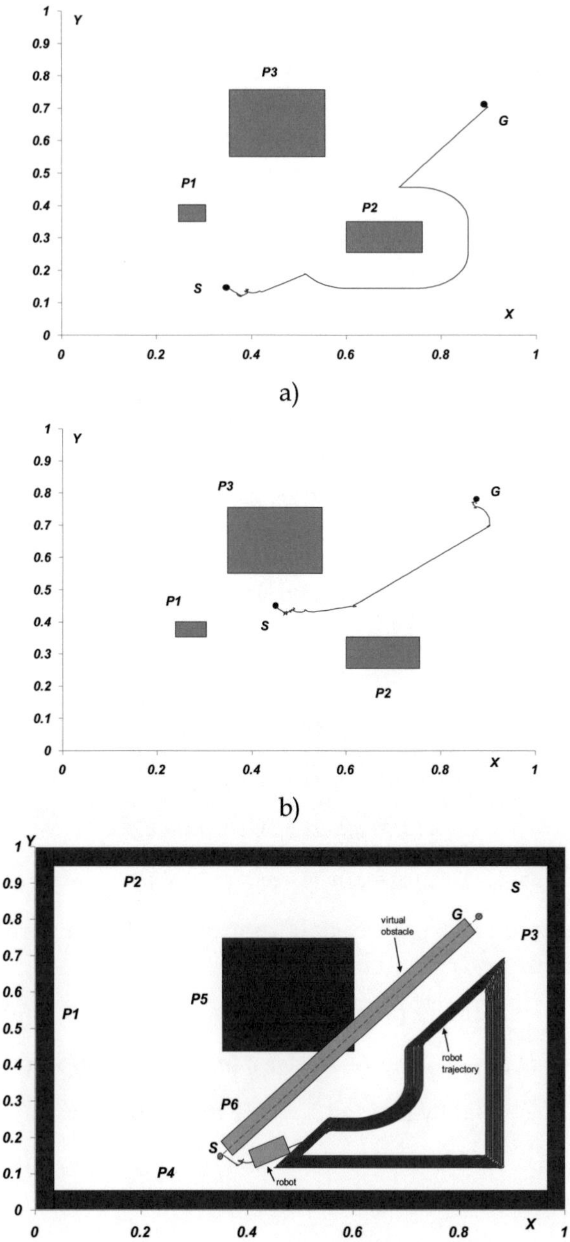

c)

Figure 12. The autonomous vehicle trajectory in xy plane for three and for five static obstacles. The autonomous vehicle is controlled by neural network clone

A neural network MKBC algorithm as an advancement according to the GoldHorm machine learning system (A. Karalič, 1991) and to the RBF is used. MKBC algorithm is applied in order to solve the situation which is given in Fig. 12a),b), c). In situation illustrated by Fig. 12a) for three obstacles the vehicle moves from the start position S(0.35,0.15) to the goal position G(0.85, 0.8) avoiding the virtual obstacle also. The experiment is successful. Like that, the experiment is successful in Fig. 12b), where the virtual obstacle is treated again and where the vehicle reaction time is quite appropriate. In Fig. 12c) is illustrated situation with five static obstacles P1 – P5, and virtual obstacle P6 which is illustrated also. After a great number of obstacles avoiding circles the vehicle attempts to move to the goal G, but some oscillations appear and the vehicle decides to avoid obstacles and reach the goal again and again, and has not the chance to stop. That process has not been time limited, but it means that experiment is quite successful: we have obstacle avoiding and the goal position reaching using the same algorithm, i.e. the same relations (R. Kulić, 2004).

4.5 Results end experiments in 3D space

4.5.1 Kinematikal 3D model

We have applied behavioral cloning and machine learning and got some results, e.g., equations, which were suspected: this results could not suit us if we had dynamical model or if we had real time dynamical system. Generally we expected, equations could be wrong in some sense, but the methodology had to be an appropriate. It is illustrated in the next section. But for some class of trajectories vehicle moving can be decomposed on moving in the vertical and moving in the horizontal plain (R. Stojić, et al., 1990), and for some class of trajectories we can take that $\psi_{desired}(n)=\theta_{desired}(n)$. In that sense the next 3D kinematical model of the autonomous vehicle is observed:

$$\psi(n)=\psi(n-1) + \Delta t\, r(n),$$

$$\theta(n)=\theta(n-1)+ \Delta t\, q(n),$$

$$x(n) = x(n-1) + \Delta t\, v \cos(\psi(n)),$$

$$y(n) = y(n-1) + \Delta t\, v \sin(\psi(n)),$$

$$z(n) = z(n-1) - \Delta t\, v \sin(\theta(n)),$$

where θ is pitch angle, q is the appropriate control variable in vertical plane. To avoid obstacle in 3D the next control rule is observed, also:

if $(n = even)$ **then** $r(n)=r(n-1)$ **else** $r(n) = (\psi_{desired}(n) - \psi(n)) / \Delta t;$

if $(n = odd)$ **then** $q(n) = (n-1)$ **else** $q(n) = (\theta_{desired}(n) - \theta(n)) / \Delta t.$

A simulated scene is illustrated by Figure 13, where the obstacles were avoided successfully.

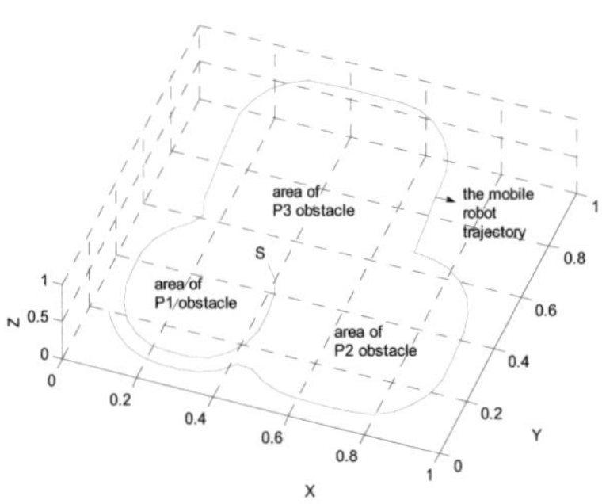

Figure 13. The autonomous vehicle 3D trajectory

It can be concluded that the proposed method gives results that can easily be applied in 3D space.

4.5.2 The complete 3D mathematical model of the vehicle

In literature (G. Campa, M. Innocenti, 1999) is presented development of a mathematical model of an vehicle and its translation under Matlab environment. At the same time the report presents and explain both the equations and the Matlab code that implements those equations. The used approach is independent on which kind of vehicle is modeled, on which environment the vehicle moves (air or water). It means that the report could be easily taken as a reference for a wide class of vehicle modeling problems. In the final chapters several examples on how to use the given nonlinear model for simulation, analysis and control synthesis purposes are given. In Figure 14 is rigid body in 6 DOF with two coordinate frames: body fixed referenced frame and earth fixed reference frame.

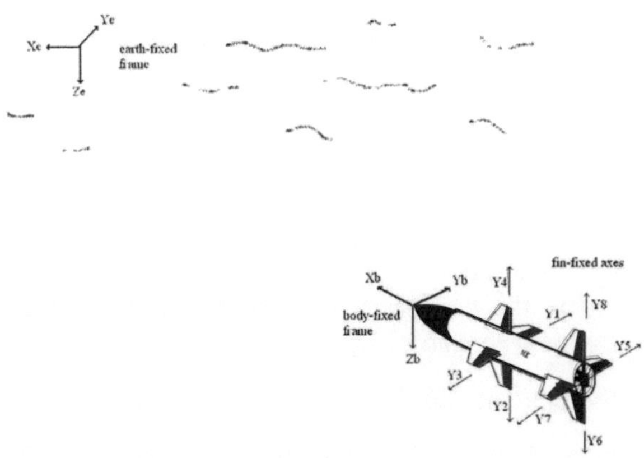

Figure 14. The observed vehicle

The considered vehicle has structure a cylindrical body, an aerodynamic prow and two sets of cross -configured mobile fins, numbered as in Figure 14. Each fin has its own fixed frame in order to rotate about its y -axis. The rotation angle is in counter clockwise. The 4 fins belonging to the front set (numbered from 1 to 4) refer as "wings", while the others 4 fins (numbered from 5 to 8) refer as "tails". Matlab function *vxdot* is the Matlab code equivalent of equation (62) (G. Campa, M. Innocenti , 1999) at equation describe a 12 state dynamic system. It takes as input a 38 - dimensional vector containing: 1) the 12 system states (generalized position and velocity); 2)fin angles vector , 8 elements; 3) external force and moment with respect to the body frame, (e. g. thrusters), 6 elements; 5) external force and torque with respect to the earth frame (e. g. attached cable or other force due to the contact with external objects), 6 elements; 6) marine current speed and acceleration with respect to the earth frame, 6 elements. Matlab *vxdot* function uses the named input and the information about vehicle structure (G. Campa, M. Innocenti,1999) stored in special global variable "veh") to compute derivatives of the system states as described in (62) . The named nonlinear model is linearized about the operating point determined by [x,y,z]=[0,0,0], [u,v,w]=[3m/s, 0, 0], the thrust is 416N, there is no sea current and there is no other external forces.

4.5.1. The classical regulator design

The Linear Quadratic Regulator is one of the regulators which is given in (G. Campa, M. Innocenti, 1999). Its appropriate scheme is given in Figure 7.

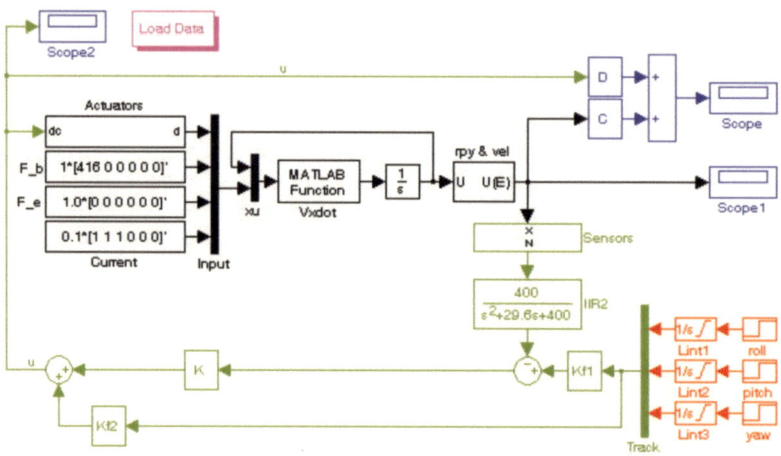

Figure 15. LQR closed loop scheme

Authors found the named regulator as a very good choice for testing the closed loop behavior of the used nonlinear model and treated it as the best in term of both performance and robustness. Having in the workspace the matrices a,b,c,d of the linear model , the feedback regulator is given in Matlab by:

```
>> Q=diag ([20 20 20 ones(1,6)]); R=eye(size(b,2));
```

That command line gives Q and R weight matrices meaning that the attitude states are weighted a value of 20 compared to the value 1 of the remaining states and inputs. Matrix K is given as:

```
>> K=lqr(a,b,Q,R);
```

The nonlinear model starts from the same input and initial conditions the previously considered for the linearization. As disturbance, a slow and constant sea current is considered, and some noice is added to each available measurements. To increase of realism of the simulation a model of the actuators with delays, position limiter and rate limiter is included as it is given by Figure 16.

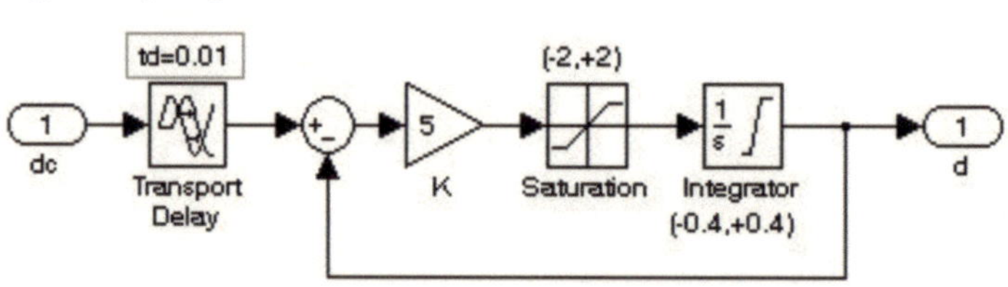

Figure 16. Actuators model

A sensors model is given in Figure 17.

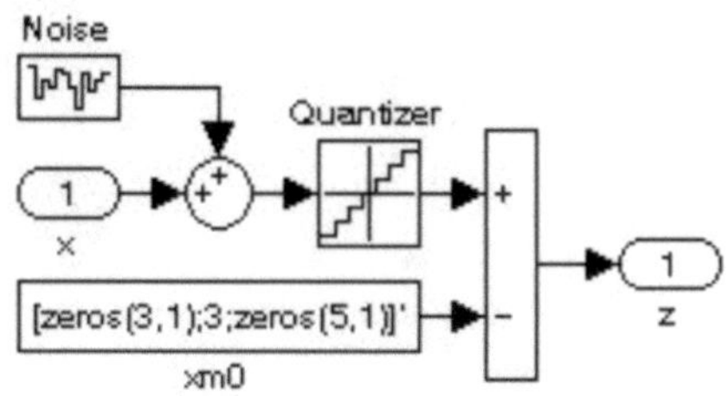

Figure 17. Sensors model

The command to be tracked consists in 3 independent steps of angle 0.1 rad and maximum amplitude 0.3 rad. The inputs or channels: yaw, pitch and roll are appear with time delay from 5s. Sample time is 0.1s. These inputs are used in experiments described in next section.

4.5.2 To avoid obstacles regulator design

As existed our good experience according learning algorithm GoldHorn (R. Kulic, Z. Vukic 2006), in training phase we decided to use it again, i.e. use it clone based on equation

$$\psi_{desired} = -0.08005 d_m - 1.54286 \quad (10),$$

$$\theta_{desired} = -0.08005 d_m - 1.54286 \quad (11),$$

which GoldHorn managed to find. In order to form training instances system in Figure 3 is changed adding Simulink subsystems containing: 1) GoldHorn clone as operator, 2) virtual simulation environments, 3) the fashion of calculating[4] d_m, 4) the fashion of detecting touch between the vehicle and the obstacles and 5) the fashion of gathering the training instances in the Matlab workspace. Part of that model is given in Figure 18.
Situation with the vehicle start position S[0,0,0] and three unmoving sphere obstacles: *O1[40,10,0,30], O2[30,40,20,20], O3[10,20,40,20]*[5] is observed. In the training phase the vehicle is controlled by GoldHorn clone in order to avoid sphere obstacle from 0s to 500s. Sample time was 0.1s. The clone did not manage to realize the trajectory which mean that the vehicle start position is equal the vehicle goal position. It was difficult: the previously operator has not been a skilled worker and GoldHorn machine learning controller was similar to him. From 5122 examples were selected 10, i.e. 500th, 1000th, 1500th, ..., 5000th, as it is given in Table 2. It is important to expose that using training instances and Matlab polyfit instruction we get some unsatisfactory results.

---

[4] *d_i is distance of the vehicle gravity center from i^{th} obstacle and d_m is defined as: $d_m = min(d_i\ , i=1,n)$, where n is number of obstacles.*

[5] *Sphere is given as: O(a,b,c,R), where a,b,c are coordinates of center and R is the radius.*

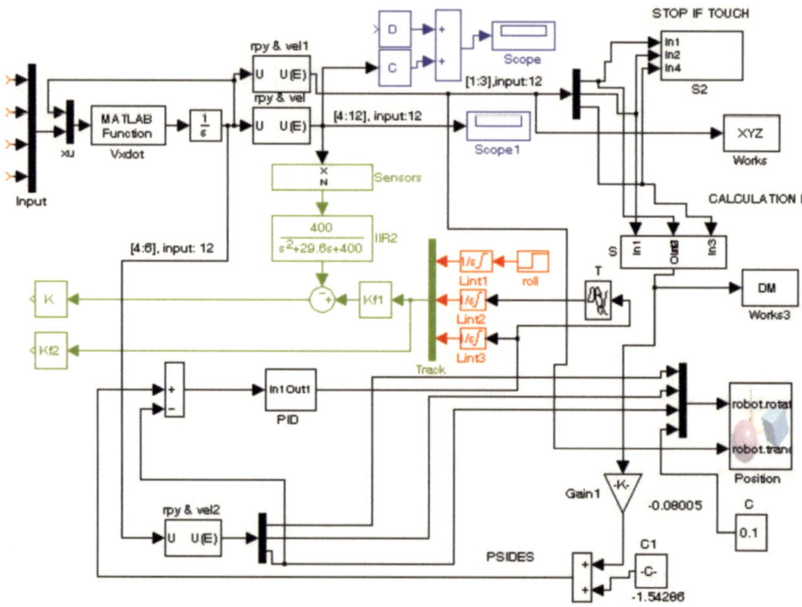

Figure 18. The part of Simulink model connected with the training phase

No.	Ψ Angle	d_m distance
1	-0.8162	2.1189
2	-1.5639	7.1298
3	-2.4173	19.0685
4	-3.4051	27.7691
5	-4.2736	28.9622
6	-4.6729	20.8607
7	-4.359	12.796
8	-3.7437	17.9315
9	-3.463	29.6801
10	-13.3025	150.2197

Table 2. The training examplse for 3D nonlinear dynamical model

4.5.3 The experiments and results

In verifying phase it is needed to connect $\psi_{desired}$ with ψ and θ inputs as it is illustrated in Figure 7. But it was not enough. Namely, it was observed that is needed to be synthesized additional regulator regarding obstacle avoiding process and to be formed the closed loop system according ψ and θ, or according minimal distance d_m between the vehicle and obstacles. To require that kind of correction is a consequence of the fact that the learner outputs the hypothesis that has the minimum error over the training examples, but has the error (T. Mitchell, 1997). So, a PID regulator was added and its founded constants are: *kp=0.005, kd=0.0009, ki=0.0001.* According to MKBC algorithm given in Section 3.8 tuned constants are: *ka=0.00005, kcte=-0.1.* Between *θ and ψ time delay was* T_θ=1s and roll angle was φ=0.1rad. Maximum amplitude was: for θ ***i*** φ (-0.3, 0.3) rad., and for ψ (-∞, +∞)rad. Part of Simulink model connected with verifying phase is illustrated by Figure 19. Situation with the vehicle start position S[0,0,0] and three unmoving sphere obstacles, with different positions and dimensions according to the training phase, are observed: *O1[20,30,40,10], O2[30,40,50,20], O3[40,10,350,10]* . The results are illustrated by Figure 20. As illustrated in figure 20b) the tendency of the regulator to guide the vehicle in such a fashion that its distance from obstacle edges increases was not observed. That fact is a good reason for the conclusion that these results, with sophisticated kinematical and dynamical model of the vehicle, quite confirm the results exposed the result the previously exposed when, when was used only kinematical model of the vehicle.

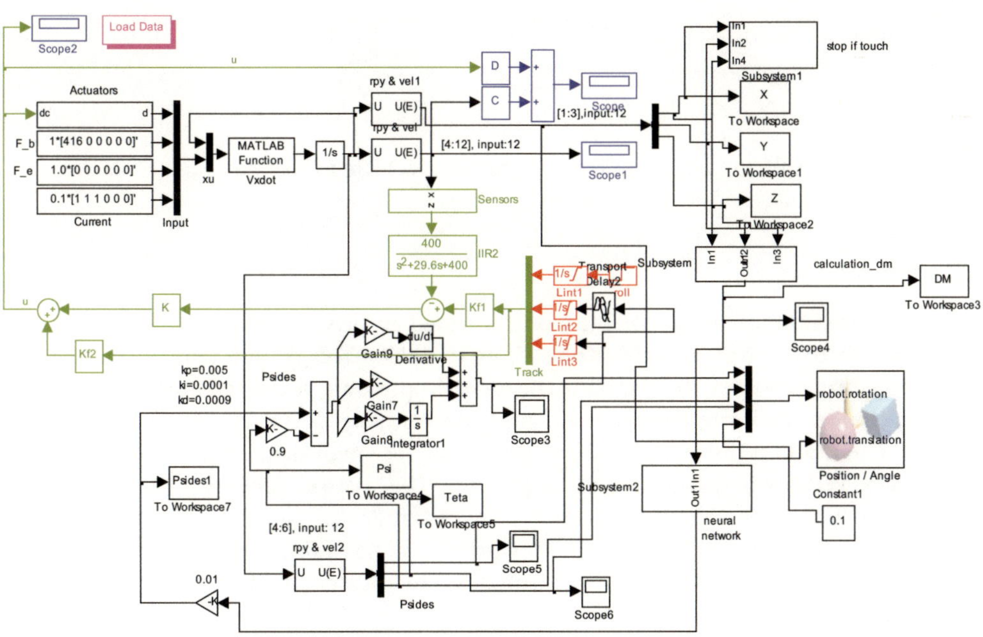

Figure 19. The part of Simulink model connected with the verifying phase

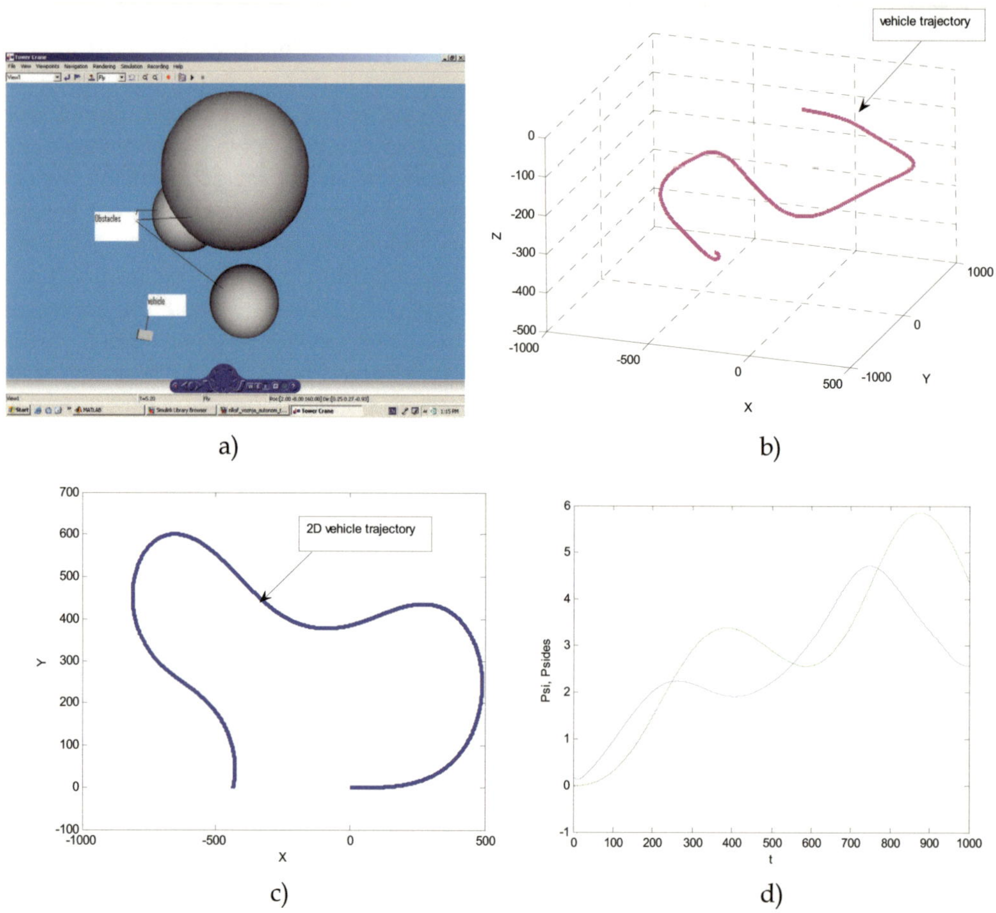

Figure 20. The simulation results of the autonomous vehicle moving and the obstacles avoiding

5. Conclusion

As it is exposed the obtained MKBC clone is tuned to produce the results that are more appropriate then those given in (R. Kulic, Z. Vukic, 2006). This is especially important when the vehicle turns around the square.The very important MKBC characteristics are the operator cloning and symplicity, the simplicity specialy according to the RBF neural network. The reaction time substantially decreases for small distance of the vehicle from the obstacle vertices and edges according to the RBF neural network. According to the different situations the vehicle managed to avoid obstacles again and again, and never did not have the chance to stop, i.e. that process is has not time limiting in 2D space specially and we observed that all experiments are quite successful. Following the given context the problem narrow passage avoiding and the goal position reaching fundamentally is observed. Namely, defining if - then rule according to the problem is observed as destroying of the consistency of the reached methodology. MKBC algorithm gives chance that the obstacle avoiding, the narrow passage avoiding and goal position reaching is enabled by the same algorithm or by the methodology which is the previously described and it means that the solution consistency is saved. The algorithm complexity is about O(n), where n is the number of obstacles. A the end, the autonomous vehicle mathematical model is given by nonlinear equations describing a 12 state dynamical system and in that case the exposed methodology is applied successfully. The advantage of this approach lies in the fact that a complete path can be defined off-line, without using sophisticated symbolical models of obstacles. By this fact the named methodology substantially differs from the others. In the next phase it is expected to confirm results in on - line simulation process.

6. References

J. C. Latombe, *Vehicle Motion Planning*, Kluwer Academic Publishers, Boston 1[1]991.

J. Reif, Compexity of the generalized mover's problem. Editors J. Schwartz, M. Sharir, and J. Hopcroft: *Planning, Geometry, and Compexity of Vehicle Motion*, Ablex, Norwood, Nj, 1987. [2]

J. Schwartz, M. Sharir, and J. Hopcroft, *Planning, Geometry, and Compexity of Vehicle Motion*, Ablex, Norwood, Nj, 1987. [3]

J. Schwartz, M. Sharir, On the piano mover's problem: I. The case of twodimensional rigid polygonal body moving amidst polygonal barriers, Editors J. Schwartz, M. Sharir, and J. Hopcroft: *Planning, Geometry, and Compexity of Vehicle Motion*, Ablex, Norwood, Nj, 1987. [4]

O. Khatib, Real - time obstacle avoidance for manipulators and autonomous vehicles, *International Journal of Vehicleics Rsearch*, 5, 1986. [5]

D. Michie, R. Camacho, Building symbolic representation of intuitive real - time skills from performance data . In: K. Furukawa, S. Muggleton (eds.) *Machine Intelligence and Inductive learning*, Oxford University Press, Oxford, 1994. [6]

C. Sammut, S.Hurst, D. Kedzier, D. Michie, Learning to Fly, D. Sleeman, P. Edwards, (eds), *Proceeding 9st International Workshop on Machine Learning* , Morgan Kaufmann, 1992. [7]

D. Michie, Knowledge, learning and Machine Intelligence In:L.S. Sterling (ed), *Intelligent System*, Plenum Press, New York 1993; [8]

R. Kulić, Z. Vukić, Methodology of concept control synthesis to avoid unmoving and moving obstacles (II), *(accepted, April 11, 2006; Online published June 14,2006) Journal of Intelligent and Robotic Systems*, Publisher Springer Netherlands, ISSN 0921-0296 (paper) 1573-0409 (Online), DOI: 10.1007/s 10846-006-9035-7, Issue: Volume 45, Number 3, Date:March 2006, Pages: 267 - 294. [9]

R. Kulić.,Trajectory design for the autonomous vehicle by operator cloning, PhD thesis, Faculty of Electrical Engineering and Computing, Zagreb, 2004. (in Croatian) [10]

A. Karalič, Regression trees learning from noice data, *Master thesis*, Univerzity of Ljubljana, Ljubljana, 1991. (in Slovenian); [11]

V. Križman, The consideration of noice data using automatic modeling of dynamical systems, *Master thesis*, Univerzity of Ljubljana, Ljubljana, 1993. (in Slovenian). [12]

L. Breiman, J.H. Friedman, R.A. Olshen, C.J Stone, *Classification and Regression Trees*, Wadsworth Int. Group, Belmont, California, USA, 1984; [13]

C, A. Michelli, Interpolation of scattered data: Distance and conditionally positive definite functions, *Constructive Approximation*, 2, 11-22, 1986. [14]

M. J.D. Powell, Radial basis functions for multivariate interpolation: A review, in *Algorithms for the Approximation of Functions and Data,* J. C. Mason and M. G. Cox, eds. Clarendon Press, Oxford, England, 1987. [15]

R. Stojić, R. Kulić, M. Živanović, Flight control for the given trajectory, *Scince-Technical observer,* Vol. XI, No. 8-9, Belgrade, 1990 (in Serbian). [16]

M..H. Hassoun, *Foundamental of Artificial Neural Networks,* MIT Press, 1995. [17]

D. S. Broomhead, D. Lowe, Multivariate functional interpolation and adaptive networks, *Complex Systems,* 2, 321-355, 1988. [18]

S. Lee, R. Kill, Multilayer feedforward potential function networks, *Proceedings of the IEEE Second International Conference on Neural Networks,* San Diego 1988, vol. I,, pp. 161-171, IEEE, New York. [19]

M. Niranjan , F. Fallside, Neural Networks and Radial Basis Functions in Classifying static Speech Patterns, *Technical Report* CUEDIF –INFENG17R22, Engineering Department, Cambridge University, 1988. [20]

T. Poggio, F. Girosi, Networks for approximation and learning, *Proceeding of the IEEE,* 78(9), 1481-1497, 1990. [21]

E. J. Hartman, at al., Layered neural networks with Gaussian hidden units as universal approximators, *Neural Computation,* 2(2), 210-215, 1990. [22]

G. Campa, M. Innocenti, *Model of an underwater vehicle,* University of Pisa October 1999. [23]

P. Vadakkepat, X. PengQ. B. Kiat, L. T. Heng, Evolution of fuzzy behaviors for multi – robotic system, *Robotics and Autonomous System,* vol 55(2). February 2007, pp 146 - 161. [24]

T. Mitchell, *Machine Learning,* McGraw – Hill Companies, Inc., 1997. [25]

15

An Immunological Approach to Mobile Robot Navigation

Guan-Chun Luh and Wei-Wen Liu
Department of Mechanical Engineering, Tatung University
Taiwan, Republic of China

1. Introduction

Autonomous mobile robots have a wide range of applications in industries, hospitals, offices, and even the military, due to their superior mobility. Some of their capabilities include automatic driving, intelligent delivery agents, assistance to the disabled, exploration and map generation for environmental cleanup, etc. In addition, their capabilities also allow them to carry out specialized tasks in environments inaccessible or very hazardous for human beings such as nuclear plants and chemical handling. They are also useful in emergencies for fire extinguishing and rescue operations. Combined with manipulation abilities, their capabilities and efficiency will increase and can be used for dangerous tasks such as security guard, exposition processing, as well as undersea, underground and even space exploration.

In order to adapt the robot's behavior to any complex, varying and unknown environment without further human intervention, intelligent mobile robots should be able to extract information from the environment, use their built-in knowledge to perceive, act and adapt within the environment. An autonomous robot must be able to maneuver effectively in its environment, achieving its goals while avoiding collisions with static and moving obstacles. As a result, motion planning for mobile robots plays an important role in robotics and has thus attracted the attention of researchers recently. The design goal for path planning is to enable a mobile robot to navigate safely and efficiently without collisions to a target position in an unknown and complex environment. The navigation strategies of mobile robots can be generally classified into two categories, global path planning and local reactive navigation. The former is done offline and the robot has complete prior knowledge about the shape, location, orientation, and even the movements of the obstacles in the environment. Its path is derived utilizing some optimization techniques to minimize the cost of the search. However, it has difficulty handling a modification of the environment, due to some uncertain environmental situations, and the reactive navigation capabilities are indispensable since the real-world environments are apt to change over time. On the other hand, local reactive navigation employing some reactive strategies to perceive the environment based on the sensory information and path planning is done online. The robot has to acquire a set of stimulus-action mechanisms through its sensory inputs, such as distance information from sonar and laser sensors, visual information from cameras or processed data derived after appropriate fusion of numerous sensor outputs. The action taken by the robot is usually an alternation of steering angle and/or translation velocity to avoid collisions and reach the desired target. Nevertheless, it does not guarantee a solution for the mission, nor is the solution the optimal one.

Reactive behavior-based mobile robot responds to stimuli from the dynamic environment, and its behaviors are guided by local states of the world. Its behavior representation is situated at a sub-symbolic level that is integrated into its perception-action (*i.e.*, sensor-motor) capacities analogous to the manifestation of the reflex behavior observed in biological systems. Some researches have focused on this kind of robot system and have demonstrated its robustness and flexibility against an unstructured world (Chang, 1996). Reactive behavior-based strategy is now becoming attractive in the field of mobile robotics (Lee, et al., 1997) to teach the robot to reach the goal and avoid obstacles. Two different kind of reactive navigation strategies have been studied. The first application task for the mobile robot is to navigate in a stationary environment while avoiding static obstacles but reaching a goal safely. A well-known drawback is that the mobile robot suffers from local minima problems in that it uses only locally available environmental information without any previous memory. In other words, a robot may get trapped in front of an obstacle or wander indefinitely in a region whenever it navigates past obstacles toward a target position. This happens particularly if the environment consists of concave obstacles, mazes, etc. Several trap escape algorithms, including the random walk method (Baraquand and Latombe, 1990), the multi-potential field method (Chang, 1996), the tangent algorithm (Lee, et al., 1997), the wall-following method (Yun and Tan, 1997), the virtual obstacle scheme (Liu et al., 2000), and the virtual target approach (Xu, 2000) have been proposed to solve the local minima problems. The second application task is to navigate mobile robot in an unknown and dynamic environment while avoiding moving obstacles. Various methods have been proposed for this purpose, such as configuration-time space based method (Fujimura and Samet, 1989), planning in space and time independently (Ferrari et al., 1998), cooperative collision avoidance and navigation (Fujimori, 2005), fuzzy based method (Mucientes et al., 2001), velocity obstacles method (Prassler et al., 2001), collision cone approach (Qu et al., 2004), and potential field method (Ge and Cui, 1989). Another approach for motion planning of mobile robots is the Velocity Obstacle (VO) method first proposed by Fiorini and Shiller (Fiorini and Shiller, 1998).

In the last decade, it has been shown that the biologically inspired artificial immune system (AIS) has a great potential in the fields of machine learning, computer science and engineering (Castro and Jonathan, 1999). Dasgupta (1999) summarized that the immune system has the following features: self-organizing, memory, recognition, adaptation, and learning. The concepts of the artificial immune system are inspired by ideas, processes, and components, which extracted from the biological immune system. A growing number of researches investigate the interactions between various components of the immune system or the overall behaviors of the systems based on an immunological point of view. Immunized systems consisting of agents (immune-related cells) may have adaptation and learning capabilities similar to artificial neural networks, except that they are based on dynamic cooperation of agents (Ishida, 1997). Moreover, immune systems provide an excellent model of adaptive process operating at the local level and of useful behavior emerging at the global level (Luh and Cheng, 2002). Accordingly, the artificial immune system can be expected to provide various feasible ideas for the applications of mobile robots (Ishiguro et al., 1997; Lee and Sim, 1997; Hart et al., 2003; Duan et al., 2005). As to mobile robot navigation problem, Ishiguro *et al.* (1995) proposed a two-layer (situation-oriented and goal-oriented) immune network to behavior control of autonomous mobile robots. Simulation results show that mobile robot can reach goal without colliding fixed or moving obstacles. Later, Lee *et al.* (2000) constructed obstacle-avoidance and goal-approach immune networks for the same purpose. Additionally, it shows the advantage of not falling into a local loop. Afterward, Vargas *et al.* (2003) developed an Immuno-Genetic Network for autonomous navigation. The simulations show that the evolved immune network is capable of correctly coordinating the system towards the objective of the navigation task. In addition, some preliminary experiment on a real Khepera II robot demonstrated the feasibility of the network. Recently, Duan et al. (2004) proposed an immune algorithm for path planning of a car-like wheeled mobile robot. Simulations indicate that the algorithm can finish different tasks within shorter time. It should be noted that, however, all of the above researches did not consider solving the local minima problems. Besides, none relative researches implement AIS for mobile robot navigating in dynamic environments.

Two different kind of reactive immune networks inspired by the biological immune system for robot navigation (goal-reaching and obstacle-avoidance) are constructed in this study. The first one is a potential filed based immune network with an adaptive virtual target mechanism to solve the local minima problem navigating in stationary environments. Simulation and experimental results show that the mobile robot is capable of avoiding stationary obstacles, escaping traps, and reaching the goal efficiently and effectively. Employing the Velocity Obstacle method to determine the imminent collision obstacle, the second architecture guide the robot avoiding collision with the most danger object (moving obstacle) at every time instant. Simulation and experimental results are presented to verify the effectiveness of the proposed architecture in dynamic environment.

2. Biological immune system

The immune system protects living organisms from foreign substances such as viruses, bacteria, and other parasites (called antigens). The body identifies invading antigens through two inter-related systems: the innate immune system and the adaptive immune system. A major difference between these two systems is that adaptive cells are more antigen-specific and have greater memory capacity than innate cells. Both systems depend upon the activity of white blood cells where the innate immunity is mediated mainly by phagocytes, and the adaptive immunity is mediated by lymphocytes as summarized in Fig. 1. The phagocytes possess the capability of ingesting and digesting several microorganisms and antigenic particles on contact. The adaptive immune system uses lymphocytes that can quickly change in order to destroy antigens that have entered the bloodstream. Lymphocytes are responsible for the recognition and elimination of the antigens. They usually become active when there is some kind of interaction with an antigenic stimulus leading to the activation and proliferation of the lymphocytes. Two main types of lymphocytes, namely B-cells and T-cells, play a remarkable role in both immunities [34]. Both B-cell and T-cell express in their surfaces antigenic receptors highly specific to a given antigenic determinant. The former takes part in the humoral immunity and secrete antibodies by the clonal proliferation while the latter takes part in cell-mediated immunity. One class of the T-cells, called the Killer T-cells, destroys the infected cell whenever it recognizes the infection. The other class that triggers clonal expansion and stimulates or suppresses antibody formation is called the Helper T-cells.

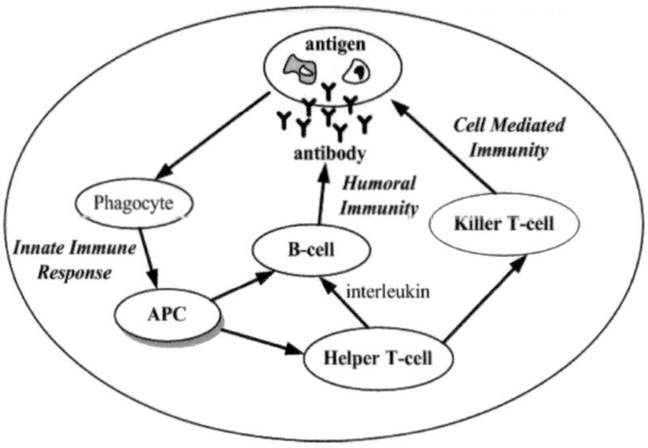

Figure 1 Illustration of the biological immune system

When an infectious foreign pathogen attacks the human body, the innate immune system is activated as the first line of defense. Innate immunity is not directed in any way towards specific invaders but against any pathogens that enter the body. It is called the non-specific immune response. The most important cell in innate immunity is a phagocyte, which internalizes and destroys the invaders to the human body. Then the phagocyte becomes an Antigen Presenting Cell (APC). The APC interprets the antigen appendage and extracts the features by processing and presenting antigenic peptides on its surface to the T-cells and B-cells. These lymphocytes will be able to sensitize this antigen and be activated. Then the Helper T-cell releases the cytokines that are the proliferative signals acting on the producing B-cell or remote the other cells. On the other hand, the B-cell becomes stimulated and creates antibodies when it recognizes an antigen. Recognition is achieved by inter-cellular binding, which is determined by molecular shape and electrostatic charge. The secreted antibodies are the soluble receptor of B-cells and these antibodies can be distributed throughout the body (Oprea, 1996). An antibody's paratope can bind an antigen's epitope according to its affinity. Moreover, B-cells are also affected by Helper T-cells during the immune responses (Carneiro et al., 1996). The Helper T-cell plays a remarkable key role for deciding if the immune system uses cell-mediated immunity or humoral immunity (Roitt et al. 1998), and it connects the non-specific immune response to make a more efficient specific immune response. The Helper-T cells work primarily by secreting substances known as cytokines and their relatives (Roitt et al. 1998) that constitute powerful chemical messengers. In addition to promoting cellular growth, activation and regulation, cytokines can also kill target cells and stimulated macrophages.

The immune system produces the diverse antibodies by recognizing the idiotype of the mutual receptors of the antigens between antigen and antibodies and between antibodies. The relation between antigens and antibodies and that amongst antibodies can be evaluated by the value of the affinity. In terms of affinities, the immune system self-regulates the production of antibodies and diverse antibodies. Affinity maturation occurs when the maturation rate of a B-cell clone increases in response to a match between the clone's antibody and an antigen. Those mutant cells are bound more tightly and stimulated to divide more rapidly. Affinity maturation dynamically balances exploration versus exploitation in adaptive immunity (Dasgupta, 1997). It has been demonstrated that the immune system has the capability to recognize foreign pathogens, learn and memorize, process information, and discriminate between self and non-self. In addition, the system can be maintained even faced with a dynamically changing environment.

Jerne (1973) has proposed the idiotypic network hypothesis (immune network hypothesis) based on mutual stimulation and suppression between antibodies as Fig. 2 illustrates. This hypothesis is modeled as a differential equation simulating the concentration of a set of lymphocytes. The concept of an immune network states that the network dynamically maintains the memory using feedback mechanisms within the network. The various species of lymphocytes are not isolated but communicate with each other through the interaction antibodies. Jerne concluded that the immune system is similar to the nervous system when viewed as a functional network. Based on his speculation, several theories and mathematical models have been proposed (Farmer et al., 1986; Hoffmann, 1989; Carneiro et al., 1996). In this study, the dynamic equation proposed by Farmer (1986) is employed as a reactive immune network to calculate the variation on the concentration of antibodies, as shown in the following equations:

$$\frac{dA_i(t)}{dt} = \left(\sum_{\ell=1}^{N_{Ab}} m_{i\ell}^{st} a_\ell(t) - \sum_{k=1}^{N_{Ab}} m_{ki}^{su} a_k(t) + m_i - k_i \right) a_i(t) \quad (1)$$

$$a_i(t) = \frac{1}{1 + \exp(0.5 - A_i(t))} \quad (2)$$

where i, ℓ, $k = 0, 1, \cdots, N_{Ab}$ are the subscripts to distinguish the antibody types and N_{Ab} is the number of antibodies. A_i and a_i are the stimulus and concentration of the ith antibody. m_{ij}^{st}, m_{ki}^{su} indicate the stimulative and suppressive affinity between the ith and the jth, kth antibodies, respectively. m_i denotes the affinity of antigen and antibody i, and k_i represents the natural death coefficient. Equation (1) is composed of four terms. The first term shows the stimulation, while the second term depicts the suppressive interaction between the antibodies. The third term is the stimulus from the antigen, and the final term is the natural extinction term, which indicates the dissipation tendency in the absence of any interaction. Equation (2) is a squashing function to ensure the stability of the concentration (Ishiguro et al., 1997).

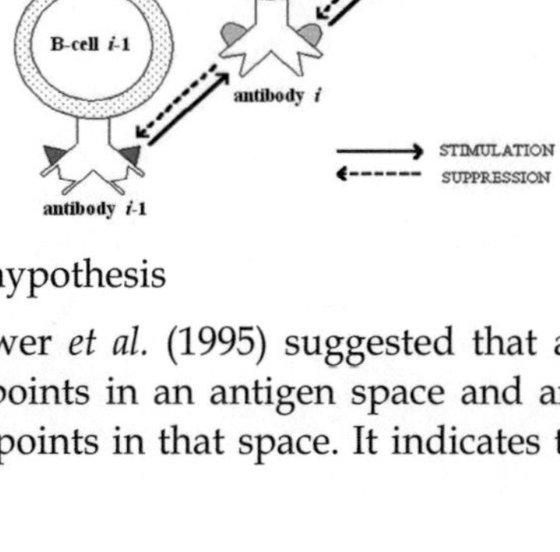

Figure 2. Idiotypic network hypothesis

On the other hand, Hightower *et al.* (1995) suggested that all possible antigens could be declared as a group of set points in an antigen space and antigen molecules with similar shapes occupy neighboring points in that space. It indicates that an antibody molecule can recognize some set of antigens and consequently covers some portion of antigen space as Fig. 3 illustrated. The collective immune response of the immune network is represented as $\sum_{i=1}^{N_{Ab}} f(Ab_i)$, where $f(Ab_i)$ indicates the immune response function between antigen and the *i*th antibody. Note that any antigen in the overlapping converge could be recognized by several different antibodies simultaneously. Afterward, Timmis *et al.* (1999) introduced similar concept named Artificial Recognition Ball (ARB). Each ARB represents a certain number of B-cells or resources, and total number of resources of system is limited. In addition, each ARB describes a multi-dimensional data item that could be matched to an antigen or to another ARB in the network by Euclidean distance. Those ARBs located in the other's influence regions would either be merged to limit the population growth or pulled away to explore new area. ARBs are essentially a compression mechanism that takes the B-cells to a higher granularity level.

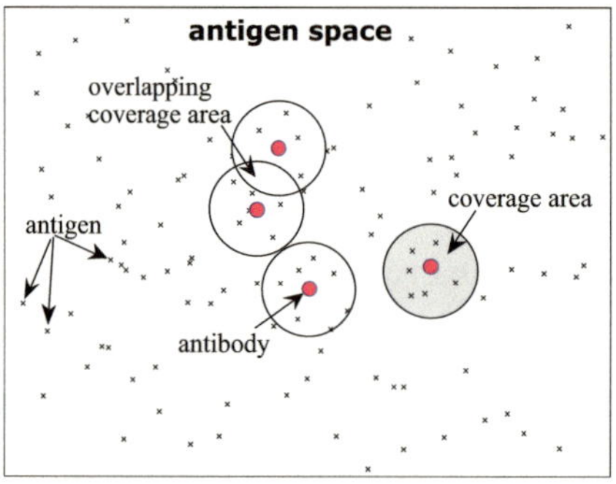

Figure 3. The antigen space

3. Motion Planning in Stationary Environments

3.1 Reactive immune network

A reactive immune network inspired by the biological immune system for robot navigation (goal-reaching and obstacle-avoidance) in stationary environments is described in this section. The architecture of the proposed navigation system is depicted in Fig. 4. The antigen's epitope is a situation detected by sensors and provides the information about the relationship between the current location and the obstacles, along with the target. This scene-based spatial relationship is consistently discriminative between different parts of an environment, and the same representation can be used for different environments. Therefore, this method is tolerant with respect to the environmental changes. The interpreter is regarded as a phagocyte and translates sensor data into perception. The antigen presentation proceeds from the information extraction to the perception translation. An antigen may have several different epitopes, which means that an antigen can be recognized by a number of different antibodies. However, an antibody can bind only one antigen's epitope. In the proposed mechanism, a paratope with a built-in robot's steering direction is regarded as a antibody and interacts with each other and with its environment. These antibodies/steering-directions are induced by recognition of the available antigens/detected-information. In should be noted that only one antibody with the highest concentration will be selected to act according to the immune network hypothesis.

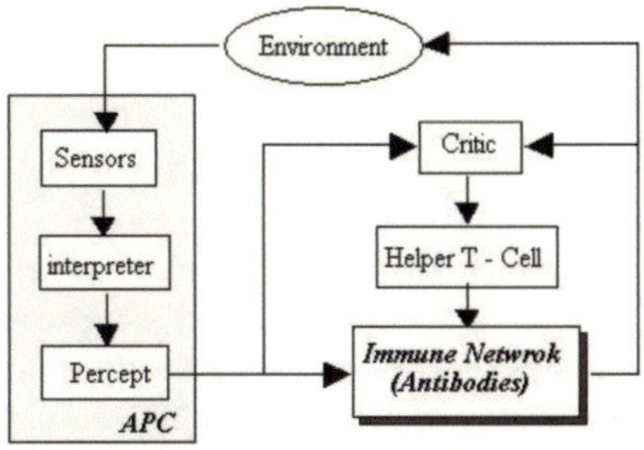

Figure 4. The architecture of the immunized network reactive system

In the proposed immune network, antibodies are defined as the steering directions of mobile robots as illustrated in Fig. 5,

$$Ab_i \equiv \theta_i = \frac{360^\circ}{N_{Ab}}(i-1) \qquad i = 1,2,\cdots,N_{Ab},$$

where N_{Ab} is the number of antibodies/steering-directions and θ_i is the steering angle between the moving path and the head orientation of the mobile robot. Note that $0° \leq \theta_i \leq 360°$.

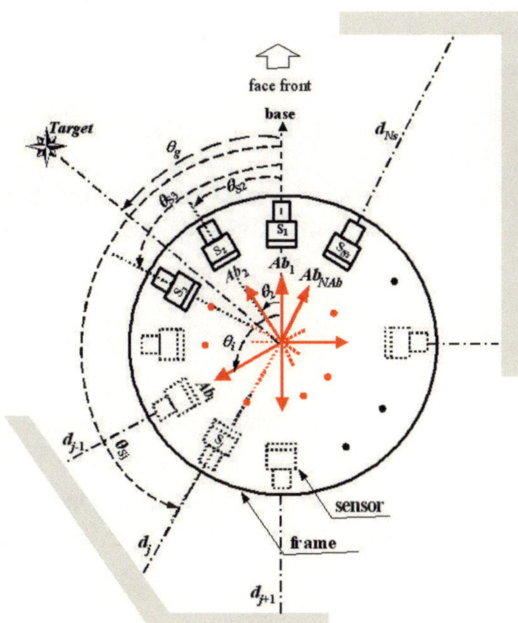

Figure 5. Configuration of mobile robot and its relatives to target and obstacles

In addition, the antigen represents the local environment surrounding the robot and its epitopes are a fusion data set containing the azimuth of the goal position θ_g, the distance between the obstacles and the *jth* sensor d_j, and the azimuth of sensor θ_{Sj},

$$\theta_{S_j} \equiv \frac{360°}{N_S}(j-1) \qquad j = 1,2,\cdots,N_s\,,$$

$$Ag_j \equiv \{\theta_g, d_j, \theta_{S_j}\} \qquad j = 1,2,\cdots,N_s$$

where N_s is the number of sensors equally spaced around the base plate of the mobile robot, $d_{min} \leq d_j \leq d_{max}$ and $0° \leq \theta_{S_j} \leq 360°$. Parameters d_{min} and d_{max} represent the nearest and longest distances measured by the range sensors, respectively. It should be noted that different antigens (local environments) might have identical epitopes (fusion information from range sensors). There is no necessary relationship between N_{Ab} and N_s since they depend on the hardware (*i.e.* motor steering angles and number of sensors installed) of mobile robot. Nevertheless, simulation results show that better performance could be derived if N_s equal to or larger than N_{Ab}.

The potential-field method is one of the most popular approaches employed to navigate the mobile robot within environments containing obstacles, since it is conceptually effective and easy to implement. The method can be implemented either for off-line global planning if the environment is previously known or for real-time local navigation in an unknown environment using onboard sensors. The Artificial Potential Field (APF) approach considers a virtual attractive force between the robot and the target as well as virtual repulsive forces between the robot and the obstacles. The resultant force on the robot is then used to decide the direction of its movements. In the proposed immune network, the resultant force on the robot is defined as m_i, the affinity value between the antigen/local environment and the *ith* antibody/steering angle,

$$m_i = w_1 F_{goal_i} + w_2 F_{obs_i} \qquad i = 1,2,\cdots,N_{Ab} \tag{3}$$

The weighing values w_1 and w_2 indicate the ratio between attractive and repulsive forces. Note that $0 \leq w_1, w_2 \leq 1$ and $w_1 + w_2 = 1$. The attractive force F_{goal_i} of the ith steering direction (*i.e.* the ith antibody) is defined as follows:

$$F_{goal_i} = \frac{1.0 + \cos(\theta_i - \theta_g)}{2.0}, \qquad i = 1,2,\cdots,N_{Ab} \tag{4}$$

Note that F_{goal_i} is normalized and $0 \leq F_{goal_i} \leq 1$. Obviously, the attractive force is at its maximal level ($F_{goal_i}=1$) when the mobile robot goes straightforward to the target (*i.e.* $\theta_i = \theta_g$). On the contrary, it is minimized ($F_{goal_i}=0$) if the robot's steering direction is the opposite of the goal.

The repulsive force for each moving direction (the ith antibody θ_i) is expressed as the following equation,

$$F_{obs_i} = \sum_{j=1}^{N_S} \alpha_{ij} \cdot \overline{d}_j \tag{5}$$

where $\alpha_{ij} = \exp(-N_s \times (1-\delta_{ij}))$ with $\delta_{ij} = [1+\cos(\theta_i - \theta_{S_j})]/2$. Fig. 6 demonstrates the relationship between α_{ij} and δ_{ij}. The parameter α_{ij} indicates the weighting ratio for the jth sensor to steering angle θ_i while $\overline{d}_j$ represents the normalized distance between the jth sensor and the obstacles. Coefficient δ_{ij} expresses influence and importance of each sensor at different locations. The equation shows that the information derived from the sensor closest to the steering direction is much more important due to its biggest δ_{ij} value. Kubota *et al.* (2001) have proposed a similar 'delta rule' to evaluate the weighting of each sensor too.

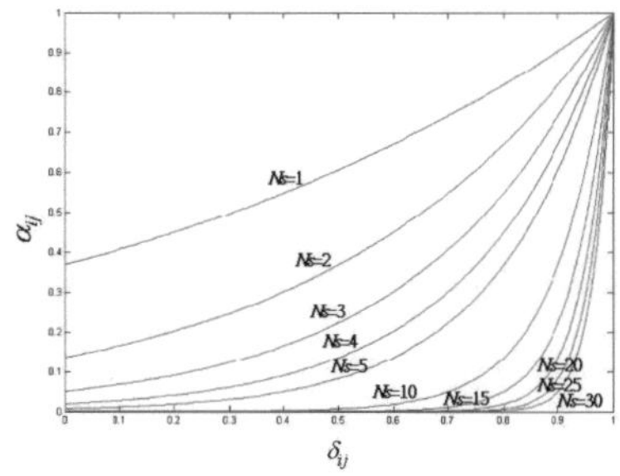

Figure 6. Relation between α_{ij} and δ_{ij}

The normalized obstacle distance for each sensor $\bar{d}_j$ is fuzzified using the fuzzy set definitions. The mapping from the fuzzy subspace to the TSK model is represented as three fuzzy if-then rules in the form of

$$\begin{array}{llll} \text{IF} & d_j \text{ is } \mathbf{s} & \text{THEN} & y = L_1 \\ \text{IF} & d_j \text{ is } \mathbf{m} & \text{THEN} & y = L_2 \\ \text{IF} & d_j \text{ is } \mathbf{d} & \text{THEN} & y = L_3 \end{array}$$

where L_1, L_2, and L_3 are defined as 0.25, 0.5 and 1.0, respectively. The input variable of each rule is the detected distance d_j of the *j*th sensor. The antecedent part of each rule has one of the three labels, namely, **s** (safe), **m** (medium), and **d** (danger). Consequently, the total output of the fuzzy model is given by the equation below,

$$\bar{d}_j = \frac{\mu_{safe}(d) \cdot L_1 + \mu_{medium}(d) \cdot L_2 + \mu_{danger}(d) \cdot L_3}{\mu_{safe}(d) + \mu_{medium}(d) + \mu_{danger}(d)} \tag{6}$$

where $\mu_{safe}(d)$, $\mu_{medium}(d)$, $\mu_{danger}(d)$ represent the matching degree of the corresponding rule. Fig. 7 illustrates the membership function and labels for measured distance d_j.

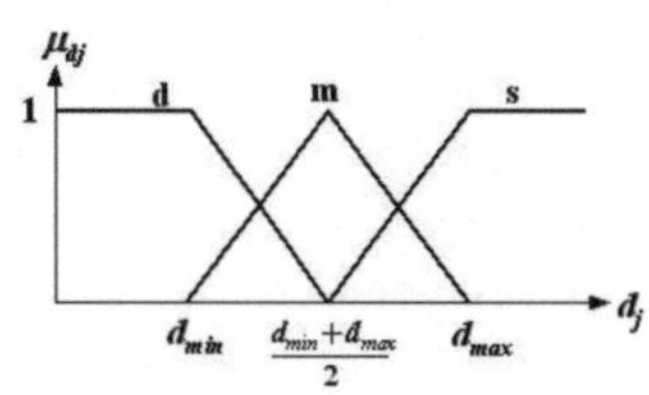

Figure 7. Membership function and labels for measured distance d_j

As to the stimulative-suppressive interaction between the antibodies/steering-directions are derived from equation (1) as follows,

$$\frac{dA_i(t)}{dt} = \left(\sum_{\ell=1}^{N_{Ab}} m_{i\ell}^{st} a_\ell(t) - \sum_{k=1}^{N_{Ab}} m_{ki}^{su} a_k(t) + m_i - k_i \right) a_i(t)$$
$$= \left[\left(m_{i1}^{st} a_1(t) + m_{i2}^{st} a_2(t) + \cdots + m_{iN_{Ab}}^{st} a_{N_{Ab}}(t) \right) - \left(m_{1i}^{su} a_1(t) + m_{2i}^{su} a_2(t) + \cdots + m_{N_{Ab}i}^{su} a_{N_{Ab}}(t) \right) + m_i - k_i \right] a_i(t)$$
$$= \left[\left(m_{i1}^{st} - m_{1i}^{su} \right) a_1(t) + \left(m_{i2}^{st} - m_{2i}^{su} \right) a_2(t) + \cdots + \left(m_{iN_{Ab}}^{st} - m_{N_{Ab}i}^{su} \right) a_{N_{Ab}}(t) + m_i - k_i \right] a_i(t)$$
$$= \left[m_{i1}^{ss} a_1(t) + m_{i2}^{ss} a_2(t) + \cdots + m_{iN_{Ab}}^{ss} a_{N_{Ab}}(t) + m_i - k_i \right] a_i(t)$$
$$= \left(\sum_{\ell=1}^{N_{Ab}} m_{i\ell}^{ss} a_\ell(t) + m_i - k_i \right) a_i(t)$$

and the stimulative-suppressive affinity $m_{i\ell}^{ss}$ between the *i*th and *j*th antibody/steering-angle is defined as

$$m_{i\ell}^{ss} = m_{i\ell}^{st} - m_{\ell i}^{su} = \cos(\theta_i - \theta_\ell) = \cos(\Delta\theta_{i\ell}), \quad i, \ell = 1, 2, \cdots, N_{Ab} \tag{7}$$

Obviously, stimulative-suppressive effect is positive ($m_{i\ell}^{ss}$ >0) if $-90° < \Delta\theta_{i\ell} < 90°$. On the contrary, negative stimulative-suppressive effect exists between two antibodies if their difference of steering angles are greater than 90° or less than -90° (*i.e.*, $\Delta\theta_{i\ell} > 90°$ or $\Delta\theta_{i\ell} < -90°$). In addition, there is no any net effect between orthogonal antibodies (*i.e.* $\Delta\theta_{i\ell} = \pm 90°$). The immune system responses to the specified winning situation that has the maximum concentration among the trigged antibodies by comparing the currently perceived situations (trigged antibodies). In other words, antibody with the highest concentration is selected to activate its corresponding behavior to the world. Therefore, mobile robot moves a step along the direction of the chosen steering angle/antibody.

3.2 Local minimum recovery

As mentioned in the previous section, one problem inherent in the APF method is the possibility for the robot to get trapped in a local minima situation. Traps can be created by a variety of obstacle configurations. The key issue to the local minima problems is the detection of the local minima situation during the robot's traversal. In this study, the comparison between the robot-to-target direction θ_g and the actual instantaneous direction of travel θ_i was utilized to detect if the robot got trapped. The robot is very likely to get trapped and starts to move away from the goal if the robot's direction of travel is more than 90°off-target (*i.e.* $|\theta_i - \theta_g|$>90°). Various approaches for escaping trapping situations have been proposed as described previously. In this study, an adaptive virtual target method is developed and integrated with the reactive immune network to guide the robot out of the trap.

In immunology, the T-cell plays a remarkable key role in distinguishing a "self" from other "non-self" antigens. The Helper-T cells work primarily by secreting substances to constitute powerful chemical messengers to promote cellular growth, activation and regulation. Simulating the biological immune system, this material can either stimulate or suppress the promotion of antibodies/steering-directions depending on whether the antigen is non-self or self (trapped in local minima or not). Different from the virtual target method proposed in [10-11], an additional virtual robot-to-target angle θ_v (analogous to the interleukine secreted by T-cells) is added to the goal angle θ_g whenever the trap condition ($|\theta_i - \theta_g|$>90°) is satisfied,

$$\theta_g(k+1) = \theta_g(k) + \theta_v(k) \tag{8}$$

with

$$\begin{cases} \theta_v(k) = \theta_v(k-1) \pm \Delta\theta_g & \text{if } |\theta_i(k) - \theta_g(k)| \geq 90° \text{ and } \theta_v(k-1) = 0 \\ \theta_v(k) = \theta_v(k-1) + sign(\theta_v(k-1)) \cdot \Delta\theta_g & \text{if } |\theta_i(k) - \theta_g(k)| \geq 90° \text{ and } \theta_v(k-1) \neq 0 \\ \theta_v(k) = \max\{0, \theta_v(k-1) - sign(\theta_v(k-1)) \cdot \theta_c(k-1)\} & \text{if } |\theta_i(k) - \theta_g(k)| < 90° \text{ and } \theta_v(k-1) \geq 0 \\ \theta_v(k) = \min\{0, \theta_v(k-1) - sign(\theta_v(k-1)) \cdot \theta_c(k-1)\} & \text{if } |\theta_i(k) - \theta_g(k)| < 90° \text{ and } \theta_v(k-1) < 0 \end{cases}$$

where $\Delta\theta_g = \theta_g(k) - \theta_g(k-1)$ and $\theta_c(k) = \theta_c(k-1) + \lambda$.

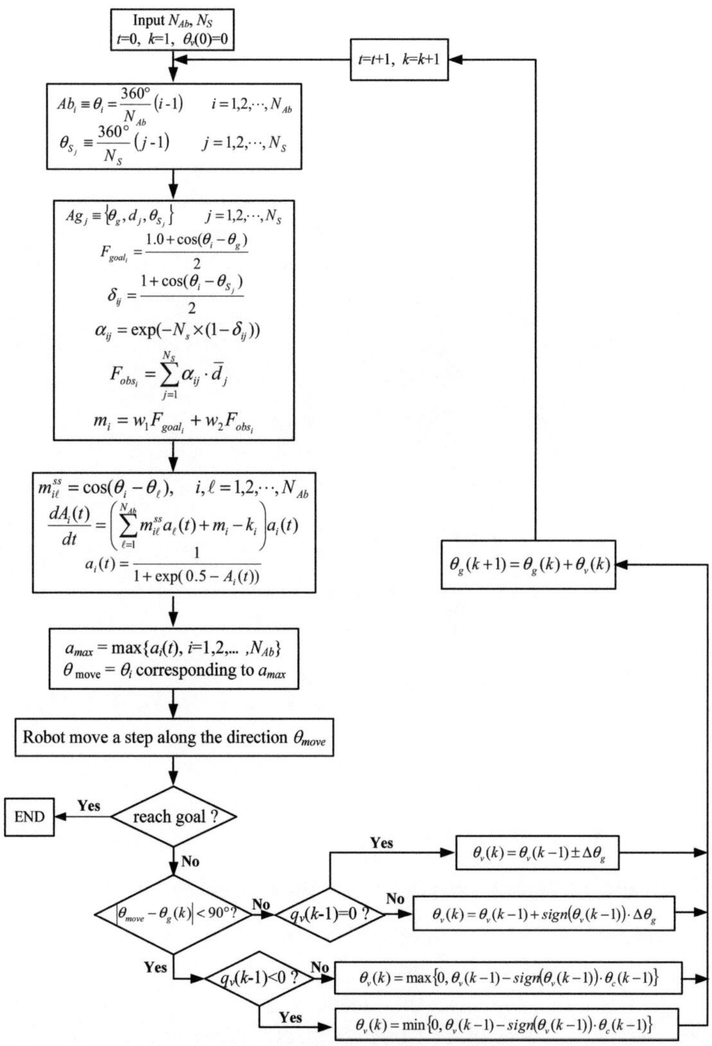

Figure 8. Flowchart of the mobile robot navigation procedure

Parameters k-1, k, and k+1 represent the previous state, the current state and the future state, respectively. Symbol "±" indicates that the location of the virtual target can be randomly switched to either the right (*i.e.* "+") or the left (*i.e.* "–") side of the mobile robot so that the robot has a higher probability of escaping from the local minima in either direction. λ is an adjustable decay angle. The bigger the value is, the faster the location of virtual target converges to that of the true one and the easier it is for the robot to get trapped in the local minima again. In this study, λ is determined after multiple simulation runs and set to 0.2. The incremental virtual angle $\theta_v(k)$ in the proposed scheme is state dependent and self-adjustable according to the robot's current state and the action it took previously. This provides powerful and effective trap-escaping capability compared to virtual target method, which keeps θ_v a constant value. θ_c is a converging angle and its initial value is 0. Fig. 8 shows the flowchart of navigation procedure for mobile robot employing the proposed reactive immune network.

For carrying out the necessary simulation and validating the efficacy of the proposed methodology, a computer program was developed using C++ language with graphical user interface. The simulation environment contains a robot and obstacle constructed by numerous square blocks 10cm in length. The environmental condition adopted in simulation is a 300cm×300cm grid. The size of the simulated robot is a circle with 10cm diameter. During each excursion, the robot tries to reach target and avoid collision with obstacle. Fig. 9 elucidates and demonstrates the performance of the proposed strategy for the robot to escape from a recursive U-trap situation, which may make the virtual target switching strategy (Xu, 2000) ineffective as Chatterjee and Matsuno (2001) suggested.

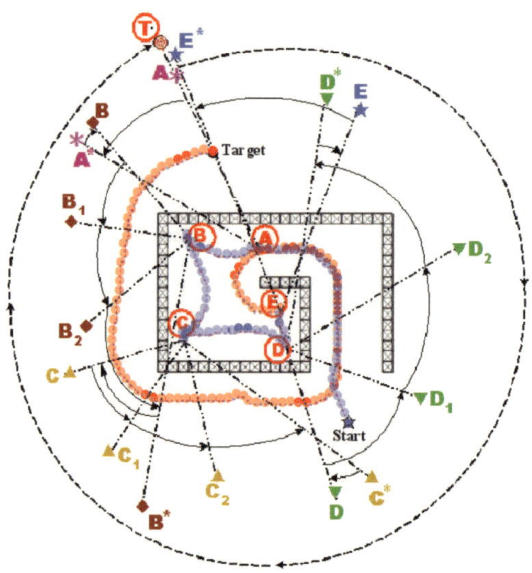

Figure 9. Robot path and state of the indices along the trajectory

The robot first enters a U-shaped obstacle and is attracted to the target due to the target's reaching behavior until it reaches the critical point Ⓐ. Clearly, the azimuth of goal θ_g is kept the same during this stage; however, the distance between the robot and the target is decreased quickly. The detection of the trap possibility because of the abrupt change of target orientation at location Ⓐ (θ_g) makes the target shift to a virtual position A* (θ_g - $\Delta\theta_g$).

$\Delta\theta_g$ is defined as 45° is this study. Note that the switch-to-left or the switch-to-right of the virtual target (*i.e.*, minus or plus $\Delta\theta_g$) is selected randomly. On the way Ⓐ→Ⓑ, $\Delta\theta_g$ is decreased gradually according to equation (8) until a new local minimal is found at location Ⓑ. Again, the location of virtual target switches from A* to B*. Fig. 9 and Fig. 10 show that there is a successive virtual target switching $A^*\rightarrow B_1\rightarrow B_2\rightarrow B^*$ when the robot moves around the left upper corner where it is tracked in a trap (to satisfy condition $|\theta_i - \theta_g|>90°$) three times. After passing through the critical point Ⓑ, the robot keeps approaching the virtual target until reaching the third critical point Ⓒ. Concurrently, the associated orientation of the virtual target is decreased from B* to C. Once more, it takes three times for the robot to escape from the trap situation in the left lower corner on the path Ⓒ→Ⓓ (orientation of the virtual target $C\rightarrow C_1\rightarrow C_2\rightarrow C^*\rightarrow D$). Similar navigation procedures take place on the way Ⓓ→Ⓔ (orientation of virtual target $D\rightarrow D_1\rightarrow D_2\rightarrow D^*\rightarrow E\rightarrow E^*$). After escaping from the recursive U-shaped trap, the mobile robot revolves in a circle and finally reaches target Ⓣ without any trapping situations (azimuth of virtual target θ_g decreases gradually from E* to T illustrated with a dashed line). The derived trajectory illustrated in Fig. 8 is quite similar to the results derived by Chatterjee and Matsuno (2001). Fig. 10 illustrates the other possible trajectory to escape the same trap situation due to the random choice of the "plus" or "minus" robot-to-target angle $\Delta\theta_g$, as shown in equation (8). Obviously, the mechanism for virtual target switching to the right or to the left (*i.e.*, $\pm\Delta\theta_g$) increases the diversity and possibility of the robot's escaping from the local minima problem.

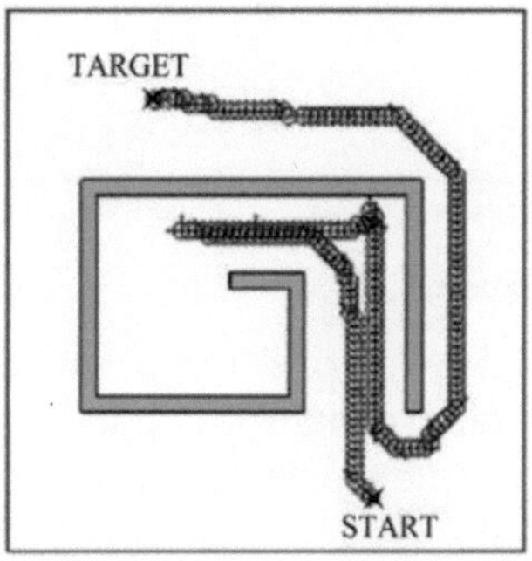

Figure 10. The other possible trajectories to escape the recursive trap situation

4. Motion Planning in Dynamic Environments

4.1 The velocity Obstacle method

This section briefly describes the velocity obstacle (VO) method for the obstacles. For simplicity, the mobile robot and moving obstacles are assumed to be approximated by cylinders and move on a flat floor. Fig. 11(*a*) shows two circular objects *A* and *B* with velocities $\mathbf{v}_A$ and $\mathbf{v}_B$ at time t_0, respectively. Let circle *A* represent the robot and circle *B* represent the obstacle. To compute the VO, obstacle *B* must be mapped into the configuration space of *A*, by reducing *A* to a point $\hat{A}$ and enlarging *B* by the radius of *A* to $\hat{B}$ as Fig. 11(*b*) demonstrates. The Collision Cone, $CC_{A,B}$, is thus defined as the set of colliding relative velocities between $\hat{A}$ and $\hat{B}$:

$$CC_{A,B} = \{\mathbf{v}_{A,B} \mid \lambda_{A,B} \cap \hat{B} \neq 0\} \tag{9}$$

where $\mathbf{v}_{A,B} = \mathbf{v}_A - \mathbf{v}_B$ is the relatively velocity of $\hat{A}$ with respect to $\hat{B}$, and $\lambda_{A,B}$ is the line of $\mathbf{v}_{A,B}$.This collision cone is the light gray sector with apex in $\hat{A}$, bounded by the two tangents λ_f and λ_r from $\hat{A}$ to $\hat{B}$ as shown in Fig. 11(*b*). Clearly, any relative velocity $\mathbf{v}_{A,B}$ outside $CC_{A,B}$ is guaranteed to be collision-free, provided that the obstacle $\hat{B}$ maintains its current shape and speed. The collision cone is specific to a particular robot/obstacle pair. To consider situation of multiple obstacles, it is better to establish an equivalent condition on the absolute velocities $\mathbf{v}_A$. This could be done simply by adding the velocity $\mathbf{v}_B$ to each velocity in $CC_{A,B}$, or equivalently, translating the collision cone $CC_{A,B}$ by $\mathbf{v}_B$, as shown in Figure 11(*b*). The velocity obstacle VO (in dark gray sector) is thus defined as:

$$VO = CC_{A,B} \oplus \mathbf{v}_{A,B} \tag{10}$$

where $\oplus$ is the Minkowski vector sum operator. The VO partitions the absolute velocities $\mathbf{v}_A$ into avoiding and colliding velocities. Selecting $\mathbf{v}_A$ outside of VO would avoid collision with *B*. Velocities $\mathbf{v}_A$ on the boundaries of VO would result in *A* grazing *B*.

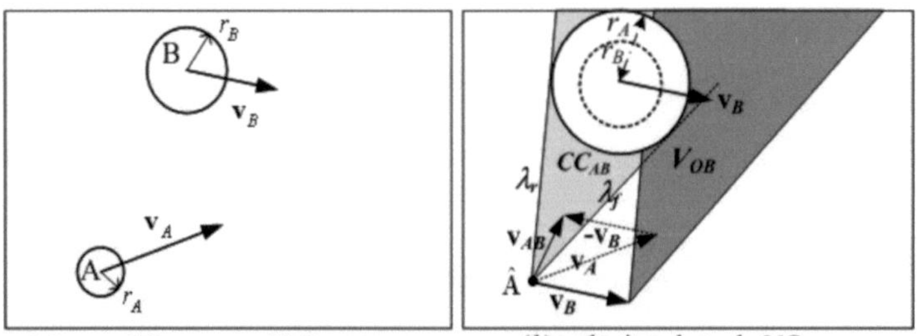

(*a*) two objects on collision course (*b*) velocity obstacle VO

Figure 11. The Velocity Obstacle approach

In the case of multiple obstacles, they are prioritized according to their danger level so that the most imminent collision obstacle is avoided first. In this study, a "collision distance index" is defined as follows to compute the danger level for each obstacle:

$$\delta = \frac{d_{r,obs_j}}{v_j \times T_s}, \quad j = 1,2,\cdots,N_{obs} \tag{11}$$

where d_{r,obs_j} represents the distance between robot and the *j*th obstacle, v_j is speed of the *j*th obstacle, and T_s is the sampling time. Obviously, the smaller the collision distance index, the more dangerous to collide obstacle.

4.2 Potential field immune network

A potential field immune network (PFIN) inspired by the biological immune system for robot navigation in dynamic environment is described in this section. For simplicity, one can make the following choices without loss of any generality:

- The mobile robot is an omni-directional vehicle. This means any direction of velocity can be produced at any time. In addition, maximum velocity and acceleration are assumed to be limited considering dynamics of robot and obstacles.

- The mobile robot and moving obstacles under consideration are approximated by cylinder with radius r_r, and r_o. This is not a severe limitation since general polygons can be represented by a collection of circles. Chakravarthy and Ghose (1998) showed that the union of all these circles can still be meaningfully used to predict collision between the irregularly shaped objects. Moreover, the resulting inexact collision cone can still be used effectively for motion planning.
- The mobile robot and moving obstacles move in a flat floor. Moving obstacles may change their velocities (amplitude and direction) at any time.
- The obstacles move along arbitrary trajectories, and that their instantaneous states (position and velocity) are either known or measurable. Prassler *et al.* (2001) have proposed such a sensor system includimg a laser range finder and sonar.

Fig. 12 illustrates the architecture of the proposed potential field immune network. The mechanism, imitating the cooperation between B-T cells, can help the robot adapt to the environment efficiently. In the immunology, the T-cell plays a remarkable key role for distinguishing a "self" from other "non-self" antigens. Resembling the biological immune system, its function is to prioritize the obstacles employing the VO method so that the obstacle with most imminent collision can be identified. In other words, T-cell in PFIN distinguishes an "imminent" from other "far-away" obstacles.

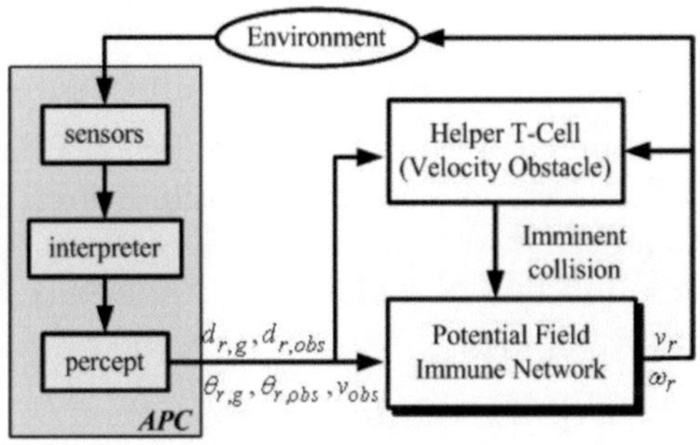

Figure 12. The architecture of the potential field immune network

In PFIN, the antigen's epitope is a situation detected by sensors and provides the information about the relationship between the robot's current states and the obstacles, along with the target as Fig. 12 depicted. Therefore, the antigen represents the local environment surrounding the robot each time interval and its epitopes are a fusion data set for each obstacle as Fig. 13 shows.

$$Ag_j \equiv \{\theta_{r,g}, d_{r,g}, \theta_{r,obs_j}, d_{r,obs_j}\} \qquad j = 1,2,\cdots,N_{obs},$$

where $\theta_{r,g}$ and θ_{r,obs_j} represent the orientations between robot and target, and the *j*th obstacle, respectively. $d_{r,g}$ and d_{r,obs_j} represent the distance between robot and target, and the *j*th obstacle, respectively. N_{obs} is the number of moving obstacles.

This scene-based spatial relationship is consistently discriminative between different parts of an environment. The interpreter is regarded as a phagocyte and translates sensor data into perception. The antigen presentation proceeds from the information extraction to the perception translation. An antigen may have several different epitopes, which means that an antigen can be recognized by a number of different antibodies. However, an antibody can bind only one antigen's epitope. In the proposed immune network, the antibody's receptor is defined as the situation between robot and the imminent collision obstacle as the following

$$Ab_1 \equiv d_{r,g}\ ;\ \ Ab_2 \equiv \theta_{r,g}\ ;\ \ Ab_3 \equiv d_{r,obs}\ ;\ \ Ab_4 \equiv \theta_{r,obs}$$

where $d_{r,obs}$ and $\theta_{r,obs}$ represent the distance and orientation between robot and the imminent collision obstacle, respectively.

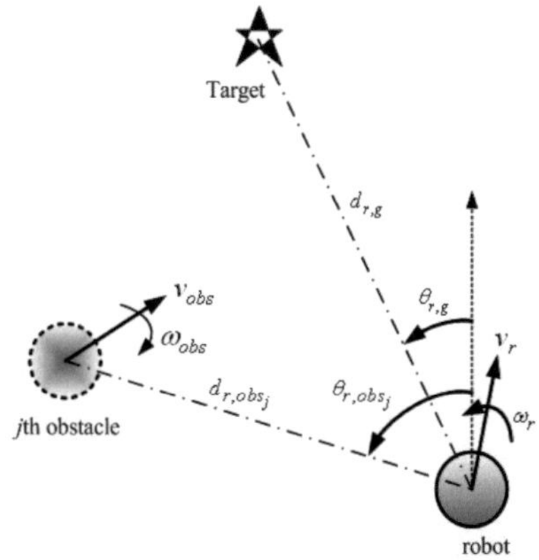

Figure 13. Configuration of mobile robot and its relatives to target and obstacles

The response of the overall immune network is thus derived by determining the set of affinities associated with the receptors and the structural similarity between antigen and antibody defined by quantification of the distance in antigen space. In this study, the collective immune response function of the immune network is defined as the following immune functions,

$$\begin{cases} v_r = f(Ab_1) + f(Ab_3) \\ \omega_r = f(Ab_2) + f(Ab_4) \end{cases} \tag{12}$$

where v_r and ω_r are the robot's velocity and angular velocity outputs, respectively. This is a kind of artificial potential field approach since it considers a virtual attractive force between the robot and the target (*i.e.* $f(Ab_1)$ and $f(Ab_2)$) as well as virtual repulsive forces between the robot and the obstacle (*i.e.* $f(Ab_3)$ and $f(Ab_4)$). The resultant force on the robot is then used to decide the velocities (*i.e.* v_r and ω_r) of its movements. Functions $f(Ab_i)$ are expressed as following,

$$f(Ab_i) = K_i \times \frac{\sum_{j=1}^{4} m_i \times m_{ij}}{\sum_{j=1}^{4} m_{ij}}, \quad i = 1,2,3,4$$

where m_i is the affinity of antigen (the most imminent collision obstacle) and the ith antibody, m_{ij} is the stimulation/suppressive affinity between the ith and jth antibody. Corresponding constant parameters are K_1= 20, K_2= 30, K_3= 15, K_4= 30, respectively. Note that these values are defined according to the velocity limitation of the robot and obstacles. The affinity of the antigen and the ith antibody m_i is fuzzified using the fuzzy set definitions as Fig. 14 illustrates.

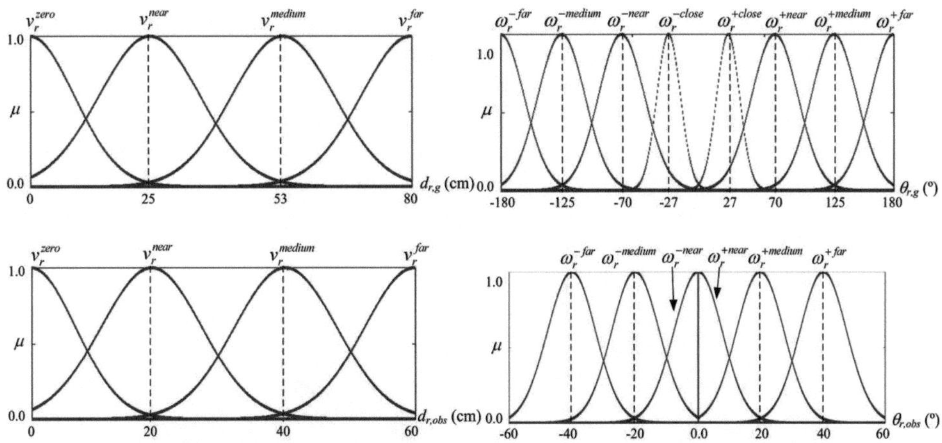

Figure 14. Membership functions of antibodies

The mapping from the fuzzy subspace to the TSK model is represented as fuzzy if-then rules in the form of

IF $d_{r,g}$	is	*zero*	THEN	$v_r = 0\text{cm/s}$
IF $d_{r,g}$	is	*near*	THEN	$v_r = 10\text{cm/s}$
IF $d_{r,g}$	is	*medium*	THEN	$v_r = 15\text{cm/s}$
IF $d_{r,g}$	is	*far*	THEN	$v_r = 20\text{cm/s}$
IF $\theta_{r,g}$	is	*-far*	THEN	$\omega_r = -30^{\circ}/\text{s}$
IF $\theta_{r,g}$	is	*-medium*	THEN	$\omega_r = -25^{\circ}/\text{s}$
IF $\theta_{r,g}$	is	*-near*	THEN	$\omega_r = -20^{\circ}/\text{s}$
IF $\theta_{r,g}$	is	*-close*	THEN	$\omega_r = -10^{\circ}/\text{s}$
IF $\theta_{r,g}$	is	*+close*	THEN	$\omega_r = 10^{\circ}/\text{s}$
IF $\theta_{r,g}$	is	*+near*	THEN	$\omega_r = 20^{\circ}/\text{s}$
IF $\theta_{r,g}$	is	*+medium*	THEN	$\omega_r = 25^{\circ}/\text{s}$
IF $\theta_{r,g}$	is	*+far*	THEN	$\omega_r = 30\,^{\circ}/\text{s}$
IF $d_{r,obs}$	is	*zero*	THEN	$v_r = -15\text{cm/s}$
IF $d_{r,obs}$	is	*near*	THEN	$v_r = -10\text{cm/s}$
IF $d_{r,obs}$	is	*medium*	THEN	$v_r = -5\text{cm/s}$
IF $d_{r,obs}$	is	*far*	THEN	$v_r = 0\text{cm/s}$
IF $\theta_{r,obs}$	is	*-far*	THEN	$\omega_r = 10^{\circ}/\text{s}$
IF $\theta_{r,obs}$	is	*-medium*	THEN	$\omega_r = 20^{\circ}/\text{s}$
IF $\theta_{r,obs}$	is	*-near*	THEN	$\omega_r = 30^{\circ}/\text{s}$
IF $\theta_{r,obs}$	is	*+near*	THEN	$\omega_r = -30^{\circ}/\text{s}$
IF $\theta_{r,obs}$	is	*+medium*	THEN	$\omega_r = -20^{\circ}/\text{s}$
IF $\theta_{r,obs}$	is	*+far*	THEN	$\omega_r = -10^{\circ}/\text{s}$

Consequently, the centroid defuzzification method is employed to calculate the weighted average of a fuzzy set,

$$m_i = \frac{\sum_{k=0}^{L} \mu_k y_k}{\sum_{k=0}^{L} \mu_k}, \quad i = 1, 2, 3, 4 \tag{13}$$

where μ_k represent the matching degree of the kth rule and y_k represent its corresponding output value. Finally, the stimulation and suppressive interaction m_{ij} between the ith and jth antibodies are optimized utilizing genetic algorithms. Hundreds of different circumstances with randomly generated moving obstacles were employed to optimize the affinity values m_{ij} of PFIN. Fig. 15 demonstrates one of the cases in which several tens of obstacles circumrotate at randomly generated positions with different radius. Fig. 15(*a*) shows that robot reaches target successfully while Fig. 15(*b*) demonstrates that robot is failed to reach target in the optimization procedure. Table 1 lists the derived optimal stimulation and suppressive affinity value m_{ij} between the ith and jth antibodies.

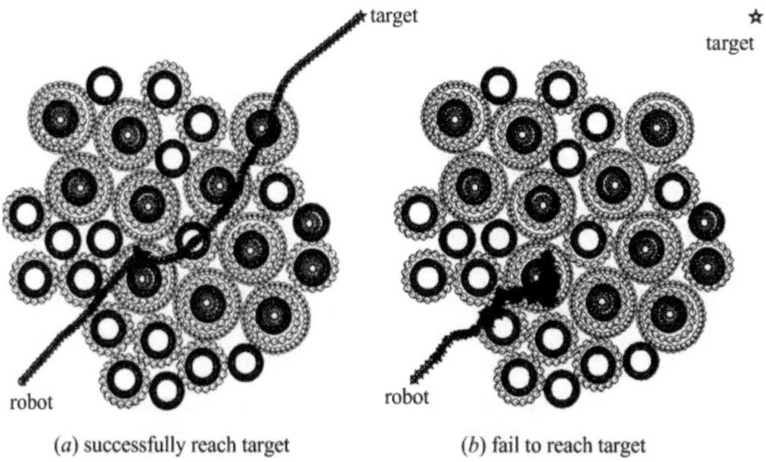

Figure 15. Randomly generated moving obstacles for optimizing m_{ij}

m_{ij}	j=1	j=2	j=3	j=4
i=1	**1**	**-0.13**	**-0.24**	**-0.04**
i=2	**-0.02**	**1**	**-0.11**	**-0.42**
i=3	**-0.37**	**-0.84**	**1**	**0.92**
i=4	**-0.21**	**-0.92**	**-0.31**	**1**

Table 1. The stimulation and suppressive interaction affinity value m_{ij}

5. Simulation and discussions

5.1 Motion Planning in Stationary Environments

Numerous simulation examples presented by researchers (Xu, 2000; Chatterjee & Matsuno, 2001; Kubota et al., 2001) were conducted to demonstrate the performance of mobile robot navigation employing RIN to various unknown environments; in particular, the capability of escaping from the traps or the wandering situations described. Assuming that the robot has eight uniformly distributed distance sensors (*i.e.* N_s=8) and eight moving directions including forward, left, right, back, forward left, forward right, back left, and back right (*i.e.* N_{Ab}=8) as Fig. 16 shows. Fig. 17(*a*) demonstrates the similar trajectory of the mobile robot to escape from loop-type and dead-end-type trapping situations in (Chatterjee & Matsuno, 2001). Again, Fig. 17(*b*) demonstrates the other possible trajectory (escaping from the left side) due to the random selection scheme ("±") mentioned previously.

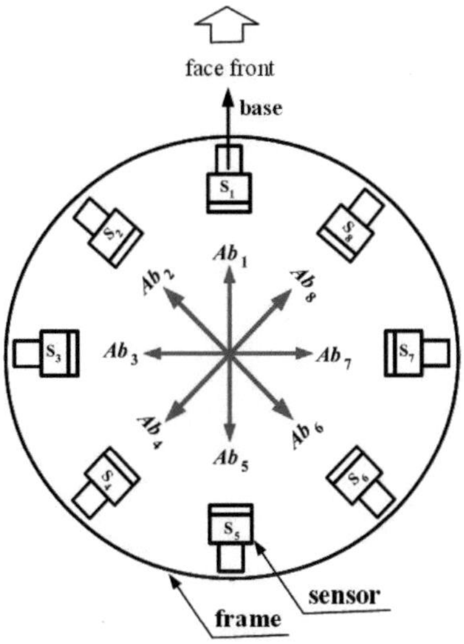

Figure 16. Configuration of mobile robot employed in simulation

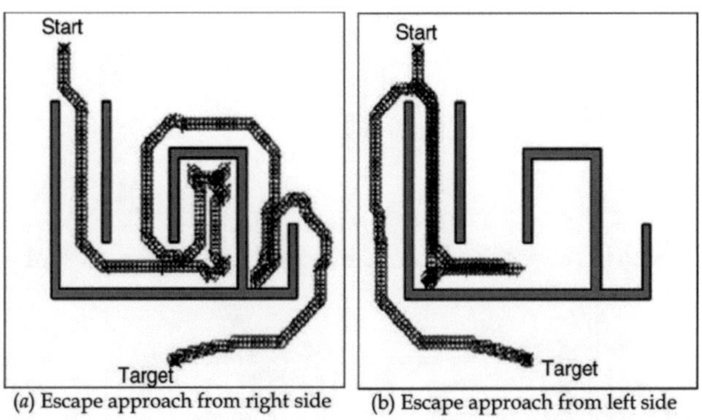

Figure 17. Robot trajectories to escape from loop type and dead-end type trap situation

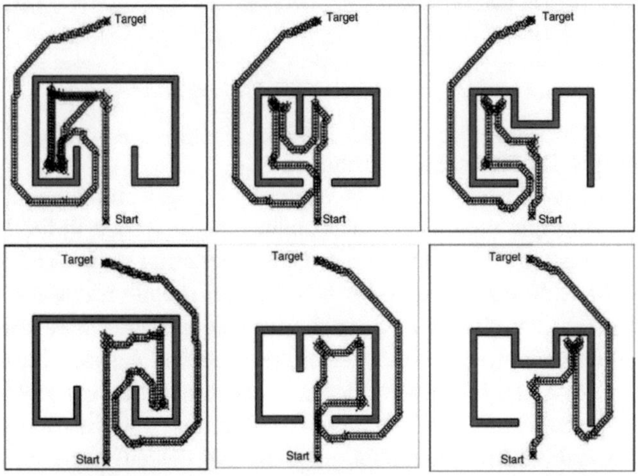

Figure 18. Robot trajectories to escape from different trapping situations

To validate the efficiency of the proposed scheme further, three trap environments adopted in (Madlhava & Kalra, 2001) were utilized in this study. Obviously, the simulation results depicted in Fig. 18 show that the robot is capable of escaping from all the traps as expected. Finally, the most famous and utilized example, U-shaped trap problem, is employed in this study. Fig. 19 shows part of the paths of the robot escaping from the U-shaped trap with different L/W (length/width) ratios. Clearly, the robot is capable of escaping from different ratio U-shaped environments. As to the double U-shaped trap environment (Kunota et al., 2001), Fig. 20 demonstrates the four trajectories for the robot to escape utilized RIN. Apparently, RIN successfully drives the robot to escape the double U-shaped trap.

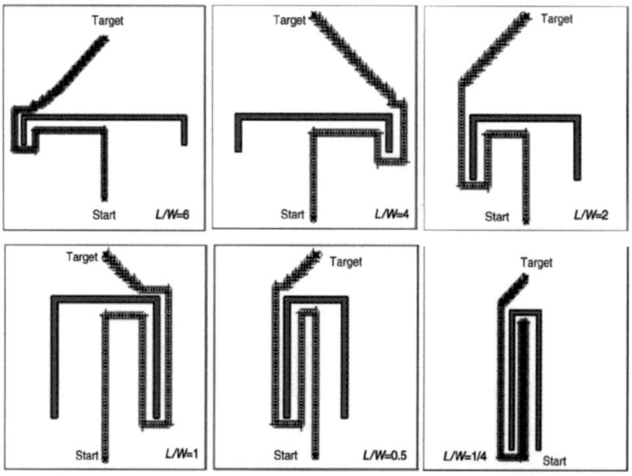

Figure 19. Robot trajectories to escape from U-shaped trap with different length/width ratio

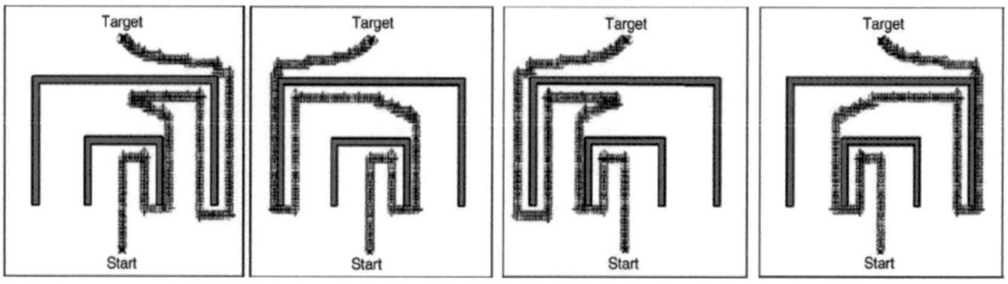

Figure 20. Robot trajectories to escape from double U-shaped trap

5.2 Motion Planning in Dynamic Environments

Numerous simulations have been utilized to evaluate the performance and effectiveness of a mobile robot among multiple moving obstacles using the proposed PFIN. In the simulations, the size of the test field is 5m ×5m, and the radius of robot and obstacles are r_r = 0.1 m and r_o = 0.1 m, respectively. In addition, the speed constraints on mobile robot and moving obstacles are $v_{r\ max}$ = 20 cm/s, $v_{o\ max}$ = 20 and $\omega_{r\ max}$ = 30°/s. The sampling time for each step is T_s = 0.03sec. To carry out these computations, a computer program was developed employing C++ programming tools with a graphical user interface. The simulation examples demonstrated in figures are given with graphical representations in which the trajectories of the moving object and the robot are described. Moreover, figures show the velocity-time history and azimuth-time history of the robot, respectively. In each figure, circles indicate the position of the robot and obstacles at each time instant when the robot executed an action. A high concentration of circles indicates a lower velocity (of the obstacle and of the robot) whilst a low concentration is a reflection of a greater velocity. In addition, the state responses (speed and orientation) of robot and obstacles are depicted in the figures. Obviously, the robot smoothly avoids the moving obstacles and reaches goal as expected for all cases.

Fig. 21 reveals that an obstacle coming from left side along a straight line cross the robot path. Within the interval of points A and D (at fifth and fourteenth sampling instant respectively), the obstacle slows down its speed in front of the robot's way to goal. Obviously, the robot appears a "hunting" behavior in this duration. Figs. 22(*a*)-22(*d*) explain this behavior employing the VO concept. At position A as Fig. 22(*a*) shown, the robot will collide with the obstacle since the relative velocity between robot and obstacle (*i.e.* $\mathbf{v}_{ro}$) is inside the velocity cone CC_{AB}. Thus the robot turns left (positive angular velocity as Fig. 21 shown) to avoid collision and reach position B. Because $\mathbf{v}_{ro}$ is outside the velocity cone in position B and there is no danger to collide the obstacle as Fig. 22(*b*) depicted, the robot turns right again due to the attraction force from the target. Once more, robot turns left to avoid collision at position C as Fig. 22(*c*) demonstrated. The "hunting" behavior (*i.e.* turn left and then turn right) is repeated until the robot reaches the position D. Subsequently, the robot finds that it may collide with the obstacle in next time step again as Fig. 22(*d*) shown. Robot decelerates its speed to stop quickly and then goes back to the position E (negative velocity from position D to E as Fig. 21 shown). Finally, the robot turns right and passed over behind the obstacle rapidly to reach the goal since the obstacle is no longer a threat.

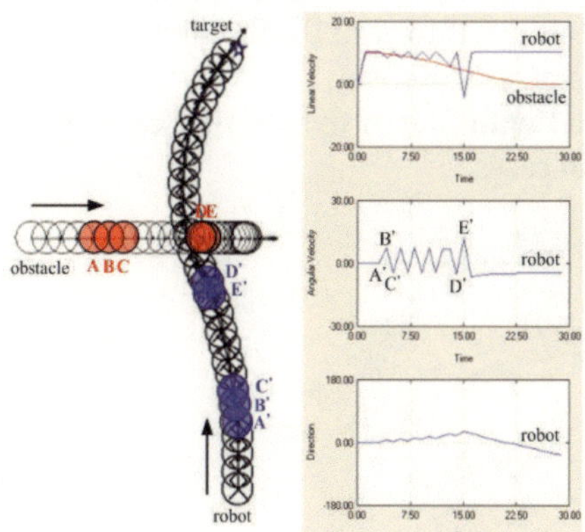

Figure 21. Trajectories and associated state responses of mobile robot and obstacle

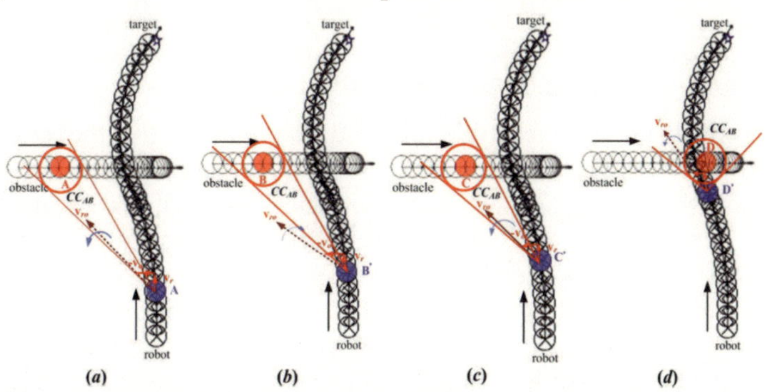

Figure 22. Velocity cone of robot at different positions

Fig. 23 shows a simulation result by which the robot can avoid two moving obstacles one after another then reach the goal. These obstacles come from different sides with arbitrary trajectory and varying speed cross the robot path. Similar to the previous simulation, the robot decelerates its speed at points A and B to avoid the first and second obstacles separately. Then it accelerates and moves towards the goal without collision

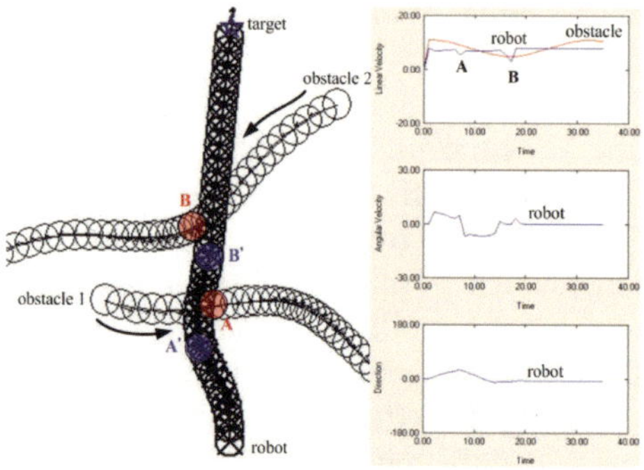

Figure 23. Trajectories and associated state responses of mobile robot and obstacles

Fig. 24 demonstrates the motion planning of a mobile robot tracking a moving goal while avoiding two moving obstacles. Obviously, mobile robot is able to reach goal and avoid moving obstacles no matter what the goal is fixed or moving employing the proposed PFIN. The robot decelerates at position A′ to wait for the first obstacle while accelerates at position C′ to exceed the second obstacle. Moreover, robot turns bigger angles at position B′ to follow the moving target compared with that of fixed target case. Note that the two obstacles have the same trajectories in both cases.

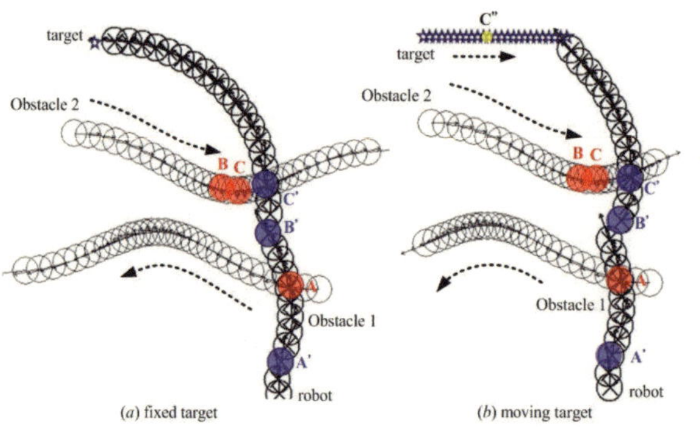

Figure 24. Trajectories of robot and obstacles for fixed/moving goals

Fig. 25 demonstrates another example of motion planning for the case of suddenly moving/stopped obstacle. Figs. 25(*a*)-25(*d*) illustrate a simulation result by which the robot successfully avoid two moving and two static obstacles. As usual, the robot exceeds the first moving obstacle at position A′ and waits for the second moving obstacle at position C′. Fig. 25(*e*) demonstrates that the robot reaches target safely even though the second static obstacle abruptly moves when the robot approaches it. Fig. 25(*f*) shows the similar result except that the second moving obstacle unexpectedly stops when it near the robot. Note that both the moving and stopping actions of the second static obstacle shown in Fig. 25(*e*) and Fig. 25(*f*) are pre-programmed to test the performance of the proposed architecture.

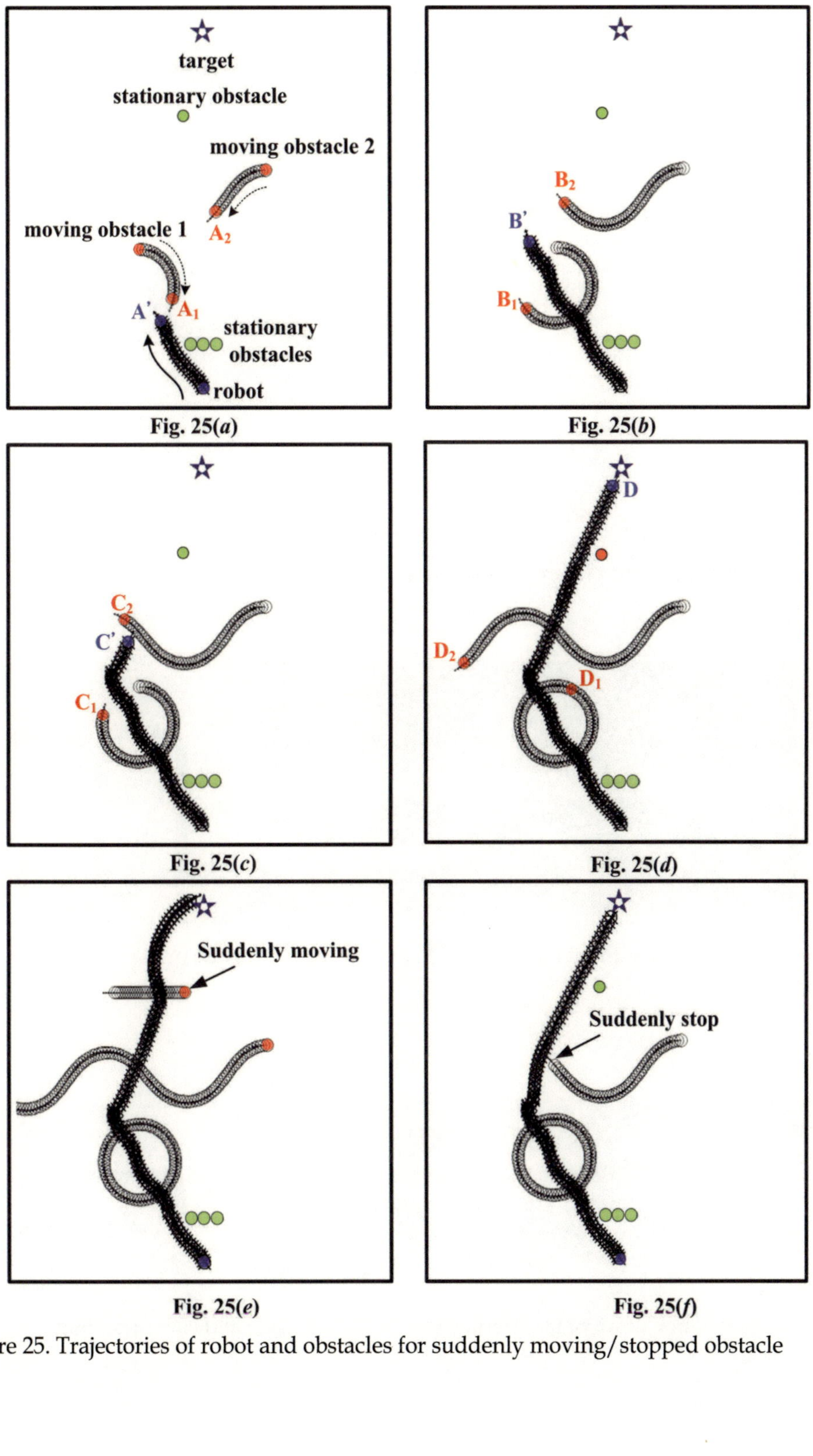

Figure 25. Trajectories of robot and obstacles for suddenly moving/stopped obstacle

6. Experimental Results

Numerous experiments were implemented to evaluate the performance in real application. Fig. 26 shows the mobile robot (with omni-directional wheel) used. Its dimension is 416*mm*×363.7*mm*×670*mm*. The robot installed with 8 ultrasonic sensors, two web-cams, and a laser range finder. Figs.27, 28 demonstrate the pictures of the robot navigate in two "U" shape obstacles (with different length and width: 160*mm*×320*mm*, and 300*mm*×100*mm*) and their corresponding trajectories, respectively.

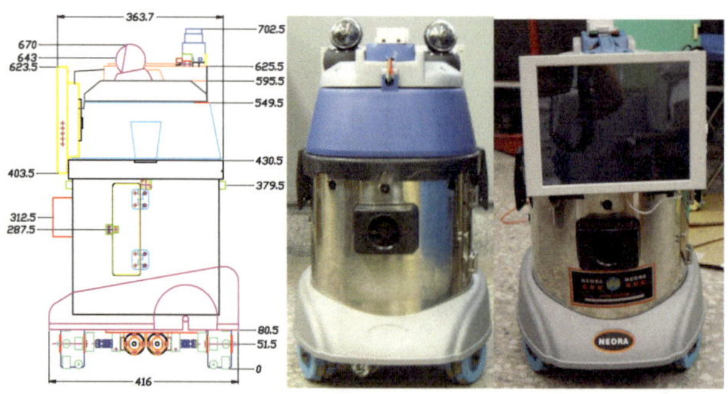

Figure 26. Dimension and pictures of the mobile robot

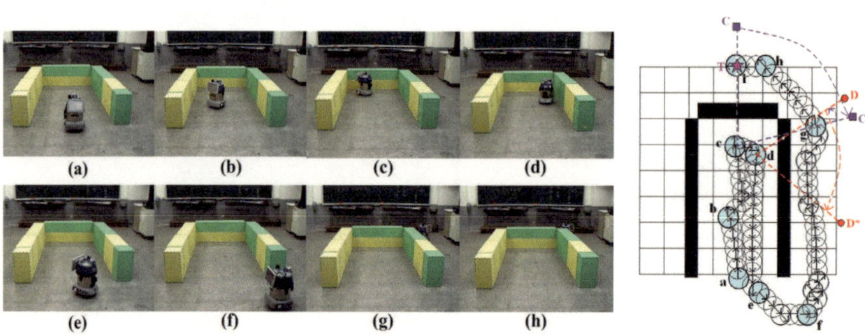

Figure 27. Navigation of robot in 160mm×320mm "U" obstacle and corresponding trajectories

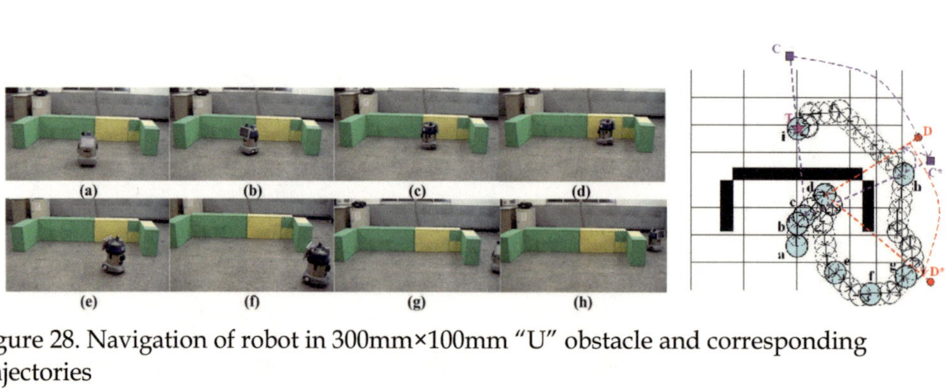

Figure 28. Navigation of robot in 300mm×100mm "U" obstacle and corresponding trajectories

Fig. 29 illustrate the pictures of the robot navigate in a "sequential-U" shape obstacle (400*mm*×190*mm* with a 90*mm* bar in middle) and its corresponding trajectory, respectively.

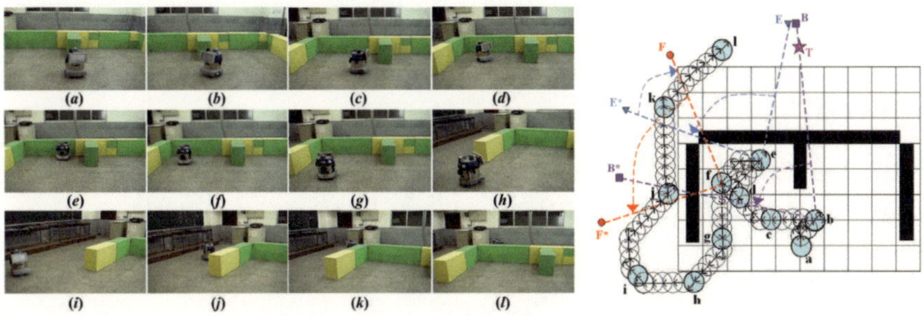

Figure 29. Trajectories in "sequential-U" shape obstacle and corresponding trajectories

All these figures show that the mobile robot is capable of navigating to the goal and escaping from local minimum traps employing the proposed reactive immune network. Note that mobile robot can approach target from both sides randomly as described in previous section.

6. Conclusion

Two different kind of reactive immune networks inspired by the biological immune system for robot motion planning are constructed in this study. The first one is a potential filed based immune network with an adaptive virtual target mechanism to solve the local minima problem navigating in stationary environments. Simulation and experimental results show that the mobile robot is capable of avoiding stationary obstacles, escaping traps, and reaching the goal efficiently and effectively. Employing the Velocity Obstacle method to determine the imminent collision obstacle, the second architecture guide the robot avoiding collision with the most danger object (moving obstacle) at every time instant. Simulation results are presented to verify the effectiveness of the proposed architecture in dynamic environment. Currently, laser range finder is utilizing to evaluate the performance of the proposed mechanism.

7. Acknowledgements

The authors would like to acknowledge the National Science Council, Taiwan, R.O.C., for making this work possible with grants NSC 95-2221-E-036-009 and NSC 96-2221-E-036-032-MY2.

8. References

Baraquand, J.; & Latombe, J.C. (1990). A Monte-Carlo algorithm for path planning with many degrees of freedom, *Proceedings of the IEEE International Conference on Robotics and Automation*, pp. 1712-1717, Cincinnati, OH, May, 1990

Carneiro, J.; Coutinho, A.; Faro, J. & Stewart, J. (1996). A model of the immune network with B-T cell co-operation I-prototypical structures and dynamics, *Journal of theoretical Biological*, Vol.182, No.4, 1996, pp. 513-529

Chakravarthy, A. & Ghose, D. (1998). Obstacle Avoidance in a Dynamic Environment: A Collision Cone Approach, *IEEE Transactions on Systems, Man, and Cybernetics – Part A: Systems and Humans*, Vol.25, No.5, 1998, pp. 562-574

Chatterjee, R. & Matsuno, F. (2001). Use of single side reflex for autonomous navigation of mobile robots in unknown environments, *Robotics and Autonomous Systems*, Vol.35, No.2, 2001, pp. 77-96

Chang, H. (1996). A new technique to handle local minima for imperfect potential field based motion planning, *Proceedings of the IEEE International Conference on Robotics and Automation*, pp. 108-112, Minneapolis, Minnesota, April, 1994

Dasgupta, D. (1997). Artificial neural networks and artificial immune systems: similarities and differences, *IEEE International Conference on Systems, Man, and Cybernetics*, pp. 873-878, Orlando, Florida, October, 1997

Dasgupta, D. (1999). *Artificial Immune Systems and Their Applications*, Springer-Verlag, ISBN 3-540-64390-7, Berlin Heidelberg

de Castro, L.N. & Jonathan, T. (1999). *Artificial immune systems: A new Computational Intelligence Approach*, Springer-Verlag, ISBN 1-85233-594-7, London

Duan, Q.J.; Wang, R.X.; Feng, H.S. & Wang, L.G. (2004). An immunity algorithm for path planning of the autonomous mobile robot, *IEEE 8th International Multitopic Conference*, pp. 69-73, Lahore, Pakistan, December, 2004

Duan, Q.J.; Wang, R.X.; Feng, H.S. & Wang, L.G. (2005). Applying synthesized immune networks hypothesis to mobile robots, *IEEE International Conference on Autonomous Decentralized Systems*, pp. 69-73, Chengdu, China, April, 2005

Farmer, J.D.; Packard, N.H. & Perelson, A.S. (1986). The immune system adaptation, and machine learning, *Physica*, Vol.22-D, Vol.2, No.1-3, 1986, pp. 187-204

Ferrari, C.; Pagello, E.; Ota, J.; & Arai, T. (1998). Multi-robot motion coordination in space and time, *Robotics and Autonomous Systems*, Vol.25, No.2, 1998, pp. 219-229

Fiorini, P. & Shiller, Z. (1998). Motion planning in dynamic environments using velocity obstacles, *International Journal of Robotics Research*, Vol.17, No.7, 1998, pp. 760-772

Fujimura, K. & Samet, H. (1989). A hierarchical strategy for path planning among moving obstacles, *IEEE Trans on Robot and Automat*, Vol.5, No.1, 1989, pp. 61-69

Fujimori, A. (2005). Navigation of mobile robots with collision avoidance for moving obstacles, *Proc Instn Mech Engrs Part I: J Systems and Control Engineering*, Vol.219, No.1, 2005, pp. 99-110

Ge, S. S. & Cui, Y. J. (1989). Dynamic motion planning for mobile robots using potential field method, *Autonomous Robots*, Vol.13, No.3, 1989, pp. 207–222

Hart, E.; Ross, P.; Webb, A. & Lawson, A. (2003). A role for immunology in "next generation" robot controllers, *Lecture Notes in Computer Science*, Vol.2787, 2003, pp. 46-56

Hightower, R.; Forrest, S. & Perelson, A. S. (1995). The evolution of emergent organization in immune system gene libraries, *Proceedings of Sixth International Conference on Genetic Algorithms*, pp. 344-350, Pittsburgh, PA, July, 1995

Hoffmann, G.W. (1989). The immune system: a neglected challenge for network theorists, *IEEE International Symposium on Circuits and Systems*, pp. 1620-1623, Portland, OR, May, 1989

Ishida, Y. (1997). The immune system as a prototype of autonomous decentralized systems: an overview, *Proceedings of Third International Symposium on autonomous decentralized systems*, pp. 85-92, Berlin, Germany, April, 1997

Ishiguro, A.; Watanabe, Y. & Uchikawa, Y. (1995). An immunological approach to dynamic behavior control for autonomous mobile robots, *IEEE/RSJ International Conference on Intelligent Robots and Systems*, pp. 495-500, Pittsburg, PA, 1995

Jerne, N.K. (1973). The immune system, *Scientific American*, Vol.229, No.1, 1973, pp. 52-60

Kondo, A. T.; Watanabe, Y.; Shirai, Y. & Uchikawa, Y. (1997). Emergent construction of artificial immune networks for autonomous mobile robots, *IEEE International Conference on Systems, Man, and Cybernetics*, pp. 1222-1228, Orlando, Florida, October, 1997

Kubota, N.; Morioka, T.; Kojima, F. & Fukuda, T. (2001). Learning of mobile robots using perception-based genetic algorithm, *Measurement*, Vol.29, No.3, 2001, pp. 237-248

Lee, D.-W. & Sim, K.-B. (1997). Artificial immune network-based cooperative control in collective autonomous mobile robots, *IEEE International Workshop on Robot and Human Communication*, pp. 58-63, Sendai, Japan, September, 1997

Lee, D.-J.; Lee, M.-J.; Choi, Y.-K. & Kim, S. (2000). Design of autonomous mobile robot action selector based on a learning artificial immune network structure, *Proceedings of the fifth Symposium on Artificial Life and Robotics*, pp. 116-119, Oita, Japan, January, 2000

Lee, S.; Adams, T.M. & Ryoo, B.-Y. (1997). A fuzzy navigation system for mobile construction robots, *Automation in Construction*, Vol.6, No.2, 1997, pp. 97-107

Liu, C.; Marcelo Jr., H.A.; Hariharan, K. & Lim, S.Y. (2000). Virtual obstacle concept for local-minimum-recovery in potential-field based navigation, *Proceedings of the IEEE International Conference on Robotics and Automation*, pp. 983-988, San Francisco, CA, April 2000

Luh, G.-C. & Cheng, W.-C. (2002). Behavior-based intelligent mobile robot using immunized reinforcement adaptive learning mechanism, *Advanced Engineering Informatics*, Vol.16, No.2, 2002, pp. 85-98

Madlhava, K. & Kalra, P.K. (2001). Perception and remembrance of the environment during real time navigation of a mobile robot, *Robotics and Autonomous Systems*, Vol.37, No.1, 2001, pp. 25-51

Mucientes, M.; Iglesias, R.; Regueiro, C. V.; Bugarín, A.; Cariñena, P. & Barro, S. (2001). Fuzzy temporal rules for mobile robot guidance in dynamic environments, *IEEE Transactions on Systems, Man, and Cybernetics – Part C: Applications and Reviews*, Vol.21, No.3, 2001, pp. 391-398

Oprea, M.L. (1996). *Antibody repertories and pathogen recognition: the role of germline diversity and somatic hypermutation*, PhD Dissertation, Department of Computer Science, The University of New Mexico, Albuquerque, New Mexico

Prassler, E.; Scholz, J. & Fiorni, P. (2001). A robotic wheelchair for crowded public environments, *IEEE Robotics and Automation Magazine*, Vol.7, No.1, 2001, pp. 38-45

Qu, Z.; Wang, J. & Plaisted, C.E. (2004). A new analytical solution to mobile robot trajectory generation in the presence of moving obstacles, *IEEE Transactions on Robotics*, Vol.20, No.6, 2004, pp. 978-993

Roitt, I.; Brostoff, J. & Male, D.K. (1998). *Immunology*, Mosby-Harcourt Publishers Ltd, ISBN 0723429189, London

Timmis, J.; Neal, M. & Hunt, J. (1999). Data analysis using artificial immune systems, cluster analysis and Kohonen networks: some comparisons, *IEEE International Conference on Systems, Man, and Cybernetics*, pp. 922-927, Tokyo, Japan, October, 1999

Vargas, P.A.; de Castro, L.N.; Michelan, R. & Von Zuben, F.J. (2003). Implementation of an Immuno-Gentic Network on a Real Khepera II Robot, *IEEE Congress on Evolutionary Computation*, pp. 420-426, Canberra, Australia, December, 2003

Xu, W.L. (2000). A virtual target approach for resolving the limit cycle problem in navigation of a fuzzy behaviour-based mobile robot, *Robotics and Autonomous Systems*, Vol.30, No.4, 2000, pp. 315-324

Yun, X. & Tan, K.-C. (1997). A wall-following method for escaping local minima in potential field based motion planning, *Proceedings of the IEEE International Conference on Advanced Robotics*, pp. 421-426, Monterey, CA, July, 1997

16

A Mobile Computing Framework for Navigation Tasks

Mohammad R. Malek[1,2], Mahmoud R. Delavar[3]
and Shamsolmolook Aliabady[2]
[1]*Dept. of GIS, Faculty of Geodesy and Geomatics Eng., KN Toosi University of Technology,* [2]*National Cartographic Center,* [3]*Dept. of Surveying and Geomatics Eng., University of Tehran*
Iran

1. Introduction

Mobile agents and movement systems have been rapidly increased worldwide. Within the last few years, we were facing many advances in wireless communication, computer networks, location-based engines, and on-board positioning sensors. Mobile GIS as an integrating system of mobile user, wireless network, and some GIS capabilities has fostered a great interest in the GIS field [14]. Without any doubt navigation and routing could be one of the most popular GIS based solution on mobile terminals. Due to this fact the mobile GIS is defined as an area about non-geographic moving object in geographic space [22].

Although the mobile computing has been increasingly growing in the past decade, there still exist some important constraints which complicate the use of mobile GIS systems. The limited resources on the mobile computing would restrict some features that are available on the traditional computing. The resources include computational resources (e.g., processor speed and memory) user interfaces (e.g., display and pointing device), bandwidth of mobile connectivity, and energy source [3], [11], [22], and [38]. In addition, one important characteristic of such environment is frequent disconnection that is ranging from a complete to weak disconnection [11] and [44]. The traditional GIS computation methods and algorithms are not well suited for such environment. These special characteristics of mobile GIS environment make us pay more attention to this topic.

In this chapter, in order to provide a paradigm that treats with mobile objects; i.e. an automatic machine that is capable of movement in a mobile information environment; a logical framework is presented. In this framework the concept of spatial influenceability is combined with well-known formal structure; i.e. network structure. In our view, influenceability which stands for both space and time domains is a primary relation. It has some exclusive properties in the mobile information context. It can be served as a basis for context-aware mobile computing because it depends on abilities of single player or agent.

Within the framework of this chapter we attempt to apply an idea to treat moving objects in mobile GIS environment based on partitioning in space and time. The idea is, to divide space-time into small parts and find solution (e.g. collision-free paths and wayfinding procedures) recursively. In this paper, finding a path without any conflict which is so-called collision-free path is highlighted. It is an important task of routing and navigation. Collision-free path and its variants find applications in robot motion planning, intelligent transportation system (ITS), and any mobile autonomous navigation system. It will be concluded that Wayfinding which is a fundamental spatial activity that people experience in daily lives, could be solved by this method.

2. Related Works

An overview of collision detection, mathematical methods, and programming techniques to find collision-free and optimal path between two states for a single vehicle or a group of vehicles can be found in [8], [19], [40] and [18] respectively. In the field of robot motion planning potential field methods introduced by Khatib, are widely used [27]. The main attraction of potential method is its ability to speed up the optimization procedure. New researches in this area can be found as well in [2] and [18]. Path planning techniques using mixed-integer linear program were developed earlier, especially in the field of aerial vehicles navigation (see e.g. [32-33], [35-36], and [39]). The reader who wants to see more related topics is referred to [12]. In almost all works it is assumed that the moving object cruises within a fixed altitude layer, with a fixed target point, and its velocity is predefined. In addition, accessibility to up-to-date knowledge of the whole mobile agents and a global time frame are prerequisite. The lack of two last conditions in distributed mobile computing environment is a well-known fact.

A method for reducing the size of computation is computation slice [13] and [30]. The computation slicing as an extension of program slicing is useful to narrow the size of the program. It can be used as a tool in program debugging, testing, and software maintenance. Unlike a partitioning in space and time, which always exists, a distributed computation slice may not always exist [13].

Among others, two works using divide and conquer idea, called honeycomb and space-time grid, are closer to our proposal. The honeycomb model [9] focuses on temporal evolution of subdivisions of the map, called spatial partitions, and gives a formal semantics for them. This model develops to deal with map and temporal map only. The concept of space-time grid is introduced by Chon et al. [5-7]. Based upon the space-time grid, they developed a system to manage dynamically changing information. In the last work, they attempt to use the partitioning approach instead of an indexing one. This method can be used for storing and retrieving the future location of moving object.

In the previous work of the first author [25-28] a theoretical framework using Influenceability and a qualitative geometry in the mobile environment with application in the relief management was presented. This article can be considered as an empirical extension of them.

3. Algebraic and Topological structure

Causality is a well-known concept. There is much literature on causality, extending philosophy, physics, artificial intelligence, cognitive science and so on (e.g. [1, 16, and 40]). In our view, influenceability stands for spatial causal relation, i.e. objects must come in contact with one another; cf. [1]. Although influenceability as a primary relation does not need to prove, it has some exclusive properties which show why it is selected. Influenceability supports contextual information and can be served as a basis for context aware mobile computing which has attracted researchers in recent years [10] and [31]. This relation can play the role of any kind of accident and collision. It is well-known that the accident is the key parameter in most transportation systems (for example see [36]). As an example the probability of collision defines the GPS navigation integrity requirement. In addition, this model due to considering causal relation is closer to a naïve theory of motion [30].

In the relativistic physics [17] based on the postulate that the vacuum velocity of light c is constant and maximum velocity, the light cone can be defined as a portion of space-time containing all locations which light signals could reach from a particular location (Figure 1). With respect to a given event, its light cone separates space-time into three parts, inside and on the future light cone, inside and on the past light cone, and elsewhere. An event A can influence (influenced by) another event; B; only when B (A) lies in the light cone of A (B). In a similar way, the aforementioned model can be applied for moving objects. Henceforth, a cone is describing an agent in mobile GIS environment for a fixed time interval. That means, a moving object is defined by a well-known acute cone model in space-time. This cone is formed of all possible locations that an individual could feasibly pass through or visit. The current location or apex vertex and speed of object is reported by navigational system or by prediction. The hyper surface of the cone becomes a base model for spatio-temporal relationships, and therefore enables analysis and further calculations in space-time. It also indicates fundamental topological and metric properties of space-time.

As described in Malek [25- 26], the movement modeling, are expressed in differential equation defined over a 4-dimensional space-time continuum. The assumption of a 4-dimensional continuum implies the existence of 4-dimensional spatio-temporal parts. It is assumable to consider a continuous movement on a differential manifold M which represents such parts in space and time. That means every point of it has a neighborhood homeomorphic to an open set in R^n. A path through M is the image of a continuous map from a real interval into M. The homeomorphism at each point of M determines a Cartesian coordinate system (x_0, x_1, x_2, x_3) over the neighborhood. The coordinate x_0 is called time. In addition, we assume that the manifold M can be covered by a finite union of neighborhoods. Generally speaking, this axiom gives ability to extend coordinate system to the larger area. This area shall interpret as one cell or portion of space-time. The partitioning method is application dependent. The partitioning method depends on application purposes [6] on the one hand, and limitation of the processor speed, storage capacity, bandwidth, and size of display screen on the other hand. It is important to note that the small portion of space and time in this idea is different from the geographical area covered by a Mobile Supported Station (MSS). This idea is similar to Helmert blocking in the least squares adjustment calculation [42].

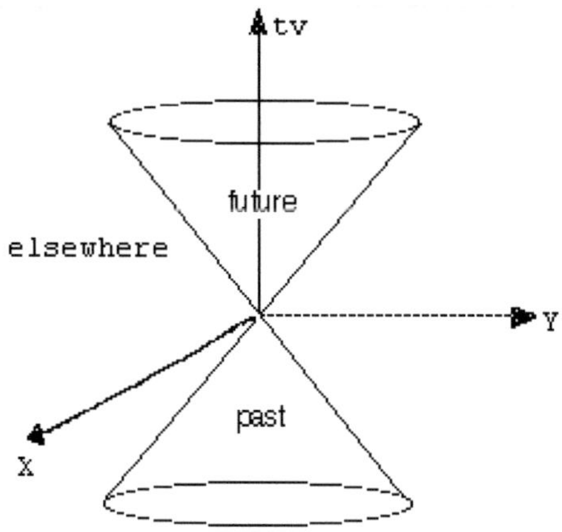

Figure 1. A cone separates space-time into 3 zones, past, future, and elsewhere

Let us take influenceability as an order relation (symbolized by $\prec$) be primitive relation. It is natural to postulate that influenceability is irreflexive, antisymmetric, but transitive, i.e.,

$$(x \prec y) \wedge (y \prec z) \Rightarrow x \prec z \tag{1}$$

Thus, it can play the role of 'after'.

Definition 1 (Temporal order): Let x and y be two moving objects with t_x and t_y corresponding temporal orders, respectively. Then,

$$(x \prec y) \Rightarrow (t_x < t_y) \tag{2}$$

Connection as a reflexive and symmetric relation [10]can be defined by influenceability as follows:

Definition 2 (Connect relation): Two moving objects x and y are connected if the following equation holds;

$$(\forall xy)C(x,y) := [(x \prec y) \vee (y \prec x)] \wedge \{\neg(\exists a)[(x \prec a \prec y) \vee (y \prec a \prec x)]\} \tag{3}$$

Consequently, all other exhaustive and pairwise disjoint relations in region connected calculus (RCC) [3], i.e., *disconnection* (DC), *proper part* (PP), *externally connection* (EC), *identity* (EQ), *partially overlap* (PO), *tangential proper part* (TPP), *nontangential proper part* (NTPP), and the inverses of the last two; TPPi and NTPPi; can be defined.

The consensus task as an acceptance of the unique framework in mobile network can not be solved in a completely asynchronous system, but as indicated by Malek [24] with the help of influenceability and partitioning concept, it can be solved. Another task in mobile network is leader election. The leader, say *a*, can be elected by the following conditions:

$$\forall x \in \{\textit{The set of moving objects}\} : a \prec x .$$

Furthermore, some other relations can be defined, such as which termed as *speed*-connection (SC) and time proper overlap (TPO) (see Figure 2):

$$SC(x,y) := \neg EQ(x,y) \wedge \\ \{[C(x,y) \wedge (\forall ab)(EC(x,a) \wedge (EC(x,b) \wedge EC(y,a) \wedge EC(y,b)] \Rightarrow C(a,b)\} \tag{4}$$

$$TPO(x,y) := \{(x \prec y) \wedge (PO(x,y) \wedge [\ \forall z\ (SC(x,z) \Rightarrow PO(y,z))]\ \} \tag{5}$$

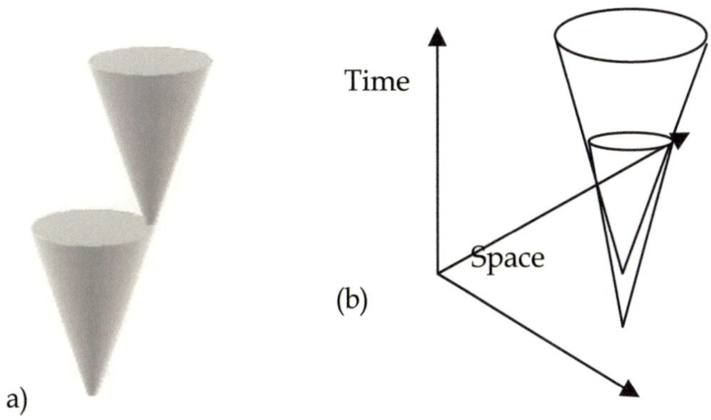

Figure 2: a) Speed-connection relation and **b)** Time-proper relation between two objects

As an example, team arrangement is considered. Team arrangement is an important task of any team coaching. Team arrangement in a mobile environment finds its applications not only in online problems such as Robocup or battlefield problem, but also in offline coaching. The main assumptions about mobile environment are valid in the usual coaching problem. Robocup is a well known application area of this problem. Team arrangement in such a mobile environment is a complex task in space and time. In this scenario players can be modeled with a cone based on their estimated speed and position (Figure 3).

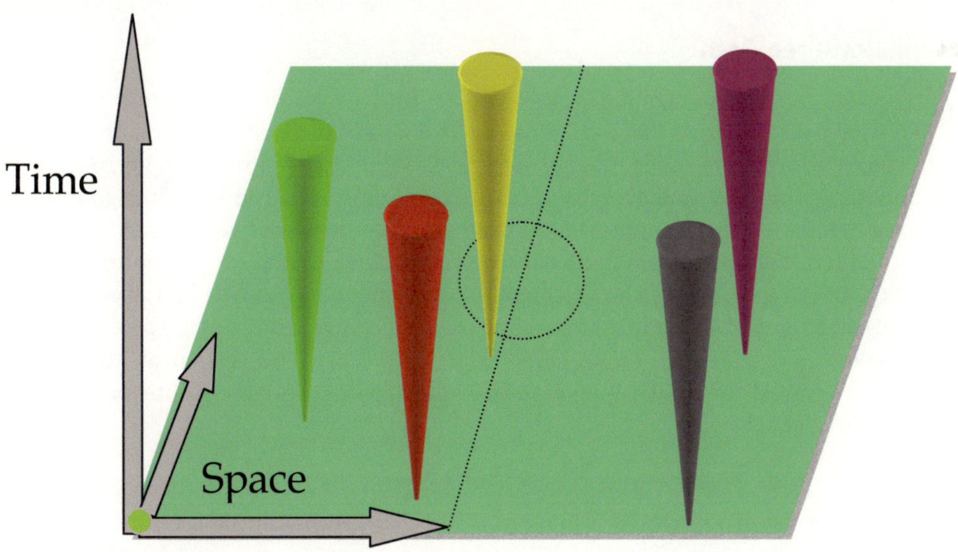

Figure 3. Robocup soccer from influenceability view point

Table 1 shows some different situations between players and their correspond relations based on our proposed model.

Presentation	Application	Relation
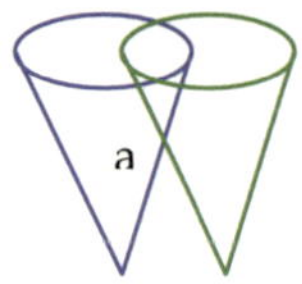	Design of the defense players	Players A and B overlap
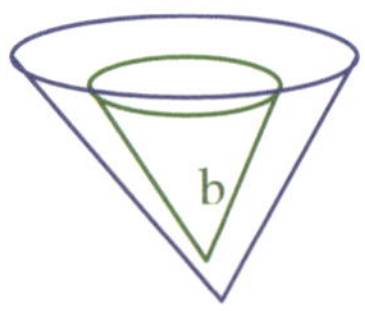	Man-for-man type of defense	Player A covers B completely
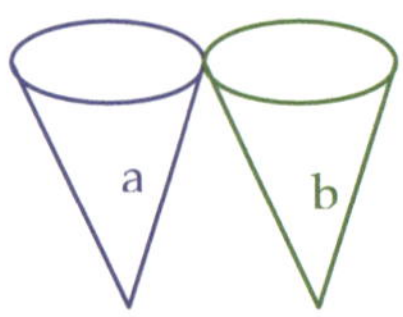	Arrangement of players to minimize empty space	Externally connection
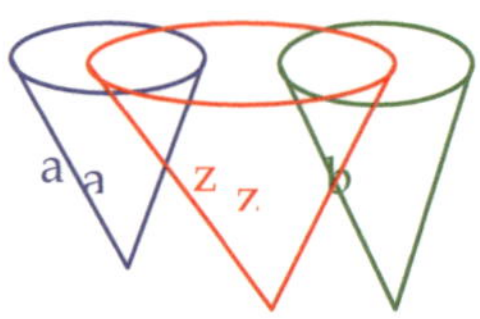	Player z can attack from gap between a and b	C(a,z) and C(b,z) but $\neg$ C(a,b)

Table 1. Some relations between players and their presentations

4. Collision-Free Path

An important task in navigation systems is to find a secure or collision-free path. A collision-free path is a route that a moving object does not have any collision or intersection with obstacles as well as other moving objects. Finding a collision-free path requires four steps, dividing the space domain into small parts, finding connected cones, computing free space, and finally solving an optimization problem. The problem discussed in this section is using a mathematical programming technique to find the optimal or near optimal collision-free path between moving objects. The details of the other steps are left for future articles.

After partitioning space-time into space-time cells, all connected cones in space-time should be calculated. Let $[\mathbf{t}]=[t_i,t_{i+1}]$ be an interval of time. The circle section of kth cone at t_j; $t_i \leq t_j \leq t_{i+1}$ is denoted by $CIR(O_k,t_j)$, where O_k is the center point. The radius of $CIR(O_k,t_j)$ is calculated by speed; v_k;

$$r_k = v_k.(t_j - t_i) \tag{6}$$

The intersection of two circles is a lens-shaped region. As it can be seen in Figure 4, the following equations can be given:

$$D^{RL} = \sqrt{[(x_R - x_L)^2 + (y_R - y_L)^2]}$$

$$a^{RL} = |r_R + r_L - D| \tag{7}$$

$$b^{RL} = r_k^2 - (r_k - a)^2$$

where x, y are the coordinates of point O.

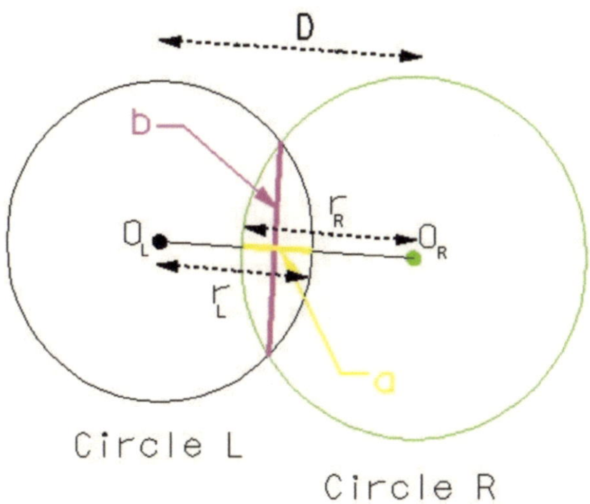

Figure 4. Intersection of two circles

Removing the barriers gives free space. It can easily be accomplished by removing all obstacles in topological structured map or by using trapezoidal map (see e.g. [8]). The trapezoidal map is an arrangement of line segments, which partitions the space into trapezoidal sections. Each trapezoidal has exactly two non-vertical boundaries and belongs to one face.

Figure 5 shows different kind of barriers. Barriers vary in size, shape and their behavior in time. The barrier (a) is constant barrier like a wall, (b) is a temporary barrier with a spatial extend, for example closing a road for a few hours, (c) is a changing size barrier in time.

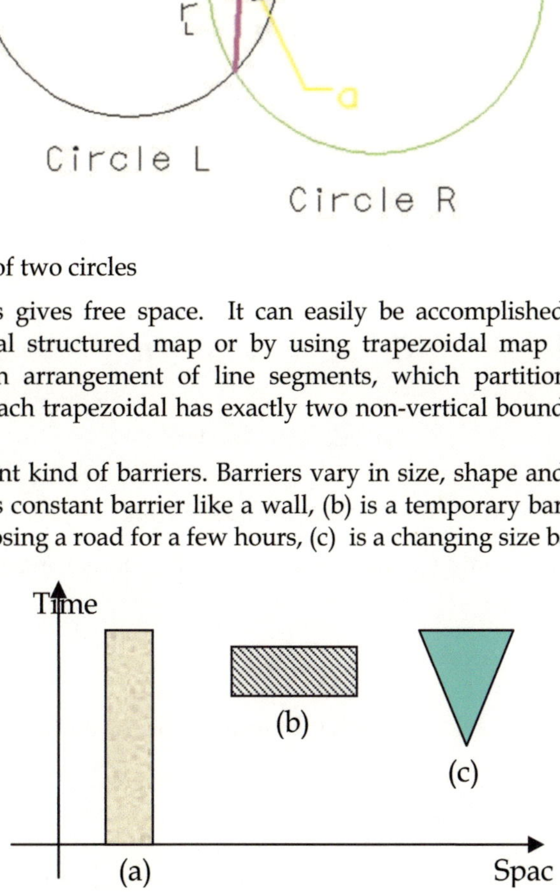

Figure 5. Different types of barriers

It will be distinguished between two network architectures, centralized and co-operative. In the centralized architecture, a control center exists which receives and sends data to moving objects. In the co-operative architecture all moving objects exchange information between themselves [21]. In the former architecture, the control variables of all nodes associates in the optimization, but in the later only variables of the active node are considered. One example of such optimization is used for unmanned aerial vehicles (UAV) or any other autonomously guided robots.

4.1. Centralized Network

As explained before, the hyper-surface of the cone indicates the fundamental topological and metrical properties of space-time. The collision-free paths, result of these cones, collectively guarantee a global collision-free route. Finding a collision-free path needs to solve the constrained optimization problem as discuss in this section.

Trying to find the new control parameters (position and speed) such that variation of parameters minimized and no collision accrued, leads to a constrained optimization problem. Consider the following model

$$\mathbf{u} = \mathbf{u}^* + \mathbf{v} \tag{8}$$

where $\mathbf{v} \in \mathbb{R}^m$ is the vector of residuals, $\mathbf{u}$ and $\mathbf{u}^*$ are the variables and estimated parameters, respectively. We select the sum of squares of residuals as the target function, which has to be minimized:

$$\Phi(\mathbf{u}) := < \mathbf{v}, \mathbf{v} > \quad \rightarrow \min \tag{9}$$

The manifold $\boldsymbol{M}_T$ can be introduced by means of target function (9) as:

$$\Phi = < \mathbf{v}, \mathbf{v} > \;\; ; \mathbf{u} \in \mathbb{U}$$

$$\boldsymbol{M}_T := \{\Phi | \; u \in \mathbb{U} \} \tag{10}$$

where $\mathbb{U} \subseteq \mathbb{R}^m$ is an open set which includes $\mathbf{u}$ and $<.,.>$ indicates inner product. Minimizing Φ means finding local extreme point in target manifold [24]. In addition, having some conditions:

$$G(u,c)=0 \tag{11}$$

the best looked for result must be satisfied in them. In the equation (11) the vector $\mathbf{c}$ includes the constants and n nonlinear equations are denoted by G. In the current problem, system equation (11) translates to

$$D^{ij} - (r_i + r_j) \geq \varepsilon_k \; ; \forall i,j : i \neq j \; , k=1,\ldots,n \tag{12}$$

where $\varepsilon = \varepsilon(u,S,E,\ldots)$ is a small quantity function of navigation parameters, shape and size of object, and other environment variables depending on specification of the problem domain. The total equations (9) and (12) results to a constrained optimization, which has to be solved by for instance one of ‘Direct substitution’, ‘Constrained variation’, ‘Lagrange multipliers’ methods [34]. The first order optimality condition leads to the system of nonlinear normal equations. Using the Lagrange multipliers and applying the first order optimality conditions the following system result [15]

$$u^{*} - (\delta G(u,c)/ \delta u^{*}).k = u$$

$$G(u,c) = 0 \quad (13)$$

where k is Lagrange multipliers. MALEK [23-24] discussed nonlinear approaches and their geometrical interpretations to solve such problems. In a centralized architecture, u consists of the parameters of all objects in the net. The result of the optimization will send to the all nodes, which have influenceability relation to each other. On the contrary, in the co-operative architecture, u consists of speed and position parameters of the active object and result will apply only to own itself. Through the following example the suggested method will be more clarified.
Example: Consider four vessels with the following properties:

	X-Coordinate	Y-Coordinate	Speed (m/sec)
V1	100	100	20
V2	200	200	16
V3	-50	-50	24
V4	500	500	20

Table 2. Coordinate and speed at start time

By $t_i=0$; $t_{i+1}=5$ and $\varepsilon=0$ the following table holds

$r_1=100$	$r_2=80$	$r_3=120$	$r_4=100$

Table 3. Radius of the CIR(O,5)
By checking (7) all connected cones can be found (Table 4).

	V1	V2	V3	V4
V1	_	Intersect	Intersect	Not Intersect
V2		_	Not Intersect	Not Intersect
V3			_	Not Intersect
V4				_

Table 4. Intersected cones

The initial position, result of system equation (13) in a centralized architecture and the final position are shown in the figure 4, table 4 and figure 6, respectively.

	X-Coordinate	Y-Coordinate	Speed (m/sec)
V1	99.224	99.224	16.944
V2	200.6	200.6	11.73
V3	-49.828	-49.828	25.214

Table 5. The result of constrained optimization

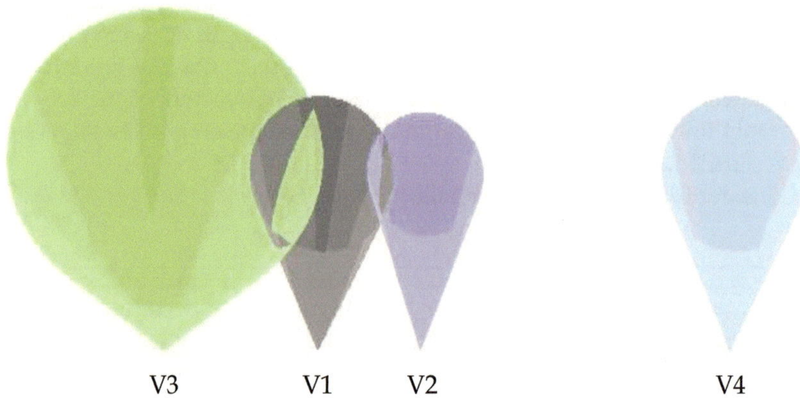

Figure 6. Four moving objects from isometric view

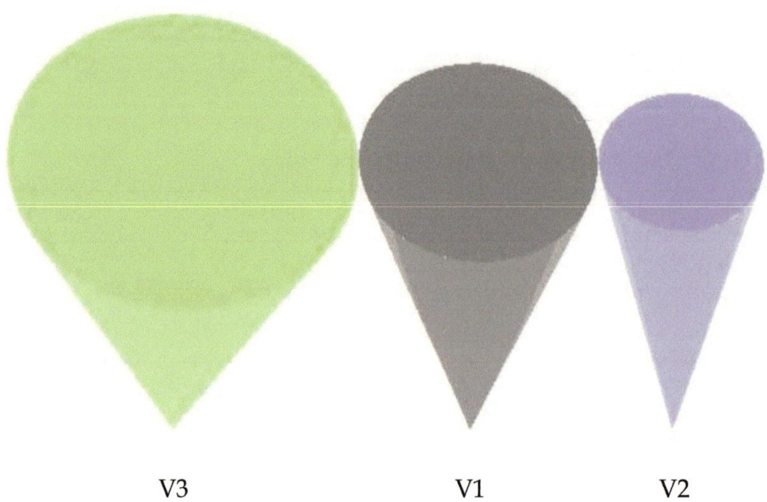

Figure 7. three moving objects after optimization from isometric view

4.2 Co-operative Network

Let us continue with the following scenario: A private company in order to attract more tourists to the lake "Wörthersee" in Austria provides an autonomous navigation system for their motorized small boats. Each boat equipped with a palm-top computer; using GPS for positioning; that can communicate via a wireless network. Based upon this capability, the system can play the role of online tourist guide at each part of the lake. The server sends the necessary information like current position and velocity of the other boats those are relevant

to the mobile host. That means, the only information of the agents that have accident possibility in the current cell will be sent. As an instance, only the information of the agent number 2 will be sent to the agent number 1 in the Figure 2. This rule can be formalized by spatial influenceability relation [25]. The mobile host will send its state information like its position and velocity once they have significant changes.
At call setup an optimal route is generated. This task can be done by the fixed host. In each cell, the preliminary route is ensured that no collision occurred. It is natural to define the target function by minimizing the distance between calculated route and the optimal one. It may be named as nearest to optimal path. The method described in this part provides a minimum distance formulation (15). It is combined with the linear collision avoidance constraints [39], turn and velocity constraints, and is extended to match with partition and conquer idea.

$$
\begin{aligned}
&Min.\ \mathbf{d}^{\mathrm{T}}\,\boldsymbol{\Sigma}\ \mathbf{d}\\
&\text{S.T.:}\\
&\textit{Collision avoidance, turn, and velocity conditions}
\end{aligned}
\tag{14}
$$

$\mathbf{d}$ is the vector of distances between optimal state parameters and the estimated control parameters in the space-time grid of interest. Finally, the linear constraint quadratic optimization problem should be solved. This part can be run in the clients and the procedure will repeat in other parts.

4.2.1 Collision avoidance condition

We shall consider for simplicity of exposition of two moving objects in a two-dimensional space. The position of agent p at time step i is given by (x_p^i, y_p^i) and its velocity by (v_{xp}^i, v_{yp}^i), forming the elements of the state vector $\mathbf{S}_{xp}^i$. The real value of the state parameter is represented by an asterisk. At every time interval the corresponding surfaces; i.e. cone; of both objects must lie outside each other. It is possible to consider one object as a point and similar to classical approach taken in robot motion planning, enlarge another object with the same size. In this case, the problem becomes easier where the point should be outside of a polygon. With this trick linear conditions introduced by Schouwenaars et al. [39] can be used:

$$
\begin{aligned}
&x_p^i - x_q^i \geq d_x - R.c_{pq1}^i \quad and\\
&x_q^i - x_p^i \geq d_x - R.c_{pq2}^i \quad and\\
&y_p^i - y_q^i \geq d_y - R.c_{pq3}^i \quad and\\
&y_q^i - y_p^i \geq d_y - R.c_{pq4}^i \quad and\\
&\sum_{k=1}^{4} c_{pqk}^i \leq 3
\end{aligned}
\tag{15}
$$

where d_l is the safety distance in direction l, c^i_{pqk} are a set of binary variables (0 or 1) and R is a positive number that is much larger than any position or velocity to be encountered in the problem.

4.2.2 Turn and Velocity Condition

It is possible to define other conditions to constrain the rate of turning ($\alpha_{\max}$) and changing velocity (Δ). Turn condition can be defined with the help of coordinates. Assuming space-time is small, linearization may apply. Other linear equations are suggested by Richards and How [36]. The velocity conditions can be derived easily as linear function from parameters.

$$\frac{x^j_p - x^i_p}{y^j_p - y^i_p} \le \alpha_{\max} \qquad (16)$$

$$v - v^* \le \Delta$$

4.2.3 Example

This example demonstrates that the suggested method forms an acceptable collision-free path for two boats. Figure 8 shows current locations of the boats, destinations, and space grids. The time axis is perpendicular to the space. Minimum distance optimality condition results to straight line paths to the destinations which clearly lead to a collision. In this example, the control parameters of the left vehicle are optimized.

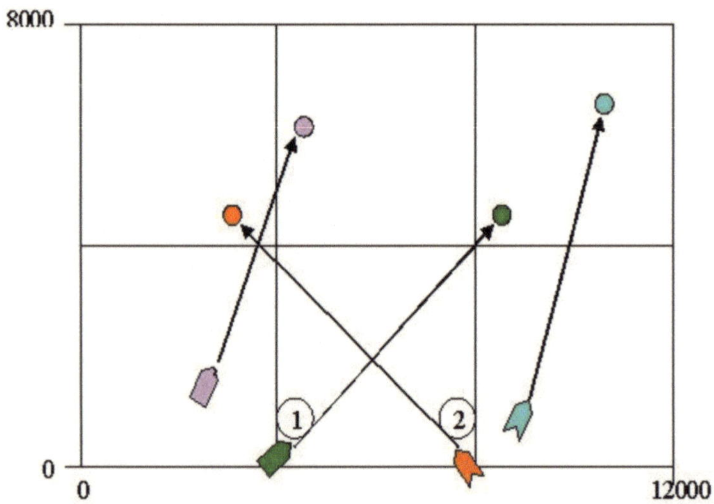

Figure 8. The trajectories of two intersecting boats. Accident will occur at fifth time interval

Let minimum speed, maximum speed, fixed time interval, and maximum deviation angle (off-route angle) be 12 m/s, 30 m/s, 20 sec., and 5 degrees, respectively. It can be easily seen that the approximate envelope of cones with that deviation angle is a rectangle in 2-dimensional and a cylinder in 3-dimensional space. Figure 3 shows the result of optimization with equal weight for all parameters. As can be seen, only velocity of the left boat at collision time is reduced without any significant change in direction. In order to reach a minimum time trajectory or maximum traveling with a fixed amount of money in our scenario, high weights for velocity are defined and the results are shown in the Figure9. By this method it is not necessary to assume that target point and the altitude are fixed. In each space-time cell some new object can appear. Due to linear formulation, this approach may be used in real or fast-time systems.

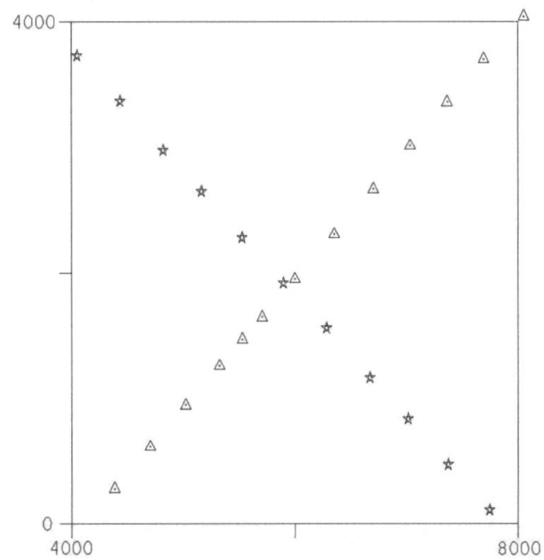

Figure 9. The designed trajectory for the left boat when all parameters are considered with equal weights

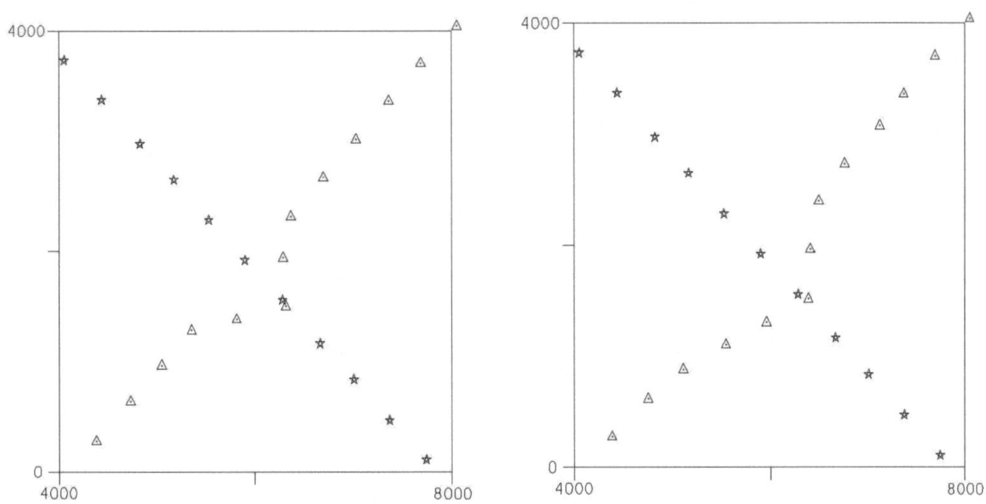

Figure 10. The designed trajectories with different turn conditions and priority of the velocity

5. Conclusion and Further Work

This chapter has demonstrated that concerns to mobile GIS theory can profitably be addressed in terms of the partition and conquer idea, and influenceability relation. Of particular significance the collision-free path problem in the context of a limited resources mobile GIS environment was addressed. We have demonstrated that concerns to mobile GIS theory can be addressed profitably in terms of the partition and conquer idea. It is based on partitioning space-time into small parts, solving the problem in those small cells and connecting the results with each other to find the final result. The reasons behind are clear. The problems can be solved easier and many things are predictable at a small part of space-time. Then, a logic-based framework for representing and reasoning about qualitative spatial relations over moving agents in space and time was derived. We provide convincing evidence of the usability of our suggested method by demonstrating how it can provide model for routing and navigation. A mathematical programming formulation has been proposed and simulated by an example to express optimal or near optimal collision-free path under the framework of such partitioning paradigm.

One important possible application of suggested methodology as our further work is mobile wayfinding services. It is based on the suggested method because wayfinding is an ordered presentation of the needed information to access an environment. It can be done in small parts as far as reaching to the desired point. A detailed uncertainty modeling for partitioning method and solving inverse problem, i.e., to determine the size and other characteristics of small parts based on the given information about needed precision, resource constraints, etc. are also among our future work.

8. Acknowledgements

Thanks to Prof. Frank (Technical University of Vienna) and Dr. Winter (University of Melbourne) for their comments.

9. References

Born, M.: Natural Philosophy of Cause and Chance, Dover Publications, New York, 1949. [1]

Buchli J. (Editor): Mobile Robotics Moving Intelligence, Pro Literatur Verlag, Germany / ARS, Austria, 2006. [2]

Caduff, D. : Sketch-Based Queries In Mobile GIS Environments, *Spatial Information Science and Engineering*, Maine, University of Main: 114, 2002. [3]

Cohn, A. G. and Hazarika, S. M.: Qualitative Spatial Representation and Reasoning:an Overview, *Fundamenta Informaticae*, 43, pp. 2-32, 2001. [4]

Chon, H., Agrawal, D. and Abbadi, A. E.: Query Processing for Moving Objects with Space-Time Grid Storage Model, Dept. of Computer Science, University of California, No: 2001-15, 2001a. [5]

Chon, H., Agrawal, D. and Abbadi, A. E.:Storage and Retrieval of Moving Objects. Proceeding of *International Conference on Mobile Data Management.*, 2001b. [6]

Chon, H., Agrawal, D. and Abbadi, A. E.: Using Space-Time Grid for Efficient Management of Moving Objects. *Proceeding of MobiDE*, 2001c. [7]

De Berg, M., Van Kreveld, M., Overmars, M. and Schwarzkopf, O.: *Computational Geometry-Algorithms and Application*,(Berlin, Springer-Verlag, 1997. [8]

Erwig, M. and Schneider, M., 1999, The Honeycomb Model of Spatio-temporal partitions. In *Spatio-Temporal database management*, (Edinburgh, Scotland, Springer), PP. 39-59. [9]

Ferscha, A. and Hoertner, H. and Kotsis, G.: Advances in Pervasive Computing, Austrian Computer Society, 2004. [10]

Forman, G. H. and Zahorjan, J., The Challenges of Mobile Computing. *IEEE Computer*, 27 (4), 38-47, 1994. [11]

GAMMA, http://www.cs.unc.edu/, Accessed: June 2006. [12]

Garg, V. K. and Mittal, N., Computation Slicing: Techniques and Theory. *Proceeding of DISC 2001*, Lisbon, Portugal, PP. 29-78, 2001. [13]

GIS-LOUNGE, http://gislounge.com/ll/mobilegis.shtml, Accessed: June 2006. [14]

Höpcke, W. : Fehlerlehre und Ausgleichrechnung, Walter de Gruyter, 1980. [15]

Lewis, D.: Causation, *Journal of Philosophy*, 70, pp. 556-567, 1973. [16]

Kaufmann, W. J.: Relativity and Cosmology, 1966. [17]

Kolski, S. (Editor): Mobile Robots Perception & Navigation, Pro Literatur Verlag, Germany / ARS, Austria, 2007. [18]

Kuchar, J. K. and Yang, L. C., A Review of Conflict detection and Resolution Modeling Methods, *IEEE Transactions On Intelligent Transportation Systems* (December), 2000. [19]

Latombe, J.-C., *Robot Motion Planning*, Kluwer academic Publishers, 1991. [20]

Laurini. R., An Introduction to TeleGeoMonitoring: Problems and Potentialities, In Atkinson P. and Martin D.:*Innovations in GIS*, Taylor & Francis, 2000 [21]

Li, L., Li, C. and Lin, Z., Investigation On the Concept Model Of Mobile GIS. *Proceeding of Symposium on Geospatial theory, Processing and Applications*, Ottawa, 2002. [22]

Malek M.R.: Nonlinear Least Squares Method from Geometric Oriented view Point, *Proceedings of The First National Optimization Conference*, Ferdowsi University, (1998). [23]

Malek M.R.:Nonlinear Least Squares Adjustment, *Proceedings of Surveying'78 Conference*, National Cartographic Center, (1999). [24]

Malek, M. R., A Logic-based Framework for Qualitative Spatial Reasoning in Mobile GIS Environment, *Lecture Note in Artificial Intelligence*, Vol. 3066, pp. 418- 426, Springer Verlag, 2004. [25]

Malek, M. R., Delavar, M. R. and Aliabady, S., A Logic-Based Foundation of Spatial Relationships in the Mobile GIS Environment, *proceeding of the 1st International Conference on Integrated Disaster Management*, Tehran, January 2006, (in Persian). [26]

Malek, M. R. and Frank, A. U. , A Mobile Computing Approach for Navigation Purposes, *Lecture Notes in Computer Science*, Vol. 4295, pp. 418-426, Springer Verlag, 2006. [27]

Malek, M. R., Delavar, M. R. and Frank A.U., A Logic-Based Foundation for Spatial Relationships in Mobile GIS Environment, In: Gartner, G. and Cartwright, William; Peterson, Michael P. (Eds.): *Location Based Services and Telecartography*, Springer, 2007. [28]

McClosky, M.: Naive theories of motion In: Gentner D. and Stevens S. (Editors): *Mental Models*, Hillsdale, New Jersey, Lawrence Erlbaum, 1983. [29]

Mittal, N., Techniques for Analysing Distributed Computations, Department of Computer Science, Austin, USA, The university of Texas, 2002. [30]

Nivala, A. M. and Sarjakoski, L. T.: Need for Context-Aware Topographic Maps in Mobile Devices, In: Virrantaus, K. and Tveite, H.: *ScanGIS'2003*, Espoo, Finland, 2003. [31]

Pallottino, L., Feron, E. and Bichini, A., Mixed Integer Programming for Aircraft Conflict Resolution. *Proceeding of Guidance, Navigation and Control Conference*, 2001. [32]

Pallottino, L., Feron, E. and Bichini, A., Conflict Resolution Problems for Air Traffic Management systems Solved with Mixed integer Programming. *IEEE Transactions On Intelligent Transportation Systems*, 3(1) (March), 3-11, 2002. [33]

Rao S.S. : Optimization: Theory and Applications, Wiley Eastern Ltd, (1978). [34]

Richards, A., How, J., Schouwenaars, T. and Feron, E., Plume Avoidance Maneuver Planning Using Mixed Integer Linear Programming. *Proceeding of AIAA 2001*, 2001. [35]

Richards, A. and How, J. P., Aircraft Trajectory Planning with Collision avoidance Using mixed Integer Linear Programming. *Proceeding of American Control Conference 2002*, 2002. [36]

Sang, J.: Theory and Development of GPS Integrity Monitoring System, *PhD Thesis*, Queensland University of Technology, 1996. [37]

Satyanarayanan, M., Fundamental Challenges in Mobile Computing. *Proceeding of ACM Symposium on Principles of Distributed Computing*, 1995. [38]

Schouwenaars, T., Moor, B. D., Feron, E. and How, J., Mixed Integer Programming For Multi-Vehicle Path Planning. *Proceeding of European Control Conference 2001*, 2001. [39]

Verma, T. S.: Causal Networks: Semantics and expressiveness, In: Sacher R. and Levitt T.S. and Kanal L.N. (ed.s):*Uncertainty in Artificial Intelligence*, Elsevier Science, 4, 1990. [40]

Van den Bergen, G., *Collision Detection in Interactive 3D Computer Animation*, Eindhoven, Eindhoven University of Technology, 1999. [41]

Wolf, H.,The Helmert block method, its origin and development, *Proceeding of Second International Symposium on Problems Related to the Redefinition of North American Geodetic Networks*, Arlington, PP. 319-326, 1978. [42]

Wolfson, O., Jiang, L., et al., Databases for Tracking Mobile Units in Real Time, *Proceeding of Database Theory- ICDT'99, 7th International Conference, LNCS 1540*, PP. 169-186, 1999.

Zaslavsky, A. and Tari, Z., Mobile Computing: Overview and Current Status, *Australian Computer Journal*, 30, 1998. [43]

Zhao, Y., *Vehicle Location and Navigation Systems*, Boston, Artech House, 1997. [44]

17

Planning with Discrete Harmonic Potential Fields

Ahmad A. Masoud
KFUPM
Saudi Arabia

1. The Harmonic Potential Field Planning Approach: A Background

Utility and meaning in the behavior of an agent are highly contingent on the agent's ability to semantically embed its actions in the context of its environment. Such an ability is cloned into an agent using a class of intelligent motion controllers that are called motion planners. Designing a motion planner is not an easy task. A planner is the hub that integrates the internal environment of an agent (e.g. its dynamics and internal representation etc.), its external environment (i.e. the work space in which it is operating, its attributes and the entities populating it), and the operator's imposed task and constraints on behavior as one goal-oriented unit. Success in achieving this goal in a realistic setting seems closely tied to the four conditions stated by Brooks which are: intelligence, emergence, situatedness, and embodiment (Brooks, 1991). While designing planners for agents whose inside occupies a simply-connected region of space can be challenging, the level of difficulty considerably rises when the planner is to be designed for agents whose inside no longer occupies a simply-connected space (i.e. the agent is a distributed entity in space). This situation gives rise to sensitive issues in communication, decision making and environment representation and whether a centralize top-down mode for behavior generation should be adopted or a decentralized, bottom-up approach may better suit the situation at hand.

To the best of this authors' knowledge, the potential field (PF) approach was the first to be used to generate a paradigm for motion guidance (Hull,1932); (Hull, 1938). The paradigm began from the simple idea of an attractor field situated on the target and a repeller field fencing the obstacles. Several decades later, the paradigm surfaced again through the little-known work of Loef and Soni which was carried out in the early 1970s (Loef , 1973); (Loef & Soni, 1975). Not until the mid-1980s did this approach achieve recognition in the path planning literature through the works of Khatib (Khatib , 1985), Krogh (Krogh, 1984), Takegaki and Arimoto (Takegaki & Arimoto, 1981), Pavlov and Voronin (Pavlov & Voronin, 1984) ,Malyshev (Malyshev, 1980), Aksenov et al. (Aksenov et al., 1978), as well as Petrov and Sirota (Petrov, Sirota ,1981); (Petrov & Sirota, 1983). Andrews and Hogan also worked on the idea in the context of force control (Andrews & Hogan, 1983).

Despite its promising start, the attractor-repeller paradigm for configuring a potential field for use in navigation faced several problems. The most serious one is its inability to guarantee convergence to a target point (the local minima problem). However, the problem was quickly solved. One solution uses a configuration that forces the divergence of the PF gradient to be zero almost everywhere in the workspace, hence eliminating the local minima problem. The approach was named the harmonic potential field (HPF) planning approach. A basic setting of this approach is shown in (1) below:

$$\nabla^2 V(X)=0 \qquad X\in\Omega \tag{1}$$

subject to: $V(X_S) = 1$, $V(X_T) = 0$, and $\frac{\partial V}{\partial \mathbf{n}} = 0$ at $X = \Gamma$,

where Ω is the workspace, Γ is its boundary, $\mathbf{n}$ is a unit vector normal to Γ, X_s is the start point, and X_T is the target point.

Although a paradigm to describe motion using HPFs has been available for more than three decades, it was not until 1987 that Sato (Sato, 1987) formally used it as a tool for motion planning (an English version of the work may be found in (Sato, 1993)). The approach was formally introduced to the robotics and intelligent control literature through the independent work of Connolly et al. (Connolly et al., 1990), Prassler (Prassler, 1989) and Tarassenko et al. (Tarassenko & Blake, 1991) who demonstrated the approach using an electric network analogy, Lei (Lei, 1990) and Plumer (Plumer, 1991) who used a neural network setting, and Keymeulen et al. (Decuyper & Keymeulen, 1990); (Keymeulen & Decuyper, 1990) and Akishita et al. (Akishita et al., 1990) who utilized a fluid dynamic metaphor in their development of the approach. Cheng et al. (Cheng & Tanaka, 1991);(Cheng, 1991);(Kanaya et al., 1994) utilized harmonic potential fields for the construction of silicon retina, VLSI wire routing, and robot motion planning. A unity resistive grid was used for computing the potential. In (Dunskaya & Pyatnitskiy 1990) a potential field was suggested whose differential properties are governed by the inhomogeneous Poisson equation for constructing a nonlinear controller for a robotic manipulator taking into consideration obstacles and joint limits.

Harmonic potential fields (HPFs) have proven themselves to be effective tools for inducing in an agent an intelligent, emergent, embodied, context-sensitive and goal-oriented behavior (i.e. a planning action). A planning action generated by an HPF-based planner can operate in an informationally-open and organizationally-closed mode; therefore, enabling an agent to make decisions on-the-fly using on-line sensory data without relying on the help of an external agent. HPF-based planners can also operate in an informationally-closed, organizationally-open mode (Masoud, 2003) ;(Masoud & Masoud, 1998) which makes it possible to utilize existing data about the environment in generating the planning action as well as illicit the help of external agents . A hybrid of the two modes may also be constructed. Such features make it possible to adapt HPFs for planning in a variety of situations. For example in (Masoud & Masoud, 2000) vector-harmonic potential fields were used for planning with robots having second order dynamics. In (Masoud, 2002) the approach was configured to work with a pursuit-evasion planning problem, and in (Masoud & Masoud, 2002) the HPF approach was modified to incorporate joint constraints on regional avoidance and direction. The decentralized, multi-agent, planning case was tackled using the HPF approach in (Masoud, 2007). The HPF approach was also found to facilitate the integration of planners as subsystems in networked controllers containing sensory, communication and control modules with a good chance of yielding a successful behavior in a realistic, physical setting (Gupta et al., 2006).

Although a variety of provably-correct, HPF-based planning techniques exist for the continuous case, to the best of this author's knowledge, there are no attempts to formally use HPFs to synthesize planners that work with discrete spaces described by weighted graphs. Planning in discrete spaces is of considerable significance in managing the complexity in high dimensions (Aarno et al., 2004); (Kazemi et al. , 2005). It is also important for dealing with inherently discrete systems such as robustly planning the motion of data packets in a network of routers (Royer & Toh, 1999).

In this work a discrete counterpart to the continuous harmonic potential field approach is suggested. The extension to the discrete case makes use of the strong relation HPF-based planning has to connectionist artificial intelligence (AI). Connectionist AI systems are networks of simple, interconnected processors running in parallel within the confines of the environment in which the planning action is to be synthesized. It is not hard to see that such a paradigm naturally lends itself to planning on weighted graphs where the processors may be seen as the vertices of the graph and the relations among them as its edges. Electrical networks are an effective realization of connectionist AI. Many computational techniques utilizing electrical networks do exist (Blasum et al., 1996);(Duffin, 1971);(Bertsekas, 1996);(Wolaver, 1971) and the strong relation graph theory has to this area (Bollabas, 1979);(Seshu, Reed, 1961) is well-known. This relation is directly utilized for constructing a discrete counterpart to the BVP in (1) used to generate the continuous HPF. The discrete counterpart is established by replacing the Laplace operator with the flow balance operator represented by Krichhoff current law (KCL) (Bobrow, 1981). As for the boundary conditions, they are applied in the same manner as in (1) to the boundary vertices. The discrete counterpart is supported with definitions and propositions that help in utilizing it for developing motion planners. The utility of the discrete HPF (DHPF) approach is demonstrated in three ways. First, the capability of the DHPF approach to generate new, abstract, planning techniques is demonstrated by constructing a novel, efficient, optimal, discrete planning method called the M* algorithm. Also, its ability to augment the capabilities of existing planners is demonstrated by suggesting a generic solution to the lower bound problem faced by the A* algorithm. The DHPF approach is shown to be useful in solving specific planning problems in communication. It is demonstrated that the discrete HPF paradigm can support routing on-the-fly while the network is still in a transient state. It is shown by simulation that if a path to the target always exist and the switching delays in the routers are negligible, a packet will reach its destination despite the changes in the network which may simultaneously take place while the packet is being routed. An important property of the DHPF paradigm is its ability to utilize a continuous HPF-based planner for solving a discrete problem. For example, the HPF-based planner in (Masoud, 2002) may be adapted for pursuit-evasion on a grid. Here a note is provided on how the continuous, multi-agent, HPF-based planner in (Masoud, 2007) may be adapted for solving a form of the sliding block puzzle (SBP).

2. HPF and Connectionism:

Learning is the main tool used by most researchers to adapt the behavior of an agent to structural changes in its environment (Tham & Prager, 1993),(Humphrys, 1995),(Ram et al., 1994). The overwhelming majority of learning techniques are unified in their reliance on experience as the driver of action selection. There are, however, environments which an agent is required to operate in that rule-out experience as the only mechanism for action

selection. Learning, or the acquisition of knowledge needed to deal with a situation, may be carried-out via hieratical, symbolic reasoning. Despite the popularity of the symbolic reasoning AI approach, its fitness to synthesize autonomous and intelligent behavior in such types of environments is being seriously questioned (Brooks, 1990); (Brooks, 1991).
Representing an environment as a group of discrete heterogeneous entities that are glued together via a hierarchical set of relations is a long standing tradition in philosophy and science. There is, however, an opposing, but less popular, camp to the above point of view stressing that representations should be indivisible, and homogeneous. Distributed representations have already found supporters among modern mathematicians, system theorists, and philosophers. Norbert Wiener said "The identity of a body is more like the identity of a flame than that of a stone; it is the identity of a structure, not of a piece of matter" (Wiener, 1950);(Wiener, 1961). In (Lefebvre,1977) Lefebvre viewes an entity or a process as a wave that glides on a substrate of parts where the relation between the two is that of a system drawn on a system. In (Campbell , 1994) Campbell argues against the hypothesis that geometrical symbols are used by creatures, to model the environment that they want to navigate. He postulate the existence of a more subtle and distributed representation of the environment inside an agent. With this in mind, the following guidelines are used for constructing a representation:

1. A representation is a pattern that is imprinted on a substrate of some kind.
2. The substrate is chosen as a set of homogeneous, simple automata that densely covers the agent's domain of awareness. This domain describes the state of the environment and is referred to as state space.
3. The representation is self-referential. A self-referential representation may be constructed using a dense substrate of automata that depicts the manner in which an agent acts at every point in state space. Self-referential representations are completely at odd with objective representations. They are a product of the stream of philosophy and epistemology (theory of knowledge) (Glasserfeld, 1986), (Lewis, 1929), (Nagel & Brandt, 1965), (Masani, 1994) which stresses that ontological (absolute or objective) reality does not exist, and any knowledge that is acquired by the agent is subjective (self-referential.)
4. In conformity with the view that objective reality is unattainable, a representation is looked upon as merely a belief. Its value to an agent is in how useful it is, not how well it represents its outside reality. Therefore, a pattern that evolves as a result of a self-regulating construction is at all phases of its evolution a legitimate representation.

A machine is a two-port device that consists of an operator port, an environment port, and a construction that would allow a goal set by the operator, defined relative to the environment to be reached. CYBERNETICS (Wiener, 1950); (Wiener, 1961), or as Wiener defined it: "communication and control in the animal and the machine," is based on the conjectures that a machine can learn, can produce other machines in its own image, and can evolve to a degree where it exceeds the capabilities of its own creator. It is no longer necessary for the operator to generate a detailed and precise plan to convert the goal into a successful motor action. The operator has to only provide a general outline of a plan and the machine will fill in the "gaps"; hence confining the operator's intervention to the high-level functions of the undergoing process. Such functions dictate goals and constrain behavior. The machine is supposed to transform the high-level commands into successful actions. CYBERNETICS unifies the nature of communication and control. It gives actions the soft nature of information. To a cybernist a machine that is interacting with its environment is an agent that is engaging in information exchange with other agents in its environment. In turn, a machine consists of interacting subagents, and is an interactive subagent in a larger machine. A controller which forces an agent to comply with the will of the operator is seen as an encoder that translates the requests of the operator to a language the agent can understand. Therefore, an action is a message, and a message is an information-bearing signal or simply information. Accepting the above paves the way for a qualitative understanding of the ability of a machine to complement the plan of the operator. Let an information theoretic approach (Shannon, 1949); (Gallager, 1968) be used to examine two agents that are interacting or, equivalently, exchanging messages. Assume that the activities of the first agent has Ix equivalence of information, and that the second has Iy. Although what is being contributed by the interacting agents is equal to Ix+Iy (self-information), the actual information content of the process is Ix+Ixy+Iy, where Ixy is called mutual information. While the measure of self information is always positive definite (Ix=-log(Px) ,Iy=-log(Py)), the measure of mutual information (Ixy= log(Px,y/(Px.Py)) is indefinite (Px and Py are the probability of x and y respectively, and Px,y is their joint probability). In an environment where carefully designed modes of interaction are instituted among the constituting agents, the net outcome from the interaction will far exceed the sum of the individual contributions. On the other hand, in non-cooperative environments the total information maybe much less than the self-information (an interaction that paralyzes the members makes Px,y≡ 0, and Ixy →-∞). It has been shown experimentally and by simulation that sophisticated goal-oriented behavior can emerge from the local interaction of a large number of participants which exhibit a much more simplistic behavior. This has motivated a new look at the synthesis of behavior that is fundamentally different from the top-bottom approach which is a characteristic of classical AI. Artificial Life (AL) (Langton, 1988) approaches behavior as a bottom-up process that is generated from elementary, distributed, local actions of individual organisms interacting in an environment. The manner in which an individual interact with others in its local environment is called the Geno-type. On the other hand, the overall behavior of the group (Phenotype, or P-type) evolves in space and time as a result of the interpretation of the Geno-type in the context of the environment. The process by which the P-type develop under the direction of the G-type is called Morphogenesis (Thorn, 1975).

To alter its state in some environment an agent (from now on is referred to as the operator) needs to construct a machine that would interface its goal to its actions. The machine (or interface) function to convert the goal into a sequence of actions that are imbedded into the environment. These actions are designed to yield a corresponding sequence of states so that the final state is the goal state of the operator. The action sequence is called a plan and it is a member of a field of plans (Action field) that densely covers state space so that regardless of the starting point, a plan always exist to propel the agent to its goal. To construct a machine of the above kind the operator must begin by reproducing itself by densely spreading operator-like micro-agents at every point in state space (Figure-1). The only difference between the operator-agent and an operator-like micro-agent is that the state of the operator evolves in time and space while the state of the micro-agent is stagnant and immobilized to one *a priori* known point in state space. The second part of machine construction is to induce the proper action structure over the micro-agent group. It is obvious that a hierarchical, holistic, centralized approach for inducing structure over the group entails the existence of a central planning agent/s that is/are not operator-like. Including such an agent in the machine violates organizational closure, i.e. the restrictions on intelligent machines receiving no influx of external intelligence to help them realize their goals. In other words, the agent must be able to lift itself from its own bootstraps. By restricting the forms of the agents constituting the machine to that of the operator, an AL approach does not require the intervention of any external intelligence to help in the construction of the machine. The AL approach, which is decentralized by definition (i.e. no supervisor is needed), requires a micro-agent to locally constrain its behavior (Genotype, or G-type behavior) using the information derived from the states of the neighboring micro-agents (Figure-1). Unlike centralized approaches where each micro-agent has to exert the "correct" action in order to generate a group structure that unifies the micro-agents in one goal-oriented unit, the AL approach only requires the micro-agent not to exert the "wrong" action that would prevent the operator from proceeding to its goal. Obviously, not selecting the wrong action is not enough, on its own, for each micro-agent to restrict itself to one and only one admissible action that would constitute a proper building block of the global structure that is required to turn the group into a functional unit. In an AL approach, the additional effort (besides that of the G-type behavior) needed to induce the global structure on the micro-agents is a result of evolution in space and time under the guidance of the environment. This interpretation or guidance is what eventually limits each micro-agent to one and only one action that is also the proper component in a functioning group structure.

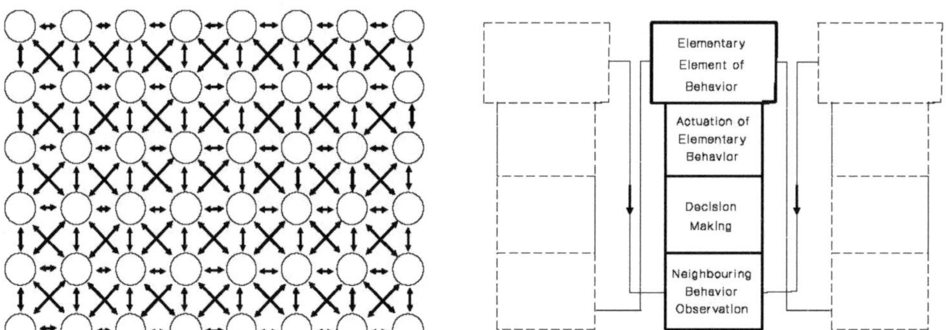

Figure 1. An interacting collective of micro-agents & Layers of functions in a micro-agent

To construct a machine that operates in an AL mode, the operator must have the means to:

1. Reproduce itself at every point in state space.
2. Clone the geno-type behavior in each member of the micro-aqent group.
3. Factor the environment in the behavior generation process.

The HPF approach lends itself to the above guidelines for the construction of an AL-driven, intelligent machine. The potential field is used to induce a dense collective of virtual agents covering the workspace. The interaction among the agents is generated by enforcing the Laplace equation. The environment is factored into the behavior generation process by enforcing the boundary conditions. More details can be found in (Masoud, 2003);(Masoud & Masoud, 1998).

3. The DHPF approach:

This section provides basic definitions and propositions that serve as a good starting point for the understanding and utilization of the DHPF approach.

Definition -1: Let G be a non-directed graph containing N vertices. Let the cost of moving from vertex i to vertex j be C_{ij} ($C_{ij}=C_{ji}$). Let a potential V_i be defined at each vertex of the graph (i=1,..,N), and I_{ij} be the flow from vertex i to vertex j defined as:

$$I_{ij}=\frac{V_i - V_j}{C_{ij}} \quad , \tag{2}$$

where $V_i > V_j$ .Note that equation-2 is analogous to ohm's law in electric circuits (Bobrow, 1981). Let T and S be the target and start boundary vertices respectively.
A discrete counterpart for the BVP in (1) is obtained if at each vertex of G (excluding the boundary vertices) the balance condition represented by KCL is enforced:

$$\sum_j I_{ij}=0 \qquad i=1,..,N,\ i \neq T,\ i \neq S$$

and

$$V_S = 1,\ V_T = 0. \tag{3}$$

Definition-2: Let the equivalent cost between any two arbitrarily chosen vertices, i and j, of G (Ceq_{ij}) be defined as the potential difference applied to the i-j port of G (ΔV) divided by the flow, I, entering vertex i and leaving vertex j (figure-2)

$$Ceq_{ij}=\frac{\Delta V}{I} \tag{4}$$

Proposition-1: The equivalent cost of a graph, G, that satisfies KCL at all of its nodes (i.e. an electric network) as seen from the i-j port (vertices) is less than or equal to the sum of all the costs along any forward path connecting vertex i to vertex j. Note that if the proposition holds for forward paths, it will also hold for paths with cycles.

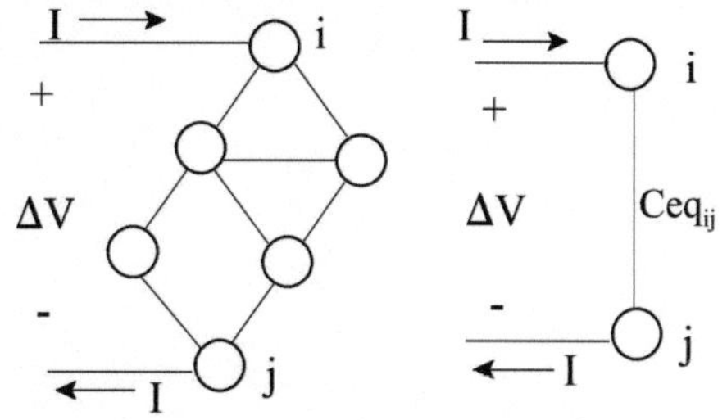

Figure 2. Equivalent cost of a graph as seen from the i-j vertices

Proof: Since the graph is required to satisfy KCL, then any internal flow in the edges of the graph is less than or equal to the external flow I (figure-3).

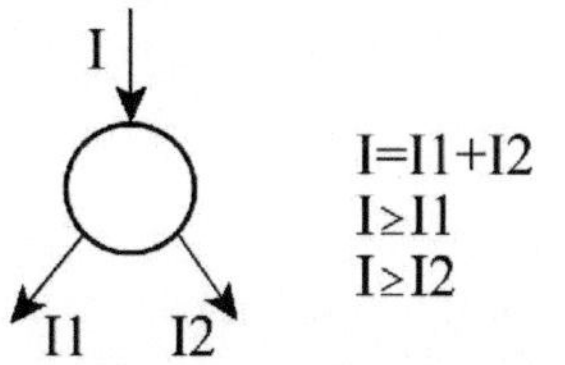

Figure 3. KCL splits input current into smaller or equal components

Consider a forward sequence of vertices connecting vertex i to vertex j: i→i+1→...→i+L-1→j. The potential difference at the k'th edge (ΔV_k) in the selected path is: $\Delta V_k = V_{i+k-1} - V_{i+k}$. The potential difference between i and j may be written as:

$$\begin{aligned}\Delta V &= \Delta V_1 + \Delta V_2 + ... + \Delta V_k + ...+ \Delta V_L \\ &= C_1 I_1 + C_2 I_2 + ... + C_K I_K + ...+ C_L I_L ,\end{aligned} \tag{5}$$

where for simplicity C_k is used to denote $C_{i+k-1,i+k}$, and I_k is used instead of $I_{i+k-1,i+k}$. Now divide both sides by the flow I:

$$\frac{\Delta V}{I} = Ceq_{i,j} = C_1 \frac{I_1}{I} + ... + C_k \frac{I_k}{I} + ... + C_L \frac{I_L}{I} \tag{6}$$

Since $I_k/I \leq 1$, we have:

$$Ceq_{i,j} \leq C_1 + ... + C_k + ... C_L . \tag{7}$$

Proposition-2: If G satisfies the conditions in equation-3, then the potential defined on the graph (V(G)) will have a unique minimum at T (V_T) and a unique maximum at S (V_S).
Proof: The proof of the above proposition follows directly from KCL and the definition of a flow (equation-2). If at vertex i V_i is a maximum, local or global, with respect to the potential at its neighboring vertices, all the flows along the edges connected to i will be outward (i.e. positive). In this case KCL will fail. Vice versa if V_i is a minimum.
Proposition-3: Traversing a positive, outgoing flow from any vertex in G will generate a sequence of vertices (i.e. a path) that terminates at T. Vice versa, traversing a negative, ingoing flow from any vertex in G will generate a sequence of vertices (i.e. a path) that terminates at S.
Proof: Assume that a hop is going to be made from vertex i to vertex i+1 based on a selected outgoing, positive flow $I_{i,i+1}$ (Figure-4).

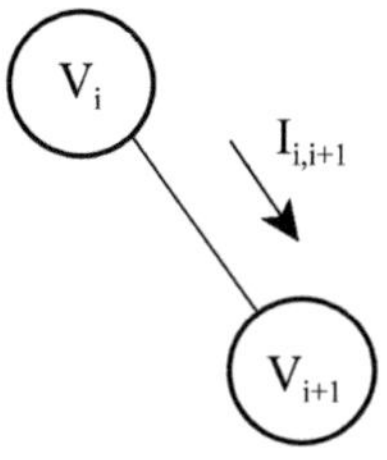

Figure 4. positive flow-guided vertex transition

From the definition of a flow we have:

$$V_i - V_{i+1} = C_{i,i+1} \cdot I_{i,i+1} . \tag{8}$$

Since we are dealing with positive costs, and the flow that was selected is positive, we have:

$$V_i - V_{i+1} > 0, \quad \text{or} \quad V_i > V_{i+1}. \tag{9}$$

By guiding vertex transition using positive flows, a continuous decrease in potential is established. This means that the unique minimum, V_T, will finally be achieved. In other words, the path will converge to T. The proof of the second part of the preposition can be easily carried out in a manner similar to the first part.

Proposition-4: A path linking S to T generated by moving from a vertex to another using a positive flow cannot have repeated vertices (i.e. it contains no loops).
Proof: If the flow used to direct vertex transition loops back to a previously encountered vertex, it will keep looping back to that vertex trapping the path into a cycle and preventing it from reaching the target vertex. Since in proposition-3 convergence to T of a positive flow driven path is guaranteed, no cycles can occur and the path cannot have repeated vertices.
Definition-3: Since a positive flow path (PFP) beginning at S is guaranteed to terminate at T with no repeated vertices in-between, the combination of all PFPs define a tree with S as the top parent vertex and bottom, offspring vertices equal to T. This tree is called the harmonic flow tree (HFT).
Proposition-5: The HFT of a graph contains all the vertices in that graph.
Proof: Since all PFPs of a graph start at S and terminate at T, the boundary vertices have to be a part of the HFT. Since at the remaining intermediate vertices the flow balance relation in equation-2 (KCL) is enforced, positive flow must enter each of these vertices; otherwise, the balance relation cannot be enforced. Therefore, every intermediate vertex has to belong to a PFP which makes it also a vertex in the HFT tree.
Proposition-6: The HFT of a graph contains the optimal path linking S to T.
Proof: If the optimum path of a graph linking S to T is not a branch of its HFT, then some of the hops used to construct that path were driven by negative flows. It is obvious that the optimum path cannot be constructed, even partially, using negative flows. If negative flows are used, convergence cannot be guaranteed. Even if countermeasures are taken to prevent a cycle from persisting, vertices will get repeated and the path will contain loops that can be easily removed to construct a lower cost path. However, a proof that negative flows cannot be used in constructing a least cost path may be obtained using the weighted Jensen inequality:

$$w_1 \cdot f(x_1)+....+w_L \cdot f(x_L) \geq f(w_1 \cdot x_1+....+w_L \cdot x_L) \tag{10}$$

where f( ) is a convex function and $w_1+...+w_L=1$, $w_i \geq 0$. Equality will hold if and only if $x_1=x_2=..=x_L$.
Assume that the optimal path connecting S to T contains L hops (L+1 vertices). The cost of that path may be written as:

$$C=C_1+C_2+....+C_L=\Delta V_1 \cdot \frac{1}{I_1}+\Delta V_2 \cdot \frac{1}{I_2}....+\Delta V_L \cdot \frac{1}{I_L}, \tag{11}$$

where ΔV_l's and I_l's are defined in the proof of proposition-1. Since KCL is enforced we have:

$$\Delta V_1+....+\Delta V_L=V_S-V_T=1-0=1. \tag{12}$$

Also, notice that $f(I_l)=1/I_l$ is a convex function of I_l. Therefore, the weighed Jensen inequality may be applied to the cost function above,

$$\Delta V_1 \cdot f(I_1)+....+\Delta V_L \cdot f(I_L) \geq f(\Delta V_1 \cdot I_1+....+\Delta V_L \cdot I_L). \tag{13}$$

or

$$\sum_{i=1}^{L} C_i \geq \frac{1}{\sum_{i=1}^{L} \Delta V_i \cdot I_i} = \frac{1}{\sum_{i=1}^{L} C_i \cdot I_i^2}. \tag{14}$$

For the case of an electric network I_k and the corresponding ΔV_k are related using ohm's law. For a positive cost having a negative ΔV_k implies a negative I_k. Therefore, the inequality will still hold for negative values of ΔV_k. This variational inequality may be used to establish a lower bound on the cost of a path connecting the start and target vertices. It is interesting to notice that the lower bound on the cost is equal to the inverse of the power consumed in traversing a path. According to the inequality, a zero difference between the two can only occur if all the flows along the path are the same. In general, the more variation in the value of the flows along the path the more is the deviation from the optimal cost will be. It is obvious that the extreme case of mixing positive and negative flows in constructing the path cannot lead to the optimum solution.
Proposition-7: The optimum path (or any PFP for that matter) must contain at most N vertices.
Proof: It is obvious that if the path contains more than N vertices, an intermediate vertex in the path is repeated. Based on the previous propositions, this cannot happen. Therefore transition from S to T must be obtained in N hops or less.

4. The M* Algorithm:

It is not hard to see that the DHPF belongs to connectionist AI. Even in its raw form, as shown by preposition-3, the approach is capable of tackling planning problems in a provably-correct manner. While connectionist AI and symbolic reasoning AI are considered to be two separate camps in artificial intelligence, there is a trend to hybridize the two in order to generate techniques that combine the attractive properties of each approach. In this section it is demonstrated that the DHPF approach can work in a hybrid mode. The reason for that is: the flow induced by the connectionist system has a structure which can be reasoned about to enhance or add to the basic capabilities of the DHPF approach. This is demonstrated in this section by suggesting the M* algorithm for finding the minimum cost path between S and T on a graph. By interfacing the connectionst layer to a symbolic layer, the quality of the path which the connectionist stage is guaranteed to find is enhanced.

4.1 The M*: direct realization:

The followings are the steps for directly realizing the M* procedure:
01. Write the KCL equations for each vertex of the graph

$$\sum_j I_{ij}=0 \quad i=1,...,N \tag{15}$$

02. From the KCL equations derive the vertex potential update equations:

$$V_i=\sum_{\substack{k=1\\k\neq i}}^{N} b_{i,k} V_k \tag{16}$$

03. Initialize the variables: $V_S=1;\ V_T=0;\ V_i=1/2 \quad i=1,..,N \quad i\neq S,\ \ i\neq T$ (17)

04. Loop till convergence is achieved performing the operations:

$$V_i=\sum_{\substack{k=1\\k\neq i}}^{N} b_{i,k}V_k \qquad i=1,..,N,\ i\neq S,\ \ i\neq T \tag{18}$$

05. Compute the flows
06. Using the flows construct the HFT of the graph
07. Starting from the last parent nodes, for each node retain the branch with lowest cost and delete the others
08. Move to the parent nodes one level up and repeat step 7.
09. Repeat step 8 till the top parent node S is reached
10. The remaining branch connected to S is the optimal path linking S to T.

4.2 An Example

Consider the weighted graph shown in figure-5. It is required that a minimum cost path be found from the start vertex S=1 to the target vertex T=5. The transition costs are: $C_{16}=1$, $C_{14}=3$, $C_{23}=4$, $C_{34}=7$, $C_{26}=1$, $C_{37}=5$, $C_{35}=2$, $C_{47}=6$, $C_{67}=9$, $C_{57}=5$. The KCL equations are:

$$\text{Vertex-1: } \frac{V_1-V_4}{C_{14}}+\frac{V_1-V_6}{C_{16}}=0 \quad \text{Vertex-2: } \frac{V_1-V_4}{C_{14}}+\frac{V_1-V_6}{C_{16}}=0$$

$$\text{Vertex-3: } \frac{V_3-V_2}{C_{23}}+\frac{V_3-V_4}{C_{34}}+\frac{V_3-V_5}{C_{35}}+\frac{V_3-V_7}{C_{37}}=0 \quad \text{Vertex-4: } \frac{V_4-V_1}{C_{14}}+\frac{V_4-V_3}{C_{34}}+\frac{V_4-V_7}{C_{47}}=0$$

$$\text{Vertex-5: } \frac{V_5-V_3}{C_{35}}+\frac{V_5-V_7}{C_{57}}=0 \qquad \text{Vertex-6: } \frac{V_6-V_1}{C_{16}}+\frac{V_6-V_2}{C_{26}}+\frac{V_6-V_7}{C_{67}}=0 \tag{19}$$

$$\text{Vertex-7: } \frac{V_7-V_3}{C_{37}}+\frac{V_7-V_4}{C_{47}}+\frac{V_7-V_5}{C_{57}}+\frac{V_7-V_6}{C_{67}}=0.$$

The update equations may be derived as:

$$V_1=\frac{1}{K_1}[\frac{V_4}{C_{14}}+\frac{V_6}{C_{16}}]=b_{1,4}V_4+b_{1,6}V_6\,, \quad V_2=\frac{1}{K_2}[\frac{V_3}{C_{23}}+\frac{V_6}{C_{26}}]=b_{2,3}V_3+b_{2,6}V_6$$

$$V_3=\frac{1}{K_3}[\frac{V_2}{C_{23}}+\frac{V_4}{C_{34}}+\frac{V_5}{C_{35}}+\frac{V_7}{C_{37}}]=b_{3,2}V_2+b_{3,4}V_4+b_{3,5}V_5+b_{3,7}V_7$$

$$V_4=\frac{1}{K_4}[\frac{V_1}{C_{14}}+\frac{V_3}{C_{34}}+\frac{V_7}{C_{47}}]=b_{4,1}V_1+b_{4,3}V_3+b_{4,7}V_7\,, \quad V_5=\frac{1}{K_5}[\frac{V_3}{C_{35}}+\frac{V_7}{C_{57}}]=b_{5,3}V_3+b_{5,7}V_7 \tag{20}$$

$$V_6=\frac{1}{K_6}[\frac{V_1}{C_{1,6}}+\frac{V_2}{C_{2,6}}+\frac{V_7}{C_{6,7}}]=b_{6,1}V_1+b_{6,2}V_2+b_{6,7}V_7$$

$$V_7=\frac{1}{K_7}[\frac{V_3}{C_{37}}+\frac{V_4}{C_{47}}+\frac{V_5}{C_{57}}+\frac{V_6}{C_{67}}]=b_{7,3}V_3+b_{7,4}V_4+b_{7,5}V_5+b_{7,6}V_6$$

where

$$\frac{1}{K_1}=[\frac{1}{C_{14}}+\frac{1}{C_{16}}] \qquad \frac{1}{K_2}=[\frac{1}{C_{23}}+\frac{1}{C_{26}}] \qquad \frac{1}{K_3}=[\frac{1}{C_{23}}+\frac{1}{C_{34}}+\frac{1}{C_{35}}+\frac{1}{C_{37}}]$$

$$\frac{1}{K_4}=[\frac{1}{C_{14}}+\frac{1}{C_{34}}+\frac{1}{C_{47}}] \qquad \frac{1}{K_5}=[\frac{1}{C_{35}}+\frac{1}{C_{57}}] \qquad \frac{1}{K_6}=[\frac{V_1}{C_{16}}+\frac{V_2}{C_{26}}+\frac{V_7}{C_{67}}]$$

$$\frac{1}{K_7}=[\frac{1}{C_{37}}+\frac{1}{C_{47}}+\frac{1}{C_{57}}+\frac{1}{C_{67}}]$$

Setting $V_1=1$, $V_5=0$ and applying the procedure described above we obtain the vertices potential: $V_1=1$, $V_2=0.74673$, $V_3=0.33753$, $V_4=0.70006$, $V_5=0$, $V_6=0.84902$, $V_7=0.41093$. The flows may be computed as: $I_{14}=0.09998$, $I_{16}=0.15098$, $I_{23}=0.1023$, $I_{47}=0.048189$, $I_{43}=0.05179$, $I_{35}=0.16877$, $I_{62}=0.1023$, $I_{67}=0.048677$, $I_{73}=0.014679$, $I_{75}=0.082186$. The graph, flows and corresponding HFT are in figure-5.

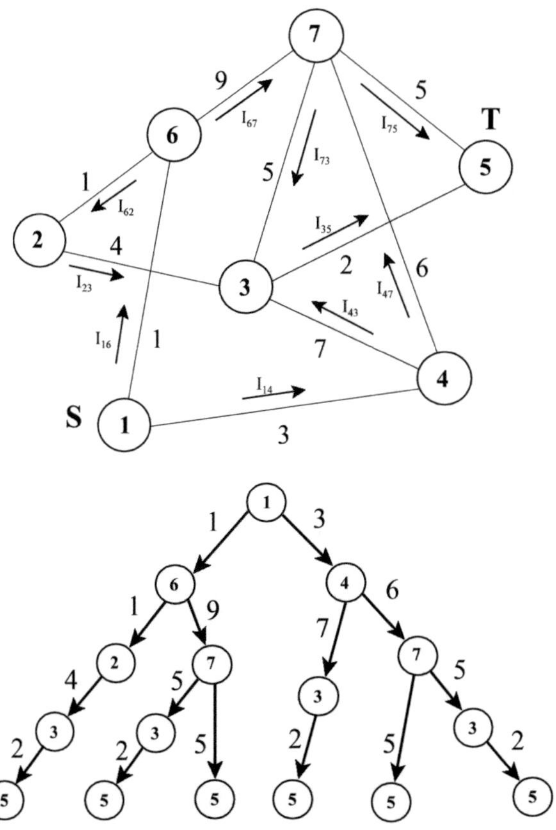

Figure 5. The graph and corresponding flows and HFT

Now start reducing the HFT,

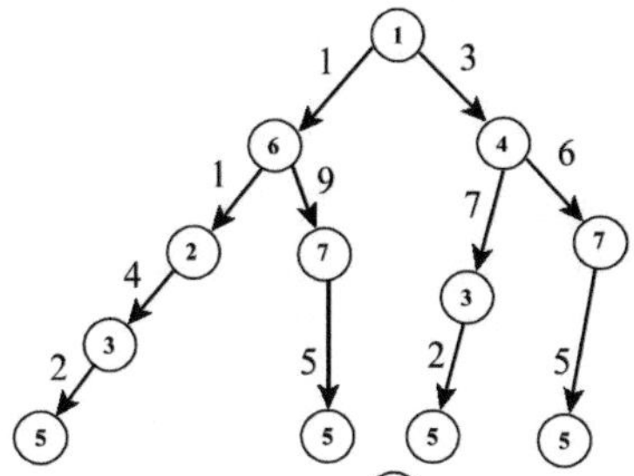

Level-1 parent node branch removal

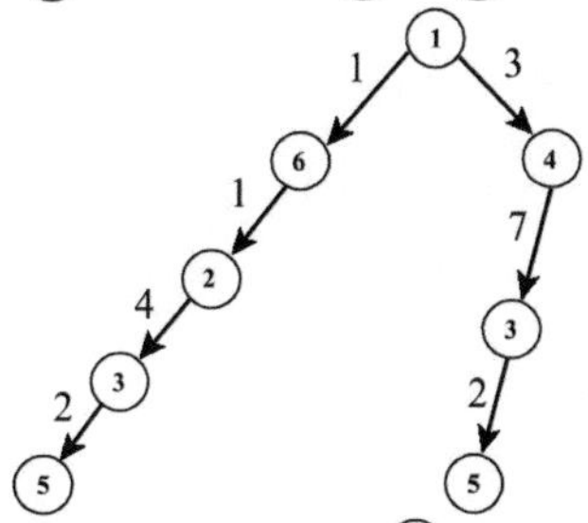

Level-2 parent node branch removal

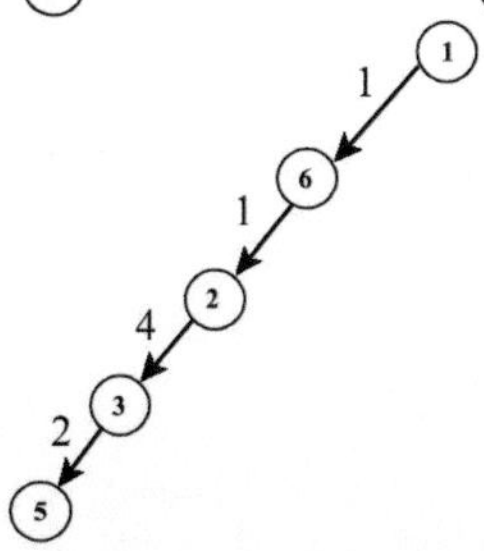

Level-3 parent node branch removal

Figure 6c. Level-3 parent node branch removal

As can be seen from figure-6 the optimum path: 5→ 3→2→6→1 with a cost 8 was obtained after only two levels of branch removal. The HFT need not be computed explicitly in order to carry out the branch removal procedure. The M* algorithm may be implemented indirectly using a procedure that successively relaxes the graph till only the optimal path is left.

4.2 The M* indirect realization:

In this subsection a successive relaxation procedure for implementing M* is suggested. The rapid growth of an HFT with the size of a graph makes it impractical to apply the algorithm on the tree directly. Here a procedure that allows us to operate on the HFT indirectly by successively relaxing the graph is suggested. In order to apply the procedure the following terms need to be defined (figure-7): positive flow index (PFI) of a vertex: number of edges connected to the vertex with outward positive flows. Negative flow index (NFI) of a vertex: is the number of edges connected to the vertex with inward negative flows.

Vertex				
PFI	0	1	2	3
NFI	3	2	1	0

Figure 7. PFI and NFI of a vertex

The procedure is:
0-compute the PFI and NFI for each vertex of the graph,
1-starting from the target vertex and using the negative flows along with the NFIs and PFIs of the vertices, detect the junction vertices and label them based on their levels,
2- at the encountered junction vertex clear NB buffers where NB=PFI of the junction vertex,
3- starting from the junction vertex, trace forward all paths to the target vertex traversing vertices with positive flows and PFIs=1,
4- excluding the lowest cost path, delete all the edges in the graph connecting the junction vertex to the first, subsequent vertices in the remaining paths,
5- decrement the PFI of the junction vertex by the number of edges removed from the graph,
6- decrement the NFIs of the first subsequent vertices from step 4 by 1,
7- if the NFI of any vertex in the graph from step 4 is equal zero and the vertex is not a start vertex, delete the edge in the graph connecting that vertex to the subsequent vertex and reduce the NFI of the subsequent vertex by 1,
8- repeat 7 till all the reaming vertices in the paths with PFIs=1 have NFIs >0,
9- go to 1 and repeat till there is only one branch left in the graph with two terminal vertices having NFI=0, PFI=1 and NFI=1, PFI=0. This is the optimum path connecting the start vertex to the target vertex.

In figure-8 the procedure is applied, in a step by step manner, to gradually reduce the graph in figure-5 to the optimum path connecting vertex 1 to 5.

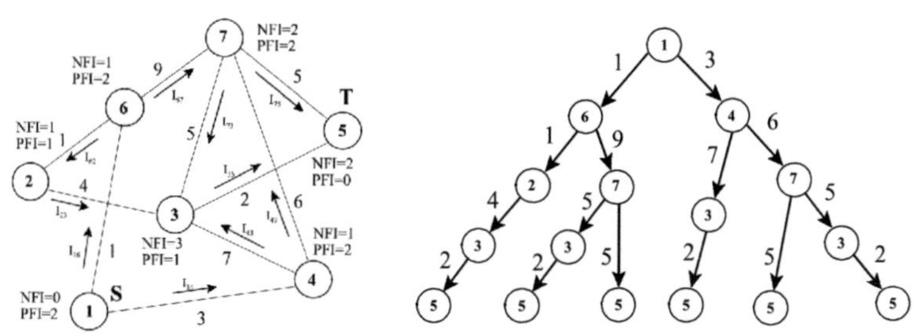

initially traced path: 5→7

constructed paths: 7→5 cost=5

7→3→5 cost=5+2=7 (eliminate edge 7→3 in graph)

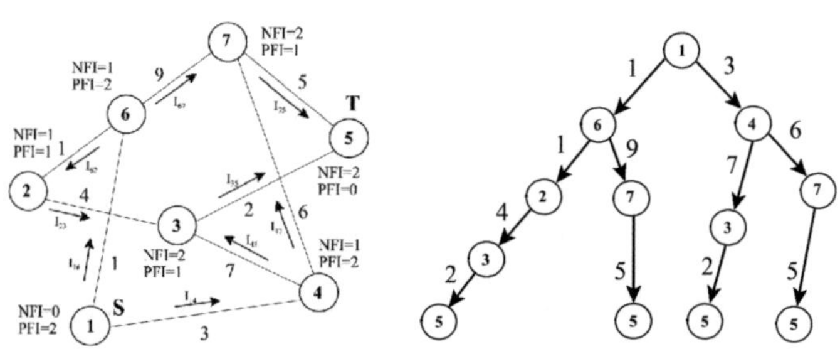

initially traced path: 5→3→2→6

constructed paths: 6→7→5 cost=9+5=14 (eliminate edge 6→7 in graph)

6→2→3→5 cost=1+4+2=7

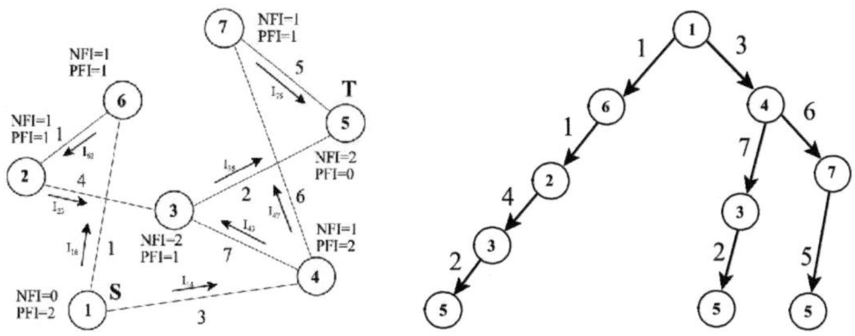

initially traced path: 5→7→4

constructed paths: 4→7→5 cost=6+5=11 (eliminate edge 4→7 in graph)

4→3→5 cost=7+2=9

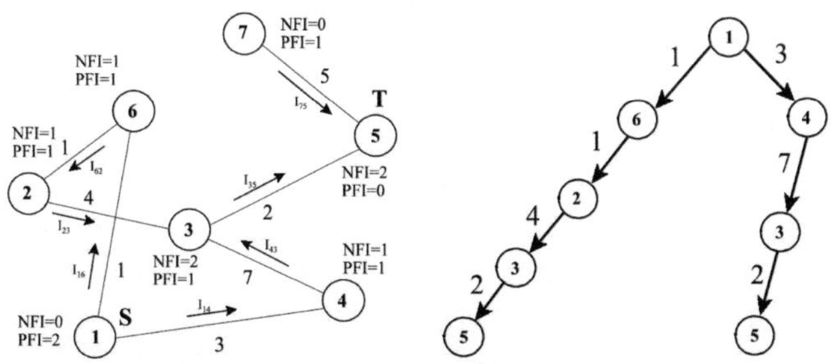

Vertex 7 NFI=0, not a start vertex (eliminate edge 7→5 in graph)

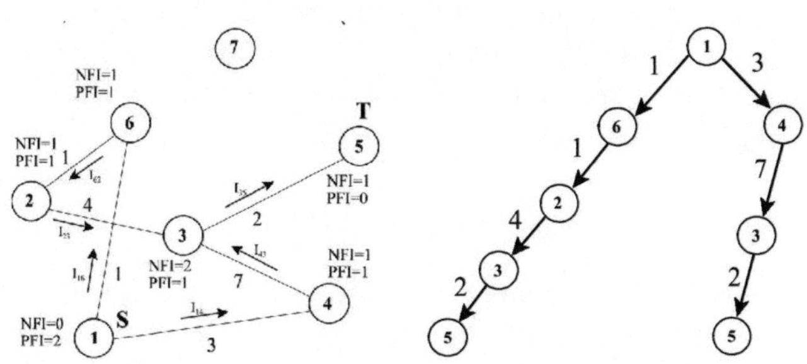

initially traced path: 5→3→2→6→1

constructed paths: 1→4→3→5 cost= 3+7+2=12 (eliminate edge 1→4 in graph)

1→6→2→3→5 cost=1+1+4+2=8

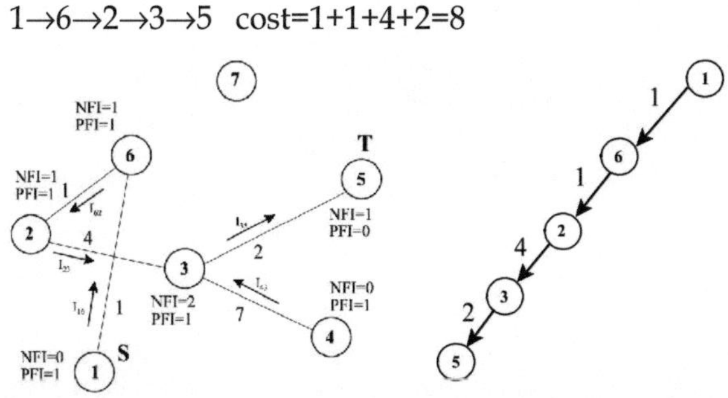

Vertex 4 has an NFI=0 and is not a start vertex (eliminate edge 4 → 3)

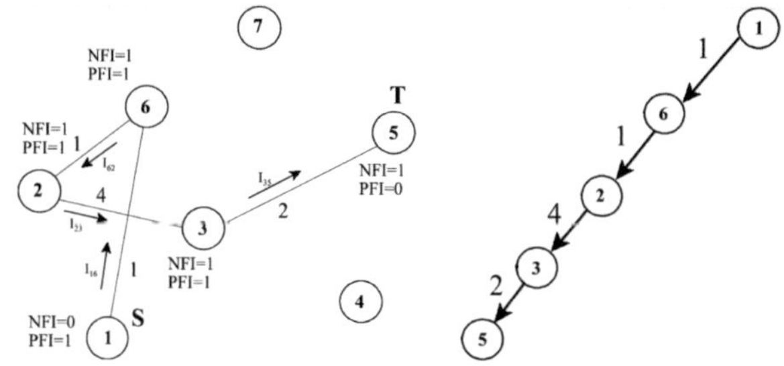

All intermediate vertices have an NFI=PFI=1 → Algorithm terminates.
Figure 8. M*-based successive relaxation.

The optimum path is: 1→6→2→3→5 having a cost of 8.

5. A Lower bound for A*

For the A* algorithm to work, a lower bound on the cost from each vertex of the graph to the target vertex has to be supplied. For spatial planning problems the Euclidian distance between the vertices provides such a bound. However, for the general case finding a lower bound may be a source of difficulties that prevents the use of the A* algorithm. In the followings it is shown that the concept of equivalent cost (resistance) from the resistive network paradigm can effectively solve this problem.
Consider the simple graph in figure-9 where S=1 and T=4. To apply the A* algorithm, the path at node 1 should be expanded towards 2 and 3. In order to sort the paths so that the next path expansion can be determined, lower estimates on the cost of moving from 2 to 4 and 3 to 4 are needed. Expansion of the path towards 2 may be achieved by simply removing all the edges of the graph that are attached to 1 leaving only the edge connected to vertex 2 (figure-9). The flows are then computed for the remaining part of the network. The equivalent cost from 2 to 4 (Ceq_{24}) may be computed as:

$$Ceq_{24}=\frac{1}{I_{12}}-C_{12} \tag{21}$$

Since in proposition-1 it is proven that the equivalent cost between two vertices in a graph is less than or equal to the least cost path connecting these vertices, the equivalent cost may be used as the lower bound estimate required by the A* algorithm. The minimum cost bounds needed for the remaining path expansions may be obtained in a similar manner.

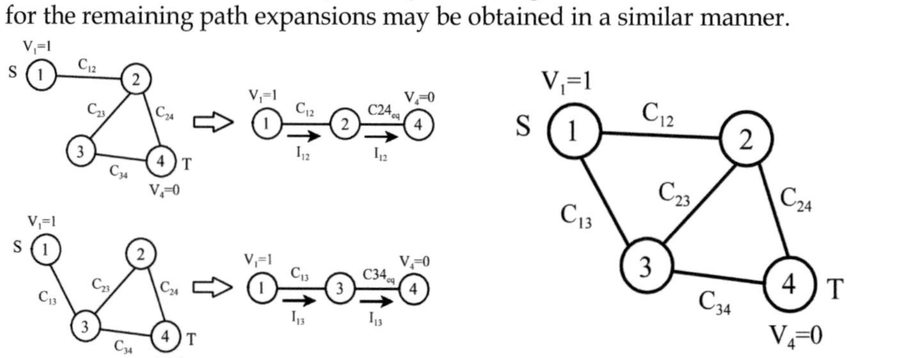

Figure 9. Path expansion, the A*

5.1 An example:

The same example in the previous section is repeated using the A* algorithm and the equivalent cost concept. The successive path expansions are shown in figure-10. The optimum path is: 1→6→2→3→5 having a cost of 8.

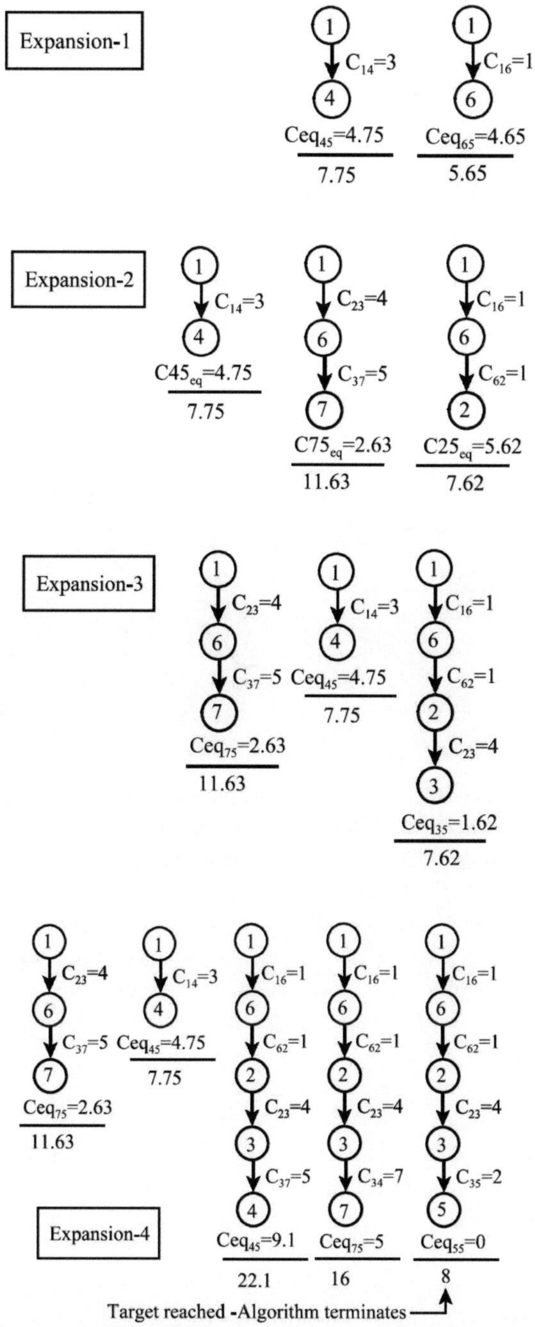

Figure 10. A* applied to the graph in figure-4

6. Routing on-the-fly

In the previous sections optimal algorithms for planning motion on a weighted graph utilizing the flow in a resistive grid are suggested. In order to apply these algorithms the graph must have a fixed structure known to the central unit that is processing the data and generating the path. While the above setting applies in many practical situations there are cases where such a scenario cannot be applied, e.g. ad-hoc networks. Also, reliability and cost issues may make it undesirable to have the whole process hinge on the success of a single, central agent. The alternative is to execute the routing process in an asynchronous, decentralized, self-organizing manner. In this case each vertex of the graph is assumed to be a router with limited sensing, processing and decision making capabilities where the immediate domains of awareness and action of a router are limited to a subset of the network with the remaining part being transparent to the router concerned. In other words, the router should sense locally, reason locally, and act locally yet produce global results (figure-11).

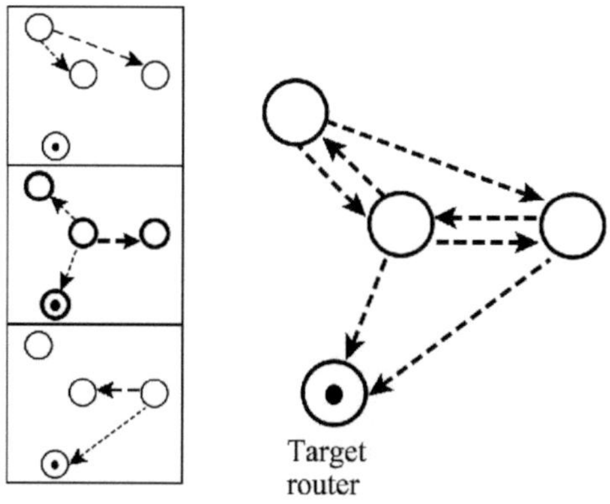

Figure 11. decentralized routing

In a centralized mode, the routers keep exchanging states till convergence is achieved. The potential is then communicated to a central agent which in a single shot lays a path to the target (figure-12). In a decentralized mode, communication of states between routers need not necessarily be sustained till a steady state is reached. Instead, during communication among the routers, whenever possible, the router with the packet attempt to pass it to a neighboring router using a simple, local, potential-based procedure that can be easily implemented on-board a router. As can be seen, under ideal situation, in a discrete HPF paradigm, the decentralized mode reduces to the centralized one.

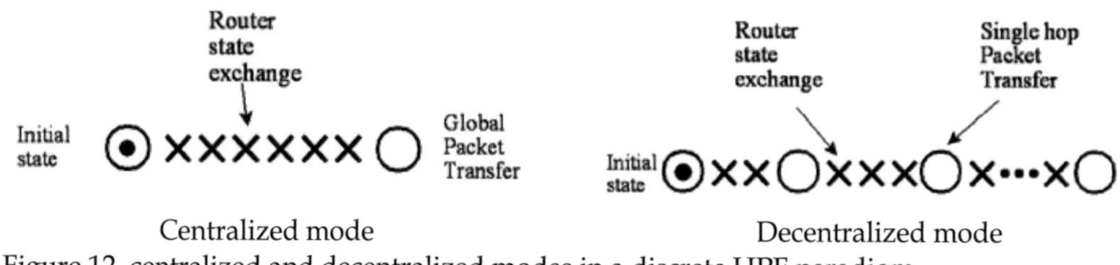

Figure 12. centralized and decentralized modes in a discrete HPF paradigm

The following is one of the decentralized procedures that may be derived from this paradigm:

0- fix the potential at the target vertex to zero,

1- each router establishes connectivity with selected neighboring routers and assigns appropriate costs,

2- fix the potential at the router that currently hold the data packet to 1,

3- excluding the routers with the packet and the target router, each router should update its potential using equation (18),

4- forward the packet from the current router to the associated router with highest positive flow,

5- if the router is not the target router go to 1,

6- target router is reached.

The procedure is simulated for the graph in figure-5. The potential field was initially set using a random number generator that is uniformly distributed between (0, 1). The output of the process is the path: $1 \rightarrow 6 \rightarrow 2 \rightarrow 3 \rightarrow 5$ having the cost 8.

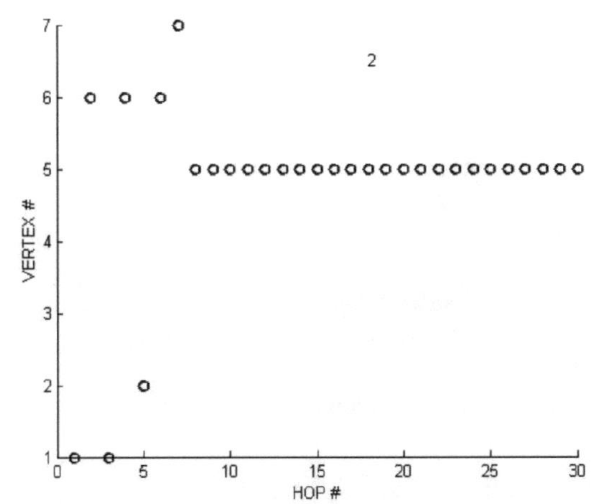

Figure 13. vertex number vs hop number under random router malfunction

To test the robustness of the procedure the example is repeated while inducing, at each hop, a malfunction in a randomly selected router (excluding the target router and the one currently holding the packet). In the following the vertex number as a function of the hop number is shown for one of the trials (figure-13). As can be seen the packet finally converges to the target vertex. Convergence was observed for all the trials that were carried out. The number of hops needed for the packet to reach the target vertex as a function of the trial number and the corresponding histogram are shown in figure-14.

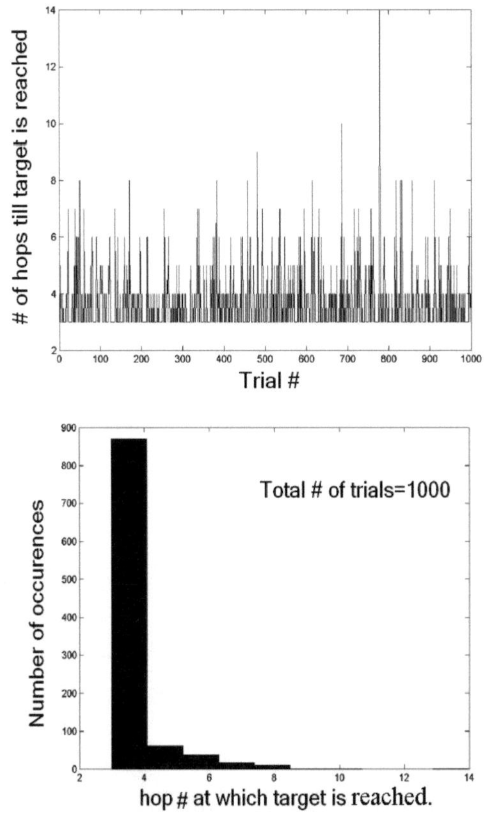

Figure 14. convergence hop number versus trial number and the corresponding histogram

7. A Note on Solving the Sliding Block Puzzle:

Designing a planer for the sliding block puzzle (Hordern, 1986) is a challenging task. It is proven that, if symbolic reasoning AI is used, the problem is PSPACE-complete. The difficulty of this problem made it an excellent choice for testing heuristic AI algorithms in order to plan the tiles' motion so that a certain target arrangement is achieved.

With little modification the continuous, multi-agent, decentralized, HPF-based planner suggested in (Masoud, 2007) may be used to tackle specific forms of the SBP problem. This HPF-based planner guarantees that convex shaped blocks can be placed in any configuration within an arbitrarily- shaped confine provided that the narrowest passage within the confine is wide enough to allow the two largest objects to pass at the same time. In other words, if the SBP has two tiles missing, it is completely solvable. The usual form of the SBP contains empty space for one block only. For this case the multi-agent planner cannot guarantee that motion of the tiles can be planned so that a final configuration, even if it belongs to the resolvable set, is attained.

The multi-agent planner uses two types of fields for controlling the motion of each tile of the SBP: a global field, called the purpose field (PRF), constructed using harmonic potential for each agent individually assuming that no other agents is sharing its workspace. This field function to drive the agent to the specified target. The second component is a local field fencing the agents. This component is called the conflict resolving field (CRF). Figure-15 shows the, evolutionary self-organizing nature of the multi-agent, HPF-based planner while attempting to lay trajectories for two sets of robots moving along opposite direction along a tight road.

The method in (Masoud, 2007) may be applied with little modification to the SBP problem. The only modification is to replace the rectangular space in which the tiles of the SBP slide with a rectangular grid (figure-16). The PRF of the individual tile may be easily computed using equation (3). The CRF components remain unchanged.

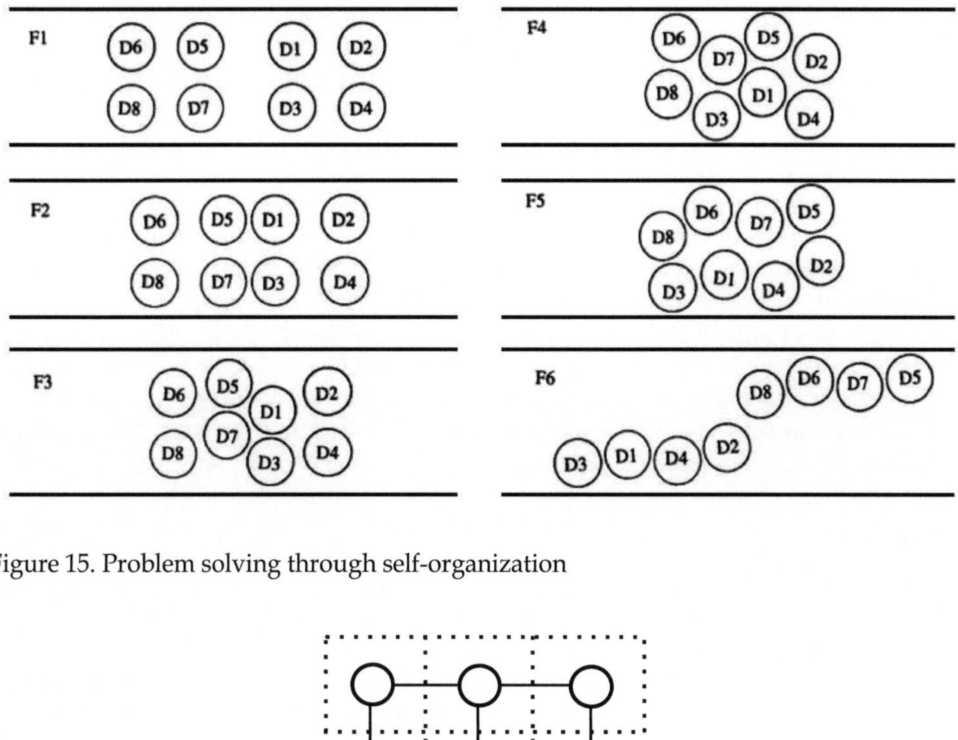

Figure 15. Problem solving through self-organization

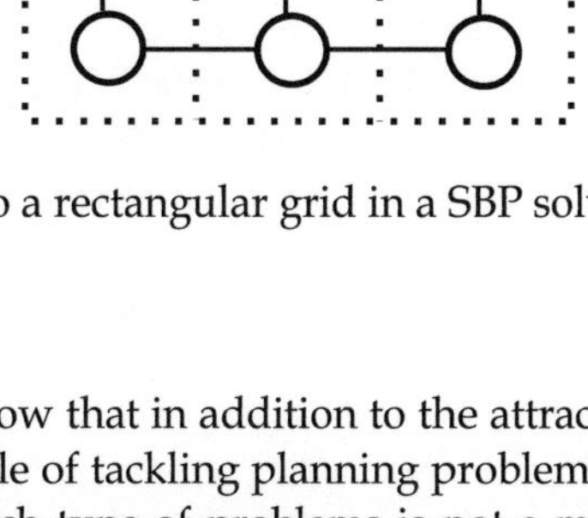

Figure 16. Restricting a PRF to a rectangular grid in a SBP solver

8. Conclusions

This work is an attempt to show that in addition to the attractive features the HPF approach already have, it is fully capable of tackling planning problems in discrete spaces. The ability to extend the approach to such type of problems is not a mere coincidence; it is rather the

product of a deeply rooted relation the HPF approach has to both AI and mathematics. This relation is what allows this approach to provide a framework able to accommodate a large variety of planning situations, both applied and abstract, as well as provide the tools needed to analyze them and understand their behavior in both qualitative and quantitative manners. Besides the abstract side, the HPF approach possess a physical intuitive side that serves as a good aid for configuring the approach to suit a desired planning situation as well as help in understanding the properties of the suggested method.
The ability of the HPF approach to work in a hybrid, symbolic-connectionist, AI mode adds more value to this method. This value is vivid in the case of data network where communication can still be maintained through local interaction even if the central entity planning traffic is no longer functioning. The fact that in this work the development for the DHPF approach is only provided for non-directed graphs does by no means reflect any shortcoming on part of the approach to deal with directed graphs. The ability of the HPF approach to enforce directional constraints for both the continuous and discrete cases has already been demonstrated in (Masoud & Masoud, 2002). The DHPF approach is even capable of making the connectivity between the vertices conditional on external, user-specified events while providing a provably-correct planning action.
This author is a firm believer in the point of view stating that a the discrete and continuous components in a hybrid system are inseparable (Wiener, 1961); (Wiener, 1950); (Lefebvre; 1977). Rather, the discrete system should appear as a pattern induced on a continuous substrate instead of a distinct discrete system module interfaced to a continuous one. This work and the associated mathematical development should be looked at only as a proof-of-principle of the ability of the HPF approach to tackle, in a provably-correct way, planning in purely discrete domains. This author views this proof of principle as an important argument against the misconception that the HPF approach is incapable of doing so.

9. References:

Aksenov G. , Voronetskaya D., Fomin V. (1978). A Construction of Program Movements of a Manipulator with the Aid of a Computer, *Engineering Cybernetics*, Vol. 16, July-August, 1978, pp. 40-45.

Andrews J. & Hogan N. (1983). Impedance Control as a Framework for Implementing Obstacle Avoidance in Manipulators, *in Control of Manufacturing Processes and Robotics Systems*, Eds. D. Hardt, W. Book, The American Society of Mechanical Engineers, pp. 243-251. (Presented at the Winter Annual Meeting of the American Society of Mechanical Engineers, Boston, MA, Nov. 13-18, 1983).

Akishita S., Kawamura S., Hayashi K. (1990). New Navigation Function Utilizing Hydrodynamic Potential for Mobile Robot, *IEEE International Workshop on Intelligent Motion Control*, , pp. 413-417, Istanbul, Turkey, August 20-22, 1990.

Aarno D., Kragic D., Christensen H. (2004). Artificial Potential Biased Probabilistic Roadmap Method, *Proceedings of the 2004 IEEE international conference on robotics and automation*, pp. 461-466, New Orleans, April 2004.

Brooks R. (1991). Intelligence Without Reason, *MIT AI Lab., Memo No. 1293*, April 1991.

Brooks R. (1990). Elephants Don't Play Chess, *Designing Autonomous Agents: Theory and Practice from Biology to Engineering*, Ed. Pattie Maes, pp. 3-15, Elsevier Science Publishers.

Blasum U., Hochstattler W., Moll C. & Rieger H. (1996). Using the Network-flow techniques to solve an optimization problem from surface physics, J. *Phys. A: Math. Gen.* 29 (1996) L459-L463.

Bollabas B. (1979), Graph Theory: An Introductory Course, Springer-Verlag, New York Heidelberg Berlin.

Bertsekas D. (1996). Thevenin Decomposition and Large-Scale Optimization, *Journal of Optimization Theory and Applications*, Vol. 89, No.1, April, 1996, pp. 1-15.

Bobrow L. (1981). Elementary Linear Circuit Analysis, second edition, OXFORD university press.

Connolly C., Weiss R. & Burns J. (1990). Path Planning using Laplace Equation, *1990 IEEE International Conference on Robotics and Automation*, pp. 2102-2106, May 13-18, 1990, Cincinnati, Ohio.

Cheng G., Tanaka M. (1991). A Parallel Iterative Routing Algorithm Based on Local Current Comparison Method, *1991 International Conference on Circuits and Systems, Shenzhen*, pp. 651-654 China, 16-17 June, 1991, Vol. 2.

Cheng G., Ikegami M., Tanaka M. (1991). A Resistive Mesh Analysis Method for Parallel Path Searching, *1991Proceedings of the 34th Midwest Symposium on Circuits and Systems*, pp. 827-830, May 14-17, Monterey, CA USA, Vol. 2.

Campbell J. (1994). "Objects and Objectivity", *Proceedings of the British Academy*, Vol. 83, Ed. Peacock C., pp. 3-45, Oxford Univ. Press.

Decuyper J., Keymeulen D. (1990). A Reactive Robot navigation System Based on a Fluid Dynamics Metaphor, *Parallel Problem Solving From Nature, 1st Workshop, PPSN1*, H. Schwefel, R. Hartmanis (Eds.), pp. 356-362, Oct. 1-3, 1990, Dortmund, FRG.

Dunskaya N. & Pyatnitskiy Y. (1990). The Objective Potential Method in Problems of Synthesizing Controls for Manipulator Robots, *Soviet Journal of Computer and Systems Sciences*, Vol. 27, Number 3, May-June 1989, English Edition Published January 1990 by Scripta Technica Inc, A Wiley Company.

Duffin R. (1971), Network Models, in *Mathematicsl Aspects of Electrical Network Analysis, Vol. III*, SIAM-AMS Proceedigns, pp. 65-91, American MAthematical Society .

Glasserfeld E. (1986). Steps in the Construction of "Others" and "Reality; *A Study in Self-regulation ,Power, Autonomy, Utopia: New Approaches Toward Complex Systems*, Ed. Trappl R.,pp. 107-116, Plenum Press, N.Y.,London.

Gallager R. (1968). *Information Theory and Reliable Communication*, John Wiely and Sons, N. Y., London, Sydney, Toronto.

Gupta R., Masoud A. & Chow M. (2006). A Network based, Delay-tolerant, Integrated Navigation System for a differential drive UGV using Harmonic Potential Field, *CDC 2006 45th IEEE Conference on Decision and Control*,pp. 1870-1875, December 13-15 2006 Manchester Grand Hyatt San Diego, California USA.

Hull C. (1932). The goal gradient hypothesis and maze learning, *Psychol. Rev.*, vol. 39, no. 1, pp. 25–43, Jan. 1932.

Hull C. (1938). The goal-gradient hypothesis applied to some 'Field-Force' problems in the behavior of young children, *Psychol. Rev.*, vol. 45, no.4, pp. 271–299, July 1938.

Humphrys M. (1995). W-Learning: Competition Among Selfish Q-learners, *April 1995 Univ. of Cambridge Comp. Lab. Tech. Rep. no. 362*.

Hordern E. (1986). *Sliding Piece Puzzles*, 1986, Oxford University Press, ISBN 0-19-853204-0.

Khatib O. (1985). Real-time obstacle avoidance for manipulators and mobile robots, *in IEEE Int. Conf. Robotics and Automation*, pp. 500–505, St. Louis, MO, Mar. 25–28, 1985.

Krogh B. (1984). A generalized potential field approach to obstacle avoidance control, in Robotics Research: The Next Five Years and Beyond, Bethlehem, PA, USA, Aug. 14–16, 1984, pp. 1–15.

Keymeulen D., Decuyper J. (1990). A Reactive Robot Navigation System Based on a Fluid Dynamics Metaphor, *AI MEMO # 90-5, Artificial Intelligence Lab.*, Vrije Universiteit Brussel.

Kazemi M., Mehrandezh M., Gupta K. (2005). An Incremental Harmonic Function-based Probabilistic Roadmap Approach to Robot Path Planning, *Proceedings of the 2005 IEEE international conference on robotics and automation*, pp. 2148-2153, April 2005 Barcelona-Spain.

Kanaya M., Cheng G., Wtanabe K. & Tanaka M. (1994). Shortest Path Searching for the Robot Walking Using an Analog Resistive Network, *1994 ISCAS'94, 1994 IEEE International Symposium on Circuits and Systems*, pp. 311-314, London UK, May 30-June 2, 1994, Vol. 6.

Loef L. (1973). An Algorithm for Computer Guidance of a Manipulator in Between Obstacles, *M.S. thesis*, Oklahoma State Univ., Stillwater, 1973.

Loef L. & Soni A. (1975). An algorithm for computer guidance of a manipulator in between obstacles, *Trans. ASME, J. Eng. Ind.*, Aug. 1975, pp. 836–842.

Lei G. (1990). "A Neuron Model With Fluid Properties for Solving Labyrinthian Puzzle", *Biological Cybernetics*, Vol. 64, pp. 61-67, 1990.

Lefebvre V. (1977). *The structure of Awareness*, Sage Publications, Beverly Hills,London, 1977.

Langton C. (1988). Artificial Life, *Artificial Life SFI Studies in the Sciences of Complexity*, Ed C. Langton , pp. 1-47, Addison-Wesley.

Lewis C. (1929). *Mind and the World Order: outline of a theory of knowledge*, Dover Publications, Inc. New York.

Malyshev V. (1980). A Method of Constructing Program Motions of a Manipulator, *Engineering Cybernetics*, Vol. 18, November-Decembert, 1980, pp. 41-45.

Masoud A. (2003). An Informationally-Open, Organizationally-Closed Control Structure for Navigating a Robot in an Unknown, Stationary Environment *2003 IEEE International Symposium on Intelligent Control*, , pp. 614-619, October 5-8, 2003, Houston, Texas, USA.

Masoud A. & Masoud S. (1998). A Self-Organizing, Hybrid,PDE-ODE Structure for Motion Control in Informationally-deprived Situations, *The 37th IEEE Conference on Decision and Control*, pp. 2535-40, Tampa Florida, Dec. 16-18, 1998.

Masoud S. & Masoud A. (2000). Constrained Motion Control Using Vector Potential Fields, *The IEEE Transactions on Systems, Man, and Cybernetics, Part A: Systems and Humans*. May 2000, Vol. 30, No.3, pp.251-272.

Masoud A. (2002), Evasion of Multiple, Intelligent Pursuers in a Stationary, Cluttered Environment: A Harmonic Potential Field Approach, *2002 IEEE International Symposium on Intelligent Control (ISIC)*, pp. 654-659, Vancouver, British Columbia, Canada, 27-30 October 2002.

Masoud S., Masoud A. (2002). Motion Planning in the Presence of Directional and Obstacle Avoidance Constraints Using Nonlinear Anisotropic, Harmonic Potential Fields: A Physical Metaphor, *the IEEE Transactions on Systems, Man, & Cybernetics, Part A: systems and humans.* Vol 32, No. 6, November 2002, pp. 705-723.

Masoud A. (2007). Decentralized, Self-organizing, Potential field-based Control for Individually-motivated, Mobile Agents in a Cluttered Environment: A Vector-Harmonic Potential Field Approach, *IEEE Transactions on Systems, Man, & Cybernetics,* Part A: systems and humans, Vol. 37, No. 3, May 2007, pp. 372-390.

Masani P. (1994). The Objectivity Axioms and Their foundation in Faith, *Kybernetes,* Vol. 23,No. 4,1994, pp.13-18.

Nagel E. & Brandt R. (1965). *Meaning and Knowledge: Systematic Readings in Epistemology,* Harcourt, Brace & World.

Pavlov V. and Voronin A. (1984). The method of potential functions for coding constraints of the external space in an intelligent mobile robot, Sov. *Automat. Contr.,* vol. 17, no. 6, 1984, pp. 45–51.

Petrov A., Sirota I. (1981). Control of a Robot-Manipulator with Obstacle Avoidance Under Little Information About the Environment, *the 8th IFAC Congress,* , pp. 54-59, Kyoto, Japan, Vol. 14, August 1981.

Petrov A., Sirota I. (1983). Obstacle Avoidance by a Robot Manipulator Under Limited Information About the Environment, *Automation and Remote Control,* Volume 44, Number 4, Part 1, April 1983, pp. 431-440.

Plumer E. (1991). Cascading a Systolic Array and a Feedforward Neural Network for Navigation and Obstacle Avoidance Using Potential Fields, *NASA Contractor Report 177575,* Prepared for Ames Research Center Contract NGT-50642, February 1991.

Prassler E. (1989). Electrical Networks and a Connectionist Approach to Path-finding, *Connectionism in Perspective,* R. Pfeifer, Z. Schreter, F. Fogelman, L. Steels (Eds.), pp. 421-428, Elsevier Science Publishers, North-Holland.

Ram A., Arkin R., Boone G. & Pearce M. (1994). Using Genetic Algorithms to Learn Receive Control Parameters for Autonomous Robotic Navigation, *Adaptive Behavior,* Vol. 2, issue 3, 1994, pp. 277-304.

Royer E. & Toh C. K. (1999). A review of current routing protocols for ah hoc mobile wireless networks. *IEEE Personal Communication,* 6(2), Apr. 1999, pp. 46-55.

Sato K. (1987). Collision Avoidance in Multi-dimensional Space Using Laplace Potential, *Proc. 15th Conf. Rob. Soc. Jpn.,* pp.155- 156, 1987.

Sato K. (1993). Deadlock-free Motion Planning Using the Laplace Potential Field, *Advanced Robotics,* Vol. 7, No. 5, 1993, pp. 449- 461.

Seshu S., Reed M. (1961). Linear Graphs and Electrical Networks, Addison-Wesley Publishing Company INC., Reading, Massachusetts, USA, London England.

Shannon C. (1949). A Mathematical Theory of Communication, *Bell System Technical Journal,* Vol.27, 1949, pp. 379-423.

Tarassenko I. & Blake A. (1991). Analogue Computation of Collision- free Paths, *1991 IEEE International Conference on Robotics and Automation,* Sacramento, pp. 540-545, California, April 1991.

Takegaki M. & Arimoto S. (1981). A new feedback method for dynamic control of manipulators, *Trans.ASMEJ. Dyn. Syst. Meas. Control,* vol. 102, June 1981, pp. 119-125.

Tham C., Prager R. (1993). Reinforcement Learning Methods for Multi-Linked Manipulator Obstacle Avoidance and Control, *IEEE Asia-Pacific Workshop on Advances in Motion Control,* , July 16-16,1993, Singapore.

Thorn R. (1975). Structural Stability and Morphogenesis, W.A. Benjamin Inc.

Wolaver D. (1971). Some Applications in Electrical Network Theory of a Linear Graph Theorem by Berge and Ghouila-Houri, in *Mathematicsl Aspects of Electrical Network Analysis,* Vol. III, SIAM-AMS Proceedigns, 1971, American MAthematical Society, pp. 149-159.

Wiener N. (1950). *The Human Use of Human Beings: Cybernetics and Society,* Houghton Mifflin Company, Boston.

Wiener N. (1961). *Cybernetics, or Control and Communication in the Animal and the Machine,* The MIT press, Cambridge, Massachusetts.

18

Mobile Robot with Preliminary-announcement and Indication of Scheduled Route and Occupied Area using Projector

Takafumi Matsumaru
Shizuoka University
Japan

1. Introduction

We are proposing approaches and equipments for preliminarily announcing and indicating to people the upcoming operation of mobile robot moving on a two-dimensional plane. We have developed four kinds of prototype robot in which our approaches are realized. This chapter focuses on the projection robot PMR-5 and the revised version PMR-5R, in which projection equipment projects a two-dimensional frame on a running surface and the frame contains the information about the robot's upcoming operation.

2. Background and Objectives

2.1 Preliminary-announcement and indication of upcoming operation

Human beings interact signaling their own and predicting others' actions and intentions nonverbally through body language, hand gestures, facial expressions, and whole body operations. Most people moving through a crowd, for example, find little trouble plotting a passage through a forest of bodies without bumping into or otherwise upsetting or needlessly distracting others. Human beings hone social and physical skills that make their movement practically second nature based on a sense of affinity, familiarity, common appearance - sharing what they like to call "common sense".

Robots call up exactly the opposite reaction - disaffinity, unfamiliarity, uncommon appearance - no sharing of the common sense that would make robot movement predictable to people - or human movement predictable to robots. If I see that you are bent forward walking fast and your eyes on a distant goal, I can predict that getting out of your way is appropriate and I can guess how fast I'll have to move to let you through. If, however, you are a robot, I probably have no idea how to deal with your movement - how fast you might walk and what direction you may take. This lack of shared knowledge and common sense between human and artificial organisms is what has made their interaction such a problem - the simplest aspect of which is to avoid the risk of contact and collision.

We are studying functions to notify people of an artificial organism's approach upcoming and intentions before it moves in order to avoid the risk of unintended contact or collision between people and artificial organism, focusing on mobile robots or transport vehicles moving on a two-dimensional plane (Fig. 1). Such robots should announce their direction and speed of movement. Such announcement should be simple and immediately understandable enough to be acceptable to people. It must also be easy for people to understand based on their common sense.

We have proposed four approaches categorized in two types to preliminarily announce and indicate the speed and direction of upcoming movement of mobile robot as shown in Table 1 (Matsumaru & Hagiwara, 2001a; Matsumaru, 2004). The first type announces the state of operation just after the present: lamp method and blowout (telescopic arrow) method, and the second type indicates the operations from the present to some future time continuously: light ray method and projection method.

The effect of preliminary-announcing and indicating the upcoming robot operation, the difference according to methods, and the appropriate timing to announce have been examined by using a software simulation before the hardware equipments that made the approaches embodiment were designed and manufactured (Matsumaru & Hagiwara, 2001b; Matsumaru et al., 2003c, 2003d; Matsumaru et al., 2004).

Figure 1. Contact and collision avoidance

(1) Announcing state just after the present		(2) Indicating operations continously	
(a) Lamp	(b) Blowout	(c) Light ray	(d) Projection
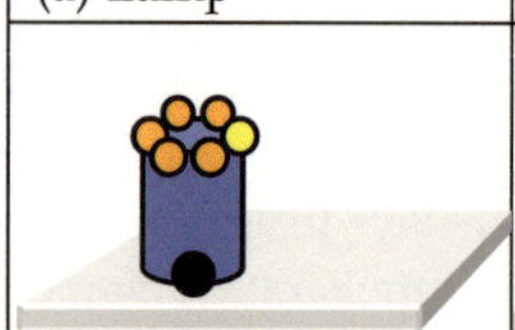			
Several lamps are set on the top. Direction of movement is by turning the lamp on. Speed of movement is by blinking speed or color of lamp.	Blowout put on a turntable is set on the top. Speed of movement is by length of blowout. Direction of movement is by tip direction of blowout.	Light ray draws scheduled route on running surface from present to some future time using pan-tilt mechanism. Situation on the way can be indicated.	Not only scheduled route but also state of operation, such as stop or going backward, can be displayed in the projected frame.

Table 1. Proposed approaches

Main content to announce and display in the second type of proposed approaches is the scheduled route on running surface from the present to some future time.

Light ray method is intended to draw only the scheduled route (Table 1 (c)). The scheduled route expressed with the afterimage of radiant irradiated on running surface was monochrome (red or green) on the developed robot PMR-1, which embodied the light ray method, since the a single colour laser pointer was used as a light source (Matsumaru et al., 2006a, 2006b). Outward appearance and main specifications of PMR-1 are shown in Fig. 2 and Table 2. Period to display the scheduled route is predefined, such as until 3-second-later. Drawn route expresses the direction of movement directly. Strong point of light ray method is the "situation on the way" can be indicated, such as whether going the shortest route directly or going via somewhere to bypass something when moving to some place. However information about time is insufficient in the route drawn in a single colour. The speed of movement can be expressed with the length of drawn route, since we decide beforehand until how many seconds after the present the scheduled route is displayed. However it is only the average speed in that period. If we decide the route is drawn until 3-second-later, it is recognized whether fast or slow as the average speed during three seconds by the length of drawn route, then a change of speed in the period is not expressed. For example, although it will be slow from the present to 1-second-later, it will become fast from 1-second-later to 2-seconds-later and it will come back to be slow after 2-seconds-later again. Moreover the problem is it is difficult for people to understand the rule at a glance since only the afterimages of radiant is the source of information. Even if people understand the afterimage of radiant expresses the scheduled route, careful observation for a while and matching between expressions and operations are required to understand both the route of movement during some predefined period and the relation between the length of drawn route and the exact amount of robot's speed.

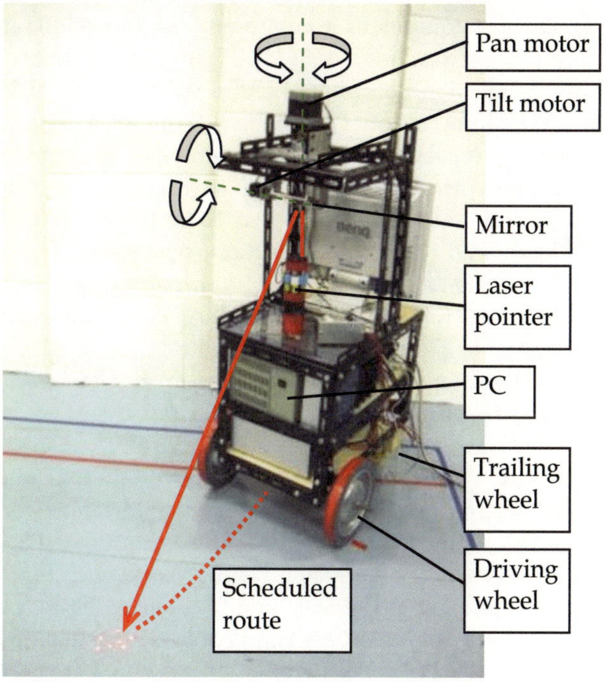

Figure 2. Outward appearance of PMR-1R

Projection method is developed from the light ray method (Table 1(d)). A projection equipment projects a two-dimensional frame on a running surface to indicate both the scheduled route and the states of operation such as stop and going backward. Internal condition of the robot such as remaining battery charge and warning on worn parts or overheating apparatus can also be displayed in the frame. If the projected frame is sufficiently large, the projector can be fixed to the robot and any moving portion for preliminary-announcement and indication function becomes unnecessary.
Toyota Motor Corp. announced "road surface depiction (laser tactile sense)" in the autumn of 2004 similar in irradiation or projection on a running surface (Toyota, 2004). However this mainly targets collision avoidance among vehicles, not intending upcoming movement of the vehicle for pedestrians.

Item	Specification
Mobile mechanism	two-wheel drive
Max. mobility	36 mm/s, 41.4 deg/s
Size	D460-W480-H910 mm
Weight	30.0 kg

Table 2. Main specifications of PMR-1R

2.2 Conventional Projection Interface

Human interface (HI) means the coupling part between people and a machine or a computer, and it indicates the scheme or structure to make artificial organisms user-friendly in a narrow meaning. Display equipment is commonly used as a human interface to show visual information through a screen, including CRT (cathode-ray tube), LCD (liquid crystal display), and projector (projection equipment). As a multifunctional interface containing some display unit for visual information, Digital Desk (Wellner, 1991) changed the previous indirect relationship between a user and a computer where a screen and a mouse device are used. On Digital Desk, three functions, visual display function (desktop is projected on a very common desk), control function (user's own fingertip is used to direct operations instead of mouse and icon), and registration function (documents and objects like doll are taken in, for example, using a camera and information is processed on desktop electronically) were considered. Researches on these three functions include Enhanced Desk (Koike et al., 2001), Sensetable (Patten et al., 2001), Illuminating Clay (Piper et al, 2002), SmartSkin (Rekimoto, 2002), Lumisight Table (Kakehi et al., 2005), Attentive Workbench (Nikaido et al., 2005), and so on. There are also some researches of projection on a wall instead of a desktop (Matsushita, 1997 and Nakanishi, et al. 2002). Those are mainly concerning human-machine interaction on visual information. Consequently reaction function (reaction force against operation is presented to the user) is studied as fourth function of Digital Desk. Researches include, for example, PSyBench (Physically Synchronized Bench) (Brave et al., 1998), Visual Haptic Workbench (Brederson et al., 2000), Actuated Workbench (Pangaro et al., 2002), and Proactive Desk (Noma et al., 2004). However those interactions are bounded between people and virtual environment even via physical force sense.
As an interface for tasks to actual objects, projector is commonly used to present information from remote supporter to on-site worker in order to support and prescribe from remote site

(Hiura et al., 2003; Machino et al., 2005). RFIG (Radio Frequency Identification and Geometry) lamps are interesting as an interaction of information in real workspace (Raskar et al., 2003). Using an active RFID tag with photo electronic sensor and a handheld projector with camera, both the ID and position of a tag within a limited area are recognized by projecting grey code pattern light from the projector, then the related information is projected on the object.

Some research uses projector as an interface between people and robot, for example, in order to teach operations to a robot (Terashima & Sakane, 1999; Sato & Sakane, 2000; Yamashita & Sakane, 2001). Those are mainly for off-line teaching, and the operation and direction to working robot are seldom taken into consideration. About the actually working robot, the information sharing function on manipulator (Wakita et al., 2001) is examined, as a method of showing people around the operational information of the robot. However this is only for ground-fixed manipulator.

3. Projection Robot, PMR-5 and PMR-5R

Projection robots PMR-5 (Matsumaru, 2006; Matsumaru, et al., 2007) and PMR-5R were developed to make the projection method embodiment. PMR-5 and PMR-5R have common system configuration which consists of mobile mechanism and announcement apparatus as shown in Fig. 3. Diameter of wheels in PMR-5 is 50 mm while that in PMR-5R is 100 mm. Liquid-crystal projector is used for the announcement apparatus. PC for control is equipped on the robot and only power source (AC 100 V) is supplied from outside.

The liquid-crystal projector is set up upward at the front of robot, and the frame reflected on the mirror just above the projector is projected on running surface. This structure makes total height of robot, about 1.0 m, lower while the shortest projection distance, 1.2 m, is secured. Size of the projected frame on running surface is 36-inch type (550 mm length by 740 mm width) with the mirror (140 mm length by 190 mm width) settled at 1.0 m heights.

Outward appearance and main specifications of PMR-5 are shown in Fig. 4 and Table 3, while those of PMR-5R are in Fig. 5 and Table 4.

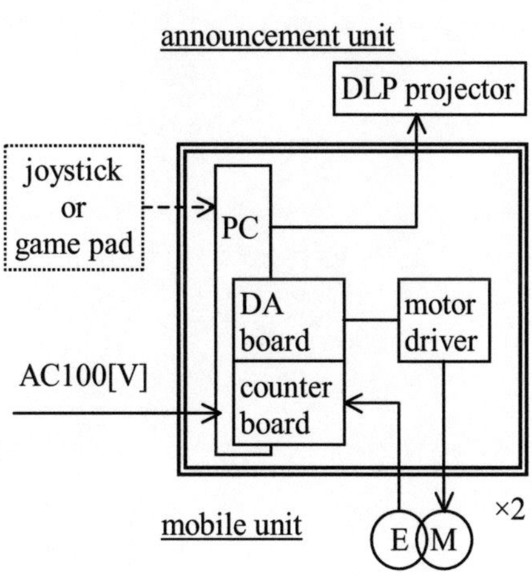

Figure 3. System configuration

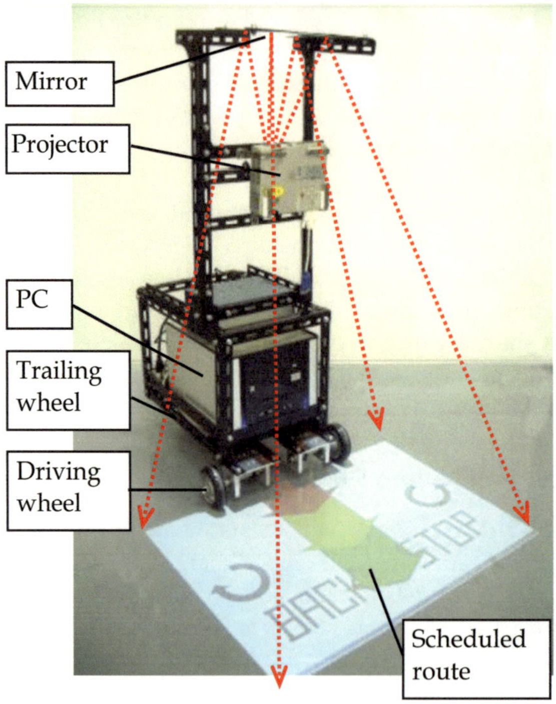

Figure 4. Outward appearance of PMR-5

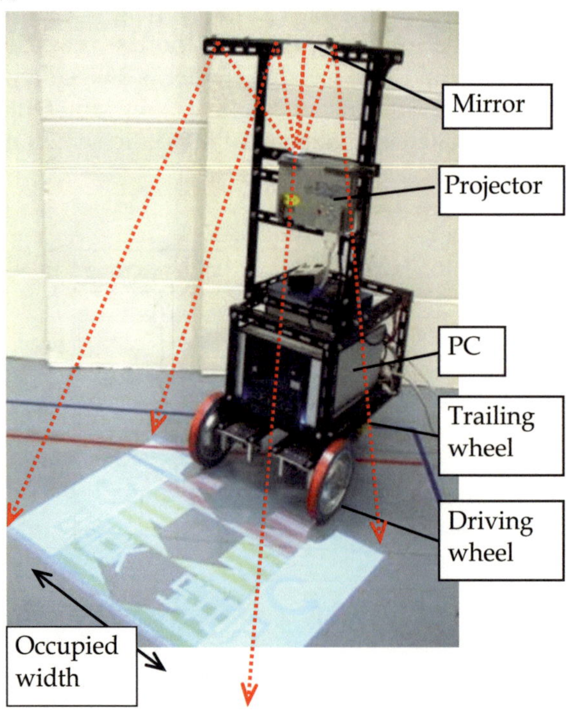

Figure 5. Outward appearance of PMR-5R

Item	Specification
Mobile mechanism	two-wheel drive
Max mobility	18 mm/s, 20.7 deg/s
Size	D500-W440-H1000 mm
Weight	25.0 kg

Table 3. Main specifications of PMR-5

Item	Specification
Mobile mechanism	two-wheel drive
Max mobility	36 cm/s, 41.4 deg/s
Size	D500-W440-H1000 mm
Weight	25.0 kg

Table 4. Main specifications of PMR-5R

4. Contents to Project in PMR-5

Main content of the projected frame is the "scheduled route", as in the light ray robot PMR-1. Scheduled route until 3-second-later can be displayed in the frame by the mounted projector and the settled mirror, considering the size of the projected frame and the mobility performance of the robot. Projected route is shown as chain of three arrows in different colours showing the time information of the route - red arrow from the present to 1-second-later, yellow arrow from 1-second-later to 2-second-later, and green arrow from 2-second-later to 3-second-later - considering that approaching robot increases risk gradually. The width of drawn arrows is also adjusted with the length depending on the speed of movement thinking that robot movement at high speed raises danger. Consequently the arrow can be curved freely and not only the length but also the width is adjusted according to the speed of movement of the robot (Fig. 5(b)). "State of operation" can also be seen in the frame (Fig. 5(c)). For on-the-spot rotation, corresponding sign is displayed at right or left in the frame depending on the direction of movement. Characters for stop or going backward are also displayed in the frame. Dark background is prepared for bright running surface to keep visibility in addition to normal white background (Fig. 6).

Straight-fast Straight-slow

(a) Speed of movement: length/width of chained arrows

Turn On-the-spot rotation

(b) Direction: curved condition of arrows or sigh

Going backward Stop

(c) State of operation: characters

Figure 5. Exact descriptions on PMR-5

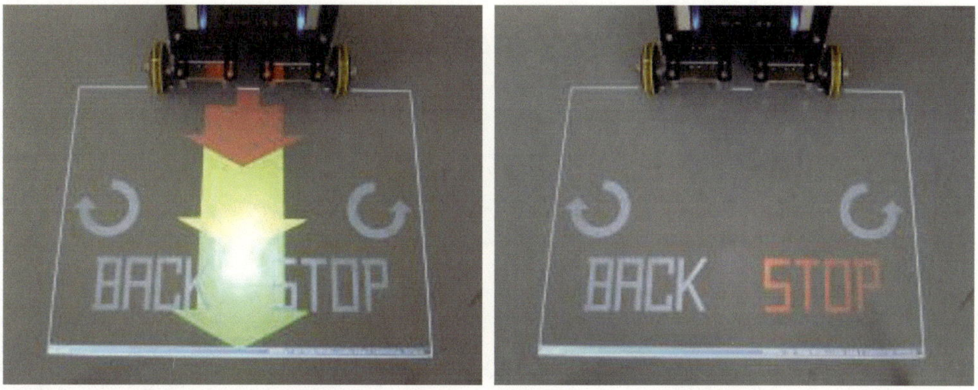

(a) Going forward (b) Stop

Figure 6. Projected frame with dark background

State of operation to display is roughly divided into four - going forward, on-the-spot rotation, going backward, and stop.

(1) Distinction of state of operation:

First, the state of operation to display is distinguished from the speed instruction, rotation factor J_x and translation factor J_x, which is taken in from some input device every 50 ms and accumulated in the instruction buffer during the latest three seconds.

- There is an instruction in which translation factor is negative. → Going forward (because the used joystick is a flight simulator type)
- There is an instruction in which translation factor is zero but rotation factor is not zero. → On-the-spot rotation
- There is an instruction in which translation factor value is positive. → Going backward
- There is an instruction in which both translation factor and rotation factor are zero at the same time. → Stop

State of operation displayed in the frame is independent each other even if two or more states are distinguished simultaneously. Accordingly multiple states may be displayed in the frame at the same time, for example, when it does stop after going forward.

(2) Contents and method to display:

Secondly, contents to display are refreshed in every 50 ms - same as command input cycle. Only one projector is equipped on the developed robot so that it can project only one frame of a fixed size to predefined position in front of the robot.

a) Going forward: Going forward is expressed with the chain of three arrows in which the scheduled route is expressed as the long axis of arrows in different colors showing the time information of the route, although in the light ray robot PMR-1 the scheduled route drawn by laser pointer is a simple line in a single color (movement afterimage of red radiant) and the length of drawn route only shows the average speed during three seconds.

Sixty polygons are prepared from sixty speed instructions acquired every 50 ms during the latest three seconds. Those are connected perpendicularly and form three arrows after transformed so as to look like the correct form. One arrow consists of twenty polygons - one triangular polygon Po_a for the tip of arrowhead, seven quadrangle polygon Po_b for the arrowhead except tip, and twelve quadrangle polygon Po_c for the rod of the arrow (Fig. 7).

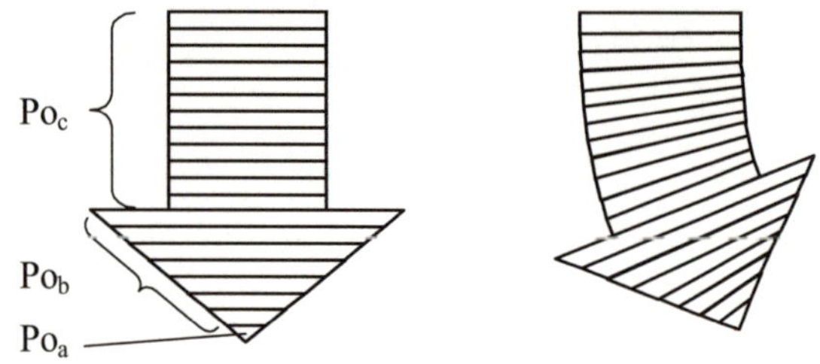

(a) Straight (b) Turn

Figure 7. Each arrow consists of twenty polygons

Three arrows are formed connecting sixty polygons from base to tip using instructions from the oldest to the latest one (Fig. 8(a)). When a new speed instruction is acquired, the polygon A shaped using the newly acquired instruction is arranged at top of green arrow, while sixty polygons are shifted to the base of arrows (toward the robot) (Fig. 8(b)). The position of polygon A (top of green arrow) shows the position where the robot will reach three seconds later. When the next speed instruction is acquired, the polygon A is shifted once to the base of arrows (toward the robot) and the polygon B which is shaped using the new instruction is arranged at the top of green arrow (Fig. 8(c)). Thus the polygon A is shifted to the base of arrows (toward the robot) one by one as time goes on. It reaches the base of arrows three seconds later, then the corresponding speed instruction is performed and the robot moves in that manner (Fig. 8(d)).

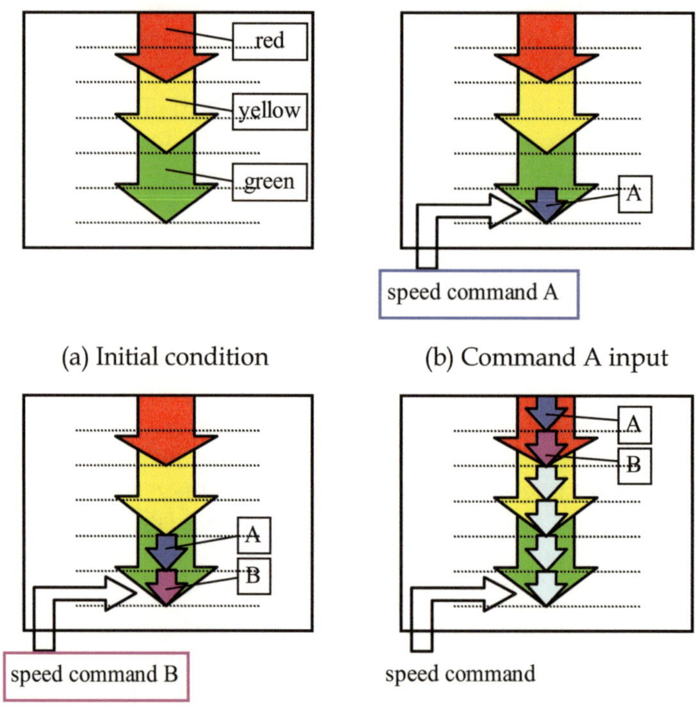

(a) Initial condition (b) Command A input

(c) Next command B (d) Three seconds later

Figure 8. Flow of polygons made from input command

The rate of the length of polygon along the direction of route against the robot's speed of movement was determined by trial and error so that the arrows might agree with the actual robot movement. Moreover the rate of the width of polygon against the amount of speed instruction was also determined by trial and error considering the visual quality of the arrows. In any case all polygons are connected perpendicularly after calculating the inclination angles and carrying out the coordinate transformation using the speed instruction, rotation factor J_x and translation factor J_y. The form of each polygon is adjusted so that the scheduled route may be expressed smoothly even though any speed instruction, not only going straight but also arbitrary rotation, is given.

Specifically the following processing is performed (Fig. 9). First, the speed of movement of wheels on both sides V_{1n}, V_{2n} mm/s is obtained from each speed instruction value J_{xn}, J_{yn} of sixty pieces for three seconds (n =1~60). The difference of movement distance S_n mm between right and left wheels during 50 ms and the rotation angle rot_n rad are computed, where Lw mm is the distance between wheels.

$$S_n = (V_{1n} - V_{2n}) \times 0.05 \tag{1}$$

$$rot_n = \tan^{-1}\left(S_n / Lw \right) \tag{2}$$

Next, the rotation angle $sumrot_n$ rad of the n-th polygon and the displacement (x_n, y_n) by the speed instruction are calculated in the coordinates on the projected frame, where the variable a is the coefficient determined by trial and error in order to make the actual movement and the projected frame in agreement.

$$sumrot_n = \sum_{n=1}^{n} rot_n \tag{3}$$

$$x_n = a \times (-J_{yn}) \times \sin(sumrot_n) \tag{4}$$

$$y_n = a \times (-J_{yn}) \times \cos(sumrot_n) \tag{5}$$

Then the central point (X_n, Y_n) of the n-th polygon from the original position (X_0, Y_0), which is shown on the bottom line in the figure, is computed on the frame.

$$X_n = X_{n-1} + x_n = X_0 + \sum_{n=1}^{n} x_n \tag{6}$$

$$Y_n = Y_{n-1} + y_n = Y_0 + \sum_{n=1}^{n} y_n \tag{7}$$

The position of both ends of the bottom line of the n-th polygon are calculated just with computing the position where it is apart from the central point only by the width d_n of arrow which is decided from the position of the polygon in the arrow (whether tip, arrowhead or rod) and the amount of speed instruction. By repeating these procedures the scheduled route can be displayed as three arrows correctly.

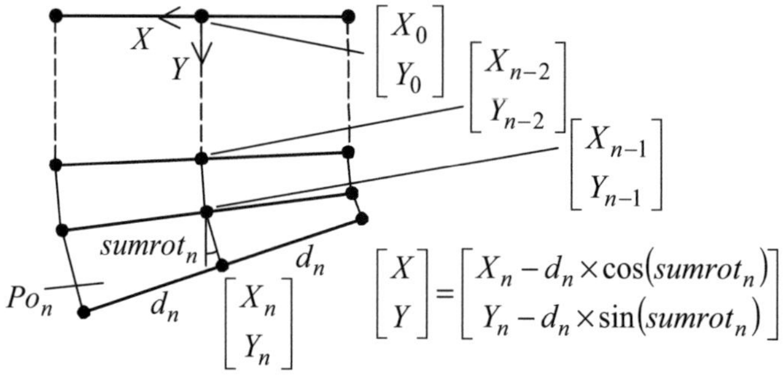

$$\begin{bmatrix} X_0 \\ Y_0 \end{bmatrix} \quad \begin{bmatrix} X_{n-2} \\ Y_{n-2} \end{bmatrix} \quad \begin{bmatrix} X_{n-1} \\ Y_{n-1} \end{bmatrix} \quad \begin{bmatrix} X_n \\ Y_n \end{bmatrix}$$

$$\begin{bmatrix} X \\ Y \end{bmatrix} = \begin{bmatrix} X_n - d_n \times \cos(sumrot_n) \\ Y_n - d_n \times \sin(sumrot_n) \end{bmatrix}$$

$$\begin{bmatrix} X \\ Y \end{bmatrix} = \begin{bmatrix} X_n + d_n \times \cos(sumrot_n) \\ Y_n + d_n \times \sin(sumrot_n) \end{bmatrix}$$

Figure 9. Position of quadrangular polygon in the frame

b) On-the-spot rotation/Going backward/Stop: In case of on-the-spot rotation, going backward, and stop, the treatment is the same in displaying the state of operation. The corresponding sign for on-the-spot rotation and the characters of BACK or STOP for going backward and stop are displayed in the frame respectively. Those are always displayed in grey, while displayed in yellow if the operation is performed within three seconds and the color turns in red when the operation is executed at that time.

5. Contents to Project in PMR-5R

Mobility performance of PMR-5R is doubled from PMR-5 with replacing the driving wheels to those of double diameters leaving the same driving mechanisms. Therefore the projected frame with the same size as on PMR-5 can inform the upcoming operation until 1.5-second-later. In PMR-5R the upcoming occupied width and area during robot's travelling are displayed on the frame by drawing a belt with the same width as the robot. People will never come into contact nor collision with robot and safety will be secured only if they do not just step on the belt. The belt is colour-coded based on the traffic signal – red part from the present to 0.5-second-later, yellow part from 0.5-second-later to 1.0-second-later, and green part from 1.0-second-later to 1.5-second-later – like arrows in PMR-5 (Fig. 10(a)). The belt can be curved freely and the length of belt is adjusted depending on the speed of movement (Fig. 10(b)). The belt has semitransparent striped pattern which moves towards robot synchronizing with robot movement. Accordingly it looks like the belt is fixed on the floor and the robot moves rewinding the belt. That makes easy to understand the meaning of the projected belt for people around. Moreover we decide to display the grey arrows along the center of belt in order to make easier for people to understand the scheduled route. The arrows are also curved freely and their width and length are adjusted depending on the speed of movement as in PMR-5. State of operation can also be seen in the frame (Fig. 10(c)). For on-the-spot rotation, corresponding sign is displayed at right or left in the frame depending on the direction of movement. Characters for stop or going backward are also displayed in the frame.

Straight-fast Straight-slow

(a) Speed of movement: length of belt and length/width of chained arrows

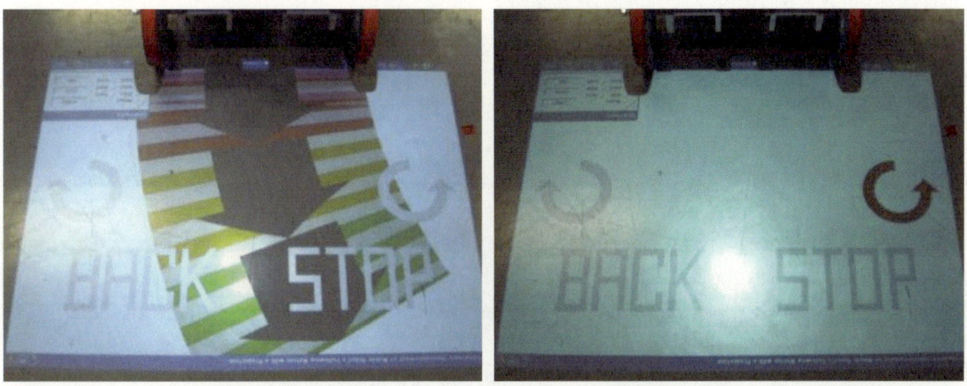

Turn On-the-spot rotation

(b) Direction: curved condition of belt/arrows or sigh

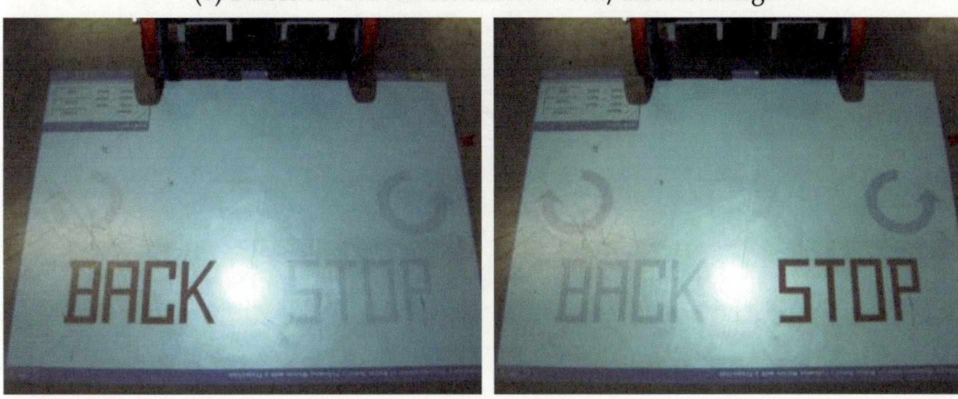

Going backward Stop

(c) State of operation: characters

Figure 10. Exact description on PMR-5R

Sixty polygons are prepared from thirty speed instructions acquired every 50 ms during the latest 1.5 seconds. That is, every speed instruction is dual partitioned from which two polygons are made to let the drawn belt look smooth (Fig. 11). Processing procedures to draw the belt and arrows are the same as in PMR-5.

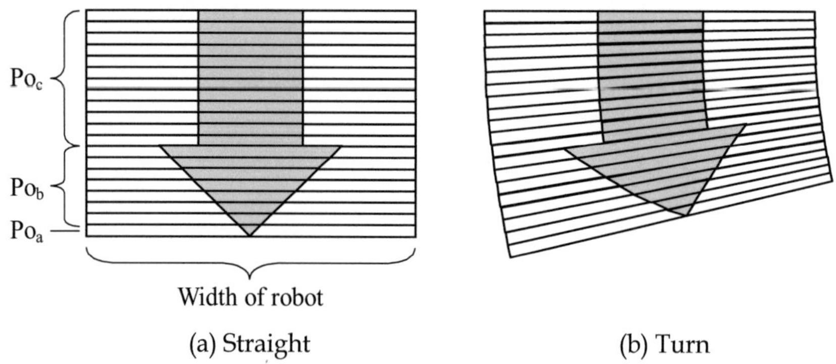

Figure 11. Each part of belt consists of twenty polygons

Three kinds of striped pattern on color coding as bright tone and dark tone of polygons are prepared. When a new speed instruction is acquired, two polygons *A* shaped using the speed instruction are arranged at tip end of the belt, while sixty polygons are sifted toward the robot (Fig. 12(a)). The corresponding polygons to form the part of arrows are also shaped as in PMR-5 at the same time, which are not shown in the figure. When the next speed instruction is acquired, the previous polygons are shifted to the robot keeping their colour tone. And new two polygons *B* shaped using the new instruction with decided colour tone are arranged at tip end of the belt (Fig. 12(b)). Two polygons *A* reach the bottom end of belt 1.5 seconds later, then the corresponding instruction is performed and the robot moves in that manner.

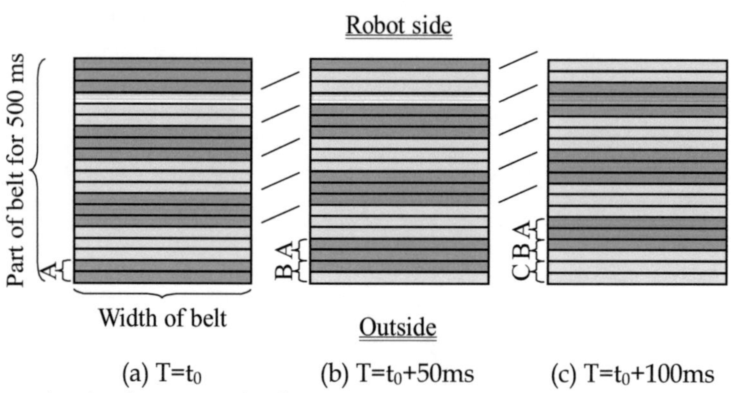

Figure 12. Flow of striped pattern of polygons

6. Discussion

In the questionnaire evaluation (Matsumaru, 2007b), PMR-5 has received the highest evaluation score comparing with other three prototype robots, eyeball robot PMR-2 (Matsumaru et al, 2003a, 2003b; Matsumaru et al, 2005), arrow robot PMR-6 (Matsumaru et al, 2001; Matsumaru, 2007a), and light ray robot PMR-1. In the experimental examination in simulated interactive situation between people and robot, the passing each other and the positional prediction, we have clarified that the method of indicating operations from the present to some future time continuously is effective when a person wants to avoid contact or collision surely and correctively owing to the feature that complicated information can be accurately transmitted of PMR-5R (Miyata et al., 2007).

Here the knowledge and consideration obtained through the research project are summarised as follows.

(1) Artificial mechatronic system which moves its body physically

This research is aimed at artificial mechatronic systems which coexists with people like a human-friendly robot. We are willing to derive some benefit from such robot moving its body physically. The research is inspired from awareness of the issues that people may feel threat and receive sense of incongruity from existing human-friendly robot because it is difficult to understand the function and performance and the upcoming operation and intention of the robot only from its outward appearance. Accordingly there is a risk of contact or collide between people and robot. There should be two approaches to solve the problem. First approach is to design the structure and movement of robot so that it is easy to understand and predict its function and upcoming operation from its appearance. Second approach is to superimpose a new function to preliminary-announce and indicate its upcoming operation. This research takes the latter. The function to preliminary-announce and indicate the upcoming operation of robot aims at providing it to those who are willing to react actively due to the information. Consequently the robot assumed here is supposed to have some safety function to detect obstacles and people around and avoid them autonomously. We do not assume that the robot comes into contact with those who have neither awareness nor interest on the robot movement. What to emphasize here is that it is insufficient even if a robot does not contact with people. It is also important that robots never pose a threat to people and people can live together with robot comfortably without anxiety.

(2) Sensory perception to be used

Each of four proposed methods complains to people's visual sense. If sound or voice is used to transmit information through auditory sense, information can be transmitted to everyone around, including those who do not want or desire it. Moreover sound or voice disturbs public peace. Accordingly only in the case that has a substantial need and requires emergency to transmit information to everyone around for safety reason, a robot might utter sound or voice, like heavy vehicles when turning at crossing. Since we decided to use visual sense here, we considered the method should be intelligible in which people can understand the shown contents immediately even at a glance, instead of taking a long looking. For that purpose it is necessary to take into consideration general common sense and customs, popular prejudice, basic knowledge stored through ordinary experiences, etc.

(3) Kind of information to transmit

Direction indicator or winker lamp of passenger cars has a serious problem by expressing several meanings only with a single sign of blinking – turn (more various when including curvature factor and selection of corner to turn), lane changing, pulling over to the kerb (following up with parking), etc. It is difficult to setup the relation between sign and meaning. Especially when we want to transmit many kinds of information, namely various robot operations, the kind of sign becomes varied. Therefore it is difficult to memorize the meaning corresponding to each sign if the difference among signs is too large. On the contrary it is easy to confuse with each other when various signs are very similar.

(4) Announcing state just after the present

Eyeball robot PMR-2 admits some interpretation between robot operation and sign for announcement, while arrow robot PMR-6 has direct connection between them. It is comparatively intelligible that the direction of movement is shown as eye positioning in the display, because the gaze is one of the keys to guess other's intention among people. Arrow is intelligible for everyone even at first sight, since it is commonly used in daily life and usually employed to express the direction. However it is difficult to show the speed of movement quantitatively by the degree of eye opening or the size of arrow. Especially on the speed of movement these two expressions might be used effectively after looking at the sign of moving robot and storing their knowledge. Accordingly these expressions will be effective in the situation and scene in which the information about the direction of movement has priority rather than that about the speed of movement. In the questionnaire evaluation, not a few people admired the eyeball expression as social and friendly due to personification and also the arrow expression as intelligible owing to straightforwardness and lucidness. Therefore those expressions will become candidates to examine for usage with greater importance to sociableness or lucidity.

(5) Indicating operations continuously

Light ray robot PMR-1 displays the scheduled route in a single colour as line expression, while projection robot PMR-5 indicates the occupied area in multiple colours as field expression. Line expression is effective in contact avoidance even for small-size robots. Field representation is valid for wide-width organisms compared with human body like automobiles. When it is difficult to project a large frame, for example, by restriction of projection equipment, drawing borderlines on both sides can be considered to indicate occupied area in which the organism is going to pass through. These two methods of drawing lines and projecting frame by light have a big advantage that information can be indicated anywhere without special maintaining. However it is supposed that the surface to draw or project is plane with little unevenness and the environment should be under the right condition in which the reflected light is clearly visible. Accordingly those expressions will be useful in mobile robot which moves indoor on a floor at first.

(6) Procedure to design and apply as interface

We have proposed and examined four methods, but we don't have to determine the best one. Each method has special and clear features respectively - indicating states of operation (subjective interpretation and clear interpretation) and displaying operations continuously (line expression and field representation). And this research has clarified their advantages and disadvantages. We think the method suitable for each application will be selected and applied referencing these clarified features. First it is necessary to declare the scene and situation where the preliminary-announcement and indication function is used, the kind and its number of required information, the contents of prior information, etc. in the intended application. Secondly candidates are chosen from four methods, then the developed method and the improved equipment are examined. The environment where the method using display or projection equipment might be effective will be restricted at first, such as indoor applications with flooring where sunlight will be never gotten directly.

7. Conclusion

This chapter presented the projection robot PMR-5 and the revised version PMR-5R in which projector projects a two-dimensional frame on a running surface which indicate the upcoming robot operation.

Robots with preliminary-announcement function of its upcoming operation are not expected to pose problems in task accomplishment when moving along predefined routes or even if the route is decided on site, like robot cleaners, because the route is announced beforehand and the robot actually moves that way for some time. In manual operation in real time, however, as with automobiles, it will cause problems. The maneuverability in operator will become worse because the time-delay will be always inserted between the instruction and the execution. In that case we have to devise the method that the robot executes the instruction from operator in real time forecasting operator's next operation from operational record and environmental condition (Akamatsu, 2003) and indicating the forecasted operation to surroundings. In addition, future work includes the study on dealing with the case that operator behaves contradictory to the forecasted operation and the case that robot does some autonomous operation suddenly to secure safety like stopping.

We will continue to study to improve the affinity between human beings and mechatronic systems to establish a standard for mutual coexistence.

8. References

Akamatsu, M. (2003): Establishing Driving Behavior Database and its Application to Active Safety Technologies, *J. of Society of Automotive Engineers of Japan*, Vol.57, No.12, (2003), pp.34-39, ISSN: 0385-7298.

Brave, S.; Ishii, H. & Dahley, A. (1998): Tangible Interfaces for Remote Collaboration and Communication, *Proceedings of the 1998 ACM conference on Computer supported cooperative work* (CSCW '98), pp.169-178, ISBN:1-58113-009-0, [Seattle, Washington, USA], (1998), ACM, New York, NY, USA.

Brederson, J.D.; Ikits, M.; Johnson, C.R. & Hansen, C.D. (2000). The Visual Haptic Workbench, *Proceedings of Fifth PHANToM Users Group Workshop* (PUG 2000), pp.46-49, ISBN:non, [Aspen, Colorado, USA], (2000), (http://www.sensable.com/openhaptics-projects-papers.htm).

Kakehi, Y.; Iida, M.; Naemura, T.; Shirai, Y.; Matsushita, M. & Ohguro, T. (2005). Lumisight Table: an interactive view-dependent tabletop display, *IEEE Computer Graphics and Applications*, Vol.25, No.1, (2005), pp.48-53, ISSN:0272-1716.

Koike, H.; Sato, Y. & Kobayashi, Y. (2001). Integrating Paper and Digital Information on EnhancedDesk: A Method for Realtime Finger Tracking on an Augmented Desk System, *ACM Transactions on Computer-Human Interaction* (TOCHI), Vol.8, No.4, (2001), pp.307-322, ISSN:1073-0516.

Hiura, S.; Tojo, K. & Inokuchi, S. (2003). 3-D Tele-direction Interface using Video Projector, *The 30th International Conference on Computer Graphics and Interactive Techniques* (ACM SIGGRAPH 2003) Sketches & Applications, ISBN:1-58113-709-5, [San Diego, California, USA], (2003), ACM, New York, NY, USA.

Machino, T.; Nanjo, Y.; Yanagihara, Y.; Kawata, H.; Iwaki, S. & Shimokura, K. (2005). Robot-augmented communication: a remote-collaboration system based on a shared field of view in real space, *2005 IEEE/RSJ International Conference on Intelligent Robots and Systems* (IROS 2005), pp.2203-2209, ISBN:0-7803-8912-3, [Edmonton, Alberta, Canada], (2005), IEEE, Piscataway, N.J., USA.

Matsumaru, T. & Terasawa, Y. (2001). Preliminary Announcement and Display for Human-Friendly Mobile Robot, *Preprints of IFAC Workshop on Mobile Robot Technology*, pp.226-231, ISBN:non, [Jejudo, Korea], (2001), IFAC, Laxenburg, Austria.

Matsumaru, T. & Hagiwara, K. (2001a). Method and Effect of Preliminary-Announcement and Display for Translation of Mobile Robot, *Proceedings of the 10th International Conference on Advanced Robotics* (ICAR 2001), pp.573-578, ISBN:9637154051, [Budapest, Hungary], (2001), IEEE Hungary Section.

Matsumaru, T. & Hagiwara, K. (2001b). Preliminary-Announcement and Display for Translation and Rotation of Human-Friendly Mobile Robot, *Proceedings of 10th IEEE International Workshop on Robot and Human Communication* (ROMAN 2001), pp.213-218, ISBN:0-7803-7222-0, [Bordeaux and Paris, France], (2001), IEEE, Piscataway, N.J., USA.

Matsumaru, T.; Iwase, K.; Kusada, T.; Akiyama, K.; Gomi, H. & Ito, T. (2003a). Preliminary-Announcement Function of Mobile Robot's Following Motion by using Omni-directional Display, *Proceeding of The 11th International Conference on Advanced Robotics* (ICAR 2003), Vol.2, pp.650-657, ISBN:972-96889-8-2, [Coimbra, Portugal], (2003), IEEE, Piscataway, N.J., USA.

Matsumaru, T.; Akiyama, K.; Iwase, K.; Kusada, T.; Gomi, H. & Ito, T. (2003b). Eyeball Expression for Preliminary-Announcement of Mobile Robot's Following Motion, *Proceedings of The 11th International Conference on Advanced Robotics* (ICAR 2003), Vol.2, pp.797-803, ISBN:972-96889-8-2, [Coimbra, Portugal], (2003), IEEE, Piscataway, N.J., USA.

Matsumaru, T.; Kudo, S.; Kusada, T.; Iwase, K.; Akiyama, K. & Ito T. (2003c). Simulation on Preliminary-Announcement and Display of Mobile Robot's Following Action by Lamp, Party-blowouts, or Beam-light, *Proceedings of 2003 IEEE/ASME International Conference on Advanced Intelligent Mechatronics* (AIM 2003), Vol.2, pp.771-777, ISBN:0-7803-7759-1, [Kobe, Japan], (2003), IEEE, Piscataway, N.J., USA.

Matsumaru, T.; Endo, H. & Ito T. (2003d). Examination by Software Simulation on Preliminary-Announcement and Display of Mobile Robot's Following Action by Lamp or Blowouts, *Proceedings of 2003 IEEE International Conference on Robotics and Automation* (2003 IEEE ICRA), Vol.1, pp.362-367, ISBN:0-7803-7736-2 (ISSN:1050-4729), [Taipei, Taiwan], (2003), IEEE, Piscataway, N.J., USA.

Matsumaru, T. (2004). The Human-Machine-Information System and the Robotic Virtual System, *Journal of the Society of Instrument and Control Engineers*, Vol.43, No.2, (2004), pp.116-121, ISSN:0453-4662.

Matsumaru, T.; Kudo, S.; Endo, H. & Ito, T. (2004). Examination on a Software Simulation of the Method and Effect of Preliminary-announcement and Display of Human-friendly Robot's Following Action, *Transactions of the Society of Instrument and Control Engineers*, Vol.40, No.2, (2004), pp.189-198, ISSN:0453-4654.

Matsumaru, T.; Iwase, K.; Akiyama, K.; Kusada, T. & Ito, T. (2005). Mobile robot with eyeball expression as the preliminary-announcement and display of the robot's following motion, *Autonomous Robots*, Vol.18, No.2, (2005), pp.231-246, ISSN:0929-5593 (ISSN:1573-7527).

Matsumaru, T. (2006). Mobile Robot with Preliminary-announcement and Display Function of Following Motion using Projection Equipment, *The 15th IEEE International Symposium on Robot and Human Interactive Communication* (RO-MAN 06), pp.443-450, ISBN:1-4244-0565-3, [Hatfield, UK], (2006), IEEE, Piscataway, N.J., USA.

Matsumaru, T.; Kusada, T. & Iwase, K. (2006a). Mobile Robot with Preliminary-Announcement Function of Following Motion using Light-ray, *2006 IEEE/RSJ International Conference on Intelligent Robots and Systems* (IROS 2006), pp.1516-1523, ISBN:1-4244-0259-X, [Beijing, China], (2006), IEEE, Piscataway, N.J., USA.

Matsumaru, T.; Kusada, T. & Iwase, K. (2006b). Development of Mobile Robot with Preliminary-announcement and Display Function of Scheduled Course using Light-ray, *Journal of the Robotics Society of Japan*, Vol.24, No.8, (2006), pp.976-984, ISSN:0289-1824.

Matsumaru, T. (2007a). Mobile Robot with Preliminary-announcement and Indication Function of Forthcoming Operation using Flat-panel Display, *2007 IEEE International Conference on Robotics and Automation* (ICRA'07), pp.1774-1781, ISBN:1-4244-0601-3 (ISSN:1050-4729), [Rome, Italy], (2007), IEEE, Piscataway, N.J., USA.

Matsumaru, T. (2007b). Development of Four Kinds of Mobile Robot with Preliminary-announcement and Display Function of its Forthcoming Operation, *Journal of Robotics and Mechatronics*, Vol.19, No.2, (2007), pp.148-159, ISSN:0915-3934.

Matsumaru, T.; Hoshiba, Y.; Hiraiwa, S. & Miyata, Y. (2007). Development of Mobile Robot with Preliminary-announcement and Display Function of Forthcoming Motion using Projection Equipment, *Journal of the Robotics Society of Japan*, Vol.25, No.3, (2007), pp.410-421, ISSN:0289-1824.

Matsushita, N. & Rekimoto, J. (1997). HoloWall: Designing a Finger, Hand, Body, and Object Sensitive Wall, *Proceedings of the 10th annual ACM symposium on User interface software and technology* (UIST'97), pp.209-210, ISBN:0-89791-881-9, [Banff, Alberta, Canada], (1997), ACM, New York, NY, USA.

Miyata, Y.; Shinbayashi, S.; Suzuki, S.; Lin, M.; Akai, K. & Matsumaru, T. (2007). Preliminary-announcement of Robot's Operation (1st report) –Experimentation of for kinds of method-, *Proceedings of the 51st Annual Conference of the Institute od Systems, Control and Information Engineers* (ISCIE), pp.159-160, ISBN:non, [Kyoto, Japan], (2007), ISCIE, Kyoto, Japan.

Nakanishi, Y.; Sato, Y. & Koike, H. (2002). EnhancedDesk and EnhancedWall: Augmented Desk and Wall Interfaces with Real-Time Tracking of User's Motion, *Proceedings of UbiComp 2002 Workshop on Collaborations with Interactive Walls and Tables*, pp.27-30, ISBN:non, [Goteborg, Sweden], (2002), (http://www.ipsi.fraunhofer.de/ambiente/collabtablewallws/).

Nikaido, M.; Sugi, M.; Tamura, Y.; Ota, J.; Arai, T.; Kotani, K.; Takamasu, K.; Yamamoto, A.; Shin, S.; Suzuki, H. & Sato, Y. (2005). Arrangement Planning for Multiple Self-Moving Trays in Human Supporting Production Cell 'Attentive Workbench', *Proceedings of 2005 IEEE/RSJ International Conference on Intelligent Robots and Systems* (IROS 2005), pp.3880-3885, ISBN:0-7803-8912-3, [Edmonton, Alberta, Canada], (2005), IEEE, Piscataway, N.J., USA.

Noma, H.; Yoshida, S.; Yanagida, Y. & Tetsutani, N. (2004). The Proactive Desk: A New Haptic Display System for a Digital Desk Using a 2-DOF Linear Induction Motor, PRESENCE: *Teleoperators & Virtual Environments*, Vol.13, No.2, (2004), pp.146-163, ISSN:1054-7460.

Pangaro, G.; Maynes-Aminzade, D. & Ishii, H. (2002). The Actuated Workbench: Computer-Controlled Actuation in Tabletop Tangible Interfaces, *Proceedings of the 15th annual ACM symposium on User interface software and technology* (UIST 2002), pp.181-190, ISBN:1-58113-488-6, [Paris, France], (2002), ACM, New York, NY, USA

Patten, J.; Ishii, H.; Hines, J. & Pangaro G. (2001). Sensetable: A Wireless Object Tracking Platform for Tangible User Interfaces, *Proceedings of the SIGCHI conference on Human factors in computing systems* (CHI '01), pp.253-260, ISBN:1-58113-327-8, [Seattle, Washington, USA], (2001), ACM, New York, NY, USA.

Piper, B.; Ratti, C. & Ishii, H. (2002). Illuminating Clay: A 3-D Tangible Interface for Landscape Analysis, *Proceedings of the SIGCHI conference on Human factors in computing systems* (CHI '02), pp.355-362, ISBN:1-58113-453-3, [Minneapolis, Minnesota, USA], (2002), ACM, New York, NY, USA.

Raskar, R.; Beardsley, P.; Baar, J.v.; Wang, Y.; Dietz, P.; Lee, J.; Leigh, D. & Willwacher, T. (2003). RFIG lamps: interacting with a self-describing world via photosensing wireless tags and projectors, *ACM Trans. on Graphics* (TOG), Vol.23, No.3, (2003), pp.406-415, ISSN:0730-0301.

Rekimoto, J. (2002). SmartSkin: An Infrastructure for Freehand Manipulation on Interactive Surfaces, *Proceedings of the SIGCHI conference on Human factors in computing systems* (CHI '02), pp.113-120, ISBN:1-58113-453-3, [Minneapolis, Minnesota, USA], (2002), ACM, New York, NY, USA.

Sato, S. & Sakane, S. (2000). A human-robot interface using an interactive hand pointer that projects a mark in the real work space, *Proceedings of 2000 IEEE Int'l Conf. on Robotics and Automation* (ICRA'00), Vol.1, pp.589-595, ISBN:0-7803-5886-4, [San Francisco, California, USA], (2000), IEEE, Piscataway, N.J., USA.

Terashima, M. & Sakane, S. (1999). A Human-Robot Interface Using an Extended Digital Desk, *Proceedings of 1999 IEEE Int'l Conf. on Robotics and Automation* (ICRA'99), Vol.4, pp.2874-2880, ISBN:0-7803-5180-0, [Detroit, Michigan, USA], (1999), IEEE, Piscataway, N.J., USA.

Toyota Motor Corp. (2004). Company - News Release - Toyota, Hino, Daihatsu to Jointly Exhibit at 11th ITS World Congress, (2004). http://www.toyota.co.jp/en/news/04/0922_2.html

Wakita, Y.; Hirai, S.; Suehiro, T.; Hori, T. & Fujiwara, K. (2001). Information Sharing via Projection Function for Coexistence of Robot and Human, *Autonomous Robots*, Vol.10, No.3, (2001), pp.267-277, ISSN:0929-5593.

Wellner, P.D. (1991). The DigitalDesk Calculator: Tactile Manipulation on a Desk Top Display, *Proceedings of the 4th annual ACM symposium on User interface software and technology* (UIST '91), pp.27-33, ISBN:0-89791-451-1, [Hilton Head, South Carolina, USA], (1991), ACM, New York, NY, USA.

Yamashita, M. & Sakane, S. (2001). Adaptive Annotation Using a Human-Robot Interface System PARTNER, *Proceedings of 2001 IEEE Int'l Conf. on Robotics and Automation* (2001 ICRA), Vol.3, pp.2661-2667, ISBN:0-7803-5886-4, [Seoul, Korea], (2001), IEEE, Piscataway, N.J., USA.

19

Occupancy Grid Maps for Localization and Mapping

Adam Milstein
University of Waterloo
Canada

1. Introduction

In order to perform motion planning it is usually necessary to define some representation of the environment and have some method of determining the robot's location in that environment. Mapping is the problem of determining the representation, while localization is the problem of finding the robot's position. Many probabilistic techniques for localization depend on the map being defined as a finite sized set of landmarks which the robot's sensors observe, giving their relative displacement from the robot. However, physical sensors do not usually detect landmarks unambiguously. Instead, they report the distance to the nearest obstacle, or return an image of the environment. In order to use a landmark based algorithm, the sensor readings must be pre-processed in a separate step to convert the raw sensor data into a set of detected landmarks, such as in [Leonard and Durrant-Whyte 1991]. The additional step introduces more error into any algorithm, as well as discarding much of the sensor information which does not detect any landmark.
One of the primary drawbacks of landmark based maps is the data association problem. Because raw sensor data is not labelled with the correct landmark detected, the sensor processing must somehow determine exactly which landmark was observed. If mistakes are made the localization and mapping algorithms which depend on the sensor data will fail. In order to compensate for the data association problem, many localization and SLAM algorithms include a method for determining the associations between the sensor data and the landmarks, however these techniques add significantly to the complexity of the solutions. Also, they do not solve the problem of actually finding landmarks in the raw sensor readings. Some examples of these algorithms include GraphSLAM [Folkesson and Christensen 2004] and Sparse Extended Information Filters (SEIF) [Thrun et al. 2004], both of which can be implemented to handle data associations in a probabilistic way as described in [Thrun et al. 2005]. Even with these integrated solutions, the data association problem adds a significant amount of error.

2. Occupancy Grid Maps

One common technique for map representation that does not suffer from data associations is to use occupancy grid maps to approximate the environment. An occupancy grid map represents the environment as a block of cells, each one either occupied, so that the robot cannot pass through it, or unoccupied, so that the robot can traverse it. Unless your

environment is composed entirely of cubes, occupancy grid maps cannot be absolutely accurate, but by choosing a small enough cell size they can provide all the necessary data. Any sensor will report the status of a set of grid cells that can be checked without reference to the rest of the map. An early implementation of occupancy grid maps was used in [Moravec 1988] to automatically generate a map of the environment. Sensor readings are compared to the map, altering the probability that observed cells are occupied. For example, a sonar sensor returns the closest object within a cone, so the cells in the volume of the cone closer than the reading are probably unoccupied. Moravec represents each cell as a probability of being traversable and initializes them to an unknown value. He describes a probabilistic technique to update cells for various types of sensors and gives a technique to allow the map to be updated as the robot moves. Unfortunately, this technique is not actually localization and does not help the robot know its own position. The map is maintained relative to the robot, rather than in a global frame of reference. In other words, the robot is assumed to be at a fixed location, while the map moves around it. As the robot moves, the map is blurred according to the motion. The robot's sensors can correct the map in its immediate area, but unobserved portions of the map must blur into uselessness. There is also no way to discover the robot's location in reference to previously visited locations.
Although the technique is problematic as a localization algorithm, it provides a very powerful way to represent the environment. Using an occupancy grid map allows the raw sensor data to be used without trying to detect and identify landmarks. Also, since raw data is used, no information is discarded because it does not correspond to a landmark. The only problem is that there are a huge number of map features, one for each grid cell. Algorithms which consider the relation of the robot to a set of distinct landmarks cannot be applied when the number of features is so large. Thus, using occupancy grid maps limits the type of localization and mapping techniques that can be used.

2.1 Mapping Technique

To create an occupancy grid map it is necessary to determine the occupancy probability of each cell. In order to do this efficiently the assumption is often made that map cells are independent. Although this is not strictly accurate, especially when considering adjacent cells representing the same physical object, it greatly simplifies the mapping algorithm without significantly increasing the error. As a result, the probability of a particular map, m, can be factored into the product of the individual probabilities of its cells. The map depends on the robot's location history, $x^t = \{x_0, \ldots, x_t\}$, and the sensor readings at each location in the path, z^t

$$p(m \mid x^t, z^t) = \prod_n p(m_n \mid x^t, z^t) \quad (1)$$

The probability of a particular cell is easy to determine given the robot's position and sensor readings, since it is determined by whether the robot observes the cell as unoccupied or occupied. Since the probability is determined by the robot's entire history, all these sensor readings must be taken into account. The mapping algorithm usually builds the cell probabilities up iteratively, considering each $\{x_t, z_t\}$ from time $t = 0$ to the most recent reading. Although these readings could be considered in any order, the iterative processing makes the most sense, allowing additional readings to be added and leading eventually to simultaneous mapping and localization (SLAM) solutions such as described in section 4.

With occupancy grid maps, the mapping step must determine the probability of each cell, as represented by equation (1). Proceeding iteratively, the map cells are updated according to the position and sensor readings. Of course, it would require significant processing to update the entire map on each step, but this is unnecessary. Only the cells which are actually observed need to be updated. Each cell that is perceived by the sensor given the robot's position is updated depending on whether the sensor indicates it is occupied or unoccupied. Although the map is defined by the occupancy probability, for simplicity the actual values for each cell, $p(m_n \mid x^t, z^t)$, are calculated in log odds form.

$$p(m_n \mid x^t, z^t) = 1 - \frac{1}{1 + e^{l_{t,n}}} \tag{2}$$

$$l_{t,n} = l_{t-1,n} + \log \frac{p(m_n \mid x_t, z_t)}{1 - p(m_n \mid x_t, z_t)} - \log \frac{p(m_n)}{1 - p(m_n)} \tag{3}$$

$p(m_n)$ is a constant prior occupancy probability, so the only important part of equation (3) is $p(m_n \mid x_t, z_t)$, which is called the inverse sensor model. Although a highly accurate inverse sensor model is difficult to determine, a simplified implementation that returns a high value if the sensors report an object in the cell and a low value otherwise is often acceptable. Occupancy grid mapping updates a map according to a sensor reading at a location so that, as evidence accumulates, the map becomes correct. Of course, the success of the mapping algorithm depends on the location x_t being correct, just as the success of localization depends on an accurate map.

3. MCL

Monte Carlo Localization uses models of various sensors, together with a recursive Bayes filter, to generate the belief state of a robot's location. In fact, MCL is a specific instance of a POMDP (Partially Observable Markov Decision Process). A standard form of MCL uses a motion model to predict the robot's motion together with a sensor model to evaluate the probability of a sensor reading in a particular location. The sensor model necessarily includes a static map of the environment. The algorithm can be applied to virtually any robot with any sensor system, as long as these two models can be created. One common implementation where MCL is very successful is on a wheeled robot using a range sensor such as a laser rangefinder. A benefit of this combination is that the map and location used by the algorithm are in a human readable format. Although I give the general algorithm in the following sections, which should be applicable to other robots, where application specific details are required, I assume the type of robot as described.

Other localization algorithms than MCL exist, but they are currently much more limited than MCL and require specific environment features in order to be effective. Most other localization algorithms require that the map be composed of discrete landmarks and often they increase in runtime with the size of the map. Extended Kalman Filter (EKF) localization [Leonard and Durrant-Whyte 1991] is an alternative technique that has both of these problems, which are exacerbated when landmarks cannot be identified exactly. Even with various optimizations to improve execution, such as using the unscented transform (UKF localization) [Julier and Uhlmann 1997] or multi hypothesis tracking (MHT), the algorithm still applies primarily to feature based maps [Thrun et al. 2005]. Since a large, indoor environment is unlikely to have discrete, unambiguous features, these techniques are ineffective for the type of problem we are considering. In order to apply them it is usually necessary to preprocess the map as in [Leonard and Durrant-Whyte 1991] to create an artificial landmark based map. The map processing step introduces additional error and discards much of the information provided by the original map. Another serious problem is that these alternative techniques cannot handle multiple hypotheses of the robot's location. Each one maintains only a single Gaussian representation of position. MCL, in contrast, can maintain multiple separate locations. Markov localization using probabilities over a grid is also possible, however it increases in runtime with the size of the state space. Since a robot in a real, dynamic environment requires a real valued state space, true Markov grid based localization is usually impossible. Because of these drawbacks, MCL is currently one of the most effective localization techniques and the most commonly used, especially for real robots operating in real environments.

3.1 Recursive Bayes Filter

MCL is an implementation of a recursive Bayes filter. The posterior distribution of robot poses as conditioned by the sensor data is estimated as the robot's belief state. A key detail of the algorithm is the Markovian assumption that the past and future are conditionally independent given the present. For a robot, this means that if its current location is known, the future locations do not depend on where the robot has been. In virtually any environment this is the case, so making the assumption is reasonable in general.
To produce a recursive Bayes filter, we represent the belief state of the robot as the probability of the robot's location conditioned by the sensor data, where sensors include odometry (u_t).

$$Bel(x_t) = p(x_t \mid z_t, z_{t-1}\dots, z_0, u_t, u_{t-1},\dots, u_0) \tag{4}$$

x_t represents the robot's position at time t, z_t the robot's sensor readings at time t and u_t is the motion data at time t. To simplify the subsequent equations we use the notation $a^t = \{a_t, \dots, a_0\}$.

$$Bel(x_t) = p(x_t \mid z^t, u^t) \tag{5}$$

While this equation is a good representation of the problem, it is not much use since it cannot be calculated as is. By applying a series of probabilistic rules, together with the Markovian assumption, equation (5) is converted into a more usable form.

$$Bel(x_t) = \eta p(z_t \mid x_t) \int_{x_{t-1}} p(x_t \mid u_t, x_{t-1}) p(x_{t-1} \mid u^{t-1}, z^{t-1}) dx_{t-1} \tag{6}$$

Obviously, $p(x_{t-1} \mid z^{t-1}, u^{t-1})$ is $Bel(x_{t-1})$, giving us the recursive equation necessary for a recursive Bayes filter. η is a normalization constant that can be calculated by normalizing over the state space. $p(z_t \mid x_t)$ is the sensor model, representing the probability of receiving a particular sensor reading given a robot's location. Finally, $p(x_t \mid x_{t-1}, u_t)$ is the motion model. It is the probability that the robot arrives at location x_t given that it started at location x_{t-1} and performed action u_t. The sensor and motion model are representations of the physical components of the robot and must be determined experimentally for each robot and sensor device.

3.2 Particle Approximation

It would appear that, given the two models, equation (6) is all that is necessary to perform localization with MCL. Unfortunately, a problem occurs with the integral. The equation requires integrating over the entire state space. Although we can evaluate the models at any point in the space, there is no closed form to the integral. Further, even a simple robot moves in a continuous, 3 dimensional state space with an x and y location together with an angle of rotation. Calculating the integral over this space is impossible, especially for a real time algorithm. In order to solve this problem, we approximate the continuous space with a finite number of weighted samples.

$$Bel(x_t) \approx \left\{x_t^i, w_t^i\right\}_{i=1,\dots,N} \tag{7}$$

The integral over the space becomes a sum over the finite number of particles.

$$Bel(x_t) = \eta p(z_t \mid x_t) \sum_N p(x_t \mid u_t, x_{t-1}) p(x_{t-1} \mid u^{t-1}, z^{t-1}) \tag{8}$$

Of course, approximating the space results in a certain amount of error when low probability locations are not represented. If the robot is really at one of these locations it can never be localized. However, if the number of particles is well chosen MCL works properly in most situations.

3.3 Resampling

One problem with using a finite set of particles to represent an infinite space is that the weight of particles representing a low probability location will quickly decrease and is unlikely to ever increase again. Similarly, if there are too few particles representing a high probability location, they will disperse and eventually lose the robot's position. What is needed is a method for relocating low probability particles to high probability locations and recalculating their probability. The method used in MCL is resampling. After the particles are weighted by the sensor model they are resampled to represent the high probability locations. N particles are chosen randomly from the list of N weighted particles, with probability according to their weight. These particles are chosen with replacement, so that after a particle is chosen it remains in the original list and has the same probability of being sampled again. A high probability particle might be selected several times and so multiple copies might occur in the new list, while a low probability particle might never be chosen at all and its location would die out. The resampled list will thus have multiple particles in high probability locations and none in low probability ones. Another effect of resampling is to set all the sample weights to 1 / N. Instead of having individual weights representing the probability of a location, the number of particles indicates the probability. A high probability location will have many particles and thus, if the robot is present, it is likely to be tracked as it moves. Of course, low probability locations will die out and be unrepresented, so localization will fail if the robot is truly at one of these positions.

3.4 Bias

Representing an infinite space with a finite number of samples necessarily introduces some error. In order to accurately represent high probability locations, the particle filter discards lower probability regions as their low likelihood particles are not selected during the resampling process. Bias is the name given to the problem that MCL tends to consider only the highest probability locations, letting others be removed. The effect is that MCL is biased towards areas that have a large number of particles, tending to converge, over time, to a single cluster in the highest probability location. However, in most situations the high probability location contains the robot and so the convergence provides the correct result.

3.5 Algorithm

As the robot moves, it reports its odometry and sensor data to the MCL algorithm. After each reported move every particle is moved according to the random motion model, based on the motion actually reported. The particles are then updated with a weight determined by the sensor model for the particle's location. Finally, the particles are resampled by repeatedly choosing samples randomly, with replacement, from the current set, according to the weights assigned by the sensor model.

1:	Create a set $\{x_t^{[n]}, w_t^{[n]}\}$ from X_{t-1} by repeating N times:
1.1:	Choose a particle $x_{t-1}^{[n]}$ from X_{t-1}. Because of the resampling step this particle may be selected either iteratively or randomly.
1.2:	Next, draw a particle $x_t^{[n]} \sim p(x_t \mid u_t, x_{t-1}^{[n]})$. This particle is the result of a random motion according to the motion model.
1.3:	Set the weight of the particle using the sensor model: $w_t^{[n]} = p(z_t \mid x_t^{[n]})$.
2:	Resample randomly according to weight from $\{x_t^{[n]}, w_t^{[n]}\}$ into X_t, which causes the particle weights to become uniform.

Table 1. MCL Algorithm

The effect of resampling is to replace the weight of the individual particles with the number of particles at that location. On the robot's next move the particles at a high probability location will spread out as they are moved randomly according to the motion model, with at least one landing in the robot's new location. Then the resampling will cause more particles to appear at the correct location, while incorrect locations die out. Assuming that the models and map are accurate, MCL will correctly track the robot's changing location. Various parameters can be tuned manually to adjust the rate of convergence and the behaviour of the models. Once the belief over the robot's location is generated, a single location for the robot can be found by looking at the mean of the particles.

3.6 Sensor Model

Corrections to the robot's location as determined by dead reckoning are made according to the robot's other sensors. The sensors, usually some type of rangefinder device, determine the weight of each particle. The weight is calculated according to $p(z_t \mid x_t)$ which represents the sensor model, the probability of getting a particular sensor reading given a suggested robot location. The sensor model depends heavily on the exact physical sensors installed on the robot, so there can be no general equation. Since the model is not sampled as is the motion model, it is often implemented as a large, precalculated table, where any particular sensor probability can be quickly looked up. A table implementation allows a more complex function to be used than could be calculated in real time. One function that is sometimes used for a laser rangefinder device gives the probability of each possible returned range value, given each possible actual distance to a wall. Such a function can be composed of a Gaussian distribution centered on the actual wall distance, since that distance is the most probable return value, together with other functions depending on the features of the physical device. Common additions are an exponential function multiplied by a linear one representing false negative values and another exponential function representing false positives. The overall probability of a laser reading, which is composed of multiple range values in different directions, is the product of the probability of each range value. Given a robot position, the distance to the wall along each sensor ray can be determined from the map and the probability of the range value returned given that distance can be looked up from the table. These probabilities are then multiplied together to get $p(z_t \mid x_t)$.

3.7 Motion Model

The motion model $p(x_t \mid x_{t-1}, u_t)$ is a critical part of MCL. Unlike the sensor model, which gives the probability of getting a specific sensor reading at a particular location, it is necessary to sample from the motion model. Given a starting location and a reported motion (x_{t-1} and u_t), MCL requires that we be able to choose a final location randomly according to the motion model. This requirement precludes us from using any motion model that is very complex. In fact, most motion models are a combination of simple Gaussian distributions. For a holonomic wheeled robot, the most common representation is with two kinds of motion leading to three kinds of error. Each movement of the robot is represented as a linear movement followed by a stationary turn. Although a particular robot probably does not follow these exact motions, if we break the robot's motion into small increments we can use them as an approximation.

These two motions are often implemented using two Normal distributions for many common robots. However, the algorithms described in this section should work for any model, provided it is possible to sample from it. In general, some collection of Gaussians works well, since they are often good approximations to a physical system while at the same time being easy to sample from and optimize.

3.8 Raytracing

Calculation of the sensor model $p(z_t \mid x_t)$ involves determining the probability of receiving a particular sensor reading given the location in the environment. For a laser rangefinder the readings are distance measurements. Given a robot's possible location in the map, the expected distance to the wall is usually determined by raytracing from the robot to the nearest wall. The robot's physical sensors determine its actual distance to the wall and, once the distance from the map and the distance from the world are known, the probability can be calculated either mathematically or by a table lookup.

4. FastSLAM

4.1 SLAM Problem

The problem of determining the map and robot position at the same time is called Simultaneous Localization and Mapping, (SLAM). It involves finding the distribution over a state space which includes both robot position and the complete map. The given data is the sensor and odometry information from the start until the current time. Even the definition of SLAM results in two different problems. Determining the map and location during operation of the robot requires finding only the current location x_t as well as the static map m. That results in the problem of online SLAM, which is intended to localize the robot during operation while also creating the map. Online SLAM is concerned with determining $p(x_t, m \mid z^t, u^t)$, which is the state at the current time. Using new information to update old estimates is not part of online SLAM, even though new information could update the map so that past localization could be corrected. The other SLAM problem is called the full SLAM problem, and it involves finding the complete pose history of the robot and the map. The probabilistic formula is $p(x^t, m \mid z^t, u^t)$, since we want all of the robot's states, instead of just the current one. The difference is that full SLAM uses current data to correct past estimates. Online SLAM is used to localize the robot dynamically while full SLAM is often an offline algorithm concerned with finding out what the robot has already done. If the robot needs to make decisions based on its location, then it is necessary to use online SLAM. If the objective is to determine where a robot controlled by some other method, for example a human driver, has been, then full SLAM is more powerful. Both the problems have an application and the various solutions are similar, although not identical. In fact, online SLAM is the result of removing the past poses from the full problem using integration.

Although SLAM is technically the definition of a particular problem, it is also the name given to the current set of solutions to the problem. These solutions all have several common elements which are shared by all effective solutions to both the online and full problems. One of the most important factors of the SLAM solutions is correspondences. Since SLAM considers the map as well as the robot pose, there must be some definition of a correct map. In SLAM, maps are defined as sets of objects and a correct map is one that has each object in the correct location. Of course, sensors do not report the location of specific objects, so it is necessary to find which object each sensor reading corresponds to. Unfortunately, it may be difficult to determine exactly which object is being observed. As we have seen, heuristic methods can be used to filter the raw sensor data into object locations, but any such technique will have a certain percentage of errors. Some SLAM algorithms explicitly take correspondence probabilities into account, adding yet another term to the posteriors. If we define c_t to be the set of correspondences between sensor readings and objects at time t, the online SLAM problem becomes $p(x_t, m, c_t \mid z^t, u^t)$, while the full problem is $p(x^t, m, c^t \mid z^t, u^t)$. Of course, increasing the size of the state space significantly increases the complexity of the problem and thus the run time of the solution. Many SLAM algorithms can be proved to eventually converge to the correct map, but only if the objects can be identified correctly. If identification is not certain, the guarantee of convergence is lost. The problem of determining which object is detected by a sensor reading is called the data association problem and it is the central drawback to SLAM algorithms. In effect, SLAM usually creates a landmark based map, rather than a pure occupancy grid map. As we have seen, correctly identifying landmarks in localization is a difficult problem, which can be overcome by using raw sensor readings in the MCL algorithm. However, the corresponding SLAM solution suffers from additional problems.

4.2 FastSLAM Derivation

Simultaneous Localization and Mapping is divided into two slightly different domains. The first, called online SLAM, is the problem of finding the robot's current pose x_t and the map m, given the sensor readings z^t and odometry u^t. The more complex problem is to find the robot's path $x^t = \{x_1, \ldots, x_t\}$ given the same data. Finding the complete path is called the full SLAM problem. Obviously, full SLAM is the more complete problem but online SLAM can be derived from full by integrating out the past poses, as shown in equation (10).

$$\text{Full SLAM}: \; p(x^t, m \mid z^t, u^t) \tag{9}$$

$$\text{Online SLAM}: \; p(x_t, m \mid z^t, u^t) = \int_{x_{t-1}} \int_{x_{t-2}} \cdots \int_{x_1} p(x^t, m \mid z^t, u^t) dx_1 dx_2 \ldots dx_{t-1} \tag{10}$$

The reduced problem is called online SLAM because it is simplified enough to be solved in real time, whereas most full SLAM solutions require offline processing.

One of the benefits of FastSLAM [Montemerlo et al. 2002] is that it simultaneously solves both the online and full SLAM problems. Of course, any solution to the full SLAM problem also produces a solution to online SLAM, but FastSLAM, although it works in real time, solves for the entire path of the robot. The key to FastSLAM is that if the robot's location is known, the locations of objects in the environment are independent. Thus, the SLAM problem can be factored into localization and mapping problems.

$$p(x^t, m \mid z^t, u^t) = p(x^t \mid z^t, u^t) \prod_n p(m_n \mid x^t, z^t) \tag{11}$$

The first term is obviously a localization problem which requires finding the robot's path x^t given its odometry and sensor data. The second term is the mapping problem which finds a particular feature's position, m_n, given the robot's path and sensor readings. Equation (11) leads to an iterative algorithm for FastSLAM where the robot's position is calculated, and then the map is updated based on that position. Unfortunately, it is unlikely that at every step a single location for the robot can be derived. Nor can the algorithm rely on a probabilistic position, since the factoring is only valid given a fixed x^t. These requirements lend themselves to a technique that represents the robot's location as a finite collection of exact states. Fortunately, such a technique, Monte Carlo Localization (MCL), exists.

4.3 Occupancy Grid FastSLAM

Unlike most SLAM algorithms, it is possible to use FastSLAM on an occupancy grid map directly, without first processing it for landmarks. Since the algorithm updates the map with reference to a given robot position, the specific representation of the map as Gaussian landmarks is not required. Instead, an occupancy grid can be used and updated according to the standard techniques. As in [Moravec 1988], observed cells are updated according to whether the sensors observed them as occupied or unoccupied. However, since each particle represents an exact robot path, there is no need to blur past readings. The grid cell implementation even overcomes the data association problem, since landmarks can no longer move, the cell being observed is exactly determined by the robot's location. An implementation called DP-SLAM, [Eliazar and Parr 2004] was able to successfully localize and map a real environment including a large loop.

The only serious problem with FastSLAM occurs with the difficult situation of loop closure. In other algorithms, when the robot re-enters known territory it becomes necessary to search a much larger set of landmarks for correspondences, possibly the entire set. However, FastSLAM represents all possible robot positions in a finite set of samples. When it closes a loop, it can only be successful if some particle has followed the true path. The longer the loop, the greater the uncertainty of the robot's position. As uncertainty increases, the number of particles necessary to represent the belief also increases. Eventually, there will not be enough particles to represent the distribution and the correct location may be lost. FastSLAM alone suffers the problem, since all other SLAM solutions use the correlations to determine the position. The problem with particle filters is that they only represent the highest probability region of a distribution, whereas the Gaussian distributions used by other techniques represent the entire distribution. Of course, particles can represent much more complex and nonlinear distributions than is possible with Gaussians. The drawback to using particle filters in FastSLAM is that the number of particles maintains the diversity of the robot's position, and as soon as the uncertainty goes beyond the number of particles, the algorithm may fail. Thus, the size of the particle set must be tuned to the environment, based on the size of the longest loop, and increasing the number of particles to this extent may make the algorithm inefficient.

Grid based FastSLAM is performed by combining the MCL particle filtering with the occupancy grid mapping algorithm using Rao-Blackwellized particle filters [Montemerlo et al. 2002; Thrun et al. 2005]. Each particle consists of both the robot's state and an occupancy grid map. Of course, particle filtering could be used over the entire state space. However, this would require a number of particles exponential in the number of state variables, in this case the number of cells in the map. Instead, the factorization in equation (11) is used to separate the robot state from the map. The particle filter is only used for the robot's state, often x, y and orientation (θ) for a terrestrial indoor robot. The map for each particle is updated according to the occupancy grid mapping algorithm, with the position fixed at the position of the particle. This separation allows the occupancy grid algorithm to work with a guaranteed position, while still allowing for uncertainty in the robot's pose. By looking at the highest probability location we can determine the current best guess of the robot's position and the map. At each step, the set of N particles X_{t-1} is updated to X_t according to the following algorithm:

for k = 1 to N

$x_t^{[k]} \sim p(x_t \mid x_{t-1}^{[k]}, u_t, m)$

$w_t^{[k]} = p(z_t \mid x_t^{[k]}, m)$

$\forall n.\ l_{t,n}^{[k]} = l_{t-1,n}^{[k]} + \log \frac{p(m_n^{[k]} \mid x_t^{[k]}, z_t)}{1 - p(m_n^{[k]} \mid x_t^{[k]}, z_t)} - \log \frac{p(m_n)}{1 - p(m_n)}$

$X_t' = X_t' \cup \langle x_t^{[k]}, m_t^{[k]}, w_t^{[k]} \rangle$

endfor

for k = 1 to N

draw i from X_t' with probability $\alpha\, w_t^{[i]}$

$X_t = X_t \cup \langle x_t^{[i]}, m_t^{[i]} \rangle$

endfor

Table 2. Occupancy Grid FastSLAM algorithm

Line 2 updates the set of particles by randomly choosing a new location for each one based on the previous location of the particle and the motion model. The sensor model is used to determine a weight for the new location based on the sensor readings in line 3. The primary difference from MCL occurs in line 4, where the occupancy probability of the map attached to the particle is updated according to the sensor readings used at the particle's new location. Of course, these maps are maintained via the log odds ratio as in section 2.1. Finally, in the loop of lines 7 - 10, the updated particles are resampled. N new particles are chosen randomly according to the weights, with replacement, to make the new particle set. Resampling has the effect of replacing the particle weight with the number of samples at a location. Thus, low probability locations die out while high probability locations gather enough particles that, on the next update, the correct location will be selected by the motion model in line 2.

5. Dynamic Maps in MCL

One drawback to localization with MCL is that it requires a static map of the environment. Sensor readings are compared with the expected values from the map and the comparison generates the probability of the robot's location. Errors in the map are partially compensated for by increasing the error that is assumed for the sensors. Another way to compensate for map errors is that the number of correct sensor readings will probably overrule incorrect ones. However, because MCL combines sensor error and map error, as map error increases, the allowable sensor error decreases until finally the algorithm fails and the map must be rescanned. Each error in the map is usually a minor matter for a localized robot; it is the combination of minor errors that can cause problems.

A localized robot rarely becomes mislocalized due to map errors, but this is not true of global localization, where the robot's initial location is unknown. Especially in symmetric environments, global localization can easily fail due to minor map errors that would be ignored by a localized robot.

The approach described in this section is based on the idea that if a robot is localized it may reasonably expect its sensor data to reflect the environment. If that is the case, then it should be possible to update the map according to the sensor data. If a known error in the map is fixed, then the robot will have a greater ability to deal with any subsequent errors. Since global localization may depend heavily on minor features, having an updated map can be a great benefit.

Violating the static map assumption and detecting changes allows localization to be more accurate and more robust to error. It also provides additional information that may be useful in planning the robot's activities. Detecting opening doors and moving objects makes path planning more reliable, because it will be based on a more accurate representation. Further, when a new opening into an unexplored are is detected, the robot can add the new region to the map. The dynamic map algorithm described here makes it far easier for a robot to be deployed long term in an environment where other agents, including humans, are present and making changes.

Dynamic maps for MCL can also be implemented by identifying binary objects, such as doors, and tracking their status using similar probabilistic methods [Avots et al. 2002]. There are several benefits of having explicit objects. Since an object consists of multiple cells that have the same probability, each scan provides more information about the object, allowing its state to be altered more quickly. Also, since most of the map is not dynamic, the probability of objects can be changed much more rapidly, since changes in the objects probably will not be able to change the map to make an invalid location match the sensors. However, explicit objects need to be manually defined before execution, adding to the work of defining maps. Since objects are binary, either present or absent, a moving object must be represented explicitly by creating a binary object at each possible location. With the dynamic maps described here, an object can appear anywhere without user interference. Finally, the method in [Avots et al. 2002] involves a different importance factor, which increases the runtime logarithmically in the number of objects, making it unsuitable for having each map cell dynamic.

Algorithms for simultaneous localization and mapping (SLAM) have the ability to localize the robot and generate the map simultaneously in real time [Montemerlo et al. 2002]. These algorithms are meant to dynamically alter the map in the same way as dynamic map MCL. Many of these methods use an algorithm which is guaranteed to converge to a correct solution. However, they suffer from the data association problem. On every sensor scan it must be possible to uniquely identify which feature of the map is responsible for each sensor reading. If this is impossible, then the guarantee of correctness does not hold. SLAM does not discover and use cell correlations, so the rate of update is slower if the map changes, since each cell must be considered independently. Further, SLAM involves significantly more processing than MCL, using up computing power that may not be necessary, especially after the map is generated. Dynamic map MCL was created specifically to provide an accurately changing map without incurring any significant overhead. Since it is a constant time addition to MCL, the map can be updated without requiring any more computing power than ordinary localization. Of course, the map cannot be generated from nothing as it can with SLAM, but once the map exists it can be kept up to date almost without cost. SLAM also, in common with ordinary MCL, makes the assumption that the map is static. Over time, the algorithm becomes more certain of the map and any changes will take longer to appear. Dynamic MCL explicitly makes the assumption that the map will change.

Algorithms that consider dynamic environments typically assume a static map with dynamic elements, such as people, which must be eliminated from consideration. In effect, these algorithms assume a static map but allow an additional form of sensor noise in the form of moving people. [Hahnel et al. 2003] describes a method for creating a map, using standard EM SLAM techniques, which can discover the static map of the environment despite dynamic elements. Similarly, [Fox et al. 1999] gives an algorithm for using MCL in an environment with many moving objects. Although both these papers give a method for handling a dynamic environment, they both assume an underlying static map. The benefit of dynamic MCL is that the static map assumption is no longer necessary. As the algorithm runs, it changes the map to correspond to the environment. Since dynamic MCL is implemented as an augmentation to ordinary MCL, there is no reason that other augmentations could not be used if warranted by the problem. For example, the algorithm described in [Fox et al. 1999] to discard readings relating to dynamic objects during MCL can coexist with my algorithm for modifying the map in accordance with changes in the environment. Dynamic MCL allows fundamental changes to be accounted for, as opposed to merely ephemeral objects that are only observed once.

5.1 Dynamic Maps

In order to alter the map, it needs to be added to the MCL formula. Consider each cell of the map to be an independent object, which can be either present or absent. Although independence is usually not entirely valid, it is an assumption that is often made. Consider $y_t = \{y_{1,t},\ldots,y_{K,t}\}$ the set of individual cells in the map. Since we are considering these cells to be independent, if the location is known, then $p(y_t \mid x_t,z_t) = \prod p(y_{k,t} \mid x_t,z_t)$.
With this background, the new state equation is $p(y_t, x_t \mid z_t, u_t)$. Unfortunately, it turns out that this equation cannot be factored, since the map state is not fully determined with only the current location. However, notice that each sample in MCL represents not only a current location, but also the history of locations that lead to that location. Since each particle is only moved according to the motion model, they may be considered as x^t instead of x_t with no change to the algorithm. If we use the equation $p(y_t, x^t \mid z^t, u^t)$, then it is possible to factor it and we can also use the MCL algorithm without significant changes. The factorization used is similar to the one in [Avots et al. 2002], which was used to add the state of doors into the MCL algorithm.

5.2 Factoring

The size of the state space of (y_t,x^t) is exponential in the size of y_t, so we need some way of factoring the posterior in order to reduce the state space.
First, Bayes rule and the Markovian property give us:

$$p(y_t,x^t \mid z^t,u^t)=\eta p(z_t \mid y_t,x_t)p(y_t \mid x^t,z^{t-1},u^t)p(x^t \mid z^{t-1},u^t) \tag{12}$$

Now, consider the 3 parts of equation (12).
Without any data we assume that all states are equally likely, and also that the probability of a random sensor scan is a constant. Therefore:

$$p(z_t \mid x_t,y_t)=\frac{p(x_t,y_t \mid z_t)p(z_t)}{p(y_t,x_t)}=\eta' p(x_t \mid z_t)\prod_k p(y_{k,t} \mid x_t,z_t) \tag{13}$$

Remembering that cells in the map change status independently in the model, and again using the Markovian assumption, we get:

$$p(y_t \mid x^t,z^{t-1},u^t)=\prod_k \sum_{y_{k,t-1}} p(y_{k,t} \mid y_{k,t-1})p(y_{k,t-1} \mid x^{t-1},z^{t-1},u^{t-1}) \tag{14}$$

Finally:

$$p(x^t \mid z^{t-1},u^t)=p(x_t \mid x_{t-1},u_t)p(x^{t-1} \mid z^{t-1},u^{t-1}) \tag{15}$$

Recombining these three equations and simplifying we get the factorization:

$$p(y_t,x^t \mid z^t,u^t)=p(x^t \mid z^t,u^t)\prod_k p(y_{k,t} \mid x^t,z^t,u^t) \tag{16}$$

Which contains the original MCL posterior and a new probability for the cells in the map. See [Avots et al. 2002] for more details about the factorization.

5.3 Binary Object Bayes Filtering

Since the method for calculating $p(x^t | z^t, u^t)$ is already known in the MCL algorithm, the only new method needed is to calculate the probability of each cell in the map. These cells are binary objects since they are either present or absent. Each $y_{k,t}$ can be either 0 or 1 with the probability of each summing to 1. Thus, the method for calculating the probabilities is the same as in [Avots et al. 2002]. Let $\pi_{k,t} = p(y_{k,t} = 1 | x^t, z^t, u^t)$. Then

$$\pi_{k,t} = \frac{p(y_{k,t}=1|x_t, z_t)\,p(z_t|x^t)}{p(y_{k,t}=1)\,p(z_t|x^t, z^{t-1}, u^t)}\pi^+_{k,t} \tag{17}$$

where

$$\pi^+_{k,t} = p(y_{k,t}=1 | y_{k,t-1}=1)\pi_{k,t-1} + p(y_{k,t}=1 | y_{k,t-1}=0)(1-\pi_{k,t-1}) \tag{18}$$

In equation (17) the only unknown probability is $p(z_t | x^t, z^{t-1}, u^t)$ in the denominator. Rather than trying to calculate it, we exploit the fact that $y_{k,t}$ is binary so $(1 - \pi_{k,t})$ can be calculated in the same way as $\pi_{k,t}$ using $y_{k,t} = 0$ instead of $y_{k,t} = 1$. The two equations are then divided to cancel the unknown quantities.

$$\frac{\pi_{k,t}}{(1-\pi_{k,t})} = \frac{p(y_{k,t}=1|x_t, z_t)}{1-p(y_{k,t}=1|x_t, z_t)}\,\frac{1-p(y_{k,t}=1)}{p(y_{k,t}=1)}\,\frac{\pi^+_{k,t}}{\pi^-_{k,t}} \tag{19}$$

The result, equation (19), consists entirely of known quantities. $p(y_{k,t}=1)$ is the prior probability that a cell is occupied. The various $p(y_{k,t} | y_{k,t-1})$ values are the transition probabilities for a cell, $\pi_{k,t-1}$ are, of course, the prior occupancy probabilities and finally, $p(y_{k,t=1} | x_t, z_t)$ is the probability of occupancy given robot location and sensor data. To get a useful value from the odds ratio, we use the equality $\pi_{k,t} = 1 - (1 + \pi_{k,t}/(1 - \pi_{k,t}))^{-1}$.

The representation of $\pi_{k,t}$ is actually in closed form, so it requires only a constant time operation to calculate. Since $p(y_{k,t}=1 | x_t, z_t)$ involves sensor values and raytraces which are already used for MCL, little additional processing should be required. It is possible to modify the importance factor, as in [Avots et al. 2002], to take into account the new map data, where each cell is not merely present or absent but has a probability of presence. Using this data results in a runtime increase at least logarithmic in the number of binary objects. The probability of a location becomes the sum of the probabilities of that location for both states of all visible objects, multiplied by the probability of the object states. While that is acceptable if there are only a small number of objects, such as doors, if the objects are the cells of a map, the number becomes unmanageable. However, most map data used for MCL is actually represented as probabilities in an occupancy grid map, but is thresholded to be either present or absent. I decided to use the same simplification for my algorithm and consider each cell as either present or absent depending on a threshold value on its probability. The processing time therefore remains unchanged, since the importance factor is calculated in the same way.

5.4 Cell Correlations

In order to perform the factorization, it is necessary to assume that map cells change independently of each other. However, this assumption is not entirely accurate. In fact, groups of adjacent cells that represent the same objects are likely to be completely dependent. To some extent ordinary MCL also assumes cells are independent, but it only becomes relevant when the cell probabilities are changed in dynamic MCL. It is easy to model correlations by annotating the map with correlation probabilities between adjacent cells, however, using this information is more difficult. Methods such as loopy belief propagation or variational methods [Jordan et al. 1999] can propagate belief through a connected graph, but they are time consuming and sometimes do not converge. Since dynamic MCL must run in real time without being much slower than ordinary MCL, these techniques are not sufficient. However, it should be noticed that the cell correlations in a map are of restricted types. Small groups of adjacent cells are highly correlated, while being uncorrelated with their neighbours. Because of the limited correlation, it is possible to use a modified variational technique in order to implement cell correlations. When a cell is updated, the update is propagated to adjacent cells along the links, but the propagation is not permitted to flow back to a cell that has already been modified. Also, the flow stops when the accumulated correlation probability falls below a threshold. In practice, only a few steps occur, but these achieve a significant improvement in the results.

The key to using cell correlations is to perform operations using two different and conflicting sets of assumptions. Each set of assumptions reduces one part of the problem to a solvable operation but makes the other part intractable. We have already seen that, by assuming cells to be independent, we can factor the belief as:

$$p(y_t, x^t \mid z^t, u^t) = p(x^t \mid z^t, u^t) \prod_k p(y_{k,t} \mid x^t, z^t, u^t) \tag{20}$$

This factorization is used to update the individual cells according to the robot's sensors. However, once the update is performed we discard both the assumption and the resulting factorization. Instead, we assume that each cell depends on its neighbours and is independent of the robot's sensors and position. According to this set of assumptions:

$$\begin{aligned} p(y_t, x^t \mid z^t, u^t) &= p(x^t \mid z^t, u^t) p(y_t \mid x^t, z^t, u^t) \\ &= p(x^t \mid z^t, u^t) p(y_t) \\ &= p(x^t \mid z^t, u^t) \prod_k p(y_{k,t} \mid y_{k-up,t}, y_{k-down,t} y_{k-left,t} y_{k-right,t}) \end{aligned} \tag{21}$$

The determination of the robot's position is unchanged, but the map cells now depend on their neighbours and not on the robot. By making this assumption any changes made to the map can be propagated to the adjacent cells and the weight of the cell correlations adjusted. Separating the algorithm into two phases with different assumptions allows the algorithm to consider additional dependencies without having to deal with the intractable problems caused by the interaction of the new dependencies with the old. In effect, during the first phase of the algorithm, as represented by equation (20), we assume that cells are influenced only by the robot, with additional effects coming from some unknown source. During the second phase, shown by equation (21), we assume that cells are only affected by their neighbours, with other changes caused by external, unconsidered, forces. Of course, two sets of contradictory assumptions cannot possibly be a reflection of reality, however, each assumption is a reasonable simplification and using both sets iteratively results in less simplification than either set exclusively.

In dynamic MCL, it is necessary to modify the cell correlation probabilities dynamically on each cycle. However, given the nature of the sensors used, it is unlikely that adjacent map cells will be observed on a single scan. The solution to the problem is to cache observed changes to each cell until an adjacent cell has also been observed. At that point, the difference in the changes of the cells can be used to adjust the correlation between them.

Adding cell correlations significantly improves the dynamic MCL algorithm since a correlated group of cells can change together whenever any member of the group is observed. The result is that although the update of individual cells must be slow to allow localization to work, if a group of cells change they will update very quickly, since each observation will correlate them, and as they become more correlated every observation of a member of the group will update the entire group. Thus, an object can appear or vanish more quickly than any single cell.

5.5 Algorithm

The preceding formulae can be used to augment an implementation of MCL in order to modify the map dynamically during processing. The MCL algorithm must raytrace along all sensor paths to calculate the probability of a particle. However, if the robot's position is known with high probability, then any differences between the sensor reading and the raytrace are more likely to be errors in the map than in the sensors. In that case, the logical action is to correct the map.

The method I used is to consider each cell of the map to be present with probability $\pi_{k,t}$. On each step of the MCL algorithm an augmented raytracer is used for the robot's most likely location. The augmented raytracer follows a ray normally, passing through each map cell along the ray. However, at each cell along the path, the probability of that cell is altered according to equation (19). Although the augmented raytracer could be run on all samples, it is more productive to determine the most likely location and use the augmented raytracer only on it. When the robot's location is not known, the new raytracer is not used.

For calculating the sensor probability of each cell, the simplifying assumption that either that cell or the existing wall is correct is used. The assumption is necessary because the normalizer for the sensor probabilities is not known, so some method must be used to normalize the values. In practice, when a new cell becomes occupied, it exceeds the threshold before any other cell, and then the assumption becomes valid again. The short period during which it is invalid for some cells does not affect the operation of the algorithm.

In order to find the robot's most likely location, the sample with the highest importance factor is used. Other locations are possible, including the weighted average of all samples. The algorithm cannot run if the robot's location is unknown.

These implementation details do not change the fundamental algorithm, which is a implementation of MCL together with the binary object formulae as described above. The only simplification to equation (19) is in the calculation of $p(y_{k,t} = 1 \mid x_t, z_t)$, a value which is at best a numerical approximation to the error in a physical sensor device.

The following pseudocode summarizes the algorithm for dynamic MCL.

```
Repeat N times
        Draw a random particle
        Move particle according to the motion model
        Annotate particle with a weight from the sensor model
Resample a new set of particles from the annotated set
Find the most probable location (mean of particles)
For each sensor reading
        Raytrace to the nearest occupied cell
        For each cell on the path
                Alter the occupancy probability of the cell
                Alter the occupancy probability of neighbouring cells according to influence
                Mark cell as observed
                If neighbouring cell marked observed
                        Adjust influence between cells
                        Unmark cells as observed
```

Table 3. Dynamic MCL algorithm

5.6 Experimental Evaluation

The dynamic map algorithm was implemented and tested using real data collected in our building. The data was created using a Pioneer 2Dxe robot equipped with a laser rangefinder. The objective of the tests was to show that the map could be updated correctly without introducing errors or causing localization to fail. Since the algorithm has an almost constant runtime there is no tradeoff necessary between the time required to update the map and the benefit obtained by doing so.

Dynamic map MCL is designed to gradually update the map of the environment used for localization. Ordinarily, MCL uses a static map which, in a dynamic environment, gradually becomes less accurate. The experiments were selected to validate the dynamic algorithm by demonstrating that over time the map becomes a more accurate representation of the environment. Obviously, localization and global localization will perform better on a more accurate map. However, the improvement is a greater tolerance for other sources of error and is not detectable from the results of localization. The experiments demonstrate that the map is updated correctly, the benefit obtained from this update depends on the specific problem.

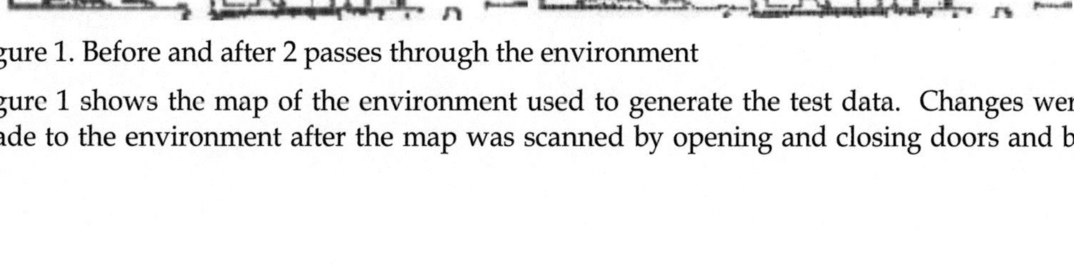

Figure 1. Before and after 2 passes through the environment

Figure 1 shows the map of the environment used to generate the test data. Changes were made to the environment after the map was scanned by opening and closing doors and by placing boxes in the corridors. After 1 pass through the changed environment the robot has mostly added the new features to the map and has correlated the changed objects, allowing them to be completed very quickly.

After 2 passes, all changes have been completely added to the map. The rate of update is slower than in [Avots et al. 2002] because each cell must be observed several times, instead of each object. However, without correlations it takes at least 5 passes to completely adapt the map. Allowing cells to become correlated permits much faster updating without compromising localization. In [Avots et al. 2002] the dynamic objects can be updated in a single pass because they are manually defined ahead of time and are known to be completely correlated. Since dynamic MCL has no predefined objects or correlations, it is necessarily slower, but because it can discover the correlations it can still update very quickly.

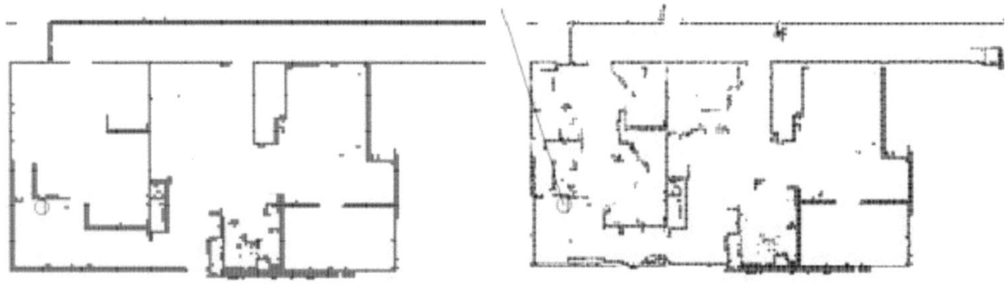

Figure 2. Before and after 5 passes through the environment using a schematic map

Another test, shown in Figure 2, was to use the same data but starting with a map consisting of the minimum possible information. From a schematic map consisting of only the walls and partitions, the algorithm was able to adapt it with all the features that were missing. Those portions of the map that were observed were corrected properly. The benefit of being able to start with a limited map is that it may not be necessary to scan a map manually with a robot. Instead, the map could be entered using blueprints of the environment and, as the robot passed through, it could correct the map until it was accurate. Usually, MCL uses the most accurate map possible, since it will lose accuracy over time, but with a dynamic map the accuracy of the map increases as the robot traverses the environment. Of course, portions of the environment that were insufficiently observed were not completely added to the map, so the result is not identical to the environment. However, observed areas have become more accurate and the map will only become a better reflection of the environment as the robot traverses it over time.

Another feature noticeable in Figure 2 is that some of the objects in the corridor are somewhat more diffuse than they appeared in Figure 1. Since the map is less accurate to begin with, localization is necessarily less accurate. As the map is corrected and localization becomes better, the location of the objects becomes clearer. After 5 passes, the objects are almost completely defined in the map, but some of them obviously require several more passes to full correct them. The benefit of dynamic MCL is that the robot can operate independently of this process. As it performs its task, the map becomes more accurate. All other data files tested exhibited similar behaviour, with the observed portions of objects being added to the map and no new errors introduced.

5.7 FastSLAM Comparison

The dynamic MCL algorithm is very similar to FastSLAM, with the major difference being that FastSLAM keeps the map state separately for each particle, while dynamic MCL maintains a single global map. The single map results in two significant changes in the behaviour of the algorithm. First, the run time over ordinary MCL is only increased by a constant in terms of the number of particles. Regardless of how many particles are necessary for localization, dynamic map MCL requires the same amount of additional processing. FastSLAM requires additional processing that is at least linear in the number of particles, disregarding the work necessary to continuously copy the maps. Dynamic MCL thus requires significantly less processing power than FastSLAM.

The per particle map is what allows FastSLAM to determine a map from nothing while localizing, since it can maintain multiple hypotheses until the robot observes a distinguishing feature. However, these map hypotheses necessarily include some unlikely maps. When the environment is mostly known these borderline maps are unnecessary. The basis of dynamic MCL is that the map is mostly known. In this case, the robot's position can be determined and the map can be altered based on the single correct position, instead of updating based on multiple hypothesized positions. In the case with a pre-existing map, FastSLAM's ability to update the map is provided by dynamic MCL without the drawback of having to consider maps based on multiple conflicting paths. Of course, if the map is unknown considering multiple paths is necessary for success, so dynamic MCL is in no way a replacement for FastSLAM, it merely uses similar ideas to apply to a situation that FastSLAM does not handle well. If the map of the environment is mostly known in advance, dynamic MCL provides an efficient solution to handling dynamic elements and previously unobserved areas, without causing additional uncertainty.

To discover if dynamic MCL provides appreciable efficiency gains over FastSLAM when the appropriate map is available, the FastSLAM algorithm was run on the same data set as in Figure 1. FastSLAM was able to generate a map, but it took 428 seconds and, of course, did not include the areas that were not visited. In contrast, dynamic MCL completed the 2 passes in 68 seconds, an 84% improvement. When only minor features need to be updated in a mostly complete map, it is unnecessary to incur the cost of FastSLAM, since in these cases dynamic MCL is far more efficient while providing the same result. Dynamic MCL also allows previously visited areas to remain in the map, even if the robot has not observed them.

6. Skeletal FastSLAM

While FastSLAM is a good solution for localization and mapping, it suffers from some problems, notably the loop closure problem. As the robot travels around a loop in the environment, it has no way to incrementally correct its position. Only once the robot arrives at the end of the loop can it realize that the correct path is the one that arrives in the right place so that the map joins up. Because particle filtering only represents the highest probability locations, over a long loop the correct path may be lost. Surviving this problem requires a number of particles relative to the size of loops in the map, which means the algorithm increases in runtime and memory with the size of the map.

SLAM is normally defined as generating a map from total uncertainty about the environment. However, there is often some information available about the map, especially in an indoor environment. Unless your robot is a bulldozer, it is constrained to follow certain paths indoors, corresponding to the building's corridors. These corridors are relatively easy to describe based on a simple floor plan, or even by observing the environment. In this section, I demonstrate how minimal information about the skeleton of the environment can be used to improve FastSLAM and reduce the loop closure problem, requiring only enough particles for the local areas of uncertainty. Skeletal FastSLAM provides an intermediate step between pure localization with a static map and pure SLAM with total uncertainty about the environment. Many problems with some preexisting knowledge might benefit from this approach.

The primary contribution of skeletal FastSLAM is to bridge the gap between the total knowledge required by MCL and the complete uncertainty that is the initial condition for SLAM. Although these two algorithms are powerful, there are many situations that are somewhere between the two conditions. For these problems, the choices are to accept the error caused by the uncertainty in MCL or to discard the initial information in SLAM. The algorithm described in this chapter allows FastSLAM to take advantage of some initial information. Similarly to the dynamic map MCL algorithm in Section 5, skeletal FastSLAM applies to problems with partial knowledge of the environment. The ability to use partial knowledge increases the usefulness of FastSLAM to situations that would ordinarily be much more difficult.

The key to using a skeleton map of the environment in FastSLAM is to realize that, especially in an indoor environment, the robot must follow certain paths. Obviously, a particle whose path corresponds to one of these corridors in the environment is more likely than one traveling at a tangent to the corridor. Of course, this only applies if the particle is close enough to the corridor, but when one of the corridors affects the robot's path, it can act as a very useful indication of the correct path.

6.1 Monte Carlo Localization with Paths

Although MCL is originally defined to solve for only the robot's current position, as in section 3 and [Dellaert et al. 1999], it is trivial, by recording the past poses of each particle, to alter it to track the robot's entire path. The derivation is similarly easy to alter, again using the Markovian assumption, producing models identical to ordinary MCL.

$$p(x^t \mid z^t, u^t) = \eta p(z_t \mid x^t) \int_{x_{t-1}} p(x^t \mid u^t, x^{t-1}) p(x^{t-1} \mid z^{t-1}, u^{t-1}) dx_{t-1} \tag{22}$$

$$\begin{aligned} p(x^t \mid u^t, x^{t-1}) &= p(x_t \mid x^{t-1}, u^t) p(x^{t-1} \mid x^{t-1}, u^t) \\ p(x_t \mid x^{t-1}, u^t) &= p(x_t \mid x_{t-1}, u^t) \\ p(x_t \mid x^{t-1}, u^t) &= p(x_t \mid x_{t-1}, u_t) \\ p(x^{t-1} \mid x^{t-1}, u^t) &= 1 \end{aligned}$$

$$p(x^t \mid z^t, u^t) = \eta p(z_t \mid x_t) \int_{x_{t-1}} p(x_t \mid u_t, x_{t-1}) p(x^{t-1} \mid z^{t-1}, u^{t-1}) dx_{t-1} \tag{23}$$

6.2 Derivation of FastSLAM with Skeleton

In order to consider a topological map in FastSLAM, we need to add it to the equations in a form that can be easily calculated. Let S be the skeletal map, then the FastSLAM factorization becomes:

$$p(x^t, m \mid z^t, u^t, S) = p(x^t \mid z^t, u^t, S) \prod_n p(m_n \mid x^t, z^t) \tag{24}$$

We assume the map is independent of the skeleton, which only affects the robot's position. Thus, the occupancy grid mapping portion of FastSLAM is unchanged. Only the localization needs to take S into account.

$$p(x^t \mid z^t, u^t, S) = \eta p(z_t \mid x_t) \int_{x_{t-1}} p(x^t \mid u^t, x^{t-1}, S) p(x^{t-1} \mid z^{t-1}, u^{t-1}, S) dx_{t-1} \tag{25}$$

Note that, because the map is independent of the skeleton, the skeleton does not affect the sensor readings z_t, which depend only on the robot's state, including the map.
The new motion model for localization is $p(x_t \mid u_t, x_{t-1}, S)$. However, it is not obvious how to sample from this model as required by MCL. Fortunately, given that the distance between x_t and x_{t-1} is small, we can factor the model into our original model and an additional term representing the motion probability of the motion given the skeleton map.

$$p(x^t \mid u^t, x^{t-1}, S) = p(S \mid x^t, u^t, x^{t-1}) p(x^t \mid u^t, x^{t-1}) / p(S \mid u_t, x^{t-1}) \tag{26}$$

Equation (26) can be greatly simplified using the Markovian assumption again. Also, notice that the same operation as in equation (23) produces the ordinary motion model in the second term, while still leaving the entire path for use with the skeleton map.

$$p(x^t \mid u^t, x^{t-1}, S) = \eta p(S \mid x^t) p(x_t \mid u_t, x_{t-1}) \tag{27}$$

Having factored the motion model from the skeleton, all that remains is to convert $p(S \mid x^t)$ into a form that can be calculated.

$$p(x^t \mid u^t, x^{t-1}, S) = \eta p(x^t \mid S) p(S) p(x_t \mid u_t, x_{t-1}) / P(x^t) \tag{28}$$

$$p(x^t \mid u^t, x^{t-1}, S) = \gamma p(x^t \mid S) p(x_t \mid u_t, x_{t-1}) \tag{29}$$

Putting equation (29) back into the localization formula of (25) results in localization which takes into account the skeleton map.

$$p(x^t \mid z^t, u^t, S) = \eta p(z_t \mid x_t) p(x^t \mid S) \int_{x_{t-1}} p(x_t \mid u_t, x_{t-1}) p(x^{t-1} \mid z^{t-1}, u^{t-1}, S) dx_{t-1} \tag{30}$$

The final equation indicates that the modification to the motion model alters the weight of each particle. Thus, the localization step continues as normal while the probability of the particle's motion based on the skeleton map is multiplied with the sensor probability to determine the likelihood of the sample. The result will be to make particles which travel according to the skeleton more likely to be resampled then those which conflict.

6.3 Creating the skeleton map

Of course, the first step in implementing this technique is to define what exactly is meant by a skeleton map. The skeleton is defined by a series of line segments marked by their endpoints. Each line segment marks a corridor that constrains the robot's direction of travel. It is not necessary for every possible path to be represented, only those composing major loops in the environment. The skeleton gives the direction and length of each corridor, including the structure of their intersections. Such a map is very easy to construct, especially in an indoor environment where a schematic diagram is often available. Also, buildings are usually constructed with corridors at right angles, making it easy to determine the intersections of the skeletal map. In such an environment, any user could easily define the necessary skeleton with minimal understanding of the underlying algorithm.

6.4 Defining the skeleton model

In order to actually implement skeletal FastSLAM as defined in equation (30), we need to create a method of calculating the model $p(x_t \mid S)$. Fortunately, this model does not need to be sampled from as does the motion model, allowing more flexibility in its creation. Since the objective is to more highly weight paths the closer they are to the corridor, some type of probability based on the difference in angle between the path and the skeleton is the obvious choice. The model used is a Gaussian distribution centered at 0 degrees mixed with a uniform distribution according to a gain value. The smaller the difference between the angle of the robot's path and the angle of the line segment of the map, the more probable the particle. Of course, this only applies as the particle travels along the particular corridor. If the robot turns away from the corridor it is probably exploring some area not represented by the skeleton. In that case, the probability is a uniform distribution. Also, the current segment of the skeleton map that the robot is following is determined by which segment is closest. However, if the distance between the robot and the line is too great, then the probability is once again uniform, since a particle must be within a corridor to be affected by it. The result is a model that increases the probability of particles traveling along the skeleton and decreases the probability of those traveling at a tangent to it, while leaving those that are not following the skeleton unchanged. The model for particles within range of a skeleton line segment is illustrated in Figure 3. However, note that the actual values depend on the parameters selected for gain, variance, and threshold angle.

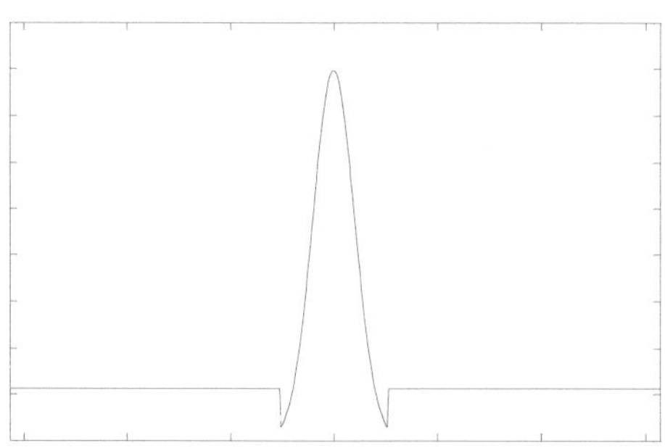

Figure 3. Skeleton map probability model

6.5 Algorithm

Given the derivation in section 6.2 and the model from 6.4, the actual implementation of skeletal FastSLAM is relatively straightforward. In order to reduce errors caused by minor corrections in the robot's heading while it follows a corridor, linear regression is used to track the line which best fits the robot's path. The regression is restarted every time the robot changes its current closest line segment. Then, the difference between the robot's course and the direction of the skeletal line segment is simply the difference between the angle of two lines, a straightforward algebraic computation. With that angle, the skeleton map model can be evaluated and the only change in the algorithm in Table 2 occurs on line 3, which becomes $w_t^{[k]} = p(z_t \mid x_t^{[k]}, m) * p(x_t^{[k]} \mid S)$.

It is, of course, necessary to provide various parameters, notably the variance and gain of the skeleton model as well as the threshold distance for the robot to be within a corridor and the threshold angle for the model. However, most of these parameters depend on the physical features of the environment and good values can be determined by examining its structure. The threshold distance depends on the corridor width, while the threshold angle depends on the relative corridor angles. Finally, the gain depends on how well the environment is represented by the skeleton map. These values probably do not need to change between different environments or robots, unless there are radical differences in the map. Even then, convergence will probably only require more particles, rather than failing.

Compared to ordinary FastSLAM, using a skeletal map adds runtime that is linear in the number of line segments in the skeleton. Since the topology of an indoor environment is usually fairly simple, the increase in processing required is minimal. If skeletal FastSLAM can reduce the number of particles required for convergence in an environment then the gains in processing time will more than offset the increase required to implement the algorithm.

6.6 Experimental Evaluation

In order to validate skeletal FastSLAM with occupancy grid maps it was tested against data sets gathered using a real robot in an indoor environment, as well as with various simulated data sets. The simulated data demonstrates the benefits of the algorithm in various situations, while the physical data sets show that it really does generate improvement in a real environment.

Loop closure is one of the major problems with FastSLAM and the skeletal algorithm is designed to reduce the processing required for loops in certain environments. The experiments were chosen to show the actual reduction provided by the skeleton in specific environments. From the results presented here we can determine that skeletal FastSLAM will provide a benefit in a wide range of circumstances where the fundamental assumption of fixed corridors applies. Since we cannot specifically test an algorithm's loop closure ability, we rely on tests of the minimum processing required to develop a map with the correct structure as determined by a human observer. Although this criteria is somewhat vague, there was no problem in making the decisions since the maps tended to either converge correctly or diverge to random nonsense. The minimum number of particles necessary to generate a correct solution was used to determine the minimum run time for each algorithm. Since skeletal FastSLAM and ordinary FastSLAM require approximately the same amount of processing the skeletal algorithm must converge on fewer particles to provide a benefit. Comparing the minimum run times for convergence proves the benefits of using the skeleton do not outweigh the extra processing required to compare particles to the skeleton map.

6.6.1 Simulated data

Data sets generated from simulated environments test the basic behaviours of the skeletal algorithm in standard situations. One of the most basic environments is a wide, straight corridor. A 40 meter long corridor is typically a very difficult situation for FastSLAM because there is no indication as to the correct direction. A straight corridor gives almost exactly the same readings as one that curves slightly. Since the robot does not turn as it traverses the corridor there is no way for FastSLAM to correct the readings. Because of this, it required a minimum of 210 particles for ordinary FastSLAM to converge correctly using the data set. Compared to that, a single corridor is a very easy environment for skeletal FastSLAM, which required only 100 particles to converge. The run time for convergence of skeletal FastSLAM was 156 seconds for this data set, an improvement of 45% over regular FastSLAM's 283 seconds. Skeletal FastSLAM provides a serious advantage when an environment provides little information about global orientation.

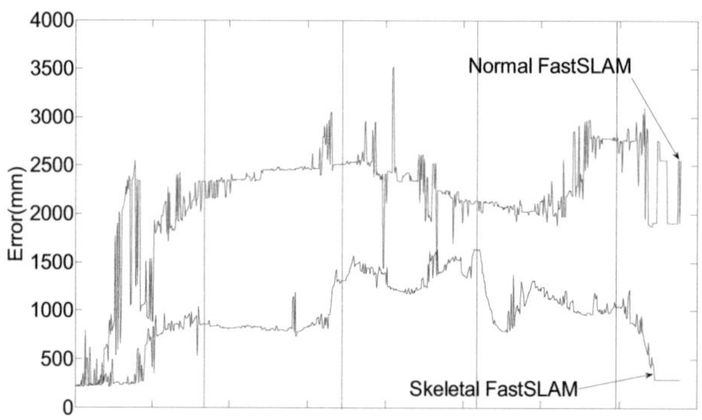

Figure 4. Error in position over time for normal FastSLAM vs. skeletal in a simple loop

The next data set was a large loop around a 40 meter square. Because the turns allow the robot to see back along its course, this environment was easier for FastSLAM to handle. Ordinary FastSLAM required 100 particles to converge with the data, while the skeletal algorithm was successful with 30. The time for convergence of 181 seconds for skeletal FastSLAM was a 67% improvement over ordinary FastSLAM's 558 seconds. In Figure **4** we can see the results of this test. For normal FastSLAM the error drifts, generally increasing over time, until the final section where the loop is closed. At that point, the error drops abruptly. In contrast, skeletal FastSLAM tends to retain a relatively constant error, changing only at the corners of the map, marked by the vertical lines, where the skeleton algorithm does not apply. The error is so much smaller when the loop is closed that it quickly decreases back to almost 0. Normal FastSLAM only managed to converge by shifting the entire map, thus retaining a larger error. By reducing the error increase in the corridors, skeletal FastSLAM is able to correct the position much more quickly when the loop is finally closed. The simulated data indicates that the algorithm provides a major benefit in the situations where it applies and leaves more leeway for handling the remaining situations, such as the corners.

6.6.2 Real data

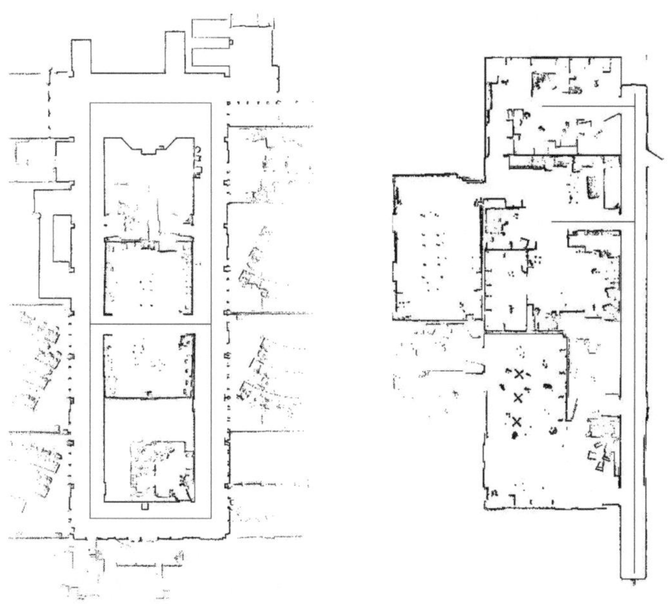

Figure 5. Two real environments with skeleton maps

Data from a 180 degree laser scanner mounted on a Pioneer 3Dxe differential drive holonomic robot was collected from two different real environments. Since there was no way to get the ground truth of the robot's position, I instead observed the minimum number of particles necessary for the map to converge to a representation which corresponded to the correct map. The primary observation about a correct map is that all of the loops are closed, connecting to the appropriate corridors. In practice, convergence was easy to determine, since, if the map did not converge, it instead diverged radically, becoming reduced to nonsense.

In the first environment in Figure 5, regular FastSLAM required at least 160 particles to converge, while using the skeleton map only required 100. The 60% larger set of particles for ordinary FastSLAM is necessary because the environment contains many long loops. Without the skeleton, more particles are necessary to allow these loops to close properly. The runtime of skeletal FastSLAM was a 58% improvement over the ordinary algorithm, converging in only 466 seconds compared to 737.

The second environment only has a single corridor and the robot travels through two rooms. Although the path into the rooms is marked by the skeleton, the path between them is not. The greater area that is not represented by the skeleton, coupled with fewer loops, results in less of an improvement. Skeletal FastSLAM needed 110 particles to converge in this environment, while ordinary FastSLAM needed 140, an increase of 27%. There was also an improvement of 23% in the runtime, with skeletal FastSLAM reducing the necessary time from 511 seconds to 391.

The improvements demonstrate that skeletal FastSLAM provides a significant improvement over ordinary FastSLAM, correctly converging with fewer particles, and thus less computation, using real data sets. Coupled with the simulated data, the results show that using a skeleton map is an effective addition to FastSLAM.

	Original		Skeletal		% improvement	
	particles	runtime	particles	runtime	particles	runtime
Simulated corridor	210	283.672s	100	156.329s	52.4%	44.9%
Simulated loop	100	558.078s	30	181.078s	60%	67.6%
Real environment 1	160	737.016s	100	466.938s	37.5%	36.6%
Real environment 2	140	511.047s	110	391.031s	21.4%	23.5%

Table 4. Experimental comparison of skeletal FastSLAM algorithm vs. original

6.7 Conclusion

Performing most tasks on a real robot, including path planning, requires knowing the configuration of the environment and the robot's position within it. One of the most adaptable representations is an occupancy grid map, which allows most types of environment to be accurately modeled. However, using occupancy grid maps limits the types of localization and mapping algorithms that can be used. Of the possible techniques, particle filter based methods are very effective. MCL provides accurate real-time localization while FastSLAM is a good solution for simultaneous localization and mapping. However, the basic implementations of these techniques have drawbacks in certain situations.

Section 5 describes an augmentation to MCL which allows the map to be updated according to the sensor measurements of a localized robot without a serious increase in running time. By considering each cell of the map to be an independent binary object and by making some simplifying assumptions, the static map required by MCL can be modified dynamically without requiring any user intervention. Instead of becoming less accurate over time, the map becomes more accurate as the robot traverses the environment. Experiments with real datasets show that the map can be updated properly without introducing errors. A change in the environment can be reflected in the map after very few passes by the robot. The result of the algorithm, having an accurate map, will always benefit the accuracy of MCL.

Dynamically correcting the map causes the largest source of error in MCL to decrease over time. Ordinarily, the best possible situation is for this error to remain constant, however in environments with dynamic elements, especially people, it is more likely that gradual changes occur. As the physical environment changes, errors build up in MCL, reducing its ability to handle any additional error. With dynamic updates the error is instead reduced over time, making localization more robust to other problems. Also, recognizing changes in the map might allow certain circumstances to be detected and considered in planning. For example, doors could be detected when they open and the robot could be sent to explore the new area. Also, new routes could be discovered as objects are moved. Removing the static map assumption greatly increases the power of MCL to handle real situations with dynamic elements. Furthermore, recognizing changes in the environment allows further improvements to be made at higher levels of control.

FastSLAM is an effective solution to both the online and full simultaneous localization and mapping problem in indoor environments where individual features are hard to determine. However, it suffers from problems in loop closure which require progressively more particles as the size of loops in the environment increase. By adding an easily created skeletal map into the algorithm, it is possible to significantly obviate this problem, allowing the FastSLAM algorithm to solve local uncertainties while aiding it in closing loops. A skeleton map indicates the direction that the robot must be taking so that, instead of wasting particles on multiple divergent trajectories, the algorithm can concentrate them around the correct path, significantly reducing the need for additional particles. As the corridors increase in length, ordinary FastSLAM requires an increasing number of particles, while skeletal FastSLAM continues to require only enough for the local uncertainties, becoming independent of the overall size of the map. Using a skeletal map is a low cost improvement to FastSLAM that is very useful in indoor environments whose overall configuration is known, even though the exact map may not be.

Skeletal FastSLAM, like dynamic map MCL, allows FastSLAM to handle situations with partial knowledge of the environment. Since initial knowledge no longer needs to be discarded, the behaviour of the algorithm is improved. By allowing additional information to be applied in the FastSLAM algorithm the technique can be very effective in specific situations where ordinary FastSLAM would require much more work. These methods of skeletal FastSLAM and dynamic map MCL lead to localization and mapping techniques that can generate a map and path from any type of starting information. Once the map and robot location are accurately known, it is possible to proceed with path planning and any other tasks the robot must perform.

7. References

Avots, D., E. Lim, R. Thibaux and S. Thrun (2002). A probabilistic technique for simultaneous localization and door state estimation with mobile robots in dynamic environments. *IEEE/RSJ International Conference on Intelligent Robots and System,* 2002. .

Dellaert, F., D. Fox, W. Burgard and S. Thrun (1999). Monte Carlo localization for mobile robots. *IEEE International Conference on Robotics and Automation.*

Eliazar, A. and R. Parr (2004). DP_SLAM 2.0. *ICRA.* New Orleans, USA.

Folkesson, J. and H. I. Christensen (2004). Graphical SLAM: A self-correcting map. *ICRA.*

Fox, D., W. Burgard and S. Thrun (1999). Markov localization for mobile robots in dynamic environments. *Journal of Artificial Intelligence Research* 11(3): 391-427.

Hahnel, D., R. Triebel, W. Burgard and S. Thrun (2003). Map building with mobile robots in dynamic environments. *IEEE International Conference on Robotics and Automation* (ICRA).

Jordan, M. I., Z. Ghahramani, T. S. Jaakkola and L. K. Saul (1999). An Introduction to Variational Methods for Graphical Models. *Machine Learning* 37(2): 183-233.

Julier, S. and J. Uhlmann (1997). A new extension of the Kalman filter to nonlinear systems. International Symposium on Aerospace/Defense Sensing, *Simulate and Controls.* Orlando, FL.

Lagarias, J. C., J. A. Reeds, M. H. Wright and P. E. Wright (1998). Convergence properties of the Nelder-Mead simplex algorithm in low dimensions. *SIAM Journal on Optimization* 9(1): 112-147.

Leonard, J. J. and H. F. Durrant-Whyte (1991). Mobile robot localization by tracking geometric beacons. *IEEE Transactions on Robotics and Automation* 7(3): 376-382.

Milstein, A. (2005). Dynamic Maps in Monte Carlo Localization. *The Eighteenth Canadian Conference on Artificial Intelligence.*

Milstein, A., J. N. Sanchez and E. T. Williamson (2002). Robust global localization using clustered particle filtering. *Proceedings of AAAI/IAAI*: 581–586.

Milstein, A. and T. Wang (2006). Localization with Dynamic Motion Models. *International Conference on Informatics in Control, Automation, and Robotics* (ICINCO).

Milstein, A. and T. Wang (2007). Dynamic motion models in Monte Carlo Localization. *Integrated Computer-Aided Engineering* 14(3): 243-262.

Montemerlo, M., S. Thrun, D. Koller and B. Wegbreit (2002). FastSLAM: A factored solution to the simultaneous localization and mapping problem. *Proceedings of the AAAI National Conference on Artificial Intelligence*: 593–598.

Moravec, H. P. (1988). Sensor Fusion in Certainty Grids for Mobile Robots. *AI Magazine* 9(2): 61-74.

Thrun, S., W. Burgard and D. Fox (2005). Probabilistic robotics, MIT Press.

Thrun, S., Y. Liu, D. Koller, A. Y. Ng, Z. Ghahramani and H. Durrant-Whyte (2004). Simultaneous Localization and Mapping with Sparse Extended Information Filters. *The International Journal of Robotics Research* 23(7-8): 693.

20

Neuro-Fuzzy Navigation Technique for Control of Mobile Robots

Dr. Dayal R. Parhi
Department of Mechanical Engineering, N.I.T.
India

Abstarct

Navigation of multiple mobile robots using neuro-fuzzy controller has been discussed in this paper. In neuro-fuzzy controller the output from the neural network is fed as an input to fuzzy controller and the final outputs from the fuzzy controller are used for motion control of robots. The inputs to the neural network are obtained from the robot sensors (such as left, front, right obstacle distances and the target angle). The neural network used consists of four layers and the back propagation algorithm is used to train the neural network. The output from the neural network is initial-steering-angle. Inputs to the fuzzy-controller are initial-steering-angle (the output from neural network) and left, front, right obstacle distances. The outputs from the fuzzy controller are the crisps values of left and right wheel velocity. From the left and right wheel velocity final-steering-angle of a robot is calculated. The neuro-fuzzy controller is used to avoid various shaped obstacles and to reach target. A Petri-net model has been developed and is used to take care of inter-robot-collision during multiple mobile robot navigation. A piece of software has been developed under windows environment to implement the neuro-fuzzy controller for robot navigation (appendix-1). Six real mobile robots are built in the laboratory for navigational purpose (appendix-2) in reality. By using the above algorithm it is visualised that, multiple mobile robots (up-to one thousand) can navigate successfully avoiding obstacles placed in the environment.

1. Introduction

Developing navigation technique for mobile robot remains on of the frontier research field since last two decades. Many researchers are focusing their thoughts in scientific manner to find out a good navigation technique for mobile robot. By increasing the number of mobile robots (i.e. multiple mobile robot navigation) the criticality of the problem is increased by many folds.

Beaufrere et al.[1,2] have discussed about navigation planning through an unknown obstacle field for mobile robot. They have used a two dimensional array to rapidly model the local free environment for the mobile robot navigation. The algorithm composed of three modules whose function were to avoid obstacles, to reach the target point and to manage direction changing of the mobile robot during path planning. For their approach they have used a method based on fuzzy reasoning and have also tested the approach by simulation. By taking Gaussian function as an activation function, a fuzzy-Gaussian neural network

(FGNN) controller for mobile robot has bin described by Watanabe et al.[3]. The effectiveness of their proposed method has been illustrated by performing the simulation of a circular or square trajectory tracking control. The paper by Racz et al.[4] has presented a neural network based approach to mobile robot localisation in front of a certain local object. Ishikawa [5] in his paper has described about a navigation method using fuzzy control. His purpose is to construct an expert knowledge for efficient and better piloting of autonomous mobile robot. He has used fuzzy control to select suitable rules i.e. tracing a path/ avoiding obstacles according to a situation, which was derived from sensor information by using fuzzy control. He has established his theory by means of simulation.

Martinez et al. [6] have considered a problem which consists of achieving sensor based motion control of mobile robot among obstacles in structured and/or unstructured environments with collision-free motion as the priority. For this they have taken fuzzy logic based intelligent control strategy, to computationally implement the approximate reasoning necessary for handling the uncertainty inherent in the collision avoidance problem. Sensor-based navigation method, which utilised fuzzy logic and reinforcement learning for navigation of mobile robot in uncertain environment, has been proposed by Boem et al. [7]. Their proposed navigator has consisted of avoidance behaviour and goal-seeking behaviour. They have designed the two behaviours independently at the design stage and then combined together by a behaviour selector at the running stage.

Kam et al.[8] have discussed about the fuzzy techniques of using sensors in robot navigation. They have also discussed about the problem in machine intelligence, including Kalmann filtering, rule-based techniques and behavior based algorithms. Wang [9] has used fuzzy logic for navigation of mobile robot. Tschicholdgurman [10] has described about fuzzy rule-net for the mobile robot navigation. Using the rule-net he has also shown the simulation result for mobile robot. Benreguieg et al.[11] have discussed about navigation of mobile robot using fuzzy logic.

Kodaira et al.[12] have described an intelligent travel control algorithm for mobile robot vehicle using neural networks. They have proposed a method that realises path planning and generation of motion command simultaneously. They have confirmed the validity of the proposed travel by computer simulation. Aoshima et al.[13] have described a simplified dynamic model of a small tunneling robot. They have constructed a dynamic model for directional correction and determined its parameter by least square method. They have used a neural network to automatically obtain four feedback gains for the directional control of both pitching and yawing. Tani et al. [14,15,16,17] have presented a novel scheme for sensory-based navigation of a mobile robot. They have shown that their scheme constructs a correct mapping from sensory inputs sequences to the manoeuvring outputs through neural adaptation, such that a hypothetical vector field that achieves the goal can be generated. Their simulation results has shown that robot can learn task of homing and sequential routing successfully in the work space of a certain geometrical complexity.

Neural network approach for navigation of indoors mobile robot has been discussed by Dubrawski [18]. His algorithm allows for an efficient search a decision space and also for a concurrent validation of the learning algorithm performance on a given data. Fiero et al.[19,20] have discussed about the navigation of mobile robot using neural network. Burgess et al.[21] have used neural network technique of navigation of miniature mobile robot. By using the sensory data and the algorithm, the robot is able to find out current heading direction. Masek et al.[22] have discussed about the mobile robot navigation using sonar data. There robot navigation task is performed by simulating a simple back-propagation neural network. Chang et al.[23] in their paper, have presented an environment predictor that provides an estimate of future environment configuration by fusing multi-sensor data in real time. They have implemented the predictor by an artificial neural network (ANN) using a relative-error-back-propagation (REBP). Their REBP algorithm enables the artificial neural network to provide output data with a minimum relative error. They have verified their result by computer simulation and navigation experiment.

In the above literature survey, it is seen that many researcher have focused their idea in finding out navigational technique for mobile robot. Still a systematic approach is needed in finding out a proper navigation technique of multiple mobile robots navigating in a highly cluttered unknown environment.

Keeping the above objectives in mind navigation technique has been developed systematically for navigation of multiple mobile robots in a highly cluttered unknown environment. A neuro-fuzzy controller has been developed for avoidance of the obstacle and for target seeking behaviour. The inputs for the neural network robots left obstacle distance, front obstacle distance, right obstacle distance and target angle(i.e. angle made by the robot with respect to. the target). The output from the neural network is initial-steering-angle. The output from the neural network along with the left-obstacle, front-obstacle and right obstacle distance is input to the fuzzy controller. The outputs from the fuzzy controller are the crisp values of left-wheel-velocity and right-wheel-velocity of the robot. From the left-wheel-velocity and right-wheel-velocity the final-steering-angle of a robot is calculated. Inter robot collision avoidance are achieved by using the Petri-net model, by which robots are prioritised. Six mobile robots are also built up in the laboratory for navigation purpose. Windows based software has been developed where the above technique has been implemented. Results achieved by the use of above technique for the navigation of multiple mobile robot shows the authenticity of the proposed technique. Navigation of multiple mobile robots in a highly cluttered unknown environment has got many application such as, automation in industry, working in hazardous conditions (where human being can not reach), space mission, or any other mass activity where there is a need of multiple mobile robots.

2. Analysis of navigation method

Navigation of multiple mobile robots in a highly cluttered environment, using neuro-fuzzy controller, has been analysed systematically in the following section. For the neuro-fuzzy controller the inputs to the neural network are left-obstacle distance, front-obstacle distance, right-obstacle distance and target angle (angle of robot with respect to target). The output from the neural network is initial-steering-angle. Again the output from the neural network along with the left-obstacle distance, front-obstacle distance and right-obstacle distance are the inputs to the fuzzy controller. The outputs from the fuzzy controller are left-wheel velocity and right-wheel velocity, which decides the final-steering-angle (Figure 1).

2.1 Analysis of neural network used in neuro-fuzzy controller

The neural network used is a four-layer perceptron. This number of layers has been found empirically to facilitate training. The input layer has four neurons, three for receiving the values of the distances from obstacles in front and to the left and right of the robot and one for the target bearing. If no target is detected, the input to the fourth neuron is set to 0. The output layer has a single neuron, which produces the steering angle to control the direction of movement of the robot. The first hidden layer has 10 neurons and the second hidden layer has 3 neurons. These numbers of hidden layer have also been found empirically. Figure 1 depicts the neural network with its input and output signals.

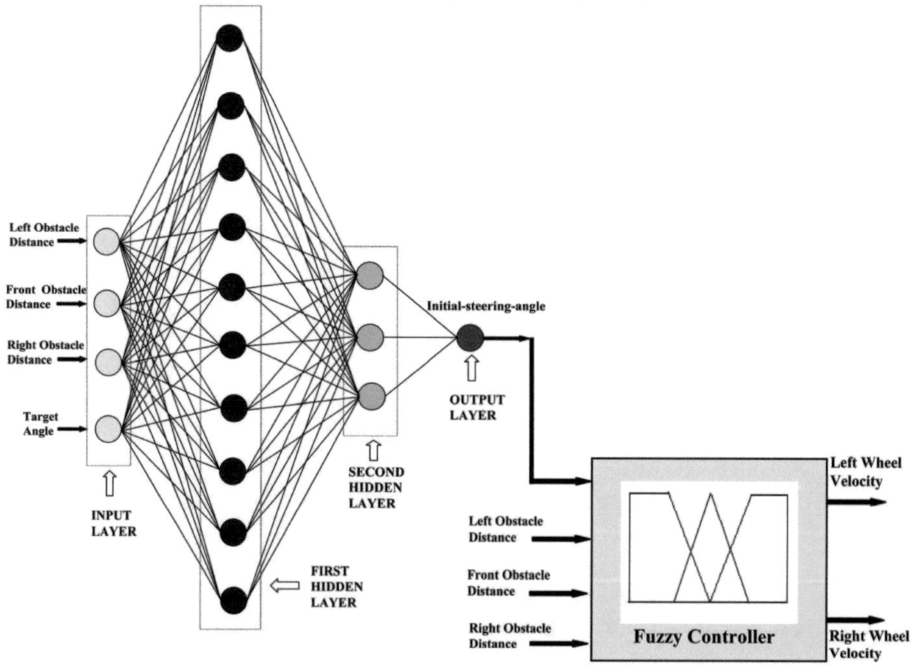

Figure 1. Neuro-fuzzy controller for mobile robots navigation

The neural network is trained to navigate by presenting it with patterns representing typical scenarios, some of which are depicted in Figure 2. For example, Figure 2a shows a robot advancing towards an obstacle, another obstacle being on its right hand side. There are no obstacles to the left of the robot and no target within sight. The neural network is trained to output a command to the robot to steer towards its left side.

During training and during normal operation, input patterns fed to the neural network comprise the following components.

$$y_1^{[1]} = \text{Left obstacle distance from the robot} \tag{1a}$$

$$y_2^{[1]} = \text{Front obstacle distance from the robot} \tag{1b}$$

$$y_3^{[1]} = \text{Right obstacle distance from the robot} \tag{1c}$$

$$y_4^{[1]} = \text{Target bearing} \tag{1d}$$

These input values are distributed to the hidden neurons which generate outputs given by:

$$y_j^{[lay]} = f\left(V_j^{[lay]}\right) \tag{2}$$

where

$$V_j^{[lay]} = \sum_i W_{ji}^{[lay]} . y_i^{[lay-1]} \tag{3}$$

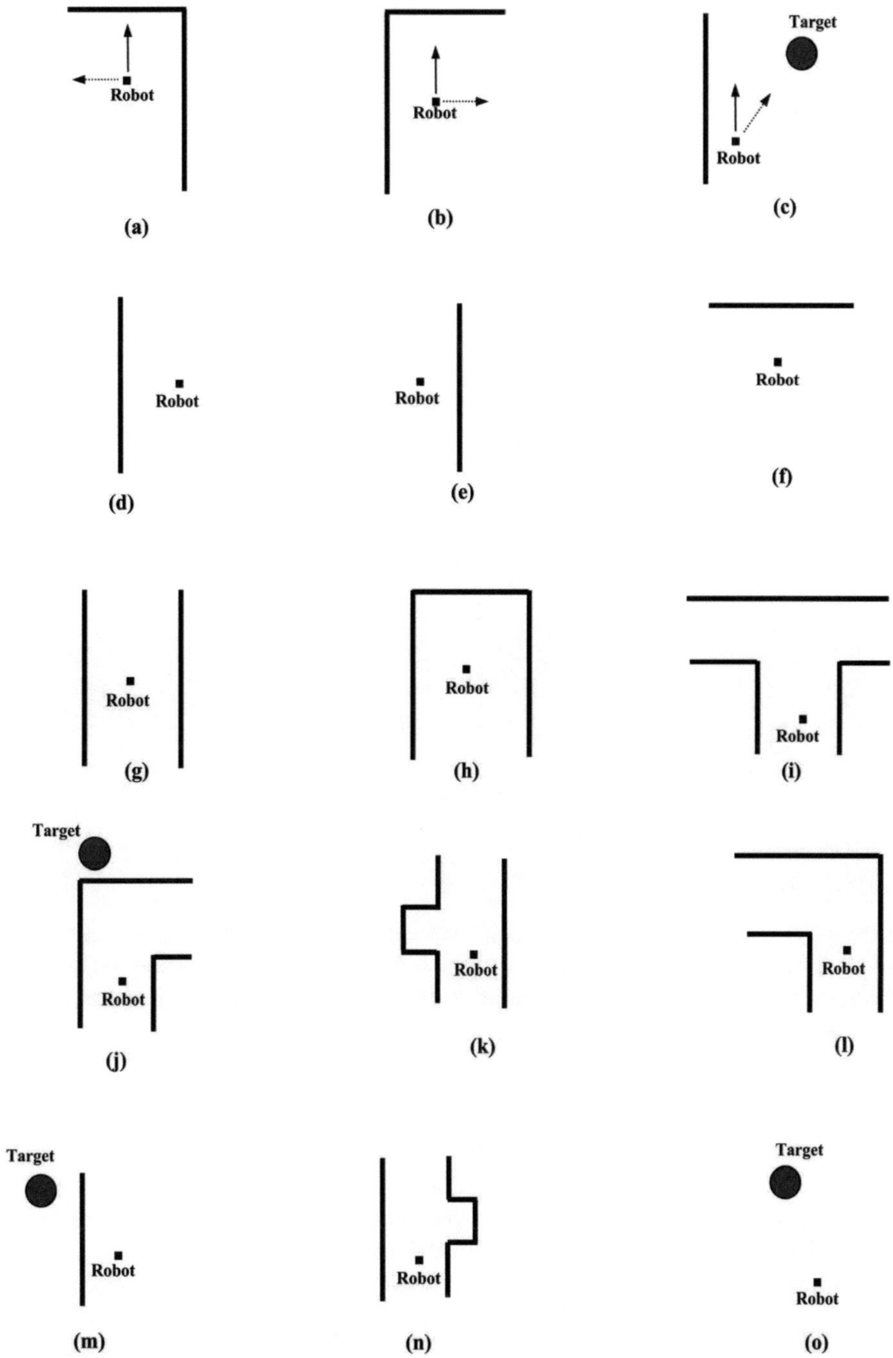

Figure 2. Example training patterns

lay = 2, 3
j = label for j^{th} neuron in hidden layer 'lay'
i = label for i^{th} neuron in hidden layer 'lay-1'
$W_{ji}^{[lay]}$ = weight of the connection from neuron i in layer 'lay-1' to neuron j in layer 'lay'
f(.) = an activation function chosen as the sigmoid function (Figure 3):

$$f(x) = \frac{e^{x} - e^{-x}}{e^{x} + e^{-x}} \quad (4)$$

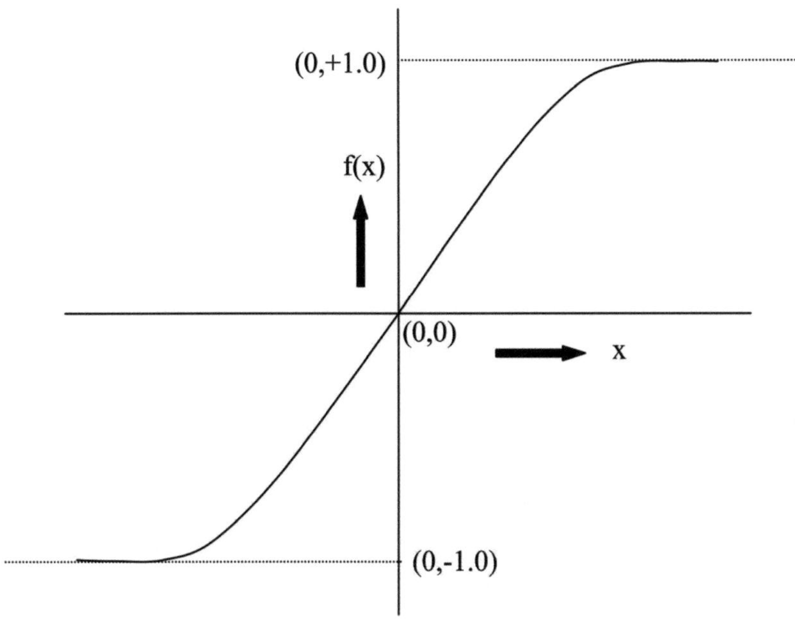

Figure 3. Hyperbolic tangent function

During training, the network output θ_{actual} may differ from the desired output $\theta_{desired}$ as specified in the training pattern presented to the network. A measure of the performance of the network is the instantaneous sum squared difference between $\theta_{desired}$ and θ_{actual} for the presented training patterns:

$$Err = \frac{1}{2} \sum_{\substack{\text{all training} \\ \text{patterns}}} \left(\theta_{desired} - \theta_{actual}\right)^2 \quad (5)$$

As mentioned previously, the error back propagation method is employed to train the network [7]. This method requires the computations of local error gradients in order to determine appropriate weight corrections to reduce Err. For the output layer, the error gradient $\delta^{[4]}$ is:

$$\delta^{[4]} = f'\left(V_1^{[4]}\right)\left(\theta_{desired} - \theta_{actual}\right) \tag{6}$$

The local gradient for neurons in hidden layer [lay] is given by:

$$\delta_j^{[lay]} = f'\left(V_j^{[lay]}\right)\left(\sum_k \delta_k^{[lay+1]} W_{kj}^{[lay+1]}\right) \tag{7}$$

The synaptic weights are updated according to the following expressions:

$$W_{ji}(t+1) = W_{ji}(t) + \Delta W_{ji}(t+1) \tag{8}$$

and

$$\Delta W_{ji}(t+1) = \alpha \Delta W_{ji}(t) + \eta \delta_j^{[lay]} y_i^{[lay-1]} \tag{9}$$

where
α = momentum coefficient (chosen empirically as 0.2 in this work)
η = learning rate (chosen empirically as 0.35 in this work)
t = iteration number, each iteration consisting of the presentation of a training pattern and correction of the weights.
The final output from the neural network is:

$$\theta_{actual} = f\left(V_1^{[4]}\right) \tag{10}$$

where

$$V_1^4 = \sum_i W_{1i}^{[4]} y_i^3 \tag{11}$$

2.2 Analysis of fuzzy logic used in neuro-fuzzy controller

The fuzzy logic used in the neuro-fuzzy controller has been discussed in the following section.
Fuzzy logic was 1st introduced in 1965, by Lotif Zadeh [24]. A fuzzy set A on the universe X is a set defined by a membership function μ_A representing a mapping

$$\mu_A : X \rightarrow \{0,1\}. \tag{12}$$

Where the value of $\mu_A(x)$ for the fuzzy set A is called the membership value or the grade of membership of $x \in X$. The membership value represents the degree of x belonging to the fuzzy set A [25]. Fuzzy-fied inputs are inferred to a fuzzy rule base. This rule base is used to characterise the relationship between fuzzy inputs and fuzzy outputs. Let us consider an example in which a simple fuzzy control rule relating the input n to the output p may be expressed in the condition-action form as follows. If x is A then z is C. Where A and C is fuzzy values defined in the universe x and z, respectively. The inference mechanism provides a set of control action according to fuzzy-fied inputs. As the outputs are in fuzzified sense, a defuzzification method is required to transform fuzzy outputs into a crisp output value, which can be applied in real sense. For this a well known defuzzification method i.e. centriod method i.e.:

$$z_0 = \frac{\int \mu_c(z)\, z\, dz}{\int \mu_c(z)\, dz} \tag{13}$$

Where $\int$ means ordinary integral . z_0 is a crisp output value of the fuzzy controller.

2.2.1. Inputs and outputs from the fuzzy controller

The inputs and outputs from the fuzzy controller are analysed in the following section.
The inputs to the fuzzy controller are left_obs (left obstacle distance), right_obs (right obstacle distance) and front_obs(front obstacle distance) and initial-steering-angle (out put from the neural network controller). Terms such as near, medium and far are used for left_obs, right_obs and front_obs (Figure 4). Terms such as pos(Positive), zero and neg(Negative) are defined for initial-steering-angle(Figure 4). The out-put from the the fuzzy controller are left_velo and right_velo. Terms such as fast, medium and slow, are defined for left_velo(left velocity) and right_velo(right velocity). The member ship functions described above are shown in Figure 4. All these membership function are triangular or trapezoidal which can be determined by three inputs, the parameters are listed in the Table-1.

Variables	Near (Meter)	Medium (Meter)	Far (Meter)
Left Obstacle (LD)	0.0	0.8	2.0
and	0.8	2.0	3.2
Right Obstacle (RD)	2.0	3.2	4.0

(a) Parameters for Left and Right Obstacle

Variables	Near (Meter)	Medium (Meter)	Far (Meter)
	0.0	0.6	2.0
Front Obstacle (FD)	0.6	2.0	3.4
	2.0	3.4	4.0

(b) Parameters for Front Obstacle

Variables	Negative (Degree)	Zero (Degree)	Positive (Degree)
	-180.0	-20.0	0.0
Heading Angle (HA)	-20.0	0.0	20.0
	0.0	20.0	180.0

(c) Parameters for Heading Angle

Variables	Slow (Meter/Sec)	Medium (Meter/Sec)	Fast (Meter/Sec)
Left Wheel Velocity (LV)	-0.35	-0.1	0.0
and	-0.1	0.0	0.1
Right Wheel Velocity (RV)	0.0	0.1	0.35

(d) Parameters for Left and Right Velocity

Table 1. Parameters of fuzzy membership functions

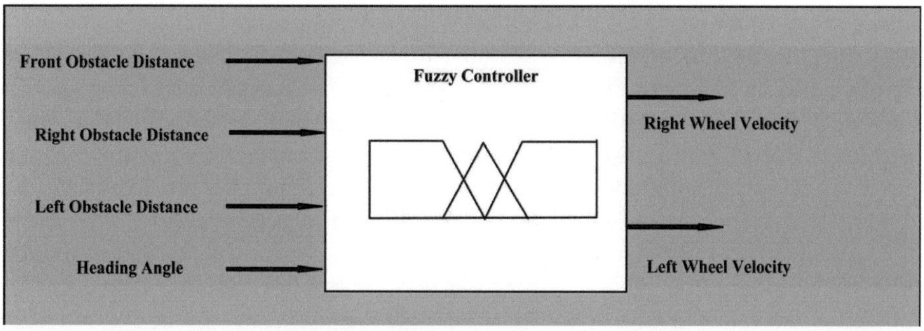

Figure 4a. Fuzzy controller for mobile robot navigation

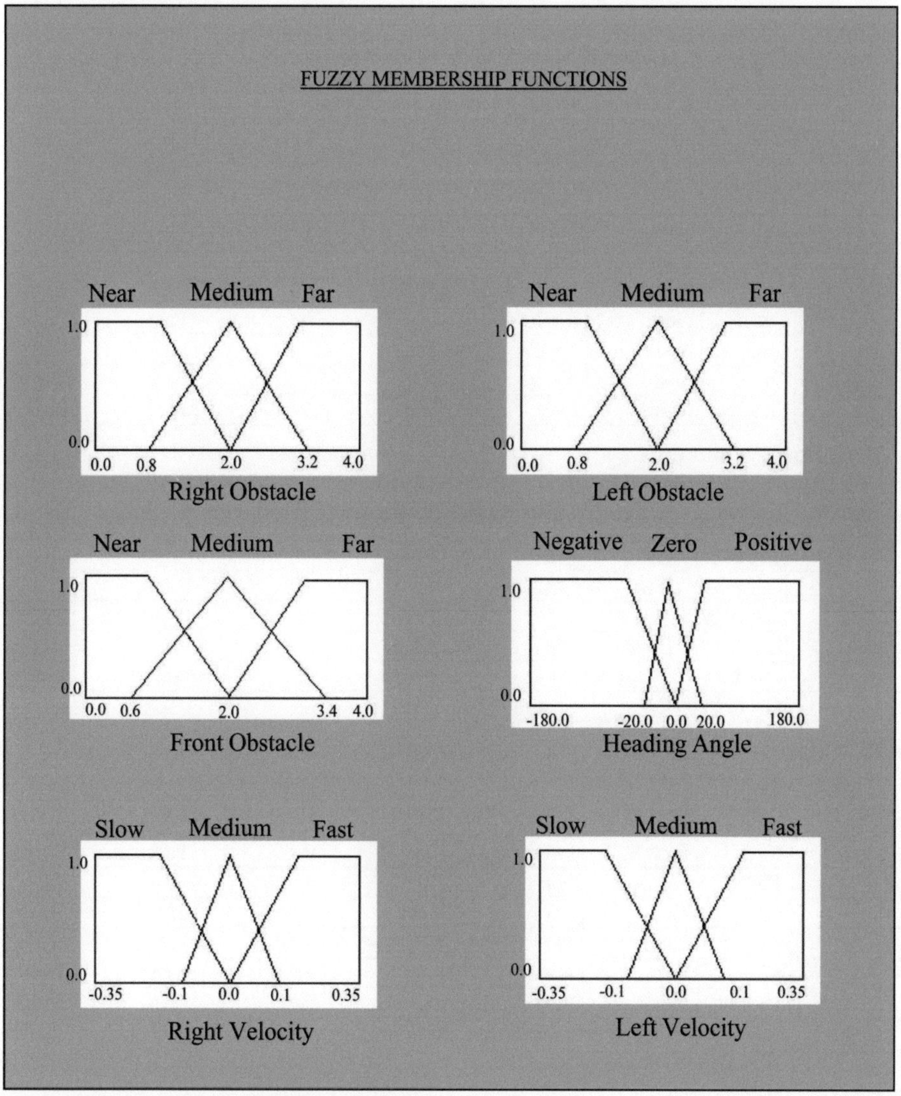

Figure 4b. Fuzzy membership functions

2.2.2 Obstacle avoidance

The fuzzy control rules for avoiding collision with obstacles are listed in Tables 2a and 2b. Rules 1-12 (Table 2a) are dealing with pure obstacle avoidance when the robot turns away from an obstacle as soon as possible. For example, if the obstacle to the left of the robot is far, the obstacles to the right and in front of the robot are near and the heading angle is negative, then the robot should immediately turn to the left. To achieve this, the left wheel velocity should be small and the right wheel velocity should be large (see Rule 3).

As with the fuzzy controller defined in Rules 13-18 (Table 2b) handle obstacle avoidance and wall following simultaneously. For instance, Rule 13 caters for the case where the left and front obstacles are far from the robot and the right obstacle is near. In this situation, the robot continues to move ahead, keeping the right obstacle (a wall) to its right hand side.

FuzzyRuleNo.	Action	Left_obs	Front_obs	Right_obs	Head_ang	Left_velo	Right_velo
1	OA	Far	Near	Med	Neg	Slow	Fast
2	OA	Far	Near	Far	Z	Fast	Slow
3	OA	Far	Near	Near	Neg	Slow	Fast
4	OA	Med	Med	Near	Neg	Slow	Fast
5	OA	Med	Near	Far	Pos	Fast	Slow
6	OA	Med	Near	Med	Neg	Slow	Fast
7	OA	Med	Near	Near	Neg	Slow	Fast
8	OA	Near	Med	Med	Z	Fast	Slow
9	OA	Near	Med	Near	Z	Fast	Slow
10	OA	Near	Near	Far	Pos	Fast	Slow
11	OA	Near	Near	Med	Pos	Fast	Slow
12	OA	Near	Near	Near	Pos	Fast	Slow

Table 2a. Obstacle avoidance

FuzzyRuleNo.	Action	Left_obs	Front_obs	Right_obs	Head_ang	Left_velo	Right_velo
13	OA & WF	Far	Far	Near	Z	Med	Med
14	OA & WF	Far	Med	Near	Z	Med	Fast
15	OA & WF	Med	Far	Near	Z	Med	Med
16	OA & WF	Near	Far	Med	Z	Med	Med
17	OA & WF	Near	Far	Near	Z	Slow	Slow
18	OA & WF	Near	Med	Far	Z	Fast	Med

Table 2b. Obstacle avoidance and wall following

FuzzyRuleNo.	Action	Left_obs	Front_obs	Right_obs	Head_ang	Left_velo	Right_velo
19	TS	Far	Far	Far	Pos	Fast	Slow
20	TS	Far	Far	Far	Neg	Slow	Fast
21	TS	Far	Far	Near	Neg	Slow	Fast
22	TS	Near	Far	Far	Pos	Fast	Slow
23	TS	Far	Near	Near	Neg	Slow	Fast
24	TS	Near	Near	Far	Pos	Fast	Slow

Table 2c. Target finding

Table 2. List of fuzzy rules for multiple mobile robot navigation

Note: Left_obs - Left Obstacle, Front_obs - Front Obstacle, Right_obs - Right Obstacle, Head_ang - Heading Angle, Left_velo - Left Velocity, Right_velo - Right Velocity, OA - Obstacle Avoidance, WF - Wall Following, TS - Target Steering, Neg - Negative, Pos - Positive, Z - Zero, Med - Medium

2.2.3 Targets finding

If a mobile robot detects a target, it will directly move towards it unless there is an obstacle obstructing its path. In that case, the robot will move around the obstacle before proceeding towards the target. The target finding rules are listed in Table 2c. An example is Rule 19 which directs the robot to turn right (high left wheel velocity and low right wheel velocity) because that is where the target is located and there are no obstacles in the vicinity.

2.3. Inter robot collision avoidance

At start robots are placed in the cluttered unknown environment, without any prior knowledge of targets, obstacles and other robots in the environment. Each robot has a aim of finding the targets avoiding obstacles and inter robot collision (Task-1) as shown in Figure 5. Once the robots have received a command to start, they will navigate in search of targets by avoiding obstacles with the help of proper steering angle, which will be decided by neuro-fuzzy controller (Task-2).

During the navigation if path of a robot is obstructed by another robot, then conflicting situation between the robots is detected (Task-3). Conflicting robots will negotiate with each other and they will be prioritised. The lower priority robot will be treated as static obstacle and higher priority robot as moving robot (Task-4). As soon as the conflicting situation is solved between the above two robots, the free robots will co-ordinate with the other conflicting robots (Task-5) to see whether there is any other conflicting situation with them.

If a robot during navigation meet with other two conflicting robots (Task-6), than the priority of the last robot will be lowest and will be treated as a static obstacle till the conflicting situation is being resolved between the first two robots. As soon as conflicting robots resolve all conflicting situations, they will again start Task-2.

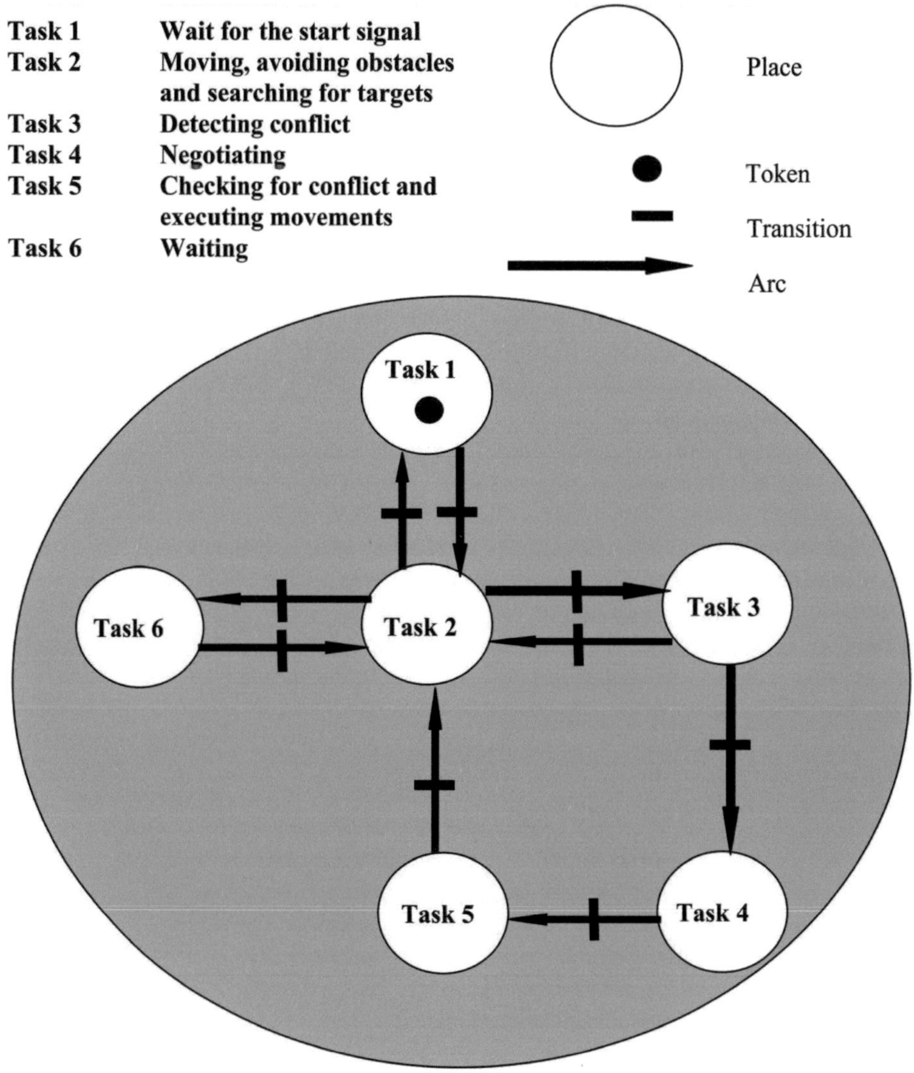

Figure 5. Petri Net Model for avoiding inter-robot collision

3. Demonstrations

3.1 Simulation Results

- Obstacle Avoidance and Target Seeking

This exercise shows that the mobile robots can navigate without hitting obstacles and can find targets in a cluttered environment. Nine robots are involved together with several obstacles including two U-shaped containers housing one target each. Figure 6a depicts the initial state when the nine robots are arranged into two groups positioned at two locations in an enclosure. From Figure 6b, it can be seen that the robots are able to locate the targets while successfully avoiding collision against the obstacles.

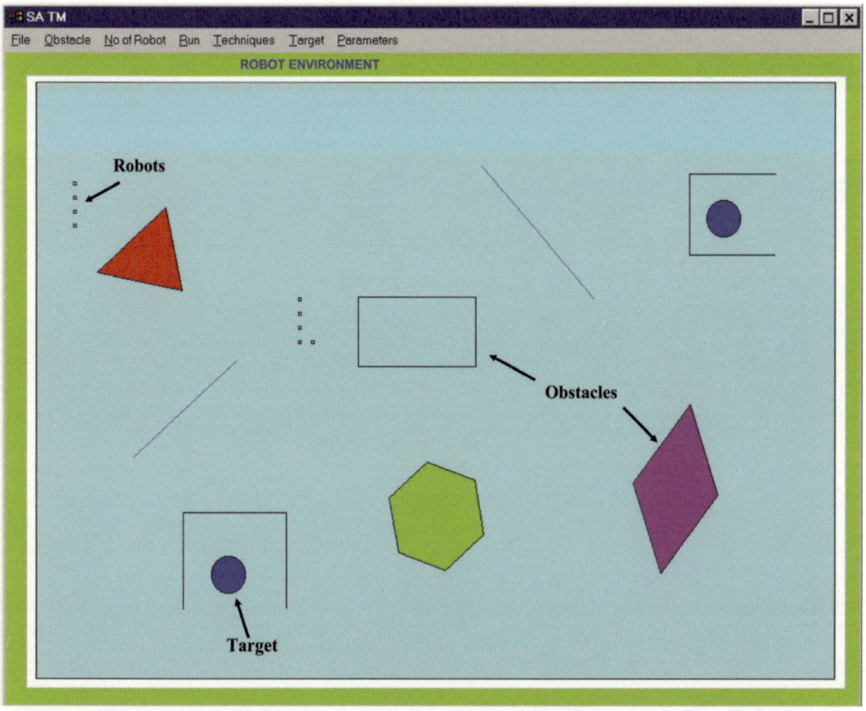

Figure 6a. Obstacle avoidance and target seeking (initial state)

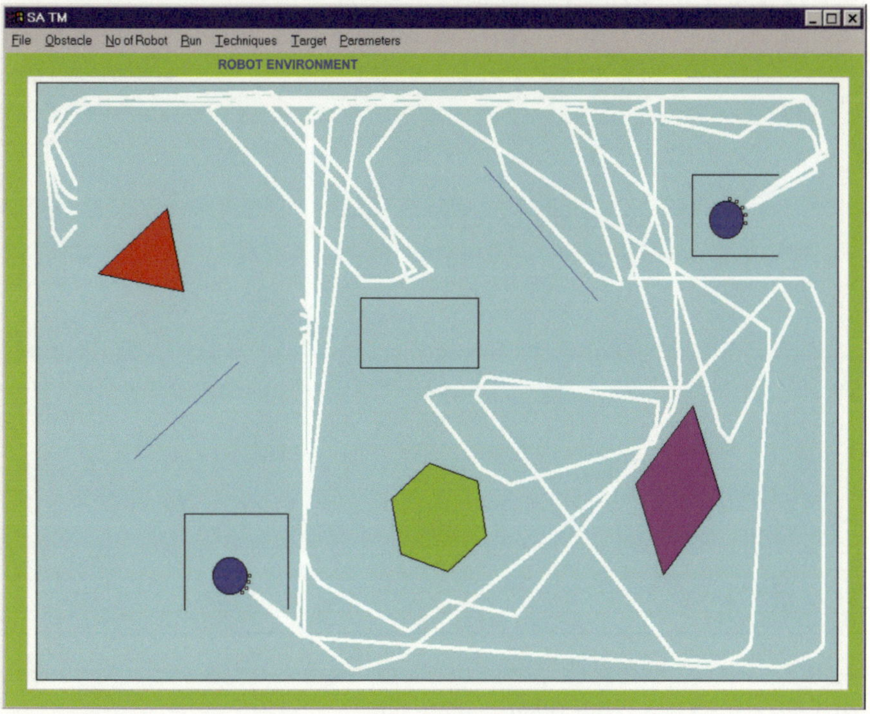

Figure 6b. Obstacle avoidance and target seeking (final state)

- Escape from Dead Ends

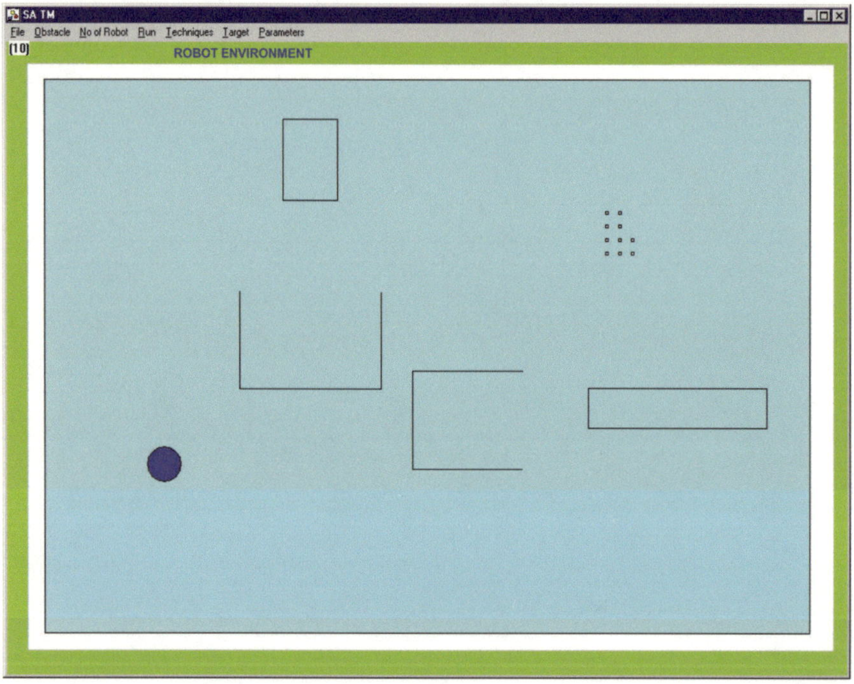

Figure7a. Escape from dead ends (initial state)

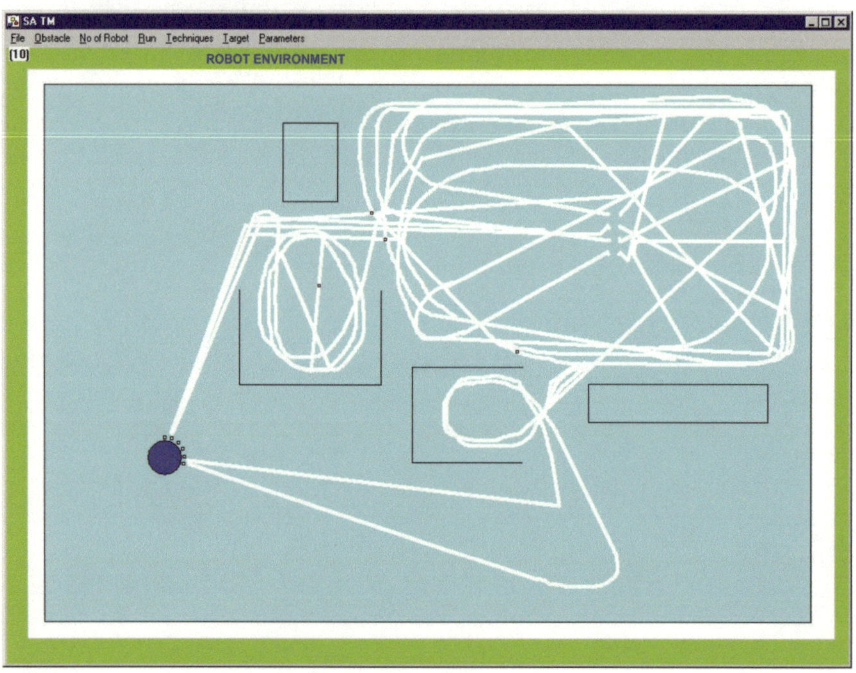

Figure 7b. Escape from dead ends (intermediate state)

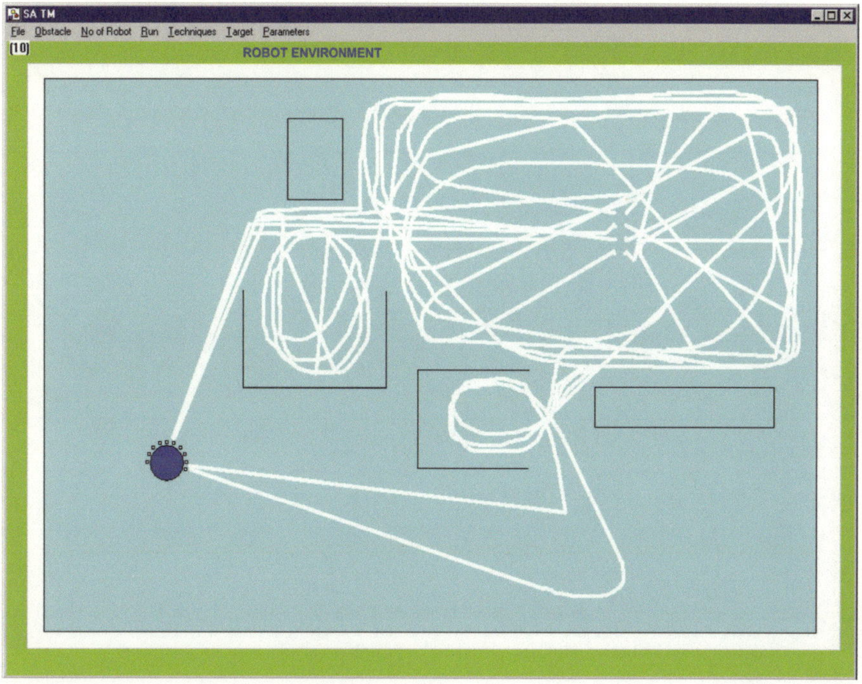

Figure 7c. Escape from dead ends (final state)

This exercise is similar to that discussed in the previous chapter (see Figure 7a) where U-shaped containers are used to simulate dead ends from which trapped robots should escape. There are 10 robots in an enclosure comprising two U-shaped containers and other obstacles. Figure 7b depicts an intermediate state. It can be seen that most of the robots are outside the containers. All of those that have strayed inside have escaped. Figure 7c shows that the other trapped robots have also escaped and all the targets have been located.

- Wall Following

Two robots are involved in this exercise. Figure 8 shows that the robots are able to follow the walls of a corridor and reach the targets successfully.

- Navigation of a Large Number of Robots

One thousand robots are involved in this exercise. Figure 9 is a snap shot of what happens. It can be noted that all the robots stay well away from the obstacles.

- Inter-robot Collision Avoidance

This exercise demonstrates that the robots do not collide with one another even in a cluttered environment. For ease of visualisation, only a small number of robots are employed. Figure 10a and 10b depict the trajectories of the robots for the neuro-fuzzy controller. It can be seen that the robots are able to resolve conflict and avoid one another.

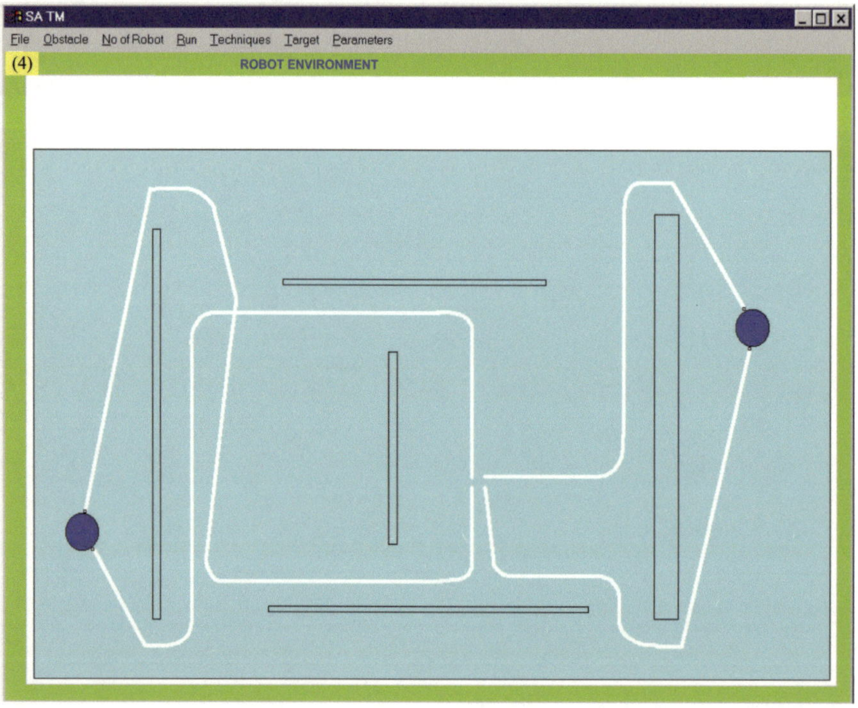

Figure 8. Wall following (final state)

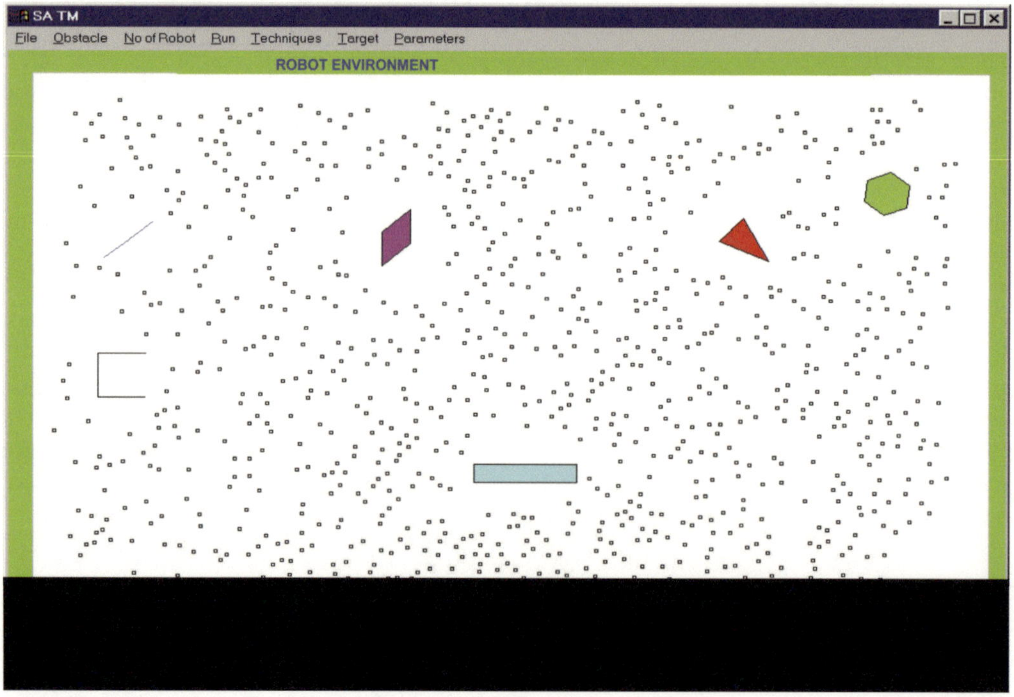

Figure 9. Navigation of a large number of robots

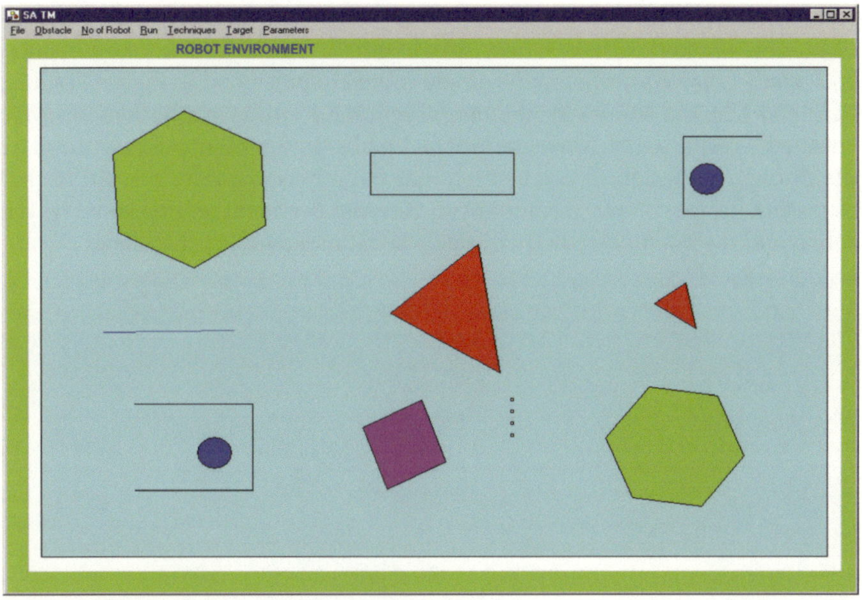

Figure 10a. Collision free movements (initial state)

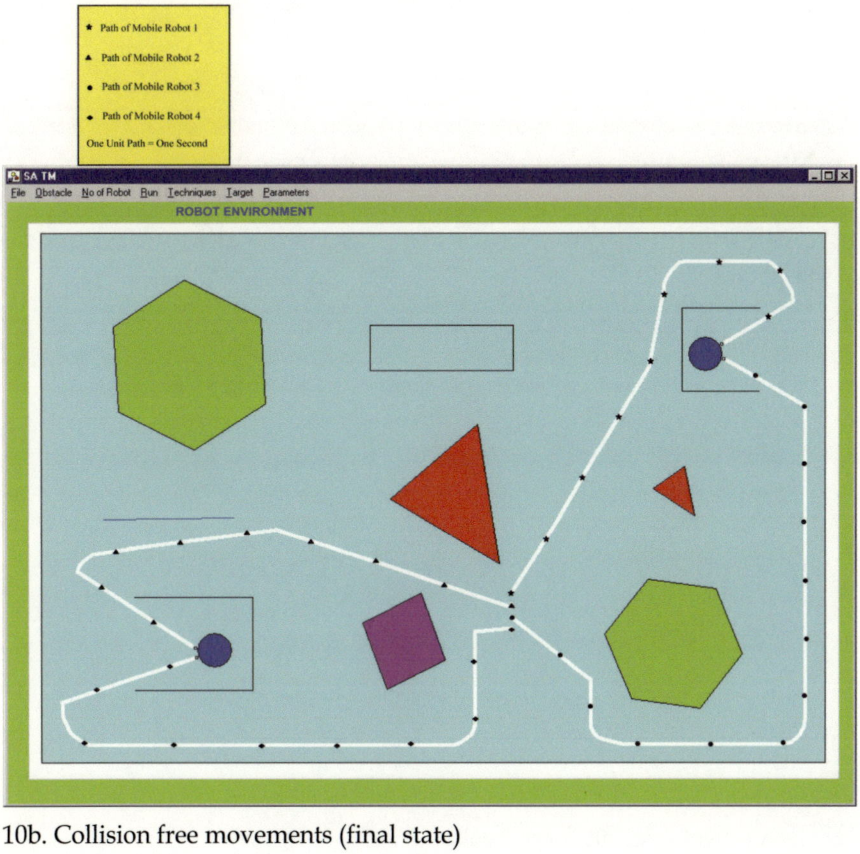

Figure 10b. Collision free movements (final state)

3.2 Experimental Results

Figures 11a, 11b, 12a and 12c show the experimental results obtained. The results for the neural and neuro-fuzzy controllers for a single robot are presented in Figure 11a and 11b. In Figures 12a and 12b, the results for the neural and neuro-fuzzy controllers for four mobile robots are shown respectively. The experimental paths drawn follow closely those traced by the robots during simulation. It can be seen that the robots are able to avoid obstacles and reach the targets. Table 3 shows a comparison between the average time taken by the robots in simulation and the practical tests for obstacle avoidance and target seeking

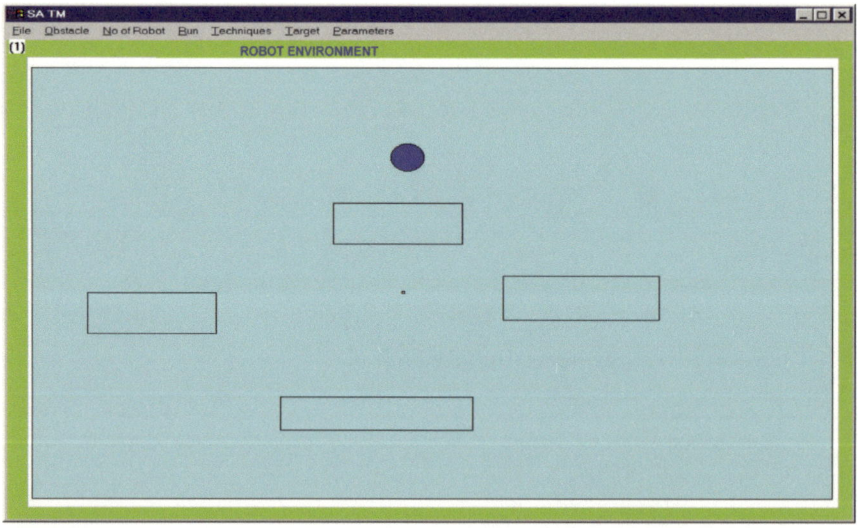

Figure 11a. Work space environment of one mobile robot (initial state)

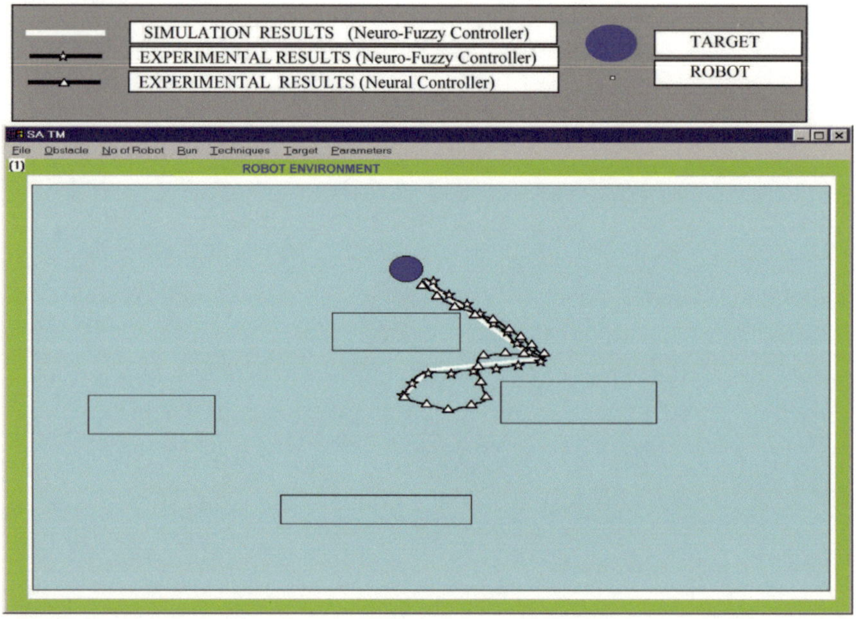

Figure 11b. Experimental results for one robot

Number of robots	Time taken (Seconds) using Neural technique	Time taken (Seconds) using Neuro-Fuzzy technique
4	11.14	10.39
8	23.37	20.25
10	22.46	19.43
16	34.42	30.49
24	54.31	51.05
40	78.12	71.05
56	123.57	111.45
70	259.01	228.54

Table 3. Times taken to reach target using different techniques

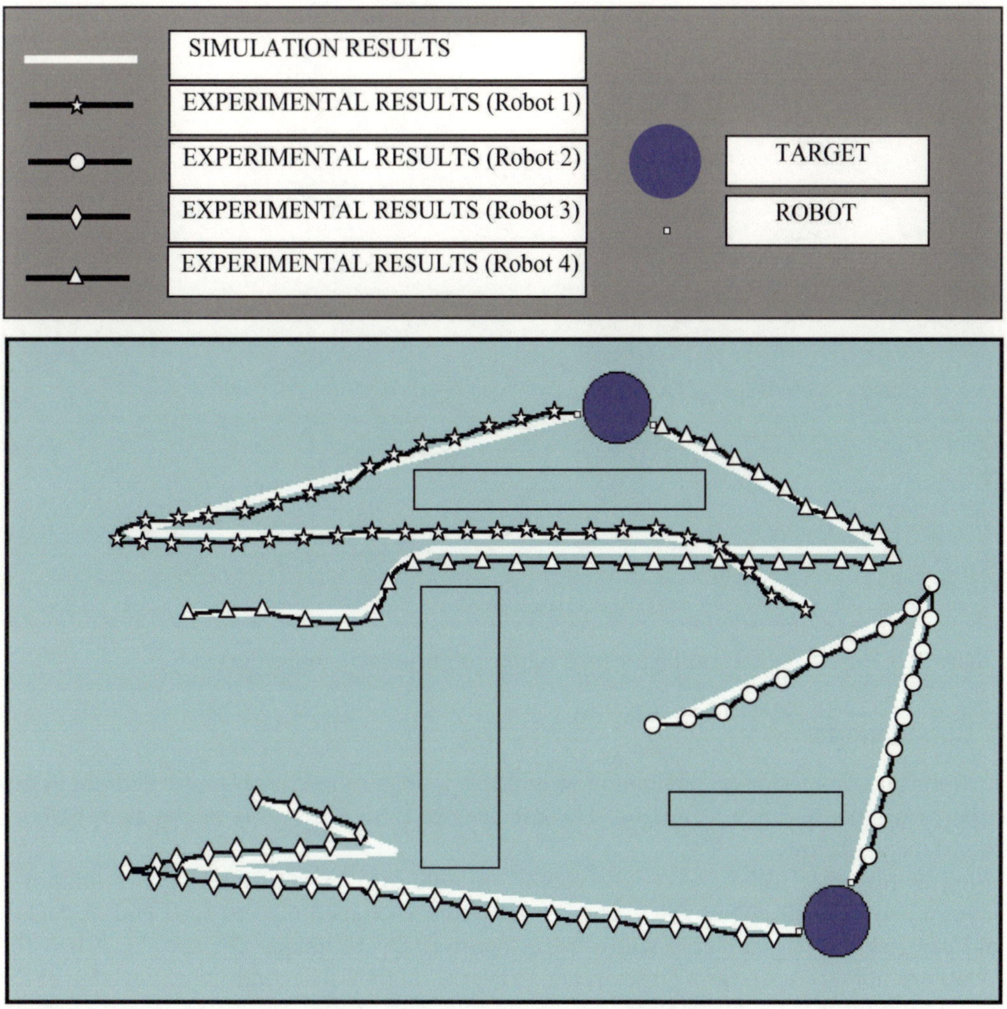

Figure 12a. Experimental results for four robots (neural controller)

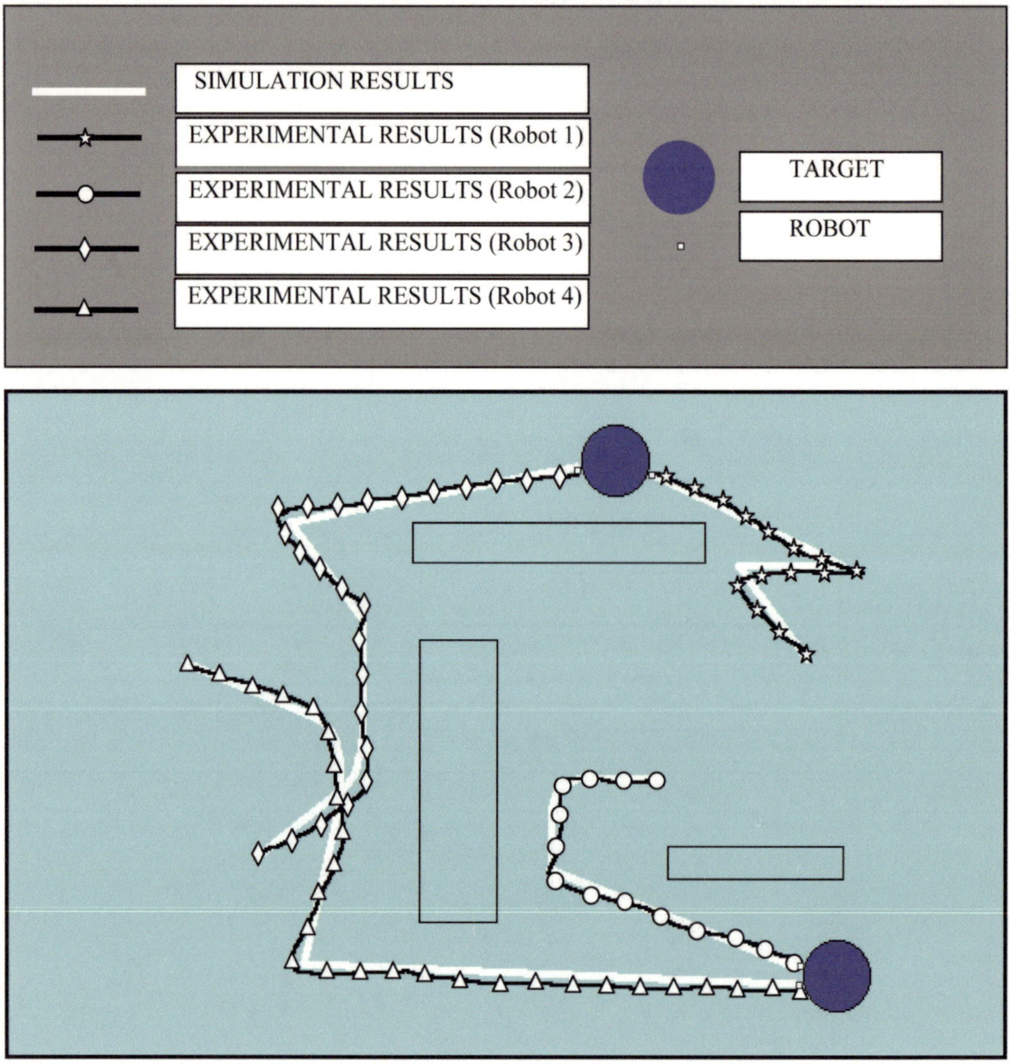

Figure 12b. Experimental results for four robots (neuro-fuzzy controller)

4. Conclusions:

Analysis and discussion on navigation of multiple mobile robots has been carried out in this paper using neuro-fuzzy technique. The conclusions drawn from the above analysis have been depicted below.

Using neuro-fuzzy technique mobile robots are able to navigate in a cluttered unknown environment. Mobile robots are able to escape from U-shaped objects(dead end obstacles) and can get the targets successfully using neuro-fuzzy technique (Figure 7.). When the robots are not able to see any target in one side of wall, they will continue at the edge of the wall assuming that the targets are lacated in other side of the wall(wall following action). The wall following action has been achieved successfully using neuro-fuzzy technique

(Figure 8.). In the present navigation method robots are able to avoid inter robot collision, which can be visualised from (Figures 10). As many as one thousand mobile robots can navigate successfully using neuro-fuzzy controller (Figure 9.).
The present research has got a tremendous application such as automation of industry, space mission, agricultural activity and mass activity where many robots are required at a time. This technique can again revised by fusing some other technique to neuro-fuzzy technique such as adaptive technique.

5. References:

Beaufrere, B. and Zeghloul, S. A mobile robot navigation method for using a fuzzy based method simulation and experimental aspects. *International journal of robotics and automation.* 1995, vol-10, no-3, 106-113.

Beaufrere, B. and Zeghloul, S. A mobile robot navigation method using a fuzzy-logic approach. *Robotica.* 1995, vol-13, no-pt5, 437-448.

Watanabe,K., Jang, J., Nakamura, M., Koga, S. and Fukuda, T., A fuzzy-gaussian neural-network and its application to mobile robot control, *IEEE transactions on contro systems technology*, Vol.4(2)(1996) 193-199.

Racz, J. and Dubrawski, A., Artificial neural-network for mobile robot topological localisation, *Robotics and autonomous systems*, Vol. 16(1)(1995) 73-80.

Ishikawa, S. A method of autonomous mobile robot navigation by using fuzzy control. *Advanced robotics.* 1995, vol-9, no-1, 29-52

Martinez, A; Tunstel, E; Jamshidi, M. Fuzzy-logic based collision-avoidance for a mobile robot. *Robotica.* 1994, vol-12, no-pt6, 521-527.

Boem, H.R. and Cho, H.S. A sensor-based navigation for a mobile robot using fuzzy-logic and reinforcement learning. *IEEE transactions on systems man and cubernetics.* 1995, vol-25, no-3, 464-477.

Kam,M, Zhu, X.X. Kalata, P. Sensor fusion for mobile robot navigation. *Proceedings of the IEEE.* 1997, vol-85, no-1, 108-119.

Wang, L.X. Modelling and control of hierarchical systems with fuzzy systems. *Automatica,* 1997, vol-33, no-6, 1041-1053.

Tschicholdgurman, N. The neural-network model rulenet and its application to mobile robot navigation. *Fuzzy sets and systems.* 1997, vol-85, no-2, 287-303.

Benreguieg, M., Hoppenot, P., Maaref, H., Colle, E., and Barret, C. Fuzzy navigation strategy: application to two distinct autonomous mobile robots. *Robotica,* 1997, Vol-15, 609-615.

M. Kodaira, T. Ohtomo, A. Tanaka, M. Iwatsuki and T. Ochuchi, Obstacle avoidance travel control of robot vehicle using neural network, *Systems and computers in Japan*, Vol. 27(12)(1996) 102-112.

S. Aoshima, K. Takeda, K. Hanari, T. Yabuta and M. Shiraishi, Dynamic simplified model and autotuning of feedback gain for directional control using a neural network for a small tunneling robot. *JSME international journal series c-dynamics control robotics design and manufacturing*, Vol.40(2)(1997) 245-252.

J. Tani and N. Fukumura, Learning goal-directed sensory-based navigation of a mobile robot, *Neural networks*, Vol. 7(3)(1994) 553-563.

J. Tani and N. Fukumura, Self-organising internal representation in learning of navigation. A physical experiment by the mobile robot YAMABICO, *Neural networks*, Vol.10(1)(1997) 153-159.

J. Tani and N. Fukumura, Embedding task-based behavior into internal sensory-based attraction dynamics in navigation of a mobile robot. *Proc. of the IEEE internatinal conf. of intelligent robots and system-94*, (1994) 886-893.

J. Tani and N. Fukumura, Embedding a grammatical description in deterministic chaos, An experiment in recent neural learning, *Biolog Cyber.*, Vol.72(1995), 365-370.

A. Dubrawski, Stochastic validation for automated tuning of neural network's hyper-parameters, *Robotics and autonomous systems.* Vol. 21(1)(1997) 83-93.

R. Fierro, F.L. Lewis, Robust practical point stabilization of a nonholonomic mobile robot using neural networks, *Journal of intelligent & robotic systems*, Vol. 20(2-4)(1997) 295-317.

R. Fierro, F.L. Lewis, Control of a nonholonomic mobile robot: Backstepping kinematics into dynamics, *Journal of Robotic Systems*, Vol.14(3)(1997)149-163.

N. Burgess, J.G. Donnett, K.J. Jeffery and J. Okeefe, Robotic and neural simulation of the hippocampus and rat navigation, *Philosophical transactions of the royal society of London series B-Biological sciences*, Vol-352(1360)(1997) 1535-1543.

Masek, V., Kajitani, M., Ming, A. and Viacic, L., Fast mobile robot obstacle detection using simultaneous firing of sonar ring sensors, *International Journal of the Japan Society for Precision Engineering*, Vol.32(3)(1998) 207-212.

C.C. Chang and K.T. Song, Environment prediction for a mobile robot in a dynamic environment, *IEEE transactions on robotics and automation*, Vol.13(6)(1997) 862-872.

Zadeh, L.A. Fuzzy SetsInformation and Control. 8, 1965, 338-353.

Tanaka, K, An introduction to fuzzy logic for practical application, 1991.

6. Appendix-1

The software ROBNAV has been developed in the laboratory under Microsoft's Windows environment.

By using the software users can have:

i. Various options and can create a navigational environment for multiple mobile robots.
ii. Many mobile robots inside the environment (multiple mobile robots).
iii. Different types of obstacles inside the environment (in order to create a cluttered environment for multiple mobile robots navigation).
iv. Many targets in the environment.

By the help of the software, neuro-fuzzy controller has been implemented for navigation of multiple mobile robots and the results obtained are shown in Figures 6 to 12. The overview outlay of the software has been shown in Figure 13.

7. Appendix-2

A prototype view of a mobile robot (out of several similar mobile robots that have been constructed in the laboratory for navigational purposes) is shown in Figure 14. Each mobile robot consists of:

i. Three wheels, of which two are driven by stepper motors and the third by caster wheel.
ii. PC-Mother board.

iii. Ultrasonic Card, six ultrasonic transmitters and receivers, for measuring the obstacle distances around the robot.
iv. Infrared card, six infrared transmitters and eight infrared receivers, used for detecting the targets.
v. Radio Modem Card, for remote transactions of commands with other robots and also with other computers.
vi. Hard-disk, for storing the software required for navigation of the mobile robot.
vii. Two touch sensors, one at the front end and one at the rear end of the robot.
viii. An onboard battery for power supply.

The components discussed above are shown in the Figure 14.

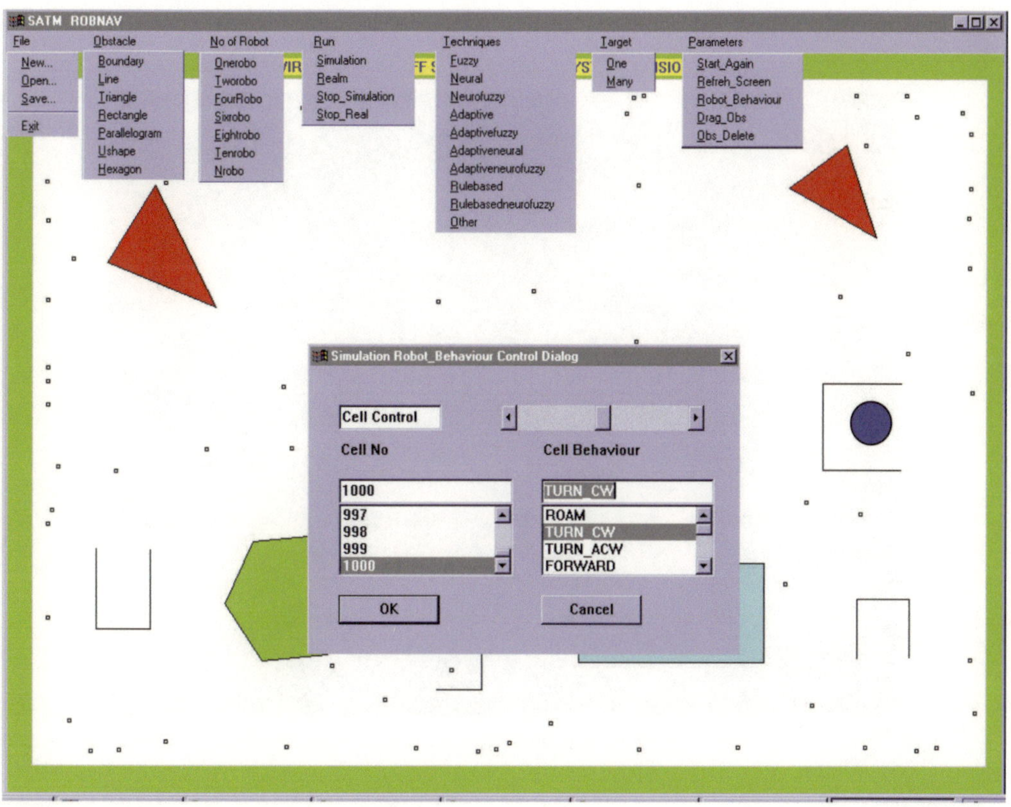

Figure 13. ROBNAV software package

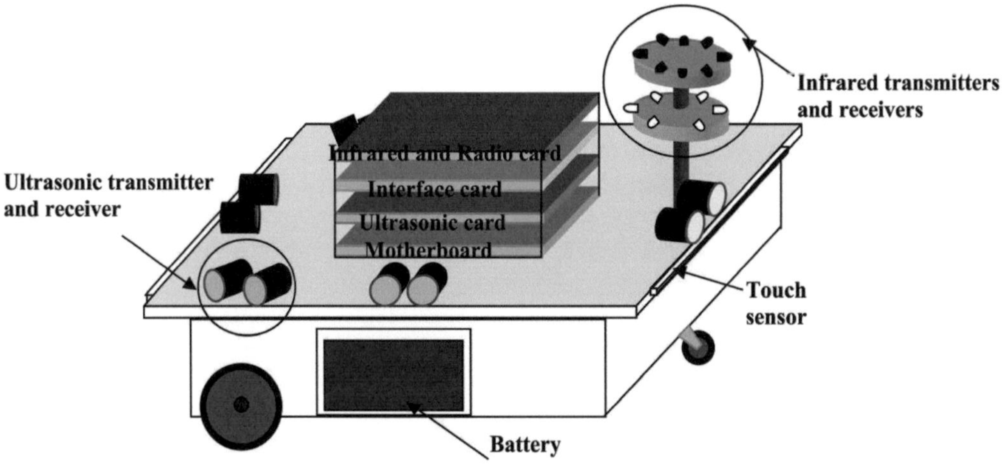

(a) Schematic view of a robot

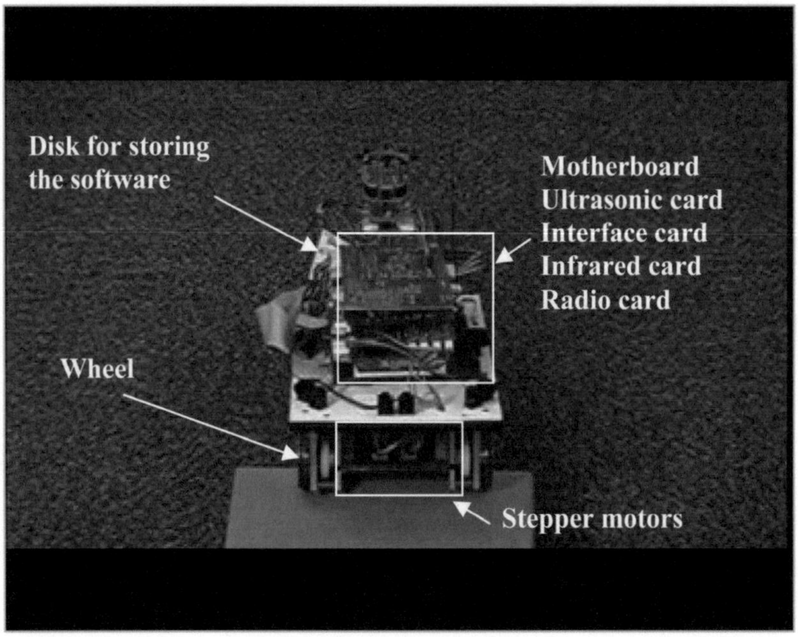

(b) Actual view of a robot

Figure 14. A mobile robot

21

Spatial Reasoning with Applications to Mobile Robotics

Lech Polkowski and Pawel Osmialowski
Polish-Japanese Institute of Information Technology
Poland

1. Introduction to Qualitative Spatial Reasoning

Qualitative Reasoning aims at studying concepts and calculi on them that arise often at early stages of problem analysis when one is refraining from qualitative or metric details, cf., [14]; as such it has close relations to the design, cf., [10] as well as planning stages, cf., [29] of the model synthesis process. Classical formal approaches to spatial reasoning, i.e., to representing spatial entities (points, surfaces, solids) and their features (dimensionality, shape, connectedness degree) rely on Geometry or Topology, i.e., on formal theories whose models are spaces (universes) constructed as sets of points; contrary to this approach, qualitative reasoning about space often exploits pieces of space (regions, boundaries, walls, membranes) and argues in terms of relations abstracted from a common-sense perception (like *connected, discrete from, adjacent, intersecting).* In this approach, points appear as ideal objects (e.g., ultrafilters of regions/solids [78]).

Qualitative Spatial Reasoning has a wide variety of applications, among them, to mention only a few, representation of knowledge, cognitive maps and navigation tasks in robotics (e.g. [39], [40], [41], [1], [3], [21], [37], [26]), Geographical Information Systems and spatial databases including *Naive Geography* (e.g., [24], [25], [33], [22]), high-level Computer Vision (e.g. [84]), studies in semantics of orientational lexemes and in semantics of movement (e.g. [6], [5]). Spatial Reasoning establishes a link between Computer Science and Cognitive Sciences (e.g. [27]) and it has close and deep relationships with philosophical and logical theories of space and time (e.g., [65], [8], [2]). A more complete perspective on Spatial Reasoning and its variety of themes and techniques may be acquired by visiting one of the following sites: [75], [83], [56].

Any formal approach to Spatial Reasoning requires Ontology, cf., [32], [70], [11]. In this Chapter we adopt as formal Ontology the ontological theory of Lesniewski (cf. [49], [50], [69], [47], [36], [18]). This theory is briefly introduced in Section 2.

For expressing relations among entities, mathematics proposes two basic languages: the language of set theory, based on the opposition element—set, where distributive classes of entities are considered as sets consisting of (discrete) atomic entities, and languages of mereology, for discussing entities continuous in their nature, based on the opposition part-whole. Due to this, Spatial Reasoning relies to great extent on mereological theories of part, cf., [4], [5], [6], [12], [15], [30], [31], [28], [71], [72], [55].

Mereological ideas have been early applied toward axiomatization of geometry of solids, cf., [45], [78]. Mereological theories dominant nowadays come from ideas proposed independently by Stanislaw Lesniewski and Alfred North Whitehead.
Mereological theory of Lesniewski is based on the notion of a part (proper) and the notion of a (collective) class, cf., [49], [51], [18], [73], [52]. Mereological ideas of Whitehead based on the dual to part notion of an extension [85] were formulated in the Calculus of Individuals [46] and given a formulation in terms of the notion of a connection [12]. Mereology based on connection gave rise to spatial calculi based on topological notions derived therefrom (mereotopology) cf. [16], [14], [20], [23], [5], [6], [15], [30], [31], [28], [71], [55].
Our approach to spatial reasoning is based on the paradigm of rough mereology, see sect. 4. Rough mereology is based on the predicate of being a part to a degree and thus it is a natural extension of mereology based on part relation.
We demonstrate that in the framework of rough mereology one may define a quasi−Čech topology [19] (a quasi− topology was introduced in the connection model of mereology [12], [5] under additional assumptions of regularity). By a *quasi-topology* we mean a topology without the null element (being the equivalent of the empty set).
Finally, we apply rough mereology toward inducing geometrical notions. It is well known, cf., e.g., [79], [8] that geometry may be introduced via notions of nearness, betweenness etc. In Section 7, we define these notions by means of a rough mereological notion of distance and we show that in this way a geometry may be defined in the rough mereological universe. This geometry is clearly of approximate character, approaching precise notions in a degree due to uncertainty of knowledge encoded in rough inclusions.
We show applications of the proposed scheme to localization and navigation by a mobile robot equipped with sonar sensors in an environment endowed with a GPS system and we discuss its implementations in the Player/Stage system.

2. Ontology of spatial objects

In reasoning with spatial objects, of primary importance is to develop an ontology of spatial objects, taking into account complexity of these objects. We propose a hierarchical ontology obtained by iterative application of the Lesniewski ontological principle. Ontology was intended by Stanislaw Lesniewski as a formulation of general principles of *being* [50], cf., also [36], [47], [69], [35]. In application-oriented spatial reasoning systems, ontology appears as typology of concepts and their successive taxonomy, cf., e.g., [54] (to quote a small excerpt: *edge* is *frontier, barrier, dam, cliff, shoreline*).
The only primitive notion of Ontology of Leśniewski is the copula " *is*".
We begin with the axiom of Ontology.
The Leśniewski Ontological Axiom is formulated in terms of the conjunctive *is* whose intuitive semantics denotes the fact that one object (individual) falls under the scope of a collective notion (a collection of individuals).
We assume a set of primitive objects S, a set of complex objects C, and we formulate a restricted ontological axiom:

$$X \text{ is } Y \text{ iff } (X \in S) \wedge (Y \in C) \wedge (\text{for all } Z \in S. Z \text{ is } X \Rightarrow Z \text{ is } Y).$$

The meaning is: X is an object from S and X belongs in Y extensively (anything in S which is X is also Y).

Example 1 *Assume a unit grid on the Euclidean d-space R^d, let S_1 be the collection of unit cubes resulting from the grid, and C_1 a non-empty collection of connected unions of finitely many cubes (one may look at this as a collection of obstacles built from single cubes). On the next level, S_2 can be taken as C_1, and C_2 can be defined as a collection of finite unions of members of C_1 which are pairwise disjoint (making an environment map of obstacles).*

Example 2 *S is a collection of closed discs, and C is a collection of finite unions of (eventually, pairwise disjoint) closed discs.*

Extending the idea just put forth, we define a Layered Ontology (LO) as a sequence $((S_i, C_i) : i = 1, 2,...)$ such that:

LO1. (S_i, C_i) are related by the Ontology Axiom for each *i;*

LO2. $S_{i+1} = C_i$ for each *i;*

LOS. $C_i = F_i(S_i)$ for each *i*, where F_i is an operator (in examples above it was taken as the union).

Ontological theories play an important role in Approximate Reasoning [11], [32], [70] witnessed with particular clearness in Spatial Reasoning [54], [22] where Ontology plays a basic role as it sets spatial concepts and their taxonomy.

3. Mereology

Yet another source of ideas and points of reference for rough mereology are mereological theories of concepts/sets. We refer here to two mainstream theories of mereology, viz., mereology due to Lesniewski [49], [51], [52], [73], [74], [18], [79], [13], [48] and mereology based on the notion of connection [12], [46], [85], [15], [16], [55], [4], [6].

Of the two theories, mereology based on connection offers a richer variety of mereotopological functors; mereology based on the notion of *part* offers a formalism of which the formalism of rough mereology is a direct extension and generalization: the latter was proposed [58], [59], [61], [62] to contain mereology as the theory of the predicate μ_1.

3.1 Mereology based on parts

We denote with the symbol *pt* the relation of part on a collection of objects, subject to conditions,

PI. X is *pt* $Y \Rightarrow X$ is $X \wedge Y$ is Y (the relation *pt* is defined for **individual entities** only);

P2. X is *pt* $Y \wedge Y$ is *pt* $Z \Rightarrow X$ is *pt* Z (*pt* is transitive);

P3. $\neg$ (X is *pt* X) (*pt* is non-reflexive).

On the basis of the notion of part, we define the notion of an *element* (an improper part; called originally in [49], an *ingredient)* as a relation *el:*

$$X \text{ is } el\ Y \Leftrightarrow X \text{ is } pt\ Y \vee X = Y.$$

The remaining axioms of mereology are related to the *class functor* which converts distributive classes (general names) into individual entities. The class operator *Cl* is a principal tool in applications, cf., [58], [59], [61], [62], [68].

We may now introduce the notion of a (collective) class via the class functor *Cl.*

3.1.1 The class operator

An individual X is the class of a non-vacuous collection M of objects, in symbols, X is $Cl\mathcal{M}$, if

Cl1. If $Y \in \mathcal{M}$ then *Y el X;*
Cl2. $\forall$ Z.(Z is *el* X $\Rightarrow \exists$ U, W.(U is Y $\wedge$ W is *el* U $\wedge$ W is *el* Z).
Let us disentangle the meaning of this definition. First, we may realize that the class operator *Cl* is intended as the operator converting names (general sets of entities) into individual entities i.e. collective classes; its role may be fully compared to the role of the union of sets operator in the classical set theory. The analogy is indeed not only functional but also formal.
By Cl1., the class contains any member of the collection; by Cl2., the class contains all objects Z with the property: each element of Z has an element in common with an object in the collection.
Thus, the class functor pastes together individuals in *Y* by means of their common elements.
A basic tool in reasoning by means of mereology, see [51], is the following inference rule,
IR. For *X, Y:* if for each Z (from Z *is el* X it follows that there is *T* such that *T* is *el* Z $\wedge$ *T* is *el* Y) then X is *el* Y.

3.2 Mereology based on connection

This approach [85], [46], [12] is based on the functor of being connected; for the uniformity of exposition sake, we will formulate all essentials of this theory in the ontology language applied above.
The requirements for a functor *Con* of connection are as follows,
Conl. *X* is *Con Y* => X is X $\wedge$ *Y* is *Y* (asserting that *Con* is defined on individuals);
Con2. *X* is *Con X* (asserting refiexivity of *Con);*
Con3. *X* is *Con Y* $\Leftrightarrow$ *Y* is *Con X* (asserting that *Con* is symmetric);
Con4. For all Z (Z is *Con X* $\Leftrightarrow$ Z is *Con Y* $\Rightarrow$ X = Y) (asserting extensionality of *Con).*
From the functor *Con*, other functors are derived,
DCon. X is *DConY* $\Leftrightarrow$ *non*(X is *ConY*) (*X is disconnected from Y*);
El_C. X is el_C Y $\Leftrightarrow$ for all Z (Z is *ConX*) $\Rightarrow$ Z is *ConY*) (X is a connection element of Y);
P. X is P Y $\Leftrightarrow$ X is el_C Y $\wedge$ *non*(Y is el_C X) (*X is a part* of Y);
Ov. *X* is *Ov Y* $\Leftrightarrow$ exists Z (Z is el_C X) $\wedge$ Z is el_C Y) (*X, Y overlap*);
ECon. X is *.ECon Y* $\Leftrightarrow$ X is *Con Y* $\wedge$ *non* (X is *Ov* Y) (X is *externally connected to Y);*
TP. *X* is *TP Y* $\Leftrightarrow$ X is *P Y* $\wedge$ exists Z (Z is *.ECon X* $\wedge$ Z is *.ECon* Y) *(X is a tangential part of Y)*;
NTP. *X* is *NTP Y* $\Leftrightarrow$ *X* is *P Y* $\wedge$ *non* (X is *TP* Y) (*X is a non-tangential part of* Y).
Connection allows for a variety of functors of topological characters (one may define a quasi-topological interior by means of *NTP*, cf., eg., [5], [6], [12], [55]).

4. Rough mereology

Rough mereology, see [59], [60], [61], [62] begins with the notion of a *rough inclusion* which is a parameterized relation μ_r such that for any pair of individual entities *X, Y* the formula Y is μ_r X means that Y is a *part of* X *to a degree r* where $r \in [0,1]$.
The following is the list of basic postulates for rough inclusions; *el* is the element relation of a chosen mereology system.
RM1. *X* is μ_1Y $\Leftrightarrow$ X is *el(Y)* (a part in degree 1 is equivalent to an element);
RM2. *X* is μ_1Y $\Rightarrow$ for all Z (Z is μ_r X => Z is μ_r Y) (monotonicity of μ);
RM3. *X* is μ_r Y $\wedge$ $s \leq r \Rightarrow$ X is μ_s Y (assuring the meaning "a part in degree *at least r*").

4.1 Rough inclusions on collections of objects

Assume that we are given two individuals X, Y being classes of (finite) names:
X=Cl(X′), Y=Cl(Y′) and that we have defined values of μ for pairs T, Z of individuals where $T$is X', $Z$is Y'.
We extend μ to a measure μ^* on X, Y by letting:

$$r = min_{Z\in Y'}\{max_{T\in X'}max\{s : Z \text{ is } \mu_s(T)\}\} \quad \text{and} \quad Y \text{ is } \mu_r^*(X).$$

It may be proved straightforwardly that the measure μ^* satisfies (RM1)− (RM4).

4.2 Transitive rough inclusions

We introduce now [58], a modification to our functors μ_r; it is based on an application of residuated implication [38] and a measure of containment defined within the fuzzy set theory (the necessity measure) [34] , [7] . Combining the two ideas, we achieve a formula for μ_r which allows for a transitivity rule; this rule will in turn allow to introduce into our universe rough mereological topologies. We therefore recall the notion of a *t-norm* $\top$ as a function of two arguments $\top : [0, 1]^2 \to [0, 1]$ which satisfies the following requirements:

2. $\top(x,y) = \top(y,x)$;
3. $\top(x, \top(y,z)) = \top(\top(x,y),z)$;
4. $\top(x,1) = x$;
5. $x' \geq x \wedge y' \geq y \Rightarrow \top(x', y') \geq \top(x, y)$.

We also invoke a notion of fuzzy containment $\subset_r$ based on *necessity,* cf., [34]; it relies on a many-valued implication Υ i.e. on a function $\Upsilon : [0, 1]^2 \Rightarrow [0, 1]$ according to the formula:

$$X \subset_r Y \Longleftrightarrow \forall Z.(\Upsilon(\mu_X(Z), \mu_Y(Z)) \geq r),$$

where μ_A is the fuzzy membership function [38] of the fuzzy set A.
We replace Υ with a specific implication, viz., the residuated implication $\vec{\top}$ induced by $\top$ and defined by the formula: $\vec{\top}(r, s) \geq t \Longleftrightarrow \top(t, r) \leq s$.
We define a predicate $\mu_{\top,r}$ where $r \in [0, 1]$, according to the formula:

$$X \text{is } \mu_{\top,r}\ Y \Longleftrightarrow \text{for all } Z \text{ (exist } t, w \text{ (}Z\text{is } \mu_t\ X\ \wedge Z\text{is } \mu_w\ Y\ \wedge \vec{\top}(t, w) \geq r.)$$

As proved in a different context in [58], $\mu_{\top,r}$ satisfies (RM1-RM4).
The rough inclusion $\mu_{\top,r}$ does satisfy a deduction rule of the form, DR. If X is $\mu_{\top,r}Y$ and Y is $\mu_{\top,s}Z$ then X is $\mu_{\top,\top(r,s)}Z$.
We now propose to synthesize basic topological and geometric constructs applied in Qualitative Spatial Reasoning based on connection, e.g., [6], [5], by means of rough mereology.

5. Mereotopology

As mentioned few lines above, topological structures may be defined within the connection framework via the notion of a non− tangential part. Interior entities are formed then by means of some fusion operators, see, e.g. [5], [55]. The functor of connection allows also for some calculi of topological character based directly on regions, e.g., *RCC* – *calculus,* see,

[31]. For a different approach where connection may be derived from the axiomatized notion of a boundary, see, [72].
These topological structures provide a mereotopological environment in which it is possible to carry out spatial reasoning. We now demonstrate that in rough mereological framework one defines in a natural way Cech topologies.
We would like to recall that a topology on a given domain *U* may be introduced by means of a closure operator *cl* satisfying the *Kuratowski axioms* [43]:
(cl1) $cl\varnothing = \varnothing$;
(cl2) $clcl\ X = cl\ X$;
(cl3) $X \subseteq clX$;
(cl4) $cl(X \cup Y) = cl\ X \cup cl\ Y$.
The dual operator *int* of *interior* is then defined by means of the formula: $int\ X = U - cl(U - X)$ and it has dual properties: $int\varnothing = \varnothing$, $intintX = intX$, $intX \subseteq X$, $int(X \cap Y) = intX \cap intY$.
The Čech topology [19] is a weaker structure as it is required here only that the closure operator satisfies the following:
(Čcl1) $cl\varnothing = \varnothing$;
(Čcl2) $X \subseteq clX$;
(Čcl3) $X \subseteq Y \Rightarrow dX \subseteq dY$.
so the associated Čech interior operator *int* should only satisfy the following: $int\varnothing = \varnothing$; $intX \subseteq X$; $X \subseteq Y \Rightarrow intX \subseteq intY$.
Čech topologies arise naturally in problems when one considers coverings induced by similarity relations [53].
In order to define Čech topologies, we first define the class $Cl_r\ X$ for any object X and $r < 1$, as the class of objects having the property $M_r\ X$ of being a part of X to a degree r: $Cl_r\ X = Cl\ M_r\ X$.
By means of the rule (DR), sect.4.2, one can establish the properties,
Mon. For $s \leq r$, $Cl_r\ X$ is *el* $Cl_s\ X$.
Her. $X$is *el* $Y \Rightarrow Cl_r\ X$is *el* $Cl_r\ Y$.
Following this, we define a new functor *int* as the class of the property $I(X)$, $int\ X = Cl\ I(X)$, where Zis $I(X) \Leftrightarrow$ exists $s < 1$ ($Cl_s\ Z$ is *el* X).
We have the following properties of *int*,
I1. $int(X)$ is *el* X;
I2. X is *el* $Y \Rightarrow int(X)$ is *el* $int(Y)$.
Properties (I1)-(I2) witness this quasi-topology is a *quasi-Čech topology*. We denote it by the symbol τ_μ.
We now study the case of mereotopology under functors $\mu_{\top,r}$ in this case, the quasi- Čech topology τ_μ turns out to be a quasi-topology.

5.1 Mereotopology in the case of $\mu_\top$

We begin with an application of deduction rule (DR). We denote by the symbol $Cl_{\top,r}X$ the set $Cl_r\ X$ in case of the rough inclusion $\mu_\top$. We assume that $\top(r, s) < 1$ when $rs < 1$. We have a new direct characterization of $Cl_{\top,r}X$: Zis *el* $Cl_{\top,r}X \Leftrightarrow$ Zis $\mu_{\top,r}X$.
This characterization implies that $Cl_{\top,r}X$ may be regarded as *"an open ball of radius r centered at X"*.

We assume now, additionally, that the t-norm $\top$ has the property that: for every $r < 1$ there exists $s < 1$ such that $\top(r, s) > r$. With this assumption, we have the following property,
INC. For $Z$is *el* $Cl_{\top,r}X$, if $s_0 = \text{arg\_min}\{s : \top(r, s) \geq r\}$ then

$$Cl_{\top,s_0}Z \text{ is } el\ (Cl_{\top,r}X).$$

We define a functor of two nominal individual variables *AND*, AND. $Z$is *el* $AND(X, Y) \Leftrightarrow Ov(X, Y) \wedge Z$is *el* $X \wedge Z$is *el* Y.
The rough mereotopology $\tau_{\mu_\top}$ has the properties:
INTER. $AND(int(X),int(Y)) = int(AND(X,Y))$ holds whenever AND(int(X),int(Y)) is non-empty.
IDEM. $int(int(X)) = int(X)$.
It follows by INTER and IDEM that the rough mereological topology induced by the rough inclusion $\mu_{\top,r}$ is a quasi-topology.

6. Connections from Rough Inclusions

In sect. 3.2, we presented basic notions related to mereological theories based on the notion of a connection. We recall that a connection is a functor which satisfies axioms (Conl)-(Con4) of sect. 3.2.
In this section we will investigate some methods for inducing connections from rough inclusions. Clearly, the presence of topology induced in the preceding section allows for a few approaches to this problem. We begin with a notion of a connection in a strong sense.

6.1 Strong connection

We define a name-forming functor $C\,on_T$ on individual entities as follows, CON(T).
$X$is $C\,on_T\,Y \Leftrightarrow non(\text{exist } r,s < 1.ext(Cl_r X, Cl_s Y))$.
Thus, X and Y are connected in the strong sense whenever they cannot be separated by means of their open neighborhoods.
We check whether $C\,on_T$ thus defined does satisfy (Conl)-(Con4). It may be clear that (Conl), (Con2), (Con3) hold irrespective of properties of μ. The status of (Con4) will clearly depend on our assumed functor μ. In case $non(X = Y)$, we have, e.g., Z is *el* X), $ext(Z,Y)$ with some Z. Clearly, Z is $C\,on_T\,X$; to prove that $non(Z$ is $C\,on_T\,Y)$, we need some assumptions about the form of μ.
We add a new property of μ,
RM4. If it is not true that X is Ov Y, then there exists $s < 1$ such that (if X is $\mu_r\,Y$ then $r < s$).
Assume (RM4) and consider $\mu_\top^s$ with a t-norm $\top$ which would satisfy the following: given $s < 1$, there exist $\alpha, \beta < 1$ with the property that $\top(\alpha, \beta) > s$. Then, $C\,on_T$ induced via $\mu_\top^s$ would satisfy (Con4).
In connection framework, the notion of an element is derived from the functor C of connection; the resulting functor of an element is denoted here by the symbol el_C We will find relationships between the original functor *el* of an element and the functor el_C. To this end, we have
ELEM1. For any functor of the form $\mu_\top$: X is *el* $Y \Rightarrow X$ is $el_{C_T}\,Y$.
ELEM2. For any functor of the form $\mu_\top^s$: X is $el_{C_T}Y \Rightarrow X$is *el* Y.
A corollary follows,
For any functor of the form $\mu_\top^s$: el_{C_T} and *el* are equivalent.

We may therefore create in the framework of rough mereology an alternative scheme of calculus of individuals based on the connection C_T inducing the same notion of an element as the original mereological one.

7. Mereogeometry

Predicates μ_r may be regarded as weak metrics also in the context of geometry. From this point of view, we may apply μ_r in order to define basic notions of rough mereological geometry.

In the language of this geometry, we may approximately describe and approach geometry of objects described by data tables; a usage for this geometry may be found, e.g., in navigation and control tasks of mobile robotics [1], [3], [21], [37], [40], [41].

It is well-known (see, [80], [8]) that the geometry of Euclidean spaces may be based on some postulates about the basic notions of a point and the ternary equi-distance functor. In [80], postulates for Euclidean geometry over a real-closed field were given based on the functor of betweenness and the quaternary equi−distance functor. Similarly, in [8], a set of postulates aimed at rendering general geometric features of geometry of finite−dimensional spaces over reals has been discussed, the primitive notion there being that of nearness.

Geometrical notions have been applied in, e. g., studies of semantics of spatial prepositions [6] and in inferences via cardinal directions [42].

7.1 Rough mereological distance, betweenness

We first introduce a notion of distance κ_r in our rough mereological universe by letting

$$\kappa_r(X,Y) \Longleftrightarrow r = \min\{max\ u, max\ w : X \text{is}\ \mu_u\ Y \wedge Y \text{is}\ \mu_w\ X\}.$$

We now introduce the notion of betweenness as a functor *T(X, Y)* of two individual names; the statement Zis *T(X, Y)* reads as '*Z is between X and Y:*

$$Z \text{is}\ T(X,Y) \Longleftrightarrow \text{for all W}\ \kappa_r(Z,W) \wedge \kappa_s(X,W) \wedge \kappa_t(Y,W) \Longrightarrow s \le r \le t \vee t \le r \le s.$$

Thus, *Z*is *T(X,Y)* holds when the rough mereological distance κ between Z and any *W* is in the non−oriented interval (i.e. between) [distance of *X* to *W*, distance of *Y* to *W*] for any *W*. One checks that *T* satisfies the axioms of Tarski [80] for *betweenness.*

Proposition 1 The following properties hold:

1. Zis T(X, X) => Z = X (*identity*);
2. Yis T(X, U) ∧ Zis T(Y, U) ⇒ Yis T(X, Z) (*transitivity*);
3. Yis T(X, Z) ∧ Yis T(X, U) ∧ X ≠ Y ⇒ Zis T(X, U) ∨ Uis T(X, Z) (*connectivity*).

Proof 1 *By means of* κ, *the properties of betweenness in our context are translated into properties of betweenness in the real line which hold by the Tarski theorem [80], Theorem 1.*

7.2 Nearness

We may also apply κ to define in our context the functor *N* of nearness proposed in van Benthem [8]:

$$Z \text{is}\ N(X,Y) \Longleftrightarrow (\kappa_r(Z,X) \wedge \kappa_s(X,Y) \Longrightarrow s < r).$$

Here, nearness means that Z is closer to X than to Y (recall that rough mereological distance is defined in an opposite way: the smaller r, the greater distance).
Then the following hold, i.e., N does satisfy all axioms for nearness in [8],
Proposition 2 1. *Zis N(X, Y) ∧ Yis N(X, W) ⇒ Zis N(X, W) (transitivity);*
2. Zis N(X, Y) ∧ Xis N(Y, Z) ⇒ *Xis N(Z, Y) (triangle inequality);*
3. *non*(*Zis N*(*X, Z*)) (*irreflexivity*);
4. Z = X ∨ *Zis N*(*Z, X*) (*selfishness*);
5. *Zis N*(*X, Y*) ⇒ *Zis N*(*X, W*) ∨ *Wis N*(*X, Y*) (*connectedness*).
Proof 2 *(4) follows by (RM1); (3) is obvious. In proofs of the remaining properties, we introduce a symbol $\mu(X, Y)$ as a value of r for which $\kappa_r(X, Y)$. Then, for (l), assume that Z is N(X,Y),Y is N(X,W) i.e. $\mu(Z,X) > \mu(X,Y)$, $\mu(X,Y) > \mu(X, W)$ hence $\mu(Z, X) > \mu(X, W)$ i.e. Zis N(X, W). In case (2), Zis N(X, Y), Xis N(Y, Z) mean $\mu(Z, X) > \mu(X, Y)$, $\mu(X, Y) > \mu(Y, Z)$ so $\mu(Z, X) > \mu(Y, Z)$ i.e. XisN(Z,Y). Concerning (v), Zis N(X,Y) implies that $\mu(Z,X) > \mu(X,Y)$ hence either $\mu(Z, X) > \mu(X, W)$ meaning Zis N(X, W) or $\mu(X, W) > \mu(X, Y)$ implying Wis N(X,Y).*
We now may introduce the notion of equi−distance as a functor *Eq(X, Y)* defined as follows:

$$Z\text{is } Eq(X,Y) \Longleftrightarrow (non(X\text{is } N(Z,Y)) \wedge non(Y\text{is } N(Z,X))).$$

It follows that
Proposition 3 *Zis Eq(X,Y) ⇔ (for allr (κ_r (X,Z) ⇔ κ_r (Y,Z)).*
We may also define a functor of equi−distance following Tarski [80]:

$$D(X,Y,Z,W) \Longleftrightarrow (\text{for all r } \kappa_r(X,Y) \Longleftrightarrow \kappa_r(Z,W)).$$

These functors do clearly satisfy the following (see, [8], [80])
Proposition 4 *1. Zis Eq(X, Y) ∧ Xis Eq(Y, Z) ⇒ Yis Eq(Z, X) (triangle equality);*
2. *Zis T(X, Y) ∧ Wis Eq(X, Y) =^> D(Z, W, X, W) (circle property);*
3. *D(X,Y,Y,X) (reflexivity);*
4. *D(X, Y, Z, Z) ⇒ X = Y (identity);*
5. *D(X, Y, Z, U) ∧ D(X, Y, V, W) ⇒ D(Z, U, V, W) (transitwity).*
One may also follow van Benthem's proposal for a betweenness functor defined via the nearness functor as follows:

$$Z\text{is } T_B(X,Y) \Longleftrightarrow [\text{for all } W \; (Z \text{ is } W \vee Z\text{is } N(X,W) \vee Z\text{is } N(Y,W))].$$

One checks in a straightforward way that
Proposition 5 The functor T_B of betweenness defined according to the above does satisfy the Tarski axioms.

7.3 Points

The notion of a point may be introduced in a few ways; e.g. following Tarski [78], one may introduce points as classes of names forming ultrafilters under the ordering induced by the functor of being an element *el*. Another way, suitable in practical cases, where the universe, or more generally, each ultrafilter F as above is finite, i.e., principal (meaning that there exists an object X such that F consists of those Y'"s for which *Xis el(Y)* holds) is to define points as atoms of our universe under the functor of being an element i.e. we define a constant name AT as follows:

$$X\text{is } AT \Longleftrightarrow non(\text{exists } Y \ (Y\text{is } el(X) \wedge non(X\text{is } = Y).$$

We will refer to such points as to *atomic points.* We adopt here this notion of a point.
Clearly, restricting ourselves to atomic points, we preserve all properties of functors of betweenness, nearness and equi−distance proved above to be valid in the universe V.

7.4 Mereogeometry: Examples

We will adopt the notion of betweenness T_B based on the nearness functor.
We give some examples of specific contexts in which this functor can be realized.
Example 3 *With reference to Example 1, we adopt as objects topologically connected unions of finitely many cubes in the unit grid on the space R^d (topological connectedness will be defined recursively: (1). a single cube is connected; (2) given a connected union C and a cube c, the union $C \bigcup c$ is connected if c is adjacent by the edge or a vertex to a cube in C). We adopt as the rough inclusion H the function μ(C,D,r) iff* $\frac{|C \bigcap D|}{|C|} \geq r$, *where* $|C \bigcap D|$ *is the number of cubes common to C and D and* $|C|$ *is the number of cubes in C. The mereological distance between C and D is then:* $\kappa(C,D) = min\{max\{r : \mu(C,D,r)\}, max\{s : \mu(D, C,s)\}\}$.
Then one checks that: the connected union E is between (in the sense of T_B) disjoint cubes c, d whenever E contains c and d and E consists of a minimal number of cubes for E being connected.
One can interpret this result as follows: when c,d are obstacles (e.g., of the size of a mobile robot) then Z minus c, d gives the space between these obstacles, free for the robot to pass.
Example 4 *In the same frame and with same μ, consider unions of grid cubes (not necessarily topologically connected). Then for disjoint cubes c, d:* a, union Z of grid cubes is between *c,d* (in the sense of) T_B whenever Z is either *c*, or *d*, or $c \bigcup d$.
One can interpret this as: when unit grid is of the size of a mobile robot, Z between c and d, can be interpreted as beacons (landmarks) between which a robot may pass safely.
Example 5 *We consider now obstacles modeled as squares on R^2 of equal size given by edge length r, but not necessarily conforming to any grid, i.e., a square can be centered at any point; we change the rough inclusion μ, to the following: μ (C,D,r) iff* $\frac{\|C \bigcap D\|}{\|C\|} \geq r$, *where* $\|C\|$ *is the area of C. In this setting, consider disjoint squares C,D whose centers are on the line e.g., x=0.*
Then one checks: any square centered on x=0 with the center on the segment joining centers of C and D of edge length r is between C, and D, in the sense of T_B.
One can interpret this as follows: a robot of size of the square of edge length r, should interpret the space formed by the rectangular space between C and D minus C,D as the eventual free space for bypassing either C or D.
Example 6 *In the setting of Example 5, we replace squares with discs of equal radius r placed arbitrarily in the space R^2, with μ, defined as therein by means of the area. Then conclusion holds with disks as with squares.*

Example 7 *The equi-distance functor Eq may be used to define spheres; for instance, admitting as Z the square* [0,1] x [0,1], *(so the edge length is 1) we have the sphere* S(Z;1/2) *to contain squares*: [0,1] x [1/2,3/2], [0,1] x [-1/2,1/2], [1/2,3/2] x [0,1], [-1/2,1/2] x [0,1]}.
A line segment may be defined via the auxiliary notion of a pattern; we introduce this notion as a functor *Pt.* We let

$$Pt(X,Y,Z) \iff Z \text{is } T_B(X,Y) \vee X \text{is } T_B(Z,Y) \vee Y \text{is } T_B(X,Z).$$

We will say that a finite sequence $X_1, X_2, \dots, X_n$ of objects *belong in a line segment whenever* $Pt(X_i, X_{i+1}, X_{i+2})$ for $i = 1,\dots, n-2$; formally, we introduce the functor *Line* of finite arity defined via

$$Line(X_1, X_2, ..., X_n) \iff \forall i < n-1. Pt(X_i, X_{i+1}, X_{i+2}).$$

Example 8 *With reference to Example 5, consider a sequence of squares* $C_1,\dots,C_k$ *all centered on x=0, and of edge length r. Then, Line*($C_1,\dots,Ck$) *holds. One can interpret this as a robot moving along a straight line when its consecutive positions are* $C_1,\dots,Ck$.
The notion of orthogonality may be introduced in a well-known way; we introduce a functor *Ortho:* for two line segments *A, B,* with an object Z common to sequences *A* and *B,* we let

$$Ortho(A,B) \iff \exists X,Y,U,W.X,Y \text{ is in } A \wedge U,W \text{ is in } B \wedge non(X \text{ is } Y)$$

$$\wedge$$

$$non(U \text{ is } W) \wedge U \text{ is } Eq(X,Y) \wedge W \text{ is } Eq(X,Y),$$

Example 9 *With reference to Example 8, consider lines A: the sequence* [0, 1] x [0,1], {[*i*/2, *i* + 1/2] x [0,1] : *i* = +/-1,+/-2,...};*B: the sequence* [0,1] x [0,1], {[0,1] x [*i*/2, *i*+ 1/2] : *i* = +/-!,+/- 2,...}. *Then Ortho(A, B) holds as witnessed, e.g., by Z* = [0,1] x [0,1], *X* = [-1/2,1/2] x [0,1], *Y* = [1/2,3/2] x [0, 1], *U* = [0, 1] x [-1/2, 1/2], *W* = [0, 1] x [1/2, 3/2].
Example 10 *Finally, we consider a more general context in which objects are rectangles positioned regularly, i.e., having edges parallel to axes in* R^2. *The measure* μ *is the one in Example 5. In this setting, given two disjoint rectangles C, D, the only object between C and D is the extent of C, D, ext(C,D), i.e., the rectangle which is the minimal rectangle containing the union* $C \cup D$. *To see this, one can consider two identical squares C, D as in Example 5 and solve the problem analytically by showing that there is no other rectangle nearer to C and D (this requires solving a set of linear inequalities); then, the general case follows by observing that linear shrinking or stretching of an edge does not change the area relations.*
Rectangles C, D and their extent ext(C, D) form then a line segment.

8. A.Szmigielski's model for localization and navigation

In his PhD disssertation [77], A. Szmigielski has proposed an approach to localization and navigation of a mobile robot using the mereological geometry. The robot was a *Pioneer P2DX* endowed with sonar emitter in the environment of sonar sensors.
The control of a robot utilized the parameters,

- length of route segment *d,*
- rotation angle α,

where the robot first traveled the distance d then rotated by α. Low level control means a sequence of pairs (d, α). Distance to a nearest obstacle was determined by robot sonars.
The system of sonar sensors can be exploited towards localization of a robot in the global reference frame. The system consists of a sonar emitter positioned on the robot and up to 16 receivers in the environment of the robot. Simultaneous measurement of times of flight between the emitter and receivers gives distances from the robot to receivers. At the same time, robot sonars detect the nearest obstacle.
As the result, two regions are determined in 2D space of the robot plane: the disk centered at the receiver with the radius equal to the distance emitter -receiver (the receiver region), and the disk centered at the robot with the radius equal to the distance to the nearest obstacle (the collision-free region). The robot position is on the boundary of the receiver region.
A relation between the two regions is expressed by the mereodistance κ.
Whereas the Euclidean distance is context—free, the mereological distance depends on the environment (which bears on radii of regions).
The goal - reaching a desired receiver, can be formulated as the requirement that radii of the receiver and the collision free regions are equal and the distance κ between them is maximal $= \frac{2}{3} - \frac{\sqrt{3}}{2\pi}$.
The stop criterion can be formulated then as the requirement that radii of two regions be equal; then, the receiver is found on the boundary of the collision—free region.

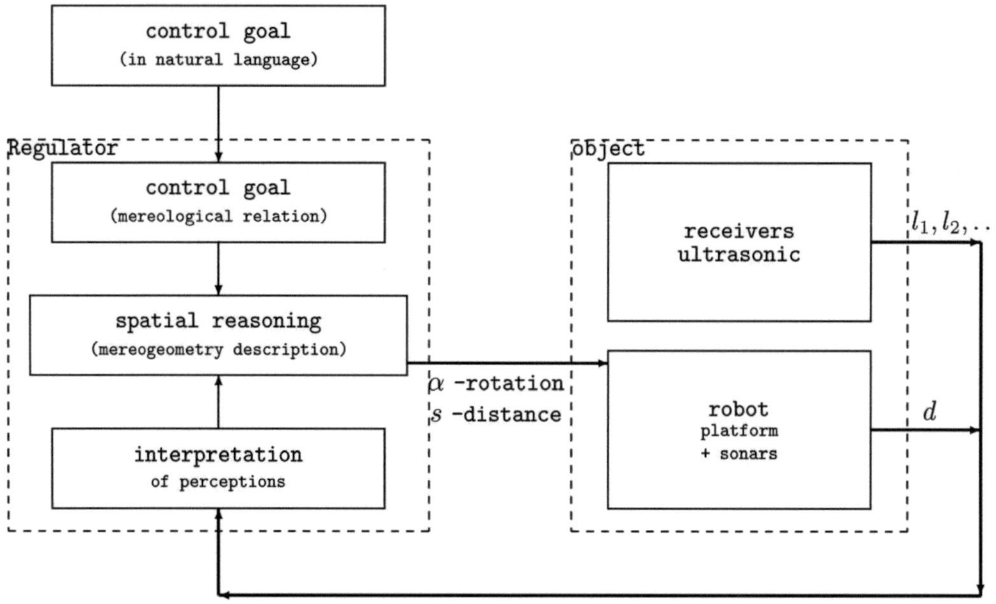

Figure 1. Spatial reasoning in closed loop

Control in dynamic setting is performed in the closed loop: perception – decision – action – perception. The inference engine gives two parameters:
s = *the distance to goal,*
α = *the rotation angle of the robot.* The input parameters are:
d = *the radius of the collision – free region,*

$l_1, l_2, \ldots, l_k$ = *radii of receiver regions.*

The block diagram of control is shown in Fig. 1.

Fig. 2 shows the result of the experiment of reaching the goal in an environment with obstacles; trajectory is shown for the robot starting at the orientation of 180 degrees with respect to obstacle.

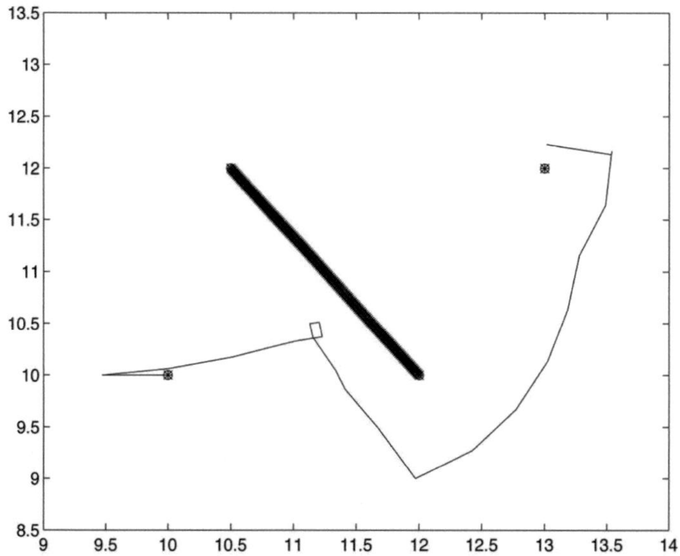

Figure 2. Robot reaches the goal

Further results can be found in [77] and [63].

9. Implementation in Player/Stage software system

Player/Stage is an Open-Source software system designed for many UNIX-compatible platforms, widely used in robotics laboratories [57]. Main two parts are Player - message passing server (with bunch of drivers for many robotics devices, extendable by plugins) and Stage - a plug-in for Players bunch of drivers which simulates existence of real robotics devices that operate in simulated 2D world. Player/Stage offers client—server architecture. Many clients can connect to one Player server, where clients are programs (robot controllers) written by a roboticist who can use Player client—side API. Player itself uses drivers to communicate with devices, in this activity it does not make distinction between real and simulated hardware. It gives roboticist means for testing programmed robot controller in both real and simulated world.

Among all Player drivers that communicate with devices (real or simulated), there are drivers not intended for controlling hardware, instead those drivers offer many facilities for sensor data manipulation, for example, camera image compression, retro—reflective detection of cylindrical markers in laser scans, path planning. One of the new features added to Player version 2.1 is the PostGIS driver: it connects to PostgreSQL database in order to obtain and/or update stored vector map layers.

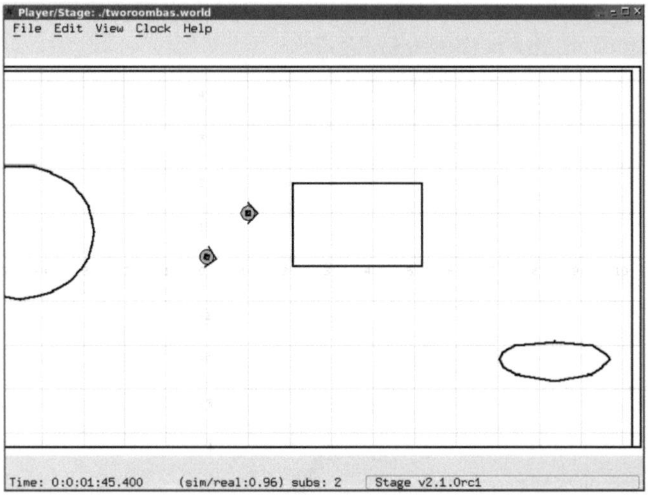

Figure 3. Stage simulator in use - two iRobot Roomba robots inside of simulated world

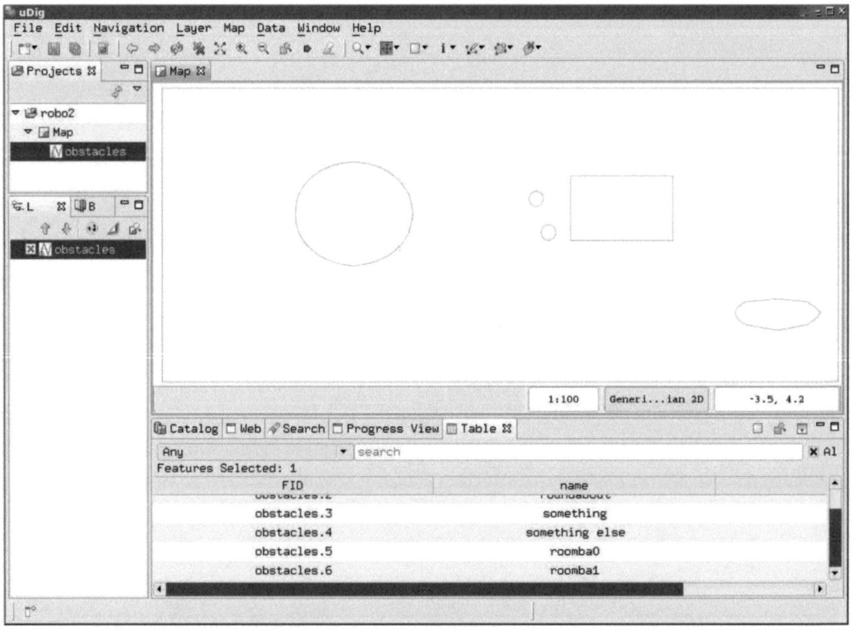

Figure 4. uDig application in use - modification of *obstacles* layer

PostGIS itself is an extension to the PostgreSQL object-relational database system which allows GIS (Geographies Information Systems) objects to be stored in the database [64]. It also offers new SQL functions for spatial reasoning. Maps which to be stored in SQL database can be created and edited by graphical tools like uDig or by C/C++ programs written using GEOS library of GIS functions. PostGIS, uDig and GEOS library are projects maintained by Refractions Research.

A map can have many named layers, for each layer a table in SQL database is created. We can assume that layer named *obstacles* is full of objects that a robot cannot walk through. Other layers can be created in which we can divide robots workspace in areas with assigned attribute which for example tells if the given area is occupied by an obstacle or not. During our experimentations, we have created a plug-in for Players bunch of drivers which constantly tracks changes of position of every robot and updates *obstacles* layer such as robots are remarked as any other obstacle. As a result, the map stored in SQL database is kept always up to date. This feature is also useful in multi-agent environments: at any time a robot controller can send a query to SQL database server regarding every other robot position.
A roboticist can write a robot controller using Player client-side API which obtains information about current situation through the *vectormap* interface. Additionally, to write such a program, PostgreSQL client-side API can be used in order to open direct connection to the database server on which our mereoge-ometry SQL functions are stored together with map database. These functions can be called using this connection, results are sent back to the calling program. This gives robot controller program ability to perform spatial reasoning based on rough mereology.
Using PostGIS SQL extensions we have created our mereogeometry SQL functions [44]. Rough mereological distance is defined as such:

```
CREATE FUNCTION meredist(objectl geometry, object2 geometry)
RETURNS DOUBLE PRECISION AS
$$
    SELECT min(degrees.degree) FROM
    ((SELECT
            ST_Area(STJntersection(extent($l), extent($2)))
            / ST_Area(extent($l))
            AS degree)
     UNION (SELECT
            ST_Area(STJntersection(extent($l), extent($2)))
            / ST_Area(extent($2))
            AS degree))
     AS degrees;
$$ LANGUAGE SQL STABLE;
```

Having mereological distance function we can derive nearness predicate:

```
CREATE FUNCTION merenear(obj geometry, ol geometry, o2 geometry)
RETURNS BOOLEAN AS
$$
    SELECT meredist($l, $2) > meredist($3, $2)
$$ LANGUAGE SQL STABLE;
```

The equi-distance can be derived as such:

```
CREATE FUNCTION mereequ(obj geometry, ol geometry, o2 geometry)
RETURNS BOOLEAN AS
$$
    SELECT (NOT merenear($l, $2, $3))
            AND (NOT merenear($l, $3, $2));
            $$ LANGUAGE SQL STABLE;
```

Our implementation of the betweenness predicate makes use of a function that produces an object which is an extent of given two objects:

```
CREATE FUNCTION mereextent(objectl geometry, object2 geometry)
RETURNS geometry AS
$$
    SELECT GeomFromWKB(AsBinary(extent(objects.geom))) FROM
    ((SELECT $1 AS geom)
    UNION (SELECT $2 AS geom))
    AS objects;
$$ LANGUAGE SQL STABLE;
```

The betweenness predicate is defined as such:

```
CREATE FUNCTION merebetb(obj geometry, ol geometry, o2 geometry)
RETURNS BOOLEAN AS
$$
    SELECT
            meredist($l, $2) = 1
            OR meredist($l, $3) = 1
            OR
                    (meredist($l, $2) > 0
                    AND meredist($l, $3) > 0
                    AND meredist(mereextent($2, $3),
                            mereextent(mereextent($l, $2), $3)) = 1);
$$ LANGUAGE SQL STABLE;
```

Using the betweenness predicate we can check if three objects form a pattern:

```
CREATE FUNCTION merepattern
    (objectl geometry, object2 geometry, objects geometry)
    RETURNS BOOLEAN AS
    $$
      SELECT merebetb($3, $2, $1)
            OR merebetb($l, $3, $2)
            OR merebetb($2, $1, $3);
    $$ LANGUAGE SQL STABLE;
```

Also having pattern predicate we can check if four objects form a line:

```
CREATE FUNCTION mereisline4
    (objl geometry, obj2 geometry, obj3 geometry, obj4 geometry)
    RETURNS BOOLEAN AS
    $$
      SELECT merepattern($l, $2, $3) AND merepattern($2, $3, $4);
    $$ LANGUAGE SQL STABLE;
```

Those predicates can be used in global navigation task. We can create additional map layer for navigational marker objects. Whenever the target is set, a robot planner should form a path across navigational markers. The path itself can be a group of objects representing areas free of obstacles. This group of objects in the path from the robot to the target should form a mereological line. A robot should follow this path by going from one area centroid to another until the goal is reached. If the changes in the world are expected (e.g. in multi-robot environments) a planner should update the path within some interval.

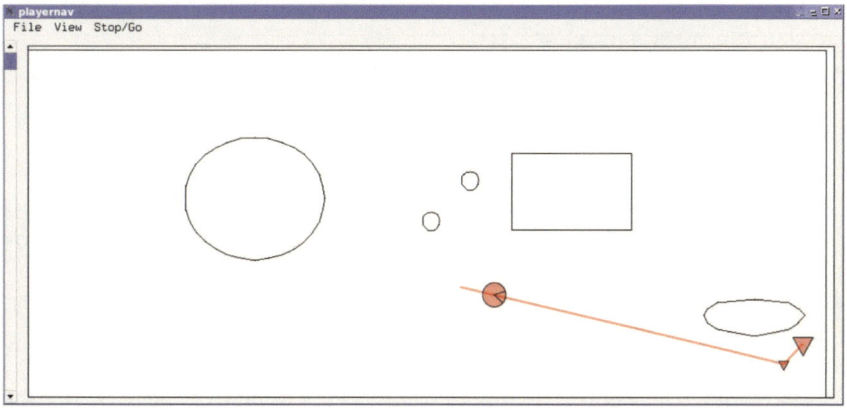

Figure 5. Playernav - a Player client-side application used to set a goal points for server-side planner driver

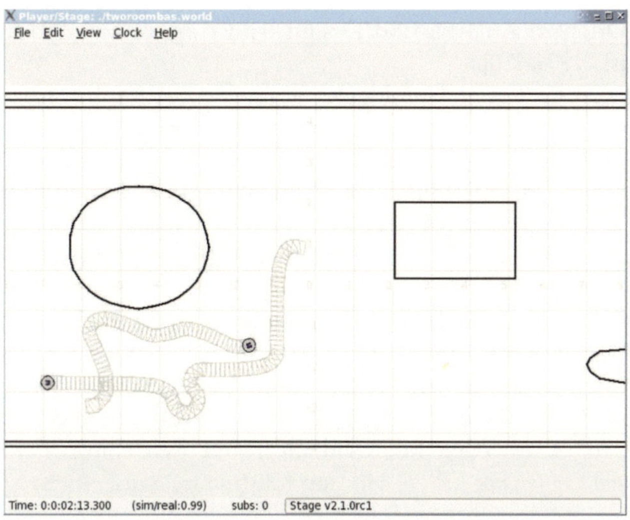

Figure 6. *Show trails* is a nice option in Stage which can be used to track robot trajectory. Here we can see two Roomba robots avoiding to hit obstacles and each other respectively. The robot controller program was using *meredist* function in order to choose free space area as a local target. This method of navigation suffers from local optima problem: a robot can start to spin around one place between obstacles

10. References

AISB-97: Spatial Reasoning in Mobile Robots and Animals, *Proceedings AISB-97 Workshop*, Manchester Univ., Manchester, 1997. [1]

J. Allen, Towards a general theory of action and time, *Artificial Intelligence* 23(20), 1984, pp. 123-154. [2]

R. C. Arkin, Behavior-Based Robotics, MIT Press, Cambridge, MA, 1998. [3]

N.Asher, L. Vieu, Toward a geometry of commonsense: a semantics and a complete axiomatization of mereotopology, in: *Proceedings IJCAI'95*, Morgan KaufTman, San Mateo, CA, 1995, pp.846-852. [4]

N. Asher, M. Aurnague, M. Bras, P. Sablayrolles, L. Vieu, De 1'espace-temps dans 1'analyse du discours, *Rapport interne IRIT*/95-08-R, Institut de Recherche en Informatique, Univ. Paul Sabatier, Toulouse, 1995. [5]

M. Aurnague and L. Vieu, A theory of space-time for natural language semantics, in: K. Korta and J. M. Larrazabal, eds., *Semantics and Pragmatics of Natural Language: Logical and Computational Aspects*, ILCLI Series I, Univ. Pais Vasco, San Sebastian, 1995, pp. 69-126. [6]

W. Bandler, L. J. Kohout, Fuzzy power sets and fuzzy implication operators, *Fuzzy Sets and Systems* 4, 1980, 13-30. [7]

J. vanBenthem, The Logic of Time, Reidel, Dordrecht, 1983. [8]

T. Bittner, On ontology and epistemology of rough location, in: C. Freksa, D. M. Mark (eds.), Spatial Information Theory. *Cognitive and Computational Foundations of Geographic Information Science*, Lecture Notes in Computer Science vol. 1661, Springer Verlag, Berlin, 1999, pp. 433-448. [9]

G. Booch, Object-Oriented Analysis and Design with Applications, Addison-Wesley Publ., Menlo Park, 1994. [10]

R. Casati, B.Smith, A.C. Varzi, Ontological tools for geographic representation, in: N. Guarino (ed.), *Formal Ontology in Information Systems*, IOS Press, Amsterdam, 1998, pp. 77-85. [11]

B. L. Clarke, A calculus of individuals based on connection, *Notre Dame Journal of Formal Logic* 22(2), 1981, pp.204-218. [12]

R. Clay, Relation of Le'sniewski's Mereology to Boolean Algebra, *The Journal of Symbolic Logic* 39, 1974, pp. 638-648. [13]

A. G. Cohn, Calculi for qualitative spatial reasoning, in: J. Calmet, J. A. Campbell, J. Pfalzgraf (eds.), *Artificial Intelligence and Symbolic Mathematical Computation, Lecture Notes in Computer Science* vol. 1138, Springer Verlag, Berlin, 1996, pp. 124-143. [14]

A. G. Cohn, N. M. Gotts, Representing spatial vagueness: a mereological approach, in: Principles of Knowledge Representation and Reasoning: *Proceedings of the 5th International Conference KR'96*, Morgan Kaufmann, San Francisco, 1996, pp. 230-241. [15]

A. G. Cohn, A. C. Varzi, Connections relations in mereotopology, in: H. Prade (ed.), Proceedings ECAI'98. *13th European Conference on Artificial Intelligence*, Wiley and Sons, Chichester, 1998, pp. 150-154. [16]

A. G. Cohn, N. M. Gotts, The "egg-yolk" representation of regions with indeterminate boundaries, in: P. Burrough, A. M. Frank (eds.), *Proceedings GISDATA Specialist Meeting on Spatial Objects with Undetermined Boundaries*, Fr. Taylor, 1996, pp.171-187. [17]

Collected Works of Stanislaw Lesniewski, J. Srzednicki, S. J. Surma, D. Barnett, V. F. Rickey (eds.), Kluwer, Dordrecht, 1992. [18]

E. Čech, Topological Spaces, Academia, Praha, 1966. [19]

Z. Cui, A. G. Cohn, D. A. Randell, Qualitative and topological relationships, in: Advances in Spatial Databases, *Lecture Notes in Computer Science* vol. 692, Springer Verlag, Berlin, 1993, pp. 296-315. [20]

M. Dorigo, M. Colombetti, Robot Shaping. *An Experiment in Behavior Engineering*, MIT Press, Cambridge, MA, 1998. [21]

M. J. Egenhofer, R. G. Golledge (eds.), Spatial and Temporal Reasoning in Geographic Information Systems, Oxford U. Press, Oxford, 1997. [22]

C. Eschenbach, A predication calculus for qualitative spatial representations, in: C. Freksa, D. M. Mark (eds.), Spatial Information Theory. Cognitive and Computational Foundations of Geographic Information Science, *Lecture Notes in Computer Science* vol. 1661, Springer Verlag, Berlin, 1999, pp. 157-172. [23]

A. U. Frank, I. Campari (eds.), Spatial Information Theory: A Theoretical Basis for GIS, *Lecture Notes in Computer Science*, vol. 716, Springer Verlag, Berlin, 1993. [24]

A. U. Frank, W. Kuhn (eds.), Spatial Information Theory: A Theoretical Basis for GIS, *Lecture Notes in Computer Science*, vol. 988, Springer Verlag, Berlin, 1995. [25]

C. Freksa, D. M. Mark (eds.), Spatial Information Theory. Cognitive and Computational Foundations of Geographic Information Science. *Proceedings COSIT'99, Lecture Notes in Computer Science* vol. 1661, Springer Verlag, Berlin, 1999. [26]

C. Freksa, C. Habel, Repraesentation und Verarbeitung raeumlichen Wis-sens, *Informatik-Fachberichte*, Springer Verlag, Berlin, 1990. [27]

A. Galton The mereotopology of discrete space, in: C. Freksa, D. M. Mark (eds.), Spatial Information Theory. Cognitive and Computational Foundations of Geographic Information Science, *Lecture Notes in Computer Science* vol. 1661, Springer Verlag, Berlin, 1999, pp. 250-266. [28]

J. Glasgow, A formalism for model—based spatial planning, in: A. U. Frank, W. kuhn (eds.), Spatial Information theory - A Theoretical Basis for GIS, *Lecture Notes in Computer Science* vol. 988, Springer Verlag, Berlin, 1995, pp. 501-518. [29]

N. M. Gotts, A. G. Cohn, A mereological approach to representing spatial vagueness,in: Working papers, *the Ninth International Workshop on Qualitative Reasoning*, QR'95, 1995. [30]

N. M. Gotts, J. M. Gooday, A. G. Cohn, A connection based approach to commonsense topological description and reasoning, *The Monist* 79(1), 1996, pp. 51-75. [31]

N. Guarino, The ontological level, in: R. Casati, B. Smith, G. White (eds.), *Philosophy and the Cognitive Sciences*, Hoelder-Pichler-Tempsky, Vienna, 1994. [32]

S. C. Hirtle, A. U. Frank (eds.), Spatial Information Theory: A Theoretical Basis for GIS, *Lecture Notes in Computer Science* vol. 1329, Springer Verlag, Berlin, 1997. [33]

M. Inuiguchi, T. Tanino, Level cut conditioning approach to the necessity measure specification, in: Ning Zhong, A. Skowron, S. Ohsuga (eds.), New Directions in Rough Sets, Data Mining and Granular-Soft Computing, *Lecture Notes in Artificial Intelligence* vol. 1711, Springer Verlag, Berlin, 1999, pp. 193-202. [34]

B. Iwanus, On Lesniewski's Elementary Ontology, *Studia Logica* 31, 1973, pp. 73-119. [35]

T. Kotarbinski, Elements of the Theory of Knowledge, Formal Logic and Methodology of Science, *Polish Sci. Publ.*, Warsaw, 1966. [36]

D. Kortenkamp, R. P. Bonasso, R. Murphy (eds.), *Artificial Intelligence and Mobile Robots*, AAAI Press/MIT Press, Menlo Park, CA, 1998. [37]

R. Kruse, J. Gebhardt, F. Klawonn, Foundations of Fuzzy Systems, John Wiley & Sons, Chichester, 1984. [38]

B. Kuipers, Qualitative Reasoning: Modeling and Simulation with Incomplete Knowledge, MIT Press, Cambridge MA, 1994. [39]

B. J. Kuipers, Y. T. Byun, A qualitative approach to robot exploration and map learning, in: *Proceedings of the IEEE Workshop on Spatial Reasoning and Multi-Sensor Fusion*, Morgan Kaufmann, San Mateo CA, 1987, pp. 390-404. [40]

B. J. Kuipers, T. Levitt, Navigation and mapping in large-scale space, *AI Magazine* 9(20), 1988, pp. 25-43. [41]

L. Kulik, A. Klippel, Reasoning about cardinal directions using grids as qualitative geographic coordinates, in: C. Freksa, D. M. Mark (eds.), Spatial Information Theory. Cognitive and Computational Foundations of Geographic Information Science, *Lecture Notes in Computer Science* vol.1661, Springer Verlag, Berlin, 1999, pp. 205-220. [42]

C. Kuratowski, Topology I, II, Academic Press and Polish Scientific Publishers, New York-Warsaw, 1966. [43]

H. Ladanyi, SQL Unleashed, Sams Publishing, 1997. [44]

T. De Laguna, Point, line, surface as sets of solids, *J. Philosophy* 19, 1922, pp. 449-461. [45]

H. Leonard, N. Goodman, The calculus of individuals and its uses, *The Journal of Symbolic Logic* 5, 1940, pp. 45-55. [46]

Cz. Lejewski, On Lesniewski's Ontology, *Ratio* 1(2), 1958, pp. 150-176. [47]

Cz. Lejewski, A contribution to lesniewski's mereology, *Yearbook for 1954-55 of the Polish Society of Arts and Sciences Abroad*, V, 1954-55, pp. 43-50. [48]

St. Lesniewski, Grundziige eines neuen Systems der Grundlagen der Math-ematik, *Fundamenta Mathematicae* 24, 192 , pp. 242-251. [49]

St. Lesniewski, Uber die Grundlegen der Ontologie, *C.R. Soc. Sci. Lettr. Varsovie* III, 1930, pp.111-132. [50]

St. Lesniewski, On the Foundations of Mathematics, (in Polish), Przeglad Filozoficzny: 30,1927, pp. 164-206; 31, 1928, pp. 261-291; 32, 1929, pp. 60-101; 33, 1930, pp. 77-105; 34, 1931, pp. 142-170. [51]

St. Lesniewski On the foundations of mathematics, *Topoi* 2, 1982, pp. 7-52 (an abridged version of the preceding position). [52]

S. Marcus, Tolerance rough sets, *Cech topologies*, learning processes, Bull. Polish Acad. Ser. Sci. Tech 42(3), 1994, pp. 471-487. [53]

D. M. Mark, M. J. Egenhofer, K. Hornsby, Formal models of commonsense geographic worlds, *NCGIA Tech. Report* 97-2, Buffalo, 1997. [54]

C. Masolo, L. Vieu, Atomicity vs. infinite divisibility of space, in: C. Freksa, D. M. Mark (eds.), Spatial Information Theory. Cognitive and Computational Foundations of Geographic Information Science, Lecture Notes in Computer Science vol. 1661, Springer Verlag, Berlin, 1999, pp. 235-250. [55]

http: /www. c s. albany. edu/ amit[56]

P. Osmialowski, Player and Stage at PJIIT Robotics Laboratory, *Journal of Automation*, Mobile Robotics and Intelligent Systems, 2, 2007, pp. 21-28. [57]

L.Polkowski, On connection synthesis via rough mereology, *Fundamenta Informaticae* 46, 2001, pp. 83-96. [58]

L. Polkowski, A rough set paradigm for unifying rough set theory and fuzzy set theory (a plenary lecture), *Proceedings RSFDGrC'03*), LNAI, vol. 2639, Springer-Verlag, Berlin, 2003, pp. 70-78. [59]

L. Polkowski, Toward rough set foundations. Mereological approach (a plenary lecture), *Proceedings RSCTC04*. LNAI, vol. 3066, Springer, Berlin, 2004, pp. 8-25. [60]

L. Polkowski and A. Skowron, Rough mereology: a new paradigm for approximate reasoning, International Journal of Approximate Reasoning 15(4), 1997, pp. 333-365. [61]

L. Polkowski, A. Skowron, Rough mereology in information systems with applications to qualitative spatial reasoning, Fundamenta Informaticae 43, 2000, pp. 291-320. [62]

L.Polkowski, A.Szmigielski, Computing with words via rough mereology in mobile robot navigation, Proceedings 2003 IEEE/RSJ Int. Conf. Intell. Robots and Systems IROS 2003, Las Vegas NV, 2003, pp. 3498-3503. [63]

P. Ramsey, PostGIS Manual, in: postgis.pdf file downloaded from Refractions Research home page. [64]

H. Reichenbach, The Philosophy of Space and Time (repr.), Dover, New York, 1957. [65]

R. Sikorski, Boolean Algebras, Springer Verlag, Berlin, 1960. [66]

P. Simons, Parts - A Study in Ontology, Clarendon, Oxford, 1987. [67]

A. Skowron and L. Polkowski, Rough mereological foundations for design, analysis, synthesis and control in distributed systems, *Information Sciences*. An Intern. J., 104, 1998, pp. 129-156. [68]

J. Slupecki, S. Lesniewski's Calculus of Names, *Studia Logica* 3, 1955, pp. 7-72. [69]

B. Smith, Logic and formal ontology, in: J. N. Mohanty, W. McKenna (eds.), *Husserl's Phenomenology: A Textbook*, Lanham: University Press of America, 1989, pp. 29-67. [70]

B. Smith, Agglomerations, in: C. Freksa, D. M. Mark (eds.), Spatial Information Theory. Cognitive and Computational Foundations of Geographic Information Science, *Lecture Notes in Computer Science* vol. 1661, Springer Verlag, Berlin, 1999, pp. 267-282. [71]

B. Smith, Boundaries: an essay in mereotopology, in: L. Hahn (ed.), The Philosophy of Roderick Chisholm, *Library of Living Philosophers*, La Salle: Open Court, 1997, pp. 534-561. [72]

B. Sobociriski, Studies in Lesniewski's Mereology, *Yearbook for 1954-55 of the Polish Society of Art and Sciences Abroad,* 5 (1954), pp. 34-43. [73]

B. Sobocinski, L'analyse de 1'antinomie Russellienne par Lesniewski, *Methodos*, I, 1949, pp. 94-107, 220-228, 308-316; II, 1950, pp. 237-257. [74]

http:/agora.leeds.ac.uk/spacenet/spacenet.html[75]

J. G. Stell, Granulation for graphs, in: C. Freksa, D. M. Mark (eds.), Spatial Information Theory. *Cognitive and Computational Foundations of Geographic Information Science, Lecture Notes in Computer Science* vol. 1661, Springer Verlag, Berlin, 1999, pp. 416-432. [76]

A. Szmigielski, A description based on rough mereology of the workspace of a mobile robot by means of a system of ultrasound sensors (in Polish), *PhD Dissertation*, Dept. Electronics and Computer Techniques, Warsaw University of Technology, 2003, L.Polkowski, supervisor. [77]

A. Tarski, Les fondements de la geometrie des corps, *in: Ksi§ga Pamiatkowa I Polskiego Zjazdu Matematycznego* (Memorial Book of the 1st Polish Mathematical Congress), a supplement to Annales de la Societe Polonaise de Mathematique, Cracow, 1929, pp. 29-33. [78]

A. Tarski, Appendix E, in: J. H. Woodger, The Axiomatic Method in Biology, Cambridge Univ. Press, Cambridge, 1937. [79]

A. Tarski, What is elementary geometry?, in: L. Henkin, P. Suppes, A. Tarski (eds.), The Axiomatic Method with Special Reference to Geometry and Physics, *Studies in Logic and Foundations of Mathematics*, North-Holland, Amsterdam, 1959, pp. 16-29. [80]

A. Tarski, Zur Grundlegung der Boolesche Algebra I, *Fundamenta Mathematicae* 24, 1935, pp.177-198. [81]

T. Topaloglou, First-order theories of approximate space, in *Working Notes of the AAAI Workshop on Spatial and Temporal Reasoning*, Seattle, 1994, pp. 47-53. [82]

http:/www.cs.utexas.edu/users/qr/[83]

http:/www.cs.utexas.edu/users/qr/robotics/argus[84]

A. N. Whitehead, Process and Reality. An Essay in Cosmology, Macmillan, New York, 1929 (corr. ed. : D. R. Griffin, D. W. Sherbourne (eds.), 1978). [85]

M. F. Worboys, Imprecision in finite resolution spatial data, *Geoinformatica* 2, 1998, pp. 257-279. [86]

22

Automated Static and Dynamic Obstacle Avoidance in Arbitrary 3D Polygonal Worlds

J.M.P. van Waveren and L.J.M. Rothkrantz
Delft University of Technology, Netherlands Defence Academy
The Netherlands

1. Introduction

In common applications, a robot will have no prior knowledge of its environment and must use sensory information to build a cognitive model for path and route finding. The process of building a good cognitive model from sensory information alone is complex and time consuming. Uncertainty of the input data can also cause the resulting paths and routes to be suboptimal or erroneous. In many of these applications, however, it is possible to augment a robot's sensory information with existing static information about the environment.

Detailed and accurate digital blueprints of buildings are a fortuitous by-product of the prevalence of computer aided design in architectural design today. These data provide some information about the layout of the environment, but typically do not include more dynamic elements such as furniture, people, or other robots. Additional records may indicate what type of furniture is in a given room, but its placement may have changed over time. A robot navigating an office building will have to deal with the uncertainties of dynamic and unexpected objects in real-time, even if the layout of the building is known. However, the situation is vastly improved if the sensory data is used to identify object types, whose precise dimensions are already known and can be retrieved as needed.

The use of existing information of the environment is important to make a robot resource-efficient, while achieving near optimal path and route finding. However, processing existing information in the form of geometric models with hundreds of thousands of polygons can be expensive if a robot system is not carefully crafted to deal with such large amounts of information.

An automated system is presented for path and route finding through arbitrary 3D polygonal environments. The system can process a polygonal representation of an environment with hundreds of thousands of polygons within a few minutes on a small grid of today's computers. The processed information allows a robot to efficiently find routes and paths through the environment in real-time. Additionally, when a robot has identified and recognized dynamic obstacles in the environment based on sensory information, the system is able to create near optimal paths in real-time around arbitrary configurations of dynamic obstacles.

The system presented here has been battle hardened in several generations of computer games, such as the triple-A titles: QUAKE III Arena, DOOM III and Enemy Territory QUAKE Wars. These computer games provide virtual environments for a robot (or artificial player), which are by no means inferior to real-life environments in that sophisticated physics simulations are used to move the robot and dynamic obstacles. These latest games show that the system can be applied to not only indoor but also outdoor environments when a polygonal representation of the terrain is available.

1.1 Previous Work

The objective of obstacle avoidance is to find collision-free trajectories between a start and a goal configuration in static and/or dynamic environments containing obstacles. A rich variety of algorithms for obstacle avoidance can be found in literature (Ribeiro, 2005).
The Certainty Grid method for probabilistic representation of obstacles in a grid-type world model has been developed at Carnegie-Mellon University (CMU) (Moravec, Elfes, 1985), (Elfes, 1987), (Moravec, 1988). This world model is especially suited to the accommodation of inaccurate sensor data such as range measurements from ultrasonic sensors. The work area is represented by a two-dimensional array of square elements, denoted as cells. Each cell contains a certainty value that indicates the measure of confidence that an obstacle exists within the cell area. The environment is scanned to update the certainty grid and local movement of the robot may be required for multiple scans. A global path-planning method is then employed for off-line calculations of subsequent paths.
Potential fields were introduced by Khatib (Khatib, 1985) for robot path-planning. Using the potential field method, every obstacle exerts a repelling force on the robot while the goal exerts an attractive force. Using this method, however, the robot might fall into local minima where it achieves a stable configuration before reaching a goal. A similar approach using repulsion vectors is used in (Johnson, 2003). A combination of the Certainty Grid method and potential fields, named Virtual Force Field, can be found in (Borenstein, Koren, 1989).
The Vector Field Histogram (VFH) approach was introduced by Borenstein (Borenstein, Koren, 1991), (Ulrich, Borenstein, 1998), (Ulrich, Borenstein, 2000). The VFH approach generates a polar histogram of the space occupancy in the vicinity of a robot. This polar histogram is then checked to select the most suitable sector out of all polar histogram sectors which have a low polar obstacle density. The steering of the robot is then aligned with that direction.
Using the Bugs algorithm (Lumelsky, Skewis, 1990), (Lumelsky, Stepanov, 1990), (Choset et al., 2005), (Kamon, 1998), also known as edge detection and wall following (Bauzil et al., 1981), (Iijima et al., 1983), (Giralt, 1984) a robot moves directly towards the goal unless an obstacle is found, in which case the obstacle is contoured until motion to the goal is again possible. However, it may be necessary to take all static and dynamic obtacles into account at the same time to predict optimal paths towards the goal. When groups of objects are contoured the path may become jagged and non-optimal. Connolly (Connolly et al., 1990), (Connolly, Grupen, 1993) presents methods for planning smooth robot paths using Laplace's equation. Techniques such as "string-pulling" also known as "line-of-sight testing" (Snook, 2000) can also be used to optimize the paths.
Most of these systems are setup to create a cognitive model from sensory information without any prior information about the environment. (Waveren, Rothkrantz, 2006) presents a system for automated path and route finding in static polygonal environments. The system uses an off-line compilation process to derive a cognitive model of the environment that is suitable for efficient route and path finding. However, the system does not deal with dynamic obstacles in real-time.
Many other automated systems that use an off-line compilation process can be found in literature such as (Farnstrom, 2006), (McAnlis, Stewart, 2008), (Hamm 2008), (Axelrod, 2008)

and (Ratcliff, 2008). Some of these systems use a discrete approach and others an analytical approach to create data structures that are suitable for route and path finding. The analytical approaches are complex or suffer from floating-point rounding errors on computers. The discrete approaches are prone to inaccuracies and errors that are inherent to discretely sampling the environment.

2. Static Obstacle Avoidance

Static obstacle avoidance deals with obstacles in the environment with a shape and position that never changes. Static obstacles may present themselves in many different forms and situations. Obvious examples of static obstacles are mountains and fences, on terrain, and the walls of a building. However, obstacles may also present themselves as gaps or pits a robot can fall into, or perhaps borders between countries that are not even physically marked. Unless some serious destruction takes place, it is safe to assume such obstacles always stay the same, or at least for an extended period of time. In many applications, pre-calculated data structures can be constructed based on existing information about the environment in order to speed up real-time route and path finding around a variety of different static obstacles.

2.1 Area System

The system presented here for dealing with static obstacles is similar to the system presented in (Waveren, Rothkrantz, 2006). The robot is assumed to live inside a simple bounding volume, and the robot has a limited number of degrees of freedom. In particular, the robot lives inside an axis-aligned bounding box that only translates through the world. This bounding box can be defined by six axis-aligned bounding planes relative to a reference point on the robot. The system can be implemented to work with different polygonal bounding volumes, but by using an axis-aligned bounding box the complexity is kept to a minimum. The system can also be extended to deal with rotations of the bounding volume of the robot. However, using more than 3 degrees of freedom significantly increases the complexity, where the movement of the robot is bounded by hyper-surfaces that are much harder to visualize and conceive.

During an off-line compilation process, the system automatically derives a cognitive model for route and path-finding, from a polygonal representation of a world with static obstacles. The first step in this off-line compilation process involves the construction of a boundary representation of configuration space (C-Space). The boundary representation of C-Space consists of one or more two-manifold triangle meshes that describe the Minkowsky sum of the polygonal world geometry and the bounding volume in which the robot resides. This boundary representation describes the extents of the complete movement freedom of the robot in an environment with only static obstacles.

The configuration space is calculated using Constructive Solid Geometry (CSG) operations on two-manifold triangle meshes, in which meshes are subtracted from each other and welded together. The system presented in (Waveren, Rothkrantz, 2006) uses a three-dimensional (3D) Binary Space Partitioning (BSP) algorithm with portalization to calculate the boundaries of configuration space. However, 3D BSP algorithms are prone to floating-point rounding errors on computers when dealing with many polygons. These rounding errors may cause the constructed boundary representation of configuration space to not accurately describe the actual extents of the complete movement freedom of the robot. The floating-point rounding errors can be localized and minimized by subdividing polygonal environments into smaller blocks that are moved into a well defined floating-point range centered about the origin. Nevertheless, 3D BSP algorithms are prone to rounding errors when dealing with a mixture of very high and very low density polygonal geometry.

Before using CSG operations to calculate the boundary representation of configuration space, each polygon that is used to describe the environment is turned into a convex polytope. Such a convex polytope is described by a two-manifold triangle mesh that represents the Minkowsky sum of a single polygon and the bounding volume of the robot.

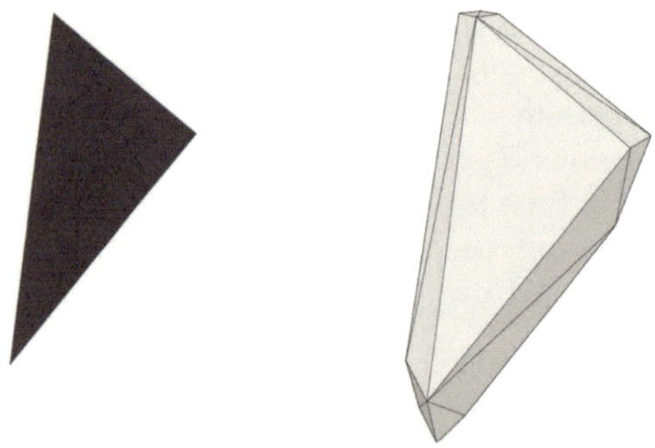

Figure 1. A triangle in 3D space and the convex polytope that describes the Minkowsky sum

There are several ways to construct such convex polytopes. One way is to place a copy of the axis-aligned bounding box in which the robot resides at each vertex position of a polygon, such that, the point of reference on the robot, relative to which the bounding box is defined, coincides with a vertex position. The convex polytope for a polygon is then constructed by calculating the convex hull of all the corners of all the positioned bounding boxes. This approach assumes the point of reference on the robot, relative to which the bounding box is defined, is centered inside the bounding box. Once a convex polytope is defined for each polygon in the environment, the two-manifold meshes that describe these convex polytopes can be welded together using CSG operations in order to construct a complete boundary representation of configuration space.

Figure 2. Boundary representation of C-Space for an outdoor environment with buildings

In the next step of the off-line compilation process, traversable surfaces are identified on the boundary representation of configuration space. First, the slope of each triangle is determined. A triangle is considered traversable if the slope is less than 45 degrees. Each triangle edge between a traversable and non-traversable triangle can now be classified and flagged as either a "wall" edge or a "ledge" edge based on whether the triangles meet at an inward or outward angle respectively. The robot is assumed to be able to cover small height differences. For instance, a robot may be able to walk up stairs. Whenever a "ledge" edge is positioned directly above a "wall" edge, the overlapping part, where the edges are not too far apart, can be considered traversable. The "ledge" and "wall" flags are removed from the parts of the edges that overlap within the height difference that is considered traversable. As such, the overlapping edges of the steps of a staircase are not flagged. By processing all edges this way, a lot of geometric detail is ignored which helps to simplify the cognitive model used for path and route finding.

The edges that are flagged as "ledge" and "wall" edges can now be used to subdivide the traversable surfaces into the least number of traversable areas, such that a robot can move in a straight line between any two points in an area. A two-dimensional (2D) Binary Space Partitioning (BSP) algorithm is used to subdivide the traversable surfaces into areas where vertical planes through the "ledge" and "wall" edges are used as splitters. This 2D BSP algorithm is significantly more robust and numerically stable than a 3D BSP algorithm with partalization because only vertical planes are used to split two-manifold meshes. The 2D BSP algorithm effectively creates convex vertical columns in space that contain one or more convex traversable surface fragments. Within a column each fragment covers the whole area of the column from a top-down perspective. However, each fragment within a single column is part of a different floor. The different floors are separated in the BSP tree by introducing additional split planes between the floors that are approximately level with the floor surface.

Figure 3. Areas for an outdoor environment with buildings

Because different floors of a building are not first separated, a floor on one level may be split by "ledge" or "wall" edges from a floor on a different level. In effect this causes unnecessary fragmentation of floors. However, convex traversable surface fragments from different columns can be trivially merged back together if they meet at an edge and the merged fragments also form a convex area in which a robot can move in a straight line between any two points. Merging traversable surface fragments together may cause multiple branches of the BSP tree to point to the same area. This is, however, not a problem because the BSP "tree"

is typically only used to quickly find the area a robot is in. The inner edges of a traversable surface fragment can now be discarded so that an area is described by a single polygon.
The calculation of the configuration space is the most expensive step in the off-line compilation process. Fortunately, the polygonal representation of the environment can be broken up into small blocks, and the boundary representation of configuration space for each block can be calculated separately. A small grid of computers can be used to process these blocks in parallel, which significantly reduces the time it takes to complete the off-line compilation process. After all the blocks have been processed, the results from the individual blocks can be merged to form one or more two-manifold meshes that describe the complete boundary representation of configuration space for the whole environment.

2.2 Avoiding Ledges

The robot movement is modelled as an axis-aligned bounding box that translates through the environment. Even if most of this bounding box is hanging over a cliff or ledge, and only a corner of the box actually rests upon an edge, the box will not tumble as far as the area system is concerned. Figure 4 shows a bounding box which is hanging over a ledge.

Figure 4. A robot partly standing over a ledge while the bounding box rests upon the edge

So called "ledge" areas are introduced to keep the legs of a robot from partly dangling over ledges. The robot may travel through such areas if absolutely necessary but when calculating routes the robot will try to avoid these areas. The ledge areas are introduced at edges that are not traversable because the height difference is too large. A vertical plane is constructed through a ledge facing away from the gap. This plane is then moved away from the ledge such that if the robot stays in front of the plane, the robot bounding volume cannot dangle over the ledge. The areas of the area system are then cut up with this plane to create areas on both sides of the plane. Areas that are behind the plane can then be marked as ledge areas and the robot can avoid these areas when calculating routes, or knows to navigate through such areas with care.

2.3 Reachability

Areas describe spaces in which a robot can travel between any two points in a straight line. To travel between areas a robot needs additional information. For this purpose so-called reachabilities are introduced that describe how a robot can travel from one area to the next. In particular, a reachability describes how a robot can travel from one area to an adjacent area or an area in close proximity. The reachabilities can then be used to calculate routes through the environment.

A reachability is stored in the area system as a link between two areas and a point in space where the robot can transition from one area to the next. Reachabilities are easy to define using the polygons that represent the areas in the area system. Many polygon edges of adjacent areas are exactly on top of each other and can be used directly to define reachabilities where there is a smooth transition from one area to the next. The robot may, however, also be able to cover small height differences such as the steps of a staircase. A lot of these small height differences are already filtered and ignored when constructing the areas, but the 2D BSP algorithm may have cut up the traversable two-manifold mesh such that there are still small height differences between areas. Because the robot is assumed to live inside an axis-aligned bounding box, these small steps always present themselves as edges from adjacent areas that are directly above each other. Areas in close proximity can be tested for edges that overlap when projected onto a horizontal plane. The vertical distance between the overlapping parts of such edges can then be calculated to find the places where a robot can reasonably be expected to cover the height distance and consequently navigate across the edges from one area to the next.

2.4 Routing

Routes are calculated between areas by following the reachabilities from one area to another. The number of areas, however, can grow quite large when the area system is employed in an environment described by hundreds of thousands of polygons. Environments that are mapped by ten thousand areas or more are not uncommon. Conventional routing algorithms can be time consuming when dealing with such large numbers of areas in real-time. All routes could be pre-calculated, but this consumes a lot of storage space and does not allow for adjustments to dynamic changes, such as areas that are temporarily disabled.

A hierarchical routing system is used to calculate routes with the same accuracy as a conventional routing algorithm. However, due to the hierarchical nature of the system, it takes significantly less time and space to calculate routes. The calculated routes are temporarily saved (cached) to avoid having to recalculate routes repeatedly. The caches can be freed and routes are recalculated, when the total size of all cached routes exceeds a threshold or when routes need to be adjusted due to dynamic changes in the environment. Routes are typically calculated and cached per goal area, because the goal area tends to stay the same over a longer period of time than the area the robot is in.

The hierarchical routing system creates a set of clusters of area. The areas within a cluster are connected through reachabilities. The areas are considered nodes of a graph and the reachabilities are considered the edges between the nodes. The clusters are separated by "cluster portals". These cluster portals are areas themselves and represent passages from one cluster to another. The clusters and portals are setup such that the only way to travel from one cluster to another is through a cluster portal. Each cluster portal separates no more and no less than two clusters.

The hierarchical routing system can save a lot of computation time and storage space by limiting routing calculations to the areas within clusters. Calculating all distances or travel times from all areas in a cluster to a specific goal area in the same cluster is significantly faster than performing routing calculations over all areas in the environment, because a cluster contains only a subset of all the areas. The upper bound for route calculations in a cluster is the square of the number of areas in the cluster; whereas the upper bound for calculating routes through the whole environment is square in the total number of areas in the environment.

The cluster portals are used to calculate routes through the whole environment. This works by interpreting the cluster portals as the nodes of a graph, and the routes through clusters as the edges between the nodes. The goal area of a route is also interpreted as a node of the same graph. The goal area is connected to cluster portals with routes through the cluster the goal area is in.

Routes through clusters are calculated with a simple breadth first routing algorithm. The breadth first algorithm starts at the goal area and uses the reversed reachabilities to flood to other areas. The algorithm never floods to areas outside the cluster, but does flood into cluster portals that touch the cluster. While flooding through a cluster, the algorithm calculates and temporarily saves travel times to the goal area for each area in the cluster. Routes through the whole environment are also calculated with a breadth first routing algorithm. The algorithm starts at the goal area and floods to all portals in the environment through the routes within clusters. For each portal the algorithm calculates and temporarily saves the travel time of the portal to the goal area.

At any given time the next area along a route can be calculated by first determining the portal of the current cluster that is closest to the goal area, and then calculating the route to this portal. If the goal area is in the same cluster as the source area, the direct route within the cluster is also considered and chosen instead if determined to be shorter.

2.5 Path Optimization

A route retrieved from the area system is a sequence of areas connected through reachabilities. The reachabilities can be used to define a path as a sequence of points in space. These intermediate points along such a path are typically at the centers of edges between areas. A line through these centers, however, may not describe the most optimal path around static obstacles. The arrows in figure 5 show a path retrieved from the area system around two wall sections. Obviously this is not the most optimal path.

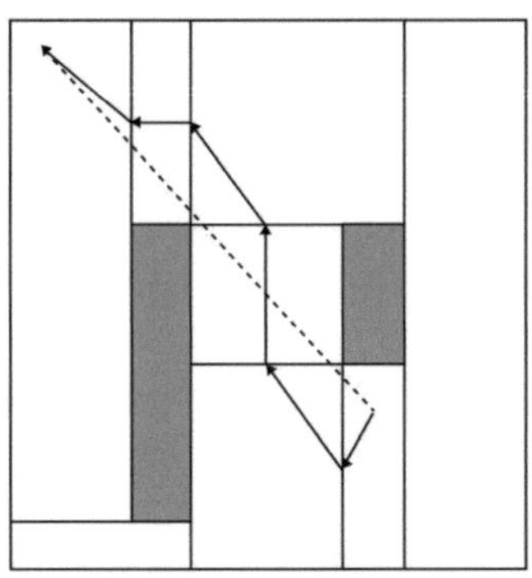

Figure 5. Top down view of path optimization around two wall sections

The path shown by the arrows in figure 5 can be optimized by plotting a line from the start position directly to one of the intermediate points that is further away, possibly several areas down the path. The robot must be able to travel along this line without being obstructed. The polygons that describe the areas from the area system can be used to test whether or not the line goes through free space without obstructions. At any given time the robot can choose the intermediate point on the path that is furthest away and to which the robot can move in a straight line. Instead of only choosing intermediate points that are on the edges between areas, the lines between successive intermediate points can be sub-sampled. A line from the start position to such a sub-sampled point can also be tested against the polygons of areas to determine if the sub-sampled point can be reached by the robot without being obstructed. Sub-sampling avoids discontinuities that may arise when a robot follows a route through areas with vastly different sizes. Optimizing paths in this manner can be done in real-time. By continuously optimizing paths while moving, the robot will follow a smooth and close to optimal path to its destination.

3. Real-Time Dynamic Obstacle Avoidance

The area system provides a robust solution for static obstacle avoidance. However, there may also be many dynamic obstacles in the environment. When the positions and dimensions of nearby dynamic obstacles are known, the areas of the area system could be recalculated for a particular configuration of such dynamic obstacles. However, recalculating the areas is typically too time consuming on today's computers to be done in real-time. Instead, the area system only handles the static part of the environment. Another system which is built on top of the area system is used to calculate paths around arbitrary configurations of dynamic obstacles. The system described here assumes the dynamic obstacles have already been identified from sensory information and records. The system is similar to the Bugs algorithm or wall following, except that the full path around obstacles can be re-calculated and optimized repeatedly.

The problem of finding paths around dynamic obstacles is simplified by using a projection which makes the obstacle avoidance system suitable for real-time use. Furthermore, the system only deals with dynamic objects represented by oriented bounding boxes (OBBs). This restriction generally does not cause any problems because all dynamic obstacles can be contained within one or more tightly fitting OBBs. The wall edges from the area system are also considered as obstacles, such that the static obstacles are also taken into account when constructing a path around dynamic obstacles.

A robot may choose to ignore very small obstacles or obstacles that do not extend more than a small distance above the floor. The dynamic obstacle avoidance system, however, will always try to find a path around dynamic obstacles, and does not try to find a path that requires the robot to move over these obstacles. In some situations a pile of stacked obstacles may present a smooth top surface that may seem suitable for robot navigation. However, it is very hard to determine whether or not a stack of obstacles will collapse if the robot tries to move over the obstacles.

Figure 6. Obstacles along the path from the camera position towards the door behind the table

Figure 6 shows a room with a pile of boxes and a door behind a table. All the boxes, the table and the chairs are dynamic and can move through the room. A robot that wants to travel from the camera position towards the door will have to navigate around these obstacles. The walls and floor of the room are not dynamic and the area system only takes these static obstacles into account. As such, the optimized path retrieved from the area system is a straight line from the camera position to the door because none of the walls block the movement towards the door.
The dynamic obstacle avoidance system searches for any obstacles in proximity of the optimized path retrieved from the area system. If no obstacles are found, the robot can just follow the optimized path. If possible obstacles are found, the nearby walls from the area system are considered as obstacles as well. Each obstacle is represented by an oriented bounding box (OBB) which is projected onto a 2D horizontal navigation plane. With this projection a polygon is created that describes the contour of the projected OBB. The polygon of a projected obstacle is expanded and beveled to create a polygon that describes the 2D Minkowsky sum of the projected OBB of the obstacle, and the axis-aligned bounding box in which the robot resides. The polygon used for the obstacle avoidance is the outline of this Minkowsky sum.

3.1 Path Tree

A path tree is built using clock-wise and counter clock-wise edge walks along the expanded polygons. Each node in the tree is a line segment that describes part of a path around obstacles. The optimized path from the area system is followed until a polygon is hit. Whenever a polygon is hit, the tree branches to follow the edges of the polygon in both a clock-wise and counter clock-wise fashion. While following the edges of a polygon, a new node is added to tree for each edge of the polygon. While following an edge of a polygon another polygon may be hit. When this happens, a new node is added to the tree to follow the polygon that is hit. Before extending the tree with a node for a new edge of a polygon, the edge is tested to see if it faces the end point of the optimized path from the area system. If an edge faces this end point, the polygon is no longer followed and a new node is added to the tree in order to continue the path directly to this end point. Any polygons hit along this new path are again used to extend the tree. This process continues until all branches of the tree either reach the end point of the optimized path from the area system, or loop back to a path already followed closer to the root of the tree. When following the edges of polygons in both a clock-wise and counter clock-wise fashion, a branch of the tree may loop back onto edges of polygons that have already been followed. Whenever that happens the branch is terminated.

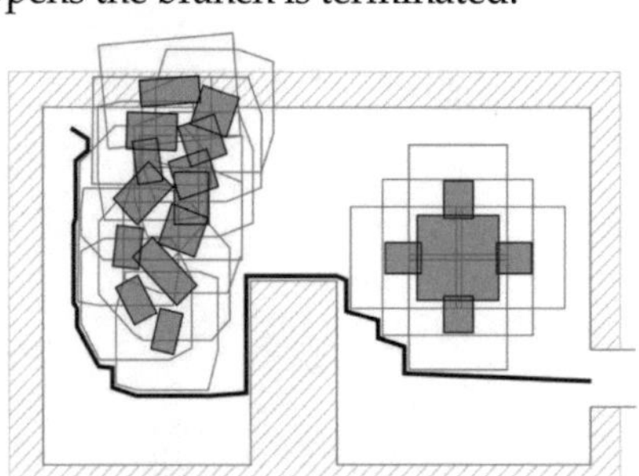

Figure 7. Top down view of one of the paths around the obstacles

Figure 7 shows a top down view of the room with obstacles that block a direct path from the camera position to the door. One of the branches of the tree is shown as a thick black line. The tree typically has many more branches, for instance, a branch that goes the opposite way around the table. These other branches are not shown for clarity.
Once the complete tree is built, the tree is traversed in a depth first manner in order to find all possible paths that may lead to the destination. Such a path is a sequence of nodes that represent line segments. A path that does not lead to the destination is discarded. A path which does lead to the destination is optimized by taking shortcuts where possible. Nodes and, as such, line segments are skipped if the new line segment that represents the shortcut does not intersect any other obstacles. Figure 8 shows the path from figure 8 that has been optimized in this way. The shortest path is chosen from all optimized paths that are found when traversing the tree. The path shown in figure 8 is the shortest path around the obstacles.

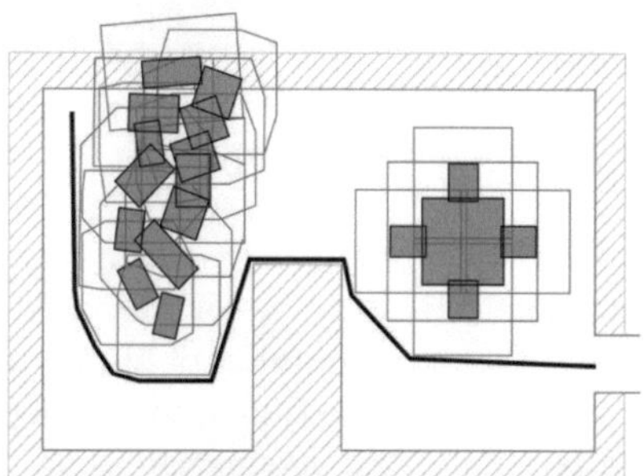

Figure 8. Top down view of optimized path around the obstacles

The start and/or end point of a path retrieved from the area system may be inside the expanded polygons of a dynamic obstacle. This can happen when the robot is standing on the edge of an obstacle near the ground, or if an obstacle covers the end point of the path from the area system. When this happens, the start and/or end point needs to be pushed out of any expanded polygons of obstacles in order to find valid paths around the obstacles. A point inside an expanded polygon is first pushed towards the closest edge of the expanded polygon. If that point is still inside other expanded polygons the point is pushed towards one of the intersection points of the expanded polygons. All intersections of the expanded polygons that contain the point are calculated and the first intersection that is outside any obstacles is chosen. If no such intersection is found, either the robot or the destination is completely surrounded by obstacles and the robot will first have to move obstacles out of the way to create a path.
The obstacle avoidance can be used to calculate paths around dynamic obstacles in real-time. The robot can recalculate paths around dynamic obstacles repeatedly such that paths will be updated in real-time while obstacles are moving. Furthermore, the system can cope with arbitrary configurations of dynamic obstacles, allowing a robot to find paths around or out of maze like configurations.
When the optimized path from the area system is recalculated repeatedly and the path around dynamic obstacles is also updated in real-time, there are cases where the area system and dynamic obstacle avoidance system may fight each other. The dynamic obstacle avoidance may change the path from the area system, which in return may cause the path from the area system to change as the robot moves along the modified path. In some

situations this may lead to confusion where the systems are alternately directing the robot to go in a different direction. The solution to this is to follow the path from the area system until dynamic obstacles are found that block this path. Once dynamic obstacles are found the path from the obstacle avoidance is followed and the optimized path from the area system is no longer updated until the destination of this optimized path has been reached or there are no longer any obstacles blocking the path. In other words, the path from the area system is not updated while avoiding dynamic obstacles.

4. Results

The robots (artificial players) in the computer game Enemy Territory QUAKE Wars use the described system for real-time route and path finding. This computer game provides a variety of challenging environments for robot navigation. The environments are largely outdoors but there are also many smaller and larger buildings the robots can enter. Each environment in the game covers approximately a square mile. Large parts of the environments never change but there are also many dynamic obstacles such as other robots, human players and vehicles.

Table 1 shows statistics for several environments from the computer game Enemy Territory QUAKE Wars. For each environment table 1 shows the number of triangles that make up the static obstacles in the environment, the number of blocks the environment is broken up into, the number of triangles used to describe the boundary representation of configuration space, the number of areas used for robot navigation, and the number of reachabilities between areas.

Name	**Obstacle triangles**	**Blocks**	**C-Space triangles**	**Areas**	**Reachabilities**
Sewer	112288	2768	156328	6089	30667
Valley	120133	4241	167638	9379	47595
Volcano	217553	2301	237680	7641	36818

Table 1. Statistics for several environments from Enemy Territory QUAKE Wars

The above table shows that the number of triangles required to describe the boundary representation of configuration space is not significantly higher than the number of triangles that are considered static obstacles for robot navigation. This is partly because the robot lives inside a simple bounding volume. When a more complex bounding volume is used the number of configuration space triangles increases.

A small grid of 10 to 20 computers was used for the off-line compilation process to construct the boundary representation of configuration space. By using this small grid of computers the off-line compile time was reduced to just a couple of minutes.

In Enemy Territory QUAKE Wars 16 or more robots use the described system simultaneously. The combined CPU consumption of all robots is no more than 20% on a Intel Pentium 2 GHz processor.

5. Conclusion

The presented system is fully automated and can be used for path and route finding through arbitrary 3D polygonal environments with both static and dynamic obstacles. A polygonal representation with hundreds of thousands of polygons describing all static

obstacles in the environment can be processed off-line within a few minutes on a small grid of today's computers. The off-line compilation process creates data structures that allow a robot to efficiently find routes and paths through the environment in real-time. Once dynamic obstacles are identified from sensory information, the system is also able to create near optimal paths in real-time around arbitrary configurations of dynamic obstacles.
The system has been successfully implemented and employed in the computer game Enemy Territory QUAKE Wars. This implementation shows that the system is resource-efficient. Multiple robots can use the system simultaneously for real-time path and route finding while only using a small percentage of all available compute power on a modest computer.

6. Future Work

For performance reasons the dynamic obstacle avoidance system uses a projection onto a 2D navigation plane. Dynamic obstacles well above a robot, or obstacles that do not extend more than a small distance above the floor can be ignored. However, imagine a thin pipe against a wall that fell over across a hallway such that it leans against the opposite wall at a 45 degrees angle. A robot may be able to navigate through the hallway by passing underneath the pipe close to the wall the pipe rest against. Unfortunately the dynamic obstacle avoidance system is unable to direct the robot through the hallway because it uses a projection of the pipe which covers the whole hallway. The pipe could be represented by multiple oriented bounding boxes where some are ignored if they are well above the robot. However, the dynamic obstacle avoidance could also be implemented as a wall following algorithm that follows the contours of 3D convex polytopes as opposed to polygons in the plane. This increases the complexity and the required compute power but will allow a robot to find paths around or through more complex configurations of dynamic obstacles.

7. References

G. Bauzil, M. Briot, P. Ribes (1981). A Navigation Sub-System Using Ultrasonic Sensors for the Mobile Robot HILARE, *1st In.. Conf. on Robot Vision and Sensory Controls*, pp. 47-58 and pp. 681-698, 1981, Stratford-upon-Avon, UK

J. Iijima, S. Yuta, Y. Kanayama (1983). Elementary Functions of a Self-Contained Robot YAMABICO 3.1, *Proc. of the 11th Int. Symp. on Industrial Robots*, pp. 211-218, 1983, Tokyo

G. Giralt (1984). Mobile Robots, *NATO ASI Series, Robotics and Artificial Intelligence*, Vol. F11, pp. 365-393, 1984, Springer-Verlag

H. P. Moravec, A. Elfes (1985). High Resolution Maps from Wide Angle Sonar, *IEEE Conference on Robotics and Automation*, pp. 116-121, 1985, Washington D.C.

Oussama Khatib (1985). Real-time obstacle avoidance for manipulators and mobile robots, *In Proceedings of the 1985 IEEE International Conference on Robotics and Automation*, pp. 500-505, March 1985

A. Elfes (1987). Sonar-based Real-World Mapping and Navigation, *IEEE Journal of Robotics and Automation*, pp. 249-265, Vol. RA-3, No 3, 1987

H. P. Moravec (1988). Sensor Fusion in Certainty Grids for Mobile Robots, AI Magazine, pp. 61-74, Summer 1988

J. Borenstein, Y. Koren (1989). Real-time Obstacle Avoidance for Fast Mobile Robots, *IEEE Transactions on Systems, Man, and Cybernetics*, Vol. 19, No. 5, pp. 1179-1187, Sept./Oct. 1989

Christopher I. Connolly, J.B. Burns, R. Weiss (1990). Path Planning Using Laplace's Equation, *Proceedings of the IEEE Conference on Robotics and Automation*, pp. 2102-2106, May 1990

V. Lumelsky, T. Skewis (1990). Incorporating range sensing in the robot navigation function, *IEEE Transactions on Systems Man and Cybernetics*, vol. 20, pp. 1058 - 1068, 1990

V. Lumelsky, Stepanov (1990). Path-planning strategies for a point mobile automaton amidst unknown obstacles of arbitrary shape, *Autonomous Robots Vehicles*, pp. 1058 - 1068, 1990, Springer, New York

J. Borenstein, Y. Koren (1991). The vector field histogram - fast obstacle avoidance for mobile robots, *IEEE Transaction on Robotics and Automation*, vol. 7, no. 3, pp. 278 - 288, 1991

Christopher I. Connolly, Roderic A. Grupen (1993). Applications of Harmonic Functions to Robotics, *Journal of Robotic Systems*, 10(7), pp. 931-946, 1993

Ishay Kamon, Elon Rimon, Ehud Rivlin (1998). Tangentbug: A Range-Sensor-Based Navigation Algorithm, *The International Journal of Robotics Research*, vol. 17, nr. 9, pp. 934-953, September 1998

I. Ulrich, J. Borenstein (1998). VHF+: Reliable obstacle avoidance for fast mobile robots, *Proc. of the IEEE Int. Conf. on Robotics and Automation*, pp. 1572 - 1577, 1998

I. Ulrich, J. Borenstein (2000). VHF*: Local obstacle avoidance with look-ahead verification, *Proc. of the IEEE Int. Conf. on Robotics and Automation*, San Francisco, pp. 2505 - 2511, 2000

Greg Snook (2000). Simplified 3D Movement and Pathfinding Using Navigation Meshes, *Game Programming Gems*, 2000, Charles River Media

Geraint Johnson (2003). Avoiding Dynamic Obstacles and Hazards, *AI Game Programming Wisdom 2*, 2003, Charles River Media

Maria Isabel Ribeiro (2005). Obstacle Avoidance, *The Robotics WEBook*, Institute for Systems and Robotics, Instituto Superior Técnico, October 2005, Lisboa, Portugal

H. Choset et al (2005). *Principles of Robot Motion: theory, algorihms and implementations*, Englewood Cliffs and New Jersey, 2005, MIT Press

J.M.P. van Waveren, Drs. dr. L.J.M. Rothkrantz (2006). Automated path and route finding through arbitrary complex 3D polygonal worlds, *Elsevier, Robotics and Autonomous Systems*, February 2006

Frederick Farnstrom (2006). Improving on Near-Optimality: More Techniques for Building Navigation Meshes, *AI Game Programming Wisdom 3*, 2006, Charles River Media

Colt McAnlis, James Stewart (2008). Intrinsic Detail in Navigation Mesh Generation, *AI Game Programming Wisdom 4*, 2008, Charles River Media

David Hamm (2008). Navigation Mesh Generation: An Empirical Approach, *AI Game Programming Wisdom 4*, 2008, Charles River Media

Ramon Axelrod (2008). Navigation Graph Generation, *AI Game Programming Wisdom 4*, 2008, Charles River Media

John W. Ratcliff (2008). Automatic Path Node Generation for Arbitrary 3D Environments, *AI Game Programming Wisdom 4*, 2008, Charles River Media

23

Reactive Motion Planning for Mobile Robots

Abraham Sánchez[1], Rodrigo Cuautle[1], Maria A. Osorio[1] and René Zapata[2]
[1]Autonomous University of Puebla, [2]LIRMM – Université Montpellier II
[1]México, [2]France

1. Introduction

Motion planning refers to the ability of a system to automatically plan its motions. It is considered central to the development of autonomous robots. In the last decade, much research effort was done on the application of probabilistic roadmaps methods (PRM) for different types of problems (Kavraki et al., 1996; Svestka & Overmars, 1997; Bohlin & Kavraki, 2000; Sánchez & Latombe, 2002).

There are two main classes of PRM planners: multiple-query and single-query planners. A multiple-query planner pre-computes a roadmap and then uses it to process many queries (Kavraki et al., 1996; Svestka & Overmars, 1997). In general, the query configurations are not known in advance and the roadmap must be distributed over the entire free configuration space (C-space). On the other hand, a single-query planner computes a new roadmap for each query (Bohlin & Kavraki, 2000; Sánchez & Latombe, 2002; Sánchez et al., 2002). Its only goal is to find a collision-free path between two query configurations. Looking for the smallest space to explore before finding a path. Planners that can answer single queries very quickly and with a little preprocessing are of particular interest. Such planners can be used to re-plan paths in applications where the configuration space obstacles can change. This occurs, for instance, when the robot changes tools, grasps an object, or a new obstacle enters in the workspace. These kinds of planners are more suitable in environments with frequent changes. The adaptation of PRM planners to environments with both static and moving obstacles has been limited so far (Jaillet & Siméon, 2004).

The planner proposed by Jaillet and Siméon (Jaillet & Siméon, 2004), uses a combination of single and multiple queries techniques. The proposed planner builds a roadmap of valid paths, considering only the static obstacles, when dynamic changes occurs, the planner uses lazy-evaluation mechanisms combined with a single-query technique as local planner to rapidly update the roadmap.

A novel real-time motion planning framework was proposed in (Brock & Kavraki, 2001). It is particularly well suited for planning problems and it decomposes the original planning into simpler sub-problems. The paradigm addresses the planning problems in which a minimum clearance to obstacles can be guaranteed along the solution path.

A method for generating collision-free paths for robots operating in changing environments was presented in (Leven & Hutchinson, 2000). The method begins by constructing a graph that represents a roadmap in the configuration space, but this graph is not constructed for a specific workspace. Later, the method constructs the graph for an obstacle-free workspace, and encodes the mapping from workspace cells to nodes and arcs in the graph. When the

environment changes, this mapping are used to make the appropriate modifications to the graph, and new plans can be generated by searching the modified graph.
A dynamic structure to enrich any non-holonomic motion planner for car-like robots with the capacity of reactiveness to environment changes was proposed by (Jaouni et al., 1998). The main advantage of the star elastic band proposed in their work, is that it allows a better reactivity than the ball band (the elastic band approach was proposed by Quinlan and Khatib in 1993 (Quinlan & Khatib, 1993)). The elastic band approach is a dynamic trajectory modification that maintains a permanent flexible and deformable path between initial and final robot configurations.
This work aims at providing a practical planner that considers reflex actions and planning with lazy techniques to account for obstacle changes. A collision-free feasible path for a mobile robot is computed using the lazy PRM method. The robot starts moving (under the permanent protection of its deformable virtual zone (DVZ)), in a free of dynamic obstacles trajectory, it does not require reflex commands and the control is performed by the lazy PRM method. If there are dynamic obstacles in its path, the reactive method takes the control and generates commands to force the robot to move away from the intruder obstacles and gives back its DVZ to the original state.

2. The DVZ principle

Artificial reflex actions for mobile robots can be defined as the ability to react when unscheduled events occurs, for instance when they move in unknown and dynamic environments. For the last seventeen years, we have been interested in the problem of reactive behaviours for collision avoidance in the domain of mobile robotics (Zapata, 1991; Zapata et al., 1994; Cacitti & Zapata, 2001). This section describes the DVZ principle. We assume that the mobile robot has not model of its surrounding space but can measure any intrusion of information (proximity-type information) at least in the direction of its own motion. The vehicle is protected by a risk zone while the deformations of the latter are directly used to trigger a good reaction.
In what follows, n will denote the dimension of the robot world (Euclidean space), $\mathbb{R}$ the real line, $\|y\|$ the Euclidean norm of vector y, and $(\partial\Xi/\partial x)$ the Jacobian of the vector-valued function Ξ. The robot/environment interaction can be described as a deformable virtual zone (DVZ) surrounding the robot. The deformations of this *risk zone* are due to the intrusion of proximity information and control the robot interactions. The *robot internal state* is defined to be a couple (Ξ, π), where the first component Ξ is called *the interaction component*, which characterizes the geometry of the deformable zone and the second component π characterizes the robot velocities (its translational and rotational velocities). In the absence of intrusion of information, the DVZ, denoted by Ξ_h is supposed to be a one-one function of π. The *internal control*, or reactive behavior is a relation ρ, linking these two components, $\Xi_h = \rho(\pi)$. In short, the risk zone, disturbed by the obstacle intrusion, can be reformed by acting on the robot velocities.

Let $\chi = \begin{pmatrix} \Xi \\ \sigma \end{pmatrix}$ be the vector that represents the internal state of the robot and let ε be the state space, which is the set of all the vectors χ. The DVZ is defined by $\Xi = \begin{pmatrix} \Xi_1 \\ \vdots \\ \Xi_c \end{pmatrix}$ and the robot velocities vector σ is defined by $\sigma = \begin{pmatrix} v \\ \dot{\theta} \end{pmatrix}$, where each component Ξ_i is the norm of the vector corresponding to the border's distance in the DVZ. These vectors belong to the straight lines that correspond to the main directions of the c proximity sensors, c_i. Generally speaking, we assume that we control the derivative ϕ of a function π for the robot velocities σ. Therefore, the control vector will be written

$$\phi = \dot{\pi} \tag{1}$$

Let H be the set of all internal states χ_h whose DVZ is not deformed. This set induces an equivalence relation in ε, defined by

$$\chi^1 \tilde{H} \chi^2 \Leftrightarrow \chi_h^1 = \chi_h^2 \tag{2}$$

where χ_h^i is the internal state corresponding to the state χ^i but without any deformation due to intrusion. In the equivalence class $[\chi]$, the vector χ_h is a one to one function for the vector π:

$$\chi_h = \rho(\pi) \tag{3}$$

which can be written as, (by separation of the two sets of variables)

$$\begin{cases} \Xi_h = \rho_\Xi(\pi) \\ \sigma = \rho_\sigma(\pi) \end{cases} \tag{4}$$

The derivative of equation (4) provides the state equation when no deformation occurs (when the state vector stays on H):

$$\dot{\chi}_h = \rho'(\pi)\dot{\pi} = \rho'(\pi)\phi \tag{5}$$

This equation is the first part of the general state equation. If we now consider deformations of the DVZ, due to intrusion, we will obtain the second part of the state equation. To do it, we denote the deformation of the state vector by Δ and study the variations of this deformation with respect to intrusion. This new vector represents the deformed DVZ, which is defined by

$$\Xi = \Xi_h + \Delta \tag{6}$$

Let $I = \begin{pmatrix} I_1 \\ \vdots \\ I_c \end{pmatrix}$ be the c-dimensional intrusion vector, where $I_i = d_{i\max} - d_i$. The sensor provides the measure $d_i = d_{i\max}$, in the absence of obstacles.

Let $\Delta = \begin{pmatrix} \Delta_1 \\ \vdots \\ \Delta_c \end{pmatrix}$ be the c-dimensional deformation vector, where

$$\Delta_i = \alpha(d_{hi}, I_i) = \begin{cases} 0 & if\ d_i > d_{hi} \\ d_{hi} - d_i & if\ d_i \le d_{hi} \end{cases} \tag{7}$$

where d_{hi} is an element of the intact DVZ (Ξ_h). Figures 1 and 2 illustrate this function.

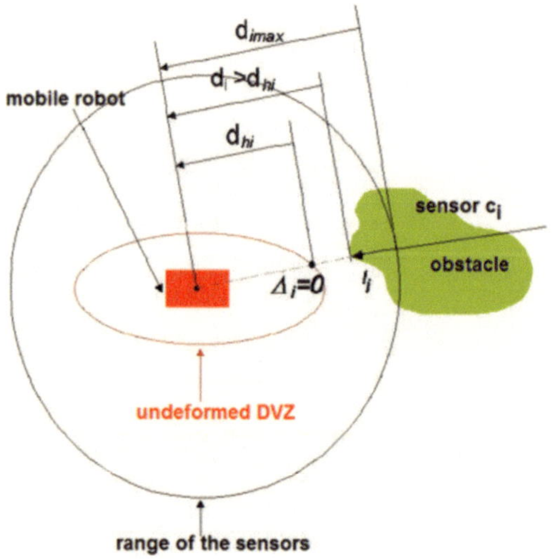

Figure 1. The deformation vector $\Delta_i = 0$

By differentiating equation (6) with respect to time, we get

$$\dot{\Delta} = \frac{\partial \alpha}{\partial \Xi_h}(\Xi_h, I)\dot{\Xi}_h + \frac{\partial \alpha}{\partial I}(\Xi_h, I)\dot{I} \tag{8}$$

By letting $\psi = \dot{I}$ and using equations (4), (5), (6) and (8), we obtain the next control equation

$$\begin{cases} \dot{\Xi} = \left(\frac{\partial \alpha}{\partial \Xi_h}(\Xi_h, I) \times \rho'_{\Xi}(\pi) + \rho'_{\Xi}(\pi) \right)\phi + \frac{\partial \alpha}{\partial I}(\Xi_h, I)\psi \\ \dot{\sigma} = \rho'(\pi)\phi \end{cases} \tag{9}$$

with

$$\begin{cases} \dot{\Xi} = \rho'_{\Xi}(\pi)\phi \\ \dot{\pi} = \phi \\ \dot{I} = \psi \end{cases}$$

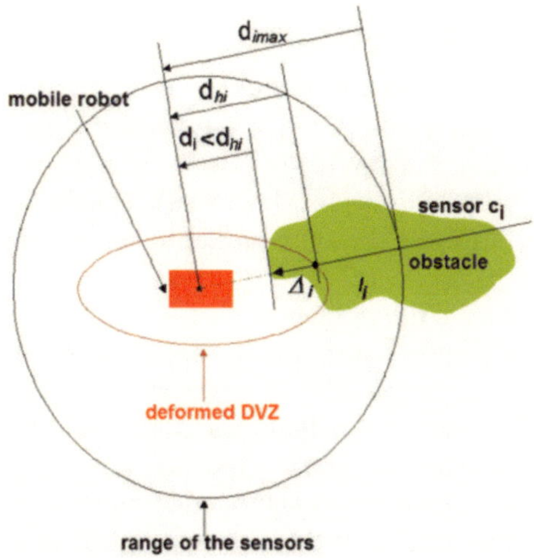

Figure 2. The obstacle deforms the DVZ

The inputs of equation (9) are the two control vectors ϕ and ψ. The first comes from the control module of the robot and the second from the environment itself.
We can consider the matrix A with dim(A)=$[c \times n]$ (c sensors, k control variables) and the matrix B with dim=(B)=$[c \times c]$ as follows:

$$A = \frac{\partial \alpha}{\partial \Xi_h}(\Xi_h, I)\rho'_{\Xi}(\pi)$$
$$B = \frac{\partial \alpha}{\partial I}(\Xi_h, I) \qquad (10)$$

By replacing equation (9) and (10) in equation (8), we obtain the evolution of the deformation

$$\dot{\Delta} = A\Phi + B\Psi \qquad (11)$$

The DVZ control algorithm consists of choosing the desired evolution $\dot{\Delta}_{des}$ of the deformation. Given $\dot{\Delta}_{des}$, the best control vector $\breve{\phi}$ in the sense of least-squares that minimizes the function $\left\|\dot{\Delta}_{des} - \dot{\Delta}\right\|^2$ is obtained by inverting equation (11):

$$\breve{\phi} = A^{+}(\dot{\Delta}_{des} - B\psi) \qquad (12)$$

where A^{+} is the pseudo-inverse of A.
A simple and efficient control law consists of choosing the desired deformation as proportional to the real deformation and its derivative:

$$\dot{\Delta}_{des} = -K_p \Delta - K_d \dot{\Delta} \tag{13}$$

where the two matrices K_p and K_d are respectively the proportional and derivative gains and are tuned in order to carry out the avoidance task. In this work, we define an ellipse as DVZ parameterized by the linear velocity and the steering angle of vehicle.

3. Lazy PRM for non-holonomic mobile robots

A Lazy PRM approach for non-holonomic motion planning was presented in (Sánchez et al., 2002). The algorithm is similar to the work presented by Bohlin and Kavraki (Bohlin & Kavraki, 2000), in the sense that the aim is to find the shortest path in a roadmap generated by randomly distributed configurations. In a later work, Sánchez et al., 2003 showed that the use of deterministic sampling improved remarkably the results obtained with random sampling.

Once a start-goal query is given, the planner performs A^* search on the roadmap to find a solution. If any of the solution edges are in collision, they are removed from the roadmap and then A^* search is repeated. Eventually, all edges may have to be checked for collisions, but often the solution is found before this happens. If no solution is found, more nodes may need to be added to the roadmap. The most important advantage of this approach, is that the collision checking is only performed when needed. In this case, all edges don't have to be collision checked as in the original PRM case (see figure 3). Experiments show that, in many cases, only a very small fraction of the graph must be explored to find a feasible path.

Planners, based on lazy strategy (Bohlin & Kavraki, 2000; Sánchez & Latombe, 2002) always use the straight-line segment (Euclidean distance) as steering method. Much research has been done on motion planning for nonholonomic car-like robots (see Laumond, 1998 for a review). Svestka and Overmars used the RTR paths as a steering method (Svestka & Overmars, 1997). An alternative choice is to use steering method that constructs the shortest paths connecting two configurations (Reeds & Shepp, 1990; Sanchez et al., 2002). Reeds & Shepp have provided a sufficient family of shortest paths for the car-like robots moving both forward and backward. Figure 4 shows the Reeds & Shepp paths and an example computed with this approach.

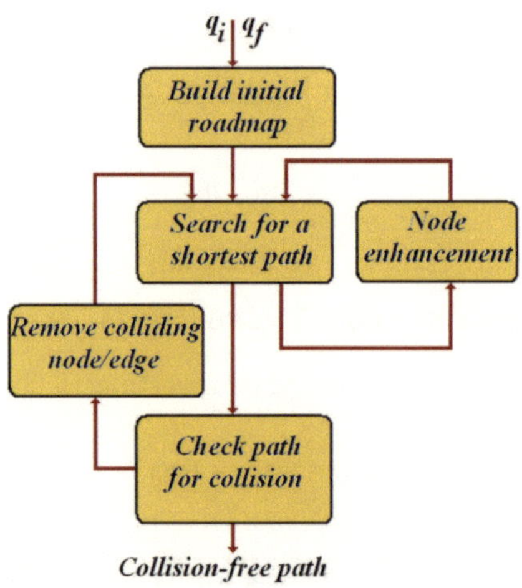

Figure 3. High-level description of the lazy PRM approach

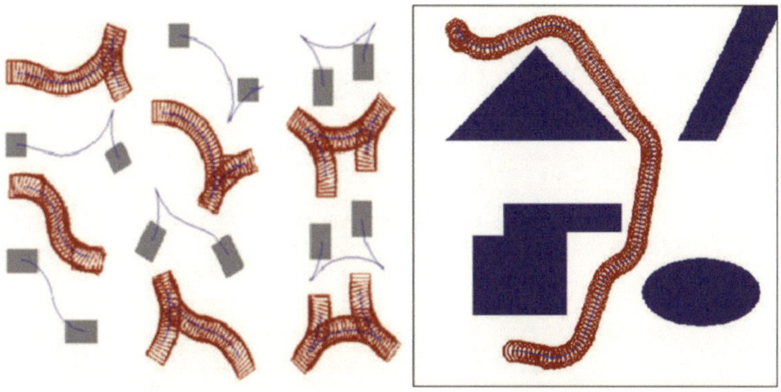

Figure 4. Shortest paths for car-like robots and a computed path in a polygonal environment

3. Reactive Lazy PRM

This section describes the proposed approach, which integrates the lazy PRM planning method and the reactive control by DVZ in the following way: a collision-free feasible path for a mobile robot is calculated by the lazy PRM method, the robot starts moving (under the permanent protection of its DVZ), in the absence of dynamic obstacles, the control is performed by the lazy PRM method and does not require reflex commands. If there are dynamic obstacles in its path, the reactive method takes the control and generates commands to force the robot to move away from the intruder obstacles and gives back its DVZ to the original state.

In this point, the robot has lost its original path, and it is necessary to search for a reconnection path to reach its goal. The new path found is a single collision-free curve of Reeds & Shepp. If the attempt of reconnection is successful, the robot executes its new path towards the goal. The new alternative path is obtained with the lazy PRM method by using the information stored in the current robot's configuration, but if a deformation appears the processes are interrupted by reflex actions that force the planner to go back to the previous state.

The algorithm can finish of three forms: i) the robot executes its path successfully, ii) the reflex action is not sufficient and a collision occurs, or iii) the robot does not find an alternative path to conclude its task. Figure 5 shows a high-level description of the proposed approach.

The lazy PRM planner for non-holonomic mobile robots is detailed in (Sanchez et al., 2000). We consider that the other components of this approach are more important and they will be detailed in the next subsections.

3.1 Reactive control by DVZ

By using the equations discussed in the section II, and the next equation. We can use the next DVZ form (see figure 6).

$$d_{hi} = K_1 V_1^2 \cos^2(\beta_i + K_2 \dot{\theta}) + d_i^{sec} \tag{14}$$

where K_1 and K_2 are constants, V_1 and $\dot{\theta}$ are the robot's velocities (see equation 14), β is the angle of the sensor c_i with respect to the transverse axis of the robot, and d_i^{sec} is a safe distance in the direction of the sensor c_i.

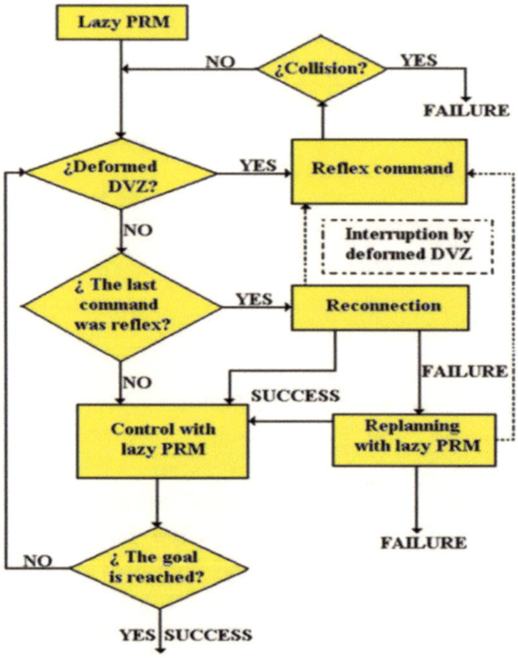

Figure 5. High-level description of our proposed approach

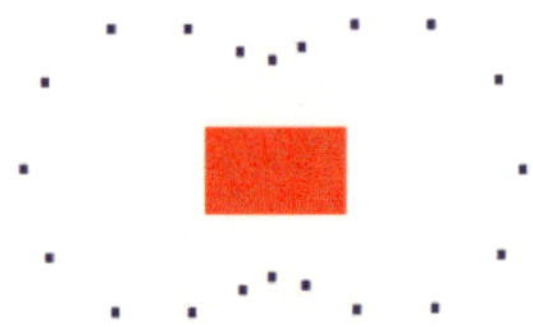

Figure 6. The obtained form of the DVZ using 20 simulated sensors

For the first case in equation (7), ($d_i > d_{hi}$), the DVZ is not deformed by the environment, the control is performed by the lazy PRM method and the reflex actions are not required. For the second case, when ($d_i \leq d_{hi}$), a re flex action is necessary, the executed path by the lazy PRM method is suspended and the robot control is taken by the DVZ method.

3.2 Generation of reflex commands

When the DVZ takes the control, it has the task of taking the robot to a free state of deformations, indicating the kinematics attitudes that should continuously have the robot. These attitudes constitute the vector π described as follows:

$$\pi = \begin{bmatrix} V_1 \\ \dot{\theta} \end{bmatrix} \tag{15}$$

We do not use the equation (12) to implement the control, the control is adapted in the following way.
Let $f_i[n]$ a vector in the direction of the sensor c_i to be defined as

$$f_i[n] = \begin{cases} \Delta_i[n] - \Delta_i[n-1] & if\, \Delta_i[n] - \Delta_i[n-1] > 0 \\ 0 & if\, \Delta_i[n] - \Delta_i[n-1] \leq 0 \end{cases} \tag{16}$$

Let $F[n]$ be the addition of the vectors $f_i[n]$

$$F[n] = \sum_{i=1}^{c} f_i[n] \tag{17}$$

then, the vector $\pi[n]$ is given by

$$\pi[n] = \begin{cases} V_1[n] = V_1[n-1] + K_v * \|F[n]\| * sign(\cos(\hat{F}[n])) \\ \dot{\theta}[n] = \dot{\theta}[n-1] + K_t * (\sin(\hat{F}[n])) \end{cases} \tag{18}$$

3.3 Kinematics of the robot

The robot learns the kinematics attitudes that it should constantly adopt through the path computed by the lazy PRM method and the reflex actions that should be taken. These attitudes are V_1 and $\dot{\theta}$ are fixed in every Δt interval. We consider a model of the car-like robot as follows:

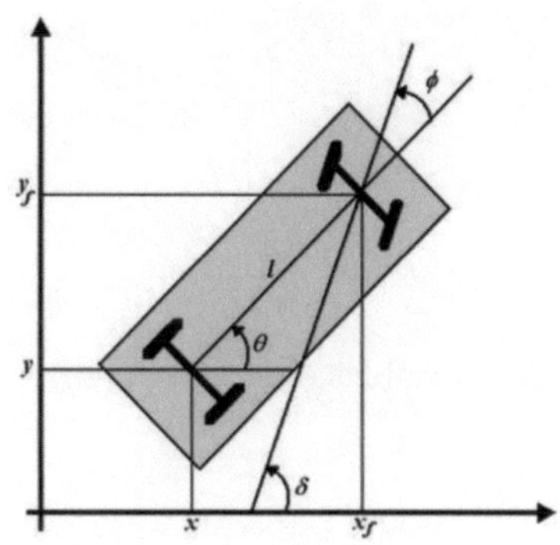

Figure 7. A car-like robot

$$\begin{bmatrix} \dot{x} \\ \dot{y} \\ \dot{\theta} \\ \dot{\phi} \end{bmatrix} = \begin{bmatrix} \cos\theta \\ \sin\theta \\ (\tan\phi)/l \\ 0 \end{bmatrix} V_1 + \begin{bmatrix} 0 \\ 0 \\ 0 \\ 1 \end{bmatrix} V_2 \tag{19}$$

Because the change of x, y and θ is constant foe every interval, the steering angle ϕ in the front wheels stays fixed in the interval (V_2=0), describing a circular path.
The last observation is useful to avoid the integration operation that may be required, otherwise, to determine the x, y and θ values for the next interval of time. Instead, it is enough to use the analytic geometry properties of the circumference.

The next equation shows how to obtain a new configuration for the robot, after the application of a specific impulse.

$$\begin{aligned} x_2 &= x_1 + r(\cos\gamma_2 - \cos\gamma_1) \\ y_2 &= y_1 + r(\sin\gamma_2 - \sin\gamma_1) \\ \theta_2 &= \theta_1 + q*(V_1/r)*\Delta t \end{aligned} \tag{20}$$

where

$$\begin{aligned} r &= \max(abs(V_1/\dot{\theta}), R_{\min}) \\ q &= sign(V_1/\dot{\theta}) \\ \gamma_1 &= \theta_1 - q(\pi/2) \\ \gamma_2 &= \theta_2 - q(\pi/2) \end{aligned}$$

$R_{\min}$ is the trajectory's radius that describes the car-like robot when its front wheels are in its maximum steering angle. The figure illustrates the relation between the variables in equation (20).

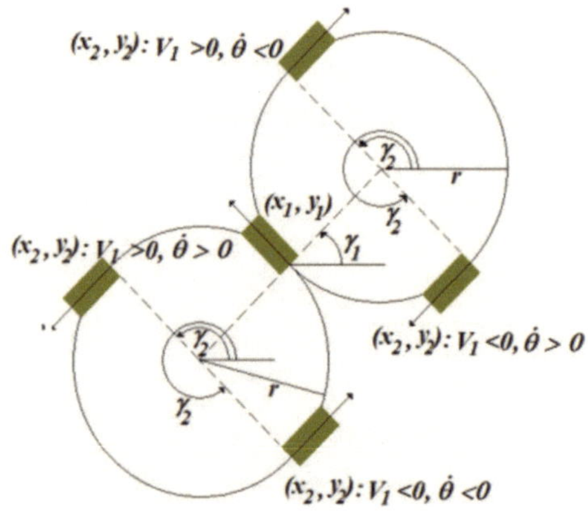

Figure 8. The relation between the variables of equation (20)

3.4 Reconnection

After a successful reflex action, the mobile robot recovers the intact state of its DVZ, but the initial planned path is lost (Fig. 9b), and the lazy PRM method needs to have a path to push the mobile robot to the goal. For this reason it is necessary to provide a path for such aim.
Since the computational cost of a complete re-planning is high, it is avoided as far as possible by executing a process that consists of a reconnection with the planned path by using a single collision-free Reeds & Shepp curve (Fig. 9c).
Initially, the algorithm tries a local path that it is interrupted by a dynamic object. The algorithm will execute a reflex action in order to reconnect with the closest point that is collision-free in the original path. If it can not reconnect after a certain number of attempts, maybe because the possible reconnection paths are blocked with obstacles, the robot will remain immovable for a certain time before executing a new attempt (see Fig. 9d).
The process will be repeated several times, but if the DVZ was deformed by an intrusion, the reconnection process will be modified and will execute the reflex commands.

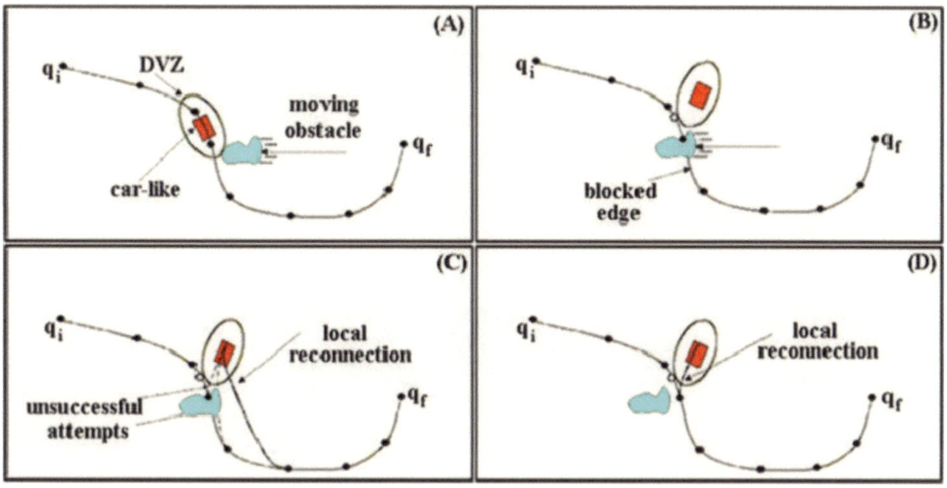

Figure 9. Cases of the reconnection process: a) to avoid a dynamic obstacle, b) after a reflex action, c) after many previous attempts, d) a successful reconnection

3.5 Re-planning

If the reconnection attempts fails, it may happen that paths are blocked by many dynamic objects, or a moving object is parked obstructing the planned path. In this case, the planner executes the lazy PRM method (the initial configuration is the current configuration in the robot). The lazy PRM will be called several times until it returns a collision-free path. If after some attempts a collision-free path can not be found, the planner reports failure.

In the case that the mobile robot is developing in a static environment (or partially static), the planned path is enough to avoid a collision. Under this assumption, there is not need to generate any reflex action when a fixed obstacle enters the DVZ.

The model cannot distinguish if an intrusion is caused by a moving or a static obstacle because the DVZ method does not use any model of the environment. To solve this problem, it is necessary to use an auxiliary image that represents the environment and it is updated every time the re-planning or reconnection procedures are called. When the sensors in the robot detect an obstacle that deforms the DVZ, the intruder object coordinates are revised to see if there was already an obstacle, registered in the auxiliary image; if this is the case, the system assumes the presence of a fixed obstacle and there is no need for a reflex action, otherwise, it will certainly assume that the object is in movement.

4. Simulation results

Some simulation results are presented in this section. The planner was implemented in Builder C++ and the tests were performed on an Intel © Pentium IV 2.4 GHz processor and 512 MB memory. After having executed our planner in different scenes, in the majority of the cases the motion planning problem is solved satisfactorily. Our planner produces a first roadmap by sampling configurations spaces uniformly. It computes the shortest path in this roadmap between two query configurations and tests it for collision.

The robot starts moving under the permanent protection of its DVZ. In absence of dynamic obstacles, the robot does not require reflex commands and the control is executed with lazy

PRM. If there are dynamic obstacles in its path, the reactive method takes the control and generates commands to force the robot to move away from the intruder obstacles and gives back its DVZ to the original state.

The moving obstacles have a square form and move at constant velocity in straight line. Whenever they collide with another object they assume in their movement a new random direction. Figure 10 shows an environment composed of narrow passages and dynamic obstacles moving randomly at the same velocity than the mobile robot.

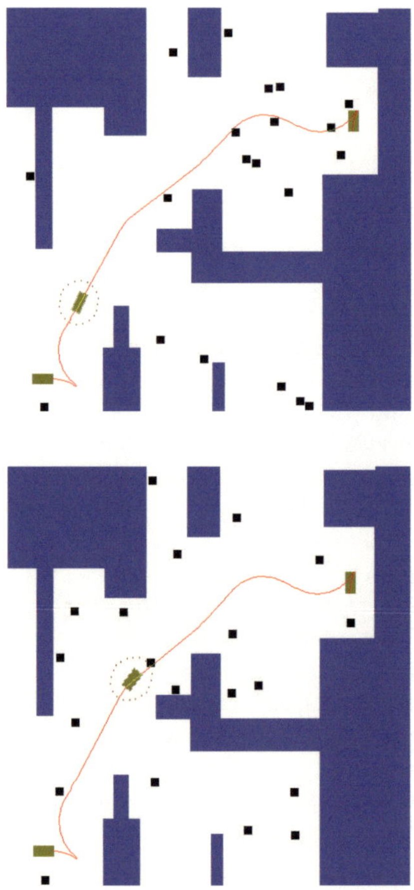

Figure 10. An example of a query and the path solution in an environment with 20 moving obstacles. The robot starts moving under the permanent protection of its DVZ

In order to evaluate the performance of the planner, we performed tests on the environment of Figure 11 for several roadmap sizes and different number of moving obstacles. The different settings are summarized in the tables 1, 2 and 3. In our case, due to the strategy of node addition, the time for the roadmap's construction is proportional to the number of nodes. The number of nodes at the beginning is a critical parameter that affects the lazy PRM's performance (Sánchez et al., 2002). To show the methodology proposed, we performed 30 trials.

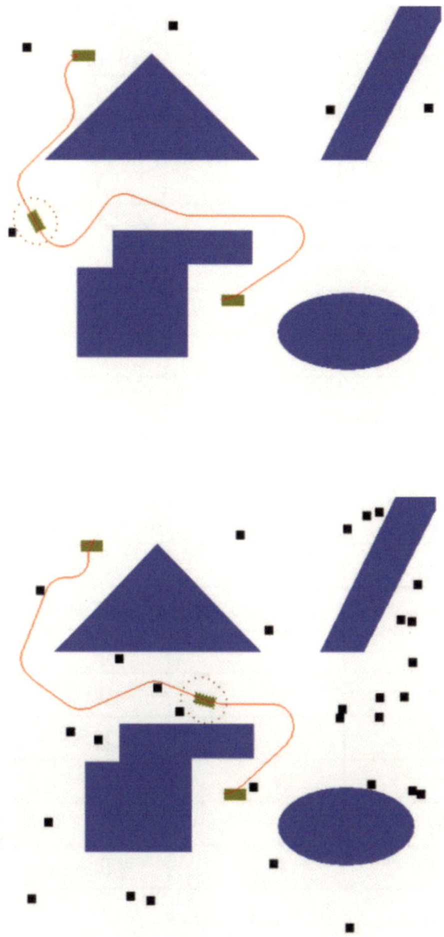

Figure 11. Trajectory execution control by the proposed planner. The environment constains 5 and 30 moving obstacles

Settings	50 nodes	100 nodes	50 nodes	100 nodes
Steering angle	45	45	70	70
Graph building	0.007	0.016	0.008	0.028
Graph searching	0.005	0.016	0.005	0.015
Coll. checking	980	1299	1093	1712
Total time (secs)	0.071	0.109	0.074	0.147

Table 1. Performance data for Lazy PRM

Reconnections	Time for reconnection	Replanning	Time for replanning	Collision	Success
43	0.027	0	0	No	Ok
12	0.025	0	0	No	Ok
3	0.025	0	0	No	Ok
65	0.030	0	0	No	Ok
97	0.027	0	0	No	Ok
84	0.031	0	0	No	Ok
12	0.026	0	0	No	Ok
16	0.026	0	0	No	Ok

Table 2. Performance data with 5 moving obstacles

Reconnections	Time for reconnection	Replanning	Time for replanning	Collision	Success
36	0.029	0	0	No	Ok
90	0.029	1	0.019	No	Ok
3	0.027	0	0	Ok	No
5	0.026	0	0	Ok	No
81	0.032	1	0.115	No	Ok
27	0.032	1	0	No	Ok
4	0.026	0	0	No	Ok
9	0.026	0	0	No	Ok

Table 3. Performance data with 10 moving obstacles

In fact, the method's performance can be considered satisfactory if it presents a fast planning phase, reflex actions based on sensors that do not require expensive algorithms, an effective process of reconnection performed in milliseconds, and a process of re-planning that is executed if the Lazy PRM and DVZ's parameters are appropriate. As mentioned in earlier sections, it can be considered that the methodology proposed here, includes these characteristics.

The planning time is reduced due to the incomplete collision detector whose work is complemented with the robot's sensors during the path execution. On the other hand, the assignation of direction angles to the nodes that conform the shortest paths obtained by the algorithm A^*, produces curves that allow the algorithm to omit the optimization process (i.e., the smoothing process). With respect to the reconnection process, the paths obtained with the planner are conformed by a single Reeds & Shepp curve and based on the incomplete collision detector, making short the time and close to optimal the curves obtained with the algorithm. Since the reflex actions are provided by the DVZ method, it is possible to interrupt the reconnection and re-planning processes if necessary, without incurring in bigger problems.

If the execution's parameters for the Lazy PRM and DVZ methods are adapted, the re-planning process will not be called very often and will be successful in the absence of narrow passages. Figure 12 presents a case where the reflex actions were not sufficient. The presence of narrow passages is an important problem to being considered.

5. Experimental results

We have implemented the approach on the Pionner-3 robot from the ActivMedia Robotics. This robot is driven by two independent wheels , it is an agile, versatile intelligent mobile robotic platform updated to carry loads more robustly and to traverse sills more surely with high-performance current management to provide power when it's needed. It has a ring of 8 forward sonar and 8 rear sonar ring. 3-DX's powerful motors and 19cm wheels can reach speeds of 1.6 meters per second and carry a payload of up to 23 kg. In order to maintain accurate dead reckoning data at these speeds, the Pioneer uses 500 tick encoders. Its sensing moves far beyond the ordinary with laser-based navigation options, bumpers, gripper, vision, stereo rangefinders, compass and a rapidly growing suite of other options.

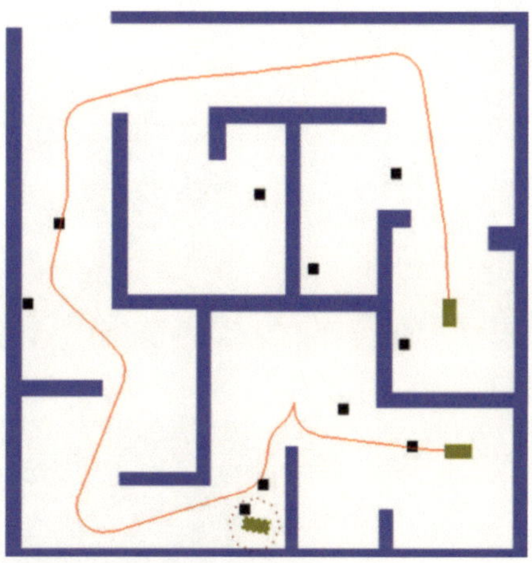

Figure 12. The reflex actions were not sufficient, the mobile robot collides with a moving obstacle

Figure 13. The mobile robot used in the experimental part

Figure 14. The mobile robot avoids an obstacle

The experimental part was done considering that the robot is able to follow a geometric trajectory previously calculated by a Lazy PRM planner, we considered a model of the environment on scale. In the absence of obstacles, the robot follows the trajectory until arriving at the goal region, if there are unknown obstacles, the robot executes reactive controls to avoid them and to return to its trajectory.

Figure 14 illustrates this single experiment, where the robot avoids an unknown obstacle. One can see that robot clearly avoids the obstacle and returns to the nominal path.

6. Conclusion

The motion planning for non-holonomic robots in moving environments is a complex problem. The results obtained in the evaluation of the reactive lazy PRM method, proposed in this work, show the importance of finding a solution for this problem.

In fact, the method's performance can be considered satisfactory if it presents a fast planning phase, reflex actions based on sensors that do not require expensive algorithms, an effective process of reconnection performed in milliseconds, and a process of re-planning that is executed if the Lazy PRM and DVZ' s parameters are appropriate.

The planning time is reduced due to the incomplete collision detector whose work is complemented with the robot's sensors during the path execution. On the other hand, the assignation of direction angles to the nodes that conform the shortest paths obtained by the algorithm A^*, produces curves that allow the algorithm to omit the optimization process (i.e., the smoothing process).

With respect to the reconnection process, the paths obtained with the planner are conformed by a single Reeds & Shepp curve and based on the incomplete collision detector, making short the time and close to optimal the curves obtained with the algorithm.

Since the reflex actions are provided by the DVZ method, it is possible to interrupt the reconnection and re-planning processes if necessary, without incurring in bigger problems.

If the execution's parameters for the Lazy PRM and DVZ methods are adapted, the re-planning process will not be called very often and will be successful in the absence of narrow passages.

A reactive lazy PRM planner for dynamically changing environments is presented in this chapter. Although some promising results are shown in its present form, the planner could be improved in a number of important ways. This approach can be extended to use real robots and to solve the problem posed by small static obstacles. Besides, some cases where the reflex action was not sufficient to avoid collisions were observed during the evaluation tests. Theses cases are difficult because they require a more intelligent behavior in order to avoid the robot to be trapped.

In those cases, it can be necessary to add a process that computes the trajectories of moving objects and corrects the path in real time.

Finally, a very interesting topic in robotics is the study of non-structured environments. This methodology can be extended to solve these cases.

7. References

Bohlin, R. & Kavraki, L. (2000). Path planning using lazy PRM, *Proceedings of the IEEE Robotics and Automation Conference*, pp. 521-528, ISBN 0-7803-5889-9, San Francisco, CA, USA, April 24-28

Brock, O. & Kavraki, L. (2001). Decomposition-based motion planning: A framework for real-time motion planning in high-dimensional configuration spaces, *Proceedings of the IEEE Robotics and Automation Conference*, pp. 1469-1474, ISBN 0-7803-6578-X, Seoul, Korea, May 21-26

Cacitti, A. & Zapata R. (2001). Reactive behaviours of mobile manipulators based on the DVZ approach, *Proceedings of the IEEE Robotics and Automation Conference*, pp. 680-685, ISBN 0-7803-6578-X, Seoul, Korea, May 21-26

Jaillet, L. & Siméon, T. (2004). A PRM-based motion planner for dynamically changing environments, *Proceedings of the IEEE Intelligent Robots and Systems Conference*, pp. 1606-1611, ISBN 0-7803-8464-4, Sendai, Japan, September 28 - October 2

Khatib, O. (1986). Real-time obstacle avoidance for manipulators and mobile robots, *International Journal of Robotics Research*, Vol. 5, No. 1, (Spring 1986) pages 90-98, ISSN: 0278-3649

Kavraki, L., Svetska, P., Latombe, J. C., & Overmars, M. (1996). Probabilistic roadmaps for path planning in high-dimensional configuration spaces, *IEEE Transactions on Robotics and Automation*, Vol. 12, No. 4, (August 1996) pages 566-580

Jaouni, H., Khatib, M., & Laumond, J. P. (1998) Elastic bands for non-holonomic car-like robots: Algorithms and combinatorial issues, *Proceedings of WAFR 98,* Houston, Texas, USA, March 5-7

Laumond, J. P (Ed.). (1998). Robot motion planning and control, Springer-Verlag, ISBN 3-540-76219-1

Quinlan, S. & Khatib. O. (1993). Elastic bands: Connecting path planning and control, *Proceedings of the IEEE Robotics and Automation Conference*, pp. 802-807, ISBN 0-8186-3450-2, Atlanta, Georgia, USA, May 93

Leven, P. & Hutchinson, S. (2000). Toward real-time path planning in changing environments, *Proceedings of WAFR 2000,* Hanover, NH, USA, March 16-18

Reeds, J. A. & Shepp, R. A. (1990). Optimal paths for a car that goes both forward and backwards, *Pacific Journal of Mathematics*, Vol. 145, No. 2, pages 367-393

Sánchez, G. & Latombe, J. C. (2002). On delaying collision checking in PRM planning: Application to multi-robot coordination, *International Journal of Robotics Research,* Vol. 21, No. 1, pages 5-26

Sánchez, L. A., Zapata, R., & Arenas B. A. (2002) Motion planning for car-like robots using a lazy probabilistic roadmap method, *Advances in Artificial Intelligence*, Springer-Verlag, pp. 1-10

Sánchez, L. A., Zapata R., & Lanzoni, C. (2003) On the use of low-discrepancy sequences in nonholonomic motion planning, *Proceedings of the IEEE Robotics and Automation Conference*, pp. 3764-3769, Taipei, Taiwan, September 14-19

Sánchez L. A., Cuautle P. R., Zapata, R., & Osorio L. M. A. (2006) A reactive lazy PRM approach for non-holonomic motion planning, *LNAI 4140 Springer-Verlag*, pp. 542-551

Svestka, P. & Overmars, M. (1997) Motion planning for car-like using a probabilistic learning approach, *International Journal of Robotics Research*, Vol. 16, No. 2, pages 119-143

Zapata, R. (1991) Quelques aspects topologiques de la planification de mouvements et des actions réflexes en robotique mobile, Thèse d'Etat (Phd Thesis in French), University of Montpellier II

Zapata, R., Lépinay, P., & Thompson, P. (1994) Reactive behaviors of fast mobile robots, *Journal of Robotic Systems*, Vol. 11, No. 1, pages 13-20

24

Integrating Time Performance in Global Path Planning for Autonomous Mobile Robots

A. R. Diéguez, R. Sanz* and J. L. Fernández
Systems Engineering and Automation Dept., University of Vigo
Spain

1. Introduction

Global path planners for autonomous mobile robots have been demonstrated to map optimal routes in structured environments. The search for the optimum path from a given starting point to a destination is usually formulated as a graph search problem.

Over the years, much progress has been made in efficient methods for global motion planning in static environments. A number of both exact and approximate approaches in this field (see, for example, Latombe, 1991; Hwang & Ahuja, 1992; LaValle, 2006). However, real world scenarios are intrinsically dynamic due to the presence of non-modelled obstacles, both static and moving. Path planning in such environments is an area of active research. Different methods have also been proposed (Heero, 2006; Jaillet & Simeon, 2004). Some approaches use robot sensors to build a model of the environment incrementally even when the real world is static. In such circumstances, a new path must be calculated every time the model is updated. The online integration of this new information in a global map is usually complex and time consuming.

Path planning with dynamic objects can be addressed by using explicit time representation to convert the problem into an equivalent static one that can then be solved using an existing static planner. However, this increases the dimensionality of the representation and requires exact motion models for moving objects (Szczerba & Chen, 1995; Hsu et al., 2002). Note that, in many circumstances (such as the presence of people near the robot in a tour-guide autonomous application), it is not possible to explicitly model the unpredictable behaviour of humans and robot movements with a time-extended representation (Philipsen et al., 2007).

In contrast to previous works, our global path planning approach is able to take into account the effect of persistent delays caused by unpredictable objects (such as people in corridors or lobbies) by considering these delays for future path planner executions.

2. Problem statement

Depending on the nature of the problem, different optimality criteria can be used to minimize the overall cost of the proposed path. Although the most frequently used cost function is based on the branch length between two nodes, in many cases it is more attractive to minimize traveling time rather than the distance traveled by the robot. Our approach focuses on problems in which this metric is of interest.

Due to the presence of non-modeled obstacles (both static and dynamic) in the environment, the robot must be able to avoid unexpected obstacles by means of a reactive behavior. This causes deviations from the original trajectory and possibly delays accomplishment of planned tasks.

The research on enhancing global path planning described in this chapter is intended for mobile robots navigating in partially known indoor environments. The method is based on a graph approach that adapts graph weights by integrating traveling time measurements from real task executions when the robot repeatedly follows the same paths.

The technique uses periodic measurements of time and the positions reached by the robot while moving towards the goal to modify the costs of the branches (Diéguez et al., 2007). These weights are good options for branch costs while the robot moves in the presence of non-modeled obstacles. Consequently, when trying to minimize the time needed to move from any given point to any given destination in dynamic environments, the search for a time-optimal path in a static global map produces better quality results than the use of a distance metric.

Another important aspect of our approach is its independence of the existence of a local motion planner or reactive guidance modules in the robot navigation architecture. This is especially attractive because it makes it possible to incorporate the time delay caused by non-modeled obstacles even in a situation where obstacles cannot be included in the map, such as people walking in crowded corridors or in a waiting room. Likewise, it makes the costs of the branches more realistic because they represent the time needed to travel in the environment.

In simulations and real experiments, results are generally better than using only the information on the obstacles stored in the global map, while the computational cost is significantly lower.

In order to illustrate the above problem statement, Figure 1 shows a simple test carried out with a mobile robot system developed in our Lab (Diéguez et al., 1998). Two possible trajectories are available to arrive at point 4 from point 1. Initially, segments 1-3-4 are chosen by the path planning algorithm because this is the shortest path, but three static non-modelled obstacles (grey boxes) found in the environment cause an increase in traveling time. Broken lines represent the actual trajectories followed to avoid the obstacles. Table 1 summarizes the results of two experiments with the robot at a constant speed of 0.5 m/sec. These non-modeled obstacles caused an important increase in the time needed to follow segment 1-3. Our global path planner recognizes this kind of situation and, as a consequence, selects segments 1-2-3-4 in further path planner executions.

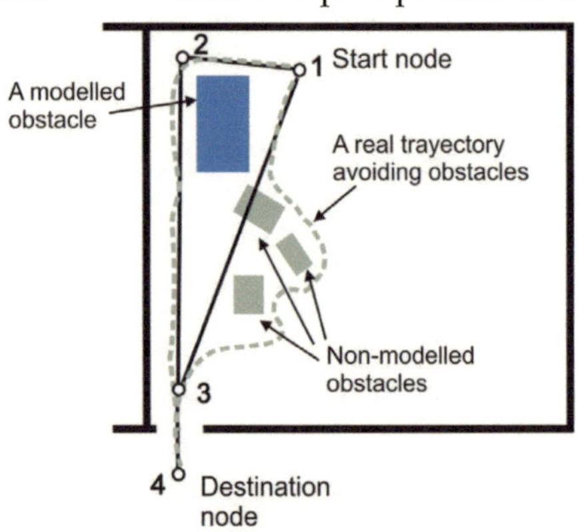

Figure 1. Mobile robot real trajectories in an experience with non-modelled obstacles

	Segment lengths	Estimated time	Actual time	Deviation
Segments 1-2-3-4	7.82 m.	15.64 sec.	16.31 sec.	4.28 %
Segments 1-3-4	6.11 m.	12.22 sec.	17.12 sec.	41.10 %

Table 1. Experimental time measurement in actual trajectories

We propose a different approach to dealing with global path planning in the presence of non-modelled obstacles that uses an aggregate time-cost function to find an optimal global path. The method uses traveling time and position measurements to calculate the time cost of each branch. We also propose a geometric method to modify the cost value, for each graph branch involved in the paths, to include the traveling time information.

Although our approach does not eliminate the need for a local motion planner in our mobile robot navigation architecture (Diéguez et al., 2003), it enables incorporation of the time delay caused by non-modelled obstacles even in the situation where these obstacles cannot be included in the map (waiting rooms or crowded corridors). It makes the cost of the branches more realistic because the time necessary to travel in the environment is represented. In many real experiments, results are generally better than using only the information on the obstacles stored in the global map, while the computational cost is significantly lower. For this reason, our method is especially attractive for incorporating the effects of mobile obstacles to the planning process.

The rest of the chapter is organized as follows: Section 3 presents the global path planning algorithm and briefly describes the skeleton (structure that represents the free space of the environment) building algorithm; the cost adaptation algorithm based on traveling time is outlined in Section 4; and some simulated results are given in Section 5. The evaluation of a real experiment with our mobile robot system is detailed in Section 6. Finally, some conclusions are given in Section 7.

3. Global path planning algorithm

The global path planning algorithm uses all currently available information about the static obstacles in a workspace to find the shortest path. In our approach, two different data structures were combined for this purpose: a cell map which containing prior knowledge of the workspace, and a path graph which is a topological representation of the free space. The cell map was created from a CAD map of the environment, whereas the path graph was created from the cell map using a wave front expansion algorithm generating a skeleton (Latombe, 1991). The skeleton is a connectivity network representing all possible routes in the free space and their intersections. The graph obtained this way allows two points anywhere in the workspace to be linked. The search for a feasible path is reduced to a graph search problem. In our method, both data structures were combined to accelerate the computation of the skeleton and to minimize search time for the optimum path.

The path resulting from a simple concatenation of the branches selected by the global path planner is called the *skeleton route* (see Table 2 for a summary of definitions). This trajectory may contain undesirable bends and detours that are unsuitable for the path. For this reason, once the skeleton route is obtained, a segment route is generated removing these imperfections and converting the route into a sequence of straight segments. Each segment is then linked to its adjacent one using smooth curves. This smoother version of the segment route is passed to the navigation module of the robot, in accordance with the model proposed in Diéguez et al. (1995).

	Definition
Skeleton	Any type of connectivity network representing the free space (e. g. a Voronoi diagram)
Node	Intersections in the skeleton
Branch	A portion of the skeleton between two intersections (nodes)
Optimum path	Fastest path between two points in the skeleton
Segment	A straight line composed of any (optimal) trajectory
Real trajectory	Route actually traveled by the robot
Segment threshold	A distance limit to a segment to compute control points
Control point	Point of the real trajectory close to a segment path

Table 2. Basic term definitions related to the global path planning algorithm

The segment route is used in our approach to adapt the weighting factors for each branch of the skeleton route. These costs are calculated from accumulated statistics of traveling times while the robot is moving through the workspace along different routes. These statistics were used to make the weights associated with each branch of the skeleton route closer to reality.

The method to calculate the optimum path consists of the following steps:

Step 1: Creating the cell map. An internal representation based on a cell map is constructed from a map of the environment stored in an external file.

Step 2: Building the skeleton of free space using an obstacle growth algorithm. A distance wave-front flows around the obstacles through all the free space in the environment. The skeleton is the set of cells where two or more wave-fronts meet (Latombe, 1991). The technique developed, unlike that used in similar algorithms, does not need to compute or store any values, making it possible to compute the skeleton rapidly and easily.

Step 3: Constructing a graph. This graph is the topological representation of the previously calculated skeleton. We have chosen this data structure because it is best suited to perform searches. A graph branch is a series of points linking two places in the free space. A graph node represents a point where two or more branches meet. Two nodes can be connected by one or more branches. A branch cost is initially defined as the estimated elapsed time for a robot to move from one node to another. The location of a node is represented by its coordinates. The graph is pruned and this reduction is transposed to the skeleton, thus eliminating irrelevant branches.

Step 4: Computing the optimum route using an exhaustive search algorithm on the graph. We implemented a breadth-first search strategy without repeating states (Russell & Norvig, 1995). Although in comparison with other non-informed methods this search strategy often requires a great amount of memory when working with cluttered environments, the memory problem does not arise in indoor structured environments due to the short number of generated states.

The path thus computed in the skeleton is called the *skeleton route*. This path is shortened by linking points of maximum visibility in the skeleton route. This is the path used to evaluate the traveling time function described in Section 4. The segments of the shortest path are then connected with arcs of circumferences and a velocity reference is assigned to each segment.

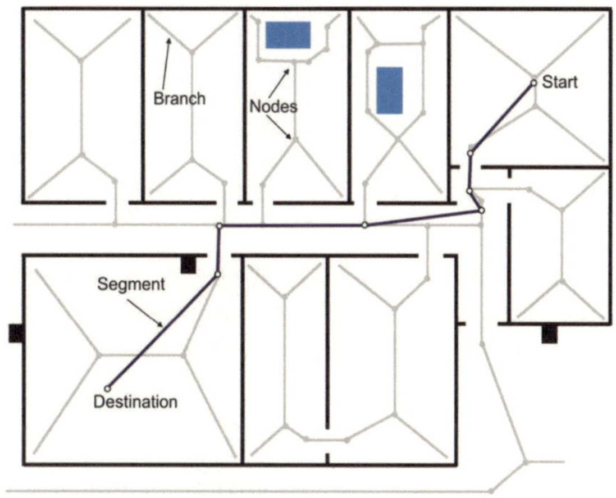

Figure 2. Workspace with rooms and corridors, with skeleton (in grey) and segment route (thin line)

Figure 2 shows a workspace describing the components used in our global path planner. The skeleton and the shortest path to arrive at the bottom right corner from the top left corner generated by the path planner are also depicted.

3.1 Description of the skeleton building algorithm

The algorithm developed to build the skeleton avoids the computation of potential field values and so dramatically reduces execution time. The algorithm iteratively grows all the obstacles (i.e. it thins the free space). The skeleton is the set of cells where these layers meet. This can be interpreted as a wave-front expansion, with each cell where two or more wave fronts meet set as a skeleton cell.

For the implementation of this algorithm, we only need the cell map. Each layer (wave front) is generated by labeling the neighboring cells of the previous layer with a temporary value. Thus, the layers use alternate cell values, it being necessary only two different temporary values. Thus, the range of values necessary can be represented by a single byte for each cell.

In order to generate each layer, a mask-based case identification method is used. This method functions on the basis of considering each cell and its eight immediate neighbors. The 3x3 resulting matrix is compared with the contents of a table with all possible cases to decide whether the central cell should be labeled as part of the layer or should be left empty. Only free (empty) cells are considered, so the 3x3 case can be represented by an 8-bit number (one bit for each cell surrounding the central one). This table only needs, therefore, to reflect 255 different cases.

When no more of the empty cells of the map can be labeled (that is, it cannot be incorporated as part of a layer), the algorithm terminates. The remaining empty cells constitute the skeleton. Figure 3 shows two examples where the central cell must remain empty and another two cases where the central cell should be labeled as part of the current growing layer.

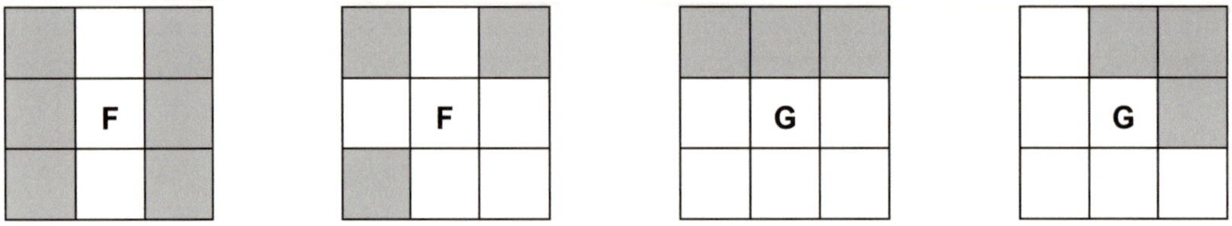

Figure 3. The first two cases must leave the central cell empty. In the other two situations, the central cell must be labeled as part of the current growing layer (G means "grow" and F means "free")

This algorithm is faster and uses less memory than the usual potential field-based wave front expansion methods. Moreover, the skeleton computed in such way is the most distant skeleton from the obstacles, leading to a lower probability of collision.

4. A geometric method for branch weighting factor calculation

This section describes the improvement incorporated in the global path planning algorithm in order to take into account statistical measurement of traveling time. The proposed method employs a geometric approach to modifying the weight of each branch according to the time spent by a robot navigating along the global path. The algorithm first maps each point of the actually performed route on the segment route. Each route segment is then associated with its corresponding branches and finally, the results are fused with the previous traveling time weights.

4.1. Mapping a real route on the segment route

Given a set of coordinate points actually performed by the mobile robot, a group of control points are selected. Control points are points common to both trajectories. Moreover, a coordinate point in the real route, which is close enough (a fixed distance threshold) to the segment path, can be regarded as a control point. Control points are employed as temporal references for traveling time assignment to the segments.

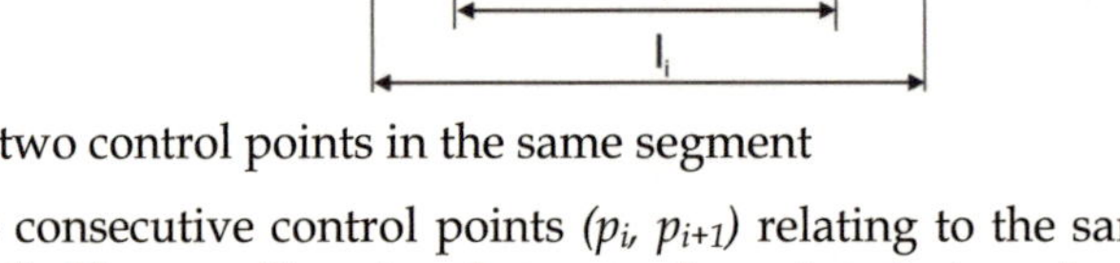

Figure 4. Mapping two control points in the same segment

Let us assume two consecutive control points *(p_i, p_{i+1})* relating to the same segment whose length is l_j (Figure 4). The traveling time between the points is t_i, and s_i is a span over this segment, and it is assumed that their contribution to the traveling time of the segment is $t_i * s_i/l_j$. This means that the contribution to the segment traveling time is the time difference between both points weighted by the share of the segment spanned. Therefore, the final traveling time of a segment is the accumulation of all the contributions made by each pair of points related to the segment.

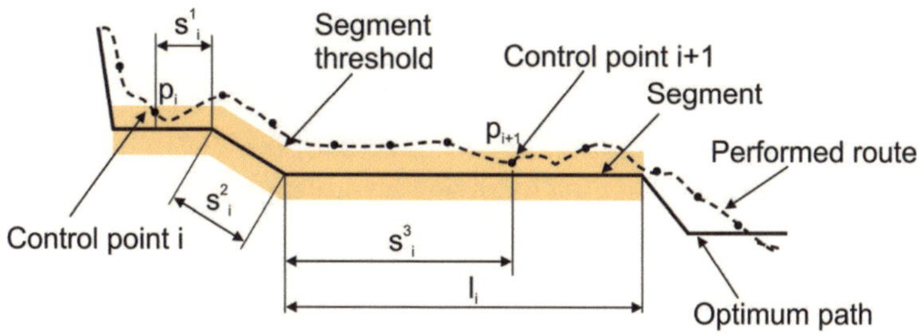

$s_i = s_i^1 + s_i^2 + s_i^3$

Figure 5. Mapping two control points in different segments

When a pair of consecutive control points spans more than one segment (or even fractions of segments), a similar procedure is implemented (Figure 5). Let us denote again by s_i the span over the segment route. Considering as l_j. the portion of the segment affected by the pair of points (note that l_j differs from the total length only for the first and the last segments), it is assumed, once more, that the contribution to the traveling time of the segment j is $t_i * s_i / l_j$. In the case of the segment being totally subtended by the pair of control points, l_j equals its total length.

Thus, once all the contributions of the control points in the segment have been established, the traveling time weight for a segment j is calculated as:

$$T_j = (1/ l_j) \cdot \Sigma_i t_i \cdot s_i \tag{1}$$

where i ranges from 1 to the total number of control points in the segment.

To summarize, the procedure for this stage is as follows:

i. Find the corresponding segment for each control point and the related information (distance, etc).
ii. Remove the control points further than a threshold from its corresponding segment.
iii. Remove all the control points corresponding to the same end of the segment, except for the first and last control points.
iv. Remove intermediate control points until only two control points in each segment remain.
v. Compute and assign the traveling time weighting for each segment.

4.2. Mapping segment route on branch route

The aim of the second phase of the algorithm is to transfer the segment traveling time to the branches of the graph that make up the skeleton route.

We only consider the control points that are the ends of segments (except for the first and the last one) because they are common points for both routes. Let us denote by sj the span of the segment on the skeleton route. The velocity v_i for the segment j is defined as S_j/T_j ; and T_i is its traveling time weight. Thus, the traveling time weight for each branch is the accumulation of the weighted contributions of each segment related to this branch. Calling l_{ij} the portion of the branch i subtended by segment j, the increase of traveling time for the branch due to the segment is v_j* l_{ij}. The total time weight for a branch j is:

$$W_j = \Sigma_i T_i{}^* S_i \tag{2}$$

where i ranges from 1 to the total number of segments related to the branch j.

4.3. Fusion of costs

The weighting factors so far computed for each branch of the skeleton route are fused with the previously existing values using the following formula:

$$W_i = \alpha(c)\, W_0 + (1 - \alpha(c))\, W_{i-1} \qquad (3)$$

The confidence measure (c) is a function of the confidences computed in the two previous phases. The factor $\alpha(c)$ has a value between 0 and 1, and takes into account deviation of the new traveling time from the existing one. The cost of each branch is calculated as a weighted sum of the costs calculated so far and a constant that is a function of the length of the branch. This approach makes the final value more stable.

5. Simulation results

This section presents the behavior of the described technique in a series of simulations. Figure 6 depicts the simulation environment perceived by the sensors of our robot. This map corresponds to a real environment in the Santiago de Compostela Conference Center (Spain) (Fernández et al., 2008), with corridors, rooms, lobbies and stands.

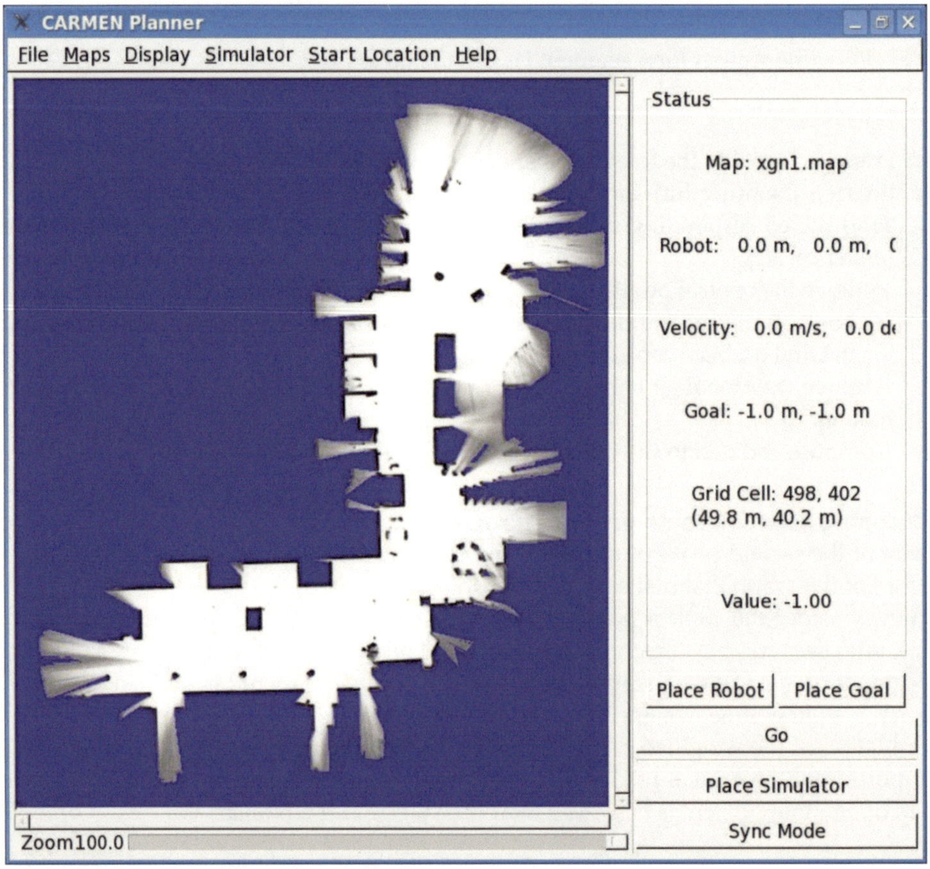

Figure 6. Environment used for the simulation

	Branch sequence	Skeleton route time	Estimated path time	Simulated path time	Deviation: estimated vs. simulated path time
Trajectory A	1→8→9→10→11→13	107.44	78.55	86.02	6.9%
Trajectory B	1→2→7→13	86.22	67.32	95.94	33.2%
Trajectory C	1→2→3→4→5→6→13	100.97	81.41	112.81	31.1%

Table 3. Estimated and simulated times (in seconds) for three alternative routes between the start and the goal

Figure 7 shows the initial position as a circle on the left, and the destination as a circle on the right. In order to reach the goal, the robot has three candidate routes, summarized in Table 3.

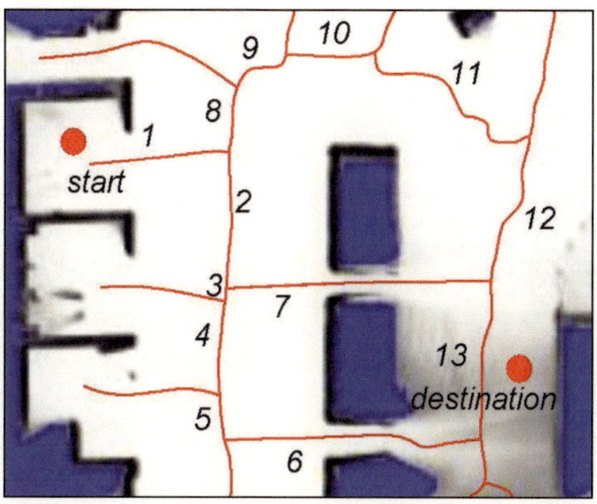

Figure 7. Simulation: skeleton of the environment, branch labels, and start and destination points for the route

	Skeleton time (sec.)	Simulated time (sec.)	Deviation: theoretical vs. simulated time
branch #1	17.73	16.73	5.6%
branch #2	16.30	11.94	26.8%
branch #7	33.02	51.99	-57.5%

Table 4. Weighting factor values for the simulated experiments

A series of tests simulating navigation by the robot in the presence of non-modeled obstacles was carried out by performing the three trajectories summarized in Table 3 and depicted in Figure 8. Each skeleton route was transformed into a segment route (straight segments) actually fed to the robot, as can be also seen in Figure 8. However branches 6 and 7 are located in narrow and busy corridors that slow down the robot run. This situation is numerically shown in Table 4. The delays are caused by both static and dynamic obstacles. The static obstacles make the robot step aside from the desired path, and the dynamic ones make the robot slow down or even stop until the obstacle (e.g. a person or another robot) has passed. In this case, there is no deviation from the segment route but, even so, a significant time delay is caused.

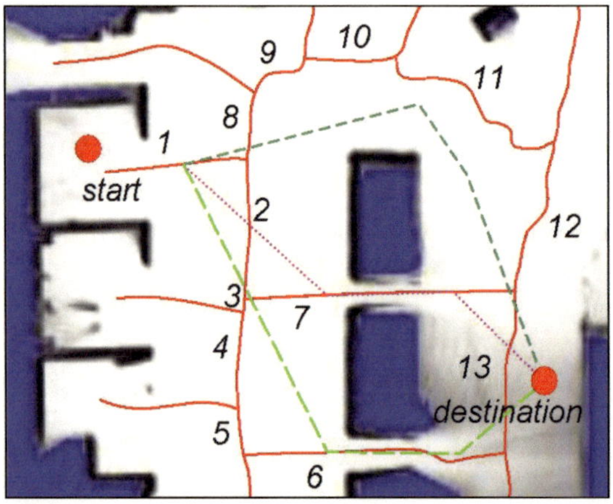

Figure 8. Skeleton and three possible segment routes

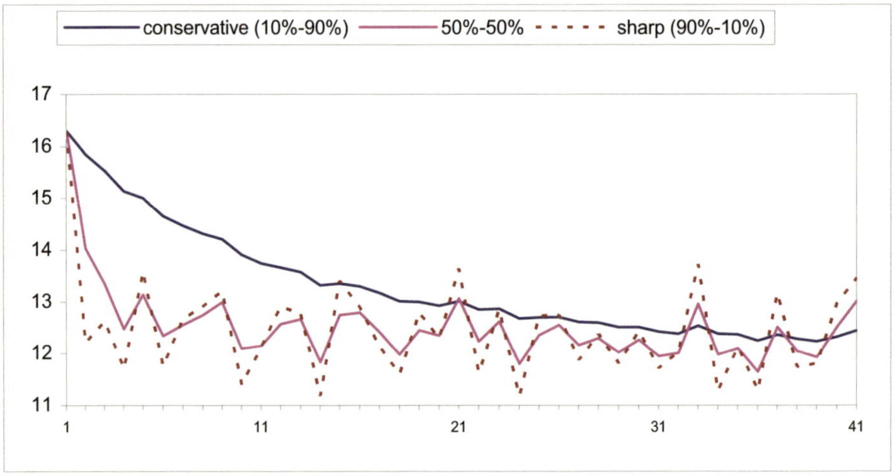

Figure 9. Branch weight convergence with different fusion ratios

After several simulations it becomes clear that the deviation from the initial cost differs depending on the branch under consideration (see Table 4). The initial weight is a function of the length of the branch. It must be pointed out that a constant speed along the whole trajectory was used in all the simulated experiments.

In order to study the relationship between fusion coefficient and branch weight convergence, several fusion ratios were used. The results can be seen in Figure 9. As expected, conservative fusion coefficients lead to slower and smoother convergence curves.

Based on this weight information, the global path planner updates the costs of the involved branches. Further executions of the global path planner choose trajectory A instead B or C, thus avoiding the aforementioned narrow, crowded corridors.

6. Experimental results with a real robot

In this section we summarize the results of a series of real-time experiments carried out with our mobile robot in the School of Engineering in the University of Vigo. The robot (called Rato) used in these experiments is based on a modified B21 RWI platform. It is equipped with a ring of 24 sonar proximity sensors and 24 bumpers on the enclosure connected with four CAN slave modules that are attached to the computer through a CAN-USB adapter card. The robot is also equipped with a SICK PLS 200 laser range finder. This sensor measures the distance to obstacles in the vicinity of the robot.

A layered software architecture is used, consisting of task scheduling, path planning, navigation, and obstacle avoidance components, each of which relies on the abstraction provided by the previous level. The robot architecture is implemented as a collection of asynchronous processes. IPC (Inter Process Communication), developed at Carnegie Mellon's Robotics Institute, is used to integrate the system and interprocess communication (Simmons, 1994). The global path planning method described here is included in this hierarchical navigation architecture (Figure 10). This navigation system combines a reactive part (BEAM module) with a deliberative one represented by the navigator and task planner modules. The advantages of both systems are thus obtained; a quick response to events that require it, and the effectiveness of the deliberative systems, as actions are regulated by references obtained by planning.

The tasks requested by the user are handled by the task planner that determines the places to visit and the order in which to do so. As long as there are tasks to be completed, this module is in charge of the selection of new destinations as the previous destinations are reached.

The goal point and the estimated position are used by the path planning module (Navigator) to determine how to travel efficiently from one location to another.

The reactive system will follow this direction avoiding any obstacles that appear. The information obtained from the sensors is integrated into a grid that represents the robot's environment. This grid is used by the localization module to estimate position.

Each module in Figure 10 is a Linux process that exchanges information with other modules using IPC, which provides a publication-subscription model. Each application registers with the the central server and specifies what types of messages it publishes and what types it listens for. Any message passed to the central server is immediately copied to all other subscribed processes.

The dynamic path planning algorithm outlined earlier is implemented as a single process.

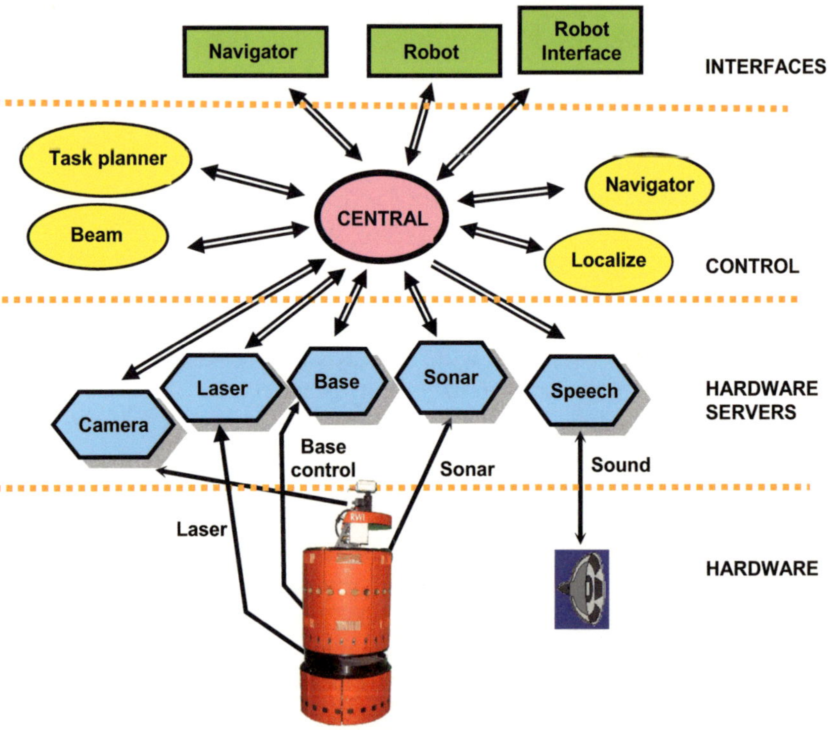

Figure 10. Graphical representation of the navigation architecture

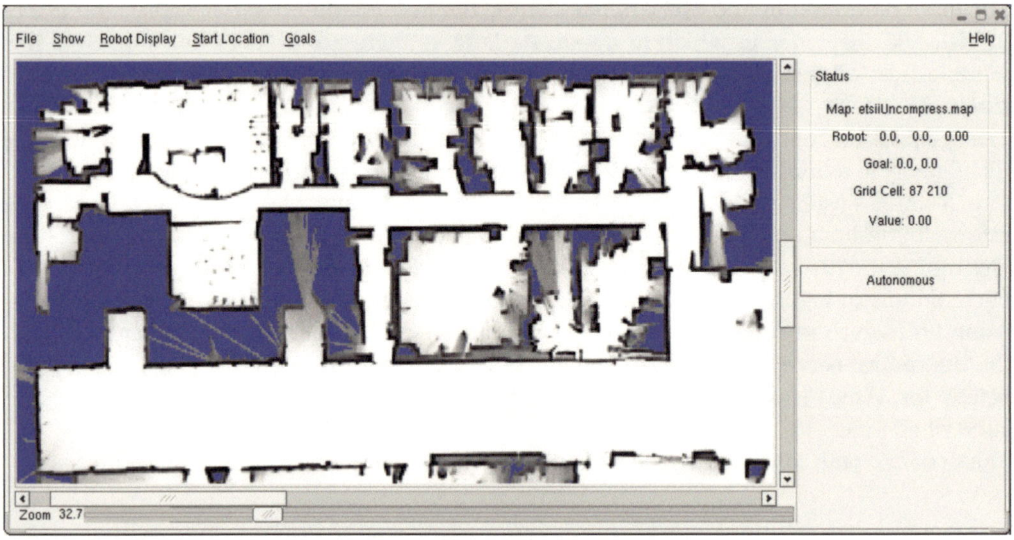

Figure 11. Environment of the experiment with the free space graph

Figure 11 shows the section of the environment (as perceived by the robot) which basically comprises offices, laboratories and corridors.

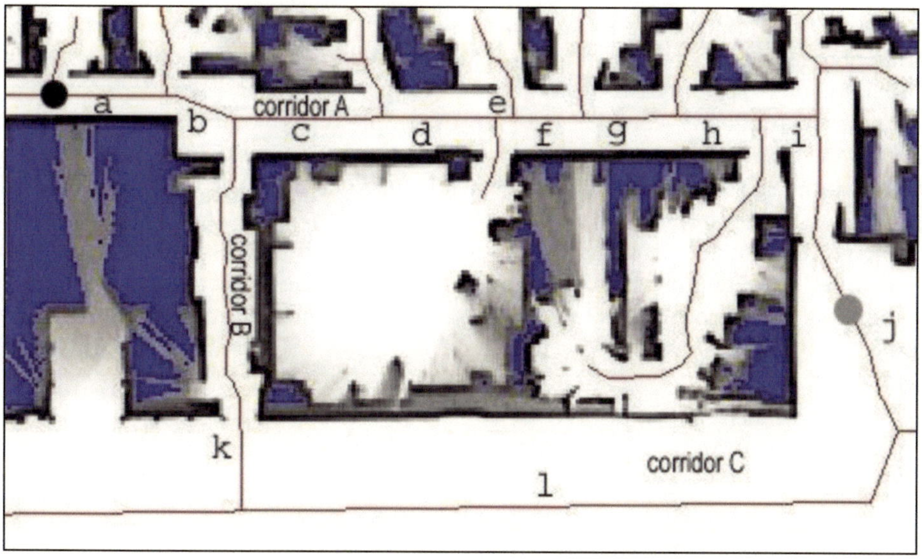

Figure 12. Branch labels, start point (black circle on the left) and destination (grey circle on the right)

Branches (route 1)	A priori cost	Real cost (measured)	New cost after fusion (fusion α = 0.8)
a	7.4	14.8	13.3
b	5.4	10.8	9.7
c	9.8	19.7	17.7
d	8.3	16.7	15.1
e	1.0	2.0	1.8
f	4.9	9.8	8.9
g	5.9	11.8	10.6
h	5.9	11.8	10.6
i	3.9	7.9	7.1
j (incomplete)	11.8	23.6	21.3
Total path cost	64.2	129.0	116.0

Table 5. Initial, measured and fused branch weights for route 1

The results under analysis come from different performances of the same global trajectory. The area of interest, along with the start and destination points (circles on the left and the right, respectively), are shown in Figure 12. To reach the goal from the initial point, two routes can be used: route 1, through corridor A, and route 2, through corridors B and C. The initial weights of the branches are computed from their lengths as the estimated time to travel them. These cost values are computed assuming a constant speed along the whole trajectory.

Thus, the initial time cost for route 1 is 64.2 seconds and 97.0 seconds for route 2. The experiments run without people in the environment led to 68.7 seconds to complete route 1 and 107.1 seconds to complete route 2.
Under normal conditions, corridor A is full of people (i. e. moving obstacles), causing a significant increase in the time it takes for the robot to traverse it. Thus, the average time needed to complete route 1 is 129.0 seconds, while for route 2 it takes 116.0 seconds, which is much closer to the a priori forecasted time.
The fusion of the previous and the new weights is shown in Tables 5 and 6. It can be seen that the final costs of the crowded branches increase. The consequence is that further planning tasks will choose route 2 over route 1 (116.3 seconds for route 1 vs. 112.6 seconds for route 2).

Branches (route 2)	a-priori cost (in secs.)	real cost (measured)	New cost after fusion (fusion $\alpha = 0.8$)
a	7.4	8.8	8.5
b	5.4	6.4	6.2
k	26.0	31.1	30.0
l	49.0	58.6	56.7
j (incomplete)	9.3	11.1	10.8
Total path cost	97.0	116.0	112.2

Table 6. Initial, measured and fused branch weights for route 2

In order to limit the influence of rare situations, a conservative fusion strategy is used in practice (see Figure 13). The progression of the fusion result for route 2 is shown in Table 7 using two different fusion ratios: conservative (20% - 80%) and sharp (75%-25%).

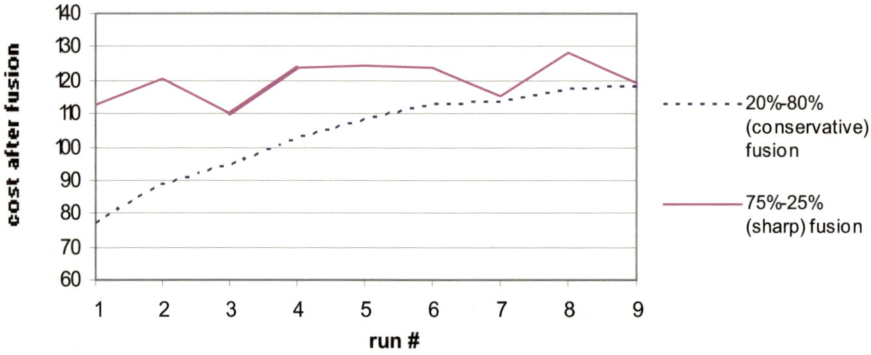

Figure 13. Graphical comparison between conservative and sharp cost fusion for route 2

Route execution	#1	#2	#3	#4	#5	#6	#7	#8	#9
Execution time (sec.)	129.0	135.2	117.4	133.9	132.1	128.7	116.1	133.2	119.5
Cost after fusion (α=20)	77.2	88.8	94.5	102.4	108.3	112.4	113.1	117.1	117.6
Cost after fusion (α=75)	112.8	120.7	110.2	124.0	124.7	123.6	115.2	128.2	118.9

Table 7. Numerical comparison between conservative and sharp cost fusion for route 2

7. Conclusions and future work

In this chapter we described a method to improve the best trajectory search for mobile robot path planning in structured environments with non-modelled obstacles. Our approach uses the experience of performed trajectories without a need to transpose information on unknown obstacles to the map for global path planning.

In the proposed algorithm, the weighting factors of the skeleton branches involved in the path are modified as a function of the traveling time. These weights are good estimators of the branch costs when the robot is moving in the presence of unknown obstacles.

Our method is especially attractive in terms of incorporating the effects of the presence of moving obstacles in the environment (for example, corridors crowded with people) in the planning process, since this obstacle information cannot be included in a map. Our approach does not eliminate the need for a local planner and its local map in our mobile robot architecture, but it reduces the number of times they will be used.

8. Acknowledgments

This work has been partially supported by the Spanish Ministry of Science and Technology, (Project DPI 2005-06210) and by the Galician Autonomous Government Department of Innovation and Industry (Program INCITE, Projects PGIDIT 07PXIB303199PR and PGIDIT06PXIC303194PN).

9. References

Diéguez, A.R. ; Raimúndez, C ; Sanz, R. ; Fernández, J.L. & Delgado, E. (1995). An intelligent supervisory model for path planning and guidance of mobile robots in non-structured environments. *Preprints of the 2nd IFAC Conference on Intelligent Autonomous Vehicles (IAV-95)*, pp. 81-86, Espoo-Helsinki (Finland), June 1995.

Diéguez, A.R. ; Sanz R. & Fernández, J.L. (2003). Deliverative On-Line Local Path Planning for Autonomous Mobile Robots. *Journal of Intelligent and Robotic Systems*, Vol. 37, 1, May 2003, pp. 1-19, ISSN: 0921-0296.

Diéguez, A.R. ; Sanz, R. & Fernández, J.L. (1998). Improving global motion planning for mobile robots by experimental measurement of traveling time. *Preprints of the 3rd IFAC Conference on Intelligent Autonomous Vehicles (IAV-98)*, 1998. pp. 403-408, Madrid (Spain), March 1998.

Diéguez, A.R. ; Sanz, R. & Fernández, J.L. (2007). A Global Motion Planner that Learns from Experience for Autonomous Mobile Robots. *Robotics & Computer Integrated Manufacturing*, Vol. 23-5, October 2007, pp. 544-552, ISSN: 0736-5845.

Fernández, J.L.; Sanz, R.; Paz, E. & Alonso, C. (2008). Using hierarchical binary Petri nets to build robust mobile robot applications: RoboGraph, Presented to *2008 IEEE International Conference on Robotics and Automation*, Pasadena, CA, May 2008.

Heero, K. (2006). *Path Planning and Learning Strategies for Mobile Robots in Dynamics Partially Known Environment*. Ph.D. Thesis, Tartu University Press, ISBN: 9949-11-308-3, Tartu, Estonia.

Hsu, D.; Kindel, R.; Latombe, J.C. & Rock. S. (2002). Randomized Kinodynamic Motion Planning with Moving Obstacles. *The International Journal of Robotics Research*, Vol. 21, No. 3, (March 2003), pp. 233-255, ISSN: 0278-3649.

Hwang, J.K. & Ahuja, N. (1992). Gross Motion Planning- A Survey. *ACM Computing Surveys*, Vol. 24, No. 3, pp. 219-291, ISSN 0360-0300.

Jaillet, J. & Simeon, T. (2004). A PRM-based motion planner for dynamically changing environments, in *Proceedings of the 2004 IEEE International Conference on Intelligent Robots and Systems* (IROS 2004), Vol. 2, pp. 1606-1611, ISBN: 0-7803-8463-6, 28 Sept.- 2 Oct. 2004.

Latombe, J. C. (1991). *Robot Motion Planning*, Kluwer Academic Pub., ISBN, Norwell, MA.

LaValle, S.M. (2006), *Planning Algorithms*, Cambridge University Press, ISBN-13: 9780521862059.

Philippsen, R.; Jensen, B.; Siegwart, R. (2007). Towards Real-Time Sensor-Based Path Planning in Highly Dynamic Environments, In: *Autonomous Navigation in Dynamic Environments, Springer Tracts in Advanced Robotics*, Vol. 35, Laugier, C.; Chatila, R., (Eds.), pp. 135-148, Springer, ISBN 978-3-540-73421-5, Berlin.

Russell, S. & Norvig, P. (1995). *Artificial Intelligence: A modern Approach*. Prentice Hall, ISBN-13: 9780131038059, Engelwood Cliffs, New Jersey.

Simmons, R. G. (1994). Structured control for autonomous robots, *IEEE Trans. on Robotics and Automation*, Vol. 10, No. 1, (February 1994), pp. 34-43, ISSN: 1042-296X.

Szczerba, R.J. & Chen, D.Z. (1995). *A Time-Sweeping Approach for Determining Shortest Paths in a Dynamic*, 2-D Environment with Multiple, Moving Goals, *Computer Science and Engineering Technical Report*, TR 1995-23., University of Notre Dame, Notre Dame, Indiana.

25

Building Internal Maps of a Mobile Robot

Branko Šter and Andrej Dobnikar
University of Ljubljana
Slovenia

1. Introduction

Classical approaches to environment modelling of mobile robots consist of geometrical space reconstruction using measurements from robot sensors. This approach could be error-prone due to noise and inaccuracy of the sensors. However, in order to perform path-planning, a mobile robot still requires some kind of representation or world-model, as pointed out in (Tani, 1996) and (Mataric, 1992).

The alternative approach consists of topological modelling of the environment. It leads to a simpler description of the environment, but still preserving essential information. Provided that a type of low-level behaviour is provided, a topological map of the space suffices to the robot to distinguish among qualitative distinct behaviours. A model of the world may be built therefore on top of the reactive behavior. This corresponds to topological navigation. A topological map of the space is usually in the form of a graph or a finite state machine (FSM). The states are typically distinctive places in the space, also called landmarks. At each node, there is a decision to be made as to where to proceed.

It was argued (Brooks, 1991) that for truly intelligent agents the type of representation must not be a designer's choice, because the abstraction is the key part of the intelligence. Human designers tend to decompose the problem to the blocks-world, i.e., they do all the abstraction and leave a supposedly intelligent agent merely to search in this simplified world. Behavior-based robotics (Brooks, 1991) consequently emerged as a paradigm, emphasizing direct embodiment of a robot in a surrounding environment, taking as the basic level a direct state-action mapping or reactive behavior that enables the robot to perform basic tasks in the real world. Further levels are to be built incrementally upon the basic level, but always having in mind strong coupling between the robot and the environment. In (Mahadevan & Connell, 1992), one of the first successful applications of reinforcement learning to a behavior-based robot is described.

Basic skills of a robot include obstacle avoidance and approaching a goal. While the obstacle avoidance skill is a reactive-type behaviour, i.e., it requires no memory, approaching a goal usually requires some knowledge about the environment. Unless the goal is observed at a given moment, the robot must have a model of the space.

The symbolic approach is simple, but also has a major drawback, namely, input symbols to the finite state machine may occasionally be overlooked or even illegal. Whenever an error occurs (due to sensor noise, for example), this symbolical representation can consequently break down.

(Tani, 1996) applied recurrent neural networks to model the environment. A robot with range sensors advances reactively according to the maximum of the smoothed range profile. When multiple maxima appear, it decides where to proceed (this is a graph node). This approach avoids a serious drawback of the methods that require localization and the position of the robot in global coordinates. The robot does not have to perform localization, since the position is implicitly contained within the current state information in the recurrent neural network (RNN). (Tani, 1996) showed that the robot, whenever it is lost, gets "situated" again, i.e., it eventually restores the correct state information (context). On the other hand, a RNN tolerates a certain amount of errors, thus overcoming the FSM approach.

The approach to handle independently low-level or reactive behavior and higher-level or planning behavior has become common in mobile robotics. The key difference between the two is the usage of memory and different granularity. Reactive behavior (Mahadevan & Connell, 1992, Krose & van Dam, 1992a, Krose & van Dam, 1992b) is responsible for reacting to current sensory information with appropriate action or motor-control. It requires no memory, therefore a purely feed-forward scheme can be applied. The first concern is always to avoid collisions with obstacles and thereby to prevent the damage on the robot. Other goals typically include approaching objects, maximizing the length of the path, etc.

The higher-level or planning behavior (Tani & Nolfi, 1999) may be therefore manipulated either by a FSM or using a model such as RNN, capable of modeling FSMs. The environment as experienced by a moving robot is treated as a dynamical system. Simple types of reactive behavior are supplemented with eventual decisions to switch among them. Switching criteria in fact define states of the FSM. Since it is embedded in the environment and dependent on the sensory flow of the robot, the notion of Embedded flow state machine (EFSM) has been introduced (Ster & Dobnikar, 2006). It is a FSM-like model, embedded in the environment and symbolically representing the sensory flow experienced by the robot when acting according to a certain type of reactive behavior and making decisions in those situations that satisfy certain pre-specified conditions or switching criteria.

An EFSM can be implemented with a RNN which is trained on a sequence of sensory contents and actions and subsequently used for planning. The EFSM is applicable to multi-step prediction of sensory information and the travelled distances between decision points, given a sequence of decisions at decision points. One of the main virtues of this approach is that no explicit localization is required, since the recurrent neural network holds the state implicitly. An important issue regards the ability of this approach to reliably predict the sensors and the traversed distance enough steps ahead.

This chapter is basically divided into two parts. The first considers learning of appropriate reactive or low-level behavior. This does not mean that a low-level behavior cannot be deterministically programmed, however, but learning is potentially more flexible, due to possible changes in the environment. The second part considers the planning behavior by building an implicit model of the environment using RNNs.

2. Learning Reactive Behaviour of a Mobile Robot

We describe an application of the reinforcement learning paradigm to learning reactive behaviour of a simulated mobile robot, equipped with proximity sensors and video color information. The value function is approximated using radial-basis functions, which are adaptively allocated in the input space. The presented approach combines multiple goals in reinforcement learning using modularity.

2.1 Q-learning in continuous space

Q-learning (Watkins & Dayan, 1992) is a variant of reinforcement learning which is appropriate for systems with unknown dynamics. A model of the system is built implicitly during learning. Q-learning is proved to converge only for discrete systems. For continuous systems it usually works by choosing an appropriate approximation architecture, which may be task-dependent. The update equation is

$$\Delta Q(s_t, a_t) = \eta_t \left[g(s_t, a_t, s_{t+1}) + \gamma \min_{a_{t+1}} (s_{t+1}, a_{t+1}) - Q(s_t, a_t) \right] \tag{1}$$

where s is state, a is action, η is learning step, and γ is discount factor. In our case the state space is continuous, while the action space is discrete and consists of three actions: turn left, forward and turn right. It must be emphasized, that our system is actually not a Markov decision process, since the sensors do not provide the complete state information.

As a function approximator, the radial-basis function (RBF) neural network is used. As a local-basis function, the Gaussian is taken. Its activation on input $x_t \in R^n$ is

$$\Phi_i(x_t) = e^{-\frac{1}{2}\|M_i(x_t - c_i)\|^2} \tag{2}$$

where n is the joint dimension of the sensory space, $c_i \in R^n$ is the center and M_i is the scaling matrix of the Gaussian. The basis functions are normalized as

$$b_i(x_t) = \frac{\Phi_i}{\sum_j \Phi_j} \tag{3}$$

which causes the approximator to be a kind of look-up table with soft transitions. The Q-values are represented as

$$Q_j(x_t) = \sum_i w_i b_i(x_t), \quad j = 1,2,3. \tag{4}$$

(Samejima & Omori, 1999) proposed adaptive state space construction method, arguing that methods using look-up tables suffer from the curse of dimensionality, while when using global-basis functions the convergence can suffer. Their adaptive basis division algorithm reportedly solved the collision avoidance problem using a smaller number of basis functions.

A similar method called Adaptive basis addition with fixed size basis (ABA-F) (Samejima & Omori, 1999, Anderson, 1993) is applied. It incrementally allocates basis functions in the observation space. A new basis function is allocated when the TD error exceeds a threshold value, $\varepsilon_{TD} > \theta_e$, and there is no basis function near the present location in the observation space, $\phi_i(x_t) < \theta_a$, $i = 1, \ldots, n$. Otherwise, the output weights are updated:

$$\Delta w_{ai} = \eta_t \left[g(x_{t+1}) + \gamma \min_{a'} Q(x_{t+1}, a) - Q(x_t, a) \right] b_i(x_t), \quad i = 1, \ldots, n, \tag{5}$$

where a is the selected action. During learning actions are selected randomly (exploration phase), while later (exploitation phase) the action with the minimal Q value is chosen at any given moment.

2.2 Reactive behavior with modular reinforcement learning

Various modular approaches have been proposed (Karlsson, 1997, Uchibe et al., 1996, Kalmar et al., 1998). In our work, two reactive or low-level modules were defined: the collision-avoidance module (named module C) and the object-approach module (named module A). For each module the activation is defined as a measure of how much a current sensory information has to do with the module. In our case the activation of the module C is 1 everywhere, where proximity sensors (*ps*) are activated, i.e., where obstacles are near the robot, and 0.01 otherwise (in order to prevent the activations of both modules to be zero).

The module A is disactivated everywhere, where the red color (from "color sensors" *cs* – this is not a video camera) is not sufficiently observed, $\sum_{i=1}^{NC} cs_i < 0.1$. Otherwise, act_A is proportional to the above sum. To bring the robot nearer to biological organisms, a function called weariness (*wr*) was defined.

We know that living beings are naturally interested in exploring the surrounding world. They are especially drawn to "interesting" objects or phenomena, i.e., those with higher levels of sensorial activation. The (colored) object approach module reflects this natural curiosity. The reward the robot gets in the vicinity of a colored object, corresponds to the pleasure of a person satisfying its curiosity. The weariness function models eventual diminishing of curiosity, due to the so-called *stimulus satiation*. Ster (2004) describes these phenomena using various theories of motivation from psychology.

The robot, therefore, while wandering around, approaches colored objects, but leaves them after a certain period, since it gets weary. When its weariness reaches a certain plateau, its activation (or motivation) decreases quite rapidly. Consequently, the collision avoidance module overcomes and takes the robot away from the object. Of course, the weariness effect requires memory, i.e., module A cannot remain purely reactive. The weariness is modeled as

$$wr \leftarrow (1-\varepsilon)wr + \varepsilon \sum_{i=1}^{NC} cs_i \tag{6}$$

and the activation is a descending function of *wr*, e.g.

$$act_a = \begin{cases} 1, & wr \le P_1 \\ \dfrac{P_2 - wr}{P_2 - P_1}, & wr > P_1 \end{cases} \tag{7}$$

We chose $\varepsilon = 0.1$, $P_1 = 2$, and $P_2 = 3$. These values depend on particular activations. How to define them, is probably not an easy issue.

The two modules are combined in the following manner

$$Q(s,a) = act_C Q_c(s,a) + act_A Q_a(s,a) \tag{8}$$

During exploration each module learns independently its action-value function Q. The activations are not used at this stage. In the exploitation phase greedy actions are chosen w.r.t. Q, the joint Q-values. However, a modular architecture is not meant to be capable in principle of achieving better solutions than a monolithic one. The actual reason for modularity is merely to facilitate learning. Simpler networks escape local minima easier than larger monolithic networks. The drawback is that a modular architecture may yield sub-optimal solutions, when modules are not chosen appropriately.

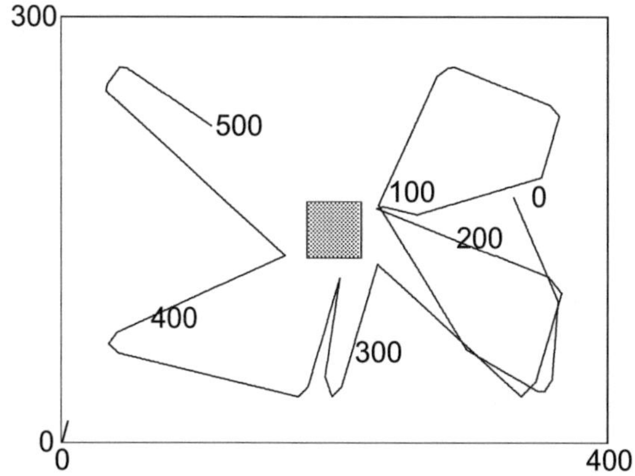

Figure 1. A test path in the environment E1, 500 steps long. The numbers indicate time

2.3 Experiments

The method was tested in two environments, E1 (Fig. 1) and E2 (Fig. 2). The learning (exploration) period lasted 5500 time steps. There was no attempt to minimize this number. In the learning period the robot selected random actions between three available actions: turn left by 0.5 rad (≈ 28.6°), forward by five units, and turn left by 0.5 rad. The actions correspond to the motor commands for the left and the right wheel: left (dl = -5, dr = 5), forward (dl = 5, dr = 5), and right (dl = 5, dr = -5). Since the radius of the robot was 10 units, angle=arc/radius=5/10 = 0.5 rad. The testing period was 500 steps long.

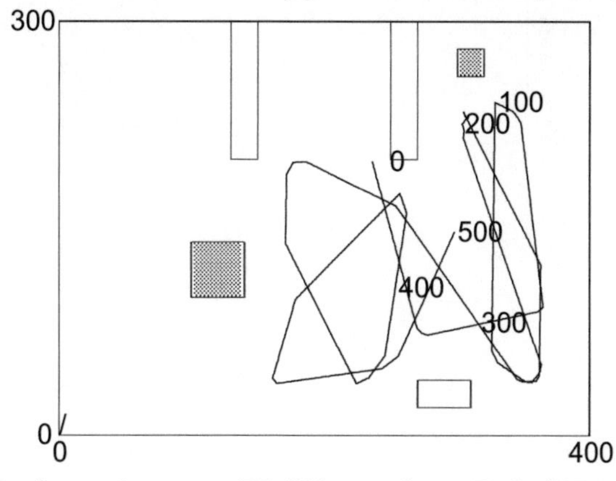

Figure 2. A test path in the environment E2, 500 steps long. Red objects are shaded, others ("pure" obstacles) are empty

The robot is assumed to have eight proximity sensors at angles 90, 45, 0, 0, -45, -90, -180, and -180, similar to the well-known miniature robot Khepera (Mondada et al., 1993). The nonlinear function of the sensors were approximated by a piecewise linear function

$$ps(d) = \begin{cases} 1, & d < 10 \\ \frac{50-d}{40}, & 10 \le d \le 50 \\ 0, & d > 50 \end{cases} \tag{9}$$

where ps is the proximity sensor activation and d is the distance.

The color information was extracted from 24 range sensors covering a 160° arc in front of the robot, yielding the "intersensorial" angle of about 7°. Each range sensor returns a color, specified in the environment file. In the presented environments a color was either default (grey) or red. This information was compressed into 8 values, each one covering three adjacent range sensors in the following manner: $cs = \frac{25}{3}\sum_{i=1}^{3}\frac{1}{d_i}$.

Parameters of the ABA-F method were: $\sigma = 0.5$, $\theta_e = 0.02$, $\theta_a = 0.1$. The discount factor γ was 0.85. Each collision was penalized by $g = 1$ and each turn action by $g = 0.1$. Seeing the red color was rewarded by a negative penalty $g = -\sum_{i=1}^{8} cs_i\left(1-\left|i-\langle i\rangle\right|/4\right)$, where <.> denotes the mean operator. The second term (in brackets) was added in order to reward the side color sensors less than the frontal color sensors.

We show here merely the results of the test run on E2 (for E1 see Ster (2003)), which show Q-values on 200 (out of 500) steps (Qc in Fig. 3, Qa in Fig. 4, and Q in Fig. 5). The actions are labeled as 0 for left, 1 for forward, and 2 for right. It can be seen that the Q-function of the two turn actions is at least 0.1 (mostly about 0.1), because the "turn penalty" was 0.1. The Q-values for forward action, Q(*s*,1), can be lower in an open space, but reach high values in the vicinity of obstacles (note that the colored object is also an obstacle besides being an object of interest), see steps around 90 and 200. At the same points a reward (g < 0) was delivered because of the colored object.

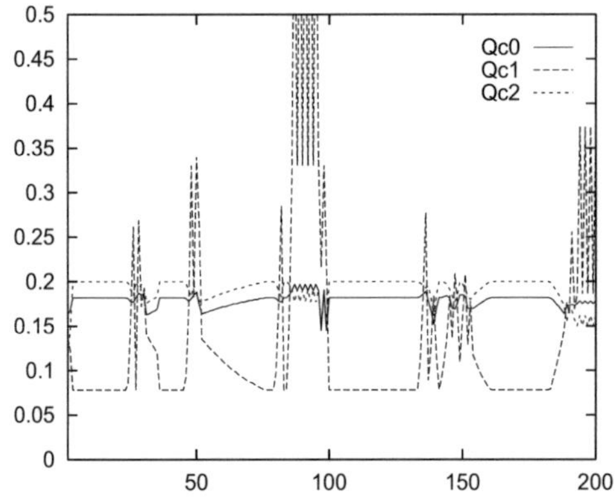

Figure 3. The Q-values of module C over 200 testing steps in E2

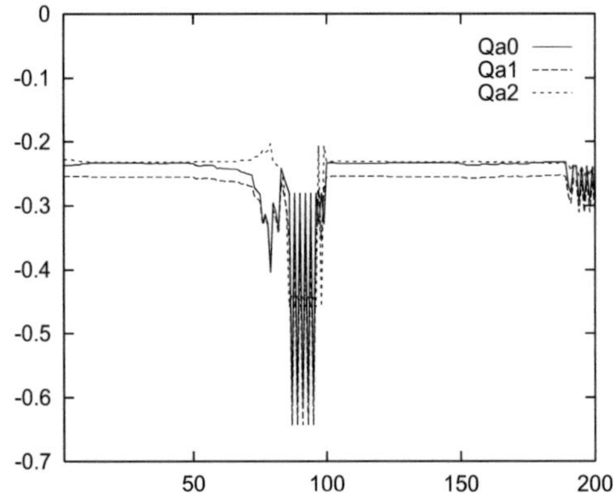

Figure 4. The Q-values of module A over 200 testing steps in E2

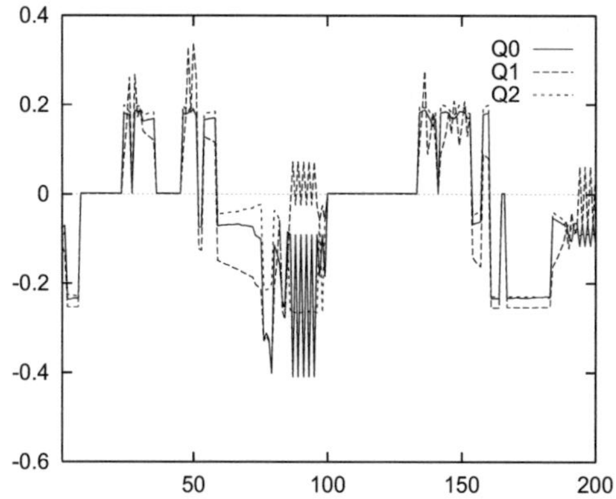

Figure 5. The joint Q-values over 200 testing steps in E2

3. Navigation of a Mobile Robot with Recurrent Neural Networks

When the robot moves according to a specified reactive behavior, a graph node may be implied according to fulfillment of a specified condition, based on sensory contents. This means that a topological graph may be defined implicitly by given switching criteria. In the exploratory or training phase, a random binary action is chosen at such branching point. In our case, one type of action will always be connected with the wall (action 0: follow the wall or return to the wall), and the other will be connected to colored objects in the environment (action 1: approach the object or look for another object). In case of different reactive behavior, of course, actions should be defined appropriately.

The Embedded flow state machine (EFSM) describes the dynamics of this process more accurately. The purpose of the EFSM is to represent a higher-level model of the sensory flow of a moving robot in an environment. The Embedded flow state machine may be formally defined as

$$EFSM = \{S, C, A, D, \delta, \lambda_1, \lambda_2\} \tag{10}$$

where S is a vector space $[0,1]^{N_s}$ of the preprocessed sensory information, C is the context vector space $[0,1]^{N_c}$, A is a finite non-empty set of possible decisions, and D is the unit interval [0,1], representing normalized distance travelled from the previous to the current decision point. N_s is dimensionality of the preprocessed sensory information and N_c is dimensionality of the context space (number of context units, as will be seen later). Functions δ, λ_1 and λ_2 provide the context, the sensors, and the travelled distance at the next decision point. From the context vector $\mathbf{c}_t \in C$, the sensory vector $\mathbf{s}_t \in S$, the distance $d_{t-1,t} \in D$, and the decision $a_t \in A$, all at time t, the next values are predicted:

$$\begin{aligned} \mathbf{c}_{t+1} &= \delta(\mathbf{c}_t, \mathbf{s}_t, d_{t-1,t}, a_t) \\ \mathbf{s}_{t+1} &= \lambda_1(\mathbf{c}_t, \mathbf{s}_t, d_{t-1,t}, a_t) \\ d_{t,t+1} &= \lambda_2(\mathbf{c}_t, \mathbf{s}_t, d_{t-1,t}, a_t) \end{aligned} \tag{11}$$

Of course, the equalities strictly hold only in the case of perfect prediction. The distance is metric information in this otherwise topological approach. It enables optimization of the predicted path in the planning phase. The preprocessed sensory information $\mathbf{s}_t$ stems, in our case, solely from a video camera.

3.1 Recurrent Neural Networks

It was shown that finite-state automata can be represented by recurrent neural networks. There has been a lot of effort to induce FSMs for regular languages with RNNs on the basis of shown examples (Cleeremans et al., 1989, Gabrijel & Dobnikar, 2003).
One of the classic algorithms for training recurrent neural networks (RNN) (Haykin, 1994) is described in (Williams & Zipser, 1989). The algorithm is gradient following and is applicable for fully connected RNNs, i.e., each unit (neuron) has a feedback connection to each unit in the network, see Fig. 6. The outputs of certain units represent outputs of the network, while the other units are called context-units, because they provide information relevant to sequence-processing problems.

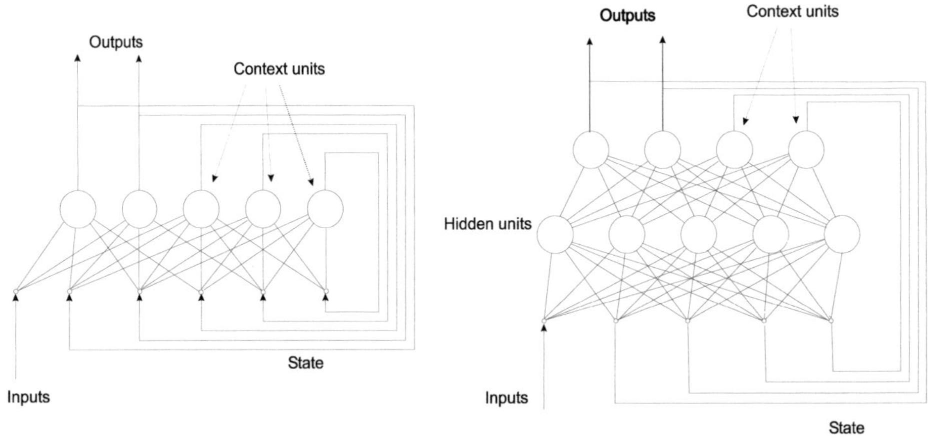

Figure 6. Fully connected recurrent neural network (left) and recurrent neural network with a hidden layer (right)

Inputs to the network are fed to each unit. Thus RNN can be trained to simulate any dynamical system, at least conceptually. The batch variant of the algorithm requires a sequence of input vectors $\mathbf{x}(t)$ and output vectors $\mathbf{y}(t)$ for $t = 1, ..., N$, where N is the length of the sequence. RNN is trained to produce the sequence of the desired outputs, if fed with the input sequence. A caution must be exercised while training the network; namely, when RNN is trained to produce desired outputs at each cycle, the functional ability of the network is restricted to a single perceptron layer, that is a linear transformation plus sigmoid activation function. To increase the functional ability of the network, inputs and outputs may change after two or more cycles of the network processing. Another possibility is, however, to apply a recurrent neural network with a hidden layer, which increases the functional ability of the network (Fig. 6 right). It is known that the 2-layer perceptron is a universal approximator (Hornik et al., 1989). The modified gradient-based algorithm is described in (Ster & Dobnikar, 2006).

3.2 Building an Internal Map of the Environment

The miniature robot Khepera (Mondada et al., 1993), see Fig. 7, with an additional video color camera was applied. Sensory information consists of infrared proximity sensors, wheel-encoders and processed image from a video camera. The task was considerably simplified using the camera solely for detection of colored objects (red, green and blue in this case). The robot is equipped with eight proximity sensors with a nonlinear activation function, which can be approximated by a piecewise-linear one, as shown in Fig. 8. The robot is able to perform three distinct actions: turn left, move forward, and turn right (by a small angle).

A video image was thresholded to extract the intensities of red, green and blue colour, which subsequently form the sensory vector. Two images at resolution 80x60, seen by the robot before the thresholding operation, are shown in Fig. 9. They correspond to branching points or EFSM states with the sensory outputs "Green reached" and "Red observed", respectively.

Figure 7. Mobile robot Khepera with an on-board video camera

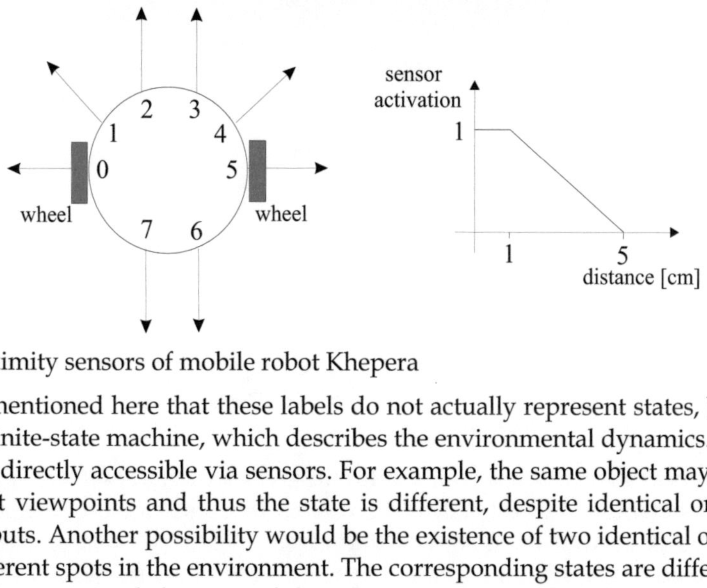

Figure 8. Proximity sensors of mobile robot Khepera

It should be mentioned here that these labels do not actually represent states, but rather the outputs of a finite-state machine, which describes the environmental dynamics. Actual states are really not directly accessible via sensors. For example, the same object may be perceived from different viewpoints and thus the state is different, despite identical or very similar sensorial outputs. Another possibility would be the existence of two identical or very similar objects at different spots in the environment. The corresponding states are different, too.

The low-level or reactive behaviour of the robot consists of following the right wall using information from proximity sensors. The low-level behaviour is deterministic and rather short to program manually, but a little longer to find experimentally. It turned out that the following portion of the low-level program (written in C language) leads to very efficient right-wall following:

Figure 9. Left: sensory output "Green reached" close to the green object, as seen by the robot. Right: sensory output "Red observed", where the red object is observed, as seen by the robot (resolution 80x60)

```
if ( proxSen[2]>100 || proxSen[3]>100 || proxSen[4]>100 )
   SetSpeed(-h/2,h/2);
else if ( proxSen[5] > 800)        // too close
   SetSpeed(-h,h);                 // left
else if ( proxSen[5] < 300)        // too far away
   SetSpeed(2*h,h);                // right and forward
else
   SetSpeed(2*h,2*h);              // forward
```

The *h* represents the speed and proxSen[0..7] is the vector of proximity sensors. Another possibility is to learn a reactive behavior by applying reinforcement learning, as described in the previous section.

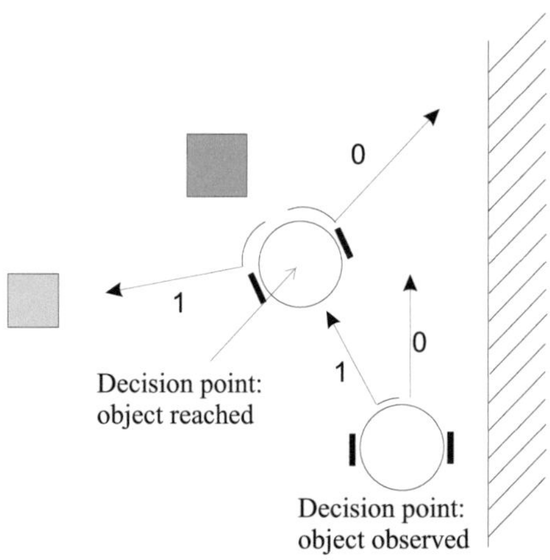

Figure 10. Decision or branching points.

The robot advances following the wall until it observes a colored object. There is a certain, experimentally determined threshold for each colour to signify the observation of an object (about 100 pixels), as well as vicinity of the object (about 600 pixels). When the robot observes a colored object, it has to decide whether to still follow the wall or to approach the object. This situation is called *branching point*. In the vicinity of a colored object, the decision is to approach another colored object or to find the wall (Fig. 10).

Looking for another object, the robot turns left until another object is observed with a specified threshold. It subsequently approaches the object. Looking for the wall, the robot turns right for approximately 90 degrees, and from then on follows Breitenberg's algorithm for low-level

behaviour, which is basically forward-motion, avoiding obstacles. The wheels' velocities are linear functions of the proximity sensors plus an offset term (Breitenberg, 1984).
In this way the whole environment can be represented as an EFSM. EFSMs corresponding to typical indoor environments are usually relatively simple (not many branching points). Therefore, in practical cases RNNs should not have problems learning the corresponding EFSMs. Of course, a complex maze would also produce a complex EFSM.

3.3 Experiments

During training the robot traveled around using random decisions at branching points. The whole path included 101 branching points. An automatic preprocessing, which dealt with situations, where the robot observed the same object again after a short period of time, was done. Thus the path with 87 branching points remained. A small part of the path is shown in Fig. 11. At the first branching point from the start the robot observes object 1 (red). Therefore, the starting sensory output is "Red observed", while the state of the EFSM might be referred to as "1 observed from the west" or O1w (Fig. 12). The training trajectory can be represented as a motor-program describing the action sequence. The vectors, extracted from images, represent the outputs of the EFSM. The outputs do not exactly correspond to the states, so there is not a one-to-one mapping from the states to the outputs. Images at distinct places may look the same, so there is an obvious need to have internal states, i.e., memory.

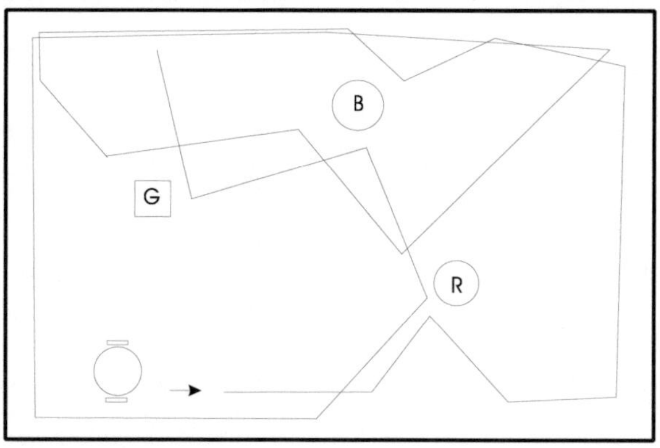

Figure 11. The path of the robot during training

RNN was trained off-line on the sensory-motor sequence for 5000 iterations (RNNs typically require large number of iterations). At each iteration the weights were updated following the gradient of the mean-squared error (MSE) on the whole training set ("batch" variant). The input vector consists of the current sensory vector (red, green and blue intensities), the distance from the previous to the current branching point, and the current decision (action). Given an input vector, a trained RNN predicts the next sensory vector and distance to the next branching point, which can be used in turn as the input at the next branching point, and so on. Given a motor sequence, RNN can thus predict the sensory flow of the EFSM.
We were interested how many steps ahead RNN was able to predict the sensory flow good enough in a closed-loop (multistep prediction). Prediction of the RNN was tested applying all possible motor programs five steps (branching points) ahead, i.e., $2^5 = 32$ programs: 00000, 00001, 00010, ..., 11111. Prediction error on the testing set showed that the first four predictions are very accurate, while the fifth is a little less reliable. MSE amounted on

average from 0.01 to 0.02 on the training set and from 0.01 to 0.25 (RMSE from 0.1 to 0.5) on the testing set. However, a few steps correspond to the whole environment, and besides, there was no effort to optimize training of the RNN.
The sense of forward-modeling of the environment is applying it in "mental" planning. The term "navigation" stands for finding the (shortest) path to a specified point. For example, we would like to find the shortest path from beginning to the green object. It is obvious from Fig. 13, that using the program 1-1-1 is the best way to do it. This corresponds to decisions: 'Approach red', 'Find another' (blue), 'Find another' (green). We need to perform a closed-loop simulation of all the motor programs, searching for the matching object, specified as the goal, and finding the least costly path to it. One possibility is proposed also in Tani (1996).

4. Conclusion

The first part of this chapter shows that using reinforcement learning with a modular architecture a mobile robot can learn to wander "sensibly", merely by applying known psychological concepts as sources of reinforcement. It is motivated by distinguishing objects in the environment, unmotivated by weariness, while at the same time avoiding collisions with obstacles and trying to maintain its course. Another possibility is to design reinforcement signals according to specific requirements.

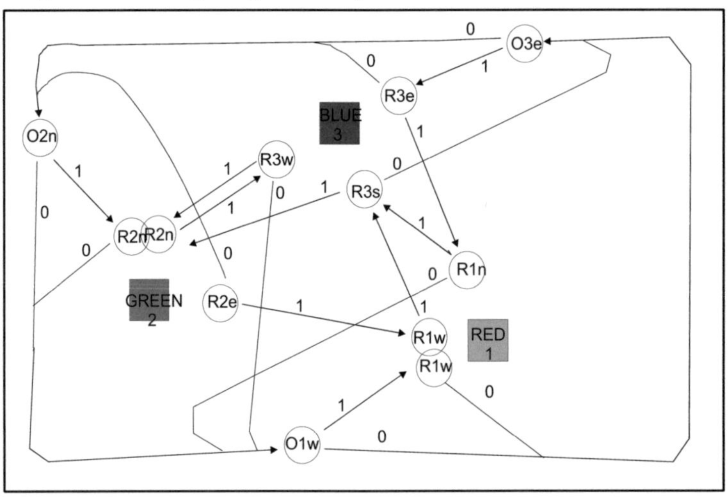

Figure 12. The EFSM, embedded in its environment. There are three landmark objects: 1 (red), 2 (green) and 3 (blue). EFSM states are denoted by circles and labelled as Oi (for 'object i observed') and Ri (for 'object i reached') with additional subscript letters signifying directions, e.g. state R2e signifies 'object 2 reached from the East'

The second part of the chapter shows how recurrent neural networks can be applied to build the internal map of the environment in a simple robot navigation task. The robot is able to induce the EFSM of a simple environment in the form of RNN and to use it for further planning. Because of the simplicity of the environment in the given task, the robot has to predict sensory flow for very short motor-programs, and it does it successfully. Further work should reveal how this behaviour scales up to larger problems and more realistic sensors, i.e., where a lower level of abstraction of sensory information is provided. Also

more reliable and possibly on-line training is required. The ultimate goal would be no human-designed representation, which seems to be still a very complex task.
There is a question whether a RNN is able to induce a finite-state automaton only on the basis of a limited number of training examples. It is obvious that the network can be trained (fitted) to a particular sequence, but it is not clear whether the complete structure of an automaton can be induced. This is not a question of functional ability, since a RNN can simulate any FSM, in principle.
There are two possible types of overtraining or overfitting here. Firstly, due to too many hidden neurons (standard overfitting in static neural networks), and secondly, due to too many context neurons. The latter might be called "dynamic overfitting", namely, the RNN uses its context units to "invent" a kind of context to lower the prediction error on the particular training sequence, especially when the latter is short. This problem would possibly be avoided with an on-line training, long enough to resolve uncertainties. For example, when turning to the left near the red object and looking for another object, the robot may overlook the blue object. The cause would be probably the video camera. In the training sequence there were four occurences of the red object followed by the action "1", i.e., looking for another object (after turning to the left.) Three times the robot observed blue and once green. In the latter case it overlooked blue, unfortunately. In case of larger sensor errors a training sequence should be much longer in order to reliably "collect statistics".

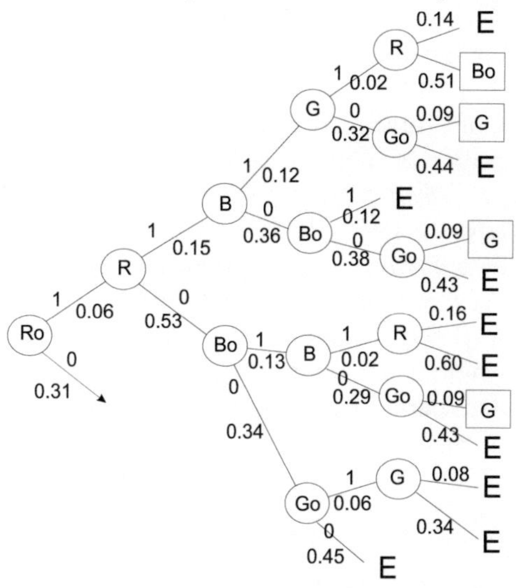

Figure 13. A half of the tree, representing motor programs of the length five. Only the programs, beginning with 1, are shown. At each arc the motor command (0 or 1) and the distance are written. Unclear predictions are labeled with E. Sensory outputs are Ro: red observed, R: red reached, Go: green observed, G: green reached, Bo: blue observed, B: blue reached

5. References

Anderson, C.W. (1993). Q-learning with hidden-unit restarting. Advances in Neural Information Processing Systems 5, San Mateo, CA: Morgan Kaufmann, 81-88.

Arleo, A.; del R. Millan, J. & Floreano, D. (1988). Learning to predict by the methods of temporal differences. *Machine Learning*, Vol. 3, No. 1, 9-44.

Breitenberg, V. (1984). *Vehicles: Experiments in Synthetic Psychology*, MIT Press, Cambridge, MA.

Brooks, R. A. (1991). Intelligence without representation. *Artificial Intelligence*, Vol. 47, 139-159.

Cleeremans, A.; Servan-Schreiber, D. & McClelland, J. (1989). Finite State Automata and Simple Recurrent Networks, *Neural Computation*, Vol. 1, No. 3, 372-381.

Gabrijel, I. & Dobnikar, A. (2003). On-line Identification and Reconstruction of Finite Automata with Generalized Recurrent Neural Networks, *Neural Networks*, Vol. 16, No. 1, 101-120.

Haykin, S. (1994). *Neural Networks*, Macmillan College Publishing Company.

Hornik, K.; Stinchcombe, M. & White, H. (1989). Multilayer feedforward networks are universal approximators, *Neural Networks*, Vol. 2, 359-366.

Kalmar, Z.; Szepesvari, C. & Lorincz, A. (1998). Module Based Reinforcement Learning for a Real Robot, In: *Proceedings of the 6th European Workshop on Learning Robots*, Lecture Notes in AI.

Karlsson, J. (1997). *Learning to Solve Multiple Goals*. PhD Thesis. University of Rochester. Rochester, New York, USA. TR 646.

Krose, B.J.A. & van Dam, J.W.M. (1992)a. Learning to avoid collisions: a reinforcement learning paradigm for mobile robot navigation, *Proceedings of the 1992 IFAC/IFIP/IMACS Symposium on Artificial Intelligence in Real-Time control, IFAC*, pp. 295-301, Delft.

Krose, B.J.A. & van Dam, J.W.M. (1992)b. Adaptive state space quantisation for reinforcement learning of collision-free navigation, *Proceedings of the 1992 IEEE/RSJ International Conference on Intelligent Robots and Systems*, pp. 1327-1332. IEEE, Piscataway, NJ, Raleigh, NC.

Mahadevan, S. & Connell, J. (1992). Automatic Programming of Behavior-based Robots using Reinforcement Learning. *Artificial Intelligence*, Vol. 55, 311-365.

Mataric, M. J. (1992). Integration of representation Into Goal-Driven Behavior-Based Robots. *IEEE Transactions on Robotics and Automation*,Vol. 8, No. 3, 304-312.

Mondada, F.; Franzi, E. & Ienne, P. (1993). Mobile robot miniaturisation: a tool for investigation in control algorithms, *Proceedings of the Third International Symposium on Experimental Robotics*, Yoshikawa, T. and Miyazaki, F. (eds.) Kyoto, Japan, October 1993.

Samejima, K. & Omori, T. (1999). Adaptive internal state space construction method for reinforcement learning of a real-world agent. *Neural Networks*, Vol.12.

Ster, B. (2004). An integrated learning approach to environment modelling in mobile robot navigation, *Neurocomputing*, No. 57, 215-238.

Ster, B. & Dobnikar, A. (2006). Modelling the Environment of a Mobile Robot with the Embedded Flow State Machine. *Journal of Intelligent and Robotic Systems*, Vol. 46, No. 2, 181-199.

Tani, J. (1996). Model-Based Learning for Mobile Robot Navigation from the Dynamical Systems Perspective. *IEEE Transactions on Systems, Man, and Cybernetics - Part B:Cybernetics*, Vol. 26, No. 3, 421-436.

Tani, J. & Nolfi, S. (1999). Learning to perceive world as articulated: an approach for hierarchical learning in sensory-motor systems. *Neural Networks*, Vol. 12, 1131-1141.

Uchibe, E.; Asada, M. & Hosoda, K. (1996). Behavior Coordination for a Mobile Robot Using Modular Reinforcement Learning, *Proceedings of the 1996 International Conference on Intelligent Robots and Systems (IROS)*, IEEE Press.

Watkins, J.C.H. & Dayan, P. (1992) Technical Note: Q-learning. *Machine Learning*, Vol. 8, 55-68.

Williams, R. J. & Zipser, D. (1989). A Learning Algorithm for Continually Running Fully Recurrent Neural Networks. *Neural Computation*, Vol. 1, 270-280.

26

Cooperative Indoor Navigation using Environment-Embedded Assistance Devices

Tsuyoshi Suzuki[1], Kuniaki Kawabata[2], Daisuke Kurabayashi[3], Igor E. Paromtchik[2] and Hajime Asama[4]
[1]*Tokyo Denki University*, [2]*RIKEN*
[3]*Tokyo Institute of Technology*, [4]*The University of Tokyo*
Japan

1. Introduction

The research in environmental robotics, ubiquitous robotics, and network robots aim to create intelligent environments for providing various services by gathering, managing, and supplying information via distributed communication, sensing, and actuation. Various applications of such robotic systems have been proposed and studied, e.g. life support (Sato *et al.*, 1996), environmental monitoring and information management (Parker *et al.*, 2003; Low, 2004; Tang, 2004), task assignment (Batalin & Sukhtame, 2003), and rescue operation (Kurabayashi *et al.*, 2001; Tadokoro *et al.*, 2003; Miyama *et al.*, 2003). These applications employ various navigation methods addressed in numerous publications (Borenstein *et al.*, 1996; Arai *et al.*, 1996b; Li *et al.*, 2003; Parker *et al.*, 2003; Nakamura *et al.*; 2003).

This chapter introduces a cooperative navigation method for multiple mobile robots operating in indoor environments, as an example of our research work in intelligent environmental robotic systems. The method relies on the information management about the environment, namely, static global information and local information. The former is represented by a topological map (Mataric, 1992) that displays the positional relation from any starting point to any goal point and is relevant for planning a route. The latter contains a map of the local environment and the traffic information for dynamic navigation.

The proposed navigation method makes use of an Information Assistant (IA) - a communication device embedded into the environment. The IA updates and manages information about the local environment and communicates with the robots. The navigation also relies on an Optical Pointer (OP) to guide robots at intersections by means of projecting a laser light onto the ground. The OP communicates with the robot via the IA, when indicating target positions to the robot. The mobile robot detects a laser beacon on the ground by means of image processing and moves towards the beacon. When the robot reaches the proximity of the beacon, the next sub-goal is indicated, and the laser beacons lead the robot along the route. The IA and OP devices assist the robot to navigate in the environment, which can be unknown to the robot.

In contrast with other navigation methods, where the robot attempts to process all the information available about the environment, e.g. simultaneous localization and mapping (Smith *et al.*, 1990; Choset & Nagatani, 2001) or an improved topological map (Tomono &

Yuta, 2001), the IA and OP devices allow us to share the environmental information among the robots and obtain more flexibility in navigation and information management.
Section 2 discusses the information management for navigation with the use of the environment-embedded IA and OP devices. The cooperative navigation is presented in section 3. Section 4 describes our experiments on robot navigation. The conclusions are given in section 5.

2. Information Management for Navigation

This section addresses the information management for robot navigation in a structured indoor environment, where passages and intersections are constrained by walls. The global environmental information is used for planning a coarse route for the robot. The local information at intersections serves for accurate navigation along the planned route.
The global information is represented by a topological map that contains the general relational data about the present location, goal, and intermediate locations. For example, the environment in Fig. 1 is represented by a topological map in Fig. 2 with the use of a graphic expression. In this map, the focus is on the topology of environment. The graphic search is used to plan a coarse route on the basis of the global map known in advance.

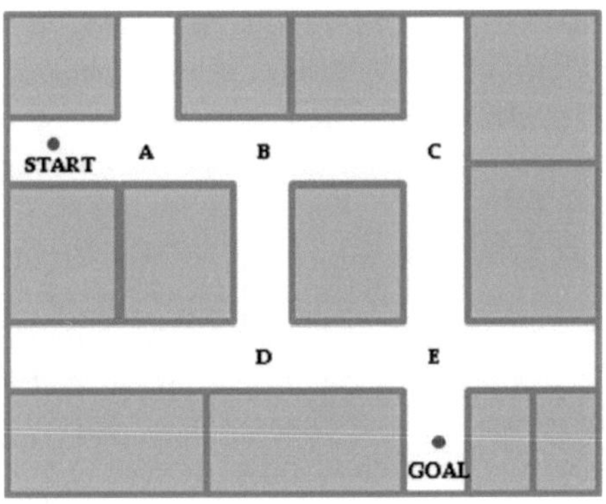

Figure 1. Example of an indoor environment: intersections are denoted by A, B, C, D, E

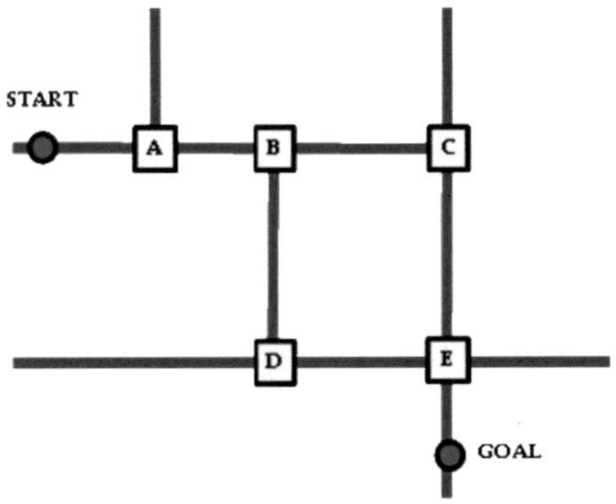

Figure 2. Topological map with passages and intersections

The processing of local information by the robot allows it to obtain a feasible path along the planned general route. While the robot can easily move in a passage along the wall, it cannot easily navigate at an intersection because the layout of any intersection is not uniform and the path cannot be supplied in advance. The traffic control is also needed at intersections to ensure collision-free motion. The information assistance at intersections facilitates the robot navigation. The local information is usually specific for each intersection and it depends on the actual state of the local environment. By means of embedding the assistance devices into the environment, the local information can be efficiently managed and made available to the robots for planning their feasible local paths.

2.1 Environment-Embedded Devices: Information Assistant and Optical Pointer

The Information Assistant (IA) is a radio device with the following functions: management of local information, robot communication, and control of interconnected devices such as sensors. The IAs are embedded into the environment at intersections and other areas to provide the relevant local information to mobile robots. The robot can retrieve or write information from/to an IA. This enables the information sharing among robots, e.g. the data written by a robot to the IA becomes available to other robots. The robot requests the local information from the nearest IA and, subsequently, plans its local path.

However, the local information is often insufficient for accurate navigation. Figure 3 shows an example of an erroneous arrival point after the robot passed the intersection because dead-reckoning was used. The following issues can also cause the navigation problems:

1. There are time variations in receiving the radio signals from the IA that results in an uncertain start position of the robot at intersections.
2. There are measurement errors of an angle between the robot's direction of motion and the wall during maneuvering.
3. The navigation errors can occur while the robot moves through intersections.
4. The presence of obstacles increases the risk of positional uncertainty, especially in the proximity of the start position at intersections.

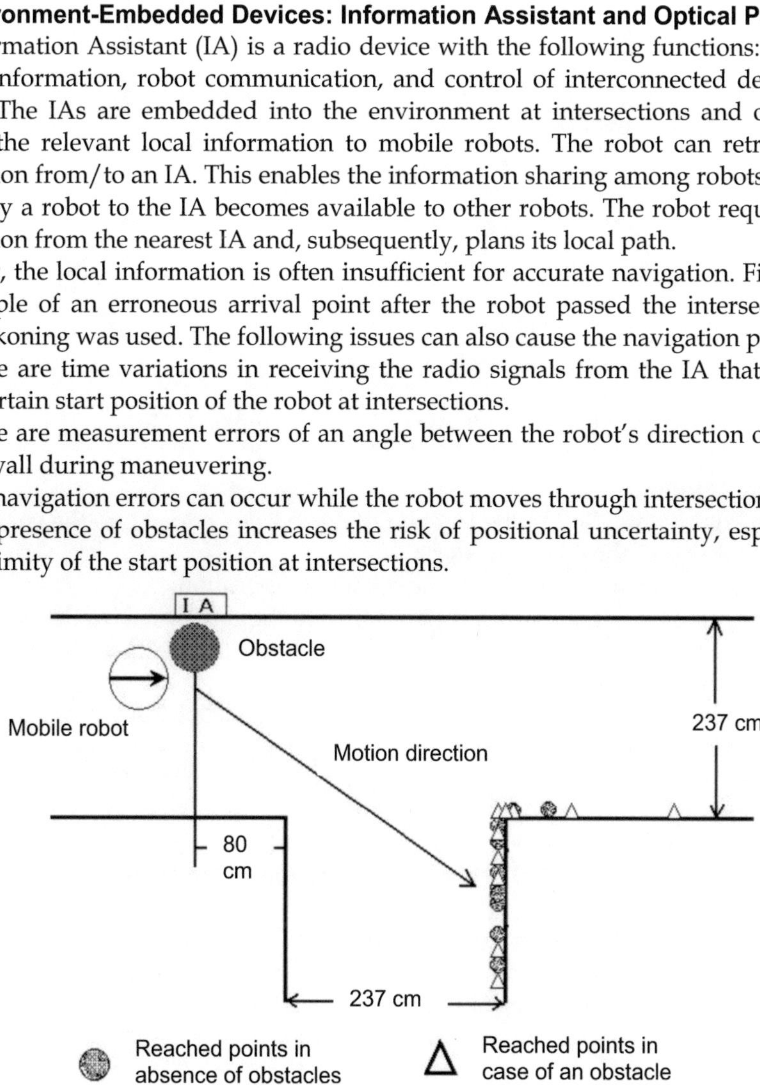

Figure 3. Example of robot navigation based solely on IA

The Optical Pointer shows a target position to the robot by means of projecting a laser light onto the ground (Paromtchik & Asama, 2001). The robot detects the laser beacon by means of image processing and moves toward it. When the robot reaches the proximity of the beacon, the OP indicates the next sub-goal along the route at intersections. Fig. 4 illustrates the concept of using IA and OP for robot navigation.

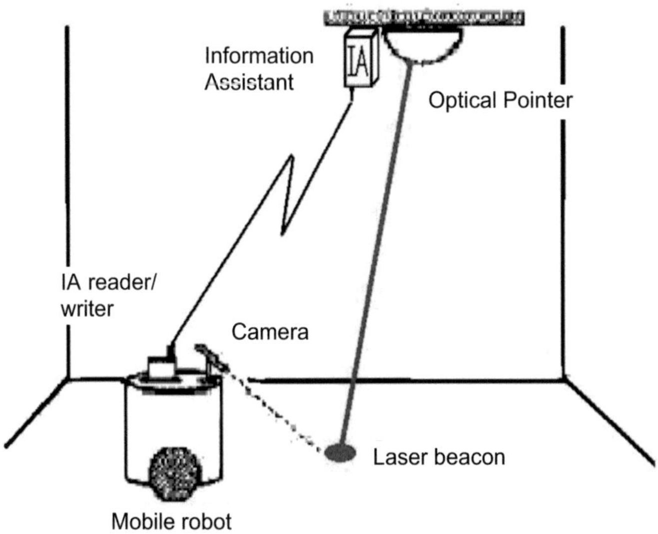

Figure 4. Concept of using the IA and OP devices for robot navigation

2.2 Navigation Algorithm

The navigation algorithm relies on the IA and OP devices embedded into the environment. The algorithm consists of the following steps:

(1) Provide a global topological map and the robot's start location.
(2) An operator specifies a goal, which is transmitted to the robot via a wireless LAN.
(3) Based on the start and goal locations, the robot plans a coarse route by means of the graphic search in the topological map.
(4) The robot moves along the planned route and monitors the local environment by means of infrared sensors to avoid collisions with obstacles.
(5) When the robot arrives at an intersection, it detects the IA's radio signal and communicates with the IA by requesting guidance at the intersection. Fig. 5 shows the communication process between the robot and the IA.
(6) The IA receives a start and goal positions from the robot, and transmits the information about the feasible local path.
(7) The OP shows a laser beacon as a sub-goal for the robot.
(8) The robot processes images to detect the laser beacon and moves toward it. The IA and OP devices guide the robot by means of the laser beacons.
(9) When the robot has entered the next passage, it sends a message to the IA to confirm its successful passing through the intersection.
(10) The robot continues its motion along a passage.
(11) When the robot reaches another intersection, the optical guidance is performed again.

Finally, when the robot receives the goal information from the IA, this signifies that the robot has completed the motion task successfully. If two or more robots arrive at an intersection, the priority is given to the robot that first requested guidance from the IA. Other robots receive a busy signal from the IA and wait in their current positions until the guidance service becomes available for them. Thus, the same navigation scheme can be used for multiple robots in the same local environment.

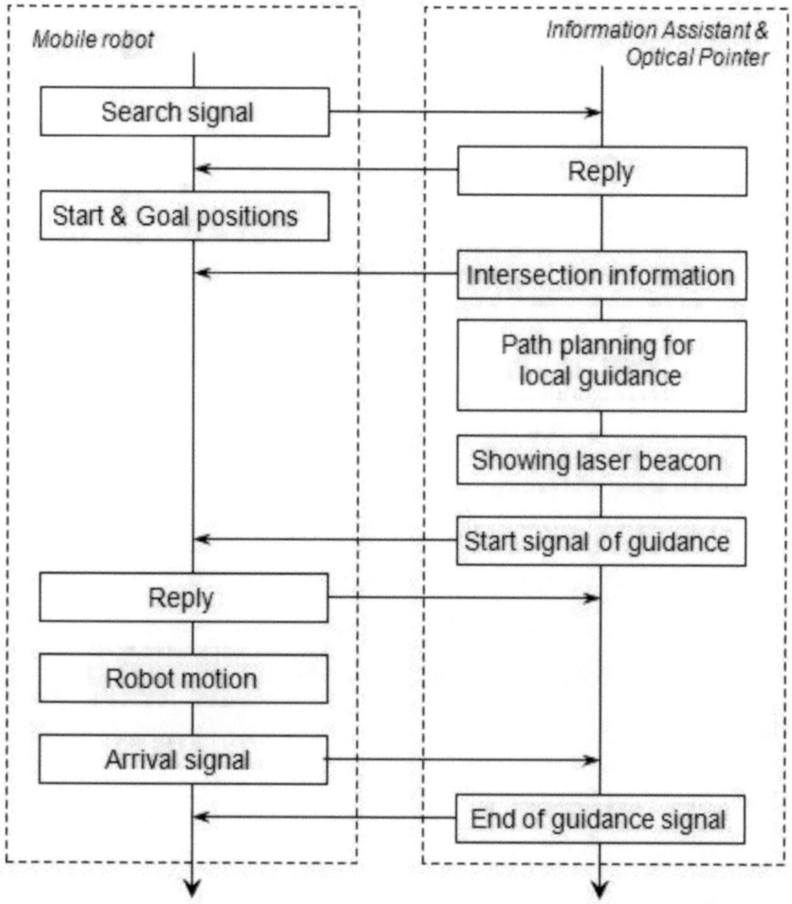

Figure 5. Communication between the mobile robot and the IA and OP devices

3. Navigation System

In order to realize the proposed navigation scheme, we have developed an experimental prototype. This section describes our experimental setup with the omni-directional mobile robot, information assistant and optical pointer.

3.1 Omni-Directional Mobile Robot

The robot is shown in Fig. 6. It is equipped with a holonomic omni-directional wheeled platform with its actuators, sensors, on-board control system and electric batteries (Asama *et al.*, 1995). Its infrared sensor system LOCISS (Locally Communicable Infrared Sensory System) is capable to distinguish a robot and an obstacle in an environment with multiple robots (Arai *et al.*, 1996a). LOCISS allows the robot to detect and avoid obstacles, follow a wall, and it also serves for local communication between robots.

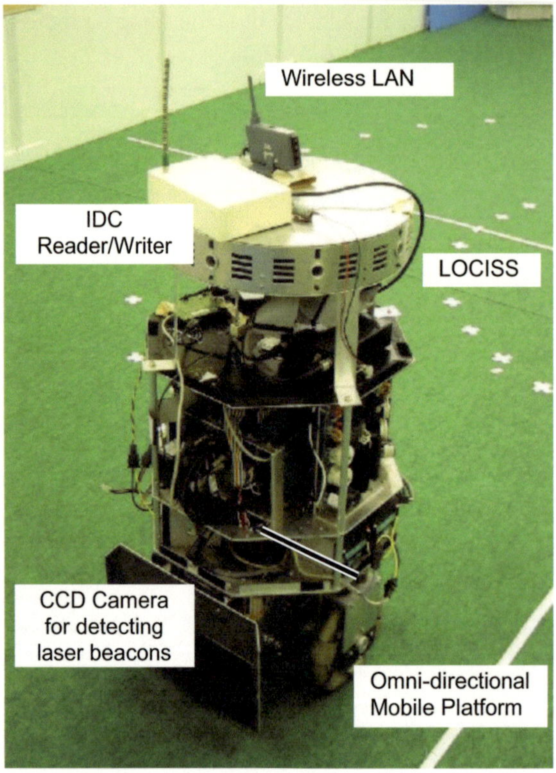

Figure 6. Mobile robot ZEN

3.2 Information Assistant and Optical Pointer

The IA device is based on the Intelligent Data Carrier (IDC) (Arai *et al.*, 1996b) and an IDC Reader/Writer. The 4th version of IDC is shown in Fig. 7. The IDC is a device with a radio communication unit, a CPU, a memory, and an electric battery. The robot communicates with the IDC through the IDC Reader/Writer.

The OP device consists of a laser pointer mounted onto a pan-tilt mechanism with two step motors. The pan-tilt mechanism directs the laser beam onto the desired positions on the ground. The Fig. 8 shows an OP which is installed at the ceiling.

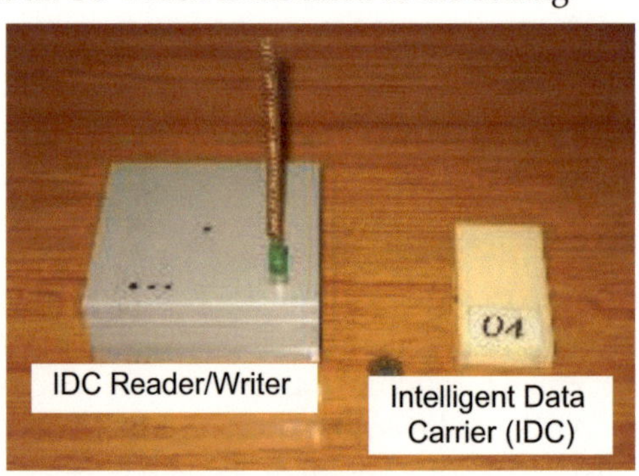

Figure 7. The IA device is based on IDC Reader/Writer and Intelligent Data Carrier

Figure 8. The OP device is mounted onto a pan-tilt mechanism at the ceiling

4. Experiments

The experiments are performed in an indoor environment of RIKEN. The floor map is shown in Fig. 9. The robot's task is to move from its start location to a goal which is set by a human operator. The IA and OP devices are placed at the intersections in the environment.

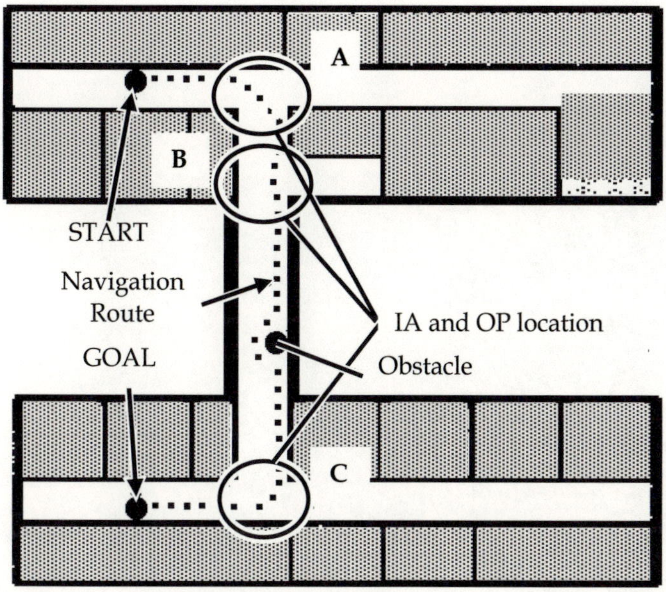

Figure 9. Floor map of our experimental environment

They contain information about their own positions and environmental conditions at the intersection. The circles and letters A, B, C in Fig. 9 denote intersections, where the IA and OP devices are placed, and the circles also indicate the communication range for local navigation. Given the start position and goal destination, the robot planned a coarse route as: START - A - B - C - GOAL by means of a graphic search on the topological map. The obstacle detection and avoidance was provided by LOCISS. After moving along a passage, the robot approached an intersection A, where it detected the IA and requested guidance through the intersection. The OP indicated the feasible path by means of laser beacons, as shown in Fig.10 and Fig. 11. The motion from START to GOAL is depicted by a dotted line in Fig. 9.

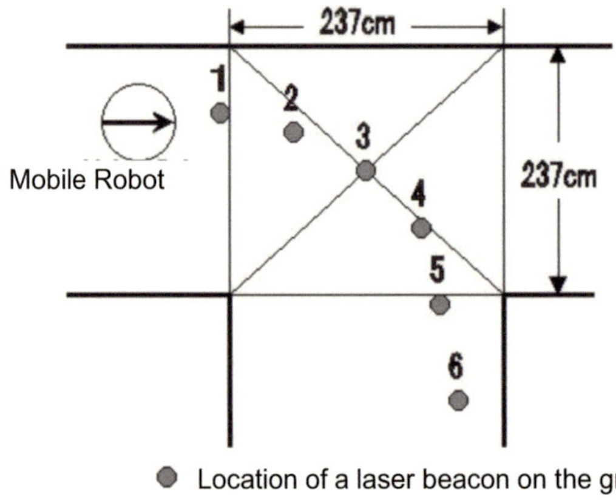

Figure 10. Location of laser beacons at intersection A, one laser beacon is shown at a time

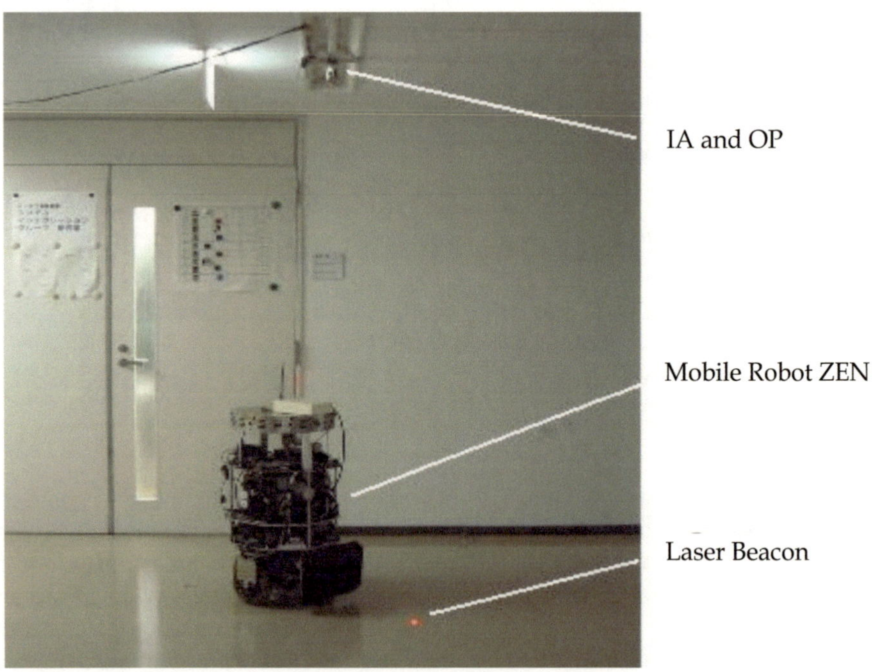

Figure 11. Navigation by means of IA and OP at an intersection

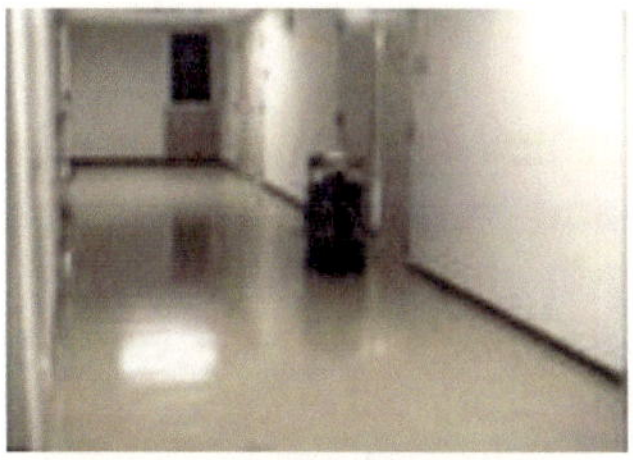

(a) The robot moves from its START location along the wall

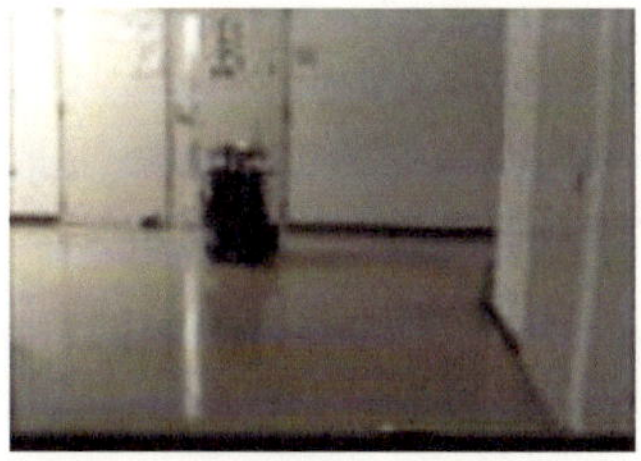

(b) The robot communicates with the IA and OP at intersection A

(c) The IA and OP guide the robot at intersection A

(d) The robot enters the passage and moves along the wall

(e) The robot communicates with the IA and OP at intersection B

(f) The robot is guided by the IA and OP at intersection B

(g) The robot moves along the wall after the guidance

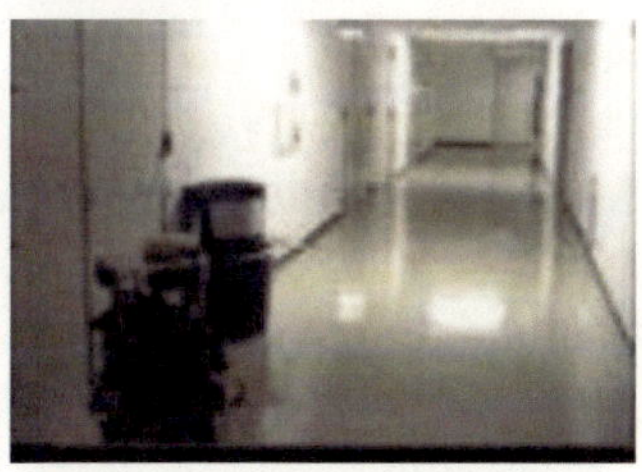

(h) The robot detects an obstacle on its route by processing range data

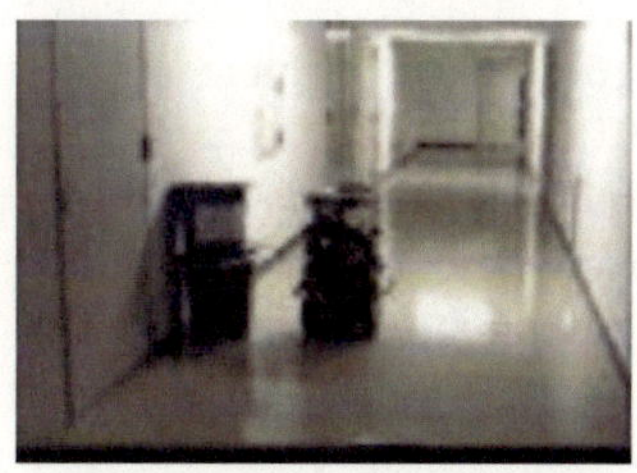

(i) The collision avoidance is performed with the use of LOCISS

(j) The robot communicates with the IA and OP at intersection C

(k) The robot is guided by the IA and OP at intersection C

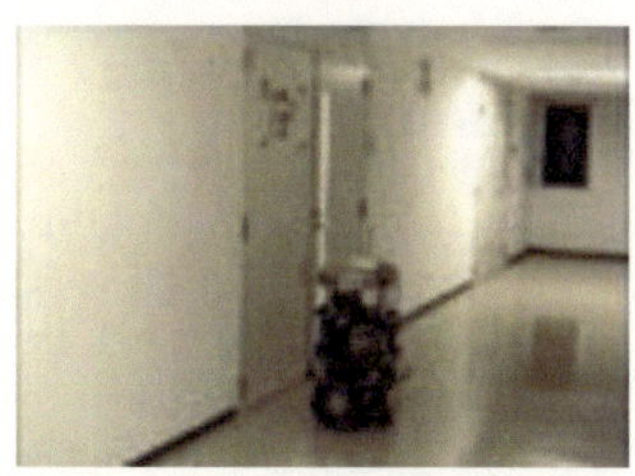

(l) The robot finally arrives at its GOAL destination

Figure 12. Robot navigation experiment with the IA and OP devices

5. Conclusion

This chapter introduced a cooperative navigation strategy for mobile robots operating in indoor environments with the embedded Information Assistant and Optical Pointer devices, as an application of an intelligent environmental robotic system. In order to provide a more flexible navigation, the management of environmental information was considered. The static global information supplies topological details such as the positional relation of any starting point to any goal point in order to create an approximate route. The dynamic local information includes a local map, obstacles and traffic information for accurate navigation. We proposed the information management and navigation algorithm based on the IA and OP devices embedded into the environment. The experimental example of navigation was described. The robot was initially provided with a coarse route to the goal, and the IA devices managed the environmental information in real-time locally. The OP device was used for guidance at intersections, and communication with mobile robots was performed through the IA device. The OP indicated target positions by means of a laser light projected from a laser pointer onto the ground. The mobile robot detected the laser beacons and followed them to reach its goal destination. The experiments have proved the feasibility of the proposed method.

6. Acknowledgment

This is a revised and extended version of our presentation at the Fourth International Conference on Field and Service Robotics (FSR'03): Suzuki, T.; Uehara, T.; Kawabata, K.; Kurabayashi, D.; Paromtchik, I. E. and Asama, H. Indoor Navigation for Mobile Robot by Using Environment-Embedded Local Information Management Device and Optical Pointer, Field and Service Robotics, STAR 24, S. Yuta, H. Asama, S. Thrun, E. Prassler and T. Tsubouchi (Eds.), pp.41-49, Springer-Verlag Berlin Heidelberg, ISBN 3-540-32801-7, Germany, 2006. The text and figures of sections 2 to 4, except Fig. 12 are reproduced with modifications from the above publication with the kind permission of Springer-Verlag Berlin Heidelberg and Springer Science+Business Media. Thanks are given to Taiki Uehara for his contribution to the development of the IA and OP devices and conducting experiments in RIKEN during his Master course study at the Toyo University.

7. References

Asama, H.; Sato, M.; Bogoni, L.; Kaetsu, H.; Matsumoto, A. & Endo, I. (1995). Development of an Omni-Directional Mobile Robot with 3 DOF Decoupling Drive Mechanism, *Proceedings of the 1995 IEEE International Conference on Robotics and Automation*, pp.1925-1930, ISBN 0-7803-1965-6, Nagoya, Aichi, Japan, May 22-27, 1995.

Arai, Y.; Suzuki, S.; Kotosaka, S.; Asama, H.; Kaetsu, H. & Endo, I. (1996a). Collision Avoidance among Multiple Autonomous Mobile Robots using LOCISS (Locally Communicable Infrared Sensory System), *Proceedings of the 1996 IEEE International Conference on Robotics and Automation*, pp.2091-2096, ISBN 0-7803-2988-4, Minneapolis, Minnesota, USA, April 22 - 28, 1996.

Arai, Y.; Fujii, T.; Asama, H.; Fujita, T.; Kaetsu, H.; Endo, I. (1996b). Self-Localization of Autonomous Mobile Robots using Intelligent Data Carriers, In: *Distributed Autonomous Robotic Systems 2*, Asama, H.; Fukuda, T.; Arai, T. & Endo, I. (Eds.), pp.401-410, Springer-Verlag, ISBN 4-431-70190-7, Tokyo, Japan.

Batalin, M. A. & Sukhtame, G. (2002). Sensor Coverage using Mobile Robots and Stationary Nodes, *Proceedings of the SPIE, Volume 4868 (SPIE'2002)* pp.269-276, Boston, MA, USA, August 2002.

Borenstein, J.; Everett, L. & Feng, L. (1996). *Navigating Mobile Robots – Systems and Techniques*, A K Peters, Wellesley, MA, USA, 1996.

Choset H. & Nagatani, K. (2001). Topological Simultaneous Localization and Mapping (SLAM): Toward Exact Localization without Explicit Localization, *IEEE Transactions on Robotics and Automation, Vol. 17, No. 2*, 2001, pp. 125-137.

Kurabayashi, D.; Asama, H.; Noda, K. & Endo, I. (2001). Information Assistance in Rescue using Intelligent Data Carriers, *Proceedings of the 2001 IEEE/RSJ International Conference on Intelligent Robots and Systems*, pp. 2294-2299, ISBN 0-7803-7398-7, Maui, Hawaii, USA, October 2001.

Li, Q.; Rosa, M. D. & Rus, D. (2003). Distributed Algorithms for Guiding Navigation across a Sensor Network, *Proceedings of the 2003 Annual International Conference on Mobile Computing and Networking (MobiCom'03)*, ISBN 1-58113-753-2, California, USA, September 2003.

Low, K. H. (2004). Reactive, Distributed Layered Architecture for Resource-Bounded Multi-Robot Cooperation: Application to Mobile Sensor Network Coverage, *Proceedings of the 2004 IEEE International Conference on Robotics and Automation, Vol.4*, pp.3747-3752, ISBN 0-7803-8232-3, New Orleans, LA, USA, April 2004.

Mataric, M. J. (1992). Integration of Representation into Goal-Driven Behavior-Based Robots, *IEEE Transactions on Robotics and Automation, Vol.8, No.3*, June 1992, pp. 304-312, ISSN 1042-296X.

Miyama, S.; Imai, M. & Anzai, Y. (2003). Rescue Robot under Disaster Situation: Position Acquisition with Omni-directional Sensor, *Proceedings of the 2003 IEEE/RSJ International Conference on Intelligent Robots and Systems*, pp.3132-3137, 0-7803-7860-1, Las Vegas, Nevada, October 2003.

Nakamura, T.; Oohara, M.; Ogasawara, T. & Ishiguro, H. (2003). Fast Self-Localization Method for Mobile Robots using Multiple Omnidirection, *Machine Vision and Applications Journal, Vol.14, No.2*, June 2003, pp.129–138, ISSN 0932-8092.

Parker, L. E.; Birch, B. & Reardon, C. (2003). Indoor Target Intercept Using an Acoustic Sensor Network and Dual Wavefront Path Planning, *Proceedings of the 2003 IEEE/RSJ International Conference on Intelligent Robots and Systems*, pp.278- 283, ISBN 0-7803-7860-1, Las Vegas, Nevada, October 2003.

Paromtchik, I. E. & Asama, H. (2001). Optical Guidance System for Multiple Mobile Robots, *Proceedings of the 2001 IEEE International Conference on Robotics and Automation*, pp.2935-2940, 2001, ISBN 0-7803-6475-9, Seoul, Korea, May 2001.

Sato, T.; Nishida, Y. & Mizoguchi H. (1996). Robotic Room: Symbiosis with Human through Behavior Media, *Robotics and Autonomous Systems 18, International Workshop on Biorobotics: Human-Robot Symbiosis*, pp.185-194, Tsukuba, Japan, May 1995, Elsevier New York.

Smith, R., Self, M. & Cheeseman, P. (1990). Estimating Uncertain Spatial Relationships in Robotics, In: *Autonomous Robot Vehicles*, Cox, I. & Wilfong, G. (Eds.), pp.167-193, Springer Verlag, 1990.

Tadokoro, S.; Matsuno, F.; Onosato, M. & Asama, H. (2003). Japan National Special Project for Earthquake Disaster Mitigation in Urban Areas, *First International Workshop on Synthetic Simulation and Robotics to Mitigate Earthquake Disaster*, pp.1-5, Padova, Italy, July 2003.

Tang, Y. (2004). Planning Mobile Sensor Net Deployment for Navigationally Challenged Sensor Nodes, *Proceedings of the 2004 IEEE International Conference on Robotics and Automation*, pp.172-179, ISBN 07803-8232-3, New Orleans, LA, USA, April 2004.

Tomono, M. & Yuta, S. (2001). Mobile Robot Localization based on an Inaccurate Map, *Proceedings of the IEEE/RSJ International Conference on Intelligent Robots and Systems (IROS'2001)*, pp.399-405, ISBN 0-7803-6612-3, Maui, Hawaii, USA, October 2001.

27

Nonlinear Motion Control of Mobile Robot Dynamic Model

Jasmin Velagic, Bakir Lacevic and Nedim Osmic
University of Sarajevo
Bosnia and Herzegovina

1. Introduction

The problem of motion planning and control of mobile robots has attracted the interest of researchers in view of its theoretical challenges because of their obvious relevance in applications. From a control viewpoint, the peculiar nature of nonholonomic kinematics and dynamic complexity of the mobile robot makes that feedback stabilization at a given posture cannot be achieved via smooth time-invariant control (Oriolo et al., 2002). This indicates that the problem is truly nonlinear; linear control is ineffective, and innovative design techniques are needed.

In recent years, a lot of interest has been devoted to the stabilization and tracking of mobile robots. In the field of mobile robotics, it is an accepted practice to work with dynamical models to obtain stable motion control laws for trajectory following or goal reaching (Fierro & Lewis, 1997). In the case of control of a dynamic model of mobile robots authors usually used linear and angular velocities of the robot (Fierro & Lewis, 1997; Fukao et al., 2000) or torques (Rajagopalan & Barakat , 1997; Topalov et al., 1998) as an input control vector. The central problem in this paper is reduction of control torques during the reference position tracking. In the case of dynamic mobile robot model, the position control law ought to be nonlinear in order to ensure the stability of the error that is its convergence to zero (Oriollo et al., 2002). The most authors solved the problem of mobile robot stability using nonlinear backstepping algorithm (Tanner & Kyriakopoulos, 2003) with constant parameters (Fierro & Lewis, 1997), or with the known functions (Oriollo et al., 2002). In (Tanner & Kyriakopoulos, 2003) a combined kinematic/torque controller law is developed using backstepping algorithm and stability is guaranteed by Lyapunov theory. In (Oriollo et al., 2002) method for solving trajectory tracking as well as posture stabilization problems, based on the unifying framework of dynamic feedback linearization was presented.

The objective of this chapter is to present advanced nonlinear control methods for solving trajectory tracking as well as convergence of stability conditions. For these purposes we developed a backstepping (Velagic et al., 2006) and fuzzy logic position controllers (Lacevic, et al., 2007). It is important to note that optimal parameters of both controllers are adjusted using genetic algorithms. The novelty of this evolutionary approach lies in automatic obtaining of suboptimal set of control parameters which differs from standard manual adjustment presented in (Hu & Yang, 2001; Oriolo et al., 2002). The considered motion control system of the mobile robot has two levels. The lower level subsystem deals with the control of linear and angular volocities using a multivariable PI controller described with a full matrix. This torque control ensures tracking servo inputs with zero steady state errors (Velagic et al., 2005). The position control of the mobile robot is a nonlinear and it is on the second level. We have developed a mobile robot position controller based on backstepping control algorithm with the extension to rapidly decrease the control torques needed to achieve the desired position and orientation of mobile robot (Lacevic & Velagic, 2005). This is important in the case if the initial position of reference robot does not belong to the straight line, determined with the robot and its initial orientation. Also, we have designed a fuzzy logic position controller whose membership functions are tuned by genetic algorithm (Lacevic, et al., 2007). The main goals are to ensure both successfully velocity and position trajectories tracking between the mobile robot and the reference cart. The proposed fuzzy controller has two inputs and two outputs. The first input represents the distance between the mobile robot and the reference cart. The second input is the angle formed by the straight line defined with the orientation of the robot, and the straight line that connects the robot with the reference cart. Outputs represent linear and angular velocity inputs, respectively. The performance of proposed systems is investigated using a dynamic model of a nonholonomic mobile robot with the friction considered. The quality of the fuzzy controller is analyzed through comparison with previously developed a mobile robot position controller based on backstepping control algorithm. Simulation results indicated good quality of both position tracking and torque capabilities with the proposed fuzzy controller. Also, noticeable improvement of torques reduction is achieved in the case of fuzzy controller.

2. Control system topology

The proposed control system with two-level controls is shown in Fig. 1. The low level velocity control system is composed of a multivariable PI controller and dynamic model of mobile robots and actuators. The medium level position control system generates a non-linear control law whose parameters are obtained using a genetic algorithm.
In the following sections the design of the control system blocks from Fig. 1 is described.

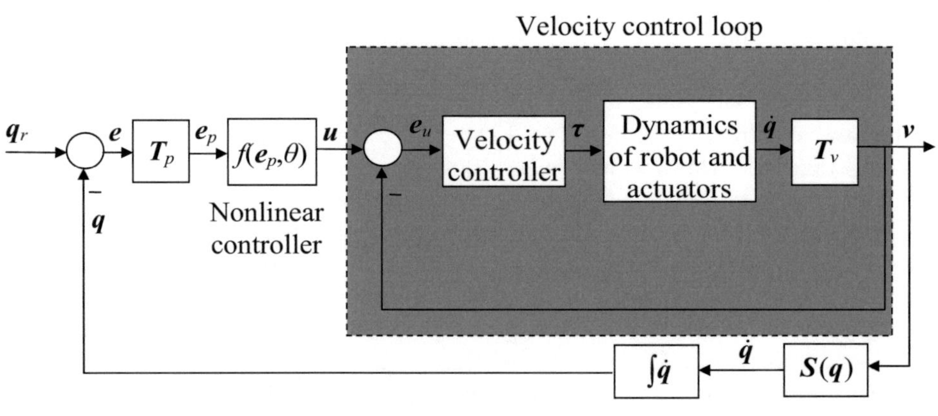

Figure 1. Mobile robot position and velocity control

2.1 Dynamics of mobile robot

In this section, a dynamic model of a nonholonomic mobile robot with the viscous friction will be derived first. A typical representation of a nonholonomic mobile robot is shown in Fig. 2.

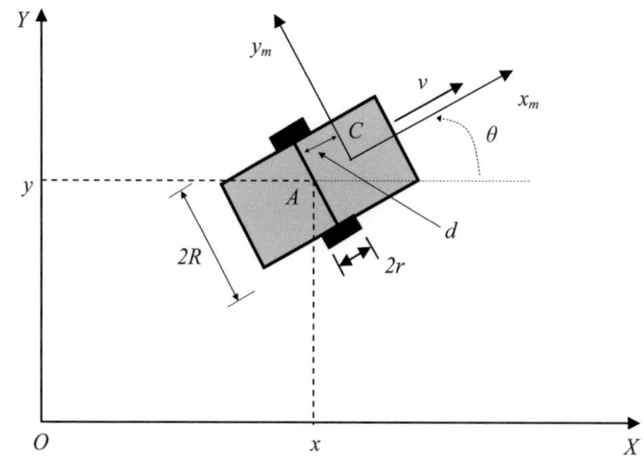

Figure 2. The representation of a nonholonomic mobile robot

The robot has two driving wheels mounted on the same axis and a free front wheel. Two driving wheels are independently driven by two actuators to achieve both the transition and orientation. The position of the mobile robot in the global frame $\{X,O,Y\}$ can be defined by the position of the mass center of the mobile robot system, denoted by C, or alternatively by position A, which is the center of mobile robot gear, and the angle between robot local frame $\{x_m,C,y_m\}$ and global frame. The kinetic energy of the whole structure is given by the following equation:

$$T = T_l + T_r + T_{kr}, \tag{1}$$

where T_l is a kinetic energy that is consequence of pure translation of the entire vehicle, T_r is a kinetic energy of rotation of the vehicle in XOY plane, and T_{kr} is the kinetic energy of rotation of wheels and rotors of DC motors. The values of introduced energy terms can be expressed by Eqs. (2)-(4):

$$T_l = \frac{1}{2}Mv_c^2 = \frac{1}{2}M(\dot{x}_c^2 + \dot{y}_c^2), \tag{2}$$

$$T_r = \frac{1}{2}I_A\dot{\theta}^2, \tag{3}$$

$$T_{kr} = \frac{1}{2}I_0\dot{\theta}_R^2 + \frac{1}{2}I_0\dot{\theta}_L^2, \tag{4}$$

where M is the mass of the entire vehicle, v_c is linear velocity of the vehicle's center of mass C, I_A is the moment of inertia of the entire vehicle considering point A, θ is the angle that represents the orientation of the vehicle (Fig. 2), I_0 is the moment of inertia of the rotor/wheel complex and $d\theta_R/dt$ and $d\theta_L/dt$ are angular velocities of the right and left wheel respectively.

Further, the components of the velocity of the point A, can be expressed in terms of $d\theta_R/dt$ and $d\theta_L/dt$:

$$\dot{x}_A = \frac{r}{2}(\dot{\theta}_R + \dot{\theta}_L)\cos\theta \,, \tag{5}$$

$$\dot{y}_A = \frac{r}{2}(\dot{\theta}_R + \dot{\theta}_L)\sin\theta \,, \tag{6}$$

$$\dot{\theta} = \frac{r(\dot{\theta}_R - \dot{\theta}_L)}{2R} \,. \tag{7}$$

Since $\dot{x}_C = \dot{x}_A - d\dot{\theta}\sin\theta$ and $\dot{y}_C = \dot{y}_A + d\dot{\theta}\cos\theta$, where d is distance between points A and C, it is obvious that following equations follow:

$$\dot{x}_C = \frac{r}{2}(\dot{\theta}_R + \dot{\theta}_L)\cos\theta - d\dot{\theta}\sin\theta \,, \tag{8}$$

$$\dot{y}_C = \frac{r}{2}(\dot{\theta}_R + \dot{\theta}_L)\sin\theta + d\dot{\theta}\cos\theta \,. \tag{9}$$

By substituting terms in (1) with expressions in equations (2)-(9), total kinetic energy of the vehicle can be calculated in terms of $d\theta_R/dt$ and $d\theta_L/dt$:

$$T(\dot{\theta}_R,\dot{\theta}_R) = \left(\frac{Mr^2}{8} + \frac{(I_A + Md^2)r^2}{8R^2} + \frac{I_0}{2}\right)\dot{\theta}_R^2 + \left(\frac{Mr^2}{8} + \frac{(I_A + Md^2)r^2}{8R^2} + \frac{I_0}{2}\right)\dot{\theta}_L^2 + \left(\frac{Mr^2}{4} - \frac{(I_A + Md^2)r^2}{4R^2}\right)\dot{\theta}_R\dot{\theta}_L \tag{10}$$

Now, the Lagrange equations:

$$\frac{d}{dt}\left(\frac{\partial L}{\partial \dot{\theta}_R}\right) - \frac{\partial L}{\partial \theta_R} = \tau_R - K\dot{\theta}_R \,, \tag{11}$$

$$\frac{d}{dt}\left(\frac{\partial L}{\partial \dot{\theta}_L}\right) - \frac{\partial L}{\partial \theta_L} = \tau_L - K\dot{\theta}_L \,, \tag{12}$$

are applied.

Here τ_R and τ_L are right and left actuation torques and $Kd\theta_R/dt$ and $Kd\theta_L/dt$ are the viscous friction torques of right and left wheel-motor systems, respectively.

Finally, the dynamic equations of motion can be expressed as:

$$A\ddot{\theta}_R + B\ddot{\theta}_L = \tau_R - K\dot{\theta}_R \,, \tag{13}$$

$$B\ddot{\theta}_R + A\ddot{\theta}_L = \tau_L - K\dot{\theta}_L \,, \tag{14}$$

where

$$A = \left(\frac{Mr^2}{4} + \frac{(I_A + Md^2)r^2}{4R^2} + I_0\right), \quad B = \left(\frac{Mr^2}{4} - \frac{(I_A + Md^2)r^2}{4R^2}\right). \tag{15}$$

In this chapter we used a mobile robot with the following parameters: M=10kg, I_A=1kgm2, r=0.035 m, R=0.175 m, d=0.05 m, m_0=0.2 kg, I_0=0.001 kgm^2 and K/A=0.5.
In the following section a design of both velocity and position controls will be established.

2.2. Velocity control of mobile robot

The dynamics of the velocity controller is given by the following equations in Laplace domain:

$$\boldsymbol{\tau}(s)=\begin{bmatrix}\tau_R(s)\\ \tau_L(s)\end{bmatrix}=\frac{1}{r}\begin{bmatrix}g_1(s) & g_2(s)\\ g_1(s) & -g_2(s)\end{bmatrix}\begin{bmatrix}e_v(s)\\ e_\omega(s)\end{bmatrix}, \tag{16}$$

where $e_v(s)$ is the linear velocity error, and $e_\omega(s)$ is the angular velocity error. This structure differs from previously used diagonal structures. Transfer functions $g_j(s)$ are chosen to represent PI controllers:

$$g_1(s)=K_1(1+\frac{1}{T_{i1}s})\cdot R,\quad g_2(s)=K_2(1+\frac{1}{T_{i2}s})\cdot R\,. \tag{17}$$

The particular choice of the adopted multivariable PI controller described by equations (16) and (17) is justified with the following theorem.
Theorem 1. Torque control (16) ensures tracking servo inputs u_1 and u_2 with zero steady state errors.
Proof: When we substitute $\dot{\theta}_R$ with ω_R, $\dot{\theta}_L$ with ω_L, and consider (16), we can write another form of (13) and (14):

$$\begin{bmatrix}As+K & Bs\\ Bs & As+K\end{bmatrix}\begin{bmatrix}\omega_R(s)\\ \omega_L(s)\end{bmatrix}=\frac{1}{r}\begin{bmatrix}g_1(s) & g_2(s)\\ g_1(s) & -g_2(s)\end{bmatrix}\begin{bmatrix}u_1(s)-v(s)\\ u_2(s)-\omega(s)\end{bmatrix}, \tag{18}$$

ω_R and ω_L can be expressed in terms of ω and v as:

$$\omega_R=\frac{v+R\omega}{r},\quad \omega_L=\frac{v-R\omega}{r}\,. \tag{19}$$

Then, equation (18) can be transformed to:

$$\begin{bmatrix}As+K & Bs\\ Bs & As+K\end{bmatrix}\begin{bmatrix}v(s)+R\omega(s)\\ v(s)-R\omega(s)\end{bmatrix}=\begin{bmatrix}g_1(s) & g_2(s)\\ g_1(s) & -g_2(s)\end{bmatrix}\begin{bmatrix}u_1(s)-v(s)\\ u_2(s)-\omega(s)\end{bmatrix},$$

and further to:

$$\begin{bmatrix}\alpha_1(s) & \alpha_2(s)\\ \alpha_1(s) & -\alpha_2(s)\end{bmatrix}\begin{bmatrix}v(s)\\ \omega(s)\end{bmatrix}=\begin{bmatrix}g_1(s) & g_2(s)\\ g_1(s) & -g_2(s)\end{bmatrix}\begin{bmatrix}u_1(s)\\ u_2(s)\end{bmatrix}, \tag{20}$$

where

$$\begin{aligned}\alpha_1(s)&=\frac{(A+B)s^2+(K+K_1T_{i1})s+K_1}{s}\\ \alpha_2(s)&=\frac{R(A-B)s^2+(RK+K_2T_{i2})s+K_2}{s}\end{aligned}\,. \tag{21}$$

Following equations could be easily derived from (20):

$$v(s)=\frac{g_1}{\alpha_1}u_1(s)=G_1u_1(s)=\frac{K_1T_{i1}s+K_1}{(A+B)s^2+(K+K_1T_{i1})s+K_1}u_1(s)$$
$$\omega(s)=\frac{g_2}{\alpha_2}u_2(s)=G_2u_2(s)=\frac{K_2T_{i2}s+K_2}{R(A-B)s^2+(RK+K_2T_{i2})s+K_2}u_2(s) \quad (22)$$

It is obvious that transfer functions G_1 and G_2 are static with gains equal to "1", which completes the **proof**.

The velocity control loop structure is shown in Fig. 1, as an inner loop. From the simulation results obtained (Figs. 3 and 4), it can be seen that the proposed PI controller successfully tracks the given linear and angular velocity profiles.

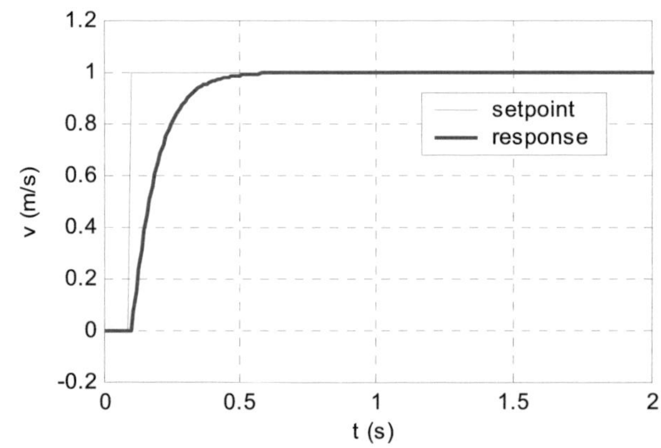

Figure 3. Linear velocity step response

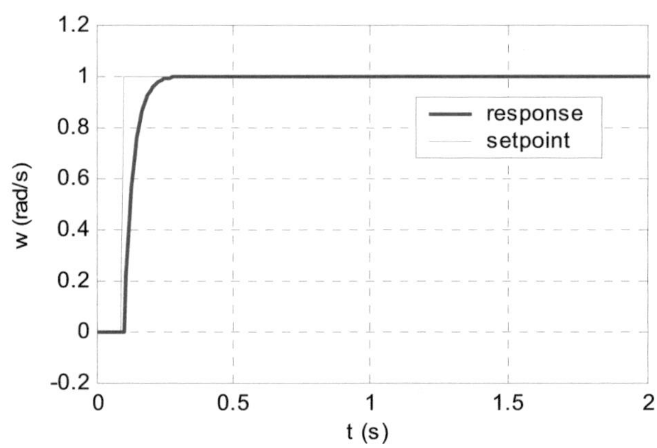

Figure 4. Angular velocity step response

The controller parameters used for this simulation are K_1=129.7749, K_2=41.0233, T_{i1}=11.4018, T_{i2}=24.1873, which are tuned using standard GA.

The design of position controls of mobile robot, backstepping and fuzzy logic controllers, will be described in the next section.

3. Position control of mobile robot

The trajectory position tracking problem for a mobile robot is formulated with the introduction of a virtual reference robot to be tracked (Egerstedt et al., 2001) (Fig. 5). The tracking position error between the reference robot and the actual robot can be expressed in the robot frame as:

$$\boldsymbol{e}_p = \begin{bmatrix} e_1 \\ e_2 \\ e_3 \end{bmatrix} = \boldsymbol{T}_p \boldsymbol{e}_q = \begin{bmatrix} \cos\theta & \sin\theta & 0 \\ -\sin\theta & \cos\theta & 0 \\ 0 & 0 & 1 \end{bmatrix} \begin{bmatrix} x_r - x \\ y_r - y \\ \theta_r - \theta \end{bmatrix}, \tag{23}$$

where $\boldsymbol{e}_q = [e_x \; e_y \; e_\theta]^T$.

The position error dynamics can be obtained from the time derivative of the (23) as:

$$\dot{e}_1 = \omega e_2 + u_1, \quad \dot{e}_2 = -\omega e_1 + v_r \sin e_3, \quad \dot{e}_3 = u_2, \tag{24}$$

where $v = v_r \cos e_3 - u_1$ and $\omega = \omega_r - u_2$.

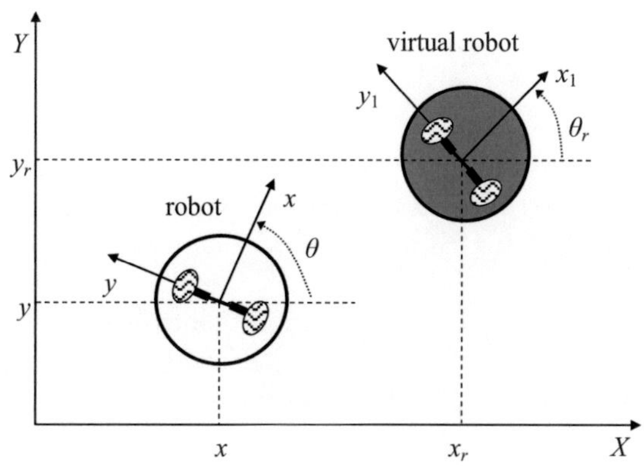

Figure 5. The concept of tracking of a virtual reference robot

3.1. Backstepping controller design

In paper (Lacevic & Velagic, 2006) we proposed the following position control law that ensures stability

$$\begin{aligned} u_1 &= -\alpha p e_1 \cdot f(\mathbf{x}) \\ u_2 &= -\frac{2\alpha p}{q}(e_1^2 + e_2^2)^{\alpha-1} v_r e_2 - q \sin e_3 \cdot g(\mathbf{x}) \end{aligned}, \tag{25}$$

where p and q are positive real constants, and $a > 1$ and f(x) and g(x) are functions of some vector $x \in \boldsymbol{R}^m$, $m \in N$, satisfying the condition: $\exists L>0$: $f(x), g(x) \geq L$, $\forall\, x \in \boldsymbol{R}^m$.

Our theorem which proved this statement is derived as follows.

Theorem 2. Control law, given in (25), provides stability of the mobile robot model, respect to the reference trajectory (i.e., $\lim_{t\to\infty}\left(\mathbf{e}_1^2(t) + \mathbf{e}_2^2(t)\right) = 0 \wedge \lim_{t\to\infty} \mathbf{e}_3(t) = k\pi,\ k \in \mathbf{Z}$)).

Proof:

Consider the Lyapunov function candidate:

$$V(e_1,e_2,e_3)=p(e_1^2+e_2^2)^\alpha+q(1-\cos(e_3)). \tag{26}$$

Deriving (26), and using the expressions from error dynamics (24), we obtain:

$$\begin{aligned}\dot V(e_1,e_2,e_3)=2\alpha p(e_1^2+e_2^2)^{\alpha-1}u_1+(q\sin e_3)u_2\\ +2\alpha p(e_1^2+e_2^2)^{\alpha-1}e_2 v_r\sin e_3\end{aligned}. \tag{27}$$

Substituting u_1 and u_2 from (25) we get:

$$\begin{aligned}\dot V(e_1,e_2,e_3)=-2\alpha^2p^2e_1^2(e_1^2+e_2^2)^{\alpha-1}\cdot f(\mathbf{x})\\ -q^2(\sin^2 e_3)\cdot g(\mathbf{x})<0\end{aligned}. \tag{28}$$

Thus, function $\dot V(e_1,e_2,e_3)$ is uniformly continuous, $V(e_1,e_2,e_3)$ tends to some positive finite value and $||e_p(t)||$ is bounded. Using Barbalat lemma, $\dot V(e_1,e_2,e_3)$ tends to zero. From (28), it is obvious that $\lim_{t\to\infty} e_1(t)=0$ and $\lim_{t\to\infty} e_3(t)=k\pi,\ k\in \mathrm{Z}$. Further, it is clear that $\lim_{t\to\infty} \dot e_3(t)=0$, and hence (using (24)) $\lim_{t\to\infty} u_2=0$. From the expression for u_2 in (25) one can conclude that $\lim_{t\to\infty} e_2(t)=0$.

The "obstacle" for global asymptotical stability of the system is the lack of guarantee that the error e_3 will converge to zero. In the worst case scenario, the robot will track the reference cart by moving backwards. This behaviour however, was not observed in any of case studies.

The parameters of both velocity and position controllers are encoded into binary chromosome (here, functions f and g are assumed as constants) is shown in Fig. 6. Each parameter is presented with 12 bits. Each individual was assigned an objective value, based on the following functional:

$$F=\sum_{i=1}^{3}\left[a_i\int_0^{t_s}\ln\left(1+\left|e_i(t)\right|\right)dt\right]+a_R\cdot\max_{t\in[0,t_s]}\left(\tau_R(t)\right)+a_L\cdot\max_{t\in[0,t_s]}\left(\tau_L(t)\right), \tag{29}$$

where

$a_1=1,\ a_2=5,\ a_3=1,\ t_s=16,\ a_R=a_L=1$.

It is obvious that the better individual has smaller value of F. The second part of the expression represents penalty for the great values of control torques. The objective value for each individual is evaluated upon the simulation run that includes the tracking of reference trifolium trajectory (Figures in simulation results section).

Simple GA with population size 51, tournament selection, uniform crossover, bit mutation, and elitism has been used.

Evolution yielded following values:

p = 1.9934, q = 0.0530, f = 9.8615, g = 2.9956, K_1 = 128.444, Ti_1 = 60.9756, K_2 = 29.048, Ti_2 = 31.4465.

p	q	f	g	K_1	K_2	T_{i1}	T_{i2}

Figure 6. Mapping the set of tunable parameters into chromosome

It has been noticed, that, at the beginning of tracking, the control torques increase rapidly if the initial position of reference robot does not belong to the straight line, determined with the robot and its initial orientation (Lacevic and Velagic, 2005) (Fig. 7).

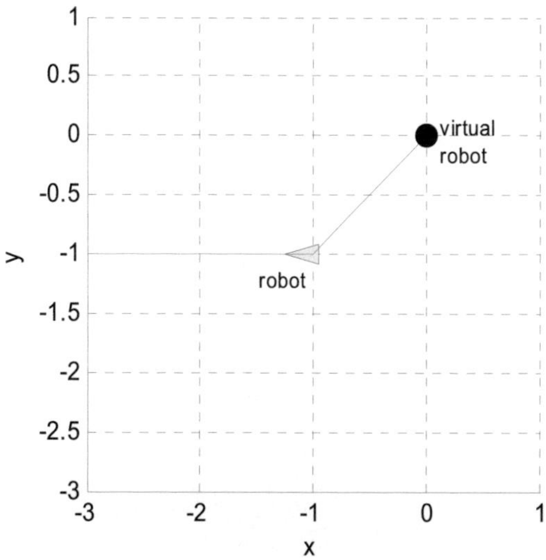

Figure 7. Tracking robot doesn't "see" virtual robot

3.2. Hybrid backstepping controller design

For improving the mentioned weakness a hybrid backstepping position controller is designed. For that purpose, the following control law, which provides velocity servo inputs, is proposed (Lacevic & Velagic, 2005):

$$\begin{aligned} \bar{u}_1(t) &= \alpha(t)u_1(t) \\ \bar{u}_2(t) &= \alpha(t)u_2(t) + (1-\alpha(t))\omega_s(t)\, ' \end{aligned} \tag{30}$$

Function $\omega_s(t)$ is produced, as the output of the following system (Fig. 8).

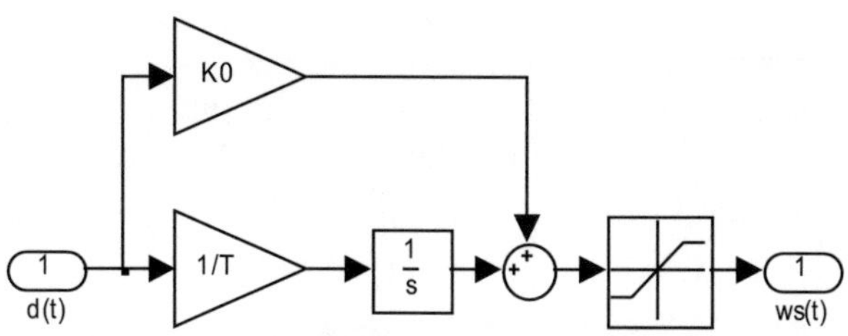

Figure 8. Producing $\omega_s(t)$

The function $d(t)$ is given with:

$$d(t) = \text{sgn}(\text{atan2}(e_y(t), e_x(t)) - \theta(t)) . \tag{31}$$

Function $a(t)$ is determined with the following differential equation:

$$b_0 \frac{d^2\alpha(t)}{dt^2} + b_1 \frac{d\alpha(t)}{dt} + \alpha(t) = z(t) , \tag{32}$$

where $z(t)$ is practically a step function given with:

$$z(t) = \begin{cases} 1, \; if \;\; \exists t_1 \in [0,t] : \theta(t_1) = \text{atan2}(e_y(t_1), e_x(t_1)) \\ \qquad 0, \; otherwise \end{cases} , \tag{33}$$

This way, the robot doesn't start tracking virtual robot instantly; it first rotates around its own axis with increasing angular velocity ω_s(t), until it "sees" the virtual robot (Fig. 9).

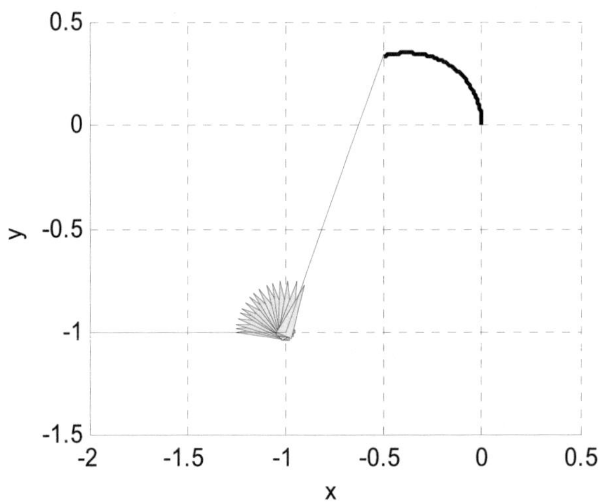

Figure 9. "Seeking" a virtual robot

In the next subsection the design procedure of proposed fuzzy position controller will be described.

3.3. Design of fuzzy controller

In order to reduce the control torques and velocity inputs, fuzzy position controller is designed. Fuzzy system based on Sugeno inference model with 2 inputs and 2 outputs is used instead of classical backstepping controller (Fig. 10).
Inputs i_1 and i_2 are following signals:

$$i_1 = \sqrt{e_1^2 + e_2^2} = \sqrt{e_x^2 + e_y^2}, \;\; i_2 = f\left(\text{atan2}\left(e_y, e_x\right) - \theta\right), \tag{34}$$

where f is given with:

$$f(x) = 2\,\mathrm{atan}\left(\tan\left(\frac{x}{2}\right)\right). \tag{35}$$

Input i_1 represents the distance between the mobile robot and the reference cart. Input i_2 is the angle formed by the straight line defined with the orientation of the robot, and the straight line that connects the robot with the reference cart. Function f ensures that variable i_2 belongs to the interval $(-\pi, \pi]$ (Fig. 11). Outputs o_1 and o_2 represent linear and angular velocity inputs respectively.

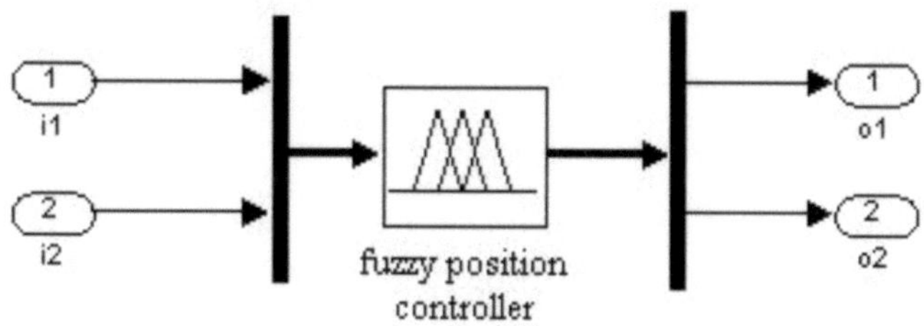

Figure 10. Fuzzy controller with 2 inputs and 2 outputs

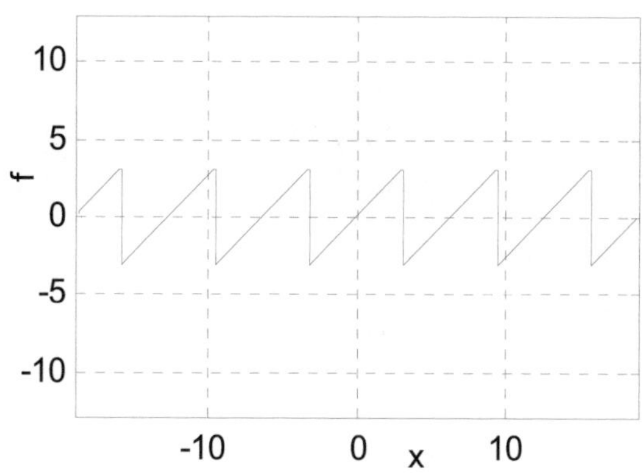

Figure 11. Diagram of the function f

Inputs i_1 and i_2 will be modeled with two trapezoidal and five triangular shape membership functions, respectively. Random initial setups of mentioned variables are shown in Figs. 12 and 13. The parameters of input variables are (a_1) and $(b_1, b_2, b_3, b_4, b_5$ i $b_6)$. Outputs o_1 and o_2 are represented by singleton functions with five and three membership values (Figs. 14 and 15). These outputs are described with parameters $(c_1, c_2, c_3, c_4$ and $c_5)$ and $(d_1, d_2$ and $d_3)$, respectively. Parameters a_i, b_i, c_i and d_i that determine the membership functions are encoded into binary chromosome in Fig 16 (the same as in a previously described algorithm), while the rules set remained invariant during the GA run. Parameters of velocity controller kept their values, obtained previously.

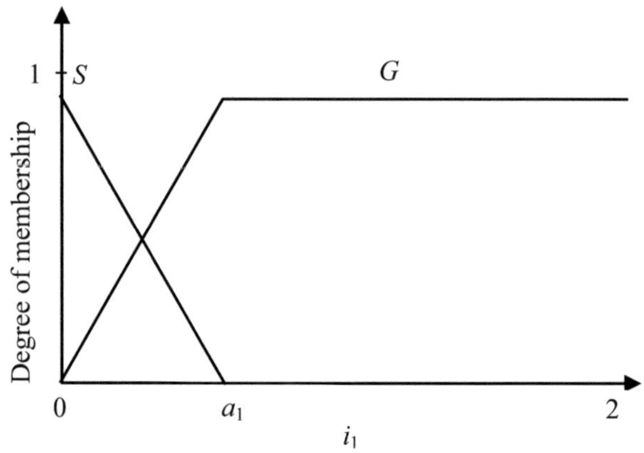

Figure 12. Random initial setup of input variable i_1

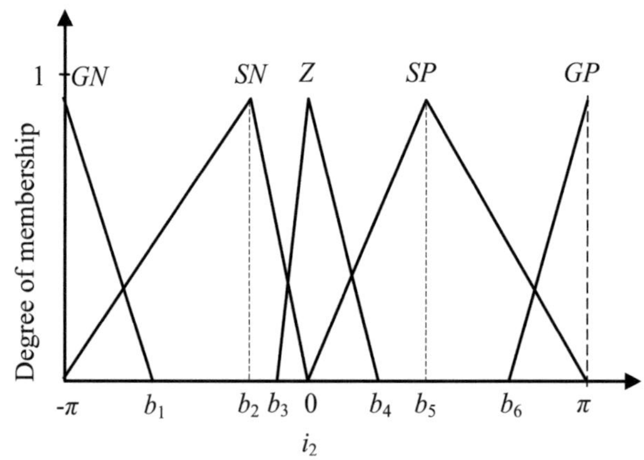

Figure 13. Random initial setup of input variable i_2

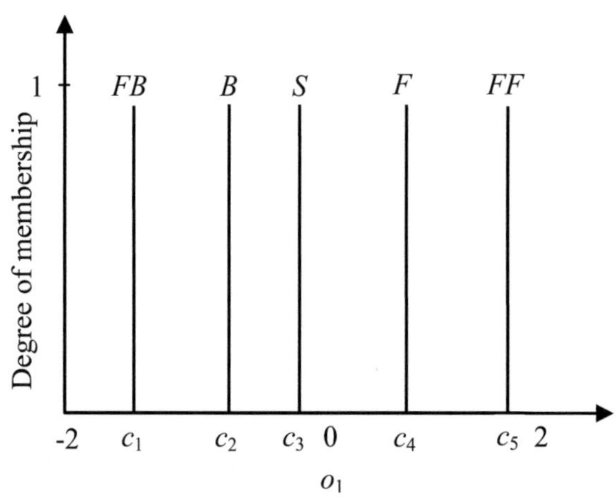

Figure 14. Random initial setup of output variable o_1

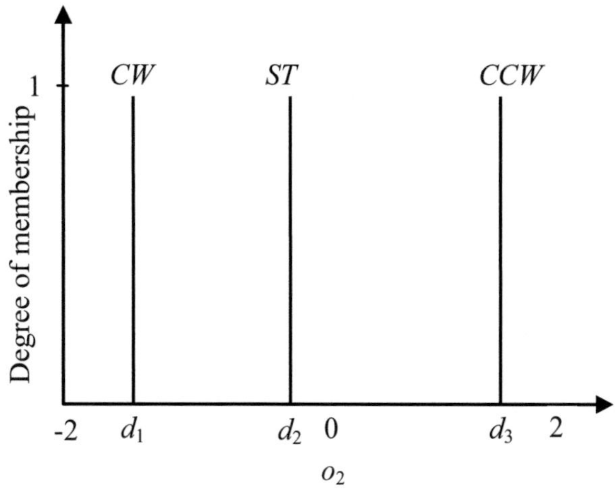

Figure 15. Random initial setup of output variable o_2

a_1	b_1	b_2	b_3	b_4	b_5	b_6	c_1	c_2	c_3	c_4	c_5	d_1	d_2	d_3

Figure 16. Representation of binary chromosome of parameters

Finally, the rules that complete the inference model are:

If (i_1 is *S*) and (i_2 is *GN*) then (o_1 is *B*) (o_2 is *CW*)

If (i_1 is *S*) and (i_2 is *SN*) then (o_1 is *F*) (o_2 is *CW*)

If (i_1 is *S*) and (i_2 is *Z*) then (o_1 is *S*) (o_2 is *ST*)

If (i_1 is *S*) and (i_2 is *SP*) then (o_1 is *F*) (o_2 is *CCW*)

If (i_1 is *S*) and (i_2 is *GP*) then (o_1 is *B*) (o_2 is *CCW*)

If (i_1 is *G*) and (i_2 is *GN*) then (o_1 is *FB*) (o_2 is *CW*)

If (i_1 is *G*) and (i_2 is *SN*) then (o_1 is *FF*) (o_2 is *CW*)

If (i_1 is *G*) and (i_2 is *Z*) then (o_1 is *FF*) (o_2 is *ST*)

If (i_1 is *G*) and (i_2 is *SP*) then (o_1 is *FF*) (o_2 is *CCW*)

If (i_1 is *G*) and (i_2 is *GP*) then (o_1 is *FB*) (o_2 is *CCW*)

Evolution of membership functions parameters is performed by identical way as in the case of backsteping controller design (subsection 3.1). The GA with population size 101, tournament selection, uniform crossover, bit mutation, and elitism has been used. Also, the objective function is same as in (29). Resulting membership functions for input and output variables are shown in Figs. 17-20.

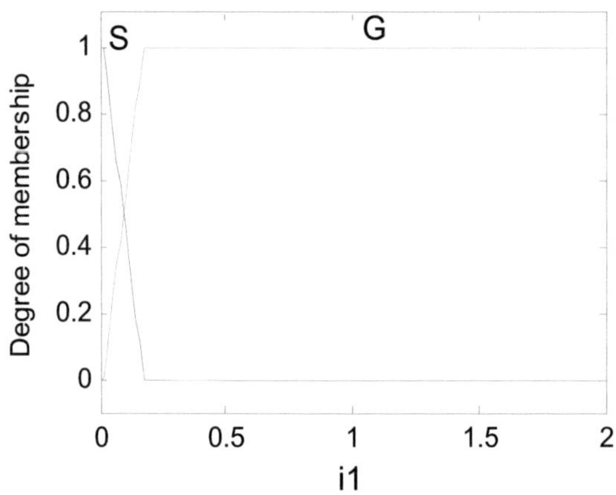

Figure 17. Membership functions for i_1

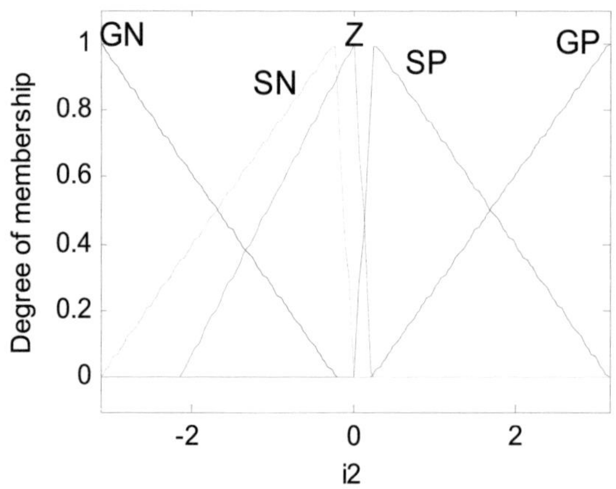

Figure 18. Membership functions for i_2

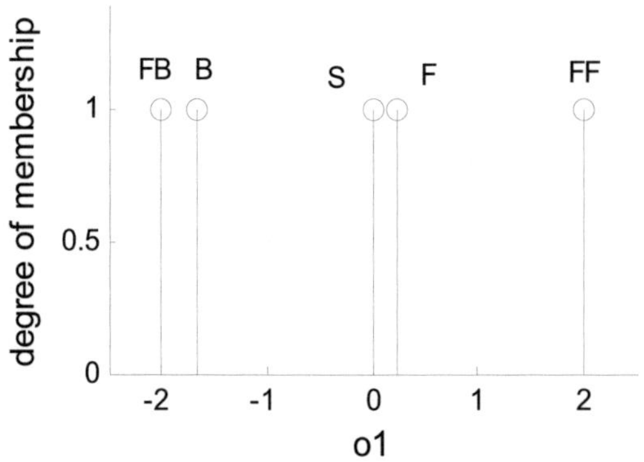

Figure 19. Membership functions for o_1 (fast backwards, backwards, stop, forward and fast forward)

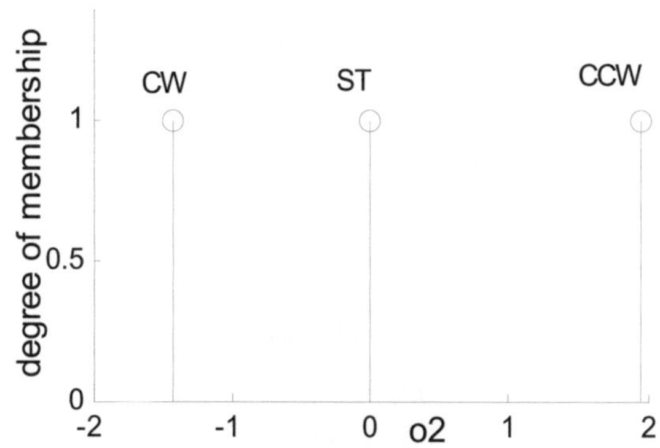

Figure 20. Membership functions for o_2 (clockwise, straight, counter clockwise)

Important characteristic of this controller is that its outputs are inherently limited. Disadvantage of this concept lies in its inavility to ensure tracking of the reference cart that has velocities which are bigger than those that fuzzy controller can "suggest". Advantage lies in the fact that control velocities (and consequently, the control torques) cannot exceed certain limits (see simulation results in the next section). Resulting control surfaces are shown in Figs. 21 and 22.

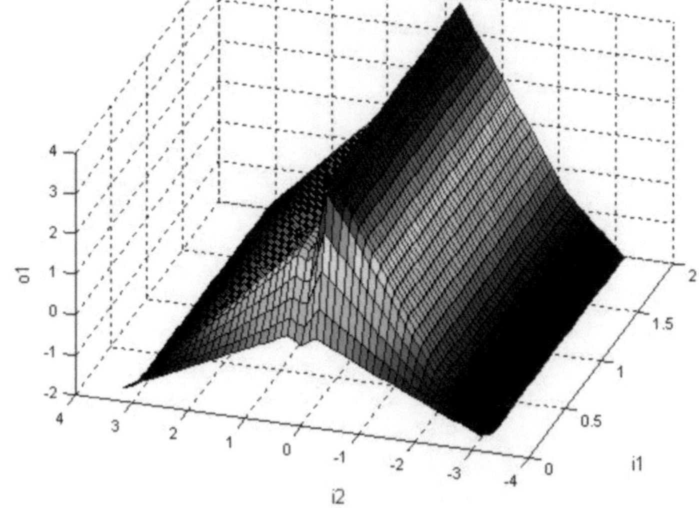

Figure 21. Control surface for o_1

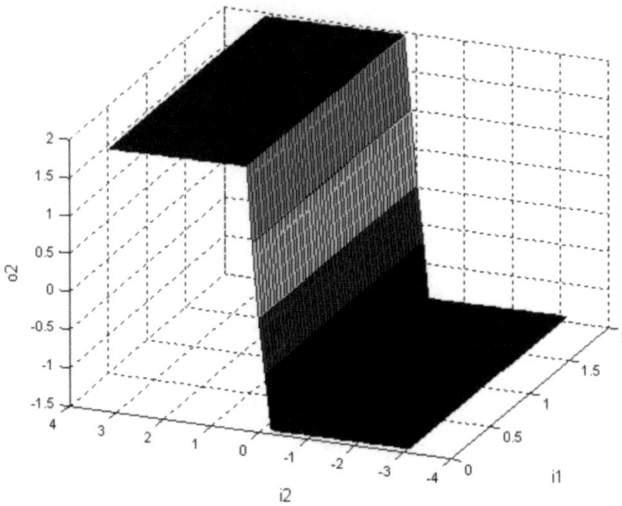

Figure 22. Control surface for o_2

4. Simulation results

The validation of proposed fuzzy controller will be tested in the comparison with backstepping control algorithm. The effectiveness of the both controllers is demonstrated in the case of tracking of a lamniscate and trifolium curves. The overall system is designed and implemented within Matlab/Simulink environment. We consider the following profiles: position, orientation, linear and angular velocities and torques.

4.1 Simulation results with backstepping controllers

The control performance of the ordinary and hybrid backstepping controllers will be illustrated in this subsection through their comparative analysis. The simulation results obtained are shown in Figs. 23-26.

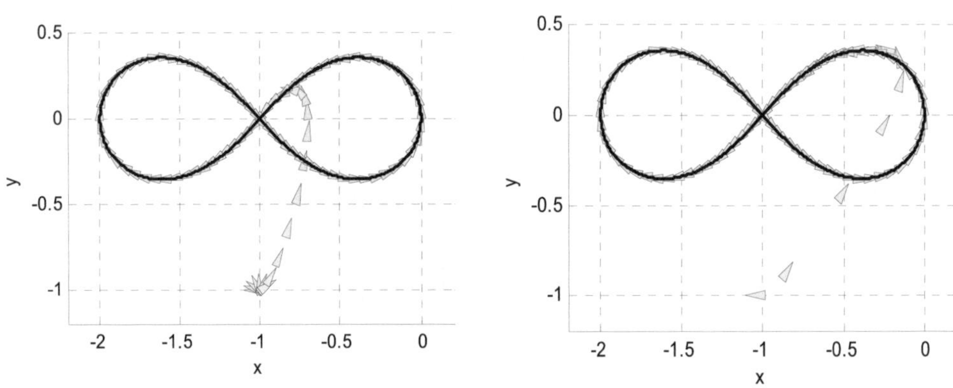

Figure 23. Tracking a lemniscate trajectory with (a) hybrid and (b) ordinary controllers

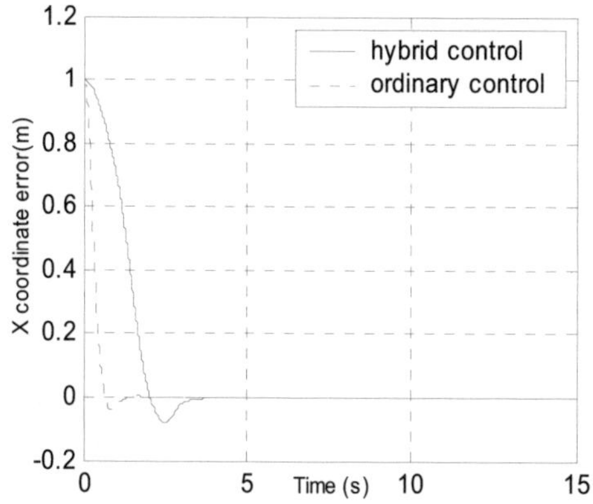

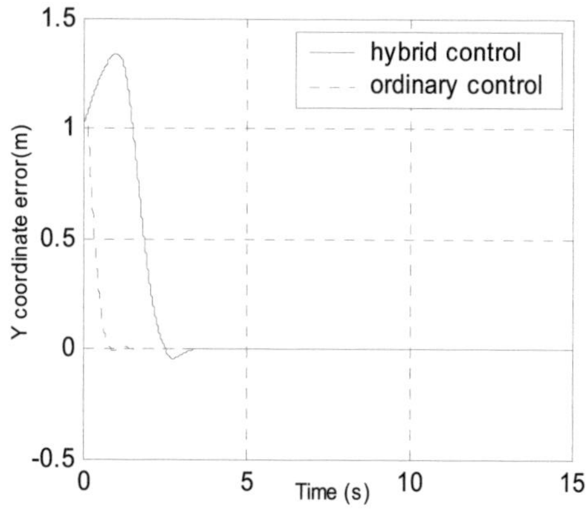

Figure 24. *X* and *Y* coordinate errors

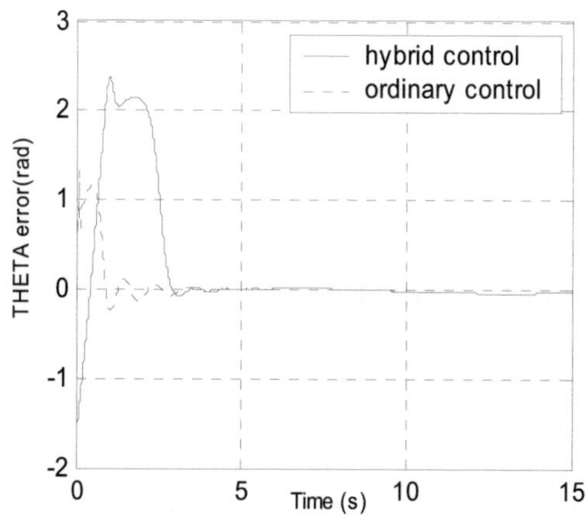

Figure 25. Orientation error

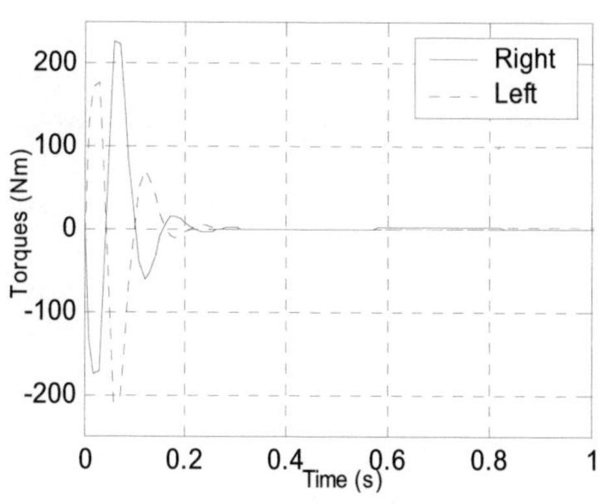

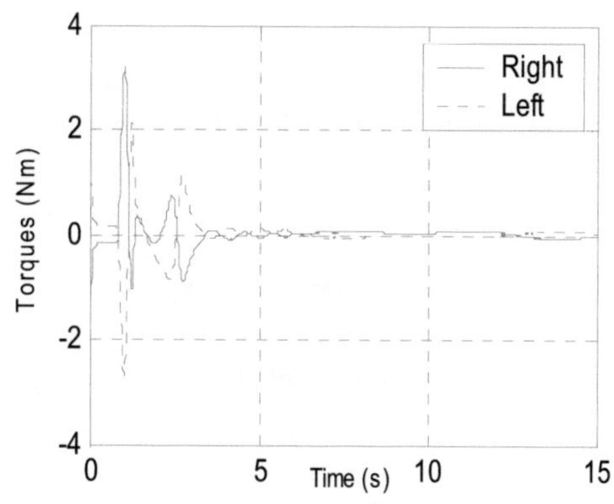

(a) (b)

Figure 26. Control torques of ordinary (a) and hybrid (b) backstepping controllers

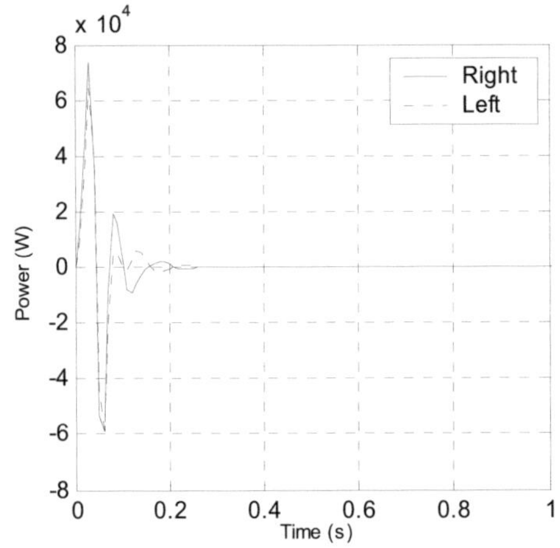

(a)

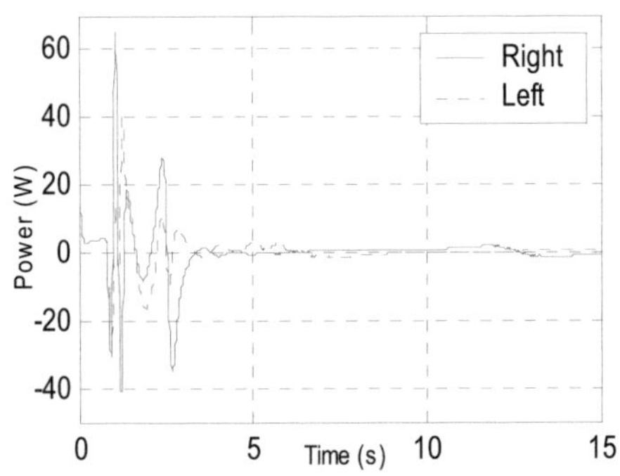

(b)

Figure 26. Requested power of DC motors: (a) hybrid and (b) ordinary backstepping controllers (ordinary controller - the first second of tracking)

These results demonstrate the good position tracking performance (Figs. 23-25), but with unsatisfactory control torques values, in the case of ordinary backstepping controller, particularly at the beginning of tracking (Fig. 25). Both torques of ordinary backstepping controller, for left and right wheels, exceed 50 Nm and they can produce the unnecessary actuators behavior. However, the hybrid controller ensures much less values of the control input torques for obtaining the reference position and orientation trajectories (Fig. 25). Consequently, the requested power of DC motors is also much less in the case of control by using the hybrid controller (Fig. 26).

4.2 Simulation results with fuzzy logic controller

The effectiveness of the fuzzy controller is demonstrated in the case of tracking of a more complex trajectory then lemniscate, such as trifolium curve. The simulation results obtained by fuzzy logic position controller are illustrated in Figs. 27-33. From figures 27-29, it can be concluded that satisfactory tracking results are obtained using this controller. Also, the fuzzy controller ensures much less values of the control input velocities (Fig. 30) then ordinary backstepping controller (Fig. 31) for obtaining the reference position and orientation trajectories in comparison with the backstepping controller. Consequently the significant decreasing of wheel torques with fuzzy control is achieved (Figs. 32). The absolute torque values of both wheels not exceed 2 Nm. These values are 30-40 times less then torque values achieved with backstepping controller. The time response of wheel torques of ordinary backstepping controller is shown in Fig. 33.

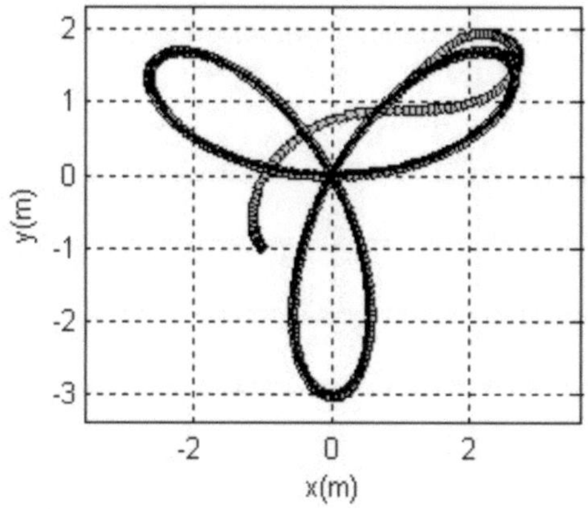

Figure 27. Tracking the trifolium trajectory

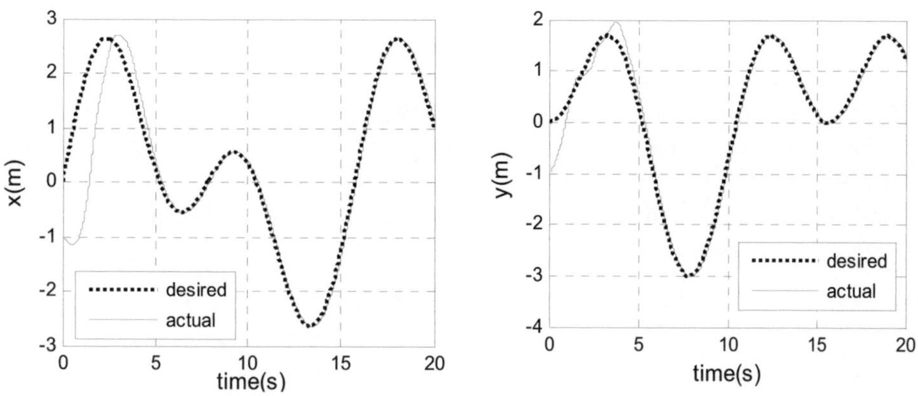

Figure 28. Time history of x and y coordinates

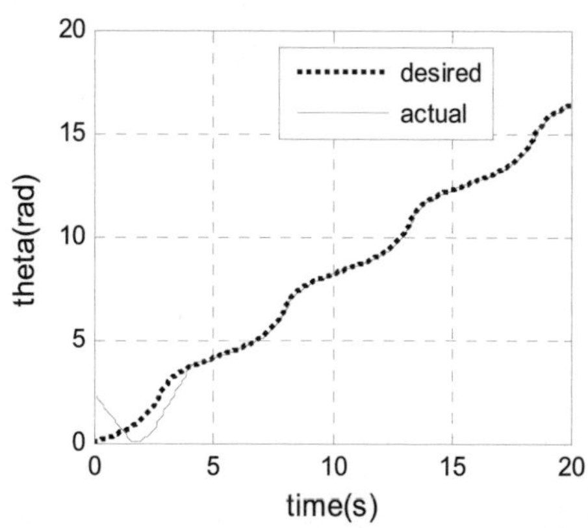

Figure 29. Orientation of the robot

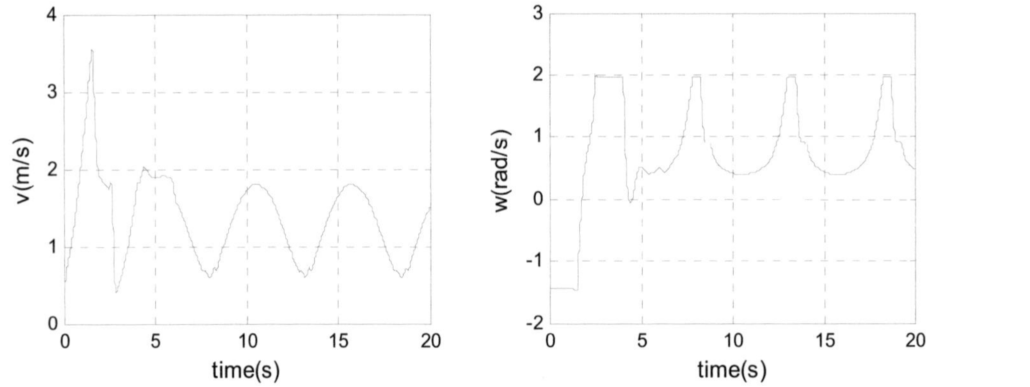

Figure 30. Linear (v) and angular (w) velocity outputs of fuzzy controller

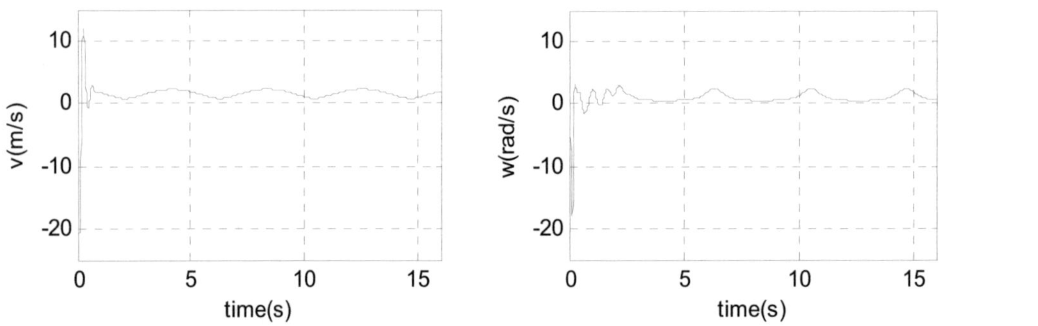

Figure 31. Linear (v) and angular (w) velocity outputs of ordinary backstepping controller

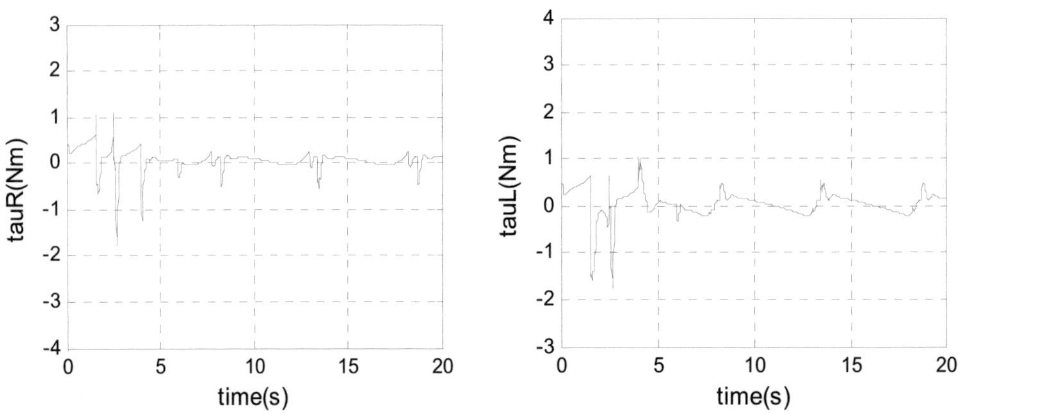

Figure 32. Right and left wheel torques of fuzzy

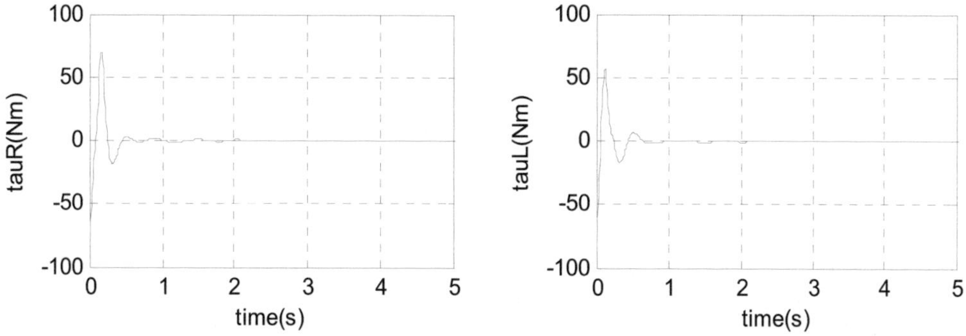

Figure 33. Right and left wheel torques of ordinary backstepping controller (first five seconds)

5. Conclusions

The experience of the design of the nonlinear position control confirmed the remarkable potential of backstepping and fuzzy logic in the development of effective decision laws capable of overcoming the inherent limitations of model-based control strategies. This paper focuses on design of hybrid backstepping and fuzzy position logic controls of mobile robot that satisfied a good position tracking performance with simultaneously satisfactory control of velocities, which has an impact on wheel torques. In our previously designed backstepping controller a good tracking performance was obtained. However, its main shortcoming is unsatisfactory control velocities values, particularly at the beginning of tracking. Control parameters of backstepping controller and membership functions of fuzzy controller are adjusted by genetic algorithms. Advantage of the proposed fuzzy controller, and also hybrid backstepping controller, lies in the fact that control velocities (and consequently, the control torques) cannot exceed certain limits. Consequently, these controllers radically decreased the control velocities without major impact on tracking performance. Finally, from the simulation results obtained, it can be concluded that the proposed hybrid backstepping and fuzzy design achieve the desired results. Future work will include the investigation of a fuzzy stability of the proposed fuzzy logic position system.

6. References

Egerstedt, M.; Hu, X. & Stotsky, A. (2001). Control of Mobile Platforms Using a Virtual Vehicle Approach. *IEEE Transaction on Automatic Control*, Vol. 46, pp. 1777-1882

Fierro, R. & Lewis, F. L. (1997). Control of a nonholonomic mobile robot: backstepping kinematics into dynamics. *Journal of Robotics Systems*, Vol. 14, No. 2, pp. 149–163.

Fukao, T.; Nakagawa, H. & Adachi, N. (2000). Adaptive Tracking Control of a Nonholonomic Mobile Robot. *IEEE Transaction on Robotics and Automation*, Vol. 16, pp. 609-615.

Hu, T. & Yang, S. X. (2001). Real-time Motion Control of a Nonholonomic Mobile Robot with Unknown Dynamics, Proceedings of the Computational Kinematics Conference, Seoul.

Lacevic, B. & Velagic, J. (2005). Reduction of Control Torques of Mobile Robot Using Hybrid Nonlinear Position Controller, *Proceedings of the IEEE International Conference "Computer as a tool"*, pp. 161-166, Belgrade, November 2005.

Lacevic, B. & Velagic, J. (2006). Stable Nonlinear Position Control Law for Mobile Robot Using Genetic Algorithm and Neural Networks, *Proceedings of the World Automation Congress, 6th International Symposium of Soft Computing for Industry (ISSCI)*, paper no. 50, Budapest, July 2006.

Lacevic, B.; Velagic, J. & Osmic, N. (2007). Design of Fuzzy Logic Based Mobile Robot Position Controller Using Genetic Algorithm, *Proceedings of the Advanced Intelligent Mechatronics (AIM2007)*, pp. 1-6, Zurich, September 2007.

Oriolo, G.; De Luca, A. & Vendittelli, M. (2002). WMR control via dynamic feedback linearization: design, implementation and experimental validation. *IEEE Transactions on Control System Technology*, Vol. 10, No. 6, pp. 835-852.

Rajagopalan, R. & N. Barakat, N. (1997). Velocity Control of Wheeled Mobile Robots Using Computed Torque Control and Its Performance for a Differentially Driven Robot. *Journal of Robotic Systems*, Vol. 14, pp. 325-340.

Tanner, H. G. & Kyriakopoulos, K. J. (2003). Backstepping for nonsmooth systems. *Automatica*, Vol. 39, pp. 1259-1265.

Topalov, A. V.; Tsankova, D. D.; Petrov, M. G. & Proychev, Th. P. (1998). Intelligent Sensor-Based Navigation and Control of Mobile Robot in a Partially Known Environment, *Proceedings of the 3rd IFAC Symposium on Intelligent Autonomous Vehicles*, IAV'98, pp. 439–444, 1998.

Velagic, J.; Lacevic, B. & Perunicic, B. New Concept of the Fast Reactive Mobile Robot Navigation Using a Pruning of Relevant Obstacles, Proceedings of the *IEEE International Symposium on Industrial Electronics (ISIE2005)*, pp. 161-166. Dubrovnik, June 2005.

Velagic, J.; Lacevic, B. & Perunicic, B. (2006). A 3-Level Autonomous Mobile Robot Navigation System Designed by Using Reasoning/Search Approaches. *Robotics and Autonomous Systems*, Vol. 54, No. 12, pp. 989-1004.

28

Planning for Unraveling Deformable Linear Objects Based on Their Silhouette

Hidefumi Wakamatsu[1], Eiji Arai[1] and Shinichi Hirai[2]
[1]Osaka University, [2]Ritsumeikan University
Japan

1. Introduction

Deformable linear objects such as tubes, cords, cables, wires, and threads are used widely for fixing, fastening, wrapping, packing, suturing, and ligating of objects including themselves. In such manipulative tasks, knotting of linear objects is required. At the same time, their raveling must be avoided. If unexpected ravel occurs, it takes much time to unravel. For example, raveling of earphone/headphone cord of a portable audio player would puzzle you sometimes. So, efficient unraveling is important as well as avoidance of such raveling.

Knotting manipulation by robots has been studied. Inoue et al. reported tying a knot in a rope with a manipulator utilizing visual feedback (Inoue & Inaba, 1984). Hopcroft et al. devised an abstract language to express various knotting manipulations and performed knot-tying tasks with a manipulator (Hopcroft et al., 1991). Matsuno et al. realized a task consisting of tying a cylinder with a rope using a dual manipulator system (Matsuno et al., 2001). Takamatsu et al. have been developing a system for knot planning from observation of human demonstrations (Takamatsu et al., 2006). Saha and Isto proposed a motion planner for manipulating ropes and realized tying several knots using two cooperating robotic arms (Saha & Isto, 2006). Yamakawa et al. proposed a new strategy for making knots with one high-speed multifingered robot hand having tactile sensors (Yamakawa et al., 2007). Unknotting manipulation, *i.e.*, the inverse of knotting manipulation, has been also studied. We have realized automatic planning and execution of knotting/unknotting manipulation (Wakamatsu et al., 2006). Ladd and Kavraki developed an untangling planner for mathematical knots represented as closed piecewise linear curves (Ladd & Kavraki, 2004).

Unraveling is equivalent to unknotting. However, the state of a raveled object can become more complex than that of a knotted object. Moreover, it is difficult to recognize the state of a reveled object completely because it may twine itself. Therefore, recognition of the object state and manipulation planning are both important for unraveling. In this paper, we propose a planning method for unraveling a linear object when 3D information about the object state is unknown. First, an unknotting process of a linear object, which is equivalent to its unraveling process, is represented as a sequence of crossing state transitions. The object state is categorized according to three properties with respect to self-crossings of the object. State transitions are defined by introducing four basic operations. Then, possible unknotting processes can be generated if the current crossing state is completely identified. Second, the crossing sequence of a linear object, which is related to its silhouette, is considered. The crossing sequence can be categorized into two types: unravelable and not-unravelable. Third, a procedure to generate efficient unraveling processes based on unravelability of the crossing sequence is explained. An object with an unravelable crossing sequence can be unraveled by pulling its both endpoints. Finally, examples of unraveling process generation with our developed system are demonstrated.

2. Unknotting Process Generation

In this section, we briefly explain a method to generate possible processes for unknotting of a linear object, which is equivalent to its unraveling. First, the state of a linear object can be topologically represented using three properties after projecting its shape on a projection plane. The first property is the *crossing sequence*. It is determined by numbering a crossing met first with tracing along the projected curve from one endpoint to the other. The *i*-th crossing is represented as symbol C_i. One endpoint where tracing starts is referred to as the left endpoint E_l and that where tracing ends as the right endpoint E_r. The second property is the *location* of a pair of points at each crossing, that is, which point is upper/lower. The upper point of *i*-th crossing is described as symbol C_i^u and the lower point of that as symbol C_i^l. The third property is the *helix* of each crossing. Let us define a crossing where the upper part overlaps first on the right side of the lower part and then overlaps on its left side as a *left-handed helical crossing*. Conversely, in a *right-handed helical crossing*, the upper part first overlaps on the left side of the lower part and then overlaps on its right side. The symbols C_i^- and C_i^+ represent the *i*-th left- and right-handed helical crossing, respectively.

Next, we introduce basic operations described in Fig.1, corresponding to state transitions. Crossing operations CO_I, CO_{II}, and CO_{IV} increases the number of crossings, while uncrossing operations UO_I, UO_{II}, and UO_{IV} decrease the number. Arranging operation AO_{III} does not change the number of crossings but permutes their sequence. Each basic operation can be applied to specific subsequences of crossings. Let us investigate subsequences to which each operation is applicable. Operation UO_I is applicable to a subsequence represented as follows:

$$\cdots - C_i^{u/l} - C_i^{l/u} - \cdots . \tag{1}$$

That is, two crossing points corresponding to one crossing C_i, should be adjacent to each other in applying UO_I. Operation UO_{II} is applicable to subsequences described as follows:

$$\cdots - C_i^{u/l} - C_j^{u/l} - \cdots - C_i^{l/u} - C_j^{l/u} - \cdots , \tag{2}$$

$$\cdots - C_i^{u/l} - C_j^{u/l} - \cdots - C_j^{l/u} - C_i^{l/u} - \cdots . \tag{3}$$

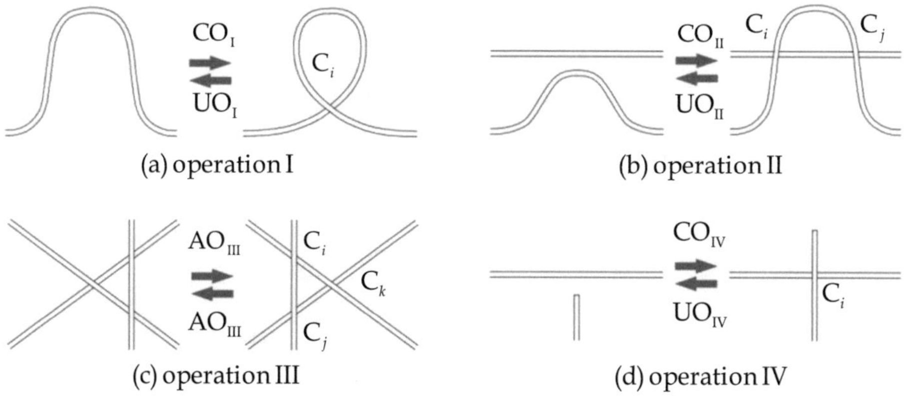

Figure 1. Basic operations

That is, two upper crossing points C_i^u and C_j^u, should be adjacent to each other and the corresponding lower crossing points C_i^l and C_j^l, should also be adjacent to each other. Operation UO$_{IV}$ is applicable to subsequences represented as follows:

$$E_l - C_i^{u/l} - \cdots - C_i^{l/u} - \cdots, \tag{4}$$

$$\cdots - C_i^{u/l} - \cdots - C_i^{l/u} - E_r. \tag{5}$$

That is, a crossing adjacent to an endpoint can be deleted by operation UO$_{IV}$. Operation AO$_{III}$ is applicable to a subsequence represented as permutation of the following three subsequences: α, β, and γ, *e.g.*, $\cdots - \beta - \gamma - \alpha - \cdots$:

$$\alpha: \ \cdots - C^u_{i/j} - C^u_{j/i} - \cdots, \tag{6}$$

$$\beta: \ \cdots - C^{l/u}_{j/k} - C^{u/l}_{k/j} - \cdots, \tag{7}$$

$$\gamma: \ \cdots - C^l_{i/k} - C^l_{i/k} - \cdots. \tag{8}$$

That is, three crossings consisting of three segments one of which overlaps the others can be permuted by operation AO$_{III}$. Uncrossing operations UO$_I$, UO$_{II}$, and UO$_{IV}$ and arranging operation AO$_{III}$ are applicable to their specific crossing subsequences indicated above. Once the initial and the objective crossing states of a linear object are given, we can generate possible sequences of crossing state transitions, that, is, possible processes of unknotting manipulation by repeating detection of applicable subsequences of individual operations and deletion/permutation of relevant crossings.

Fig.2 shows an example of automatic generation of possible unknotting processes by our developed system. Required manipulation corresponds to untying a slipknot. Assuming that only uncrossing operations can be used, *i.e.*, without operation AO$_{III}$, 14 crossing states and 39 state transitions are derived as shown in Fig.2. Including operation AO$_{III}$, we can derive 21 crossing states and 68 state transitions.

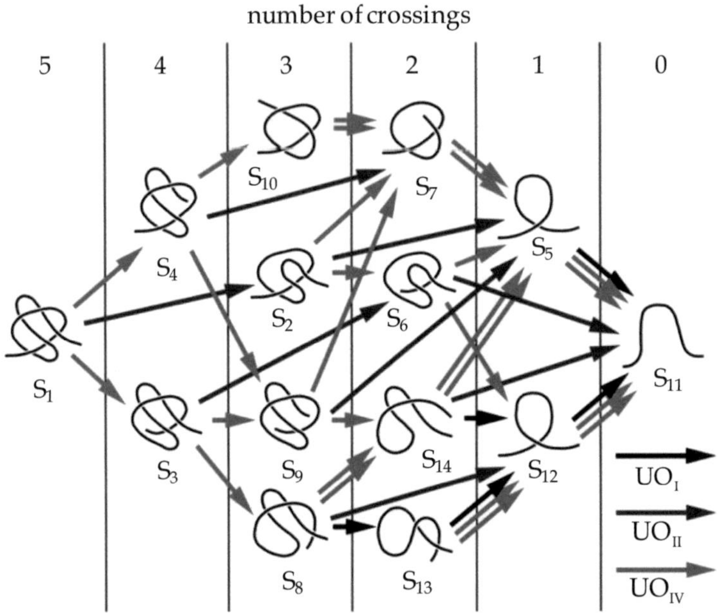

Figure 2. Example of state transition graph generation -untying slipknot-

3. Introduction of Unravelable Crossing Sequence

Once the current crossing state of a linear object is identified, we can unravel the object using the method proposed in the previous section. To identify the crossing state completely, the location at each crossing should be known. Morita recognized the state of a linear object with a 9-eye stereo camera (Morita et al., 2003) and Matsuno identified it utilizing variance of luminance at crossings (Matsuno et al., 2005). Now, let us assume that only silhouette, that is, 2D information about the object state is available. It means that the location and the helix of any crossing can not be identified. Then, we can perform operation UO_I even if the location at crossing C_i shown in Fig.1-(a) is unknown. Operation UO_{IV} can also be realized regardless of the location at crossing C_i shown in Fig.1-(d). Contrary, whether operations UO_{II} and AO_{III} can be applied depends on the location of crossings. Fig.3 shows examples of crossings with a subsequence to which operations UO_{II} and AO_{III} are applicable but with locations to which they can not be applied. Any knot can be unknotted by applying operations UO_{IV} alone (Wakamatsu et al., 2006). Note that the state transition graph shown in Fig.2 includes unknotting processes consisting of only UO_{IV} operations. This implies that a raveled linear object can be unraveled by applying operations UO_{IV} alone regardless of the location at each crossing. Recall that we often search for an endpoint and manipulate it to unravel a self-entwined rope. However, such manipulation may be not efficient when the object is raveled intricately, *i.e.*, it has many crossings. In this section, we propose a method for generating efficient unraveling processes of a linear object based on its crossing sequence, *i.e.*, its silhouette.

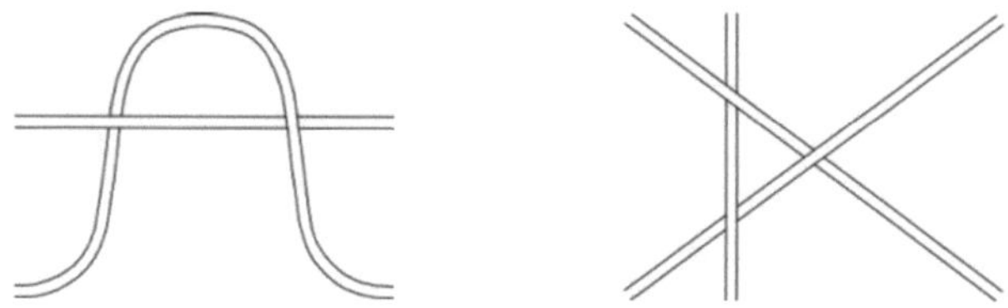

(a) with 2 crossings (b) with 3 crossings

Figure 3. Crossings not applicable uncrossing operations

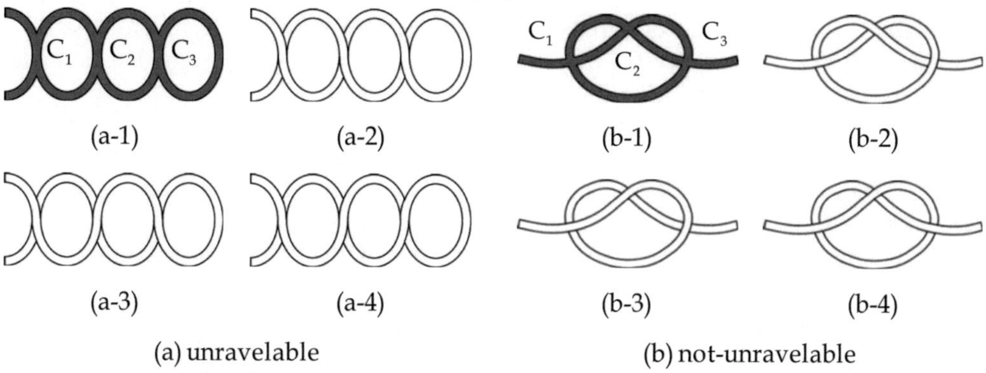

Figure 4. Crossing sequences

First, we define a knot in which some crossings remain even if all possible operations UO_I, UO_{II}, and AO_{III} are applied as a *tightenable knot*. For example, an overhand knot and a figure-of-eight knot are tightenable knots. Contrary, a knot which can be unknotted completely by applying operations UO_I, UO_{II}, and/or AO_{III} is defined as an *untightenable knot*. The untightenable knot is unknotted when its both endpoints are pulled away from each other. We can check whether a knot is tightenable or untightenable from its crossing state description (Wakamatsu et al., 2006).

Fig.4-(a-1) illustrates the silhouette of a knot with 3 crossings. Its crossing sequence is described as follows:

$$E_l - C_1 - C_2 - C_3 - C_3 - C_2 - C_1 - E_r . \tag{9}$$

Knots shown in Fig.4-(a-2) through (a-4) have the same crossing sequence. They include subsequence $\cdots - C_3 - C_3 - \cdots$ to which operation UO_I can be applied. When crossing C_3 is deleted, it is found that crossing C_2 can also be deleted by application of operation UO_I. After deletion of crossing C_2, we can delete crossing C_1 by applying operation UO_I once more. This means that knots shown in Fig.4-(a-2) through (a-4) are untightenable. Any knot with the crossing sequence described by eq.(9) can be unraveled by pulling its both endpoints regardless of the location at each crossing. In this paper, we define such crossing sequence as an *unravelable crossing sequence*. An untighenable knot has an unravelable crossing sequence.

A knot shown in Fig. 4-(b-1) also has 3 crossings, sequence of which is as follows:

$$E_l - C_1 - C_2 - C_3 - C_1 - C_2 - C_3 - E_r . \quad (10)$$

It is equivalent to that of knots shown in Fig. 4-(b-2) through (b-4). Knots in Fig. 4-(b-3) and (b-4) are both untightenable, but the knot in Fig. 4-(b-2) corresponds to an overhand knot, that is, it is tightenable. This implies that a tightenable knot with the crossing sequence described by eq.(10) exists. Consequently, such crossing sequence is not unravelable. Note that knots in Fig.4-(b-3) and (b-4) can be unraveled, but they can not be distinguished from the knot in Fig.4-(b-2) when 3D information, *i.e.*, the location at each crossing is not given. Thus, we can categorize the crossing sequence of a knot into two types: unravelable and not-unravelable. The former can be unraveled by pulling its both endpoints regardless of the location at each crossing, while the latter may be tightened according to the location when its both endpoints are pulled.
Fig.5 shows looped prime knots in knot theory. We can not reduce the number of crossings of these knots even if any operation corresponding to Reidemister move (Adams, 1994) is applied. They are closely related to tightenable knots. Let us discuss the relationship between looped prime knots and unravelable crossing sequences. If the looped prime knot with 3 crossings is cut as shown in Fig.5-(a), its crossing state is described as follows:

$$E_l - C_1^{l+} - C_2^{u+} - C_3^{l+} - C_1^{u+} - C_2^{l+} - C_3^{u+} - E_r . \quad (11)$$

If the crossing state of an unlooped linear object is described by eq.(11), it is equivalent to an overhand knot. If the object has 3 crossings but their sequence differs from eq.(11), it can be unknotted by applying operation UO_{I}, UO_{II}, and/or AO_{III}. Consequently, a linear object with 3 crossings can be unraveled by pulling both endpoints if and only if it does not have a not-unravelable crossing sequence: $E_l - C_1 - C_2 - C_3 - C_1 - C_2 - C_3 - E_r$.
Fig.5-(b) shows the looped prime knot with 4 crossings. Cutting the knot as shown in Fig.5-(b-1) and tracing it counterclockwise from one endpoint, the crossing sequence is described as follows:

$$E_l - C_1 - C_2 - C_3 - C_1 - C_4 - C_3 - C_2 - C_4 - E_r . \quad (12)$$

In the case of Fig.5-(b-2) and (b-3), the crossing sequence is described as follows:

$$E_l - C_1 - C_2 - C_3 - C_4 - C_2 - C_1 - C_4 - C_3 - E_r . \quad (13)$$

A figure-of-eight knot has this crossing sequence. A knot with the crossing sequence described by eq.(12) or (13) may be tightened. This implies that a linear object with 4 crossings is unraveled if it does not have the above two sequences.
There are two types of the looped prime knot with 5 crossings as shown in Fig.5-(c). One type illustrated in Fig.5-(c-1) has the crossing sequence as follows:

$$E_l - C_1 - C_2 - C_3 - C_4 - C_5 - C_1 - C_2 - C_3 - C_4 - C_5 - E_r . \quad (14)$$

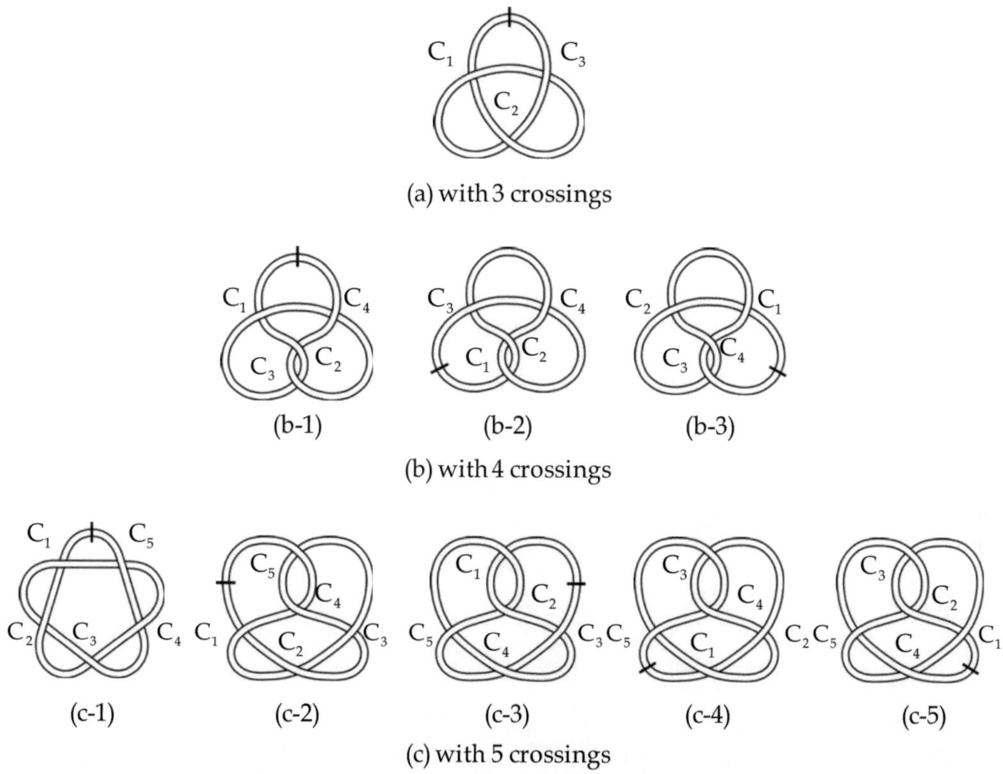

Figure 5. Looped prime knots

This sequence corresponds to that of a double overhand knot. The other type illustrated in Fig.5-(c-2) through (c-5) has the following crossing sequences:

$$E_l - C_1 - C_2 - C_3 - C_4 - C_5 - C_3 - C_2 - C_1 - C_4 - C_5 - E_r, \tag{15}$$

$$E_l - C_1 - C_2 - C_3 - C_4 - C_5 - C_1 - C_2 - C_5 - C_4 - C_3 - E_r, \tag{16}$$

$$E_l - C_1 - C_2 - C_3 - C_4 - C_2 - C_1 - C_5 - C_3 - C_4 - C_5 - E_r, \tag{17}$$

$$E_l - C_1 - C_2 - C_3 - C_1 - C_4 - C_5 - C_2 - C_3 - C_5 - C_4 - E_r. \tag{18}$$

Then, a linear object with 5 crossings but without the crossing sequence described by eqs.(14) through (18) is unravelable. Thus, we can derive not-unravelable crossing sequences from looped prime knots in knot theory. If the crossing sequence of a linear object with n crossings does not include not-unravelable sequences with 3 through n crossings, it can be unraveled by pulling its both endpoint instead of applying n UO_{IV} operations.

4. Procedure to Generate Efficient Unraveling Processes

In this section, we explain a procedure to generate unraveling processes. Let us assume that the silhouette of a linear object shown in Fig.6-(a-1) is given. Its crossing sequence is described as follows:

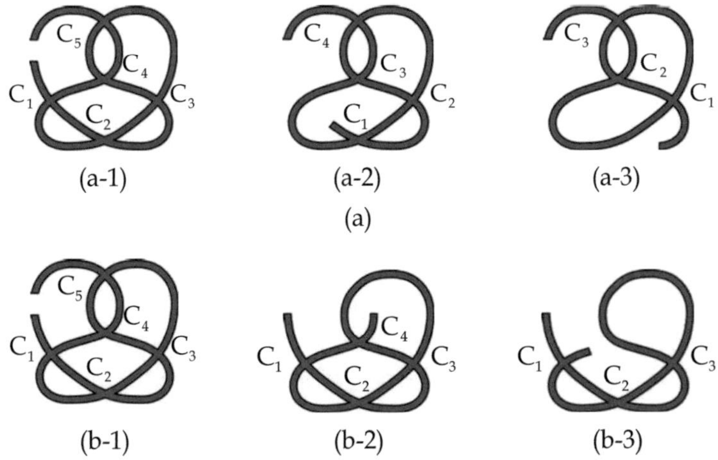

Figure 6. Unraveling processes

$$E_l - C_1 - C_2 - C_3 - C_4 - C_5 - C_3 - C_2 - C_1 - C_4 - C_5 - E_r . \tag{19}$$

The above sequence corresponds to a not-unravelable sequence with 5 crossings. It means that the object may be raveled and tightened if its both endpoints are pulled. Then, let us consider application of operation UO_{IV} so that the object does not include any not-unravelable sequence. If we apply operation UO_{IV} to the left terminal segment, the object state changes into the state shown in Fig.6-(a-2). Its crossing sequence is described as follows:

$$E_l - C_1 - C_2 - C_3 - C_4 - C_2 - C_1 - C_3 - C_4 - E_r . \tag{20}$$

A set of closed regions surrounded by a linear object is defined as the *inner region*, and the other region in the projection plane as the *outer region*. Moreover, segments touch the outer region are referred to as *outer segments*, and segments do not touch as *inner segments* (Wakamatsu et al., 2006). In Fig.6-(a-2), the left terminal segment is an inner segment. When one of terminal segments is inner, we can not pull both endpoints sufficiently without changing the crossing sequence. So, we apply another operation UO_{IV} to the left terminal segment. Then, the following sequence is derived:

$$E_l - C_1 - C_2 - C_3 - C_1 - C_2 - C_3 - E_r . \tag{21}$$

The above sequence is equivalent to the not-unravelable sequence with 3 crossings. This implies that additional UO_{IV} operations are required to unravel the object. Contrary, if we apply 2 consecutive UO_{IV} operations to the right terminal segment as shown in Fig.6-(b), the crossing sequence becomes as follows:

$$E_l - C_1 - C_2 - C_3 - C_3 - C_2 - C_1 - E_r . \tag{22}$$

As this sequence differs from the not-unravelable sequence, the knot shown in Fig.6-(b-3), which is equivalent to that in Fig.4-(a-1), can be unraveled by pulling its both endpoints. Consequently, we can conclude that unraveling process shown in Fig.6-(b) is more efficient than that shown in Fig.6-(a). Thus, we can generate efficient unraveling processes of a linear object based on only its crossing sequence, *i.e.*, its silhouette. This indicates that we may unravel a linear object without a stereo camera.

Not-unravelable sequences can be extracted from the list of looped prime knots in knot theory (Rolfsen, 1976). Let us define the following subsequence as a *not-unravelable subsequence* with 3 crossings:

$$\cdots - C_i - \cdots - C_j - \cdots - C_k - \cdots - C_i - \cdots - C_j - \cdots - C_k - \cdots \quad (i < j < k). \tag{23}$$

If the crossing state includes the above subsequence, the object has a part which may be tightened. The not-unravelable sequence described by eq.(10) is a kind of this subsequence. We can also define not-unravelable subsequences with n crossings referring to not-unravelable sequences. When such not-unravelable subsequence is detected from the crossing sequence, we delete a crossing included in the subsequence and nearest to one endpoint by applying operations UO_{IV} repeatedly. Let C_{li} and C_{rj} be i-th and j-th crossing met first when we trace an object from the left and the right endpoint, respectively. When the object has n crossings, we assume that only operation UO_{IV} is applied to delete k (k=1, ⋯, n) crossings. Then, we check the number of remaining not-unravelable subsequences after deleting crossings C_{li} (i=k, k-1, ⋯, 1, 0) and C_{rj} (j=k-i). If the crossing sequence does not include any not-unravelable subsequence by n-3 crossings are deleted, the rest can be uncrossed by applying one pulling operation instead of some UO_{IV} operations. This implies that the object can be unraveled efficiently. For example, the crossing sequence described by eq.(19) is equivalent to the not-unravelable sequence with 5 crossings and includes three not-unravelable subsequences with 3 crossings:

$$\cdots - C_1 - \cdots - C_4 - C_5 - \cdots - C_1 - C_4 - C_5 - \cdots , \tag{24}$$

$$\cdots - C_2 - \cdots - C_4 - C_5 - \cdots - C_2 - \cdots - C_4 - C_5 - \cdots , \tag{25}$$

$$\cdots - C_3 - C_4 - C_5 - C_3 - \cdots - C_4 - C_5 - \cdots . \tag{26}$$

In this case, crossing $C_{l1} = C_1$ or $C_{r1} = C_5$ can be deleted by operation UO_{IV}. If crossing C_5 is uncrossed, all these subsequences are deleted. Then, the object becomes unravelable. Contrary, subsequences described by eqs.(25) and (26) remain even if crossing C_1 is deleted. Consequently, we select application of operation UO_{IV} to crossing C_5 as the first process for unraveling. After that, the object is completely unraveled by pulling its both endpoints. Thus, efficient unraveling processes of a linear object can be derived even if only its crossing sequence is identified.

5. Case Study

In this section, we discuss the effectiveness of our proposed method for efficient unraveling with some examples. Fig.7 shows two examples of a raveled object. They correspond to not-unravelable sequences with 8 crossings. The crossing sequence in case-1 shown in Fig.7-(a) is described as follows:

$$E_l - C_1 - C_2 - C_3 - C_4 - C_5 - C_6 - C_2 - C_1 - C_7 - C_8 - C_6 - C_5 - C_4 - C_3 - C_8 - C_7 - E_r \,. \quad (27)$$

In case-2, the crossing sequence shown in Fig.7-(b) is represented as follows:

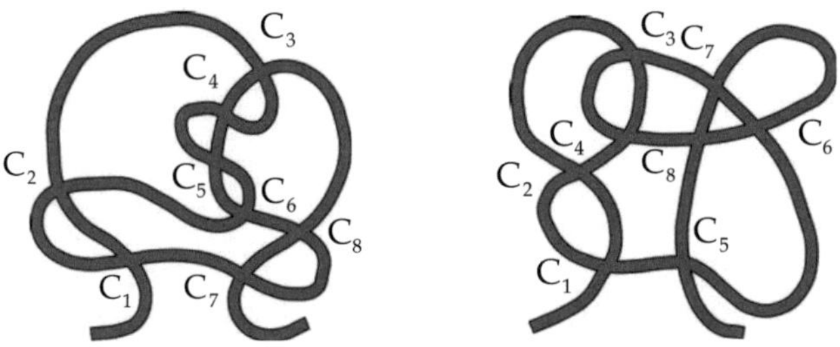

Figure 7. Silhouette of raveled objects

$$E_l - C_1 - C_2 - C_3 - C_4 - C_2 - C_1 - C_5 - C_6 - C_7 - C_3 - C_4 - C_8 - C_6 - C_7 - C_8 - C_5 - E_r \,. \quad (28)$$

We developed a system to detect not-unravelable subsequences from a given crossing sequence. Using this system, it was found that the crossing sequence in case-1 include 49 not-unravelable subsequences and that in case-2 includes 19. Table 1 and 2 show the number of remaining not-unravelable subsequences after deleting crossings C_{li} and C_{rj} in case-1 and case-2, respectively. As shown in Table 1, when 3 crossings C_1, C_2, and C_7 or C_1, C_7, and C_8 are deleted, the crossing sequence in case-1 does not include any not-unravelable subsequences. So, we can delete the rest crossings, *i.e.*, we can unravel the object at once pulling away its both endpoints. This indicates that we can perform unraveling with less operations than unraveling in which all 8 crossings are deleted by operation UO_{IV}. In case-2, 5 crossings C_4, C_5, C_6, C_7, and C_8 must be deleted to exclude not-unravelable subsequences from the crossing sequence as shown in Table 2. After that, we can unravel the object with one pulling operation. This is a more efficient unraveling process than that consisting of 8 UO_{IV} operations. But, we have to delete more crossings in case-2 to change the crossing sequence into the unravelable one than in case-1. Thus, it is found that the effectiveness of our proposed method depends on the crossing sequences. Efficient unraveling processes are derived from some crossing sequences but they are not from others. However, we can determine whether 3D information is needed for its efficient unraveling from its crossing sequence, *i.e.*, its silhouette. For more efficient unraveling in case-2, for example, we have to identify the location of some crossings. If crossings C_3 and C_4 can be deleted by applying operation UO_{II}, the object can be separated into two parts. Moreover, the left part is unravelable and it can be unraveled independently of the right part. Consequently, we should indentify the location of these crossings first. If their location satisfies eq.(2) or (3), the object can be separated actually. Thus, we can rank crossings the location of which should be identified in order of importance for efficient unraveling. 3D information of not all crossings have to be identified when the unravelability, which is determined from the silhouette of a linear object, is considered.

deleted crossings	remaining not-unravelable subsequences
none	49
C_1	21
C_7	21
C_1, C_2	7
C_1, C_7	7
C_7, C_8	7
C_1, C_2, C_3	3
C_1, C_2, C_7	0
C_1, C_7, C_8	0
C_3, C_7, C_8	3

Table 1. Unraveling process for case-1

deleted crossings	remaining not-unravelable subsequences
none	19
C_1	11
C_5	11
C_1, C_2	7
C_1, C_5	7
C_5, C_8	8
C_1, C_2, C_3	2
C_1, C_2, C_5	5
C_1, C_5, C_8	5
C_5, C_7, C_8	4
C_1, C_2, C_3, C_4	1
C_1, C_2, C_3, C_5	2
C_1, C_2, C_5, C_8	4
C_1, C_5, C_7, C_8	2
C_5, C_6, C_7, C_8	2
C_1, C_2, C_3, C_4, C_5	1
C_1, C_2, C_3, C_5, C_8	1
C_1, C_2, C_5, C_7, C_8	1
C_1, C_5, C_6, C_7, C_8	1
C_4, C_5, C_6, C_7, C_8	0

Table 2. Unraveling process for case-2

6. Conclusions

A planning method for unraveling deformable linear objects based on their silhouette was proposed. First, an unknotting process of a linear object, which is equivalent to its unraveling process, was represented as a sequence of crossing state transitions. It can be generated on a computer if 3D information about the current crossing state is given. Second, the crossing sequence of a linear object, which corresponds to its 2D information, was categorized into two types: unravelable and not-unravelable. Third, a procedure to generate efficient unraveling processes based on unravelability of the crossing sequence was explained. An object with an unravelable crossing sequence can be unraveled by pulling its both endpoints. Finally, examples of unraveling process generation with our developed system were demonstrated. The crossing sequence is not sufficient information for deriving efficient unraveling processes, but it is useful for that.

7. References

Adams, C. C. (1994). The Knot Book: An Elementary Introduction to the Mathematical Theory of Knots, Henry Holt & Co.

Hopcroft, J. E., Kearney, J. K., and Krafft, D. B. (1991). A Case Study of Flexible Object Manipulation. *Int. J. Robotics Research*, Vol.10, No.1, pp.41-50.

Inoue, H. and Inaba, M. (1984). Hand-eye Coordination in Rope Handling, *Robotics Research: The First International Symposium*, MIT Press, pp.163-174.

Ladd, A. M. and Kavraki, L. E. (2004). Using Motion Planning for Knot Untangling. *Int. J. Robotics Research*, Vol.23, No.7-8, pp.797-808.

Matsuno, T., Fukuda, T., and Arai, F. (2001). Flexible Rope Manipulation by Dual Manipulator System Using Vision Sensor, *Proc. of Int. Conf. Advanced Intelligent Mechatronics*, pp.677-682.

Matsuno, T., Tamaki, D., Arai, F., and Fukuda, T. (2005). Rope Structure Recognition for Manipulation Using Topological Model and Knot Invariant, *J. SICE*, Vol.41, No.4, pp.366-372, (in Japanese).

Morita, T., Takamatsu, J., Ogawara, K., Kimura, H., and Ikeuchi, K. (2003). Knot Planning from Observation, *Proc. of IEEE Int. Conf. Robotics and Automation*, pp.3887--3892.

Rolfsen, D. (1976). Knots and Links, Publish or Perish, Inc.

Saha, M. and Isto, P. (2006). Manipulation Planning for Deformable Linear Objects, *J. IEEE Transactions on Robotics*, Vol.23 No.6, pp.1141-1150.

Takamatsu, J., Morita, T., Ogawara, K., Kimura, H., and Ikeuchi, K. (2006). Representation for Knot-Tying Task, *J. IEEE Transactions on Robotics*, Vol.22 No.1, pp.65-78.

Wakamatsu, H., Arai, E., and Hirai, S. (2006). Knotting/Unknotting Manipulation of Deformable Linear Objects, *Int. J. Robotics Research*, pp.371-395.

Yamakawa, Y., Namiki, A., Ishikawa, M. , and Shimojo, M. (2007). One-handed Knotting of a Flexible Rope with a High-speed Multifingered Hand having Tactile Sensors, *Proc. of IEEE/RSJ Int. Conf. Intelligent Robots and Systems*, pp.703-708.

29

Smoothing of Piecewise Linear Paths

Michel Waringo and Dominik Henrich
Lehrstuhl Angewandte Informatik III (Robotik und Eingebettete Systeme),
Universität Bayreuth
Germany

1. Introduction

The pointwise traversal of a given path is a popular task in the area of robotics, e.g. in mobile, industrial or surgical robotics. The easiest method to describe paths is by a sequence of linear segments, and for many tasks the precision of a path approximated by linear segments is sufficient. The movements to be accomplished by a mobile robot or the robot's end-effector are described by a sequence of points in space, that have to be traversed by the robot. However, in practice, a pre-computed path unfortunately often consists of more path points than are necessary for sufficiently accurate execution. An excessive number of path points renders the movement jerky if the path points are dispersed around the optimal path, leading to unnecessary mechanical stress of both robot and tool. A second problem is that many path points lying close to one another can lead to high computational cost when traversing the path and can reduce traversal speed.

Paths described by teach-in methods are one example where the path can consist of too many path points. In these methods, the desired movement is recorded while the operator moves the robot's arm, either directly, through a master device or by giving instructions through a control panel. Because of the rather intuitive input of the human operator, the path suffers from deficiencies and frequent unnecessary changes of direction.

The taught-in path can be traversed better after smoothing the path. Another example is voxel-based path planning. Here, only space points with discrete coordinates can be traversed, which may lead to a stair shaped approximation of diagonal paths. Just as with smoothed taught-in paths, smoothed voxel-based paths can be traversed more efficiently, because there are fewer changes of speed and direction, and the total path length is reduced.

The remainder of the text first provides a problem description (Section 2), after which the state of the art is presented (Section 3). Then, the proposed procedure is described in detail (Section 4), and different specializations of the proposed method are shown for points with fixed orientation (Section 5) as well as with variable orientation (Section 6). Finally, experiments are described (Section 7), and open issues and further enhancements are discussed (Section 8).

2. Problem description

The problem of path smoothing can be described as follows. A path $\mathbf{P} := < \mathbf{p}_1, \mathbf{p}_2, \dots, \mathbf{p}_n >$ of n points is given, represented by an ordered list of m-dimensional Cartesian path points

$\mathbf{p}_i := (p_{i1}, p_{i2}, \ldots p_{im})^T$. All path points $\mathbf{p}_i$ have the same dimensionality $m := m_p + m_o$, depending on the degrees of freedom of the robot or the requirements of the task to be performed. The list is sorted in the natural point order, assigning the index 1 to the path's start point and index n to its end point. The parameters $m_p, m_o \in \mathbb{N}$, $m_p, m_o \geq 0$ designate the dimension of the two coordinate types position and orientation of a path point.

The *neighbourhood* of a path point $\mathbf{p}_i$ is defined as the sequence of points in $\mathbf{P}$ between and including the nearest neighbours of that point in the path $\mathbf{P}$ to the left and and right of $\mathbf{p}_i$.

The *deviation* d_i between a smoothed path $\mathbf{P}'$ and the original path $\mathbf{P}$ in the neighbourhood of the path point $\mathbf{p}_i$ can be computed according to various error functions, such as the standard deviation or the area spanning both paths. The deviation must be computed differently depending on which coordinate type, position or orientation, is considered.

The *error function* K represents the criterion used to compute d_i. Its input are the two paths $\mathbf{P}$ and $\mathbf{P}'$ as well as the index i, and its output is the deviation d_i between them. Finally, we need a threshold value d_{lim} indicating the maximum allowed d_i.

If the path points consist of both coordinate types, either position or orientation may, but need not, be used as a constraint in addition to the optimization criterion which is mandatory. If not only an optimization criterion but also a constraint is being used, a second threshold value c_{lim} is needed. In that case, we compute a second deviation c_i for each path point which may not exceed c_{lim}.

Thus, we search for a path $\mathbf{P}'$ whose deviation d_i from $\mathbf{P}$ does not exceed d_{lim} at each individual path point according to K. The number of path points of the path $\mathbf{P}'$ is minimized under the given optimization criteria and optionally a constraint (Figure 1).

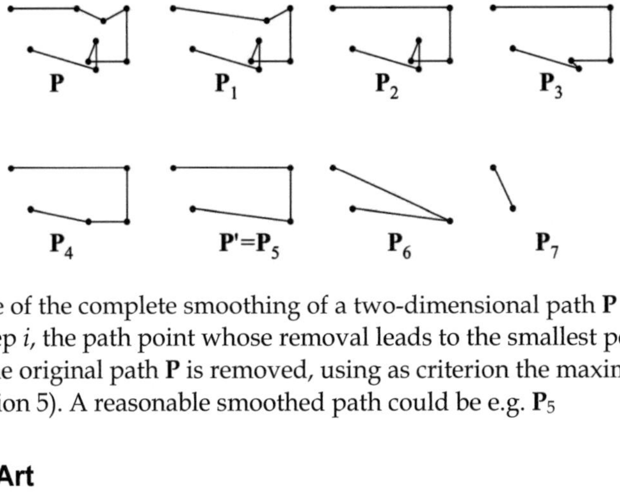

Figure 1. Example of the complete smoothing of a two-dimensional path $\mathbf{P}$ with nine path points. In each step i, the path point whose removal leads to the smallest possible deviation between $\mathbf{P}_i$ and the original path $\mathbf{P}$ is removed, using as criterion the maximum Euclidean distance (see Section 5). A reasonable smoothed path could be e.g. $\mathbf{P}_5$

3. State of the Art

We can distinguish two main categories of problems where a reduced number of path points is required: path planning and path shortening.

In collision-free motion planning, e.g. for mobile robots, the start and the end points are given, and a path connecting them is sought. There may be obstacles and narrow passages like doors or corridors. A good path avoids all obstacles and is short. In a first step, e.g. using a stochastic approach, in a *path planning procedure* (Subsection 3.1), a path of possibly poor quality is created, containing many superfluous segments and being much longer than necessary. It is improved in a second step by a path shortening procedure. There is no need

for any similarity between the original and the collision-free shortened path apart from having the same start and end point.
In other approaches, the improved path must remain similar to the original path. Not only the start and end point, but also the path between them is given. However, the quality of this path may be unsatisfying, the path being jerky or consisting of too many segments. In this *path smoothing* procedure (Subsection 3.2), small deviations from the original path are allowed as long as the number of path segments can be reduced noticably and both paths stay close enough. Figure 2 shows the difference between path shortening and path smoothing.

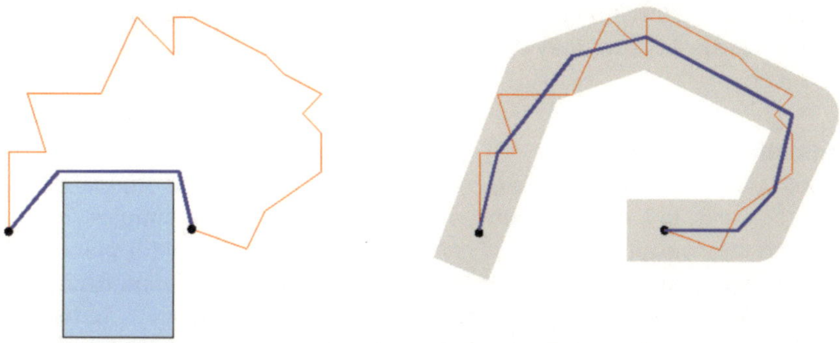

Figure 2. Left: path shortening procedure where obstacles must be avoided, right: path smoothing procedure where a path must stay within a given vicinity of the original path

3.1. Path planning procedures

In collision-free motion planning, planning is usually performed in the configuration space. The problem of finding a path between a start and an end point is PSPACE-hard in the degrees of freedom and in the number of obstacle surface patches. Therefore, most algorithms in this category are stochastic. Two main classes can be distinguished. Probabilistic roadmap (PRM) approaches (Amato, N.M. & Wu, Y., 1996), (Geraerts, R. & Overmars, M. H., 2002) proceed in two steps. First, a collision-free path is constructed as a graph in robot configuration space. In a second step, pairs of promising vertices are chosen and a simple local planner is used to find a better collision-free connection between them. Approaches based on Rapidly-exploring Random Trees (RRTs) (Kuffner, J. J.; LaValle S. M., 2001), (LaValle, S. M., 1998) use a collision-free path tree that is grown incrementally. In each iteration, a random configuration is chosen, and an attempt is made to find a path to it from the nearest RRT vertex.
Other path planning algorithms exist which do not belong to one of these two categories. One example are potential field based methods which can be used for path planning if there are only few obstacles. Another example is the Randomized Path Planner (RPP) (Barraquand, J. & Latombe, J.-C., 1991) where the path is planned according to potential fields and random walks are used to escape from local minima. Another group of path planning algorithms is based on the elastic-band method (Quinlan, S.; Khatib, Q., 1993), where contractive and repulsive forces emanating from obstacles determine the deformation of an original path.

If path segments are to remain piecewise linear, other strategies can be used. Path modifications may be performed by shifting or splitting path segments. Furthermore, path segments can be merged by removing their point of connection (Baginski, B., 1998). A similar procedure is used in (Berchtold, S. & Glavina, B., 1994), where path points are removed based on a heuristic to locally reduce the path length. Another example of collision-free paths for robots can be found in (Urbanczik, C., 2003). Here, path points can be shifted or removed, and segments can be split. After each step the list is re-sorted by pairs of neighbouring segments. Path shortening can be performed using a divide-and-conquer algorithm which removes all path points between the first and the last point in one recursion step if the direct path between them is allowed, and otherwise bisects the path points list (Carpin, S.; Pillonetto G., 2006). However, these approaches are not valid for the application we envisage because they do not guarantee a similarity between the original and the smoothed path.

3.2. Path smoothing procedures

In applications where the shortened path must remain similar to the original path, similar strategies can be used, but optimization criteria are different. The simplest form of path smoothing is the removal of superfluous collinear path points, i.e. path points lying on a straight line between their two immediate neighbours. Here, no deviation from the original path arises, but only a few path points can be removed in general. A reduction in the number of path points can also be achieved by approximating the path by curves of a higher degree consisting of nonlinear path segments (e.g., defined by quadratic or cubic functions) (Hein, B., 2003) or non-uniform rational B-Splines (NURBS) (Aleotti, J.; Caselli, S., 2005).

In (Engel, D., 2003) a smoothing procedure for piecewise linear paths is described that removes path points $\mathbf{p}_j$ not exceeding a given deviation from a path segment $<\mathbf{p}_i, \mathbf{p}_k>$ with $i < j < k$. A disadvantage is that the path point list is treated only once and thus some smoothing steps are not executed which are only possible when the path was already smoothed in a previous step.

A well-known point reduction method is linear regression (Bronstein, I. N.; Semendjajew, K. A; Musiol, G.; Mühlig, H., 2001), but it does not guarantee an upper limit for the deviation. Here, a path defined by scattered points is replaced by a path consisting of one straight-line segment placed as close as possible to the scattered points.

3.3. Conclusions

Although a smoothed path slightly deviates from the original path, it can be better suited for a specific application as long as the deviations are not too big.

A downside of the discussed methods is that they are not able to handle paths with both positions and orientations. They are either restricted to one coordinate type (usually positions) or they work in the configuration space. In that space, there is no differentiation into two coordinate types either. Although algorithms working in the configuration space can smooth paths having position and orientation coordinate components, they need a robot model and a forward kinematics.

The path smoothing method we propose offers some advantages which make it particularly suited for time-critical applications working either in configuration space or work space. Due to the order in which the path points are removed, our method has anytime ability, i.e. it can be aborted prematurely and still returns a valid smoothed path, with the result quality increasing monotonically over time. The optimization criterion can be easily exchanged (depending on the application), and an upper bound for the deviation between the original path and the smoothed one can be guaranteed. Furthermore, the algorithm is efficient, as both the computation time and storage space are linear in the number of path points. The algorithm is very versatile due to its capability to handle points with both coordinate types.

4. Path smoothing procedure

The path points $\mathbf{p}_i$ are stored in a list ordered by index. Their description (i.e., the Cartesian coordinates) is extended by three components:

- The flag $r_i \in \{true, false\}$ indicates whether the path point has been removed while smoothing.
- The variable $d_i \in \mathbb{R}$ indicates the deviation of the path in the neighbourhood of the path point, which will be explained in detail in Section 5. This variable holds the optimization value used to decide which point has to be removed next so that the path deviation stays as small as possible.
- The variable $c_i \in \mathbb{R}$ stores the deviation of the path in the neighbourhood of the path point according to a second coordinate type. This variable holds the constraint value. The use of c_i is detailed in Section 6. If the path has only one coordinate type, c_i is not used [1].

The path points removed during path smoothing are not deleted from the list, but are instead only marked as removed, since the procedure must be able to access the original path at any time. When smoothing is complete, all path points not marked as removed are copied into a target path point list containing only the path points of the smoothed path.

The algorithm for removing path points works as follows:

(1) For all path points $\mathbf{p}_i$, set $r_i = false$ and $c_i = 0$.

(2) For all path points $\mathbf{p}_k$ not yet removed ($r_k = false$), compute the arising deviation d_k between the smoothed path $\mathbf{P}'$ and the original path $\mathbf{P}$, assuming that $\mathbf{p}_k$ is removed from the path (in addition to the path points removed so far). If a constraint is being used, compute c_k.

(3) Select among all path points with $c_k < c_{\lim}$ the path point p_k with the smallest d_k.

  (3a) If the deviation d_k is smaller than the specified value $d_{\lim}$, then mark $\mathbf{p}_k$ as removed ($r_k = true$) and go to (2).

  (3b) Otherwise, no further path points can be removed from the path, and the path $\mathbf{P}' := < \mathbf{p}_i \mid r_i = false, i = 1, \ldots, n >$ is returned.

If no constraints are being used, no computations of c_i are performed and c_i stays zero, being without any effect.

The path point whose elimination leads to the smallest deviation from the original path while not violating the constraints is removed during each iteration. In this way, it is ensured that a (locally) maximum number of path points can be removed before the deviation locally exceeds the threshold $d_{\lim}$ and the algorithm terminates.

---

[1] This is realised by initially setting $c_i = 0$ and not modifying it any more and setting $c_{\lim}$ to an arbitrary value >0.

In order to prevent gradual drifting of the path in each iteration, the current path must not be compared with the path from the previous iteration step, but with the original path.
A certain computational speed improvement can be obtained by using an efficient implementation. Given a path with n path points, the maximum smoothing of the path would require $n - 2$ iterations, as the first and last path point are not removed. For a given iteration step j, the number of local deviation computations is $n - 2 - j$. This belongs to the complexity class O (n^2), since the total number of computation steps is

$$\sum_{i=1}^{n-2}(n-2-i) = \left(\sum_{i=1}^{n-2} n\right) - \left(\sum_{i=1}^{n-2} 2\right) - \left(\sum_{i=1}^{n-2} i\right)$$

$$= n\cdot(n-2) - 2\cdot(n-2) - \frac{n-2}{2}(n-1) = \frac{1}{2}n^2 - \frac{5}{2}n + 3 \tag{1}$$

The procedure can be accelerated by considering that the path deviation only changes in the proximity of a path point that is removed. Thus only in the first iteration step, the deviation needs to be computed for all path points, and in the further steps only for the path points in the neighbourhood of the last removed path point.
For this purpose, the component d_i of all path points $\mathbf{p}_i$ is required for buffering the corresponding deviation. After a path point has been removed, the deviation can change only for the two neighbouring path points without considering all path points already removed. Therefore, only two instead of $n - 2 - i$ deviations need to be determined per iteration step. This results in a complexity of

$$n + \sum_{j=2}^{n-2} 2 = n + 2(n-3) = 3n - 6 \tag{2}$$

computation steps, with complexity dropping from O (n^2) to O (n). The same holds true for the computation of the constraint c_i, which, if used, is computed whenever d_i is computed.
In the following, we describe how the interval of path points needed for the computation of the deviation d_i is determined. We are looking for two path points $\mathbf{p}_{min}$ and $\mathbf{p}_{max}$ that border on the interval in question: $I := [\mathbf{p}_{min} ; \mathbf{p}_{max}] = < \mathbf{p}_{min}, \dots , \mathbf{p}_i, \dots , \mathbf{p}_{max} >$.
In the first iteration no path points have been removed yet. Trivially, only three path points need to be regarded: the path point $\mathbf{p}_i$ as well as its neighbours p_{i-1} and $\mathbf{p}_{i+1}$, and the path segment $< \mathbf{p}_{i-1}, \mathbf{p}_i, \mathbf{p}_{i+1} >$ must be compared with $< \mathbf{p}_{i-1}, \mathbf{p}_{i+1} >$ in order to compute d_i. The manner in which this comparison is performed depends on the error function used and is described in Section 5.
In the subsequent iterations we must consider which path points are removed and must extend the path interval of interest beyond the previously removed path points so that its borders again consist of the next two non-removed path points $\mathbf{p}_{min}$ and $\mathbf{p}_{max}$. Thus $\mathbf{p}_{min}$ and $\mathbf{p}_{max}$ are the direct neighbours of $\mathbf{p}_i$ that have not yet been removed. The deviation d_i is computed by comparing a path segment of the original path $\mathbf{P}_{i,o} = [\mathbf{p}_{min}, \mathbf{p}_{max}] = < \mathbf{p}_{min}, \dots , \mathbf{p}_i , \dots , \mathbf{p}_{max} >$ and the corresponding path segment on the smoothed path $\mathbf{P}_{i,s} = < \mathbf{p}_{min}, \mathbf{p}_{max} >$.
The following table shows exemplarily the deviation computations necessary for smoothing a path with six path points $\mathbf{P} = < \mathbf{p}_1, \mathbf{p}_2, \mathbf{p}_3, \mathbf{p}_4, \mathbf{p}_5, \mathbf{p}_6 >$ in the order $\mathbf{p}_4$, $\mathbf{p}_3$, $\mathbf{p}_2$ and $\mathbf{p}_5$.

	$\mathbf{p}_1$	$\mathbf{p}_2$	$\mathbf{p}_3$	$\mathbf{p}_4$	$\mathbf{p}_5$	$\mathbf{p}_6$
compute	d_1	d_2	d_3	d_4	d_5	d_6
remove point				r_4=true		
recompute			d_3		d_5	
remove point			r_3=true			
recompute		d_2			d_5	
remove point		r_2=true				
recompute	d_1				d_5	
remove point					r_5=true	
recompute	d_1					d_6

Table 1. Steps performed during the smoothing of a path. In each iteration other than the first one only two deviations need to be determined

Prior to each iteration step, the deviation d_i is known for all remaining path points $\mathbf{p}_i$ not yet removed (thus all path points with r_i = false). Based on this information, the path point whose removal leads to the least deviation from the original path can reliably be removed.
Fig. 3 and Fig. 4 illustrate the two steps marked in gray from Table 1 based on a two-dimensional geometry. For example, the area between a path segment of the smoothed path and the appropriate original path is defined as the error function K. Using this K, the deviations d_i are areas, which are shown in gray.
The path point $\mathbf{p}_4$ is already marked as removed (represented by a white dot in Figure 3 (a)). $\mathbf{p}_3$ is now removed additionally, resulting in the modification of the smoothed path that can be seen in Figure 3 (b) and (c).

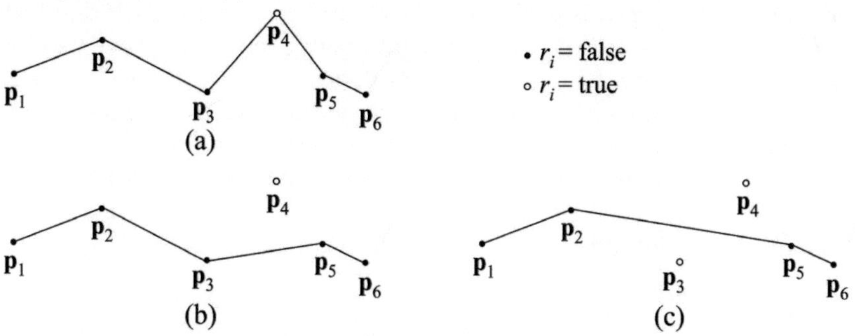

Figure 3. Example for the removal of a path point. Black dots represent still existing path points and white dots represent removed path points. (a) shows the entire path and (b) and (c) the smoothed path before and after removal of $\mathbf{p}_3$, respectively

With the removal of $\mathbf{p}_3$, the deviations d_2 and d_5 occurring upon removal of $\mathbf{p}_2$ and $\mathbf{p}_5$ also change and d_2 and d_5 must therefore be updated. Figure 4 (a) and (b) clarify why the deviation must be recomputed for the path point $\mathbf{p}_2$. Before the removal of $\mathbf{p}_3$, the removal of $\mathbf{p}_2$ only affected the path segment $< \mathbf{p}_1, \mathbf{p}_3 >$. Now, it affects the path segment $< \mathbf{p}_1, \mathbf{p}_5 >$. In the smoothed path, $\mathbf{p}_2$ now has $\mathbf{p}_1$ and $\mathbf{p}_5$ as direct neighbors rather than $\mathbf{p}_1$ and $\mathbf{p}_3$, thus its deviation has changed.
Similarly, in Figure 4 (c) und (d), the deviation for the path point $\mathbf{p}_5$ determined in an earlier iteration is now invalid and must be recomputed. No new calculations need to be performed

for the other path points, since with linear interpolation the removal of $\mathbf{p}_3$ only affects the segments that were direct neighbors of that path point before it was removed.

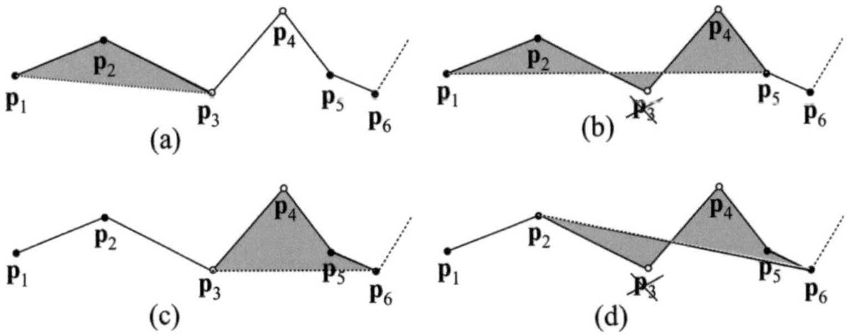

Figure 4. Example of the computational cost reduction. Black dots represent still existing path points and white dots represent removed path points. The deviations are shown as gray areas. In the upper two figures, the deviation d_2 occurring if $\mathbf{p}_2$ is removed is shown, both before removal of $\mathbf{p}_3$ (a) and after its elimination (b). (c) and (e) similarly show the deviation d_5 before and after removal of $\mathbf{p}_3$

In the following we describe how the interval of path points is determined that is needed for the computation of the deviation d_i. We are looking for two path points $\mathbf{p}_{\min}$ and $\mathbf{p}_{\max}$ that border the interval in question: $I = [\, \mathbf{p}_{\min} ; \mathbf{p}_{\max} \,] = \langle \mathbf{p}_{\min}, \ldots, \mathbf{p}_i, \ldots, \mathbf{p}_{\max} \rangle$.

In the first iteration no path points have been removed yet. Trivially, only three path points need to be regarded: the path point $\mathbf{p}_i$ as well as its neighbors $\mathbf{p}_{i-1}$ and $\mathbf{p}_{i+1}$, and the path segment $\langle \mathbf{p}_{i-1}, \mathbf{p}_i, \mathbf{p}_{i+1} \rangle$ must be compared with $\langle \mathbf{p}_{i-1}\ \mathbf{p}_{i+1} \rangle$. The manner in which this comparison is performed depends on the error function used and is described in Section 5.

In the subsequent iterations we must consider which path points are removed and we must, as shown in Figure 4, extend the path interval of interest beyond the previously removed path points so that its borders again consist of two non-removed path points. Let i be the index of the path point for which the deviation of the deletion is to be computed and $\mathbf{p}_i$ be the corresponding path point. Let n be the number of path points in the original path $\langle \mathbf{p}_1, \ldots, \mathbf{p}_i, \ldots, \mathbf{p}_n \rangle$.

We obtain the following algorithm:

```
min := i – 1
while min > 1 and r_min = true
    min := min – 1

max := i + 1
while max < n and r_max = true
    max := max + 1
```

Thus $\mathbf{p}_{\min}$ und $\mathbf{p}_{\max}$ are the direct neighbors of $\mathbf{p}_i$ that have not yet been removed. The deviation d_i is computed by comparing a path segment of the original path $\mathbf{P}_{i,o} = [\, \mathbf{p}_{\min}, \mathbf{p}_{\max} \,] = \langle \mathbf{p}_{\min}, \ldots, \mathbf{p}_i, \ldots, \mathbf{p}_{\max} \rangle$ and the corresponding path segment on the smoothed path $\mathbf{P}_{i,s} = \langle \mathbf{p}_{\min}, \mathbf{p}_{\max} \rangle$.

5. Path deviation functions for a fixed orientation

In this section, we consider only paths with positional coordinates and no orientation and we do not use any constraints (Waringo, M.; Henrich D., 2006). Depending on the application, different error functions K can be used. We investigated three error functions:

- K_1: d_i is the maximum of the Euclidean distances of all path points of the interval $\mathbf{P}_{i,o}$ from the corresponding interval $\mathbf{P}_{i,s}$ in the smoothed path. This criterion can be used in applications where motion is constrained to a safety corridor, e.g. in a master-slave robotic guidance system, car manufacturing, or robotic endoscope holding systems.
- K_2: d_i is the root-mean-square deviation of the distances (i.e. the square root of the mean of the squares of the shortest distances) of all path points of the interval $\mathbf{P}_{i,o}$ from the corresponding interval $\mathbf{P}_{i,s}$ in the smoothed path. This criterion is useful mainly for theoretical analysis.
- K_3: d_i is the area between the smoothed paths segment $\mathbf{P}_{i,s}$ and the corresponding segment $\mathbf{P}_{i,o}$ on the original path. K_3 is the best choice for sweeping applications, e.g. bones milling or cleaning robots.

The first two error functions can be determined quickly and work for path points with any dimensionality. Error function K_3 is useful and intuitively plausible for paths defined in a plane, i.e. two-dimensional paths. However, K_3 can also be used for more dimensions.

The error function K_1 guarantees that the smoothed path never deviates by more than the distance d_{lim} from the original path. One disadvantage involves the computation of each deviation d_i: Only one path point $\mathbf{p}_i \in [\,\mathbf{p}_{\min}\,;\,\mathbf{p}_{\max}\,]$ is used and the distance from the other path points in that interval is neglected. Path points far away from $\mathbf{P}_{i,s}$ are rated strongly, whereas a constant slight deviation of the path across all path segments under consideration leads to a smaller deviation.

This drawback can be avoided by using the error function K_2 as this function uses all path points $\mathbf{p}_i \in [\,\mathbf{p}_{\min}\,;\,\mathbf{p}_{\max}\,]$ for the computation. Additionally, path points far away from $\mathbf{P}_{i,s}$ are considered because the distance to the smoothed path $\mathbf{P}_{i,s}$ is squared. The computation of K_2 is also quite fast.

The computation of K_3 is more costly, however unlike K_1 and K_2 it also considers the distance between path points $\mathbf{p}_i \in [\,\mathbf{p}_{\min}\,;\,\mathbf{p}_{\max}\,]$ on the original path, not just the distances between paths points from the original path and the smoothed path $\mathbf{P}_{i,s}$.

Figure 5 illustrates the three error functions.

The algorithm uses the heuristic of always removing the point yielding the smallest deviation. Although this provides good results in practice, the path obtained is not necessarily globally optimal. Because the algorithm does not look ahead to try to remove more than one path point at a time and does not allow the deviation to exceed the limit d_{lim} in any iteration step, opportunities to shorten the path can be missed. Consider for example Figure 6 (a). When using criterion K_1 and a maximum allowed deviation $d_{\text{lim}} = 0.6 \cdot \|\, \mathbf{p}_2\, \mathbf{p}_3 \,\|$ the optimal path (b) cannot be obtained. The removal of either $\mathbf{p}_2$ or $\mathbf{p}_3$ temporarily leads to a deviation that is close to $1 \cdot \|\, \mathbf{p}_2\, \mathbf{p}_3 \,\|$, whereas by removing $\mathbf{p}_2$ and $\mathbf{p}_3$ at the same time, d_{lim} would not be exceeded. The smoothing procedure aborts without being able to remove $\mathbf{p}_2$ or $\mathbf{p}_3$.

Nevertheless, the paths created are valid because the deviation does not exceed the maximum allowed. An advantage of the proposed method is that the algorithm is anytime capable, i.e. it can be aborted prematurely and still delivers a valid result. The quality of the path increases monotonically until termination. This is an important feature for time-critical applications, such as sensor-based motion planning.

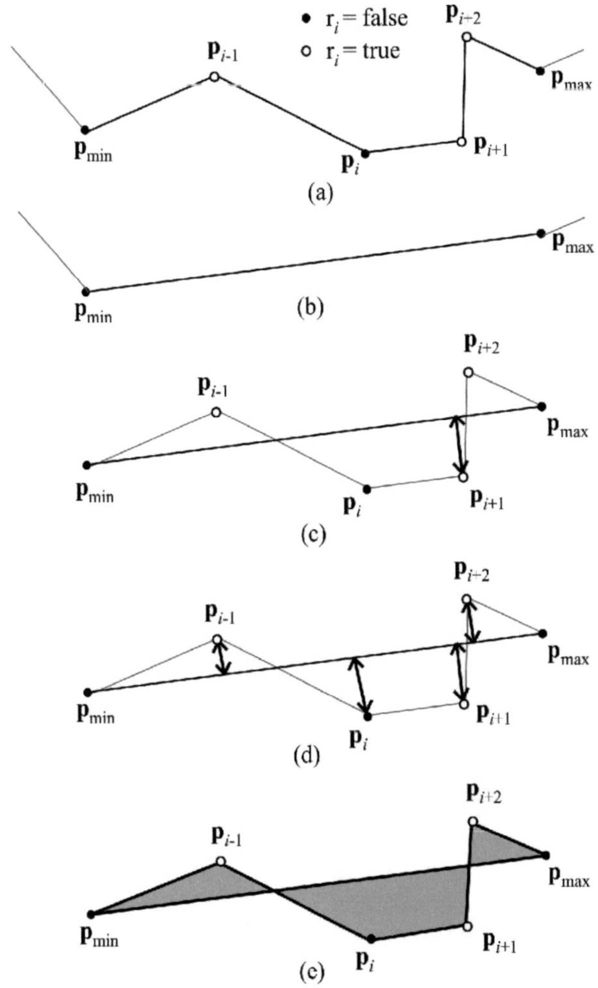

Figure 5. Sketch describing the determination of path deviation d_i. The linear path segments to be compared are the original path (a) and the smoothed path (b). The error functions K_1, K_2, and K_3 are illustrated in (c), (d), and (e)

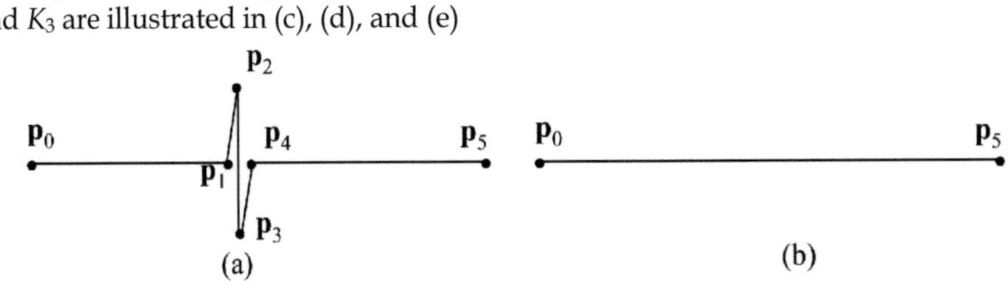

Figure 6. Example for the non-optimality of the proposed algorithm. (a) A path consisting of five points, (b) the optimal path with a maximum allowed deviation of $0.6 \cdot \|\mathbf{p}_2\,\mathbf{p}_3\|$

6. Path deviation functions for variable orientations

In Section 5 paths with arbitrary dimensionality, but just one coordinate type have been treated. However, Cartesian paths whose points contain information on both coordinate types (position as well as orientation) cannot be handled reasonably with this approach, as positions and orientations can not be treated in the same way. For example, a positional value is unique, whereas an orientation is unique in a range of 360°.

Therefore, we choose to compute position and orientation separately and combine positional and orientational deviation in an optimization procedure as optimization criterion and/or constraint, respectively. At that point, both are real numbers which are comparable again.

We use quaternions for representing the orientation of a path point, following the representation in (Maillot, P., 1990). This representation overcomes severe disadvantages of a vector angle representation like e.g. Euler angles.

Eq. (3) shows the representation of an orientation $o_i = \varphi_i$ as a quaternion.

$$\varphi_i = a_i + b_i \cdot j + c_i \cdot k + d_i \cdot l \text{ with } a_i, b_i, c_i, d_i \in \mathbb{R}. \tag{3}$$

Just like the distance between two positions $\mathbf{p}_1$, $\mathbf{p}_2$ can be computed, we can easily obtain the distance between two orientations $\mathbf{o}_1$, $\mathbf{o}_2$. It corresponds to the angle between the orientations (Eq. 5):

$$d_p(\mathbf{p}_1,\mathbf{p}_2) = \sqrt{(p_{2,x} - p_{1,x})^2 + (p_{2,y} - p_{1,y})^2 + (p_{2,z} - p_{1,z})^2} \tag{4}$$

$$d_o(\mathbf{o}_1, \mathbf{o}_2) = \operatorname{acos}(a_1 \cdot a_2 + b_1 \cdot b_2 + c_1 \cdot c_2 + d_1 \cdot d_2) \tag{5}$$

From Eq. (5), it is obvious that the distance between two orientations can not exceed the range [–180°, +180°] whereas the distance between two points is unrestricted, Eq. (4).

The criteria K_1 and K_2 are directly applicable for orientations if we replace Eq. (4) by Eq. (5) in the computation. For criterion K_3, we need to obtain the cumulated orientational deviation. We solve that problem by numerically integrating the orientational deviation along the path between two path points. We need to interpolate orientations. Quaternion interpolation can be performed using either LERP or SLERP interpolation (Maillot, P., 1990). Although SLERP interpolation is slightly more computationally expensive, we chose to use SLERP, as LERP interpolation yields only an approximated result. Not interpolating at all but only computing the angle difference between path points $\mathbf{p}_k \in \mathbf{P}_{i,o}$ and the corresponding path points of the smoothed path $\mathbf{P}_{i,s}$ would also only give an approximation.

SLERP interpolation works as follows. Let $p = 0, \ldots 1$ be a control parameter, $\mathbf{q}_1$ and $\mathbf{q}_2$ two orientations given as quaternions and the angle θ between $\mathbf{q}_1$and $\mathbf{q}_2$, as computed in Eq. (6), (7) and (8).

We obtain the control variables

$$r_1 = \sin((1-p) \cdot \theta) / \sin(\theta) \tag{6}$$

$$r_2 = \sin(p \cdot \theta) / \sin(\theta) \tag{7}$$

and the SLERP-interpolated orientation

$$\mathbf{q}' = r_1 \cdot \mathbf{q}_1 + r_2 \cdot \mathbf{q}_2. \tag{8}$$

By summing the orientational deviations of a smoothed path segment from the corresponding unsmoothed segment, we obtain the orientational deviation using criterion K_3. For this, we do not directly use the unsmoothed path, but the segmentwise projection of the unsmoothed path onto the smoothed path. To illustrate this idea, we show the interpolation of the orientation for the orientational dimensionality $m_o = 1$ and the obtained deviations as well as the sum of the deviation (Figure 7).

For $m_o = 2$ or $m_o = 3$, the procedure works in exactly the same way, as Eq. (5) and Eq. (8) are handling 3D orientations. Orientations of higher dimensionality, although untypical in practical applications, can also be easily used if the path deviation functions are adapted accordingly, just like positions of higher dimensionality are usable.

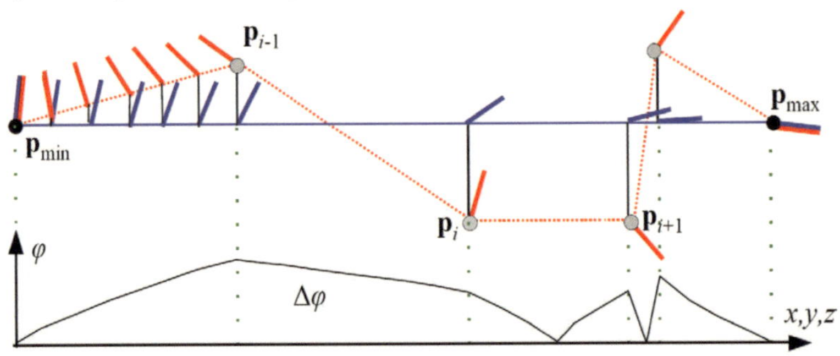

Figure 7. Sketch visualizing the computation of the angle sum $\Delta\varphi$. For simplification, the angle is represented only in 1D. Top: Curve progression of the orientation of a smoothed path $[\mathbf{p}_{min}, \mathbf{p}_{max}]$ (blue) and the original path $[\mathbf{p}_{min}, \dots, \mathbf{p}_{i-1}, \mathbf{p}_i, \mathbf{p}_{i+1}, \dots, \mathbf{p}_{max}]$ (red). Bottom: Curve progression of the angle deviation φ between the two paths

In a simple optimization procedure, the computed positional and orientational deviations p_i and o_i are assigned to the optimization and constraint value c_i and d_i. If not both coordinate types are used, all c_i stay zero and the algorithm works without any constraint. On the other hand, the optimization value d_i is indispensable. We obtain five different cases, as shown in Table 2.

	Case				
	(a)	(b)	(c)	(d)	(e)
Optimization value d_i	p_i	o_i	p_i	o_i	$\frac{p_i}{p_{max}} + \frac{o_i}{o_{max}}$
Max. optimization deviation d_{lim}	p_{lim}	o_{lim}	p_{lim}	o_{lim}	2
Constraint value c_i	/	/	o_i	p_i	/
Max. constraint deviation c_{lim}	/	/	o_{lim}	p_{lim}	/

Table 2. Enumeration of the five possible cases when combining both coordinate types. The table entries indicate the assignments of the path points positional and orientational deviation p_i and o_i to the algorithm's values c_i and d_i, as well as the assignments of the indicated positional and orientational deviation limit values p_{lim} and o_{lim} to the algorithm's values c_{lim} and d_{lim}

Case (a) is the case seen in Section 5. No orientational information is evaluated, and there is no constraint. The computed positional deviation p_i is used as d_i in the path smoothing procedure. Similarly, in Case (b), we are smoothing orientations without using constraints. The criterion function used to compute d_i is the version adapted to orientations (Eq. (5)), and the deviation d_i used in the optimization procedure is the orientational deviation o_i.
In Case (c), the positional deviation is used as in (a) as optimization value, but in addition we use the orientation as constraint. As long as the limit is not exceeded ($c_i < c_{lim}$) in any point with the lowest positional deviation, the path is smoothed as in Case (a). Points whose orientational deviation o_i exceeds o_{lim} are not removed, no matter how low p_i is. Case (d) is analogue to Case (c), the roles of position and orientation being exchanged.
In Case (e), we have to optimize the position and orientation simultaneously. Because they are incompatible, we normalize p_i and o_i to p_{lim} resp. o_{lim}, giving an indication on how close both deviations are to the limit and bringing them into the same domain. Now, we can simply sum up both values to obtain d_i. The ratio between p_{lim} and o_{lim} correlates linearly with the relation of the influences of position and orientation. As both p_i / p_{max} and o_i / o_{max} are positive numbers and their sum may not exceed 2, both positional and orientational deviations are restricted to twice the maximum indicated.

7. Experiments and Results

In this section, we present and analyze a practical surgical application of our algorithm. First, we compare the results yielded when applying the three error functions described previously. Then, we briefly describe the original paths used in the surgical application and their drawbacks. Finally, we describe the improvement achieved by applying the smoothing algorithm.
In the first rather theoretic experiment, a perturbed linear path consisting of 1000 points, positioned equidistantly on the x axis in the interval [0; 1000] with increasing x coordinate and distributed uniformly on the y axis with y values between -10 and +10, is smoothed using criterion K_1 (maximum deviation). The y coordinate of the first and the last point is 0 (Figure 8).

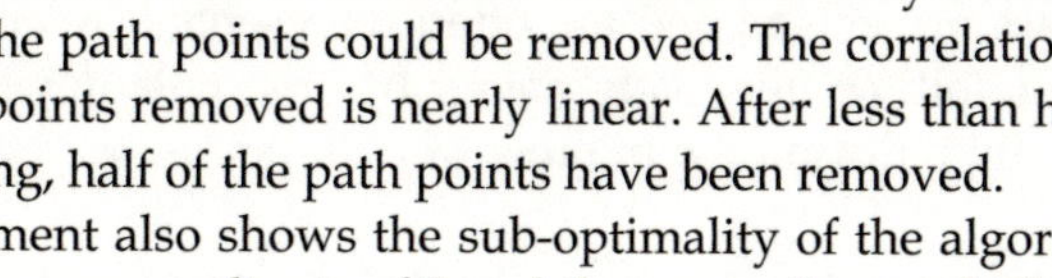

Figure 8. Experimental perturbed path. The units on both axes are millimetres

The experimental results in Figure 9 demonstrate the anytime property of the algorithm. It can be aborted at any time. With a maximum deviation of only 1 mm, reached after 20 ms, already one third of the path points could be removed. The correlation of computation time and number of path points removed is nearly linear. After less than half of the time needed for complete smoothing, half of the path points have been removed.
However, this experiment also shows the sub-optimality of the algorithm. Because the first and the last point have a y coordinate of 0 and the y coordinate of all other points lies in the interval [-10; 10], the path could be reduced to its start and end point with a maximum deviation d_{lim} =10 mm. Yet, in general the algorithm finds that solution only at d_{lim} =20 mm.

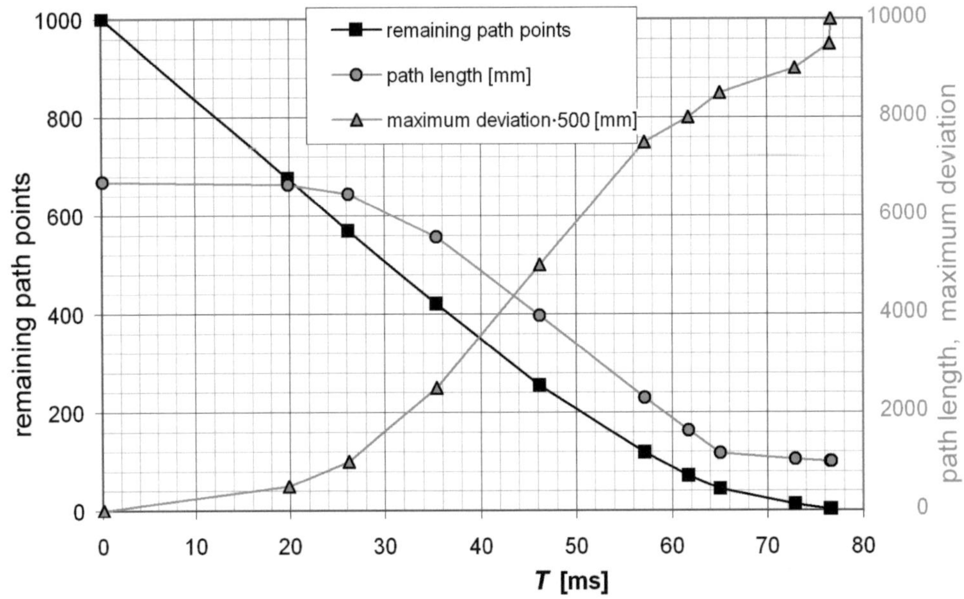

Figure 9. Remaining path points, path length and maximum deviation against computation time *T* [ms] for the perturbed linear path segment of Figure 8

When milling cavities in workpieces, problems with overly fragmented and angulated milling paths arise, cf. the RONAF project (Robot-based Navigation for Milling at the Lateral Skull Base (Federspil, P. A.; Geisthoff, U. W.; Henrich, D. & Plinkert, P. K., 2003)) (Figure 10). The goal of the RONAF project is the development and examination of a system for navigation on the lateral skull base with the purpose of an interactive supervision of a surgical robot during interventions. Modular navigation and control procedures are being used. The operation is planned on a preoperatively acquired 3D dataset, e.g. computed tomography (CT) or magnetic resonance tomography (MRT). The robot and its attached tool are moved relative to this dataset.

Milling in the skull bone demands high precision (with tolerances below one millimeter) in spite of the high force required to remove larger quantities of bone, a combination that is very straining for a surgeon and poses little problem to a robot. Therefore an important increase of processing quality is expected.

In the RONAF system, three-dimensional path planning (Waringo, M.; Henrich D., 2004) is used in order to mill a given implant volume with a robot-controlled miller. The paths planned in a voxel space are angled and are often represented by an excessive number of path points. The robot follows the path points by interpolating linearly between two successive path points. By reducing the number of path points, we can significantly reduce the milling duration.

Path smoothing was applied to milling paths planned in a voxel space for milling a hearing aid implant volume (Figure 11). The milling time for the non-smoothed path is unsatisfactory long. The traversal speed is strongly reduced in regions where the path points are close to each other or where the directional change between two consecutive path segments is high. This drawback is due to robot dynamics restrictions such as the maximum acceleration in the robot joints and restrictions involving computing time (i.e. the maximum number of path segments that can be processed per second).

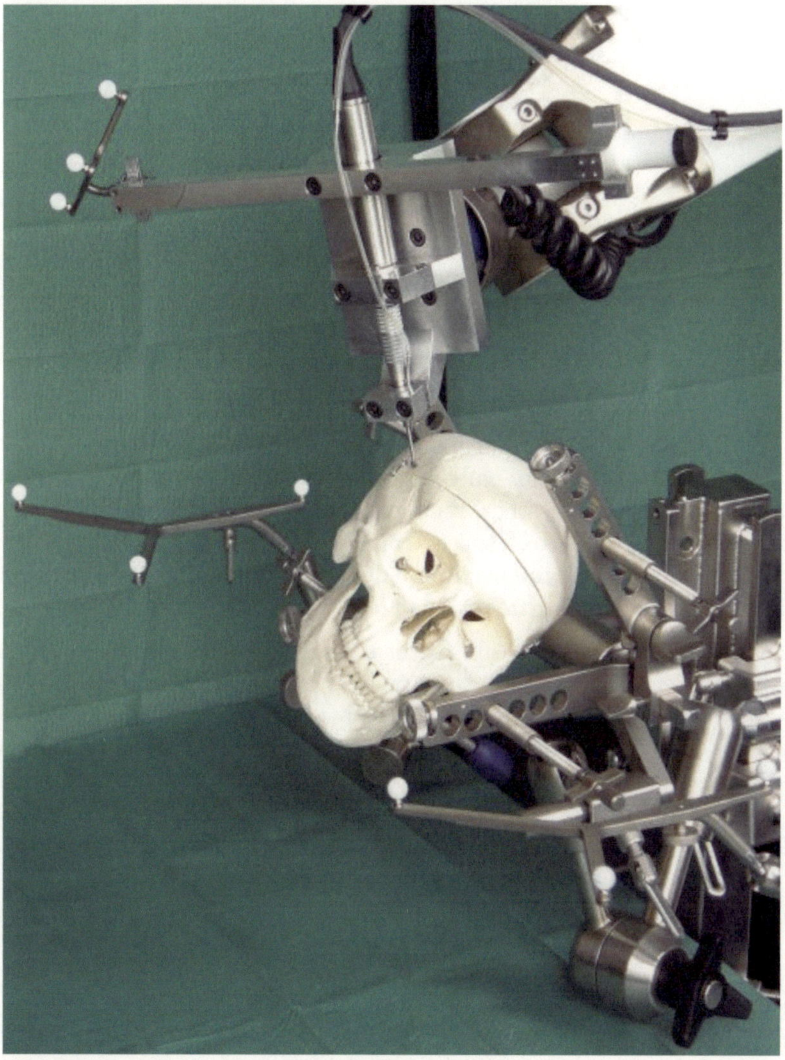

Figure 10. Experimental setup from the RONAF project

Part of the robot motions occurs above the workpiece (the long straight segments in Figure 11), where the tool moves above the bone without touching it. These segments serve to change the currently processed region. They segments can not be removed, since otherwise the miller would mill bone at forbidden locations. Therefore, even with a maximum allowed deviation of infinite in the path smoothing process, such a milling path can not be reduced to its start and end point. The path smoothing only applies to the horizontal path segments located in the bone. The smoothing algorithm was applied to the entire path, with all vertical and horizontal path segments needed for changing a region marked as non-removable.

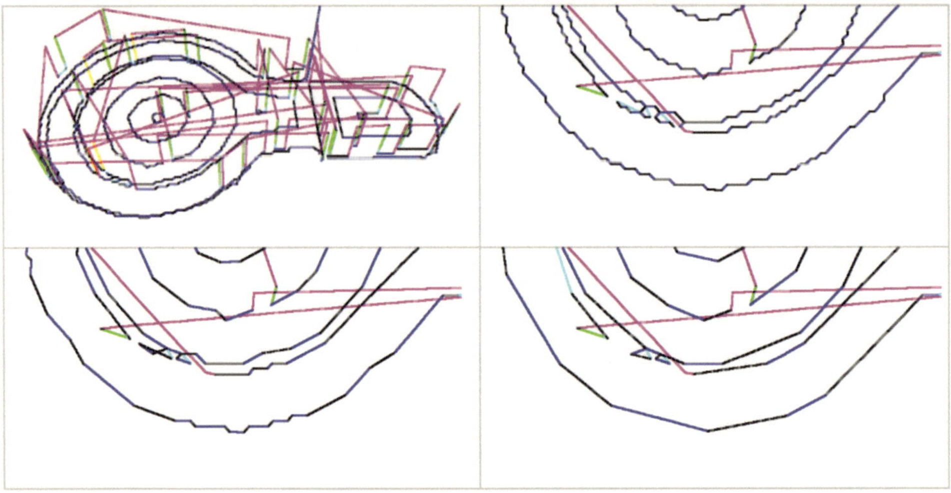

Figure 11. Milling paths for the hearing aid implant Vibrant Soundbridge (Siemens/Symphonix). The circular paths lie in the horizontal plane, and path segments perpendicular to that plane are vertical segments which connect these circular paths. Upper left: original path planned in a voxel space; Upper right: close-up of the original path (4403 path points); Bottom left: path with maximum deviation of 0.18 mm (2788 path points); Lower right: path with maximum deviation of 0.35 mm (1405 path points)

While keeping the modification to the path at a non-critical level so that no noticeable changes occur in the milled geometry [2], the number of path points can be reduced by more than 50%, as shown in Table 3. With an acceptable tolerance of 0.35 mm, it is possible to eliminate about 46% of the milling duration, 69% of the path points and 65% of all changes normally arising in the non-smoothed path. The computation time for path smoothing was measured on an AMD Athlon XP 2600+ PC with 512 MB of RAM. The computation time is nearly exactly linear with the number of points removed, in this example about 0.6 ms per point.

Table 4 shows a comparison of paths smoothed using the three error functions and evaluated according to the three error functions. As expected, path planned with error function K_i, $i \in \{1,2,3\}$ ranked best when the evaluation was performed according to the same error function K_i. No clear advantage can be determined and no error function is made redundant by the other two.

[2] As the miller's diameter is 4.5 mm and the robot's repeatability accuracy is 0.35 mm, a maximum deviation in the path of 0.35 mm is acceptable.

maximum deviation d_{lim} [mm]	# path points	time required for path traversal [min:sec]	computation time [sec]	path length [mm]	angular integral [°]
0	4403	12:35	0.00	5545	239,417
0.10	4355	12:24	0.05	5527	234,048
0.13	2788	09:22	0.94	5460	150,501
0.18	2107	08:06	1.32	5427	118,268
0.25	1629	07:12	1.58	5403	96,645
0.35	1405	06:44	1.77	5367	82,650
0.60	1207	06:19	1.80	5334	74,520
1.00	1098	06:05	1.88	5306	72,273
2.00	997	05:49	1.90	5232	69,016
Infinite	832	04:21	2.04	4171	58,320

Table 3. Number of path points remaining, necessary time requirement for traversal and for path smoothing, path length and angular integral in the function of the maximum allowed deviation. In order to avoid damage to the patient, areas of the path are not allowed to be modified, as described previously. Therefore, the path can not be reduced to its start and end point with an infinite maximum deviation. Without this restriction, the effects of the path shortening would be even more noticeable

		Error function for path evaluation					
		K_1 [mm]		K_2 [mm²]		K_3 [mm²]	
		max	avg.	max.	avg.	max.	avg.
Error function for path computation	K_1	0.281	0.171	0.078	0.015	2.125	0.257
	K_2	0.478	0.216	0.046	0.019	1.531	0.358
	K_3	0.884	0.197	0.297	0.020	0.393	0.213

Table 4. Comparison of the three error functions K_1, K_2 and K_3 when reducing the milling path of the implant Vibrant Soundbridge from 4403 to 1500 path points. K_1: maximum deviation, K_2: root-mean-square deviation, K_3: spanned area. The error function used for path planning is noted in the rows and the error function used for evaluation is noted in the columns. In the cells, the deviations are noted, with both the maximum and the average value per path segment

Another example of path smoothing in the RONAF project is shown in Figure 12. In order to record an ultrasound image of the patient's skull, the skull is sampled manually with the tip of an infrared marker whose spatial position is sampled at equidistant times. This path is then traversed by the robot, an ultrasound head being mounted on the effector. Between recording and traversing of the path, smoothing is performed. This way, in addition to rendering the path less jerky, there are also agglomerations of path points being removed which appear when the surgeon interrupts the movement of the marker.

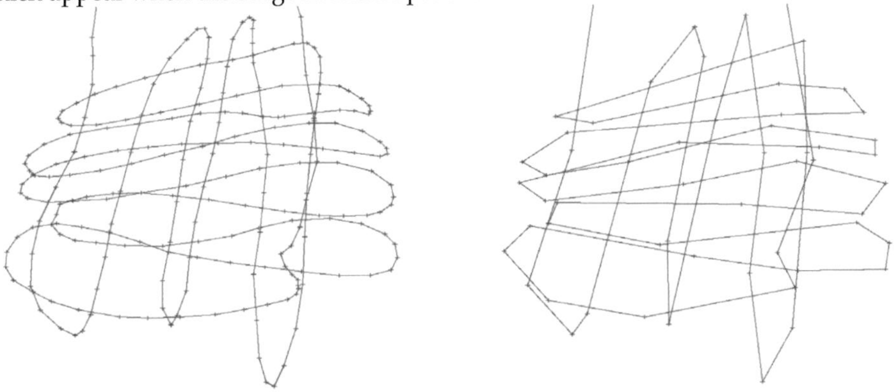

Figure 12. Scanline path for ultrasound recording of the human skull. Left: original path (307 path points), right: smoothed path using K_1 and $d_{\text{lim}} = 1$ mm (90 path points)

8. Conclusions

We have presented a method that smooths paths of any dimensionality consisting of linear segments until the deviation between the smoothed path and the original path locally exceeds a given threshold. The error function for deviation determination can easily be exchanged and adapted for diverse applications. The computational requirement has been reduced from quadratic to linear in the number of path points used. Our method is anytime-capable, i.e. it can be aborted at any time and returns a valid path for which the maximum deviation increases monotonically and the number of path points decreases monotonically in the allowed computation time.

Possible extensions of the algorithm include the consideration of forbidden regions that may not be crossed by the path and a distance computation that varies depending upon the position on the path. Additionally, if applied to motion planning in a cluttered environment, the algorithm does not handle collisions with obstacles close to the unsmoothed path. In this case, further conditions are required which are evaluated in addition to the error functions and which avoid the smoothed path getting too close to the obstacles. In this scenario, the geometry of the actuator must be considered too.

For a path smoothing with using both positions and orientations as optimization criterion (Case (e) in Section 7), one could in addition use one of them as constraint, giving even more control over the maximum allowed deviation.

If a globally optimal path has to be found, the presented method is not suitable, as it is a local method and can get stuck in local optima, as shown in the first experiment. In order to overcome this disadvantage, a global method has to be used.

9. Acknowledgments

This work has been supported by the German Research Foundation (DFG) under the project name "Robotergestützte Navigation zum Fräsen an der lateralen Schädelbasis" (RONAF) with the identifier 227100. Further information can be found at *http://ai3.inf.uni-bayreuth.de/projects/ronaf*.

10. References

Aleotti, J.; Caselli, S. (2005). Trajectory clustering and stochastic approximation for robot programming by demonstration, *IEEE/RSJ Int. Conf. On Intelligent Robots and Systems*, pp. 2581-2586

Amato, N.M. & Wu, Y. (1996). A randomized roadmap method for path and manipulation planning, *Proceedings of the IEEE Int. Conf. Robot. & Autom.*, pp. 113-120.

Baginski, B. (1998). Motion planning for Manipulators with Many Degrees of Freedom - The BB Method, *Doktorarbeit*, TU München

Barraquand, J. & Latombe, J.-C. (1991). Robot motion planning: a distributed representation approach, *Int. Journal of Robotics Research*, 10 (6), pp. 628-649, 1991

Berchtold, S. & Glavina, B. (1994). Kosten-Nutzen-optimale Verbesserung kollisionsfreier Roboterbewegungen mittels Polygon-Manipulation, in: *10. Fachgespräch Autonome mobile Systeme*

Bronstein, I. N.; Semendjajew, K. A; Musiol, G.; Mühlig, H. (2001). *Taschenbuch der Mathematik*, Verlag Harri Deutsch, ISBN 3-8171-2005-2

Carpin, S.; Pillonetto G. (2006). Motion planning using adaptive random walks, *IEEE Transactions on Robotics and Automation*, 21 (1)

Engel, D. (2003). Sensorgestützte Robotersteuerung für den Einsatz in der Chirurgie, Dissertation, *Fakultät für Informatik der Universität Fridericiana zu Karlsruhe (TH)*

Federspil, P. A.; Geisthoff, U. W.; Henrich, D. & Plinkert, P. K. (2003). Development of the First Force-Controlled Robot for Otoneurosurgery, *Laryngoscope* 113

Geraerts, R. & Overmars, M. H. (2002). A comparative study of probabilistic roadmap planners. In *Proc. Int. Workshop on Algorithmic Foundations of Robotics (WAFR)*

Hein, B. (2003). Automatische offline Programmierung von Industrierobotern in der virtuellen Welt, Dissertation, *Fakultät für Informatik der Universität Fridericiana zu Karlsruhe (TH)*, 2003.

Kuffner, J. J.; LaValle S. M. (2001). RRT-Connect: An efficient approach to single-query path planning, in *Proceedings of the IEEE Conference on Robotics and Automation*, San Francisco, pp. 995–1001.

Maillot, P. (1990). Using quaternions for coding 3d transformations, in *Graphics Gems I*, Academic Press Inc., Boston pp. 498-515

Quinlan, S.; Khatib, Q. (1993). Elastic bands: Connecting Path Planning and Control, in *Proceedings of the IEEE Conference on Robotics and Automation*, Atlanta, pp. 802-807.

LaValle, S. M. (1998). Rapidly-exploring random trees: A new tool for path planning, TR 98-11, *Computer Science Dept., Iowa State University*

Urbanczik, C. (2003). SIMERO: Bildbasierte Kollisionserkennung und Bahnglättung im Konfigurationsraum, Diploma Thesis, *Fakultät für Informatik, Universität Kaiserslautern*, D-67653 Kaiserslautern

Waringo, M.; Henrich D. (2004). 3-Dimensionale schichtweise Bahnplanung für Any-Time-Fräsanwendungen, *Robotik*, München/Germany

Waringo, M.; Henrich D. (2006). Efficient smoothing of piecewise linear paths with minimal deviation, *Proceedings of the IEEE/RSJ International Conference on Intelligent Robots and Systems*, Beijing/China, pp. 3867-3872

30

A Novel Feature Extraction Algorithm for Outdoor Mobile Robot Localization

Sen Zhang[1,3], Wendong Xiao[2,3] and Lihua Xie[1]
[1]*Nanyang Technological University,* [2]*Institute for Inforcomm Research,*
[1, 2]*Singapore,* [3]*China*

1. Introduction

Navigation is one of the basic problems for autonomous mobile robots. Its history can be traced back to long time ago. Today, navigation is a well-understood quantitative science, used routinely in maritime and aviation applications [Adams, 1999]. Given this, the question must be asked as to why robust and reliable autonomous mobile robot navigation remains such a difficult problem. The core of the problem is the reliable acquisition or extraction of information about navigation beacons from sensor information and the automatic correlation or correspondence of these with some navigation map [Guivant et al., 2000].

Many navigation systems use artificial beacons to realize their navigation task, but the approach may not be realistic in applications such as exploration of jungles or other unknown environments. In this situation, one needs to utilize naturally occurring structure of typical environments to achieve a similar performance. Hence, fast and reliable algorithms capable of extracting features from a large set of noisy data are important in such applications. Some of the early efforts in this direction have focused on extracting line features in an indoor environment based on the information provided by sonar and laser sensors. In [Crowley, 1985], a least-squares line fitting technique was applied to extract edges from ultrasonic sensor data. In [Taylor & Probert, 1996], a recursive line fitting system is used to extract line segments under polar coordinates and an ellipse fitting method is also implemented for data from a laser sensor. In [Vandorpe et al., 1996], line segments are detected using a regression least-squares parameter estimation method whereas the center and radius of a circle feature are estimated based on the average value of the measurements of the circle from a 2D range scanner. Later, a two-layer Kalman filter was used to calculate the parameters of a line by an on-line method in [Roumeliotis & Bekey, 2000]. Observe that the aforementioned articles are focused on indoor applications and are mainly concerned with line extraction.

For an outdoor environment, the problem of feature selection and detection is more challenging. In our view, in most typical semi-structured outdoor environments, such as campuses, parks and suburbs, tree trunks and tree-like objects, such as pillars, are relatively stable, regular and naturally occurring features that can provide very useful information for mobile robot navigation. Recently, some research on the use of these kinds of geometrical features has been carried out in [Guivant et al., 2002]. Also, [Guivant et al., 2002; Bailey, 2002] addressed the problem of extracting tree trunks from laser scan data where the centre and radius of a circle are estimated by averaging the measurements. This method can be

susceptible to outliers which can significantly affect the accuracy of the center and radius estimates.
In this paper, we shall address the problem of extracting edge and circle features for semi-structured outdoor mobile robot navigation. We classify features into edges, circles and random clutter and propose an approach for their extraction. First, a model based data segmentation method is applied which divides the collected data into groups that are possibly associated with different features of the environment. The extended Kalman filter or other filtering techniques can be applied for segmentation. Edges are also detected during segmentation. We then give a procedure to identify the type of features with which a given group of data is associated. For a circle feature, a modified Gauss-Newton optimization is proposed to obtain estimates of its centre and radius. Several experiments are carried out to demonstrate the feasibility and effectiveness of the proposed feature extraction method. In the experiments, the data association method proposed in [Zhang et al. 2005] is used to enhance the robustness of features. The results show that our method for feature extraction is implementable in real-time and outperforms existing methods such as that in [Bailey, 2002].
The structure of the paper is as follows: Section 2 presents our feature extraction algorithm, and section 3 shows the experimental results using the proposed algorithm in several outdoor environments. Conclusions are drawn in Section 4.

2. Feature Extraction Algorithm

We observe that in many semi-structured outdoor environments, planes such as building walls and cylindrical surfaces such as tree trunks or tree-like objects are often encountered. We consider two kinds of features for these semi-structured environments. Observe that in most outdoor environments, trees or tree trunks can be very useful features for mobile robot navigation. In [Guivant et al., 2002; Bailey, 2002], the problem of extracting circle features was addressed by averaging their measurements. Here, we shall propose an algorithm which is able to extract edges and tree trunks with a higher accuracy. The essential components of this algorithm include two parts: the first is the segmentation of the scan data and the second is the parameter acquisition.

2.1. Segmentation and Edge Detection

Segmentation is a process of aiming to classify a set of scan data into several groups, each of which possibly associates with different structures of the surroundings. The segmentation process is realized through the EKF [Adams, 1999; Zhang et al. 2003; Zhang et al. 2004a] or other filtering techniques. At each time instant the range estimate is compared to the range measurement based on their statistics in order to decide if an edge has been detected. When the difference between the measured range and the predicted range is beyond a certain threshold, we consider that an edge has been detected. This can be achieved by using a validation gate during the prediction process with the EKF.

2.1.1. Planar Model

Let us first introduce a mathematical framework for a planar surface. Consider a vertical plane shown in Fig. 1 and the corresponding sensed data points from a perfect 3D line of the sight sensor. Similar to the description in [Adams, 1999], we have:

$$d_{i+2} = \frac{d_i d_{i+1}}{2d_i \cos\gamma - d_{i+1}} \tag{1}$$

where γ is the constant angle between successive samples of the sensor as it rotates about its vertical axis. Note that the relationship given in equation (1) is independent of the elevation angle α.

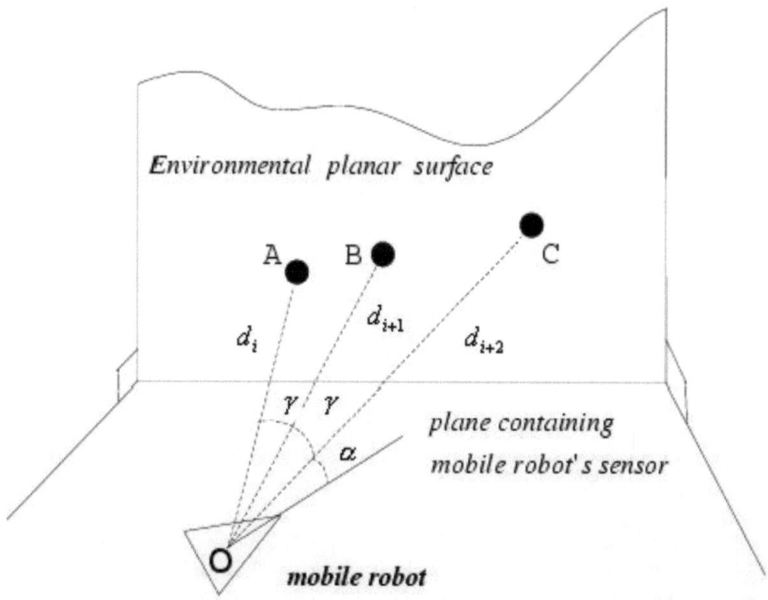

Figure 1. The relationship between successive range readings when scanning a planar surface

2.1.2. System Model

Equation (1) is clearly a second order difference equation with respect to time. Define $x_1(k+1) = d_{i+2}$ and $x_2(k+1) = x_1(k) = d_{i+1}$, where $x_1(k)$ and $x_2(k)$ are the state variables at time instant k. Therefore equation (1) can be fully defined by the state space equations:

$$\mathbf{x}(k+1) = \mathbf{f}_1(\mathbf{x}(k)) + \mathbf{v}(k) \tag{2}$$

where $\mathbf{x}(k+1) = [x_1(k+1)\ \ x_2(k+1)]^T$,

$$\mathbf{f}_1(\mathbf{x}(k)) = \begin{bmatrix} \dfrac{x_1(k)x_2(k)}{(2x_2(k)\cos\gamma - x_1(k))} \\ x_1(k) \end{bmatrix}$$

and $\mathbf{v}(k)$ is the process noise which reflects possible imperfection of the surface. We assume that $\mathbf{V}$ is a white noise with variance $\mathbf{Q}(k)$. Clearly, a small variance $\mathbf{Q}(k)$ implies that the surface is close to be perfect. In the experiments in this paper, we set $Q(k) = 10^{-4}$. Equation

(2) represents a system model which will be used to predict the next range value from the sensor before the actual range measurement is recorded.
Similar to [Adams, 1999], our observation model is:

$$\mathbf{z}(k) = \mathbf{H}_1 \begin{bmatrix} x_1(k) \\ x_2(k) \end{bmatrix} + \mathbf{w}(k) \tag{3}$$

where $\mathbf{H}_1 = [1 \quad 0]$ and $\mathbf{w}(k)$ is a zero mean Gaussian noise with a known variance σ_r^2. The EKF is used to realize the prediction and validation process.
Note that if the degenerate case (almost parallel) is detected, we suggest that these measurements can be rejected.

2.1.3. Extended Kalman Filter and Validation Gate

Based on the above system model, an extended Kalman filter is used to implement the prediction and update. In order to identify if a measurement is associated with a new edge (large discontinuity), certain criterion needs to be established. Use the innovation $\nu(k+1)$ and the innovation variance $\mathbf{s}(k+1)$ to define:

$$d(k+1) = \nu^T(k+1)\mathbf{s}^{-1}(k+1)\nu(k+1) \tag{4}$$

Note that since ν is a Gaussian random variable, d is a random variable following the χ^2 distribution. The smaller the $d(k+1)$, the higher the probability that the measurement $z(k+1)$ is obtained from the same planar surface. Thus, a validation gate, δ, is used to decide whether the measurement $z(k+1)$ is a close enough match to the predicted data point to continue the filter update. If the measurement is such that $d(k+1) > \delta$, a discontinuity is found. From the χ^2 distribution table, we know that if the observation is from the same planar surface, then $d(k+1) < 6.63$ with a probability of 0.99. If a small δ is selected, there will be more edges found. Here we set $\delta = 6.63$.
After the data segmentation process, we need to decide if each segment of data is associated with a line or a circle (note that the laser sensor data points only form an arc which is part of a circle, here we call it a circle feature) or a clutter. For a line, the average error between the observation and the EKF prediction at each point should be very small. Note that the prediction error (innovation) sequence $\{\nu\}$ of (5) is a Gaussian white noise and its covariance is given by $\mathbf{s}(k)$. Assume that the number of points of the segment is M. Then the sequence $\{\nu\}$ is of the length $M-2$ (note that the first two pints are used to initialize the filter). The average prediction error and its covariance are then given by

$$\bar{\nu} = \frac{\sum_{k=3}^{M} \nu(k)}{M-2}, \quad \bar{\mathbf{s}} = \frac{\sum_{k=3}^{M} \mathbf{s}(k)}{(M-2)^2} \tag{5}$$

Hence, $P\left\{|\bar{\nu}| \le 3\sqrt{\bar{\mathbf{s}}}\right\} = 0.997$. A threshold for the average prediction error can be chosen as $3\sqrt{\bar{\mathbf{s}}}$. The threshold is used to distinguish a line from a circle or a clutter. If the average prediction error is smaller than the threshold, we consider that this segment of data is associated with a line, otherwise, it is associated with a circle or a clutter. It is noted that if a circle shape clutter is detected as a circle feature after several successive scans, the circle shape clutter is the same as a circle feature. If after several successive scans, we can't detect the circle feature that is found in the previous scans, the feature detected should be a circle shape clutter.

For a circle, we shall need to estimate its parameters such as the center and the radius of the circle so that future measurements of the circle may be used for robot navigation. In the following, the modified Gauss-Newton method [Dennis & Schnabel, 1983] is applied.

2.2. Parameter Acquisition

A circle can be defined by the equation $(x-x_0)^2+(y-y_0)^2=r^2$ where (x_0, y_0) and r are the center and the radius of the circle, respectively. For a circle fitting problem, a data set of (x, y) is known and the circle parameters (x_0, y_0, r) need to be estimated. Assume that we have obtained M measurements (x_m, y_m), $m=1,2,\ldots M$, of the circle. Our objective is to find $p=(x_0, y_0, r)$ that minimizes

$$E(p)=E(x_0,y_0,r)=\sum_{m=1}^{M}[(x_m-x_0)^2+(y_m-y_0)^2-r^2]^2 \tag{6}$$

This is equivalent to performing the least-squares process using the equations

$$g_m(x_0,y_0,r)=(x_m-x_0)^2+(y_m-y_0)^2-r^2=0, \quad m=1,2,\ldots,M \tag{7}$$

The equation (7) is not linear about the unknown parameters x_0, y_0, and r, therefore it is a nonlinear least-squares problem. We propose to use the modified Gauss-Newton optimization method [Dennis and Schnabel, 1983] to solve the problem. In our case the Jacobian matrix for the modified Gauss-Newton algorithm is

$$A=\begin{bmatrix} \frac{\partial g_1}{\partial x_0} & \frac{\partial g_1}{\partial y_0} & \frac{\partial g_1}{\partial r} \\ \frac{\partial g_2}{\partial x_0} & \frac{\partial g_2}{\partial y_0} & \frac{\partial g_2}{\partial r} \\ \vdots & \vdots & \vdots \\ \frac{\partial g_M}{\partial x_0} & \frac{\partial g_M}{\partial y_0} & \frac{\partial g_M}{\partial r} \end{bmatrix} \tag{8}$$

Let $\bar{g} = (g_1 \ g_2 \ \cdots g_M)^T$ with g_m as defined in (7).

At the k-th step, using the modified Gauss-Newton method to search the solution according to the following equation:

$$(A_k^T A_k + \lambda_k I)\triangle p_k = -A_k^T \bar{g}_k \tag{9}$$

where $\triangle p_k = p_{k+1} - p_k$ and p_k is the estimate of $p=[x_0 \ \ y_0 \ \ r]^T$ at the k-th iteration. We set the initial value $\lambda_0 = 0.01$ and carry out the following iterations for calculating a suboptimal p:

Step 1: Calculate $\triangle p_k$ using equation (9);

Step 2: Calculate the sum error $E(p_k + \triangle p_k)$ by equation (6);

Step 3: Compare with the sum error of last step $E(p_k)$. If $E(p_k + \vartriangle p_k) > E(p_k)$, increase λ_k by a factor of 10, and go back to Step 1;
Step 4: If $E(p_k + \vartriangle p_k) < E(p_k)$, decrease λ_k by a factor of 10, update the trial solution, i.e. replace p_k by $p_k + \vartriangle p_k$ and go back to Step 1 until the algorithm converges.
The convergence condition can be defined by the sum of the error square and the number of iterations.
Observe that a starting guess for these parameters is required. We use the first three points (x_i, y_i) $i = 1,2,3$ and (7) to compute an estimated initial value of (x_0, y_0, r). The more accurate the initial value is, the faster the algorithm converges.
Remark 1. Note that the Hough transform is commonly used for parameter acquisition and segmentation [Iocchi & Nardi, 2002]. The transform is implemented by quantizing the Hough parameter space into finite accumulator cells. As the algorithm runs, each point is transformed into a discretized curve and the number of intersections of the accumulator cells is counted. However, the problem of how to decide the number of the cells in the parameter space remains unsolved. If the Hough transform is applied for fitting a circle, the parameter space is of three dimensions, which makes the problem more difficult. And with the increased dimension of the parameter space, the Hough transform method becomes more complex and slower. Hence, we use the modified Gauss-Newton method instead of the Hough transform for parameter acquisition.
Remark 2. In our algorithm, since each group of data is formed after data segmentation, any outlier of measurements has been removed since an outlier produces a large discontinuity in segmentation. On the other hand, in complex outdoor environments, features extracted by the above proposed method may become unstable. In order to use these features for navigation, the correspondence between a current feature extracted by the above method and a feature in the map built thus far has to be established. This is the so-called data association problem. In this paper, we apply the data association algorithm proposed in [Zhang et al. 2005; Zhang et al. 2004*b*] where the problem is formulated as a (0,1) integer programming one and solved by a combined linear programming and iterative heuristic greedy rounding (IHGR) method. The details can be found in [Zhang et al. 2004*b*] and will not be repeated here.
Remark 3. For the proposed algorithm, if the angular uncertainty is considered, the feature detection algorithm results will be improved.

3. Experimental Results

The laser sensor used in the following experiments is Sick PLS200. The field of view is 180 degrees in front of the robot and up to 50 meters of distance. To obtain a 360 degree scene, we use 2 back to back Sick sensors. The range samples are spaced every half a degree, all within the same plane.
In the first experiment, data is collected outdoors as shown in Fig. 2 (a) where there are 10 pillars labeled from a to j, and the surroundings are building walls and low balusters with small shrubs at a long distance. In this figure, the six cross points represent the six positions at which the robot scans the surroundings. The laser scanner is placed on the top of a mobile robot at approximately 1.2 meters above the ground. At this level, the sensor can see the object outside the balusters. In Fig. 2 (b), we show the real data from one scan of the

environment. The origin is the place at which the robot is located. Because the distance is very near to the sensor around the origin, the points here are not very regularly distributed. We have done the feature extraction for all the 6 scans. Here we show the feature extraction results at two different positions. The results are given in Fig. 3 to Fig. 4.

Fig. 3 (a) and Fig. 3 (b) show the feature extraction results at position 2 using the proposed method of the last section. Zoomed views of the regions inside the dashed box of Fig. 3(a) are given in Fig. 3(b) where the extracted features can be seen clearly. In these figures, the detected edges are denoted by crosses. Similarly, the feature detection results at position 4 are shown in Fig. 4.

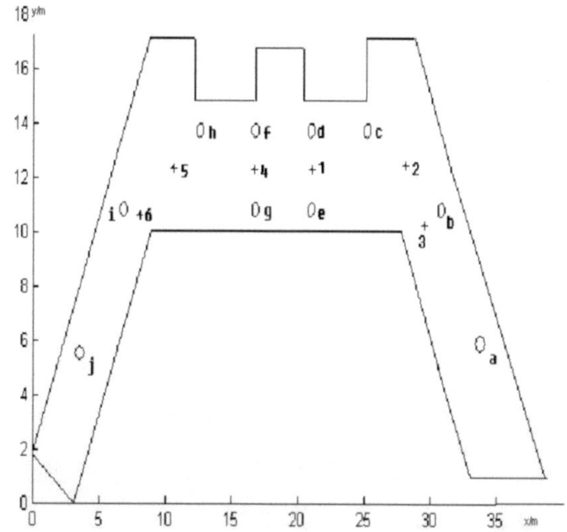

Figure 2. (a) The place to be explored by the robot

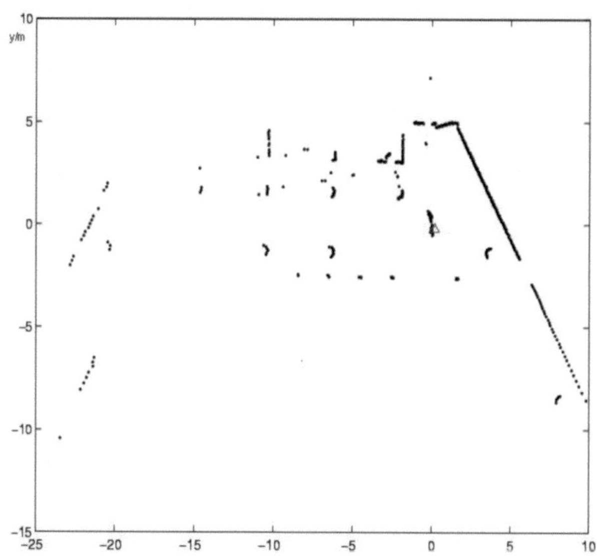

Figure 2. (b)Data from one whole

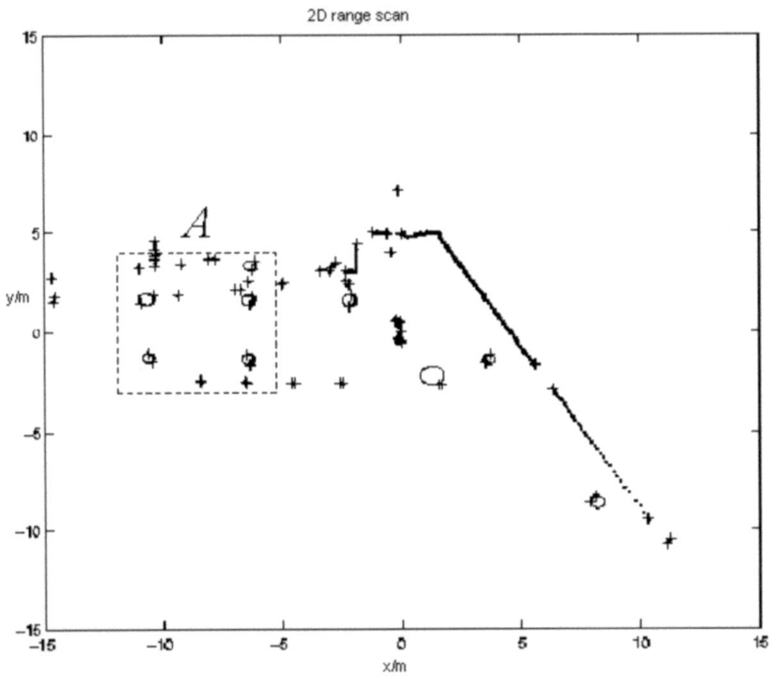

Figure 3 (a). Circles and edges extracted from data scanned at position 2 (the normal view)

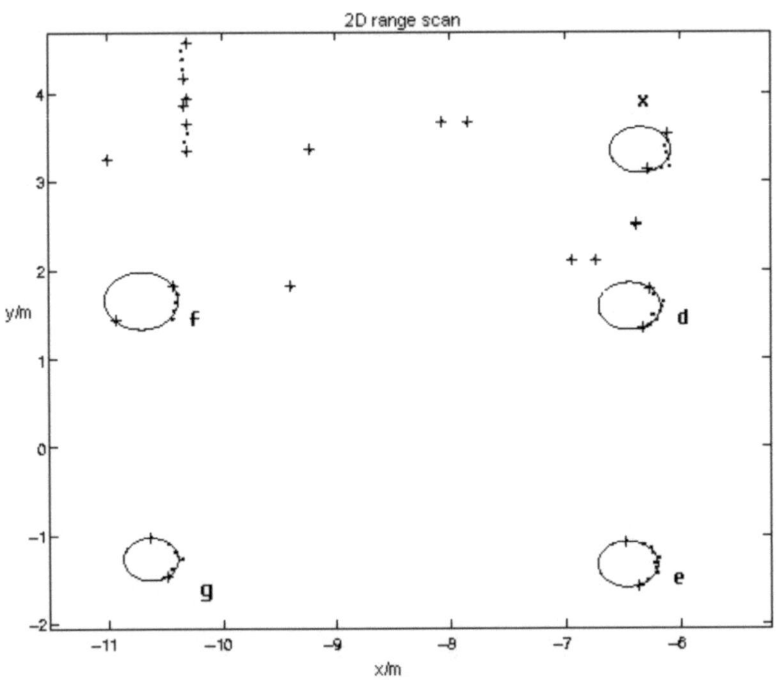

Figure 3(b). A zoomed view of the region inside box A in Fig. 3 (a)

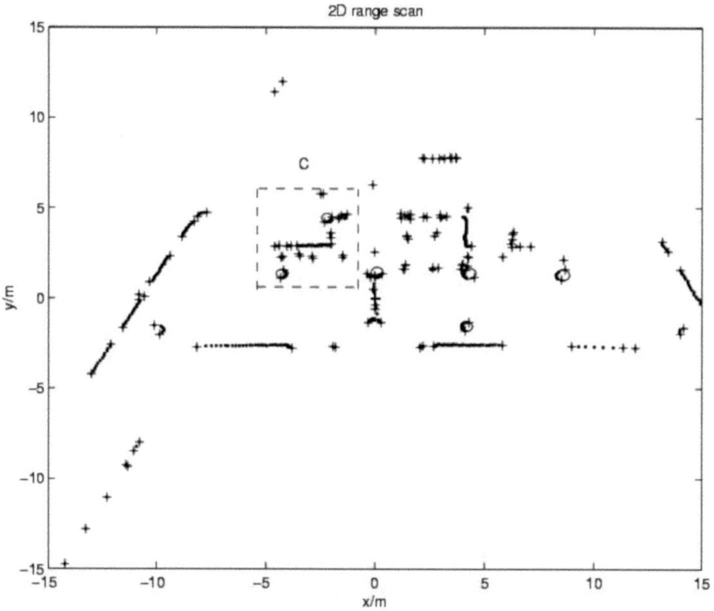

Figure 4(a). Circles and edges extracted from data scanned at position 4 (the normal view)

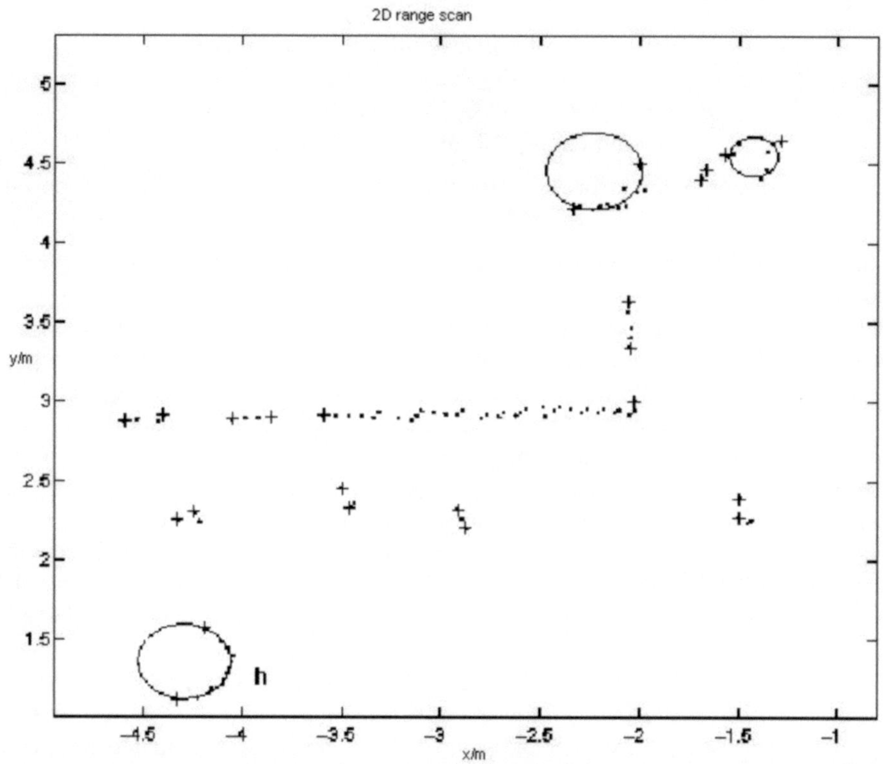

Figure 4 (b). A zoomed view of the region inside box C in Fig. 4(a)

To give an indication of the accuracy of the algorithm, we compare our method with some existing method. In the work by Bailey [Bailey, 2002], navigation methods are presented which use circular features from trees. We calculate the relative errors of the estimated center coordinates and radius of each pillar as follows:

$$CE = \frac{\sqrt{(x_{true} - x_{estimate})^2 + (y_{true} - y_{estimate})^2}}{\sqrt{x_{true}^2 + y_{true}^2}} \tag{10}$$

$$RE = \frac{| r_{true} - r_{estimate} |}{r_{true}} \tag{11}$$

where x_{true}, y_{true} and r_{ture} are the actual coordinates and the actual center of the circle feature which are obtained from hand measurements and $x_{estimate}, y_{estimate}$ and $r_{estimate}$ are their estimated values. The results are shown in Table 1 and Table 2.

	pillar d	pillar e	pillar f	pillar g
Proposed method	0.0179	0.0158	0.0136	0.0137
The method in [4]	0.0251	0.0244	0.0271	0.0166

Table 1. A comparison of the error CE of the four circle features between the proposed method and the method in [Bailey, 2002]

	pillar d	pillar e	pillar f	pillar g
Proposed method	0.0373	0.0011	0.1057	0.0588
The method in [4]	0.2096	0.2326	0.2626	0.1384

Table 2. A comparison of the error RE of the four circle features between the proposed method and the method in [Bailey, 2002]

From the above tables, we can see that the proposed method is more accurate than the method in [Bailey, 2002].

In order to test the feature extraction method for localization, the outdoor experiment has also been carried out for simultaneous localization and map building using the proposed feature extraction algorithm. The experimental environment is shown in Figure 6. There are 8 tall trees and building walls and some bushes which constitute the semi-structured outdoor environment. For this semi-structured environment, the main features for localization are tree trunks. The proposed feature extraction algorithm is applied for extracting the features. The vehicle used in the experiment is Cycab, a car-like vehicle, as shown in Figure 5. It is equipped with a laser range sensor, Sick LMS 200, with dead reckoning capabilities. There are four encoders fixed on the wheels of the vehicle. A DGPS with up to 2cm accuracy is used as a reference to give the ground truth of the vehicle pose to get the estimation error.

In the experimental environment, the vehicle moves along the path as shown in Figure 7 where the stars denote the trees of the environment which are detected, the dashed line indicates the real pose of the vehicle and the solid line means the estimated path using the simultaneous localization and mapping algorithm with the proposed feature extraction method. The data association method in the implementation is the same as that in [Zhang et al. 2005]. Figure 8 shows a typical laser scanner frame. The dashed line box A indicates the region whose clearer view is shown in Figure 9. In these two figures, we can find that there are lines, arcs and edges (point features). However, in the experiment, we only use circle features from the tree trunks for localization. The 8 features are all detected during the SLAM process after the continuous observation. It should be noted that there are additional features that are detected in some scans, but they haven't been used for the SLAM more than 3 times, we didn't draw them in the map.

To make a comparison on feature extraction performance, we also implement the method in [Bailey, 2002]. Figure 10 shows the range and bearing innovations of the measurements when we apply the feature extraction method in [Bailey, 2002] and our method during the SLAM and their 3σ bounds. The dash-dot line in the middle of each sub-figure is the result of the localization using our proposed feature detection method whereas the solid one in the middle is the result of the localization by the feature detection method in [Bailey, 2002]. Figure 11 shows the vehicle's position and orientation errors in the prediction and their 3σ error bounds. Further, we calculate the average absolute estimation error as defined by

$$\Delta x = \frac{\Sigma|\Delta x_i|}{N}; \ \Delta y = \frac{\Sigma|\Delta y_i|}{N}; \ \Delta \theta = \frac{\Sigma|\Delta \theta_i|}{N}$$

where Δx_i, Δy_i and $\Delta\theta_i$ are the vehicle pose errors at each time instant and N is the time horizon of the whole localization process. The comparison of the "average absolute error" for the two methods is given in Table 3 below.

	Proposed method	Method in [Bailey, 2002]
Δx	0.0575	0.0823
Δy	0.0571	0.0732
$\Delta\theta$	0.0353	0.0528

Table 3. The Δx , Δy $\Delta\theta$ of the vehicle pose when using different feature detection methods

In the table, the unit for Δx and Δy is meter and that for $\Delta\theta$ is radian.

Figure 5. The Cycab, a car-like vehicle in our experiment

We also test the detectability of the features by these two methods. We calculated 4 different scan sets' feature extraction false detection rate (the ratio of the number of false features to the total number of detected features in a scan) as shown in Table 4.

	scan 10 to 40	scan 50 to 80	scan 90 to 120	scan 130 to 160
Proposed method	0.067	0.034	0.067	0.1
The method in [4]	0.034	0.067	0.034	0.067

Table 4. A comparison of the false detection rate between the proposed method and the method in [Bailey, 2002]

Figure 6. The experimental environment (the whole scene)

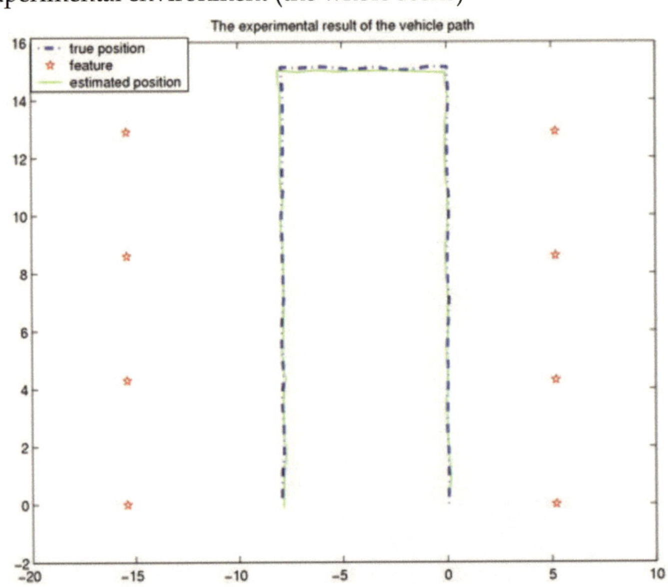

Figure 7. The estimated path and the true trajectory of the vehicle during the SLAM

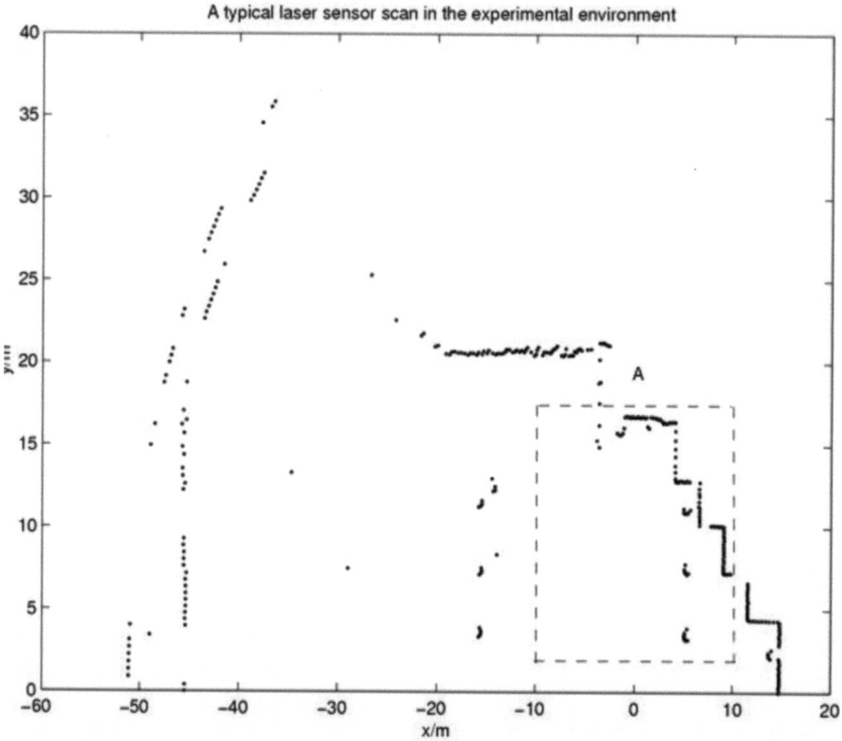

Figure 8. A whole scan data of the experiment environment corresponding to figure 5

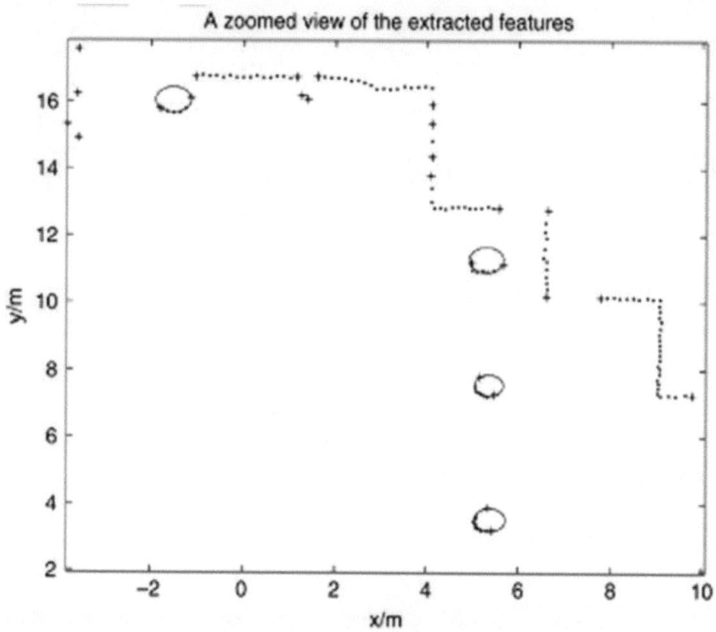

Figure 9. The circle features (trees) and edges extracted from the environment in figure 5 using Gauss-Newton algorithm

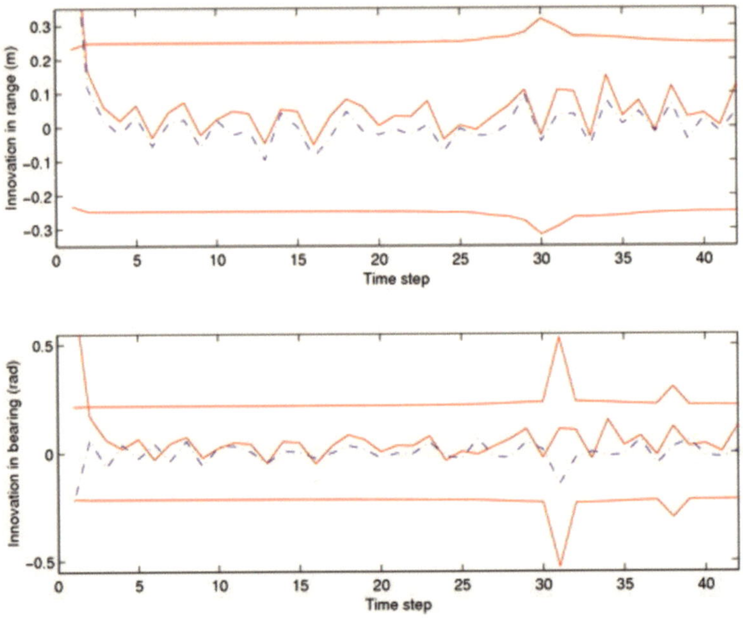

Figure 10. A comparison on range and bearing innovations during the localization when using the proposed method (dash-dot line) and the method in [Bailey, 2002] (solid line in the middle)

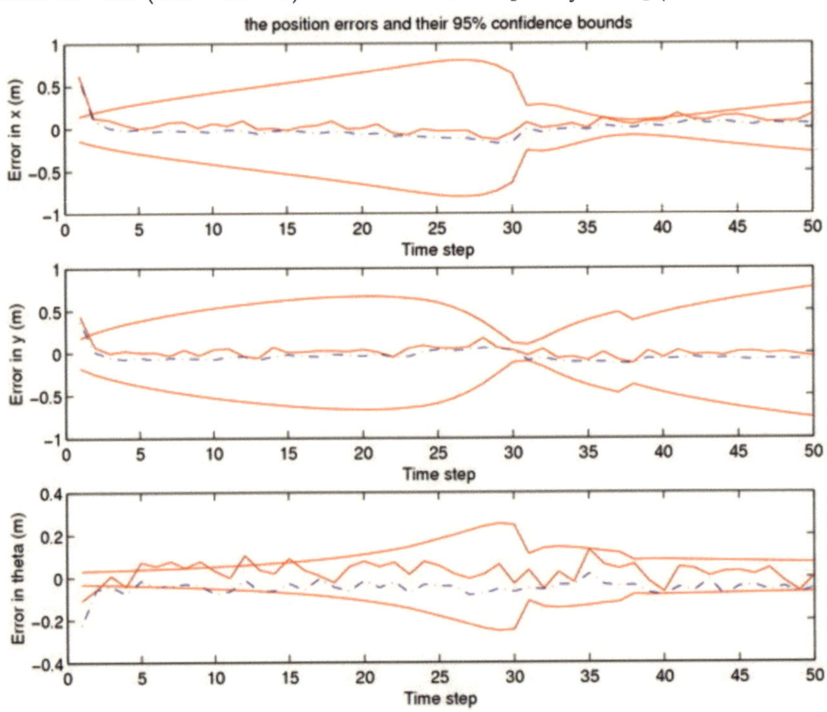

Fig. 11. The error and 3σ error bounds of the vehicle when using different feature detection methods. The dashed line is the result of the localization using the proposed feature detection methods

From the table, we know that the detectability of the method in [Bailey, 2002] is a little better than our method. However, in both algorithms, the features can be mostly detected for the localization purpose. From this point of view, Bailey's approach is more general than ours for the irregular circles' detection in some cases. However, the false detection rates of both the two algorithms are considered to be low.

It should be noted that there are false features that are detected in some scans, but they have not been used for the SLAM more than 3 times. Hence, we did not draw them in the map.

4. Conclusions

In this paper a new algorithm for feature detection in semi-structured outdoor environments has been presented. It can be used for the extraction of planar surfaces, tree trunks or tree-like objects and edges in semi-structured outdoor environments for mobile robot navigation. Experimental results show that the proposed method can extract features for navigation purposes successfully.

5. References

Adams, M. D. (1999). Sensor Modeling, Design and Data Processing for Autonomous Navigation, *World Scientific*, Singapore.

Bailey, T. (2002). Mobile robot localization and mapping in extensive outdoor environments, Universsity of Sydney *Ph.D Thesis*, 2002.

Crowley, J. L. (1985) Navigation of an intelligent mobile robot, *IEEE Journal of Robot Automation*, RA-1 No.1 , pp. 31-41.

Dennis, J. E., and Schnabel, R. B. (1983). Numerical Methods for Unconstrained Optimization and Nonlinear Equations. Prentice-Hall, 1983.

Guivant, J. E., Masson, F. R., and Nebot, E. M. (2002). Simultaneous localization and map building using natural features and absolute information, *Robotics and Autonomous Systems*, Vol. 40 , pp. 79-90.

Guivant, J. E., Nebot, E. M., and Baiker, S. (2000). Localization and map building using laser range sensors in outdoor applications, *Journal of Robotic Systems*, Vol. 17, pp. 565-583.

Iocchi, L., and Nardi, D. (2002) Hough localization for mobile robots in polygonal environments, *Robotics and Autonomous Syst.*, Vol. 40, pp.43-58.

Roumeliotis, S. I., and Bekey, G. A. (2000). Segments: a layered, dual Kalman filter algorithm for indoor feature extraction, *Proceedings of IEEE/RSJ International Conference on Intelligent Robots and Systems*, pp. 454-461, Japan, 2000.

Taylor, R. M., and Probert P. J. (1996). Range finding and feature extraction by segmentation of images for mobile robot navigation, *Proceedings of the IEEE International Conference on Robotics and Automation*, pp. 95-100, Minnesota, USA, 1996.

Vandorpe, J., Brussel, H. V., and Xu, H. (1996). Exact dynamic map building for a mobile robot using geometrical primitives produced by a 2D range finder, *Proceedings of the IEEE International Conference on Robotics and Automation*, pp. 901-908, Minnesota, USA, 1996.

Zhang, S., Xie, L., and Adams, M. D. (2003). Geometrical feature extraction using 2d range scanner, *Proceedings of the Fourth International Conference on Control and Automation.* pp. 901–905, Montreal, Canada, June 2003.

Zhang, S., Xie, L., and Adams, M. D. (2004a). Gradient model based feature extraction for simultaneous localization and mapping in outdoor applications, *Proceedings of the Eighth International Conference on Control, Automation, Robotics and Vision* (ICARCV 2004), pp.431-436, Kunming, China, Dec. 2004.

Zhang, S., Xie, L., and Adams, M. D. (2004b). An efficient data association approach to simultaneous localization and map building, *Proc. of IEEE Int. Conf. Robot. Automat.*, pp. 1493–1498, New Orleans, USA, April 2004.

Zhang, S., Xie, L., and Adams, M. D. (2005). An efficient data association approach to simultaneous localization and map building, *The International Jounal of Robotics Research* Vol. 24 pp. 49-60.